AF606161

ENCYCLOPEDIA OF APPLIED PHYSICS

ENCYCLOPEDIA OF APPLIED PHYSICS

UPDATE 2

Edited by

GEORGE L. TRIGG

Associate Editors

EDUARDO S. VERA

WALTER GREULICH

Managing Editor

EDMUND H. IMMERGUT

Weinheim · New York · Chichester · Brisbane · Singapore · Toronto

George L. Trigg
18 Duzine Road
New Paltz, NY 12561, U.S.A.

Edmund H. Immergut
Christopher Thomas Moran
2 Sidney Place
Brooklyn, New York 11201, U.S.A.

Walter Greulich
Walter Greulich Verlagsservice
Bergstraße 46
D-69469 Weinheim, Germany

Eduardo S. Vera
Science and Technology Information Center (ICT)
University of Chile
Beaucheff 850
Santiago, Chile

Library of Congress Card No. applied for.

A catalogue record for this book is available from the British Library.

Deutsche Bibliothek Cataloguing-in-Publication Data:
Encyclopedia of applied physics / ed. by George L. Trigg. Managing ed. Edmund H. Immergut. – [Aktualisierte Neuausg.]. – Weinheim ; New York ; Chichester ; Brisbane ; Singapore ; Toronto : Wiley-VCH
Update 2 2000
ISBN 3-527-29308-6

Printed on acid-free and chlorine-free paper.

Composition: Asco Typesetters, Hong Kong. Printing: Strauss Offsetdruck GmbH, D-69503 Mörlenbach. Bookbinding: Wilhelm Osswald & Co., D-67433 Neustadt.

Printed in the Federal Republic of Germany.

ADVISORY BOARD

EDITORIAL CONSULTANTS

ARTICLES

ADDENDA

The subject matter in the *Encyclopedia of Applied Physics* is presented in approximately 500 individual articles, arranged alphabetically. The topics can be classified into 20 sections, similar to the AIP Physics and Astronomy Classification Scheme (PACS):

01 General Aspects: Mathematical, Computational, and Information Techniques
02 Measurement Science, General Devices and/or Methods
03 Nuclear and Elementary Particle Physics
04 Atomic and Molecular Physics
05 Electricity and Magnetism
06 Optics (classical and quantum)
07 Acoustics
08 Thermodynamics and Properties of Gases
09 Fluids and Plasma Physics
10 Condensed Matter A: Structure and Mechanical Properties
11 Condensed Matter B: Thermal, Acoustic, and Quantum Properties
12 Condensed Matter C: Electronic Properties
13 Condensed Matter D: Magnetic Properties
14 Condensed Matter E: Dielectrical and Optical Properties
15 Condensed Matter F: Surfaces and Interfaces
16 Materials Science
17 Physical Chemistry
18 Energy Research and Environmental Physics
19 Biophysics and Medical Physics
20 Geophysics, Meteorology, Space Physics, and Aeronautics

Each article has been assigned a code number consisting of two digits which denotes the section, and a letter which gives the type of article. There are six types: A = Devices, Equipment; B = Materials; C = Methods, Processes; D = Phenomena, Effects; E = Scientific or Technological Fields; F = Institutions, Companies, Societies and other organizations.

CONTRIBUTORS

R. A. Baartman, TRIUMF, 4004 Wesbrook Mall, Vancouver, British Columbia, Canada V6T 2A3
Cyclotrons

H. L. Berk, Institute for Fusion Studies, C 1500, The University of Texas at Austin, Austin, TX 78712
Fusion, Magnetic Confinement

Evan G. Colgan, IBM Watson Research Center, P. O. Box 218, MS10-204, Yorktown Heights, NY 10598
Silicides

Daniel T. Crane, Salt Lake Technical Center, Occupational Safety and Health Administration, U.S. Department of Labor, 1781 South 300 West, Salt Lake City, UT 84165
Safety and Health in the Physical Science Laboratory

Katherine Creath, Creath Optineering Services, 2247 E. La Mirada Street, Tucson, AZ 85719
Interferometric Techniques

Philippe M. Fauchet, Department of Electrical Engineering, University of Rochester, C. S. Building, Room 514, Rochester, NY 14627
Silicon, Porous

Shaun Fischer, School of Physics and Chemistry, Lancaster University, Lancaster, LA1 4YB, Great Britain
Superfluidity: Liquid Helium Systems

Jeffrey P. Gambino, IBM Microelectronics, Hopewell Junction, NY 12533
Silicides

Kohji Hohkawa, Kanagawa Institute of Technology, 1030 Shimo-Ogino, Atsugi, Kanagawa 243-02, Japan
Modulators and Demodulators, Electrical

M. Hohage, Materials Research Group, University of Wisconsin–Madison, Madison, WI 53706
Surfaces and Interfaces of Solids

Shun-Ichiro Karato, Dept. of Geology & Geophysics, University of Minnesota, 108 Pillsbury Hall, Minneapolis, MN 55455
Earth, Internal Structure of the

Michael J. Kelly, Sandia National Laboratories, P.O. Box 5800, MS0343, Albuquerque, NM 87185-0755
Capillary Electrophoresis

Jürgen Kiefer, Strahlenzentrum der Justus-Liebig-Universität, Leihgesterner Weg 217, 35392 Giessen, Germany
Biological Effects of Electromagnetic and Particle Radiation

Gunther Kürbitz, Zeiss Optronik GmbH, Carl-Zeiss-Str. 22, 73447 Oberkochen, Germany
Photographic Imaging: Special Techniques

M. G. Lagally, Materials Research Group, University of Wisconsin–Madison, Madison, WI 53706
Surfaces and Interfaces of Solids

R. E. Laxdal, TRIUMF, 4004 Wesbrook Mall, Vancouver, British Columbia, V6T 2A3, Canada
Cyclotrons

Feng Liu, Materials Research Group, University of Wisconsin–Madison, Madison, WI 53706
Surfaces and Interfaces of Solids

G. H. Mackenzie, TRIUMF, 4004 Wesbrook Mall, Vancouver, British Columbia, V6T 2A3, Canada
Cyclotrons

Jürgen Mangelsdorf, Rollei Fototechnic Braunschweig, Salzdahlumer Str. 196, 28136 Braunschweig, Germany
Photography: Physics and Technology

Alice C. Mignerey, Department of Chemistry and Biochemistry, University of Maryland, College Park, MD 20742
Dating Techniques

Curtis D. Mowry, Sandia National Laboratories, P.O. Box 5800, MS0755, Albuquerque, NM 87185-0755
Capillary Electrophoresis

Wesley L. Nyborg, Physics Department, Cook Physical Sciences Bldg., University of Vermont, Burlington, VT 05405
Biological Effect of Sound and Ultrasound

Eiji Ohtani, Inst. of Mineralogy, Tohoku University, Sendai, Japan
Earth, Interior Structure of the

George Pickett, School of Physics and Chemistry, Lancaster University, Lancaster, LA1 4YB, Great Britain
Superfluidity: Liquid Helium Systems

Marc H. Pinsonneault, Department of Astronomy, Ohio State University, 174 W. 18th Avenue, Columbus, OH 43210-1106
Sun, Structure of

Barrett H. Ripin, The American Physical Society, One Physics Ellipse, College Park, MD 20740
Fusion, Inertial Confinement

Raymond G. Roble, High Altitude Observatory, National Center for Atmospheric Research, Box 3000, Boulder, CO 80307-300
Atomspheric Structure

Kenneth A. Rubinson, The Five Oaks Research Institute, Cincinnati, Ohio, USA
Chemical Analysis

Joanna Schmit, Veeco Instruments Incorporated, 2650 E. Elvira Road, Tucson, AZ 85706
Interferometric Techniques

Norbert Schuster, Rollei Fototechnic Braunschweig, Salzdahlumer Str. 196, 28136 Braunschweig, Germany
Photography: Physics and Technology

John P. Sibilia, Sibilia Associates, Inc., 12 Balmoral Drive, Livingston, NJ 07039
Characterization and Analysis of Materials

Dietmar Theis, Siemens AG, Dept. ZFE TWI, 81730 Munich, Germany
Display Technology

J. H. Whealton, Oak Ridge National Laboratory, P.O. Box 2009, Building 9108, Oak Ridge, TN 37831-8088
Negative Hydrogen-Ion Sources

Articles

CAPILLARY ELECTROPHORESIS

Curtis D. Mowry and Michael J. Kelly, *Sandia National Laboratories, Albuquerque, New Mexico, U.S.A.*

INTRODUCTION

In 1809, the Russian physicist Reuss performed the first recorded experiment of what would come to be known as electrophoresis (Reuss, 1809). Electrophoresis is defined as the migration, or movement, of electrically charged particles or dissolved species (solutes) in a conductive liquid or gel under the influence of an applied electric field. Reuss observed the electrophoretic migration of colloidal clay particles at an anode placed in a glass tube containing a bottom layer of wet clay, a second layer of wet sand, and a top layer of water. At the cathode, constructed in a similar fashion and ionically connected to the anode by a glass tube filled with water, the water remained clear and increased in volume, demonstrating the principle of electroosmotic flow (EOF). Also known as electroendosmotic flow or electroendosmosis, EOF is defined as the transport of neutral molecules or particles by a conducting solution under the influence of an electric field.

Numerous other experiments advanced the science of electrophoresis. Quincke (1861) showed that the migration rate of particles in an electric field is a linear function of potential gradient, while Hardy (1899) demonstrated the principle of the isoelectric point, which is defined as the pH at which a zwitterionic solute has zero electrophoretic mobility in an electric field. It was Michaelis (1909) who actually coined the phrase "electrophoresis" to describe the migration of ionic species in electric fields, and determined the isoelectric points of some enzymes in a U-shaped electrophoresis cell by measurements of electrophoretic mobilities at various pH values. Tiselius (1937) performed free-solution electrophoresis for protein separations and was the first to show that serum is composed of albumin and α-, β-, and γ-globulins, for which he won the Nobel prize in 1948. In Tiselius's experiment, the arms of a U-tube cell were filled with a pH buffer and a small amount of the protein under study, and then connected to nonpolarizable electrodes. The motion of the protein–buffer boundaries

ISBN 3-527-29308-6

was observed under the influence of a known potential gradient using a schlieren optical detection method to observe the boundaries between proteins.

Despite some significant instrumental improvements in electrophoresis experiments conducted with U-tube cells or on rectangular-shaped gel slabs, electrophoresis techniques were cumbersome and time-consuming, and produced separations with somewhat poor resolution among the separated components. Dramatic improvements were on the horizon, beginning in 1967 with the construction of the first experimental capillary electrophoresis (CE) system by Hjerten (1967), who then used the instrument to achieve high-efficiency separations of proteins, nucleic acids, and inorganic ions. In CE, the glass tubes or gel slabs used previously to separate solutes based on their electrophoretic mobilities were replaced by narrow-bore capillaries as the separation device. In 1979, Mikkers *et al.* (1979) performed electrophoresis in 200-μm inside diameter (i.d.) capillaries, and Jorgenson and Lukacs began electrophoresis experiments in 75-μm i.d. glass capillaries that resulted in seminal papers (Jorgenson and Lukacs, 1981, 1983) that showed the full potential of the technique for high-resolution separations. An intense period of research and development on CE science, separation methods, and instrumentation ensued, culminating in the introduction of the first commercial instrument in 1988. The first international scientific symposium devoted to CE was held in 1989, and the field has experienced exponential growth to the present day.

Today, capillary electrophoresis is one of the fastest growing techniques in analytical chemistry. Approximately 1500 scientific papers are published each year on CE, covering not only advances in instrumentation and methodology, but also applications to important problems in a variety of scientific disciplines. Among the important applications of CE are high-resolution separations in life science [DNA, polymerase chain reaction (PCR) products, oligonucleotides, purines, proteins, polypeptides, amino acids, carbohydrates, catecholamines, vitamins, chiral isomers, pharmaceuticals], environmental science (polyaromatic hydrocarbons, pesticides, pollutants), analytical chemistry (polymers, inorganic and organic ions, organic acids and bases, dyes and their precursors) and forensic and anti-terrorist science (illicit drugs, gunshot residues, DNA, explosive residues). Clearly, it is not possible to do justice to all areas of such a large field in a short article such as this. Therefore, in the sections that follow, the emphasis is on the most important theoretical, chemical, and procedural aspects of capillary electrophoresis.

The outline of the remainder of the article is as follows. The first section is devoted to a description of basic concepts that apply to all embodiments of capillary electrophoresis. Next, the instrumentation section describes the five essential components of a CE instrument, also known as a capillary electropherograph: the sample introduction system, the capillary (where the separation of complex mixtures into individual components occurs), the buffer system, the high-voltage power supply, and the detector. In the section entitled "Modes of Separation", a discussion of separation principles (*i.e.*, the physical and chemical mechanisms for resolving a mixture into its individual components) is combined with a description of diverse applications from the current scientific literature to illustrate the usefulness and limitations of the different CE separation techniques.

1. BASIC CONCEPTS

1.1 Electrophoresis

The objective in an electrophoresis experiment is to move individual solutes in a mixture through a separation medium at different velocities so that they exit from the medium at different times or, alternatively, reach different locations on the medium at a given time. In one popular embodiment of conventional electrophoresis, the medium is a gel slab either in the form of a free-standing membrane or as a gel supported on a planar stationary support, *e.g.*, a rectangular glass plate. Typical gels used in this application are two-phase colloidal systems (paper, starch, cellulose acetate, agarose) in an aqueous solvent or polymeric materials such as polyacrylamide. The slab is immersed in a buffer solution in an electrophoresis chamber and an electrical potential is applied across the length of the gel to set up an electric field (*i.e.*, a gradient of electrical potential). At the end of the experiment, the individual components will have migrated to different points along the direction of the gradient by processes that are described in Sec. 3. The locations of the components on the slab can be determined by a number of methods, including staining (followed by optical detection), radiography (for radioactive or radioisotopically tagged solutes), or detection with immunoreagents. This electrophoresis method is slow, labor intensive, not amenable to full automation, and prone to relatively poor reproducibility. The last results from a number of factors, including difficult-to-control solute adsorption effects on the gel and thermal contributions to solute migration (due to poor dissipation of Joule heat) that vary across the slab and result in broadening of the zones of the separated components. Improvements in the separation of mixtures are achieved by two-dimensional techniques, where successive electrophoresis experiments are performed on the same gel slab in perpendicular directions *via* two different separation modes.

Despite the problems described above, gel slab electrophoresis is an extremely useful and versatile method that is widely used today, principally in the separation and identification of complex mixtures of biological macromolecules. However, these problems have been addressed by capillary electrophoresis, and dramatic improvements in separations have been achieved. CE grew out of a remarkable convergence of scientific and technological advances in analytical chemistry. The principal advance came in the recognition that performing electrophoretic separations in small i.d., open-tubular capillaries, such as those used in gas chromatography (GC; see CHROMATOGRAPHY), reduces the problems associated with adsorption on gel media and thermally induced convective dispersion (Mikkers *et al.*, 1979). Separations in small i.d. capillaries also increases the degree of interaction between solutes and the capillary surface, which we will see has a large effect on separation efficiency. Further, detectors developed for chromatography instrumentation (principally liquid chromatography, LC) could be adapted, with certain modifications, to automatic detection of components eluting from the electrophoresis capillary. With these advances, it became possible to develop extremely fast, high-resolution separations for small quantities of materials. Capillary electrophoresis complements chromatography because separation selectivities are based on different chemical and physical principles and because it provides high-resolution separations of materials that are not amenable to chromatographic separations (*e.g.*, some large biological macromolecules). In fact, one approach for certain difficult samples takes advantage of their complementary nature by coupling CE to LC in a serial fashion to get two-dimensional separations with resolutions that exceed the capabilities of either method alone.

CE has a number of dramatic advantages compared with LC. These include faster analyses, higher efficiencies (defined in Sec. 1.3), smaller sample size requirements ($\leq \approx 10\,\mathrm{nL}$ *versus* $\approx 10\,\mu\mathrm{L}$), greater mass sensitivity (attomole *versus* picomole), lower flow rates ($\approx 100\,\mathrm{nL\,min^{-1}}$ *versus* $\approx 1\,\mathrm{mL\,min^{-1}}$), which reduce reagent consumption, and fewer column inventory requirements (one CE capillary suits a variety of samples so that different columns for different sample types are not required as in LC). These are among the key factors that account for the increasing popularity of CE for separations in liquid media.

1.1.1 Electrophoretic Mobility Electromigration is defined as the movement of a charged species under the influence of an electric field. In electrophoresis, however, it is more generally used to describe the movement of all solutes, whether they be neutral or charged. In this article, the term migration is used for brevity. In the case of a charged solute, a major contributor to migration results from its electrophoretic mobility, μ_{EP}, defined as

$$\mu_{\mathrm{EP}} = q/6\pi\eta r, \tag{1}$$

where q is the solute's electrical charge, η is the viscosity of the medium through which the solute is migrating, and r is the solute radius. Electrophoretic mobility is essentially a measure of the velocity of a charged species in an electric field of unit strength, in units of $\mathrm{cm^2\,V^{-1}\,s^{-1}}$. The linear dependence of μ_{EP} on the charge/size ratio is exploited in many modes of CE separations. Further, η is a strong function of such experimental parameters as pH, temperature, and ionic strength (a quantity related to the sum of the products of concentration and squared charge for all the ionic species in solution, expressed in terms of concentration units, *e.g.*, molarity). This provides additional experimental control of μ_{EP} in capillary electrophoresis. The effect of some parameters can be predicted *a priori*, *e.g.*, the dependence of the electrophoretic mobility of a solute with a known acid dissociation constant (K_a) as a function of pH. In the case of a weak acid, for example, the equilibrium concentration of the charged (unprotonated) and neutral (protonated) forms of the solute as a function of pH will control its mobility.

1.1.2 Electroosmotic Flow As described earlier, electroosmotic flow (EOF) is the transport of neutral molecules by a conducting solution under the influence of an electric field. In capillary electrophoresis, EOF results directly from the physical and chemical properties of the materials and chemicals used in the experiment. In the most common CE configuration, a fused-silica capillary is used with an aqueous buffer at $p\text{H} > 3$ (see Secs. 2.2 and 2.3). In this case, the surface of the capillary is composed of ionized silanoate groups (SiO^-). In the buffer, the layer closest to the capillary wall contains specifically adsorbed (tightly held) buffer cations, while a second layer consists of nonspecifically adsorbed (loosely held), hydrated buffer cations. This arrangement is known as an electrical double layer, and the electrical potential difference across the shear plane between these layers is termed the zeta potential (ζ). Under the influence of an applied electric field, the loosely held cations migrate toward the negatively charged cathode and drag other buffer constituents (water, cations, anions, solutes) along with them. This is known as EOF, and it results in the transport of neutral molecules and cations in the direction of the cathode. In addition, even though anions have a migration component in the direction of the anode due to their electrophoretic mobilities, they too will be transported toward to the cathode if the electroosmotic mobility (defined below) is greater than their electrophoretic mobility. This is typically the case at alkaline pH for most solutes of interest.

The electroosmotic mobility μ_{EOF} is defined as

$$\mu_{EOF} = \varepsilon\zeta/4\pi\eta, \tag{2}$$

where ε is the dielectric constant of the buffer and ζ is the zeta potential. Electroosmotic mobility is essentially a measure of the velocity of a neutral species in an electric field of unit strength, in units of $cm^2\,V^{-1}\,s^{-1}$.

There are CE applications where it is desirable to eliminate EOF or reverse its direction. Coating the capillary with a hydrophobic material that prevents access of the buffer to the surface silanoate groups would eliminate the electrical double layer described above that is a requirement for EOF. In this ideal case, there is no EOF, and only charged species in the capillary will migrate in the electric field, and they will do so according to their electrophoretic mobilities. The simplest way to reverse EOF is to reverse the polarity of the two electrodes; however, this does not reverse the relative directions of EOF and the electrophoretic mobility of a given solute. To accomplish this type of EOF reversal, a flow modifier such as cetyltrimethylammonium bromide (CTAB) or cetyltrimethylammonium chloride (CTAC) is added to the buffer at concentrations above a threshold required to effect flow reversal. These molecules attach to the silanoates on the capillary in a manner that creates a modified capillary surface with a net positive charge. The electrical double layer is then composed of anions and hydrated anions from the buffer, in contrast to the example described earlier, and EOF will be in the direction of the anode. In general, the magnitude of μ_{EOF} is dependent on experimental parameters such as pH, temperature, ionic strength, and viscosity. In special cases (*e.g.*, ampholytic flow modifiers, see Sec. 3.2), μ_{EOF} magnitude and EOF direction are both strong functions of pH.

1.1.3 Separation in Capillary Electrophoresis In a capillary electrophoresis separation, the net mobility of a solute being analyzed (an analyte) is the sum of its electrophoretic and electroosmotic mobilities. The velocity of the analyte (in $cm\,s^{-1}$) through the capillary is simply the product of the mobility and the electric field strength. Therefore,

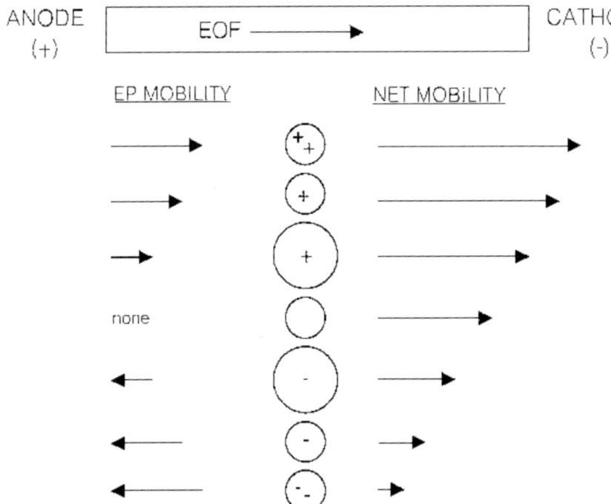

FIG. 1. Schematic representation of the electrophoretic migration of a mixture of anions (−), cations (+), and neutral molecules in capillary electrophoresis. The sizes of the molecules are represented by the sizes of the circles. The magnitude of the ionic charge is represented by the number of + and − symbols in the circles. In this example, the mixture was introduced at the anode end of the capillary and the direction of EOF is towards the cathode. The net mobility of a species is the sum of its electrophoretic and electroosmotic mobilities, and the magnitude and direction of each of these are indicated by the length and direction of the arrows, respectively.

the fastest separations can be performed with the highest field strengths. In practice, however, separations are performed at moderate field strengths to reduce Joule heating of the buffer, thereby preventing such undesirable consequences as vaporization of the buffer and thermal degradation of temperature-sensitive samples.

In the absence of specific analyte interactions with the capillary (*e.g.*, adsorption on silanoate groups), with materials added to the capillary (*e.g.*, a gel or a chromatographic stationary phase), or with a buffer modifier, a mixture of analytes will exit from the capillary (elute) at the cathode in the order of their net mobility. As an example, consider the electrophoretic separation of a mixture of anions and cations (of a variety of sizes and charges) and neutral molecules in an uncoated capillary. The analytes will elute at the cathode in the following order (see Fig. 1):

1. small, multiply charged cations;
2. small, singly charged cations;
3. large, singly charged cations;
4. neutral molecules (unresolved, *i.e.*, all elute at the same time);
5. large, singly charged anions;
6. small, singly charged anions;
7. small, multiply charged anions.

If the mixture being analyzed contains only anions, the direction of EOF can be reversed so that EOF and anion electrophoretic mobility are in the same direction. This causes the anions to migrate through the capillary more quickly and results in a faster separation. In this case, small, multiply charged anions would elute prior to large, singly charged anions.

In Sec. 3, a variety of CE separation modes that include special capillary and buffer arrangements (designed to promote specific analyte interactions with the capillary or the buffer modifier) are described. It will be seen that additional levels of selectivity in CE separations can be added to electrophoretic mobility and electroosmotic flow.

1.2 Resolution

The objective in a capillary electrophoresis experiment is to move individual analytes in a mixture through the capillary at different velocities so that they elute at different times. The analytes then move into an on-line detector that provides a real-time plot of detector response as a function of migration time. This plot is known as an electropherogram, and is a direct analog of the chromatogram (see CHROMATOGRAPHY).

The statistical treatment of electropherograms and that of chromatograms to determine important qualities of separations (efficiency, resolution, separation factor, and capacity factor) are fairly similar. The statistical methods for modes of CE that involve partitioning of the analyte between the aqueous phase and another phase (gel, micelle, chromatographic stationary phase, *etc.*) are strictly analogous to those used for chromato-

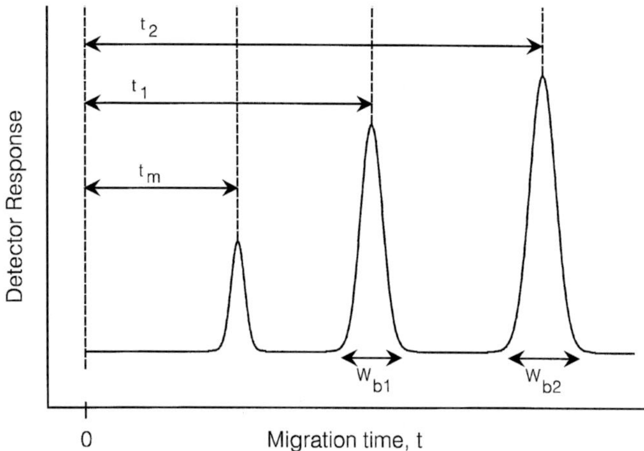

FIG. 2. Hypothetical electropherogram showing the response of a detector as a function of migration time t. The migration times for analyte 1, analyte 2, and a neutral marker are t_1, t_2, and t_m, respectively. Electrophoretic peak with is W_b.

graphic data (see CHROMATOGRAPHY), and are not included in this article. However, the treatment for modes of CE based on electrophoretic and electroosmotic mobilities is briefly discussed here.

The hypothetical electropherogram shown in Fig. 2 displays the response of a detector as a function of migration time t. The sample (composed of analytes 1 and 2) is introduced into the capillary at time zero as a narrow zone. The molecules of analytes 1 and 2 migrate through the capillary and elute at their characteristic migration times t_1 and t_2. As a way to normalize the time axis of the electropherogram, the elution time t_m of a "neutral marker", a molecule that is transported only by EOF, is used. While the components in the sample are being separated and eluted, the once-narrow zones broaden. This is primarily the result of diffusion under the influence of both temperature and concentration gradients. As a result of this zone spreading, the components elute as Gaussian-like peaks rather than as sharp bands. The degree of spreading is indicated by the magnitude of the peak width W_b. A fundamental chromatographic parameter that encompasses migration time and band spreading is the number of theoretical plates N:

$$N = 16(t_1/W_{b1})^2. \tag{3}$$

In terms of electrophoretic variables, N is expressed as

$$N = (\mu_{EP1} + \mu_{EOF})V/2D_1, \tag{4}$$

where V is the applied electrical potential (commonly referred to as the applied voltage) and D_1 is the diffusion coefficient of the analyte. Theoretical plates were first used to describe separation mechanisms in distillation processes, and were later adapted to chromatographic and electrophoretic separations. Briefly, a theoretical plate is an imaginary unit in the capillary wherein one complete mechanistic separation process step occurs. For example, in micellar electrokinetic capillary chromatography (MECC, Sec. 3.4), this would be a single equilibrium partitioning of the analyte between the micellar phase and the aqueous phase. In the entire column, there are N of these imaginary units. CE practitioners routinely calculate N values from electropherograms because it is a good indication of separation quality and the ability of a method to provide narrow peaks (high values of N being more desirable). This measure of CE efficiency is frequently normalized to the capillary length L to give the height equivalent of a theoretical plate,

$$H = L/N. \tag{5}$$

In the CZE of large biomolecules, wall adsorption can adversely affect separation efficiency, and models have been developed to predict its effects on plate height (Minarik *et al.*, 1995).

Other fundamental parameters are used as measures of the quality of a separation scheme. The separation factor α is a measure of the ability of a particular separation scheme to separate two components, and is defined in terms of electrophoretic variables as follows (Karger and Foret, 1993):

$$\alpha = \mu_{EP1}/\mu_{EP2}. \quad (6)$$

Therefore, the separation of two analytes depends on their relative electrophoretic mobilities. The extent of separation depends on the difference in migration times and the sharpness of the peaks, and is commonly described in terms of the resolution R:

$$R = 2(t_2 - t_1)/(W_{b1} + W_{b2}). \quad (7)$$

Narrow peaks with very different migration times are said to be separated with high resolution. A well-resolved peak arising from a pure substance is very desirable from both an analytical and a preparative viewpoint. The resolution can also be expressed in terms of electrophoretic variables (Jorgenson and Lukacs, 1983):

$$R = 0.177(\mu_{EP2} - \mu_{EP1})\{V/[(\mu_{EP} + \mu_{EOF})D]\}^{1/2}, \quad (8)$$

where μ_{EP} is the average electrophoretic mobility and D the average diffusion coefficient of the two analytes. It can be seen that resolution increases as $\mu_{EP} + \mu_{EOF}$ approaches zero, but this requires increasingly longer separation times. For example, if μ_{EP} equaled $-\mu_{EOF}$, resolution would be infinite; however, the solutes would be moving at equal but opposite velocity to the EOF and therefore would never elute. Therefore, a compromise between resolution and analysis time is usually required.

CE separations are typically faster and more efficient than LC separations. This advantage arises in great part from the characteristics of EOF and the fast migration times. Capillary EOF is characterized by a flat flow profile across the cross-sectional diameter of the capillary, with smaller flow rates at locations only in extremely close proximity to the capillary wall. This flat flow profile results in a narrow electrophoretic zone for a given solute, and gives narrow, well-resolved peaks. In contrast, pumped flow in LC is characterized by the parabolic profile of laminar flow, with the flow velocity at the column walls much smaller than the flow at the center of the column. This results in a broadening of the chromatographic zone for a given solute, and gives rise to broader peaks with lower resolution compared with CE.

CE separations with $N = 400\,000$–$2\,700\,000$ have been reported on 80–100-cm-long capillaries for separations that require only a few minutes. This is approximately 100 times better than the efficiencies of LC separations (as defined by the number of theoretical plates, N), which typically take about 10 times as long. For comparison, CE efficiencies are also approximately 2–10 times better than that of capillary GC separations. CE has the advantage that, with minimized injection and detection volumes and properly dissipated Joule heat, the largest contributor to band broadening is longitudinal diffusion of the analyte. The effects of longitudinal diffusion are reduced by keeping analysis times short.

A fundamental knowledge of the physical and chemical processes occurring in the capillary is critical to the successful design and implementation of a separation. The experimental variables that control N, α, and R are different for the various modes of CE, and are discussed in Sec. 3.

1.3 Qualitative and Quantitative Analysis

Qualitative analysis is the identification of a component responsible for a peak in the electropherogram. It is not possible to predict migration times exactly on the basis solely of the characteristics of the sample and the CE separation scheme. Several empirical expressions have been developed to calculate effective mobilities better on the basis of operational parameters. One such expression allows the effective mobilities of singly and multiply charged ions to be predicted over a wide range of buffer ionic strength (Friedl *et al.*, 1995). However, from an operational standpoint, the ability to predict migration times is generally not required. Correct peak assignments for an unknown analyte can be made by comparison with a literature electropherogram run under identical conditions, but this is not a very reliable approach. More reliable peak assignments are made by comparison with an electropherogram from injection of a known standard, by the use of an internal standard, or by identification using an on-line or off-line analytical method. Identification of unknown materials can sometimes be done on line with detectors based on ultraviolet–visible absorbance, fluorescence, mass spectroscopy, or electrochemistry (Sec. 2.5). Alternatively, it can be done by collection of effluent fractions (corresponding to individual components) followed by off-line spectroscopic, chemical, or biochemical analysis.

Quantitative analysis of analyte concentration is usually based on measurement of the height or area of the electrophoretic peak and comparison with peaks resulting from CE of a known standard. Modern CE instrumentation has data-acquisition and -processing equipment (*e.g.*, computers with integration and calibration software) for this purpose. However, CE peak areas are a function not only of analyte concentration, but also of analyte migration time. Since the analyte zone velocity in CE is a function of both μ_{EP} and μ_{EOF}, the separated components do not all migrate to the detector at the same velocity. The peak areas for two components of equal concentration and detector response will therefore not be equal, as the later eluting component travels more slowly through the detector. In contrast, for high-performance liquid chromatography (HPLC) the later eluting peak would be shorter and broader, but the two peak areas would be the same (see CHROMATOGRAPHY). This velocity effect in CE must be factored into the calibration calculations if the goal is the determination of analyte concentrations based on peak areas. High accuracy and precision in quantitative analysis by CE requires precise control of many factors, including EOF, capillary surface chemistry, temperature, buffer composition (*p*H, ionic strength, concentration of modifiers), and sample introduction volumes.

2. INSTRUMENTATION

Modern CE instrumentation (see schematic diagram in Fig. 3) has reached a very sophisticated state of development. It is highly automated, with computer control of the sample introduction, separation, and detection parameters and of the data acquisition and analysis. Special software for issues specific to CE separations, such as shifting migration times, correction of peak areas for analyte velocities, and estimation of analyte isoelectric points and molecular weights is also commercially available. At present, there are more than ten suppliers of capillary electropherographs in the U.S.

2.1 Sample Introduction

There are several sample introduction methods commonly used in capillary electrophoresis. The specific method used depends upon parameters such as separation mode and sample matrix. In each method, however, the inlet end of the capillary is physically moved from the buffer inlet vial to a sample vial for sample introduction, then returned to the inlet vial for electrophoresis. The length of the sample plug is usually more important than the actual volume, and only picoliters or nanoliters of sample solution are actually injected or consumed. Such a small volume requires care in handing, and it is typically contained in microliter or milliliter sample vials. The sample is often dissolved in the same buffer solution used during CE separation (Sec. 2.3). The sample concentration should be significantly lower than the buffer concentration, since concentrated samples may distort the electric field within the capillary, reducing separation efficiency and resolution. Sample introduction methods can be divided into three types: hydrodynamic, electrokinetic, and sample concentration.

Hydrodynamic sample introduction methods use flow generated by pressure, vacuum, or gravity to transport the sample solution into the capillary. In all cases, important factors for reproducible injections include buffer viscosity, sample viscosity, capillary radius, capillary length, and injection time. Sample temperature is also important because it affects buffer and sample viscosities. When these factors are reproducibly controlled, the volume of sample introduced is reproducible even if solution composition or *p*H is not well controlled.

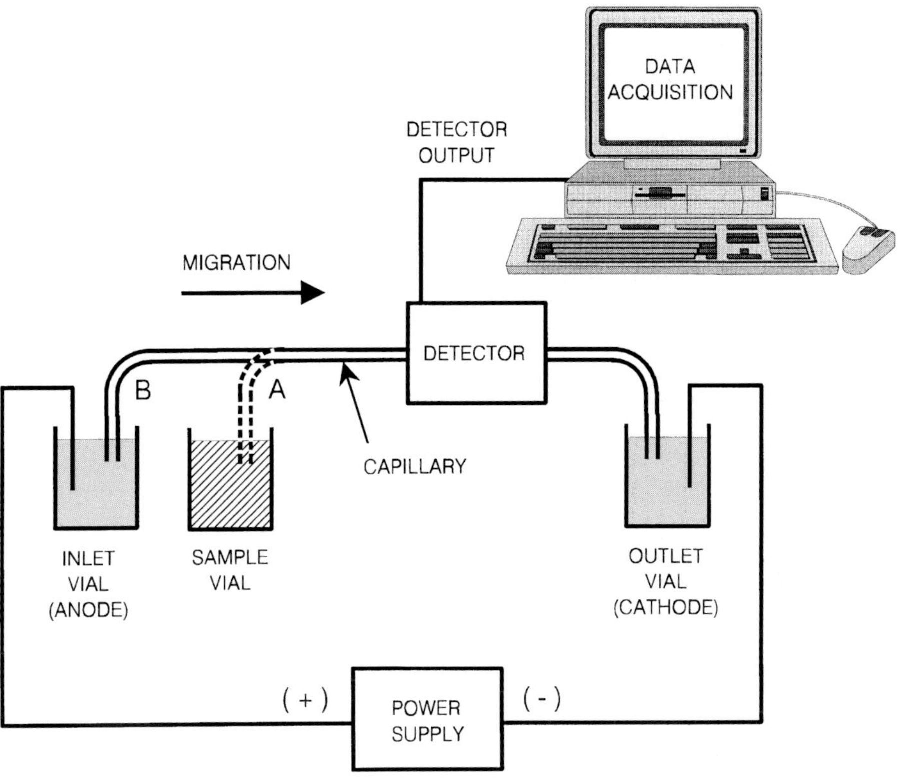

FIG. 3. Schematic diagram of capillary electrophoresis instrument, also known as a capillary electropherograph. During sample introduction (A), the capillary is immersed in the sample vial. During electrophoresis (B), the capillary is immersed in the inlet vial. This schematic diagram shows the use of an on-column detector (see Sec. 2.5).

Hydrodynamic methods require more instrumentation than electrokinetic methods and do not work well with packed capillaries (*e.g.*, capillary gel electrophoresis, CGE), yet have the advantages that sampling bias is not created and matrix effects are minimal. For aqueous sample solutions, the sample plug volume or length can be easily calculated from the experimental parameters. Pressure-based hydrodynamic injection requires that pressure be applied to the sample vial or, alternatively, a vacuum applied at the detector side of the capillary. Pressure leaks affect reproducibility, and the time period that pressure is applied must be carefully controlled. This type of sample introduction works particularly well with capillary isoelectric focusing (CIEF, Sec. 3.2), since the entire capillary can be filled in a short time. Gravity-based hydrodynamic injection, also known as siphoning injection, is performed by inserting the inlet end of the capillary into the sample vial and then raising this vial above the capillary outlet vial. The height and time required are on the order of millimeters and seconds, both of which must be carefully controlled, yet this type of sample introduction remains instrumentally simpler than pressure-based methods. Electrophoresis is initiated after the sample has been introduced and the capillary is returned to the inlet vial. It is important to keep the inlet and outlet vials at the same level to prevent siphoning during electrophoresis, which would induce laminar flow in the capillary and resultant degradation of separation efficiency and resolution.

Electrokinetic (or electromigration) sample introduction uses an electric field to transport analytes into the capillary. The amount of analyte injected depends upon the applied voltage, capillary radius, μ_{EOF}, analyte concentration, analyte μ_{EP}, viscosity, and injection time. Sample solution temperature, buffer temperature, conductivity, and *p*H will affect μ_{EOF} and μ_{EP} and therefore injection reproducibility. In some cases, it is necessary to adjust the ionic strength of the sample to obtain good injection precision. Electrokinetic introduction is instrumentally simpler than pressure-based methods, as the high-voltage supplies and controls are already in place on the instrument. The main disadvantage of electrokinetic introduction is that because analyte transport onto the column is based upon both μ_{EP} and μ_{EOF}, a sample loading bias is created (*i.e.*, a higher analyte μ_{EP} results in greater loading). However, this bias can be corrected for mathematically and can actually be used for an additional level of selectivity. For example, a capillary treated to eliminate electroosmotic flow injects cations or anions exclusively, and neutrals will not be injected.

Sample concentration methods for sample introduction include techniques known as sample stacking and field-amplified sample injection. These methods are similar to electrokinetic sample introduction and also concentrate the analyte in the capillary, often leading to signal enhancements of 100–1000-fold. Sample stacking utilizes a sample solvent (*e.g.*, water) with lower conductivity than the electrophoretic buffer. This results in higher electric fields when the capillary is placed in the sample vial and voltage is applied. Under the higher electric field, analytes migrate to the boundary between the low-conductivity sample plug and the high-conductivity buffer. At this interface the migration is slowed on account of the (relatively) lower electric field. This serves to focus or decrease the width of the analyte plug; however, it must be noted that neutrals are not focused. During sample introduction, cations are stacked at the front of the lower-conductivity plug, whereas anions are at the back. Laminar flow and Joule heat can degrade performance, however, and so typical voltages are lower than those used during CE separations. Field-amplified sample injection utilizes similar principles to focus the analytes; however, a small plug of very low-conductivity solution that does not contain analyte is injected first. This results in even higher electric fields than generated in sample stacking and smaller injected sample volumes.

Modern CE instrumentation includes autosampling accessories for automatic sample introduction by the techniques just described. Automatic sample introduction typically increases the precision of CE analyses, and also increases the efficiency of laboratory analyses since it does not require the constant attention of an operator.

2.2 Capillaries

The capillaries used in CE are made of fused silica, Teflon, or Pyrex, with fused silica being used for the vast majority of separations. They range in length from 3 to 100 cm (with most separations performed using 30–100-cm-long capillaries), in i.d. from 25 to 200 µm (with 50- and 75-µm i.d. capillaries being fairly typical) and in outside diameter (o.d.) from 200 to 375 µm (with 350–375 µm being typical). The fused-silica capillaries have a 10–20-µm-thick external polyimide cladding to impart strength and flexibility.

Fused-silica capillaries are commonly available in untreated, surface-modified (coated), and gel-filled forms. Capillary choice is affected by the CE separation mode used, compatibility with the buffer or reagent choices, and cost. Careful control of the surface of untreated capillaries is important for reproducible control of electrophoretic and electroosmotic mobilities and can be easily achieved by selection of a buffer *p*H that sets the silanol/silanoate equilibrium chemistry at the desired point. The capillary surface can also be modified to reduce or eliminate EOF and to reduce analyte adsorption or interaction with the column. This was first demonstrated by Hjerten (1985). Analyte–column interactions can cause significant problems such as band broadening, irreproducibility, and irreversible analyte binding, especially in the analysis of large biopolymers such as peptides, proteins and DNA. Modifiers include covalently bonded molecules, polymers, and groups with either anionic or cationic functionality. Non-covalently bonded modifiers can also be added to the buffer system for temporary, removable surface modification. Modifier characteristics to consider include stability with the buffer during long-term use and inertness to analytes and buffers. Commonly used modifiers include bonded alkyl chains, polymers such as polyacrylamide (PA) or poly(vinyl alcohol) (PVA), and zwitterionic buffer additives. PA is the most common polymer used in this application and is also frequently used as the packing material in gel-filled capillaries for CGE. These CGE capillaries are prepared by performing the polymerization directly in the capillary to produce a cross-linked polymeric gel with a porous network having narrow, well-defined pore size distributions. Finally, surface-modified capillaries ordinarily used in gas chromatography and cyclodextrin-modified capillaries (for chiral separations) have also found limited use.

Precise control of capillary temperature is essential for reproducible CE, since migration time, electrophoretic and electroosmotic mobilities, injection volume, and separation efficiency and resolution are all functions of temperature. In addition, temperature control helps dissipate the Joule heat generated by the electrical power applied to the capillary, thereby preventing such undesirable consequences as vaporization of the buffer and thermal degradation of temperature-sensitive samples. Temperature control of the capillary (at temperatures near ambient, to approximately $\pm 0.1\,^{\circ}\mathrm{C}$) is achieved by a variety of methods, including forced-air cooling, thermostatted liquid cooling, and the use of Peltier cooling devices. Finally, pressurization of the capillary at the inlet and outlet is sometimes required to prevent solvent outgassing (especially in capillary electrochromatography, CEC), which can perturb the flat flow profile required for high-resolution CE separations.

In capillary zone electrophoresis (CZE) and MECC, uncoated capillaries are used; in CZE, capillary ion electrophoresis (CIE), CIEF, and dynamic sieving capillary electrophoresis (DSCE) applications, coated capillaries are used; and in CGE and CEC, capillaries packed with sieving gels and LC stationary phases, respectively, are used. More details on these separation modes are presented in Sec. 3.

2.3 Buffers and Other Reagents

CE separations can be performed over a wide pH range (approximately 2–10.5). A variety of buffer systems are used in CE, including phosphate, tetraborate, carbonate, phthalate (for indirect detection methods, discussed below), and tris(hydroxymethyl) aminomethane–boric acid ("Tris–borate buffer"). Buffers are used individually and in combination, at typical concentrations of 0.01–0.1 M. Precise control of the buffer composition in the inlet vial, outlet vial, and capillary is essential for reproducible CE. In practice the buffer solution is replaced after a few experiments to prevent ionic depletion, which causes irreproducibility. The primary purposes of the buffer are to provide a conducting medium for electrophoretic and electroosmotic movement of analytes, to control the surface chemistry of fused silica capillaries, and to control the ionization and partitioning equilibria of analytes. These are key factors in adjusting the resolution of a CE separation and can be achieved by controlling the pH or by the use of buffer modifiers.

Increasing buffer pH will increase μ_{EOF}, since the zeta potential (with silica capillaries) increases with increasing pH. In contrast, increasing the ionic strength of the buffer will decrease μ_{EOF}, since the zeta potential decreases with increasing ionic strength. It is also important to know the isoelectric point (pI) or the acid dissociation constant K_a (typically expressed as $pK_a = -\log K_a$) of the analytes being separated, since they play important roles in determining their electrophoretic migration properties. Many CE separations are designed around the fact that at high pH, most biological molecules are anionically charged and are repelled from the negatively charged walls (in the case of an uncoated capillary); this reduces sample–wall interactions and sample losses due to adsorption on the wall. This approach allows uncoated capillaries to be used in the separation of proteins, as first shown by Lauer and McManigill (1986).

A variety of additives can be added to the buffer to modify the resolution of CE separations. For example, organic solvents are added to improve the solubility of analytes with limited solubility in water. In some cases, even 100% nonaqueous buffers have been successfully used in CE separations. Other materials (*e.g.*, potassium sulfate, ammonium sulfate) are added to prevent analyte adsorption on the capillary. Cyclodextrins, which are donut-shaped, water-soluble oligosaccharides with hydrophobic cavities and hydrophilic exteriors, can be added as additional partitioning phases to help separate analytes on the basis of their relative sizes and hydrophobic/hydrophilic characteristics. Another important cyclodextrin application is in the separation of stereochemical isomers, or enantiomers (see Sec. 3.1). In MECC (Sec. 3.4), detergents (*e.g.*, sodium dodecyl sulfate, SDS) are added as micellar partitioning phases. Soluble ion exchangers and charged colloidal particles have also been used in partition-assisted CE separations.

Additives can also be used to measure important CE parameters or to improve detector sensitivity. A small molecule, soluble and neutral at the *p*H used for analysis, can be added as a neutral marker to provide a measure of EOF velocity, since it has zero electrophoretic mobility. In ultraviolet–visible absorbance detection (Sec. 2.5.1), highly absorbing modifiers can be added to the buffer to perform indirect detection, where the decrease in measured absorbance due to the elution of a nonabsorbing analyte is detected as a negative peak. A similar approach can also be used in fluorescence (Sec. 2.5.2) and electrochemical (Sec. 2.5.4) detection.

In CZE, MECC, and CGE, the same buffer is used in the inlet and outlet vials; in CIEF, different solutions are used in the inlet and outlet vials, while an ampholyte is used in the capillary; and in capillary isotachophoresis (CITP), leading and trailing buffers are used to segregate the analytes into discrete zones. More details on these separation modes are presented in Sec. 3.

2.4 High-Voltage Power Supplies

High electric fields are required to generate the electrophoretic and electroosmotic velocities needed for high-resolution CE separations. Electric field strengths of 100–700 V cm^{-1} are generated across capillaries using high-voltage power supplies rated for use between 10 and 70 kV, with a typical experiment being performed at 30 kV. The current requirements on the power supply are fairly low, typically $<300\,\mu A$, and so the power ratings for these supplies are on the order of 10 W. The voltage and current outputs should be very stable for reproducible separations. Capillary electropherographs have interlocks to prevent electrical shocks to operators *via* accidental exposure to high voltages. Most CE modes are performed with constant-voltage power supplies, with the exception of CITP (Sec. 3.5), which is done in a constant-current mode.

Additional levels of efficiency and sensitivity in CE separations can be obtained with programmable power supplies. ac voltages, dc voltage pulses, and voltage ramps ("gradients") have all been used to modify analyte velocities in the capillaries in ways that improve resolution. The use of voltage gradients, where the applied voltage is changed linearly with time during the electrophoresis experiment, has been particularly successful in high-resolution separations of DNA fragments by CGE. Reversible-polarity options can allow the direction of migration to be reversed for applications where this is desired. Improved sensitivity in radiometric detection (Sec. 2.5.5) can be realized by increasing analyte residence time in the detector by voltage control of the analyte migration velocity. Finally, the use of a second power supply to apply a radial voltage to the capillary has been used to change the zeta potential and, therefore, change μ_{EOF} [Eq. (2)].

2.5 Detectors

A wide variety of detectors are available for use with CE. Many of these are common to LC (see CHROMATOGRAPHY) and have been modified to accommodate the unique constraints of CE. Requirements common to all CE detectors include issues related to connection to the capillary, compatibility with the buffer chemicals, low flow rates, high electrical fields at the capillary outlet, and sensitivities for the small amounts of analytes used in CE ($\leq 10^{-15}$ mol). Real-time, on-line detection is the preferred approach because of the problems inherent in handling small amounts of analytes for post-CE detection by an off-line analysis method.

The choice of a suitable detector from among the many available types is based on the requirements of the separation problem as well as detector availability and cost. General principles and CE applications of the most common detectors are discussed briefly in the rest of this section. Nondestructive detectors do not chemically or physically destroy the sample and are well suited to preparative applications where it is desirable to collect pure fractions of an individual component eluting from the capillary. Preparative CE is often performed to collect a pure sample of an analyte for subsequent off-line chemical, biochemical, or spectrometric analysis. In addition, nondestructive detectors can be used as the first detector in a tandem detector arrangement to improve the selectivity of on-line analyte detection.

In the sections that follow, detection limits (the minimum amount of detectable material) are given for each detector, although it must be emphasized that these are approximate values that depend on many variables, particularly the chemical and physical properties of the material being detected. The detection limits are given in terms of moles of analyte in the injected sample.

2.5.1 Ultraviolet–Visible Absorbance Ultraviolet–visible (UV–vis) absorbance detectors simply measure the absorption of light (in a particular wavelength range) by the analytes, and are the most widely used detectors in CE. This is primarily a result of two factors. First, optical absorbance detectors are amenable to integration with optically transparent fused-silica capillaries. Second, most compounds of interest absorb 190–800-nm radiation, while many other compounds that do not absorb can be chemically complexed or derivatized (*e.g.*, in a post-capillary column reactor) to form highly absorbing species. An example of this is the absorbance detection at 520 nm of the complexes formed between 4-(2-pyridylazo)resorcinol (PAR) and transition-metal ions such as Fe^{II}, Fe^{III}, Cu^{II}, Ni^{II}, Zn^{II}, Co^{II}, and Pb^{II}. If a nonabsorbing analyte is not amenable to complexation, CE can be performed in a buffer containing a highly absorbing modifier or in a light-absorbing buffer (*e.g.*, phthalate), and the analyte can be detected as a negative peak in a high-absorbance background (Small and Miller, 1982). This is sometimes referred to as indirect UV–vis absorbance detection. Monochromator-based scanning spectrometers and photodiode array detectors can provide an added level of identification, as the UV–vis absorbance spectrum can sometimes be used to confirm the identity of an analyte giving rise to a CE peak. The current generation of fast-scanning monochromator detectors typically have better signal-to-noise ratios than photodiode array detectors, and are preferred for applications requiring greatest sensitivity.

The capillary itself is typically used as the flow-through absorbance cell ("on-column" detection). Here, a section of the external polyimide cladding is removed from the capillary, and the UV–vis radiation is directed through the capillary. The absorbance A of an analyte is given by Beer's law:

$$A = \varepsilon b C, \tag{9}$$

where ε is the analyte molar absorption coefficient, b is the optical path length, and C is the analyte concentration. The small path lengths offered by the capillary i.d. have been largely offset by operation at low wavelengths (190–220 nm) where absorption coefficients are typically greatest, by sample-concentration injection methods (Sec. 2.1), which increase analyte concentrations in the capillaries by 100–1000-fold, and by the use of extended–path-length absorbance cells. An extended–path-length capillary known as a bubble cell is formed by a proprietary process that increases capillary i.d. (and therefore optical path length) in the detector by 3–5-fold. A second type of extended–path-length cell ("Z-cell") is formed by making two right-angle bends in the capillary and then directing the sample beam down the length of the middle leg of the "Z" instead of through the cross-sectional diameter of the capillary. This results in approximately a 20-fold improvement in sensitivity. Finally, optical detector designs are very efficient at reducing the amount of scattered light that reaches the photodiode, which also helps sensitivity.

UV–vis absorbance detectors are nondestructive, and for strongly absorbing analytes the detection limit is about 10^{-13}–10^{-16} mol. Further details on UV–vis absorbance can be found in OPTICAL SPECTROMETERS and MOLECULAR SPECTROSCOPY.

2.5.2 Fluorescence Fluorescence detectors measure the light emitted by the analyte after optical excitation by either a lamp (deuterium, quartz–tungsten–halogen, xenon, mercury) or a laser (laser-induced fluorescence, LIF). When a broad-band radiation source such as a lamp is used, the excitation of the analyte is limited to a narrow wavelength range by use of a monochromator. Detection of the fluorescent emission is performed with a photodiode or a photomultiplier tube located at a right angle from the excitation source. As in UV–vis absorbance detection, the capillary itself can be used as a flow-through detection cell, but improvements in sensitivity have been achieved by coupling the capillary to a cell with planar quartz windows, which reduces the amount of light scattering (Chen and Dovichi, 1994).

Fluorescence detectors are more selective than UV–vis absorbance detectors because not all compounds that absorb light subsequently fluoresce. Since only a small percentage of compounds natively fluoresce, analytes are usually derivatized with fluorescent tags. Derivatization can be performed either prior to the separation, in the capillary using a fluorescent tag added as a buffer modifier, or after elution from the capillary in a post-column reactor. Novel, analyte-specific fluorescent tags form an intense area of research, and have resulted in tremendous improvements in sensitivity for important analytes. For example, the fluorogenic reagent 3-(4-carboxybenzoyl)-2-quinoline-carboxaldehyde produces highly fluorescent isoindole derivatives with amino acids and peptides (Liu *et al.*, 1991). Nonfluorescing analytes can also be detected by indirect detection, whereby the CE separation is performed in a buffer containing a highly fluorescing modifier and the analyte is detected as a negative peak in a high-fluorescence background.

LIF is used when low detection limits are required, since the intensity of the emitted fluorescence is proportional to the number of excited molecules, which in turn is proportional to the intensity of the excitation radiation. LIF has been particularly useful for DNA analyses with intercalating dyes, which has eliminated the need for derivatization of the DNA. The increasing number and types of semiconductor and solid-state laser sources has increased the popularity of fluorescence detection and made it applicable to a wider range of analytes. Fluorescence detectors can be nondestructive (*e.g.*, if the analyte is not derivatized) and they have a detection limit of about 10^{-14}–10^{-21} mol. Further details on molecular fluorescence can be found in OPTICAL SPECTROMETERS and MOLECULAR SPECTROSCOPY.

2.5.3 Mass-Spectrometric Mass-spectrometric detectors measure the abundance and mass-to-charge (m/z) ratio of ions produced from analytes eluting from the capillary. For this measurement, the molecules are ionized (singly or multiply) after they elute and then introduced into the vacuum chamber of a mass analyzer. Ionization and subsequent mass measurement are accomplished with a variety of instrumentation. In order to do on-line coupling of CE with mass spectrometry (MS), the mass spectrometer must accommodate conditions unique to CE, including high–salt-content buffer systems, high electrical fields on the capillary outlet (which is interfaced to the inlet of the MS), and low flow rates. Although the low flow rates characteristic of CE make coupling to MS easier than is the case for LC, the other conditions constrain the ionization methods and mass analyzers that are used in CE.

Ionization methods coupled on-line with CE include electrospray ionization, ion-spray ionization, atmospheric-pressure ionization, and inductively coupled plasma ionization. Matrix-assisted laser desorption ionization (MALDI) has been used successfully off line in CE analyses. Electrospray ionization interfaces were first developed by Smith *et al.* (1988) and are the most popular devices for coupling capillaries to mass spectrometers. Typically, electrospray ionization uses buffer flow through a steel needle (on the end of the capillary) held at a high electrical potential to ionize analytes that are amenable to either negative or positive ionization modes. For optimum ionization efficiency, and therefore optimum detection limits, the design of the interface between the capillary outlet and the electrospray needle is critical. Combinations of sheath flow liquids (Wahl and Smith, 1994) with capillary outlet tip modifications have been used to enhance detection sensitivity.

Mass analysis brings an additional level of information to CE analysis. The mass spectrum of a detected peak, which is a plot of ion intensity versus m/z, contains molecular-weight, formula, and structural information. Mass analyzers utilized include quadrupole ion trap, quadrupole filter, time-of-flight, and ion-cyclotron-resonance (ICR) mass spectrometers. Each of these can also be used to perform tandem experiments ("MS–MS") if the analyzers are properly equipped. MS–MS experiments provide detailed structural information about an analyte and have been applied to polypeptide sequence determination. Time-of-flight analyzers are useful for the measurement of very high-mass analytes (>3 000 atomic mass units, u) and for rapid CE (in which eluted peaks may be only a few seconds wide), while ICR instruments are used in applications requiring the highest mass resolution. Mass analyzers can be operated either as universal detectors (by monitoring the intensity of all m/z regions of the mass spectrum) or as selective detectors (by monitoring selected m/z regions). Mass-spectrometric detectors are destructive and have typical detection limits of about 10^{-15}–10^{-18} mol. Further details on mass spectrometers can be found elsewhere in this Encyclopedia (see MASS SPECTROMETERS).

2.5.4 Electrochemical Electrochemical detectors measure changes in current or conductivity between a set of electrodes in a detection cell. It must be noted that these are not the same electrodes that are used to generate the high electric fields needed for electrophoretic migration. Electrochemical detectors are seeing increasing use in biochemical analyses because many biochemical compounds are electroactive since they are easily oxidized. Also, several novel detector designs have been reported that are free from electrophoretic voltage interference, making this a reliable detection method.

Amperometric detectors measure the current produced by a controlled-potential electrolysis of the analyte in the buffer. The analytes from the column elute past a working electrode (one of three electrodes in the detector cell) set at a specific potential with respect to a reference electrode. If an analyte can be oxidized or reduced at this potential, the transfer of electrons at the electrode–buffer interface serves as the amperometric detection signal. Some recent designs use on-column detection with 5–25-μm-diameter working electrode wires either placed over the capillary outlet (Zhong and Lunte, 1996) or inserted directly into the end of the capillary (Swanek *et al.*, 1996). An additional level of information can be obtained with voltammetric detectors, wherein the potential of the working electrode is changed with respect to time as the analyte passes through the cell. Voltammetric detectors provide additional measures of the thermodynamic and kinetic properties of the electron-transfer reactions of the analytes, which can be used to aid in the identification of unknown analytes. As with UV–vis and fluorescence detectors, indirect methods of detection are possible with amperometric detectors, whereby a nonelectroactive analyte is detected as a negative peak when the CE separation is performed in a buffer containing a high concentration of an electroactive modifier.

In conductivity detectors, the electrical conductivity of the eluting buffer is measured by applying an alternating voltage between two electrodes and measuring the resulting current. Conductivity is related to the current/voltage ratio and represents the contribution of the individual equivalent conductivities of all ions in solution, including the analyte ions. Measuring the small conductivity due to an eluting ion in the presence of the high background conductivity of the buffer is difficult. Suppression of the conductivity of the buffer (Small *et al.*, 1975) dramatically improves the sensitivity. For example, consider the case where a sodium carbonate buffer is used in anionic separation by CE. As the buffer exits from the capillary, it passes through a cation exchanger (suppressor) loaded with hydrogen ions. Sodium ions from the buffer displace hydrogen ions from the ion-exchange sites on the suppressor, with two beneficial results. First, carbonic acid is formed by the ion-exchange reaction, so that detection is carried out with a background characterized by the low conductivity of carbonic acid, as opposed to the higher conductivity of sodium carbonate. Second, the analyte anions are paired with hydrogen ions instead of sodium ions, so that detector sensitivity is increased because the equivalent conductivity of the hydrogen ion is approximately seven times as large as that of the sodium ion. Several on-column conductivity-cell designs have been developed that incorporate the electrodes and the suppressor directly on the end of the capillary. Conductivity detection is usually described as suppressed or nonsuppressed to indicate whether ion exchangers are incorporated.

Amperometric detectors are destructive and have typical detection limits of about 10^{-18}–10^{-19} mol. Conductivity detectors are destructive if a suppressor is used, and have typical detection limits of about 10^{-15}–10^{-16} mol. Further details on electrochemistry can be found in ELECTROCHEMISTRY.

2.5.5 Other Detectors Other detectors use optical, radioactive, or other analyte properties as the basis of detection, but are not as widely used as those already discussed. Detectors based upon analyte optical properties (or on the properties of the products of post-separation reactions involving the analyte) include refractive-index, thermooptical, and chemiluminescence detectors. Radiometric detectors detect radioactive decay from commonly used isotopes of phosphorus, sulfur, or hydrogen that have been incorporated into the analyte. Detectors based on other analyte characteristics (*e.g.*, nuclear magnetic resonance spectrum, Raman spectrum, proton-induced x-ray emission, particle light scattering) and capillary characteristics (laser-induced capillary vibration) have found special applications as on-line detectors in capillary electrophoresis.

2.6 Integrated Capillary Electrophoresis Microdevices

An exciting development in capillary electrophoresis is the fabrication of miniaturized capillary electropherographs on planar substrates such as silicon, quartz and glass. These CE instruments were first introduced in 1991 (Fan and Harrison) with CE separation channels (the planar analog to conventional capillaries) built on the surface of silicon substrates. These channels were fabricated by use of a combination of masking, photolithographic, and etching techniques that had been primarily developed for the fabrication of silicon-based integrated circuits and microelectromechanical structures (MEMS). Since then, sample-introduction, separation-channel, and detector components have been fabricated on glass and quartz by modifications of silicon micromachining methods. While some detectors, such as electrochemical detectors, have been integrated into these devices (Ewing and Gavin, 1996), other detection schemes, such as laser-induced fluorescence, are being performed with some integrated components (*e.g.*, post-column reactors) and some external components (*e.g.*, the laser and photodiode) not integrated onto the substrate (Fluri *et al.*, 1996). On account of the extremely small size of the separation channels, which are typically $5 \times 50\,\mu m^2$ in cross-section, the efficiency of these separations is excellent, with 70 000–160 000 theoretical plate separations reported on 5–50-mm-long channels with 0.1-nL sample volumes and 10–20-s migration times (Manz *et al.*, 1993). In addition, these devices can perform multichannel separations of a single sample by multiple, complementary modes of separation. This is achieved by voltage control of the direction and speed of sample electrophoretic migration in a grid array of perpendicular channels, each of which can be tailored to provide a particular mode of separation (Sec. 3).

3. MODES OF SEPARATION

A wide range of separation selectivity can be obtained by separating analytes not only on the basis of differences in their electrophoretic mobilities, but also on differences in their isoelectric point, molecular size, hydrophobic and hydrophilic nature, molecular structure, polarity, and stereochemistry. In this section, CE modes of separation are discussed that exploit these analyte properties. In addition, applications from the current scientific literature are used to illustrate the usefulness and limitations of the different CE separation techniques. It will be seen that CIEF, CGE, and CITP are capillary analogs of slab gel electrophoresis, while CZE, CIE, MECC, and CEC are uniquely capillary techniques.

3.1 Capillary Zone Electrophoresis (CZE)

Capillary zone electrophoresis (CZE), also known as free-solution capillary electrophoresis (FSCE), separates analytes primarily on the basis of differences in their electrophoretic migration rates. When CZE is performed in untreated capillaries, migration is related to the sum of the electrophoretic and electroosmotic mobilities. In treated (*i.e.*, wall-coated) capillaries, migration is due solely to the electrophoretic mobility of the analytes. The capillaries are not packed with a gel or a sta-

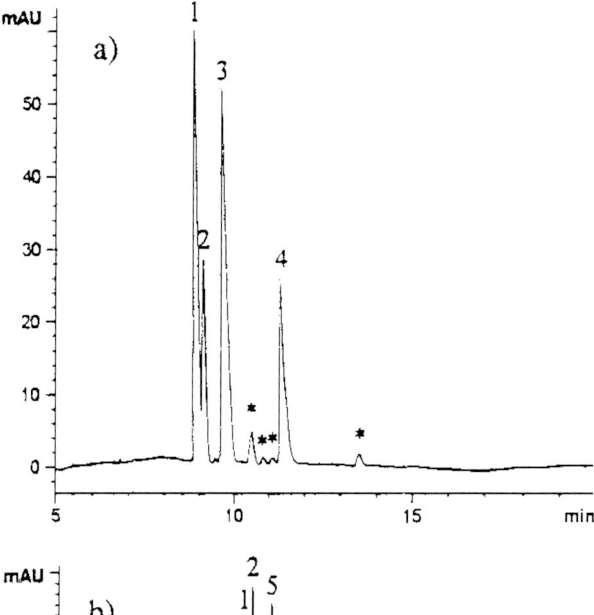

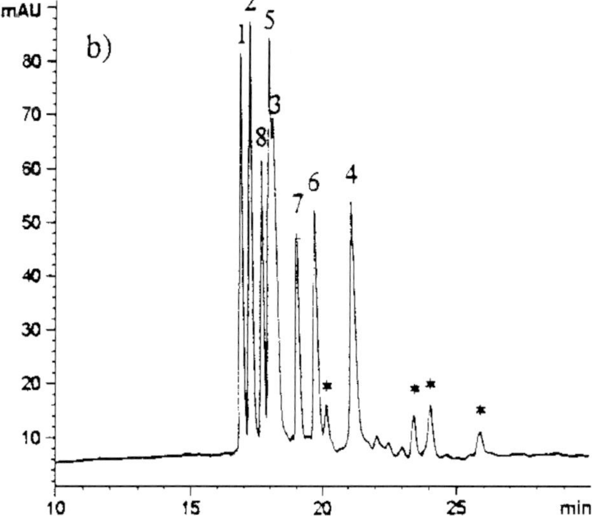

FIG. 4. CZE separation of all eight stereoisomers of Tyr-Lys-Trp (a) without and (b) with 18-crown-6 tetracarboxylic acid added to the buffer as a chiral selector. Sample introduction: hydrodynamic (pressure) injection, 50 mBar, 2 s. Capillary: fused silica, untreated, 50 cm long × 75 μm i.d., 30 °C, 30 kV applied voltage. Buffer: *p*H 2 Tris-borate buffer, with or without 0.01 *M* chiral selector. Detection: UV–vis absorbance at 210 nm. Elution order of analytes in (a): 1, LLL/DDD; 2, LDL/DLD; 3, LLD/DDL; 4, DLL/LDD. Elution order of analytes in (b): 1, LLL; 2, LLD; 3, LDL; 4, DLL; 5, DDL; 6, DLD; 7, LDD; 8, DDD. Asterisks indicate peaks due to side products of the crude peptides. Reprinted (in part) from Riester *et al.* (1996). Copyright (1996) American Chemical Society.

tionary phase material, and the same buffer is used in the capillary and the inlet, outlet, and sample vials. This mode of CE has found particularly good applications in the determination of amino acids, peptides, proteins, nucleosides, nucleotides, and vitamins.

Chiral compounds can be separated by addition of an appropriate modifier (*e.g.*, cyclodextrin) to the buffer. These compounds are molecules that are not superimposable with their mirror images; *e.g.*, in organic chemistry, they are molecules that contain a carbon atom that is covalently bonded to four different functional groups. They are also known as stereoisomers, optical isomers, optically active isomers, diastereomers, and enantiometers. In many pharmaceutical, agrochemical, and biological applications, where optical purity can influence effectiveness, stereoisomers can be difficult to separate in HPLC. The ability of CZE to separate stereoisomers is demonstrated in Fig. 4, showing separation of the eight stereoisomers produced in the solid-phase synthesis of the (target) tripeptide D-Tyr-L-Lys-L-Trp (Riester *et al.*, 1996). The prefixes D and L refer to the relative configuration of the atoms on the enantiomeric forms of each peptide group. In solid-phase peptide synthesis the polypeptide is produced stepwise from individual amino acids in an automated reactor. Side reactions, deletion errors due to an incomplete coupling cycle, and isomerization can all lead to undesired stereoisomers. The eight different tripeptide stereoisomers are separated into four pairs that each elute together in the absence of a chiral selector [Fig. 4(a)], but are completely resolved when a chiral selector is added as a buffer modifier [Fig. 4(b)]. This CZE method was used to investigate the rate of racematization (a reaction whereby a pure L or D form of a stereoisomer is converted into a mixture of equal parts of L and D forms). This work demonstrated the ability of CZE to measure the optical purity of the derivatized amino acid starting materials used in solid-phase peptide synthesis and was the first separation of all stereoisomers of a three–chiral-center tripeptide by CZE.

Figure 5 shows the electropherogram for the separation of 5-(dimethylamino)naphthalene-1-sulfonic acid (Dns-OH) and four L-amino acids using a capillary electropherograph microfabricated on quartz and Pyrex glass substrates (Fluri *et al.*, 1996). The analytes were derivatized with *o*-phthaldialdehyde in a microfabricated post-column reactor prior to fluorescence detection. The remarkable speed and efficiency of separations performed on integrated CE microdevices is demonstrated by the fact that all five analytes elute within 20 s, with N ranging between 23 000 and 94 000 theoretical plates based on analysis of the peaks (Sec. 1.2).

The speed and efficiency of CZE illustrated in the determination of small polypeptides and amino acids has also been applied in the determination of proteins. For large molecules such as proteins, the secondary and tertiary structures of the molecule have an effect on the net charge and its size or hydrodynamic radius. The effect of these parameters on the electrophoretic mobility of some proteins has been studied (Ma *et al.*, 1995). Protein modifications such as degradation, glycosylation, deamidation, or sequence changes will result in altered CZE migration times. These effects have allowed CZE to be successfully applied to confirmation of protein identities and detection of purity, heterogeneity and post-translational modifications in both native and synthetic proteins (see Sec. 3.2). Alternatively, protein sequence can be determined by peptide mapping, as reviewed for recombinant proteins by Ganzler *et al.* (1995). Peptide mapping involves enzymatically breaking a protein into smaller polypeptides. Analysis by CZE provides a map or fingerprint of the constituent polypeptides. Different enzymes are used to produce a variety of maps, and each map can be used for comparison or as a quality check for recombinant methods. Further, using a mass spectrometric detector, molecular weight can be measured and in many cases the sequence of each individual polypeptide can be determined.

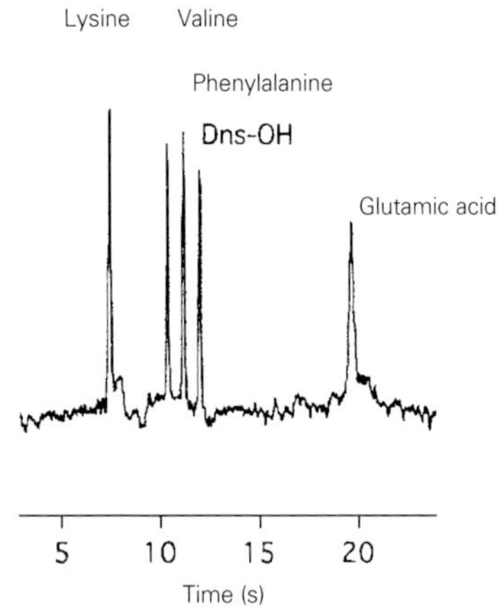

FIG. 5. CZE separation of a mixture of of 5-(dimethylamino)naphthalene-1-sulfonic acid (Dns-OH) and four L-amino acids using a capillary electropherograph microfabricated on quartz and Pyrex glass substrates. Sample introduction: electrokinetic injection, −1.8 kV, 5 s. Separation channel: Pyrex glass, 4.5 cm long × 45 μm wide × 15 μm deep, −11 kV. Buffer: 0.027 *M* sodium hydrogen carbonate and 0.012 *M* sodium carbonate, adjusted to *p*H 9.7. Detection: LIF following post-column reaction with 0.004 *M* *o*-phthaldialdehyde, He/Cd excitation (9.3 mW, 325 nm), photomultiplier tube detection. Reprinted (in part) from Fluri *et al.* (1996). Copyright (1996) American Chemical Society.

Capillary ion electrophoresis (CIE), also sometimes referred to as capillary ion analysis (CIA), is a special name given to CZE separations that are optimized for the separation of small ionic species. UV–vis absorbance (Sec. 2.5.1) is the most common detection mode used in CIE; however, since many of these ions do not absorb UV–vis radiation, indirect de-

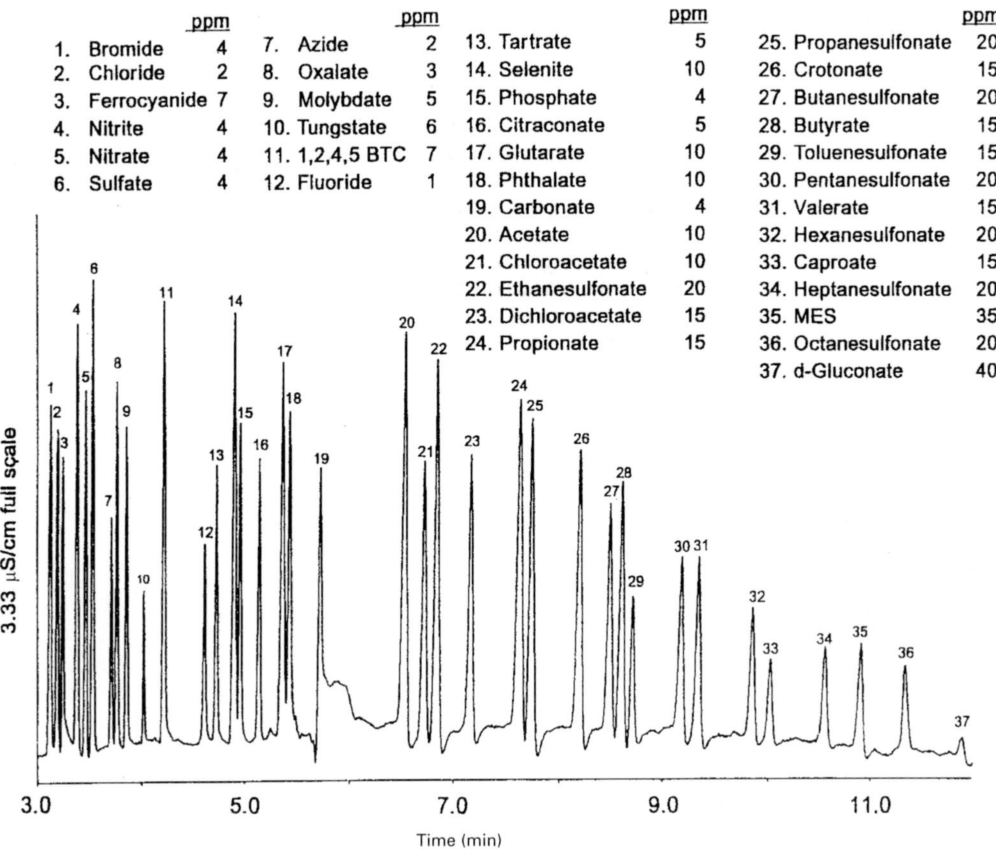

FIG. 6. CIE separation of 37 anions. Sample introduction: hydrodynamic (pressure) injection, 25 mbar, 12 s. Capillary: fused silica, 60 cm long × 50 μm i.d., room temperature, 30 kV applied voltage. Buffer: 0.05 *M* 2-{*N*-cyclohexylamino}-ethanesulfonic acid, 0.02 *M* lithium hydroxide, 0.03% Triton® X-100 surfactant (Rohm & Haas, Philadelphia, PA, U.S.A.). Detection: nonsuppressed conductivity, with detector response shown as 33 μS cm^{-1}. Cetyltrimethylammonium bromide (CTAB), 0.001 *M*, was flushed through the capillary prior to analysis. Peak assignments for the 37 anions are shown along with the concentrations of the individual anions in the mixture in parts per million (ppm). MES represents 2-[*N*-morpholino]ethanesulfonic acid. Reprinted (in part) from Jones *et al.* (1996) with permission from *American Laboratory.* Copyright (1996) International Scientific Communications, Inc.

tection schemes are typically used. Conductivity detectors, both suppressed and nonsuppressed, are also finding increasing use in CIE. When cations and anions are both present in the sample mixture, untreated fused-silica capillaries are used. However, when the sample contains only anions, wall-coated capillaries can be used to decrease the migration time for the analyte anions.

In Fig. 6, the CIE separation of 37 anions is shown (Jones *et al.*, 1996). To decrease the electrophoretic migration time of the anions, the direction of electroosmotic flow was reversed by treating the fused-silica capillary with CTAB prior to analysis. The elution order is determined by the electrophoretic mobility and, therefore, the charge/size ratio of the anions. As an example, consider the order of elution for the Group VIIA anions in the sample (bromide, chloride and fluoride). These anions are singly charged and their hydrated ionic radii increase in the order bromide < chloride < fluoride. Therefore, bromide elutes first, followed in order by chloride and fluoride. The excellent speed and resolution of capillary electrophoresis is also demonstrated by this example, as nearly all of the peaks are separated with baseline resolution in a separation that takes less than 12 min and uses only 12 nL of sample solution.

3.2 Capillary Isoelectric Focusing (CIEF)

In capillary isoelectric focusing (CIEF), analytes are separated on the basis of differences in their isoelectric points. As opposed to the other CE zonal modes of separation, CIEF is, as the name implies, a focusing technique. Further discussion of CIEF requires some definitions of terms. An ampholyte (or amphotere) is a compound that can exist as either a cation or an anion depending on the buffer *p*H. Ampholytes can take many forms, but the type most commonly used in CIEF comprises organic compounds with both amino (basic) and carboxylic acid functional groups. In its basic form, an ampholyte is an anion, while in its acidic form, it exists as a cation. The buffer *p*H at which the ampholyte is neutral is the isoelectric point, *p*I.

CIEF is performed by a three-step process on capillaries that have been treated to eliminate EOF. First, the sample is injected into the capillary along with a mixture of ampholytes. This is typically done by hydrodynamic injection of relatively large sample volumes, since the sample can fill the entire capillary. Second, a focusing step is performed with high electric fields (typically 400–700 V cm^{-1}) across the length of the capillary. The inlet (anode) vial contains a dilute, acidic anolyte (*e.g.*, phosphoric acid), and the outlet (cathode) vial contains a dilute, basic catholyte (*e.g.*, sodium hydroxide). Focusing is performed until the current drops to a steady value, and usually takes 3–5 min. During this step, a *p*H gradient is established along the length of the capillary, and sample analytes migrate to their individual isoelectric points in the capillary and are focused into sharp zones. These zones are the narrowest when there are a large number of ampholytes with evenly spaced isoelectric points introduced during the first step. In the third and final step, the analyte zones are mobilized so that they move to the CE detector. Mobilization can be achieved by an electrophoretic process, whereby the high voltage is first turned off, the compositions of the anolyte and/or the catholyte are changed (*e.g.*, by addition of a salt), and the voltage is then turned back on. When the voltage is turned on, a gradual shift in the *p*H gradient in the capillary occurs, the analytes acquire electric charges, and the analytes then move towards the detector as a result of electrophoresis. Adding a salt containing a cation other than the hydrogen ion to the anolyte will produce mobilization towards the anode. Conversely, adding a salt containing an anion other than the hydroxyl ion to the catholyte will produce mobilization to the cathode. Mobilization can also be performed by hydrodynamic means. Here, hydrostatic pressure caused by pressurization of the catholyte or anolyte vial, or by changing the level of either the catholyte or anolyte vial, causes buffer flow through the capillary.

CIEF is used for a variety of separation problems, such as proteins, monoclonal antibodies, and hemoglobins. In Fig. 7, the CIEF separation of recombinant human growth hormone (rhGH) variants is shown (Shimura and Karger, 1994). The analytes are detected as a complex formed by post-column reaction with a reagent (TR-Fab') synthesized by the reaction of a purified mouse monoclonal antibody with tetramethylrhodamine-5-iodoacetamide. CIEF was a particularly good mode of separation for this analysis on account of the fairly large sample volume that could be loaded into the capillary. In CIEF, the order of elution is related to the *p*I's of the individual analytes. With appropriate *p*I standards, a calibration curve of *p*I versus mobilization time can be used to determine the *p*I of an unknown analyte with an absolute accuracy of 0.05 pI units.

3.3 Capillary Gel Electrophoresis (CGE)

In capillary gel electrophoresis (CGE), separation is enhanced by size exclusion (*i.e.*, sieving) in polymeric gels containing a network of pores with narrow, well-defined size distributions. Small analytes permeate through the pores in the polymer and appear early in the electropherogram. Large analytes are slowed down by the gel; therefore, they have longer migration times and appear later in the electropherogram. This elution order is the

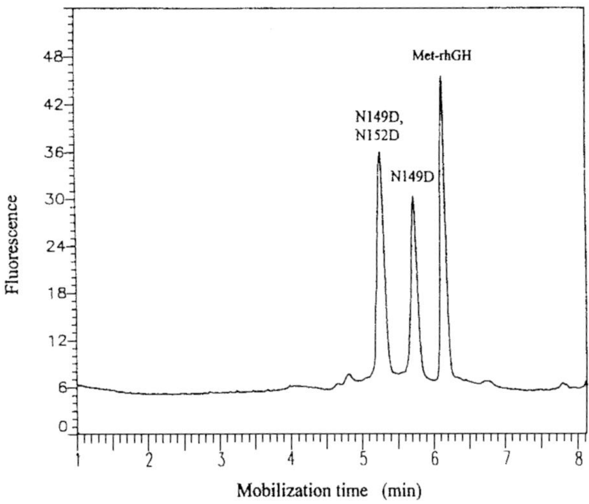

FIG. 7. CIEF separation of a mixture of recombinant human growth hormone (rhGH) variants. Mixture contains met-rhGH, the monodeamidated variant (N149D), and the dideamidated variant (N149D, N152D), each at a concentration of 10 μg L^{-1}. Sample introduction: hydrodynamic (vacuum) injection, 10 μL syringe. Capillary: polyacrylamide-coated fused silica, 15 cm long × 75 μm i.d. Ampholyte: 6% Pharmalyte 3–10 (Pharmacia LKB Biotechnology, Piscataway, NJ, U.S.A.). Catholyte: 0.02 *M* sodium hydroxide. Anolyte: 0.01 *M* phosphoric acid. Focusing: 1 min at 200 V cm^{-1} and 4 min at 400 V cm^{-1}. Mobilization: change anolyte to 0.01 *M* phosphoric acid containing 0.1 *M* sodium chloride, 400 V cm^{-1}. Detection: LIF at 580 nm following post-column reaction with TR-Fab' complexing reagent, Ar^{+} excitation (2 mW, 488 nm). Reprinted (in part) from Shimura and Karger (1994). Copyright (1994) American Chemical Society.

opposite of that in size-exclusion chromatography (SEC; see CHROMATOGRAPHY), where molecules that are too large to permeate the pores in the stationary-phase particles flow around the particles and appear first in the chromatogram, while small molecules that permeate the pores travel a long and tortuous path through the particles and elute last. In CGE, capillary walls are typically treated to enhance gel adhesion and stability. EOF is therefore eliminated and the analytes migrate through the gel as a result of their electrophoretic mobilities, which can add a second degree of separation selectivity to that based on size exclusion.

Typical gels in CGE include polyacrylamide (PA), agarose, and poly (ethyleneoxide), with PA being the most common. CGE with PA gels is frequently referred to as capillary polyacrylamide gel electrophoresis (C-PAGE). The gel is usually either a rigid, cross-linked polymer that is attached to the capillary wall or a linear polymer solution that is not cross-linked and is not fixed to the capillary. CGE capillaries with cross-linked gels occasionally have problems with gel stability (which can limit their usefulness to 100–200 analyses) and with bubble formation during electrophoresis (which can be addressed by careful control of experimental parameters). Also, electrokinetic injection is typically used with these types of gels because hydrodynamic injection is not usually capable of introducing the sample into the gel-filled capillary. In contrast, when the gel is a polymer solution composed of non-cross-linked polymers (*e.g.*, a linear polyacrylamide gel), either hydrodynamic or electrokinetic injection can be used. Also, capillary lifetimes are much longer than for capillaries with cross-linked polymers, as it is possible to introduce a new polymer solution into a capillary by a simple pressure loading technique prior to a given set of CGE experiments. With this mode of CGE, sometimes referred to as dynamic sieving capillary electrophoresis (DSCE), the capillary wall is treated to prevent EOF transport of the sieving gel out of the capillary. The polymer type, chain length, and concentration are all selected to give the desired size-exclusion properties for the separation problem at hand.

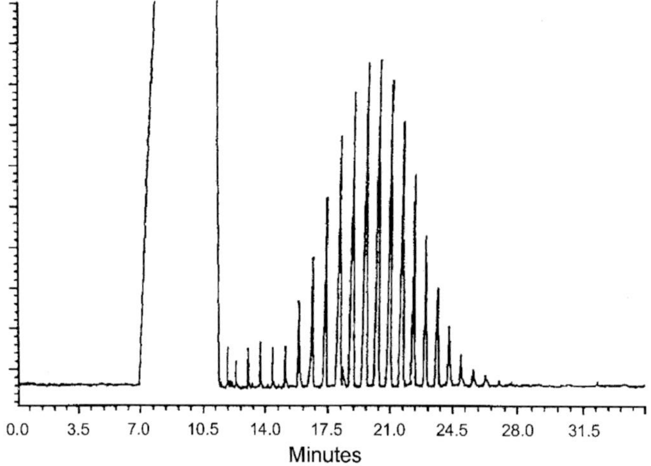

FIG. 8. CGE separation of phthalate-derivatized poly (ethylene glycol) oligomers. Sample introduction: electrokinetic injection, −10 kV, 60 s, sample concentration 1300 mg L^{-1}. Capillary: fused silica, 45 cm long× 75 μm i.d., filled with cross-linked polyacrylamide gel, −10.9 kV applied voltage. Buffer: Tris–borate, pH 8.3. Detection: UV–vis absorbance at 275 nm. Reprinted (in part) from Wallingford (1996). Copyright (1996) American Chemical Society.

In the CGE electropherogram shown in Fig. 8, the separation of phthalate-derivatized poly(ethylene glycol) oligomers is shown (Wallingford, 1996). An oligomeric set is a collection of polymers, each composed of one more monomer unit than the previous polymer. It is important to analyze these because the distribution of polymers within the set and the average molecular weight of the set determine key physical properties of the polymer. The oligomers elute in order of increasing size, and a smooth curve drawn so as to connect the individual oligomeric peak maxima can be related to the molecular-weight distribution (polydispersity) of the oligomers. With appropriate molecular-weight standards, a calibration curve of log(molecular weight) *versus* migration time can be used to estimate the molecular weight of an unknown polymer by CGE analysis. The benefits of CGE in the example shown in Fig. 8 were higher-accuracy measurements relative to gel permeation chromatography (GPC) and higher separation efficiencies (100 000–300 000 theoretical plates) than is possible with HPLC. Differences in the average weight and polydispersity between two vendors of the "same" polymer were also observed by the author.

Gel-filled capillaries were introduced by Cohen and Karger (1987), who first used them for the separation of polynucleotides by CGE. Today, CGE is widely used for many important separations in biochemistry, including nucleic acids, oligonucleotides, PCR products, DNA molecules (with as many as 5000 base pairs), DNA sequencing reaction products, and proteins. Native proteins can be separated by CGE as uncomplexed species, but molecular size determination becomes difficult because of mixed electrophoretic and size-exclusion contributions to the net mobilities of the proteins. In one very popular CGE method, proteins are treated with sodium dodecyl sulfate (SDS). The SDS denatures and interacts with each protein to produce adducts with virtually the same charge/size ratio, yet the size differences are retained. Similarly, some oligonucleotides and DNA fragments differ from one another in size but not in their charge/size ratios. None of these samples can be resolved by CZE (since they all have the same charge/size ratio), but the size discrimination properties of CGE allow separation with high resolution. CGE can also therefore be used to determine molecular weights.

The application of CGE to DNA sequencing analyses, first demonstrated by Swerdlow and Gesteland (1990), has provided significant advances in terms of speed and resolution compared with traditional slab gel electrophoresis. Standard DNA sequencing methodologies are used to produce single strands of DNA that terminate after each nucleotide type (deoxyadenosine, deoxycytidine, deoxyguanosine, and deoxythymidine, or A, C, G, T). For a DNA strand that contains G at base

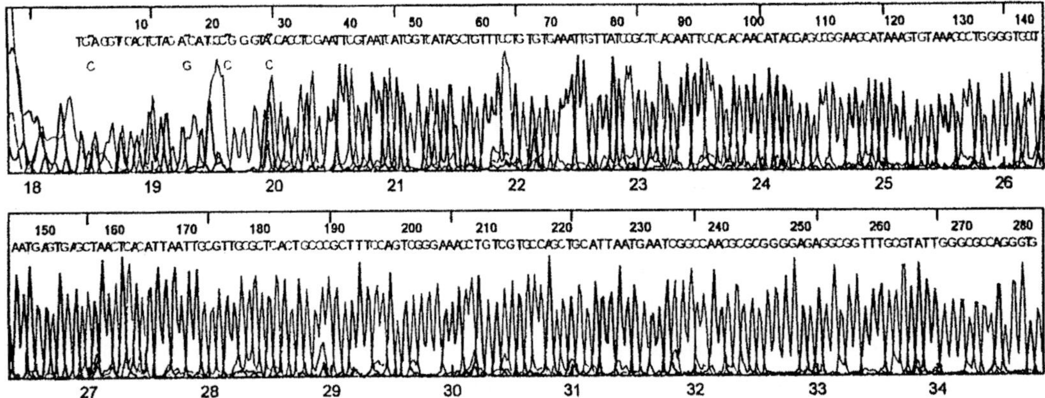

FIG. 9. Capillary gel electrophoresis of DNA sequencing products: partial electropherograph (fluorescence intensity *versus* time) showing separation of products of length 1 through 282 bases. Upper numbers show length (in bases), letters indicate terminal base. Sample introduction: electrokinetic injection, $-200\,\mathrm{V\,cm^{-1}}$, 5 s. Capillary: fused silica, poly (vinyl alcohol) coated, filled with 2% T (T = total monomer concentration) linear polyacrylamide, 30 cm long × 75 μm i.d., $-150\,\mathrm{V\,cm^{-1}}$ applied voltage. Buffer: 100 m*M* Tris-50 m*M* TAPS-2 m*M* EDTA-7 *M* urea. Detection: LIF, argon-ion laser (5 m*W*, 514 nm) excitation, photodiode array detection. Reprinted (in part) from Carrilho *et al.* (1996). Copyright (1996) American Chemical Society.

number 54 and 87, for example, two individual strands (sequencing products) of lengths 54 and 87 are produced with a terminal G. Ordinarily, the four groups of sequencing products would be labeled with radioactive isotopes and determined individually in a single slab gel experiment, with the analysis alone taking many hours. With CGE, however, each group of sequencing products is labeled with a different fluorescent tag and the groups are combined and analyzed by CGE with a significantly shorter analysis time.

An example of this type of analysis, using a replaceable linear polyacrylamide, low voltages, and laser-induced fluorescence detection, is shown in Fig. 9 (Carrilho *et al.*, 1996). By identifying the terminal base on each sequenced product, the sequence of the entire polynucleotide was determined. The experiment shown in Fig. 9 was used to sequence a polynucleotide (ssM13mp18 template) with greater than 1000 bases (only a portion is shown) in less than 90 min, whereas comparable gel-slab analysis would take from 8 to 18 h. The ability to sequence large polynucleotides by CE provides significant improvements relative to slab gel electrophoresis. CE is faster, uses smaller reagent and sample quantities, generates less waste, and is easier to automate.

3.4 Micellar Electrokinetic Capillary Chromatography (MECC)

As the name implies, micellar electrokinetic capillary chromatography (MECC) takes advantage of both electrophoretic and chromatographic separation mechanisms. The detector output is therefore called an electrokinetic chromatogram. MECC can separate charged and neutral analytes, but its most important application is for separation of neutral species, which cannot be separated by CZE. MECC is performed in untreated capillaries and analytes are partially solubilized by micelles added to the buffer, *i.e.*, they partition between the micellar and buffer phases. The degree to which each analyte partitions between the micelles and the buffer is the basis for separation. Electroosmotic mobility and micellar electrophoretic mobility are typically controlled so that the time an analyte spends in the micelle phase is directly related to its migration time. The micelles are analogous to the stationary phase in reversed-phase LC (see CHROMATOGRAPHY) and are referred to as a "pseudophase" because they migrate through the capillary. The use of a micellar phase to enhance electrophoretic separation was first demonstrated by Terabe *et al.* (1984). Drugs, vitamins, and

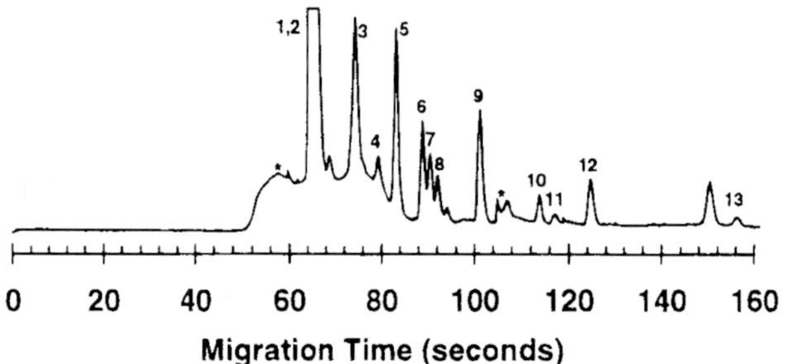

FIG. 10. MECC analysis of fifteen amino acids, allowing quantitation of ten. Sample introduction: electrokinetic injection, −100 V, 2 s. Capillary: fused silica, untreated, 10 cm long × 25 μm i.d., −12 kV applied voltage. Buffer: 0.175 *M* 2-(*N*-cyclohexylamino) ethanesulfonic acid adjusted to pH 9; containing 0.10 *M* sodium dodecyl sulfate (SDS) as the micellar additive. Detection: LIF following pre-column on-line reaction with 0.110 *M* *o*-phthaldialdehyde and 0.220 *M* ß-mercaptoethanol in 0.025 *M* borate buffer at pH 9.5, He/Cd excitation (2 mW, 354 nm), photomultiplier tube detection. Elution order of analytes: 1 and 2, glutamine and serine and threonine; 3, alanine and glycine; 4, tyrosine; 5, taurine; 6, valine; 7, glutamate; 8, methionine; 9, aspartate; 10, leucine; 11, phenylalanine; 12, isoleucine; 13, lysine. Asterisks indicate peaks due to impurities and/or reagents. Reprinted (in part) from Lada and Kennedy (1996). Copyright (1996) American Chemical Society.

large peptides are among the compounds commonly determined by MECC.

Micelles are created in the buffer solution by the addition of surfactant molecules. These straight-chain molecules have a hydrophilic end and a hydrophobic end, and, above a critical micelle concentration (typically 10^{-3} *M*), they aggregate to form a micelle. The number of molecules that form a micelle is called the aggregation number and can vary from 2 to over 100. Different surfactants will have different critical concentrations and aggregation numbers. Surfactant molecules can be nonionic, anionic, cationic, or zwitterionic (*e.g.*, having both cationic and anionic functional groups). SDS, the most common surfactant used in MECC, is an anionic molecule that forms micelles with an aggregation number of 162, an average charge of −63, and an average molecular weight of 18 000. In aqueous buffers, SDS molecules orient their hydrophobic ends inward and hydrophilic ends outward. Surfactant concentration and structure determine the degree of partitioning for each analyte and therefore affect resolution and migration time. The charge/size ratio determines micelle electrophoretic mobility, and buffer *p*H control is used to keep the surfactant molecules in the correct ionic state. Hydrophilic analytes that remain in the buffer will elute first, while hydrophobic analytes remaining in the micelle will elute last. Surfactant molecules are chosen for their hydrophilic/hydrophobic balance, solubility, purity, and cost. The buffer is then chosen so that the electrophoretic mobility of the micelles is opposite to, and of lower magnitude than, the electroosmotic flow. In this manner the micelles are attracted to the inlet vial (based on electrophoretic mobility) but ultimately migrate to the detector. Finally, it should be noted that separation of chiral compounds can be achieved by the use of a chiral surfactant (such as deoxycholate or taurodeoxycholate).

Separations can be enhanced in MECC by the use of buffer modifiers. Molecules added to the buffer can change the charge of the micelle, change the solubility of the analytes in the buffer, or act as a secondary pseudophase into which the analytes partition. These can all affect analyte solubility in the micelle and therefore MECC separation characteristics. Common modifiers include organics such as methanol and isopropanol, and metal ions, which attach to the surface of the micelle. Cyclodextrins are also used as modifiers, and analytes can partition into cyclodextrin molecules on the basis of the relative sizes of the analyte and the cyclodextrin cavity and on hydrophobic/hydrophilic interactions.

Figure 10 shows an electrokinetic chromatogram of a mixture of amino acids detected in a microdialysis fluid (Lada and Kennedy, 1996). Continuous dialysate flow was mixed on line with a derivatization reagent and introduced into a flowing buffer solution (orthogonal to the capillary inlet). The buffer flow was paused periodically to allow electrokinetic injection into the capillary. Most of the derivatized analytes were unresolved under CZE conditions. In contrast, MECC using SDS micelles allowed the separation and quantitation of ten amino acids, including leucine and isoleucine. The rapid analysis time allowed for greater temporal resolution during *in vivo* monitoring. Compounds were identified by matching migration times to known standards, while unidentified peaks and the broad peak early in the electropherogram were products of side reactions or minor products detected as a result of *in vivo* conditions. The rapid analysis and high number of theoretical plates for the MECC method permitted a fast sampling rate and simultaneous quantitative monitoring of ten analytes.

3.5 Capillary Isotachophoresis (CITP)

Capillary isotachophoresis (CITP) can be used either as a separation method or as a sample preconcentration method prior to another separation mode. In CITP, the sample solution is sandwiched between two buffers, a leading buffer and a trailing (or terminating) buffer. The leading buffer contains an ion with a higher electrophoretic mobility than any analyte ions in the sample. Similarly, the trailing buffer contains an ion with a lower mobility than all those in the sample. A constant current is passed through the capillary from the high-voltage power supply, and the analyte ions segregate into individual, discrete zones. Analyte ions having the highest mobilities form a zone at the boundary near the leading buffer, whereas those with the lowest mobilities form a zone nearest the trailing buffer. In this way the analytes are "stacked" according to their mobilities, with the length of each zone determined by the total amount of each analyte present. An analyte present in a large amount in the sample, for example, will form a wide zone, while a trace component will form a narrow zone. In addition, the concentration of each analyte in its particular zone is proportional (based upon mobility factor) to the concentration of the leading buffer. Thus, CITP can be used to calibrate many analytes with only a single component concentration standard to determine the relationship between zone length and the amount of analyte.

Once the analyte zones come to equilibrium, both buffers and all analyte zones begin to migrate together to the detector, and they do so at the migration velocity of the leading buffer. The zones do not separate as in other CE modes, but remain adjacent to one another. The result is that a CITP electropherogram has an unusual step-function appearance (each step a different analyte zone) rather than the series of Gaussian-like peaks observed in the electropherograms for other CE modes. CITP is usually carried out in coated capillaries to eliminate electroosmotic flow; thus, CITP is typically used to determine anions and cations. Migration time cannot be used for analyte identification in CITP because it is a function of the number of analyte ions in the sample.

The ability of CITP to focus a trace component into a narrow, concentrated zone at a concentration level near that of the leading buffer makes it an ideal candidate for analysis of dilute samples, since they are, in effect, preconcentrated by this method. Once the analytes have been focused, additional leading buffer is loaded behind the terminating buffer to enable electrophoretic migration and analyte separation to occur. The time at which electrophoresis is initiated is important because the analytes begin to migrate after focusing occurs, and enough capillary length must remain for adequate separation. This method of focusing followed by normal electrophoresis is termed transient CITP. Focusing the analytes in this way allows very large sample volumes ($>$100-fold larger than other methods, filling 20–50% of the capillary volume) to be injected without loss of electrophoretic performance.

The use of transient CITP to focus and separate analytes is illustrated in Fig. 11, where the separation of various length oligomers of poly(deoxyadenylic acid) is demonstrated under varying CITP focusing conditions (Auriola *et al.*, 1996). The decrease in migration time and resolution observed in Fig. 11(d) occurred because the focusing voltage was applied beyond the point in time at which the analytes were focused. As a result, the focused analytes migrated through the capillary prior to sepa-

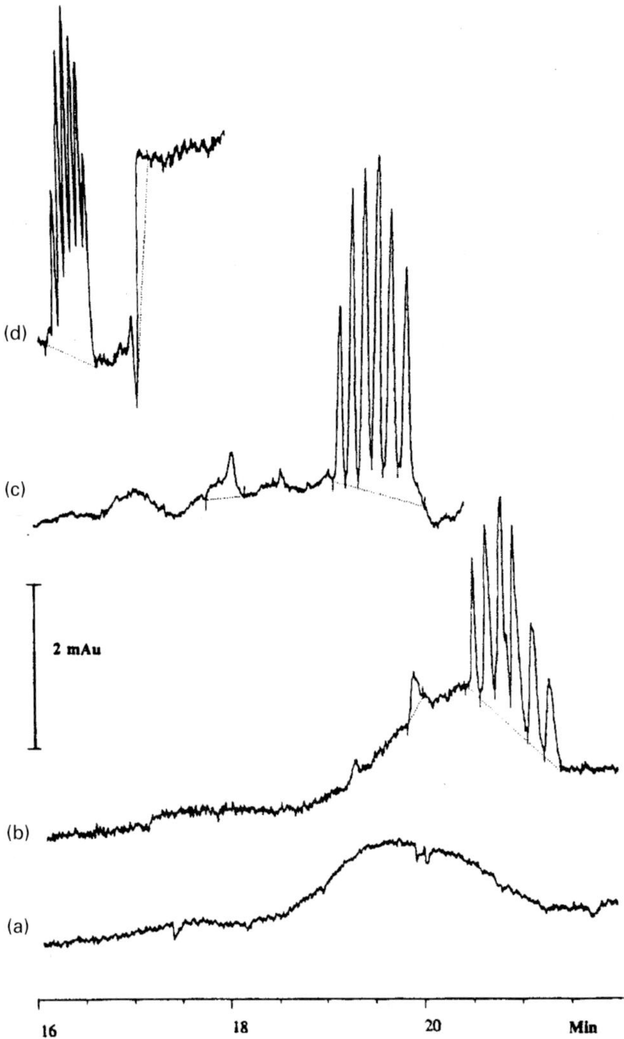

FIG. 11. Separation of poly(deoxyadenylic acid) pd(A)$_{19-24}$ oligomers with lengths of 19 through 24 units, using transient CITP. The effects of the following focusing times and voltages are shown: (a), no focusing; (b), 0.25 min; −15 kV; (c), 0.5 min, −30 kV; (d), 1.5 min, −30 kV. Sample introduction: hydrodynamic (pressure) injection, 48 s (10% of column length). Capillary: fused silica, untreated, 60 cm long ×50 μm i.d., 23 °C, −30 kV applied voltage for electrophoresis. Buffer: leading buffer 0.100 *M* ammonium formate, *p*H 4.5, containing 12% polyethylene glycol (average molecular weight 20 000) as a molecular sieving agent and to reduce EOF; terminating buffer 2-(*N*-morpholino)ethanesulfonic acid (MES) adjusted to *p*H 4.5. Detection: UV–vis absorbance at 254 nm. Elution order of analytes: pd(A)$_{19}$ through pd(A)$_{24}$ in order of increasing molecular weight. Reprinted (in part) from Auriola *et al.* (1996). Copyright (1996) American Chemical Society.

ration. This reduced the effective electrophoretic capillary length, adversely affecting electrophoretic efficiency.

3.6 Capillary Electrochromatography (CEC)

Capillary electrochromatography (CEC) is another separation method that utilizes both electrophoretic and chromatographic separation mechanisms. Similar to LC, separation in CEC is provided by the partitioning of analyte molecules between a mobile phase and a stationary phase, and it is well suited to neutral compounds. CEC uses electroosmotic flow rather than pressure to drive the mobile phase through the column. This results in better resolution than LC because it eliminates zone broadening due to parabolic flow profiles. Further, the small stationary-phase particles (1.5 μm diameter) required in LC to achieve resolutions approaching those obtained in CE require extraordinarily high pumping pressures, which are difficult to obtain in practice. Currently, CEC is only slightly (5–10 times) superior to traditional pumped-flow LC in terms of separation efficiency (N), and the expected narrow peaks are not observed on a routine basis. This has been theorized (Wan, 1997) to result from the effects of overlapping electrical double layers (on neighboring silica particles), which disturb the EOF profile. The magnitude of this perturbation decreases with increasing interparticle porosity, increasing particle diameter, and increasing buffer concentration.

In CEC, stationary phases consist of untreated or surface-modified silica particles that are typically 1.5 µm (or greater) in diameter. Untreated silica particles can contain ionizable surface groups that contribute to generation of EOF, while surface-modified particles reduce EOF in interparticle regions. The most common stationary phases are those based on siloxanes (see CHROMATOGRAPHY), including particles that have been treated with long-chain–hydrocarbon functional groups to form $Si{-}O{-}Si(CH_3)_2{-}(CH_2)_{17}{-}CH_3$. These octadecylsilica particles, also known as C_{18} particles, are the most common stationary phase in CEC, although other particles where the C_{18} hydrocarbon group has been replaced with a more polar functional group (*e.g.*, phenyl, cyano, amino) can be used in the separation of more polar molecules. Particles can be loaded into the capillary as a dry powder or as a slurry, using either high-pressure liquid, high-pressure gas, electroosmosis, or electrophoresis. The stationary phases are retained in the capillaries by column-end frits of various designs. Common problems encountered with frits include bubble formation and zone broadening. Some research has been directed toward creating packing material by *in situ* polymerization, and such stationary phases do not require column-end frits. Pressurization of the capillary at the inlet and outlet is sometimes required to prevent solvent outgassing in CEC, which can perturb the EOF flow profile. Finally, electrokinetic injection is typically used in CEC because hydrodynamic injection is not usually capable of introducing the sample into the packed capillary.

Mobile phases commonly used in CEC are similar to those used in reversed-phase LC (see CHROMATOGRAPHY), such as acetonitrile–water mixtures; however, a buffer salt is added to support electroosmosis and electrophoresis. Solvent gradients can be created by pressure programming, or by using a T-shaped inlet capillary in combination with two programmable power supplies to combine two different mobile phases dynamically (Yan *et al.*, 1996).

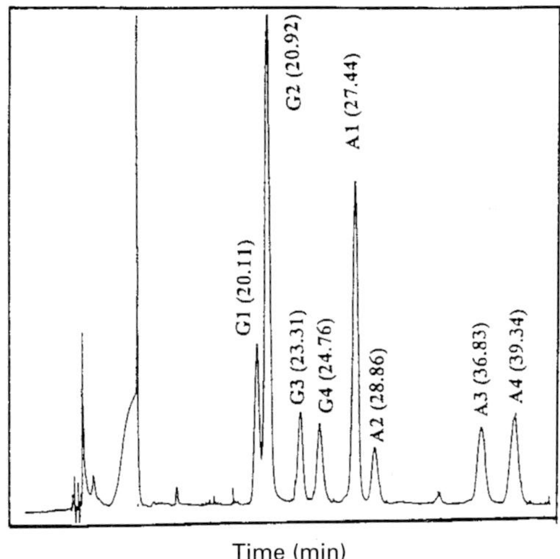

FIG. 12. CEC separation of eight diastereoisomeric deoxyguanosine and deoxyadenosine adducts of *anti*-benzo[g]chrysene. Sample introduction: electrokinetic injection. Capillary: fused silica, untreated, 40 cm long × 75 µm i.d., packed with 5 µm octadecylsilica particles, 15 kV applied voltage. Buffer: sample buffer 0.005 *M* ammonium acetate; chromatographic buffer 0.0085 *M* ammonium acetate in 29:71 acetonitrile:water. Detection: UV absorbance at 261 nm. Elution order of analytes: deoxyguanosine adduct diastereomers G1–G4 followed by deoxyadenosine adduct diastereomers A1–A4. Numbers in parentheses next to analyte peak labels are migration times in minutes. Reprinted (in part) from Ding and Vouros (1997). Copyright (1997) American Chemical Society.

CEC is often used to separate analytes that are difficult to resolve by LC, for neutral analytes not resolved by CZE, and for hydrophobic analytes not separated well by MECC. Here, CEC separations performed without surfactants (as are needed in MECC) make interfacing the capillary to some detectors, such as a mass spectrometer, less troublesome. Figure 12 shows the application of CEC to the separation of enantiomeric adducts of deoxynucleosides and anti-dihydrodiol epoxides of benzo[g]chrysene (Ding and Vouros, 1997). The relative amounts of the stereoisomers produced in the reaction between the deoxynucleoside and the epoxide provides information useful for mutagenic analysis. In this example, CEC separation results in nearly baseline resolution of all eight possible stereoisomers in the standard mixture.

ACKNOWLEDGMENTS

This work was supported by the U.S. Department of Energy under Contract DE-AC04-94AL85000. Sandia is a multiprogram laboratory operated by Sandia Corporation, a Lockheed Martin Company, for the United States Department of Energy.

GLOSSARY

Ampholyte, Amphotere: A compound that can exist as either a cation or an anion depending on buffer *p*H. In its basic form, an ampholyte is an anion, while in its acidic form, it exists as a cation.

Analyte: In capillary electrophoresis, the solute being determined.

Anode: In capillary electrophoresis, a positively biased electrode used in applying high electrical potentials across capillaries; typically immersed in the inlet buffer vial of a capillary elelctropherograph.

Anolyte: In capillary isoelectric focusing, buffer solution in the inlet (anode) vial of a capillary electropherograph.

Buffer: A solution whose composition is selected to minimize changes in *p*H that would otherwise occur as a result of chemical reactions. The *p*H is controlled by the relative concentrations of acidic and basic forms of ionizable solutes.

Capillary, Capillary Column: In capillary electrophoresis, the narrow-bore column of fused silica, Pyrex, or Teflon in which separation occurs.

Capillary Electrochromatography (CEC): Capillary electrophoresis mode whereby analytes are separated on the basis of differences in both electrophoretic migration rates and partitioning between the buffer and a chromatographic stationary phase.

Capillary Electropherograph: Instrumentation used in capillary electrophoresis.

Capillary Electrophoresis (CE): Technique for high-resolution separations of complex mixtures based on electrophoresis within small-bore capillaries.

Capillary Gel Electrophoresis (CGE): Capillary electrophoresis mode whereby analytes are separated on the basis of differences in both electrophoretic migration rates and molecular size.

Capillary Ion Analysis (CIA), Capillary Ion Electrophoresis (CIE): Capillary zone electrophoresis mode that is optimized for the analysis of small ionic species.

Capillary Isotachophoresis (CITP): Capillary electrophoresis mode used to separate analytes in dilute solutions on the basis of differences in their electrophoretic migration rates.

Capillary Isoelectric Focusing (CIEF): Capillary electrophoresis mode used to separate analytes on the basis of differences in isoelectric points.

Capillary Zone Electrophoresis (CZE): Capillary electrophoresis mode performed in open-tubular, uncoated capillaries whereby analytes are separated on the basis of differences in electrophoretic migration rates.

Cathode: A negatively biased electrode used in applying high electrical potentials across capillaries; typically immersed in the outlet buffer vial of a capillary electropherograph.

Catholyte: In capillary isoelectric focusing, the buffer solution in the outlet (cathode) vial of a capillary electropherograph.

Chromatography: Chemical separation and analysis technique based on differences in partitioning of an analyte between a solid or liquid stationary phase and a gas, liquid, or supercritical fluid mobile phase.

Dynamic Sieving Capillary Electrophoresis (DSCE): Capillary electrophoresis mode whereby analytes are separated on the basis of differences in both electrophoretic migration rates and molecular size.

Electroendosmosis, Electroendosmotic Flow, Electroosmotic Flow (EOF): Transport of neutral molecules or particles by a conducting solution under the influence of an electric field.

Electromigration: Movement of a charged species under the influence of an electric field. In electrophoresis, however, more generally used to describe the movement of all solutes, whether they be neutral or charged.

Electroosmotic Mobility: A measure of the velocity of a neutral species in an electric field of unit strength.

Electropherogram: In capillary electrophoresis, a plot of detector response as a function of migration time.

Electrophoresis: Migration, or movement, of electrically charged particles or solutes in a conductive liquid or gel under the influence of an applied electric field.

Electrophoretic Migration: See **Migration**.

Electrophoretic Mobility: A measure of the velocity of a charged species in an electric field of unit strength.

Elution: In capillary electrophoresis and chromatography, a term used to describe the process whereby the analyte exits from the separation column in a flowing medium.

Free-solution Capillary Electrophoresis (FSCE): See **Capillary Zone Electrophoresis**.

Gas Chromatography (GC): Chromatography with a gaseous mobile phase.

Gel: In slab gel electrophoresis, a two-phase colloidal system or a polymer used as a partitioning media. In capillary electrophoresis, a polymer containing a network of pores with a narrow, well-defined size distribution that is incorporated into the capillary either as a fixed, cross-linked polymer or as a non-cross-linked polymer solution.

Gel Electrophoresis: Separation and analysis technique, performed on gel slabs, where analytes are separated on the basis of differences in electrophoretic migration rates and then determined as separated zones on the slab.

Height Equivalent of a Theoretical Plate: In capillary electrophoresis and chromatography, a quantity used to describe the efficiency of a separation.

Ionic Strength: A term equal to one-half the sum of the products of concentration and squared charge for all the ionic species in solution. Denoted by the symbol I and expressed in terms of concentration units, *e.g.*, molarity.

Isoelectric Point: The pH of a solution in which an ampholyte or some other ionizable species is neutral and has zero electrophoretic mobility in an electric field. Denoted by the symbol pI.

Liquid Chromatography (LC): Chromatography with a liquid mobile phase. Also referred to as high-performance liquid chromatography.

Micellar Electrokinetic Capillary Chromatography (MECC): Capillary electrophoresis technique in which the analytes are partially solubilized by micelles added to the buffer, *i.e.*, they partition between the micellar and buffer phases. The degree to which each analyte partitions between the micelles and the buffer is the basis for separation.

Micelle: Aggregation of straight-chain molecules, each of which has both a hydrophilic end and a hydrophobic end. The aggregation occurs above a certain concentration, called the critical micelle concentration. Used in micellar electrokinetic capillary chromatography to add separation selectivity based on the degree of analyte partitioning between the buffer and micellar phases.

Migration: In electrophoresis, an abbreviated version of the term electromigration, used to describe the movement of both neutral and charged species under the influence of an electric field. Also referred to as electrophoretic migration.

Migration Time: In capillary electrophoresis, the length of time it takes an analyte to travel through the capillary.

Neutral Marker: In capillary electrophoresis, a molecule that is transported only by electroosmotic flow and can therefore be used to measure EOF mobility.

Number of Theoretical Plates: In capillary electrophoresis and chromatography, a measure of the efficiency of a particular separation scheme.

Resolution: In capillary electrophoresis and chromatography, a measure of the extent of separation achieved between two analytes.

Separation Factor: In capillary electrophoresis, the ratio of the electrophoretic mobilities of the two analytes. In chromatography, the ratio of the capacity factors (or partition coefficients) of the two analytes.

Size Exclusion Chromatography (SEC): Liquid chromatography technique that separates chemical compounds on the basis of their physical size.

Slab Gel Electrophoresis: See **Gel Electrophoresis**.

Solute: A species dissolved in a liquid.

Theoretical Plate: In capillary electrophoresis and chromatography, an imaginary unit in the capillary or column wherein one complete mechanistic separation process step occurs. For example, in micellar electrokinetic capillary chromatography, this would be a single equilibrium partitioning of the analyte between the micellar phase and the buffer phase.

Zeta Potential: In capillary electrophoresis, the potential difference across the shear plane between the layers of the electrical double layer formed at the capillary/buffer interface.

Zone: In capillary electrophoresis, a term used to describe the width and profile of an analyte band as it migrates through the capillary.

Zwitterion: A molecule containing both cationic and anionic functional groups.

Works Cited

Auriola, S., Jääskeläinen, I., Regina, M., Urtti, A. (1996), *Anal. Chem.* **68**, 3907–3911.

Carrilho, E., Ruiz-Martinez, M. C., Berka, J., Smirnov, I., Goetzinger, W., Miller, A. W., Brady, D., Karger, B. L. (1996), *Anal. Chem.* **68**, 3305–3313.

Chen, D. Y., Dovichi, N. J. (1994), *J. Chromatogr. B: Biomed. Appl.* **657**, 265–269.

Cohen, A. S., Karger, B. L. (1987), *J. Chromatogr.* **397**, 409–417.

Ding, J., Vouros, P. (1997), *Anal. Chem.* **69**, 379–384.

Ewing, A. G., Gavin, P. F. (1996), *J. Am. Chem. Soc.* **118**, 8932–8936.

Fan, Z., Harrison, D. J. (1991), *Anal. Chem.* **66**, 177–184.

Fluri, K., Fitzpatrick, G., Chiem, N., Harrison, D. J. (1996), *Anal. Chem.* **68**, 4285–4290.

Friedl, W., Reijenga, J. C., Kenndler, E. (1995), *J. Chromatogr. A.*, **709**, 163–170.

Ganzler, K., Warne, N., Hancock, W. S. (1995), in: P. G. Rhigetti (Ed.), *Capillary Electrophoresis in Analytical Biotechnology*, Boca Raton, FL: CRC Press.

Hardy, W. B. (1899), *J. Physiol.* **24**, 288.

Hjerten, S. (1967), *Chromatogr. Rev.* **9**, 122.

Hjerten, S. (1985), *J. Chromatogr.* **347**, 191–198.

Jones, W. R., Soglia, J., McGlynn, M., Haber, C., Reineck, J., Krstanovic, C. (1996), *Am. Lab.* **28** (5), 25–33.

Jorgenson, J. W., Lukacs, K. D. (1981), *Anal. Chem.* **53**, 1298–1302.

Jorgenson, J. W., Lukacs, K. D. (1983), *Science* **222**, 266–272.

Karger, B. L., Foret, F. (1993) in: N. A. Guzman (Ed.), *Capillary Electrophoresis Technology*, New York: Marcel Dekker, Inc., p. 17.

Lada, M. W., Kennedy, R. T. (1996), *Anal. Chem.* **68**, 2790–2797.

Lauer, H. H., McManigill, D., (1986), *Anal. Chem.* **58**, 166–170.

Liu, J. P., Hsieh, Y. Z., Wiesler, D., Novotny, M. (1991), *Anal. Chem.* **63**, 408–412.

Ma, S., Hodel, A., Fox, R. O., Horvath, C. (1995), *Electrophoresis* **16**, 595.

Manz, A., Verpoorte, E., Effenhauser, C. S., Burggraf, N., Raymond, D. E., Harrison, D. J., Widmer, H. M. (1993), *J. High Resolut. Chromatogr.* **16**, 433–436.

Michaelis, L. (1909), *Biochem. Z.* **16**, 81.

Mikkers, F. E. P., Everaerts, F. M., Verheggen, Th. P. E. M. (1979), *J. Chromatogr.* **169**, 11–20.

Minarik, M., Gas, B., Rizzi, A., Kenndler, E. (1995), *J. Cap. Electrophor.*, **2**, 89–96.

Quincke, G. (1861), *Ann. Phys. (Leipzig)* **113**, 513.

Reuss, F. F. (1809), *Mem. Soc. Imp. Natur. Moskau* **2**, 327.

Riester, D., Wiesmuller, K. H., Stoll, D., Kuhn, R. (1996), *Anal. Chem.* **68**, 2361–2365.

Shimura, K., Karger, B. L. (1994), *Anal. Chem.* **66**, 9–15.

Small. H., Miller, T. E. (1982), *Anal. Chem.* **54**, 462–469.

Small, H., Stevens, T. S., Bauman, W. C. (1975), *Anal. Chem.* **47**, 1801–1809.

Smith, R. D., Barinaga, C. J., Udseth, H. R. (1988), *Anal. Chem.* **60**, 1948–1952.

Swanek, F. D., Chen, G., Ewing, A. G. (1996), *Anal. Chem.* **68**, 3912–3916.

Swerdlow, H., Gesteland, R. (1990), *Nucl. Acids Res.* **18**, 1415.

Terabe, S., Otsuka K., Ichikawa, K., Tsuchiya, A., Ando, T. (1984), *Anal. Chem.* **56**, 111.

Tiselius, A. (1937), *Trans. Faraday. Soc.* **33**, 524.

Wahl, J. H., Smith. R. D. (1994), *J. Cap. Electrophor.* **1**, 62–71.

Wallingford, R. A. (1996), *Anal. Chem.* **68**, 2541–2548.

Wan, Q. H. (1997), *Anal. Chem.*, **69**, 361–363.

Yan, C., Dadoo, R., Zare, R. N., Rakestraw, D. J., Anex, D. S. (1996), *Anal. Chem.* **68**, 2726–2730.

Zhong, M., Lunte, S. M. (1996), *Anal. Chem.* **68**, 2488–2493.

Further Reading

Baker, D. R. (1995), *Capillary Electrophoresis*, New York: John Wiley and Sons.

Camilleri, P. (Ed.) (1993), *Capillary Electrophoresis: Theory and Practice*, Boca Raton, FL: CRC Press.

Coleman, D. (Ed.) (1994), *Directory of Capillary Electrophoresis*, Amsterdam: Elsevier.

Foret, F., Krivankov, L., Bocek, P. (1994), *Capillary Zone Electrophoresis*, New York: VCH Publishers, Inc.

Guzman, N. A. (Ed.) (1993), *Capillary Electrophoresis Technology*, New York: Marcel Dekker, Inc.

Hartwick, R. A. (1994), *Introduction to Capillary Electrophoresis*, Boca Raton, FL: CRC Press.

Landers, J. P. (Ed.) (1997), *Handbook of Capillary Electrophoresis*, 2nd ed., Boca Raton, FL: CRC Press.

St. Claire, R. L. (1996), "Capillary Electrophoresis," *Anal. Chem.* **68**, 569R–586R.

Weinberger, R. (1993), *Practical Capillary Electrophoresis*, Boston: Academic Press.

Yang, Q., Hidajat, K., Li, S. F. Y., (1997) "Trends in Capillary Electrophoresis," *J. Chromatogr. Sci.* **35**, 358–373.

DATING TECHNIQUES

ALICE C. MIGNEREY, *Department of Chemistry and Biochemistry, University of Maryland, College Park, Maryland, U.S.A.*

INTRODUCTION

The ability to assign a time to an object's formation or to an event on historic, prehistoric, or geologic timescales relies on a set of techniques that are commonly termed dating. In practice, this can mean a variety of things depending on the field of investigation, but it usually refers to the chronological date before the present. The most widely used dating tool is the transformation of elements via radioactive decay. Depending on the choice of radioactive nuclides, times from a few days or years to the age of the solar system can be explored. Techniques based on chemical tracers and atomic transformations have also proven important as dating tools in specific applications.

The science of dating techniques is central to a broad range of disciplines, including geomorphology, geochemistry, tectonics, cosmology, art history and archaeology, paleontology, hydrology, oceanography, environmental

ISBN 3-527-29308-6

science, and even criminology. The field has expanded rapidly in recent years with advances in the ability to quantify very small amounts of even the rarest of nuclides. This article attempts to give a broad overview of the field as it is currently practiced, with specific references to some of the more recent work.

The first section contains brief discussions of the basic principles of radioactive decay, sources of radionuclides, and the techniques of mass spectrometry, thermoluminescence, and electron-spin resonance. The subsequent sections describe specific applications in a variety of fields. Section 2 is devoted to the uses of the primordial radionuclides. The text briefly presents each nuclide or system of nuclides with regard to its origin, uses, and limitations. Section 3 focuses on the applications of cosmogenic and anthropogenic nuclides and is aligned according to application, since a number of nuclides can be applied to a given problem. Each section has been augmented with suggested reading to provide more in-depth treatments in specific areas. Summary tables are provided to correlate nuclides with various areas of research. The interested reader is encouraged to explore the titles in the Further Reading list for discussions on the wide variety of applications.

1. BASIC PRINCIPLES

1.1 Radioactive Decay

The general term radioactive decay is applied to any process by which one nuclide transforms into another via the emission of an α particle or a β particle (positive or negative), fission, electron capture, and even, in rare cases, emission of a proton (see RADIOACTIVITY). This transformation follows a first-order rate law, with the rate of change of the number of radionuclide atoms N given by

$$\frac{\mathrm{d}N}{\mathrm{d}t} = \lambda N, \tag{1}$$

where the decay constant λ is ln2 divided by the nuclide half-life, $t_{1/2}$. In Eq. (1), $\mathrm{d}N/\mathrm{d}t$ is termed the activity and expressed in disintegrations per unit time. Integrating Eq. (1) with respect to time gives the relationship between the number of initial atoms N_0 and the number remaining after time t,

$$N = N_0 \mathrm{e}^{-\lambda t}. \tag{2}$$

Knowing the number of atoms N and with an estimate of the initial N_0, the time t over which the system has been decaying can be calculated. For example, after a time $t = t_{1/2}$, N will be equal to $\frac{1}{2}N_0$.

Alternatively, if the daughter nuclide is stable, the number of daughter atoms present at time t, D_t, is given by the number of parent atoms which have decayed in time t, $P_0 - P_t$, plus any initial daughter atoms D_0 present at time $t = 0$:

$$D_t = D_0 + (P_0 - P_t) = D_0 + P_t(e^{\lambda t} - 1). \tag{3}$$

The time t can then be obtained as

$$t = \frac{1}{\lambda}\ln\left(1 + \frac{D_t - D_0}{P_t}\right). \tag{4}$$

Equations (3) and (4) are valid only for closed systems (with no loss of parent or daughter or intermediate nuclides.)

Sometimes the parent is connected to the stable daughter through a series of shorter-lived radionuclides, such as in the ^{238}U–^{206}Pb series shown in Fig. 1. Here the parent ^{238}U decays to ^{234}Th via the emission of an α particle. The ^{234}Th goes on to decay to ^{234}Pa via β^- emission, and so on until the stable end product ^{206}Pb is reached. This chain involves a total of eight α-particle and six β^- decays. Provided the decay time of the parent is long compared with those of the intermediate products, Eqs. (3) and (4) can still be used to connect the initial parent and final stable daughter nuclides.

In the case where the measured daughter is itself decaying, the parent–daughter relationship becomes

$$D_t = D_0 \mathrm{e}^{-\lambda_2 t} + \frac{\lambda_1}{\lambda_2 - \lambda_1} P_0(e^{-\lambda_1 t} - e^{-\lambda_2 t}), \tag{5}$$

with λ_1 and λ_2 being the parent and daughter decay constants, respectively. If the parent half-life is longer than the daughter half-life, an equilibrium situation is reached after a

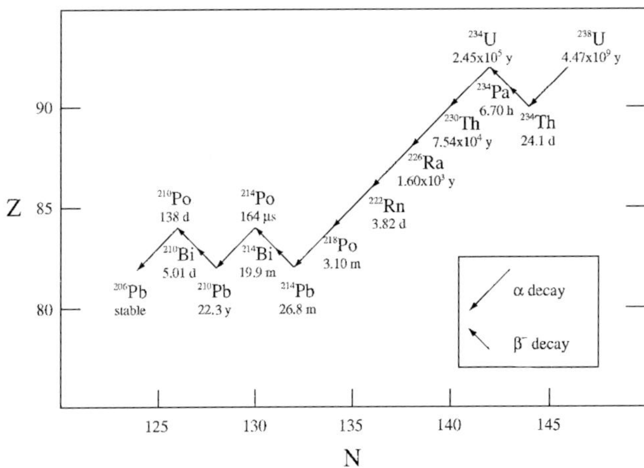

FIG. 1. Schematic diagram of the the ^{238}U decay series, displayed on a grid of proton number Z *versus* neutron number N.

few daughter half-lives, where the activity ($A = \lambda N$) ratio of the parent to the daughter, A_1/A_2, is given by $(\lambda_2 - \lambda_1)/\lambda_2$. In the case of a very long-lived parent (such as ^{238}U compared with ^{234}U in Fig. 1), $\lambda_2 \gg \lambda_1$ and $A_1 = A_2$. This is called secular equilibrium and can be extended to the activity of any nuclide in the chain, $A_1 = A_2 = A_3 \ldots$.

1.2 Sources of Radionulides

1.2.1 Primordial Radionuclides At the condensation of the solar system from the solar nebula there were a large number of unstable isotopes of the elements that are present on earth today, and even a few higher-Z elements no longer present. Most of these have since decayed to the stable nuclides that make up our natural isotopic abundances. The few exceptions are those with half-lives on the order of, or greater than, the age of the Earth (about 4.5×10^9 y). These long-lived radionuclides form the group called the primordial or primary radionuclides, because they have been around since the beginning of the solar system. Those commonly used for dating are given in Table 1, along with their half-lives.

1.2.2 Cosmogenic Radonuclides Cosmogenic nuclides are constantly being formed through nuclear reactions between the cosmic-ray flux and various constituent atoms of the Earth's atmosphere, hydrosphere, and lithosphere. Cosmic rays consist mostly of high-energy protons, about 10% α particles and 1% heavier nuclides. They originate from the Sun and from outside our solar system. The portion from the Sun shows a definite modulation, which has been correlated with sunspot activity. The Earth's magnetic field tends to focus the cosmic rays at the poles, giving a latitudinal variation in intensity. The energy distribution of the incident or primary cosmic rays is very broad and peaks around 300 MeV. This is sufficient energy to produce numerous secondary products from a single primary collision, in what is commonly termed a shower. These secondaries also include neutrons, which can interact even at very low energies via capture reactions. Lal and Peters (1967) contains an in-depth characterization of the cosmic-ray flux, its spatial distribution in height and latitude, and the resulting reactions.

The atmosphere is quite efficient at reacting with cosmic rays, with about $\frac{2}{3}$ of the atmospheric cosmogenic nuclides produced in the stratosphere and $\frac{1}{3}$ in the troposphere. Those produced in the stratosphere must first make it to the troposphere before they are removed via both wet and dry deposition.

The cosmic-ray flux that reaches the Earth's surface is greatly reduced in intensity by interactions with the atmosphere. Nevertheless, significant cosmogenic nuclides are produced by interactions of cosmic-ray components with target nuclei in the oceans, and in

Table 1. Primordial radionuclides: parent–stable daughter systems.

Parent–daughter	Parent half-life[a] (y)	Parent natural abundance (%)[a]	Applications
^{40}K–^{40}Ar(10.72%)–^{40}Ca(89.28%)	1.277×10^9	0.0117	Micas, volcanic rocks, glauconite
^{87}Rb–^{87}Sr	4.75×10^{10}	27.835	Sedimentary rocks, lunar rocks, volcanic and plutonic rocks, authigenic clays, meteorites
^{147}Sm–^{143}Nd (^{148}Sm–^{144}Nd) (^{144}Nd–^{140}Ce)	1.06×10^{11} 7×10^{15} 2.29×10^{15}	15.0 11.3 23.80	Meteorites, lunar rocks, archean volcanites, zircons, mantle evolution
^{176}Lu–^{176}Hf	3.73×10^{10}	2.59	Meteorites, apatites, garnets, monazites
^{187}Re–^{187}Os	4.35×10^{10}	62.60	Mantle separations, sulfides, iron meteorites
^{232}Th–^{208}Pb ^{235}U–^{207}Pb ^{238}U–^{206}Pb	1.405×10^{10} 7.04×10^8 4.468×10^9	100 0.720 99.2745	calcite, zircon and rutile granites, meteorites, crustal evolution

[a] From Nuclear Wallet Cards, July 1995, National Nuclear Data Center, Brookhaven National Laboratory, P. O. Box 5000, Upton, NY 11973–5000, U.S.A.

surface soils and rocks. Since cosmic rays are strongly absorbed in the surface, these *in situ* produced nuclides provide important means of studying erosion rates and exposure histories for a variety of geomorphic surfaces (Cerling and Craig, 1994; Bierman *et al.*, 1995). Figure 2 shows a representative depth profile for the neutron flux generated by cosmic-ray interactions in concrete (Liu *et al.*, 1994). Note that the flux peaks slightly below the surface. Only reactions induced by muons are important at significant depths.

1.2.3 Noncosmogenic *In Situ* Production A low flux of slow neutrons can also be generated by the reaction of α particles from U and Th with light elements such as Ca and K in the Earth via (α, n) reactions. These neutrons can be captured by ^{35}Cl to produce ^{36}Cl or by ^{235}U to induce fission. These are *in situ* processes that are not cosmogenic in origin, and are important in U- and Th-bearing deposits.

1.2.4 Anthropogenic Radionuclides The advent of the nuclear age has provided a new source of radionuclides. The atmospheric nuclear tests of the late 1950s and early 1960s spread neutron-activation and fission products around the globe that are now serving as tracers for processes on the order of 30–40 y. Examples include tritium (^3H), ^{36}Cl, and ^{129}I. This source is termed the bomb pulse and produced concentrations of tritium and ^{36}Cl in rainwater and ice cores several orders of magnitude above the prebomb background. This effect is clearly shown in Fig. 3. Man's continued activities with nuclear power reactors and nuclear fuel reprocessing have significantly elevated the global inventory of ^{129}I, producing point sources for tracing modern groundwaters and ocean circulation patterns (Rucklidge, 1995; Raisbeck *et al.*, 1994). There is also evidence for local elevated concentrations of ^{36}Cl in surface and shallow ground water from nuclear fuel reprocessing plants and reactor emissions.

1.3 Detection Techniques

1.3.1 Traditional Decay Counting The ability to determine the number of parent and daughter nuclides at a given time, or the activity due to their decay, is a necessary requirement of most nuclear dating methods. This can be achieved by measuring the nuclide directly by mass spectrometric methods or by counting the number of β or α particles emitted (or the associated γ-ray emission) in the decay and using Eq. (1) to obtain the activity. Activity can then be converted back to the number

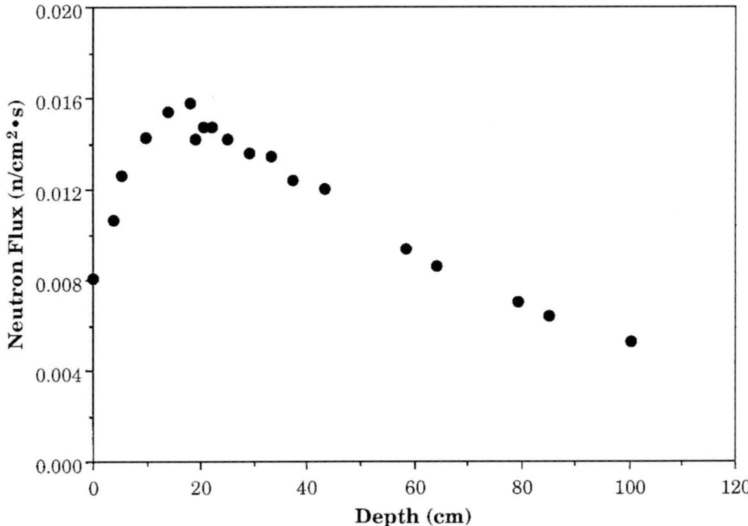

FIG. 2. Measured depth profile of neutrons produced in concrete (data from Lui *et al.*, 1994).

of atoms N or, in some cases, used directly. For example, the widely used radionuclide ^{14}C ($t_{1/2} = 5730\,y$) is traditionally quantitated by counting its β^- decays. This is possible because the average activity of "modern" prebomb carbon in the biosphere is about 13.6 decays per minute per gram carbon; counting rates above background are obtained when sufficient carbon is available. If the nuclide is very long-lived, but has a measurable abundance, then the activity is also measurable by virtue of the much greater N. This is the case with many of the nuclides in the uranium and thorium decay series. However, decay counting becomes less feasible, if not impossible, for much rarer nuclides with longer half-lives. Reviews of nu-

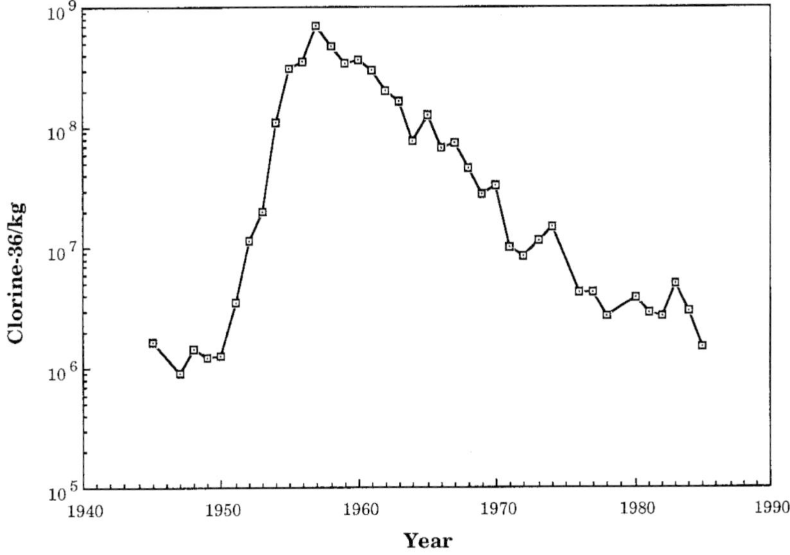

FIG. 3. Measured ^{36}Cl concentration in ice cores from Dye-3 (data from Synal *et al.*, 1990).

clear detection techniques can be found in this Encyclopedia and in Knoll (1979).

1.3.2 Traditional Mass Spectrometry Modern mass spectrometry has revolutionized the field of nuclear dating. While the ability to measure time comes from the radioactive decay clock, it is no longer necessary to count these decays to obtain an accurate estimate of the number N in Eq. (2). Only a brief overview of the technique is given here. For a more complete discussion of traditional mass spectrometric techniques the reader is referred to MASS SPECTROMETRY.

The basic principle of a mass spectrometer is the fact that accelerated ions in a uniform magnetic field will bend with radii of curvature squared, r^2, proportional to their mass to charge ratio m/q. If a sample of an element is ionized to a single charge state, then the individual isotopes of that element will appear at different positions in the focal plane of the magnet and can be individually identified. Dickin (1995) discussed the requirements as applied to measurements of geological samples, including chemical separations necessary prior to analysis and preferred means of ionization and detection. Thermal-ionization mass spectrometry (TIMS) is the most widely used means of ionization today.

The analytical quantification of the individual isotope concentration is usually achieved through isotope dilution. In this technique the sample is measured with a spike of enriched isotope of known concentration. Comparison of the results for the pure sample, the pure spike and mixtures of the two allows the initial isotope concentration to be determined. This has the advantage that any matrix effects cancel because two isotopes of the same element are compared directly.

1.3.3 Accelerator Mass Spectrometry The technique of accelerator mass spectrometry (AMS) follows the basic principles of mass spectrometry, except that in AMS the ions are accelerated to much higher energies, as high as tens of MeV. This is most often carried out in the two stages of a tandem accelerator, similar to that depicted in Fig. 4. In the first stage

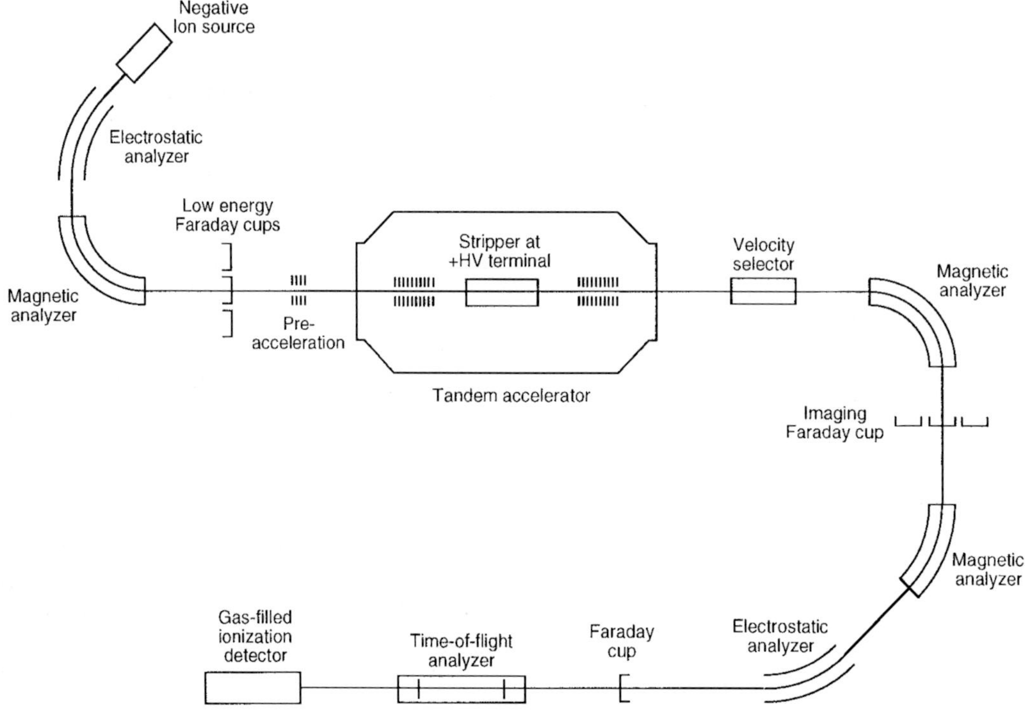

FIG. 4. Schematic diagram for an accelerator system used for accelerator mass spectrometry (AMS).

negative ions from the ion source are electrically and magnetically selected for the desired mass (for example, 14 for ^{14}C determination). In some applications the negative-ion injection has an advantage in that interfering isobars (nuclides with the same mass but different nuclear charge Z) may not make negative ions (as is the case for ^{14}N in the ^{14}C determination) However, this restriction can also prove problematic for the determination of very electropositive species, such as ^{41}Ca, or noble gases such as Ar.

The negative ions are accelerated through a potential gradient to energies of 3–10 MeV (depending on the terminal voltage) at the central terminal of the accelerator. At this point the ions are passed through a stripper foil or gas and transformed into positive ions. The result is a distribution of positive charge states, which is a function of the initial ion identity and ion velocity. At this stage any residual molecular species which may have survived the initial ionization process are destroyed. The positive ions are then further accelerated back to ground potential at the exit of the accelerator.

The resultant ions exiting the accelerator have a spread in energies due to multiple charge states produced in the central stripper. These are separated in an electrostatic analyzer and/or velocity filter. An additional magnetic analyzer further reduces the background, selecting only those ions of a given mass (which now ideally have the same charge state.) This set is not necessarily unique and may still have a mix of isobars. The next major advantage of AMS is the detection of the product ions using nuclear detection techniques. These techniques are capable of counting individual atoms and separating them by their elemental Z. For example, modern AMS systems can separate ^{36}Cl from the much more common ^{36}S and measure 1 atom of ^{36}Cl in the presence of 10^{15} atoms of stable chlorine atoms (^{35}Cl and ^{37}Cl). A more thorough discussion of the technique of AMS can be found in Elmore and Phillips (1987) and Tuniz *et al.* (1998).

Accelerator mass spectrometry has opened up the area of cosmogenic nuclides to a broad range of applications, making possible studies with isotopes which are too rare or in concentrations too low to have been measured by decay counting. In the case of the widely used ^{14}C, AMS now allows measurements of much smaller samples and extends the technique to much older dates. The main disadvantage of AMS is the cost involved. Historically developed on machines built for basic nuclear physics research, the field has now come into its own with many laboratories acquiring much smaller accelerators that have been optimized for the purpose of AMS. One such facility is the National Oceans Science AMS facility (NOSAMS) at Woods Hole Oceanographic Institute, Massachusetts, U.S.A., dedicated to the measurement of ^{14}C. There are currently over 30 facilities worldwide with AMS capabilities.

1.4 Thermoluminescence and Electron-Spin Resonance

Nuclear radiation is typically called ionizing radiation because it is of sufficient energy to eject bound electrons from substances, producing free electrons and positive ions. The techniques of thermoluminescence (TL) and electron-spin resonance (ESR) rely on the trapping of these electrons by defects in the crystal structures of nonconducting solids. These defects can be vacancies, displacements (sometimes caused by the ionizing radiation itself), or impurities in the lattice, which provide sites for free electrons. The electrons usually remain trapped until dislodged by thermal vibrations (heating) or photoexcitation by ultraviolet light (as in solar bleaching). Once dislodged the electrons can diffuse through the crystal until they either become retrapped in a stable (deeper) site or recombine with a positive ion. If the recombination is radiative, light is emitted whose wavelength is characteristic of the type of luminescence center or crystal impurity. This process is called thermoluminescence (TL).

Potassium-, uranium-, and thorium-bearing minerals in the surrounding soils, or in the sample itself, provide a relatively constant flux of ionizing radiation. Therefore, the object to be dated has been accumulating dose over the period of time since it was last heated to a temperature (or exposed to the Sun's UV radiation) that released the electrons from any trapping sites and reset the TL clock. If we assume that the TL light obtained is propor-

Table 2. Primordial radionuclides: disequilibrium systems.

Nuclides	Half-lives[a]	Applications
^{238}U–^{234}U–^{230}Th	4.468×10^9 y–2.455×10^5 y–7.538×10^4 y	Carbonates: corals, shells; sediments; ferromanganese nodules/crusts
^{235}U–^{231}Pa	7.038×10^8 y–3.276×10^4 y	Carbonates, sediments
^{210}Pb–^{210}Bi–^{210}Po–^{206}Pb	22.3 y–5.013 d–138.376 d–stable	Deposition rates: snow, glaciers, sediments

[a] From Nuclear Wallet Cards, July 1995, National Nuclear Data Center, Brookhaven National Laboratory, P.O. Box 5000, Upton, NY 11973–5000, U.S.A.

tional to the accumulated paleodose dose D_P, the time t since the last resetting of the TL clock can be calculated by

$$t = D_P / D_R. \qquad (6)$$

The annual dose rate D_R is a sum of the doses due to the α, β, and γ radiations from the nuclear decays. If the sample has been near the surface for any length of time, there will also be a background flux from cosmic-ray reactions. The specific contributions of each component in a sample depend on a number of factors, including burial history and soil and sample water content (which attenuates the radiation.) When possible the γ flux is measured at the excavation site.

In practice, a material to be TL dated is sequentially heated. Different types of traps release their electrons at different temperatures, termed glow peaks, those with the highest temperatures being the most stable. By measuring the number of photons of each type emitted as a function of temperature, the total radiation dose D_P can be determined if the trapping efficiencies are known. This is usually calibrated by making TL measurements with γ-ray sources at known doses.

Since TL requires heating the sample to high temperatures, it is unsuitable for a number of materials which decompose upon heating. An alternative is the method of electron-spin resonance (ESR), which measures the microwave absorption by unpaired electrons in the material when it is placed in a varying magnetic field. Although not as sensitive as TL, ESR has the advantage of measuring the trapped electrons in their trapping centers, and thus does not require luminescence centers or annealing the material. This means that the same sample can be measured numerous times and used to calibrate the dose by adding additional radiation through exposure to γ sources. Applications include tooth enamel, bone, and calcite (Aitken, 1990). The reader is referred to PHYSICS IN ARCHAEOLOGY for additional discussions of TL and ESR.

2. TECHNIQUES BASED ON PRIMORDIAL RADIONUCLIDES

This section contains an overview of the major applications of primordial radionuclides in use today. The parent–daughter systems are summarized in Tables 1 and 2. Many of these utilize what is called the isochron method for age determination. This method is presented in Sec. 2.1.1 in relation to the Rb–Sr system. Since it is not possible to discuss each system in great depth, the reader is referred to Faure (1986), Titayeva (1994), and Dickin (1995) for details on applications of interest.

2.1 Equilibrium Techniques

2.1.1 Rb–Sr and the Isochron Method The element rubidium has two naturally occurring isotopes, ^{85}Rb and ^{87}Rb, with abundances of 74.17 and 27.83%, respectively. The dating capabilities come from the radioactive ^{87}Rb, which decays with a half-life of 4.75×10^{10} y by β^- emission to ^{87}Sr (natural abundance 7.00%). While rubidium has no minerals of its own, the ionic radius of the Rb^+ ion is close to that for the K^+ ion. Since both are alkali metals, Rb^+ replaces K^+ in all potassium-bearing minerals. The product Sr isomorphously replaces Ca in some minerals, but differs in coordination number in others.

This results in considerable variation in the Rb/Sr ratio between formations with different histories and even between different minerals in the same formation. The differentiation of Rb and Sr in magmatic melts produces Rb/Sr ratios which vary from 0.06 in basalts to 0.53 in granite, with an average value of 0.23 for the Earth's crust (Titayeva, 1994).

In principle, the time since the formation of a rock or mineral can be found by Eq. (3). In the Rb–Sr case, D_t = number of ^{87}Sr atoms in the sample, D_0 = number of initial ^{87}Sr atoms and P_t = number of ^{87}Rb atoms in the sample. However, measuring absolute abundances is very difficult. A way around this problem is to normalize everything to a nearby isotope with the same approximate abundance, and measure all three by mass spectrometry. In this case, ^{86}Sr (natural abundance of 9.86%) is chosen as the normalizing isotope, and Eq. (3) becomes

$$\left(\frac{^{87}\mathrm{Sr}}{^{86}\mathrm{Sr}}\right)_{\mathrm{sample}} = \left(\frac{^{87}\mathrm{Sr}}{^{86}\mathrm{Sr}}\right)_{\mathrm{initial}} + \frac{^{87}\mathrm{Rb}}{^{86}\mathrm{Sr}}(e^{\lambda t} - 1) \quad (7)$$

The use of Eq. (7) requires knowledge of the initial ^{87}Sr/^{86}Sr ratio. This is difficult to estimate on its own, and today determination through the construction of isochrons is the method of choice. Inspection of Eq. (7) shows that it has the form of a straight line. If a number of samples with the same age but different Rb/Sr ratios are plotted on a graph of ^{87}Sr/^{86}Sr versus ^{87}Rb/^{86}Sr, the slope $e^{-\lambda t} - 1$ will give the age and the intercept gives the initial ^{87}Sr/^{86}Sr ratio of the sample, a quantity of geological interest in itself. In this case the resulting line is a true isochron. Because of the large natural abundance of the decay product ^{87}Sr and the long half-life of the parent ^{87}Rb, this technique works best when applied to samples with high Rb/Sr ratios.

Just because the data plot on a line does not necessarily mean that they are derived from samples which are the same age. One test of a true isochron is to calculate the mean square weighted deviates (described in Dickin, 1995). The MSWD, as it is commonly referred to, gives a measure of the scatter of the data points about the straight line. If the scatter is larger than predicted by the analytical uncertainties of the measurements, then the MSWD > 1. The larger the MSWD, the less likely it is that the line formed is a true isochron. This pitfall is demonstrated in Fig. 5 from Innocent *et al.* (1997), showing Rb/Sr isochrons from the dating of celadonites, green clays close to glauconite. The data in Fig. 5(a) are derived from a group of whole minerals, giving an effective age for the group of 96 ± 1 million years. However, the large MSWD of 44 suggests that there are large variations in the individual ages of the minerals, even though the group defines a single line. An alternative method is to construct an internal isochron, derived from the leachates and residues of a single mineral. Figure 5(b) is an example where the ratios for two leachates (enriched in ^{86}Sr) and the residue (depleted in ^{86}Sr, but enriched in radiogenic ^{87}Sr) are plotted along with the whole-mineral ratios. The

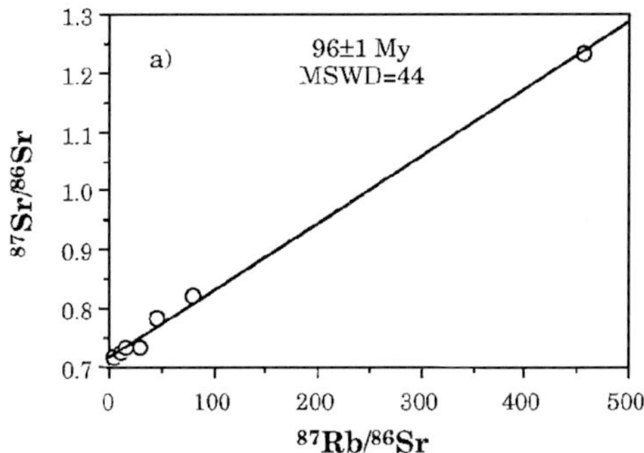

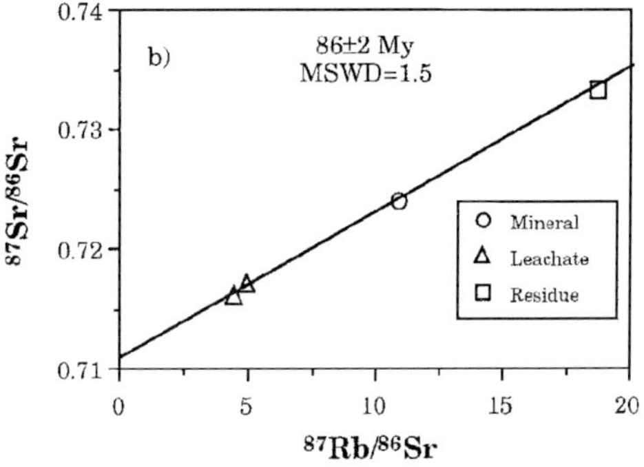

FIG. 5. Rb/Sr isochrons for (a) whole-mineral samples and (b) leachates, whole mineral, and residues for a single sample (data from Innocent *et al.*, 1997).

data clearly define a line with an extracted age of 86 ± 2 millions years for the sample (with a corresponding MSWD of only 1.5).

2.1.2 Sm–Nd There are seven naturally occurring isotopes of samarium, three of which are radioactive (see Table 1). The shortest-lived ^{147}Sm is the only one useful for dating, with a half-life of 1.06×10^{11} y and a natural abundance of 15.0%. It decays by α-particle emission to ^{143}Nb (abundance 12.18%). The isotope of choice for normalization in an equation analogous to Eq. (7) is ^{144}Nd, with a natural abundance of 23.80%. (Although ^{144}Nd is itself radioactive, its half-life is so long that is can be considered stable in this context.) Since both the parent and daughter are lanthanides, they have very similar chemistries and are not differentiated by secondary processes. This gives much more uniform natural Sm/Nd ratios in samples than is seen in Rb–Sr dating, and isochrons tend to cover a very small range in ^{143}Nd/^{144}Nd ratio. The Sm–Nd technique is particularly well suited to dating the origin of very old rocks and sediments and meteorites.

2.1.3 Lu–Hf Lutetium exists in nature as two isotopes, the stable ^{175}Lu (natural abundance 97.41%) and ^{176}Lu (natural abundance 2.59%), which has a half-life of 3.78×10^{10} y and decays by β^- emission to ^{176}Hf (natural abundance 5.206%). Standard isochron analysis can be performed using 18.606% abundant ^{177}Hf as the normalizing isotope.

One of the principal applications is in dating zircon. Since hafnium is a chemical analogue of zirconium, it is enriched in zircons and the chemical resistance of zircon to weathering makes it a good candidate for dating ancient sediments. However, ^{176}Lu/^{177}Hf is strongly fractionated in sedimentary systems. This may actually prove useful in differentiating between sediment reservoirs and recycling of basalts to the mantle.

2.1.4 Re–Os Rhenium-187 has a natural abundance of 62.60%. It decays with a half-life of 4.35×10^{11} y via β^- emission to stable ^{187}Os (natural abundance 1.6%). The two elements are strongly fractionated between mantle and crustal material, the Re being partitioned into magmatic liquids and the Os staying with the mantle material. This has the potential for studying such processes as ore formation (Freydier *et al.*, 1997), magma genesis, and mantle evolution (Esperanca *et al.*, 1997). The Re–Os system is also well suited to the study of iron meteorites, which contain relatively high concentrations of Re and Os and have large variations in the Re/Os ratios, making it easy to construct isochrons with high precision (Smoliar *et al.*, 1996). The normalizing isotope of choice is ^{188}Os, with a natural abundance of 13.3%.

The potential of Re–Os dating has only begun to be realized through advances in ionization techniques in mass spectrometry, most notably negative thermal ionization mass spectrometry (NTIMS) (Creaser *et al.*, 1991). Rhenium–osmium isochrons in iron meteorites have also been measured with AMS by Ding *et al.* (1997). The two methods compare favorably, with AMS having the advantage of being able to make measurements on single grains with no chemical preparation of the sample.

2.1.5 K–Ar, Ar–Ar The element potassium has three naturally occurring isotopes: ^{39}K, ^{40}K, and ^{41}K. The radioactive ^{40}K has a natural abundance of only 0.0117% and a half-life of 1.227×10^9 y. However, since potassium is prevalent in minerals, there is sufficient ^{40}K in the Earth's crust to be a useful dating tool. The decay of ^{40}K has two branches; β^- decay produces ^{40}Ca in 89.33% of the decays and electron capture (EC) leads to ^{40}Ar in the other 10.67%. The product ^{40}Ca is of limited use because it comprises 96.94% of natural calcium, which is very abundant in most rocks. This makes identifying the small radiogenic component essentially impossible. Argon, on the other hand, is a noble gas and rarely found naturally in rock formations. The main contaminant is atmospheric Ar, which is 99.60% radiogenic ^{40}Ar. This is overcome by using the 0.337% abundant, non-radiogenic ^{36}Ar as an indicator of atmospheric contamination, since purely radiogenic ^{40}Ar in rocks should be free of ^{36}Ar. Isochrons can also be constructed for the K–Ar system, in this case adding a separate term for the atmospheric contribution and accounting for the branching ratio for the EC decay:

$$\left(\frac{^{40}\mathrm{Ar}}{^{36}\mathrm{Ar}}\right)_{\text{sample}} = \left(\frac{^{40}\mathrm{Ar}}{^{36}\mathrm{Ar}}\right)_{\text{atm}} + \left(\frac{^{40}\mathrm{Ar}}{^{36}\mathrm{Ar}}\right)_{\text{initial}} + \frac{^{40}\mathrm{K}}{^{36}\mathrm{Ar}}\frac{\lambda_{\mathrm{EC}}}{\lambda_{\mathrm{tot}}}(e^{\lambda_{\mathrm{tot}}t} - 1). \quad (8)$$

The analytical techniques commonly used to obtain the individual concentrations of Ar and K are discussed in Dickin (1995). The potassium content of the sample can be found by a number of means, including x-ray fluorescence, atomic absorption, and isotope dilution mass spectrometry. The argon is determined by isotope dilution mass spectrometry using ^{38}Ar as the known spike.

A variation on the K–Ar dating technique is Ar–Ar dating (sometimes referred to as ^{40}Ar/^{39}Ar dating). In this application the sample is placed in a fast-neutron flux and ^{39}K (natural abundance 93.258%) is converted into ^{39}Ar via the (n, p) reaction. The amount of potassium in a sample can then be determined by mass spectrometry of ^{39}Ar, at the same time as the other Ar determinations are being made. Since the number of ^{39}Ar atoms formed is a function of the reactor flux Φ and the energy distribution of the neutrons, standards of known age are used as flux monitors and irradiated along with the sample. One problem with this technique is the possibility of other nuclear reactions occurring during irradiation that also produce argon isotopes, the principal ones being ^{40}K(n, p)^{40}Ar, ^{40}Ca(n, nα)^{36}Ar and ^{40}Ca(n, α)^{39}Ar. Loss of ^{39}Ar from mineral grains due to recoil during the irradiation is also a concern.

The assumption of a closed system is often difficult to fulfill when the species is a trapped noble gas. The Ar–Ar method allows the sample to be heated incrementally and the ages determined from the argon released in each step. This is an important test of alteration in the ages due to loss of argon, with the first ages often younger than those obtained later in the heating process (Ortega-Rivera *et al.*, 1997).

Another advance in Ar–Ar dating is the application of laser microprobe ablation. This allows single grains to be analyzed and selection of specific inclusions can then be made. A portion of the September 1997 issue of *Geochimica et Cosmochimica Acta* (Vol. 61, No. 18) is devoted to papers on Innovative Applications of ^{40}Ar/^{39}Ar Microanalytical Research.

2.1.6 U–Pb, Th–Pb, Pb–Pb Lead has four naturally occurring isotopes: ^{204}Pb, ^{206}Pb, ^{207}Pb, and ^{208}Pb. Only the ^{204}Pb is non-radiogenic. The other three are the final stable decay products of ^{238}U, ^{235}U, and ^{232}Th, respectively (see Fig. 1 and Table 1). If the time scale is long enough, each parent–daughter system satisfies Eq. (3), with ^{204}Pb as the normalizing isotope. For example, the equation for the ^{238}U–^{206}Pb system is

$$\left(\frac{^{206}\mathrm{Pb}}{^{204}\mathrm{Pb}}\right)_{\text{sample}} = \left(\frac{^{206}\mathrm{Pb}}{^{204}\mathrm{Pb}}\right)_{\text{initial}} + \frac{^{238}\mathrm{U}}{^{204}\mathrm{Pb}}(e^{\lambda t} - 1), \quad (9)$$

with analogous equations for ^{235}U–^{207}Pb and ^{232}Th–^{208}Pb. In principle, isochrons can be constructed for each of these three systems, but in practice the requirement of a closed system is usually difficult to meet on account of the mobility of Pb, Th, and U. Nevertheless, direct isochron dating has been successful in certain applications, such as marine sediments (Smith and Farquhar, 1989) and ore-stage calcite (Brannon *et al.*, 1996).

Lead usually exists as Pb^{2+} with an ionic radius of 0.12 nm, which makes it the wrong charge and too large to fit into the lattice of zirconium, uranium, or thorium minerals. This means that the only lead present in these minerals can usually be considered radiogenic. [Corrections for any initial lead are made by determining the ^{204}Pb content (natural abundance 1.4%).] For these cases the ^{235}U–^{207}Pb and ^{238}U–^{206}Pb decay equations then simplify to

$$^{207}\mathrm{Pb}_{\text{radiogenic}} = {}^{235}\mathrm{U}(e^{\lambda_{235}t} - 1) \quad (10)$$

and

$$^{206}\mathrm{Pb}_{\text{radiogenic}} = {}^{238}\mathrm{U}(e^{\lambda_{238}t} - 1) \quad (11)$$

The applicability of U–Pb dating is greatly

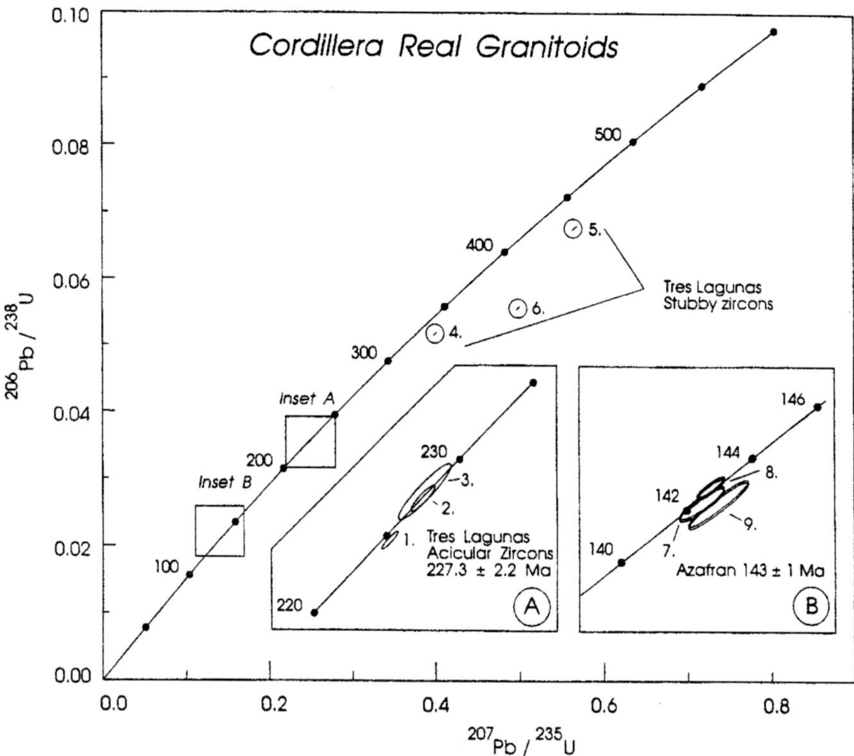

FIG. 6. Concordia plot of $^{206}Pb/^{238}U$ as a function of $^{207}Pb/^{235}U$ (from Noble *et al.*, 1997, reproduced by permission of Geological Society of America). The ages along the curve are indicated by the numbers, in units of Ma (My). Samples labeled 4, 5 and 6 are clearly discordant.

expanded by creating what are termed concordia plots. Since the decay constants are known, plots of $^{206}Pb/^{238}U$ as a function of $^{207}Pb/^{235}U$ for varying times should give a unique curve called a concordia. If there has been loss of radiogenic lead, the data will be discordant and will fall off the curve. Since the mechanisms for lead loss should be the same for the two uranium isotopes, the discordant points fall on a straight line below the concordia. Extrapolation of this discordant line to the concordia curve gives the "true" age of the sample. Figure 6 is an example of a concordia for zircons taken from Noble *et al.* (1997).

Alternatively, the ratio of Eqs. (10) and (11) gives

$$\left(\frac{^{207}\mathrm{Pb}}{^{206}\mathrm{Pb}}\right)_{\mathrm{radiogenic}} = \frac{^{235}\mathrm{U}(e^{\lambda_{235}t}-1)}{^{238}\mathrm{U}(e^{\lambda_{238}t}-1)}, \qquad (12)$$

with the $^{235}U/^{238}U$ ratio well known to be 1/137.88. Application of Eq. (12) is called Pb–Pb dating, since it only requires knowledge of the ratio of two isotopes of lead. The Pb–Pb ratios are a reflection of the Pb/U ratios; however, they have the advantage of being quite insensitive to recent conditions that might produce uranium loss, since it is the radiogenic daughter lead isotope which is measured. Taking into account any initial ^{206}Pb and ^{207}Pb and using the isochron technique of normalizing to the nonradiogenic ^{204}Pb gives

$$\frac{\left(\frac{^{207}\mathrm{Pb}}{^{204}\mathrm{Pb}}\right)_{\mathrm{sample}} - \left(\frac{^{207}\mathrm{Pb}}{^{204}\mathrm{Pb}}\right)_{\mathrm{initial}}}{\left(\frac{^{206}\mathrm{Pb}}{^{204}\mathrm{Pb}}\right)_{\mathrm{sample}} - \left(\frac{^{206}\mathrm{Pb}}{^{204}\mathrm{Pb}}\right)_{\mathrm{initial}}} = \frac{1}{137.88}\frac{(e^{\lambda_{235}t}-1)}{(e^{\lambda_{238}t}-1)}. \qquad (13)$$

While this cannot be solved in closed form for

the time t, computer integration yields isochrons as a function of time, which can be matched to the data. This technique is used when recent loss of uranium is suspected.

The Pb–Pb technique was first applied to meteorites. The Diablo Canyon meteorite produced an age of 4.55×10^9 y. Subsequent studies using Pb–Pb dating have shown that some deep ocean sediments lie on the Diablo Canyon isochron. This isochron is considered characteristic of the age of the Earth and now called the geochron. Numerous other special applications of U–Pb dating have been used to study the evolution of the Earth. For further variations and applications the reader is referred to Titayeva (1994) and Dickin (1995).

2.2 Nonequilibrium Techniques and the U–Th Decay Chains

The methods discussed in Sec. 2.1 all utilize simple parent–daughter decay, where the daughter is stable. However, inspection of Fig. 1 for the decay of ^{238}U shows that the final stable product ^{206}Pb is the result of a complex sequence of decays. Similar decay schemes connect ^{235}U to ^{207}Pb and ^{232}Th to ^{208}Pb. If conditions are right to remove and isolate one of the intervening long-lived decay products, this will then be the parent for its own decay chain. If an unstable daughter nuclide is monitored, the daughter formation and decay follows Eq. (5). Deviations from secular equilibrium also provide opportunities to gain information by monitoring the intermediate radionuclides. There are numerous possibilities, a few of which will be described here. The reader is referred to Faure (1986) and Dickin (1995) for more extensive treatments.

2.2.1 ^{238}U–^{234}U–^{230}Th System: Dating of Corals and Ocean Sediments One example of a non-equilibrium situation is the ^{238}U–^{234}U–^{230}Th parent–daughter system. In a marine environment, uranium exists as soluble carbonates which are readily incorporated into corals and shells along with calcium carbonate. In contrast, thorium is insoluble and is absorbed onto detrital particles, precipitates, and is preferentially incorporated into sediments. Therefore, the ^{232}Th/^{238}U ratio in seawater is very low (on the order of 10^{-5}) and the ^{230}Th/^{238}U ratio in seawater is about 10^{-5} times what would be obtained from ^{238}U equilibrium decay (Edwards *et al.*, 1986). Growing coral incorporates ^{234,238}U along with the carbonate, but will have very little of the daughter ^{230}Th initially. Monitoring the subsequent decay of ^{234}U and in-growth of ^{230}Th provides an important means of dating coral formations as old as 500 ky, a timescale inaccessible using radiocarbon dating.

In the ^{238}U–^{234}U–^{230}Th system, with the assumption of no initial ^{230}Th, Eq. (5) leads to

$$1 - \left[\frac{^{230}\mathrm{Th}}{^{234}\mathrm{U}}\right]_{\mathrm{act}} = \mathrm{e}^{-\lambda_{230}t} - \left[\frac{\delta^{234}\mathrm{U}}{1000}\right] \times \left(\frac{\lambda_{230}}{\lambda_{230}-\lambda_{234}}\right)[1 - e^{(\lambda_{234}-\lambda_{230})t}], \quad (14)$$

where the subscript "act" denotes activity ($A = \lambda N$). The δ^{234}U is a measure of the deviation of the ^{234}U/^{238}U activity ratio from unity (secular equilibrium) and is defined as

$$\delta^{234}\mathrm{U} = \left[\frac{\left(\frac{^{234}\mathrm{U}}{^{238}\mathrm{U}}\right)}{\left(\frac{^{234}\mathrm{U}}{^{238}\mathrm{U}}\right)_{\mathrm{equil}}} - 1\right] \times 1000, \quad (15)$$

and

$$\delta^{234}\mathrm{U}(t) = \delta^{234}\mathrm{U}(0)\mathrm{e}^{-\lambda_{234}t}. \quad (16)$$

If the ^{234}U and ^{238}U were in secular equilibrium is seawater, the δ^{234}U would be zero and the second term on the right of Eq. (14) would vanish. However, this is not the case, and in practice isochrons are constructed of the activity ratios of ^{234}U/^{238}U as a function of ^{230}Th/^{234}U.

The dating of corals has been important in studies of sea-level stance and climatic fluctuation of the Pleistocene era. The use of TIMS has made coral dating possible with samples as small as 200 mg and analysis of growth rings. This now allows extension of the ^{14}C calibration curve to ages older than accessible by dendochronology (Bard *et al.*, 1990). Applications can also be made to shells, but in this case the possibility of secondary impurities is dealt with by successive stripping and measuring the individual fractions.

A second simplifying case occurs for ocean sediments and ferromanganese nodules (thorium coprecipitates with the Fe and Mn hydroxides), where there is essentially no parent ^{234}U to feed the ^{230}Th; this is called unsupported ^{230}Th. For example, the decay of the ^{230}Th can be followed as a measure of sedimentation rate R, with $^{230}Th = {}^{230}Th_{initial}e^{-\lambda_{230}t}$. A graph of log($^{230}Th$ activity) versus depth should give a straight line with a slope $-\lambda_{230}/R$. This is usually true for shallow depths, but the effect of U-supported ^{230}Th usually shows up as depth increases. This can be corrected for by determining the ^{234}U content (which is a measure of the ^{230}Th activity in secular equilibrium with any uranium) and subtracting this from the total ^{230}Th activity.

2.2.2 ^{235}U–^{231}Pa The radionuclide ^{231}Pa is the first long-lived nuclide in the ^{235}U decay chain

$$^{235}U(t_{1/2} = 7.04 \times 10^6\,y)$$
$$\rightarrow {}^{231}Th(t_{1/2} = 25.5\,h)$$
$$\rightarrow {}^{231}Pa(t_{1/2} = 3.28 \times 10^4\,y) \rightarrow \ldots .$$

Hence, in a closed system ^{231}Pa can be considered in equilibrium with its parent ^{235}U. However, in seawater ^{231}Pa is similar to ^{230}Th and is absorbed onto particulate matter and rapidly incorporated into sediments. This leaves the seawater rich in ^{235}U, but almost devoid of the daughter ^{231}Pa, allowing dating of corals by following the ingrowth and decay of ^{231}Pa (Edwards *et al.*, 1997). Under conditions where the system had no initial ^{231}Pa and was closed with respect to ^{231}Pa and ^{235}U, the time elapsed since closure is given by

$$t = \frac{-\ln\left(1 - \left[\frac{^{231}Pa}{^{235}U}\right]_{act}\right)}{\lambda_{231}}, \tag{17}$$

where "act" denotes the ratio of the activities of ^{231}Pa and ^{235}U. It has been proposed that since both ^{231}Pa and ^{235}U are daughters of uranium isotopes, concordia could be constructed and used in a manner similar to the U–Pb dating of minerals. (See the discussion in Sec. 2.1.6.)

2.2.3 ^{210}Pb Lead-210 is a short-lived ($t_{1/2} = 22.3\,y$) intermediate in the decay chain of ^{238}U (see Fig. 1). It is produced in the atmosphere by the decay of ^{222}Rn, which is a noble gas and escapes from the earth at a rate of about 42 atoms/min cm^2 (Faure, 1986). With a residence time in the atmosphere of only 10 d, the ^{210}Pb is quickly removed by precipitation and becomes an excellent dating tool for snow pack and glaciers. In water columns ^{210}Pb has a residence time of 1–2 y before it becomes attached to sediment particles. Since the lead is now removed from its radioactive precursors, it decays away unsupported with a half-life of 22.3 y, making it useful for dating sediments over a period of several hundred years. Under the assumption of a constant deposition rate, the age of a sample can be found by solving Eq. (2) as applied to ^{210}Pb in units of atoms or activity,

$$^{210}Pb = {}^{210}Pb_{initial}e^{-\lambda_{210}t}, \tag{18}$$

where $^{210}Pb_{initial}$ is the value at the surface. In a plot of log (^{210}Pb) versus depth, the slope gives the sedimentation or accumulation rate. The major limitation is the contribution from ^{210}Pb generated from ^{226}Ra in the sediments. This contribution is termed supported, since it should be in equilibrium with the long-lived ^{226}Ra and any intermediate decay products. The supported ^{210}Pb can be corrected for by the measurement of ^{226}Ra (or the short-lived intervening daughters).

While the ^{210}Pb itself emits a low-energy β particle that is difficult to detect, it has traditionally been quantitated by measuring the β decay of its daughter ^{210}Bi or the α decay of the ^{210}Po daughter. However, advances in γ-ray spectroscopy have made the direct measurement of ^{210}Pb possible through detection of its 46.5-keV γ ray (Böllhofer *et al.*, 1994). This has the advantage of also allowing detection of any ^{226}Ra ($E_\gamma = 186.1\,keV$) or its daughters ^{214}Pb ($E_\gamma = 352.0\,keV$) and ^{214}Bi ($E_\gamma = 609.3\,keV$), important in correcting for any supported ^{210}Pb. Lead-210 dating is now

Table 3. Extinct radionuclides (adapted from Dickin, 1995).

Parent–daughter	Parent half-life[a]	Daughter natural abundance (%)[a]
^{244}Pu–fission products including ^{132}Xe, ^{134}Xe, ^{136}Xe	8.08×10^7 y	26.9, 10.4, 8.9
^{129}I–^{129}Xe	1.57×10^7 y	26.4
^{107}Pd–^{107}Ag	6.5×10^6 y	51.839
^{53}Mn–^{53}Cr	3.74×10^6 y	9.501
^{60}Fe–(^{60}Co)–^{60}Ni	1.5×10^6 y (1.9251×10^3 d)	26.223
^{26}Al–^{26}Mg	7.4×10^5 y	11.01
^{41}Ca–^{41}K	1.03×10^5 y	6.7302

[a] From Nuclear Wallet Cards, July 1995, National Nuclear Data Center, Brookhaven National Laboratory, P.O. Box 5000, Upton, NY 11973–5000, U.S.A.

an established method for dating marine and lake sediments. It has recently become important in supplying dates that can be correlated with anthropogenic activities, such as lead and other heavy-metal deposition and radioactive fallout.

2.3 Extinct Radionuclides

The nucleosynthetic processes that produced the naturally occurring nuclides found on Earth today also produced numerous unstable nuclides that have long since decayed away to their stable daughter products. In fact, many of our stable nuclides are the ultimate end points of the decay of short-lived radioactive species. Only the longest-lived ones are still around for us to study today. However, the evidence for now extinct radionuclides can be found in meteorites, which preserve remnants of the original solar nebula from which our solar system condensed. This evidence is in the form of the anomalous isotopic abundances of daughter nuclides. If the radiogenic component can be isolated from that which was directly produced, limits can be set on the time between the nucleosynthesis event and the formation of the meteoritic material in which the isotope was found. However, this requires a number of assumptions on the initial nucleosynthesis-produced isotopic ratios, the time over which the production occurred, and the possible mixture of sources.

Table 3 (adapted from Dickin, 1995) lists a number of extinct radionuclide systems that have been studied in meteorites. They cover a range of half-lives from 10^5 to 10^7 y. Combining time estimates for a number of systems provides constraints on production scenarios that cannot be obtained with a single system alone. For example, times derived from measuring ^{129}Xe and xenon isotopes in ^{244}Pu fission fragments are consistent with a delay of on the order of 100 My between nucleosynthesis and closure of the system to xenon loss. However, studies of the shorter lived radionuclides in Table 3 also provide evidence for a late injection of material into the solar nebula between 2 and 20 My before condensation.

2.4 Fission-track Methods

The kinetic energy released when a heavy nucleus such as uranium undergoes fission is on the order of 200 MeV. This large amount of energy is divided between the two fission products (and any neutrons) according to the law of conservation of momentum. If the fission were symmetric (giving two equal-size fragments), then each would carry away about 100 MeV of kinetic energy. These heavy, high-Z, energetic fragments are highly ionizing and produce damage tracks in their stopping material.

The only naturally occurring radionuclide that has a significant spontaneous fission branch is ^{238}U. Even this is very small, accounting for only 0.0001% of all ^{238}U decays. However, given the moderately large concentrations of uranium in many minerals, the effects of the stopping of fission products in the material are clearly observable. If the damage

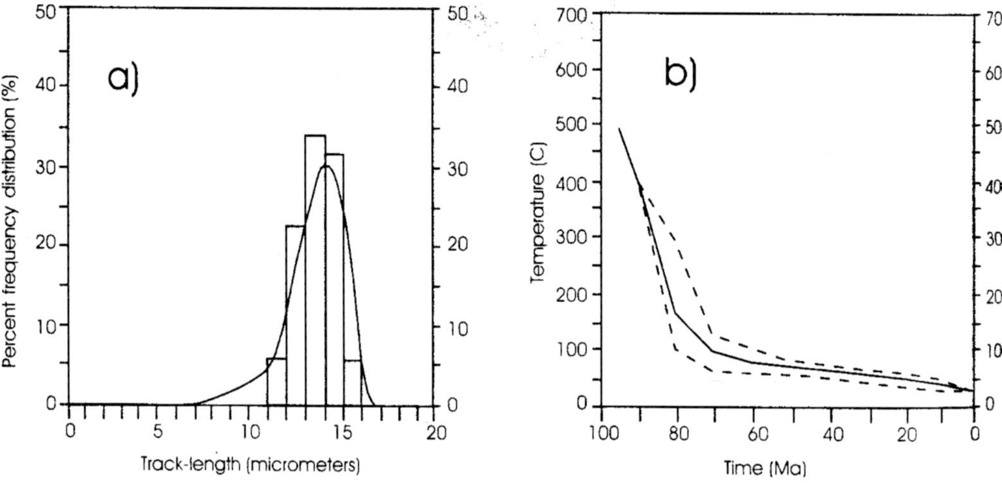

FIG. 7. Results of modeling fission track lengths (from Ortega-Rivera *et al.*, 1997, reproduced by permission of Geological Society of America). (a) Comparison of measured track-length distributions (histogram) and model length distributions (solid curve). (b) Time–temperature ranges for 250 acceptable solutions generated by the model (dashed curve.) The solid line represents the exponential-mean solution.

tracks are enhanced by etching with acid or alkaline solution, they can be counted and used as a measure of the integrated dose received by the material. The dose can be calibrated by exposing like materials to known fluxes of induced fission fragments and comparing the number of etched tracks. A thorough discussion of the analytical techniques can be found in Dickin (1995) and Ravenhurst *et al.* (1994).

How long fission tracks stay preserved depends on the material type and the thermal history. In general, the annealing process can be described by a simple Boltzmann distribution, where the annealing time can be written as a function of the material specific activation energy U and the absolute temperature T:

$$t = Ae^{U/kT}, \tag{19}$$

with A a calibration constant and k the Boltzmann constant.

Considerable work has been done on the effects of temperature on track annealing in apatites (Laslett *et al.*, 1987). As tracks anneal they decrease in length, producing a length distribution that can be modeled in terms of a time and temperature profile. The partial annealing zone (PAZ) is sensitive to temperatures as low as 60–110 °C. Figure 7(a), from Ortega-Rivera *et al.* (1997), shows a track length profile from an apatite from the Sierra San Pedro Martir pluton in Baja California. A model fit to the track distribution gives the time–temperature relationship constraints shown in Fig. 7(b). The fission-track technique is well suited to studies of cooling, exhumation, and denudation histories of tectonic formations and the effects of erosion on these formations (Johnson, 1997).

2.5 (U–Th)/He Thermochronometry

The α decays in the uranium and thorium parent–daughter systems produce radiogenic helium, which can be retained in rock masses. Recent work in apatite is providing a new means of tracking thermal histories of low-temperature processes by following the diffusive loss of He. The closure temperature in apatite is around 75 °C, with partial retention occurring between 40 and 75 °C (Wolf *et al.*, 1997). This is analogous to the partial annealing zone in fission-track dating (see Sec. 2.4) and can provide a means of extending studies on apatites to lower temperatures than accessible via fission-track analysis. The analytical details of the technique can be found in Wolf *et al.* (1996).

2.6 Thermoluminescence (TL) Dating Applications

Dating artifacts via TL has been an important part of art history and archaeology. The technique was first pioneered for dating pottery and other baked clays, being applied to both the clays and glazes. In this usage the TL light typically comes from the quartz and feldspars in the clay, with the clock reset to zero upon firing of the pot. Both bulk pottery fragments and mineral grains can be dated. In alkali feldspars with a high potassium content, the paleodose is mainly internal, coming from the decay of ^{40}K. This has the advantage of being less sensitive to the moisture content of the surroundings than the external dose.

The widespread availability of fired clay at most archaeological sites makes TL a particularly useful dating tool. While TL dating is not as accurate as radiocarbon dating for samples with ages between 1000 and 10 000 y, it is often the method of choice for ages older than 10 000 y, where the radiocarbon time scale is poorly calibrated, or for samples younger than 1000 y, where radiocarbon can give ambiguous ages (see Sec. 3.1). The ability of TL to extend to ages older than those accessible by radiocarbon is particularly important for paleontological artifacts such as burnt flint and cooking stones.

The TL dating limit depends on the lifetimes of the trapping sites and the saturation light sum. Quartz has proven to be an excellent paleodosimeter for Quarternary deposits, since it does not reach its saturation light sum until a total dose of 10^4 Gy, which means that it can be sensitive to times as long as 10^6 y, depending on the dose rate. This is in contrast to radiocarbon's practical limit of around 50 000 y. Geological applications of TL include lunar material and meteorites, volcanic eruptions, sediments, and stalagmitic calcite. Specific applications and techniques can be found in Aitken (1985) and Aitken (1990).

3. TECHNIQUES BASED ON COSMOGENIC NUCLIDES

Cosmogenic nuclides provide access to time scales much shorter than the primordial radionuclides. Since they are constantly being produced, radionuclides with half-lives as short as 53 d (as in the case of ^{7}Be) can prove useful. Table 4 lists the main cosmogenic nuclides used as dating tools today, along with their production modes, half-lives, and relevant applications. Until the advent of accelerator mass spectrometry (AMS), relatively large sample sizes were necessary to obtain nuclide concentrations, using low-level decay counting. However, not all of the nuclides in Table 4 are currently measurable by AMS at levels low enough to be useful, and decay counting is still an important analytical technique. Today cosmogenic nuclides are finding uses in a broad range of research areas, from archaeology and paleontology, to paleoclimate and ocean circulation studies, to hydrology and geomorphology, and even cosmology. The following subsections highlight a few of these, with a separate section devoted to ^{14}C.

3.1 Carbon-14

The most widely used cosmogenic nuclide is probably ^{14}C, commonly termed radiocarbon. Carbon-14 is produced in the atmosphere primarily through the reaction ^{14}N(n,p). It is rapidly incorporated into atmospheric CO_2 and becomes thoroughly mixed with the global carbon pool, showing little variation with latitude. When a plant or animal dies it is removed from active exchange with the carbon pool, and its radiocarbon content will steadily decrease as the ^{14}C decays. Carbon-14 has been a mainstay of archaeology since Libby first proposed using it as a dating tool in the 1950s (Libby, 1955), and many laboratories still specialize in radiocarbon dating through decay counting. The archaeological applications of ^{14}C are discussed in PHYSICS IN ARCHAEOLOGY.

One of the biggest challenges to obtaining accurate radiocarbon dates is the calibration of the time dependence of the atmospheric equilibrium ^{14}C/C ratio. This varies with the cosmogenic production rate through the effect of solar cycles and the Earth's magnetic field on cosmic ray flux. An additional effect on atmospheric ^{14}C has been the increasing use of fossil fuels. This has been diluting the natural radiocarbon reservoir with nonradioactive

Table 4. Cosmogenic nuclides.

Nuclide	Production modes	Half-life[a]	Applications
^{3}H	$^{14}N(n, ^{3}H)^{12}C^{b,c}$, $^{6}Li(n, \alpha)^{3}H^{d}$	12.33 y	Post-bomb groundwater dating
^{3}He	^{14}N, $^{16}O(spallation)^{c,e}$ $^{3}H(\beta^{-} decay)^{b,e}$	Stable	Exposure dating
^{7}Be	$^{14}N(n, 3p5n)^{c}$, $^{14}N(p, 4p4n)^{c}$ $^{16}O(p, 5p5n)^{c}$	53.29 d	Soils and sediments
^{10}Be	$^{14}N(p, 4pn)^{c}$ $^{16}O(spallation)^{c,e}$	1.51×10^{6} y	Soils and sediments, erosion rates and exposure dating, glacial ice
^{14}C	$^{14}N(n, p)^{b,c}$ $^{16}O(p, ^{3}H)^{c,e}$ $^{17}O(n, \alpha)^{d}$	5730 y	Erosion rates and exposure dating, ocean studies, groundwater studies, soil and sediments, glacial ice, paleodating, archaeological artifacts
^{21}Ne	$^{24}Mg(n, \alpha)^{e}$, $^{23}Na(n, p2n)^{e}$	Stable	Exposure dating
^{26}Al	$^{40}Ar(spallation)^{c}$ $^{26}Mg(p, n)^{e}$ $^{28}Si(p, 2pn)^{e}$	7.4×10^{5} y	Soils and sediments, erosion rates and exposure dating, meteorites
^{32}Si	$^{40}Ar(spallation)^{c}$	140 y[f]	Sedimentation rates, glacial ice
^{36}Cl	$^{40}Ar(p, 2p3n)^{c}$ $^{35}Cl(n, \gamma)^{b,d,e}$, $^{40}Ar(n, 4np)^{b}$ ^{39}K, $^{40}Ca(spallation)^{e}$ $^{39}K(n, \alpha)^{d}$, $^{39}K(\mu, p2n)^{e}$ $^{40}Ca(\mu, \alpha)^{d}$	3.01×10^{5} y	Erosion rates and exposure dating, groundwater studies, glacial ice, paleodating of organic fossil material
^{39}Ar	$^{40}Ar(n, 2n)^{c}$ $^{39}K(n, p)^{d,e}$	269 y	Erosion rates and exposure dating, groundwater studies
^{53}Mn	$^{53}Fe(p, 2p)^{e}$, $^{53}Fe(p, \alpha)^{e}$	3.74×10^{6} y	Meteorites
^{59}Ni	$^{59}Co(p, n)^{e}$, $^{60}Ni(p, pn)^{e}$	7.6×10^{4} y	Meteorites
^{81}Kr	$^{82}Kr(n, 2n)^{c}$, $^{80}Kr(n, \gamma)^{c,d}$ $^{238}U(spontaneous\ fission)^{d}$	2.29×10^{5} y	Groundwater dating
^{85}Kr	$^{238}U(spontaneous\ fission)^{d}$ $^{235}U(n, fission)^{d}$, $^{84}Kr(n, \gamma)^{c}$, $^{86}Kr(n, 2n)^{c}$	3934.4 d	Groundwater dating
^{129}I	$^{129}Xe(n, p)^{c}$ $^{131}Xe(spallation)^{c}$ $^{235}U(n, fission)^{b,e}$ $^{238}U(spontaneous\ fission)^{e}$	1.57×10^{7} y	Oil formation and salt brine dating, groundwater studies, ocean circulation

[a] From Nuclear Wallet Cards, July 1995, National Nuclear Data Center, Brookhaven National Laboratory, P.O. Box 5000, Upton, NY 11973-5000, U.S.A.
[b] Anthropogenic.
[c] Atmospheric cosmogenic.
[d] Subsurface.
[e] *In situ.*
[f] Silicon-32 half-life from Morgenstern *et al.*, 1996.

carbon (termed dead carbon), causing a steady decrease in the equilibrium atmospheric $^{14}C/C$ ratio, called the Suess effect after the individual who first observed it. Even more recently, nuclear bomb tests have elevated the atmospheric $^{14}C/C$ ratio.

The unit usually used to express ^{14}C in a sample is $\Delta^{14}C$(‰), which is defined using a

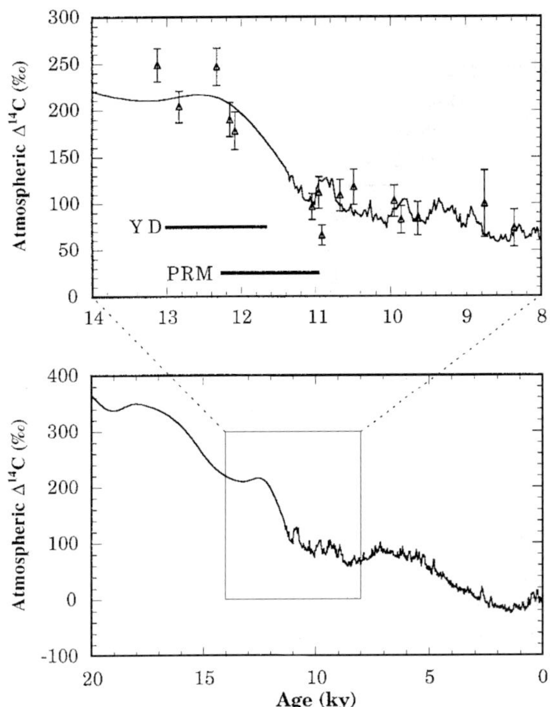

FIG. 8. Atmospheric radiocarbon expressed as Δ^{14}C [actually plotted as the negative of Eq. (20)] as a function of age, derived from bidecadal tree-ring data and U/Th coral data (obtained from CALIB program as described in Stuiver and Reimer, 1993). The triangles are the coral data from Edwards *et al.* (1993), plotted against ^{230}Th date.

prebomb-era oxalic acid carbon standard as

$$\Delta^{14}\text{C‰} = \frac{\left(\frac{^{14}\text{C}}{^{12}\text{C}}\right)_{\text{sample}} - 0.95\left(\frac{^{14}\text{C}}{^{12}\text{C}}\right)_{\text{standard}}}{0.95\left(\frac{^{14}\text{C}}{^{12}\text{C}}\right)_{\text{standard}}} \times 1000. \tag{20}$$

The 0.95 is a correction factor to modern ratios. A Δ^{14}C value of zero is defined at 1950 as modern carbon. The more negative the value, the older the radiocarbon age. Positive values reflect input from bomb-produced carbon and are termed post-modern. [Note that in some cases the negative of Eq. (20) is defined as Δ^{14}C, as in Fig. 8.]

The field of dendochronology (tree-ring dating) has been a key to understanding the variation in Δ^{14}C and its correlation with absolute dates, but is limited to around 10 000 years. Calibration dates have recently been extended by the study of corals (which grow in layers like trees), correlating radiocarbon dates with those obtained from independent U–Th dating (Bard *et al.*, 1990; Edward *et al.*, 1993). (See Sec. 2.2.1.) Figure 8 shows Δ^{14}C as a function of age as derived from bidecadal tree-ring data and U–Th coral dates (Stuiver and Reimer, 1993). It is clear that there are regions of the curve where a single ^{14}C value does not correspond to a unique age. As the expanded region shows, there can be considerable variation over both long and relatively short times due to fluctuations in the cosmic-ray flux (see Sec 1.2.2). In cases where a series of Δ^{14}C measurements have been made at what are interpreted as regular time intervals, the ambiguity in the age determination can be reduced by matching the trend in the sequence of Δ^{14}C values. This technique is referred to as wiggle matching.

Applications of ^{14}C have expanded tremendously since AMS made measurements of small samples feasible. For example, the coral records in the Huon Peninsula, Papua, New Guinea, in conjunction with the ^{230}Th dating, show interesting signatures of changes in the atmospheric–oceanic exchange of ^{14}C during

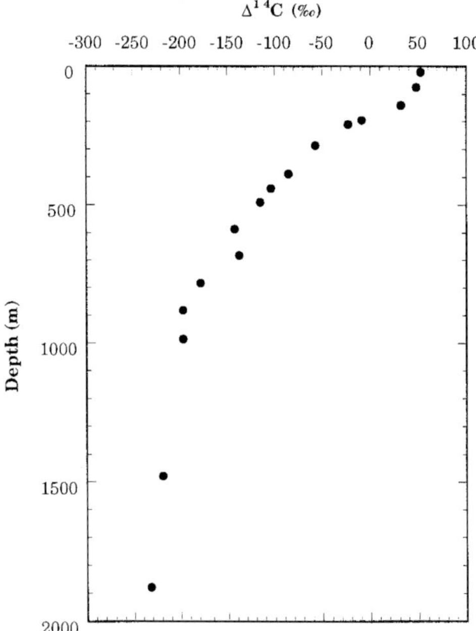

FIG. 9. Depth profile of the $\Delta^{14}C$ in ocean water, measured for a site in the Pacific Basin collected as part of the World Ocean Circulation Experiment (WOCE) (data from von Reden *et al.*, 1997).

the Younger Dryas (YD) climate event and the period of reduced melting (PRM) since the last major ice age (Edward *et al.*, 1993). These data are shown as the triangles in the expanded graph in Fig. 8. The times of the YD and PRM are marked by the horizontal bars and correspond to a rapid change in the inferred atmospheric $\Delta^{14}C$.

Studies of radiocarbon in the oceans were carried out long before AMS was available, using large-volume samples. Now small-volume samples are being measured at the National Ocean Sciences AMS facility (NOSAMS) at Woods Hole Oceanographic Institute with a precision of better than 0.4%, rivaling that obtained using decay counting (von Reden *et al.*, 1997). Figure 9 shows a ^{14}C depth profile for a site in the Pacific basin, which is part of the data set collected in the World Ocean Circulation Experiment (WOCE). The surface waters clearly show the influence of bomb carbon. The deep waters have reached an equilibrium value. Similar results have been obtained for the Canadian Basin of the Arctic Ocean (Schlosser *et al.*, 1997). The differences in radiocarbon content of surface and deep waters provide a tool for studying ocean currents and mixing.

Another application of ^{14}C dating is the ages of sediments in lakes and peat bogs. Figure 10 shows the results of dating a sediment core from Germany and its correlation to the pollen analysis showing the history of the prevailing vegetation at the site (Kretschmer *et al.*, 1997). The time scale for ^{14}C also makes it perfect for dating many groundwater aquifer systems. This has been attempted with varying degrees of success using both the dissolved CO_2 (inorganic carbon, DIC) and the dissolved organic carbon (DOC) (Murphy *et al.*, 1989; Drimmie *et al.*, 1991). In the case of DIC, corrections must be made for any dissolution of carbonate minerals. Using DOC to obtain an age requires knowledge of the radiocarbon age of the initial input and the assumption of no addition of dead carbon by subsurface bacterial activity producing organic compounds from kerogen sources. The input function can be either contaminated with anthropogenic ^{14}C or much older than modern by reason of the long residence time of organic matter in soils. Dating of soil organic matter has been shown to be quite complex, and a detailed understanding of the maturation process at a given site is needed before radiocarbon ages can be interpreted (Y. Wang *et al.*, 1996).

3.2 *In Situ* Dating of Geological Formations

The production of radionuclides inside mineral grains from cosmic-ray interactions has been used to estimate erosion rates and exposure ages of surface land forms. This is a multidimensional problem, with the radionuclide concentration changing with the time exposed to cosmic rays and the flux of cosmic rays. The cosmic-ray flux is a function of depth (see Sec. 1.2.2) and can change over time by either surface erosion (which will increase the flux) or burial (which decreases the flux). An additional effect is the change in tilt of the surface, which changes the path length of the cosmic rays and the attenuation. Add to this the fact that the flux of cosmic rays is a function of latitude and can vary as a result of solar and geomagnetic factors, and the complexity of the problem becomes apparent.

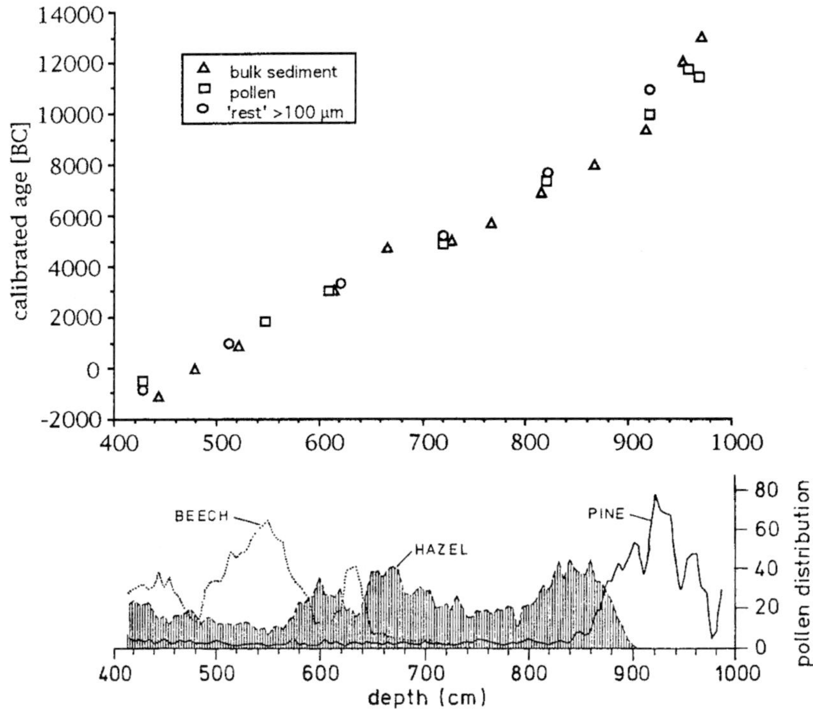

FIG. 10. Calibrated radiocarbon ages of sediment constituents as a function of depth, compared with the vegetation type determined from pollen analysis, for a site in Germany (data from Kretschmer *et al.*, 1997, drawing reproduced by permission of North-Holland Publishing Co.).

In the simplified case of a flat surface exposed to a constant flux, the concentration of the cosmogenic nuclide N_i in atoms g^{-1} is given by

$$N_i = \frac{P_i}{\lambda_i + \varepsilon/k}\left\{1 - \exp\left[-\left(\lambda_i + \frac{\varepsilon}{k}\right)t\right]\right\} + N_0 e^{\lambda t}, \qquad (21)$$

where P_i = production rate, atoms $(\text{g-ky})^{-1}$; λ_i = decay constant, ky^{-1}; ε = actual erosion rate, cm ky^{-1}; k = scale depth, calculated as cosmic-ray attenuation coefficient Λ divided by the density ρ; t = actual time, ky; and N_0 = initial concentration, atoms g^{-1} (notation after Gillespie and Bierman, 1995). In Eq. (21) the attenuation of the cosmic-ray flux is assumed exponential with depth. However, this is not strictly true of the neutron flux generated, which peaks at a depth of about 20 cm before displaying exponential behavior (see Fig. 2).

In the case of negligible erosion, with no initial concentration N_0, Eq. (21) reduces to

$$N_i = (P_i/\lambda_i)(1 - e^{-\lambda_i t}). \qquad (22)$$

In this limiting case, the concentration of a single isotope can give the exposure age. The solutions to Eq. (22) for some commonly used cosmogenic isotopes are displayed graphically in Fig. 11. Alternatively, if an independent estimate can be found for either erosion rate or exposure age, other can be found. In either case, a steady state is eventually reached. This is shown in Fig. 12(a) for the buildup of ^{10}Be in quartz for various erosion rates. When the cosmogenic isotope is stable, a steady state is still reached as a result of the erosion rate, as shown in Fig. 12(b).

If both erosion rate and exposure time need to be determined, this can be achieved by measuring two isotopes in the same sample. Isotopes usually paired are ^{10}Be, ^{26}Al, and ^{36}Cl. Readers are referred to Lal (1991), Cerling and Craig (1994), Gillespie and Bierman

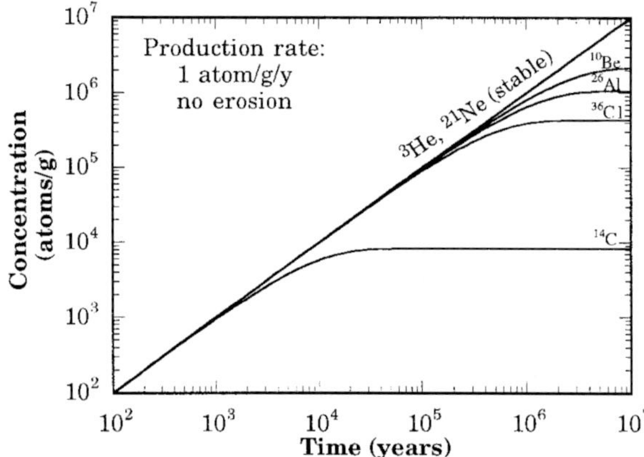

FIG. 11. Buildup of various nuclides with time for the case of no erosion [Eq. (22)].

(1995), and Clark *et al.* (1995) for discussion of the many considerations needed to obtain accurate results from this technique. An interesting application is the study of the Canyon Diablo meteorite. Surface-exposure dating of the surrounding crater using ^{10}Be/^{26}Al (Nishizumi *et al.*, 1991) and ^{36}Cl (Phillips et al., 1991) agreed with the thermoluminescence date of 49 000 ± 3000 y (Sutton, 1985). Measurements of the *in situ* ^{10}Be, ^{26}Al, and ^{36}Cl in the meteorite itself were used to estimate the production rates before the meteorite's fall to Earth. Because of the relatively short half-lives of the radionuclides, it could be assumed that

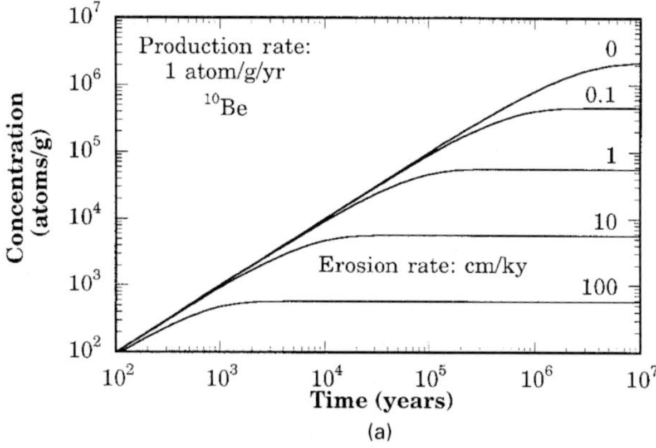

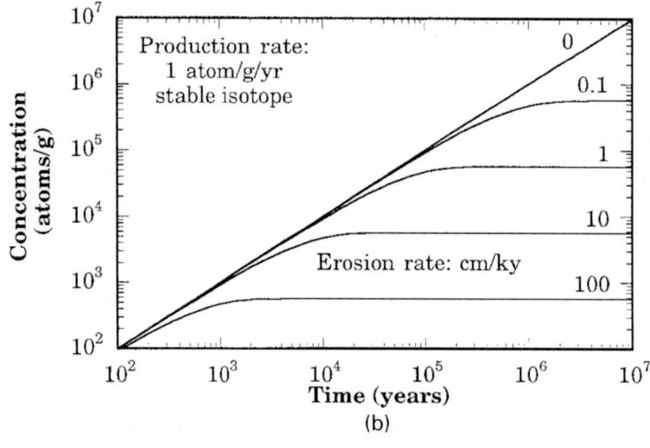

FIG. 12. Build-up of (a) ^{10}Be and of (b) a stable nuclide with time for different erosion rates, calculated with Eq. (21) using $k = 57$ cm.

they had reached equilibrium at the time of the fall, with production equaling decay. After correcting for the time on Earth, the production rate as a function of preatmospheric depth could be found. Using the ratio of the radionuclide activity to a stable noble-gas concentration ($^{10}Be/^{21}Ne$, $^{26}Al/^{21}Ne$, and $^{36}Cl/^{36}Ar$) as an integrator of the exposure period, the cosmic-ray exposure ages of six samples were found to give means of 560 ± 130 My for $^{10}Be/^{21}Ne$, 550 ± 170 My for $^{26}Al/^{21}Ne$, and 470 ± 120 My for $^{36}Cl/^{36}Ar$ (Michlovich *et al.*, 1994). Differences among samples were attributed to variations in production rates due to matrix effects (elemental composition) and irradiation geometries.

3.3 Ice Cores, Ocean and Soil Studies

3.3.1 Atmospheric Beryllium-10 in the Lithosphere and Hydrosphere Beryllium-10 has a very short residence time in the atmosphere, being strongly adsorbed onto aerosol particles and removed by wet and dry deposition. Both processes provide pathways for entrance into the hydrosphere. Profiles of ^{10}Be in ice cores and snow packs have been used as a measure of cosmic-ray flux variations and correlated with other cosmogenic isotopes, such as ^{14}C and ^{36}Cl. Arctic and Antarctic ice cores show clear evidence of the effects of the 11 y sunspot cycle (Beer *et al.*, 1990; Steig *et al.*, 1996). The influence of geomagnetic variations is less evident, with conflicting reports on the importance of latitudinal mixing (Raisbeck *et al.*, 1985; Mazaud *et al.*, 1994; Steig *et al.*, 1996).

Under the right conditions, ^{10}Be can be used as a dating tool for sediments and soil cores. Beryllium becomes attached to surface soils and can produce ^{10}Be-rich soils. If the mobility and initial deposition rate can be approximated, then a depth profile can give an age estimate (Graham *et al.*, 1997). More often, however, the concentration profile is affected by climatic variations, which can produce decreases by increased dissolution (wetter) or enrichment from glacial melt or erosion runoff. Dry climates produce more windblown soils, which are usually enriched in ^{10}Be (Gu *et al.*, 1996). Graham *et al.* (1997) studied two cores from the Wanganui region of New Zealand that had been previously characterized by magnetic analysis (Turner and Kamp, 1990). A large sixfold enhancement in ^{10}Be during the Matuyama–Brunhes magnetic transition has been interpreted as only partly due to the reversal. Since the date coincides with the beginning of an interglacial period of warming, the enhancement could be due to increased erosion and soil input and/or melting of adjacent ice sheets.

In the oceans ^{10}Be both attaches to particles and has a soluble phase, which lengthens the residence time to hundreds of years. Beryllium-10 concentration profiles in deep ocean waters are consistent with that of deep sea sediments. L. Wang *et al.* (1996) have shown that the direct atmospheric contribution (as separated by leaching with NaOH) is relatively constant, while that derived from the lithosphere is influenced by climate factors that increase erosion and runoff and by changes in ocean circulation patterns.

3.3.2 Aluminum-26 in the Oceans Another potential cosmogenic nuclide for ocean studies is ^{26}Al. The majority of the ^{26}Al in the oceans is atmospheric in origin. Once it enters the ocean, it is rapidly scavenged and locally deposited. This is in contrast to ^{10}Be, which can travel considerable distances from its source before being deposited. While the atmospheric $^{26}Al/^{27}Al$ ratio is readily measured with AMS, dilution from ^{27}Al derived from eroded crustal material has put determination of ^{26}Al in bulk sediment cores close to the AMS detection limit. A new leaching technique using NaOH now allows separation of the ^{26}Al-enriched atmospheric component and brought the $^{26}Al/^{27}Al$ ratios well within the capabilities of AMS (Wang *et al.*, 1996). When compared with ^{10}Be it may be possible to distinguish the effects of ocean dynamics and climate from variations in the cosmogenic input.

3.4 Groundwater Dating

3.4.1 Natural Sources Naturally occurring (nonanthropogenic) cosmogenic radionuclides that have the potential of dating groundwaters include ^{14}C, ^{36}Cl, ^{39}Ar, ^{81}Kr, and ^{129}I. All of these are produced in the atmosphere and enter groundwaters through wet and dry deposition, either dissolved in water droplets or attached to aerosol particles. The half-life of ^{14}C (5730 y) makes it well suited for dating waters up to around 50 000 y. The potential uses of ^{14}C and their associated complications have been discussed in Sec. 3.1. Argon-39 is relatively short-lived ($t_{1/2} = 269$ y) and can be used as an indicator of more modern water. Krypton-81, with a half-life comparable to that of ^{36}Cl, has not found widespread use on account of the analytical difficulties of measurement. Should these be overcome, ^{81}Kr may prove to be a nuclide with the fewest interfering contributions as a tool for groundwater dating (Lehmann *et al.*, 1993). The remaining nuclides, ^{36}Cl and ^{129}I, have the potential to provide ages for groundwaters that are well beyond the reach of ^{14}C. They are readily measured via AMS as ratios to their stable isotopes, $^{36}Cl/Cl$ and $^{129}I/I$. Background atmospheric levels are well above the detection limits of 1×10^{-15} for $^{36}Cl/Cl$ and 1×10^{-14} for $^{129}I/I$, providing the ability to follow their decays through several half-lives. With half-lives of 3.01×10^5 y for ^{36}Cl and 1.57×10^7 y for ^{129}I, these two nuclides could provide a range of ages from 0.1 to 80 My (Fabryka-Martin *et al.*, 1987; Frölich *et al.*, 1991).

Chloride is soluble in aqueous systems and is generally considered to be a conservative tracer in groundwaters, although there may be evidence of retention of the bomb pulse in surface soils. Aside from radioactive decay, factors that influence the $^{36}Cl/Cl$ ratio are changes in the chloride and cosmogenic ^{36}Cl of the recharge water, introduction of salt from ancient trapped brines or formation waters, diffusion of salts from the aquifer material, and subsurface production of ^{36}Cl (Fabryka-Martin *et al.*, 1989; Nolte *et al.*, 1991). Much the same is true for ^{129}I, except that the atmospheric input is expected to vary less with time by reason of an additional atmospheric contribution from volcanoes emitting ^{129}I from ^{238}U fission and the large buffering capacity of the oceans for ^{129}I. The residence time of ^{129}I in the oceans is long enough for mixing to occur with little decay. This is in contrast to ^{36}Cl, which is negligible in the oceans, because of its shorter half-life, and which has a very pronounced temporal dependence on the cosmic-ray production.

The subsurface production mechanism for ^{129}I is directly as a fission product of ^{238}U. In the case of ^{36}Cl, the main subsurface production mode is from the $^{35}Cl(n, \gamma)$ reaction on chloride in the water itself. The neutrons come from the interaction of α particles, emitted in the U–Th decay chain, with light elements such as K and Ca in the surrounding minerals, via (α, n) reactions. In some settings, as in the Stripa mine in Sweden, these subsurface sources dominate and the $^{36}Cl/Cl$ and $^{129}I/I$ ratios increase with depth, limiting their direct use as dating tools (Andrews *et al.*, 1986, 1989). It may actually be possible to use the buildup of these isotopes to obtain residence-time estimates.

While subsurface production will increase isotopic ratios, the introduction of old (dead) chloride or iodide from the aquifer material or from salt in the initial recharge water will decrease the ratio. It is well documented that the chloride concentration in rainwater decreases with distance from the coast as the influence of marine chloride decreases (Bentley *et al.*, 1986; NADP/NTN, 1990). In a study of the Aquia aquifer in southern Maryland, U.S.A., the chloride profile was found to vary with flow distance in a manner consistent with the amount of marine chloride in the recharge water, as governed by changing distance from the sea in the period covering the last glacial maximum (Purdy *et al.*, 1996). The variation in total chloride concentration can be corrected for by multiplying by the $^{36}Cl/Cl$ ratio to obtain the concentration of ^{36}Cl directly, usually expressed as atoms L^{-1}. However, this quantity is now sensitive to dilution and/or concentration effects caused by changing climatic conditions at recharge (changes in rainfall and/or evapotransporation), as well as variations in the cosmic-ray flux. Unfortunately, the time scale of interest is outside the range of good flux monitors; otherwise, ^{36}Cl concentration in confined aquifers might be a good measure of paloeclimate.

The initial promise of a simple technique for dating groundwater has not been fulfilled. However, cosmogenic isotopes have provided another experimental dimension to test models of groundwater systems and to provide constraints on ages within a given set of conditions.

3.4.2 Anthropogenic Sources The activities of man in the nuclear age have produced additional sources of nuclides commonly identified with cosmogenic processes. The bomb tests in the late 1950s and 1960s produced large pulses of tritium (^{3}H), ^{14}C, ^{36}Cl, and ^{129}I. Atmospheric levels of tritium and ^{36}Cl have mostly returned to prebomb levels (Bentley *et al.*, 1982 and Fig. 3), with the notable exception of nuclear reactor and fuel reprocessing sites (Milton *et al.*, 1994). This is not true of ^{129}I, which is being continually released by the nuclear power industry in quantities sufficient to have maintained elevated levels. All except tritium have half-lives long enough to serve as tracers of modern input for the far future. Even tritium, with its 12-y half-life, is still readily detectable (directly or as its decay product ^{3}He), and its short half-life can be used for modern age determinations (Schlosser *et al.*, 1988). The so-called bomb pulse can therefore serve as a dating tool for modern input. This might better be called a tracer, but its detection puts a definite time frame to a process, be it the advance of a contaminant plume in groundwaters (Solomon *et al.*, 1995) or the movement of currents in the ocean (Santschi *et al.*, 1996).

3.5 Silicon-32: The Potential

The widespread use of ^{32}Si as a dating tool has been hampered by the difficulty in achieving $^{32}Si/Si$ ratios below modern levels through AMS (Morgenstern *et al.*, 1996). This has limited studies to events that have elevated ratios or to the use of laborious processing for decay counting, requiring large sample sizes. The half-life of ^{32}Si is also poorly known, but recent work using AMS may improve this. The great lure of ^{32}Si is that, with a half-life of around 140 y, it fills a gap in dating that is left between tritium and ^{14}C. Exploratory studies (Morgenstern *et al.*, 1995, 1996) show promise as a dating tool in glacial ice, groundwaters, and sediments, and as a tracer for ocean circulation (Somayajulu *et al.*, 1987).

4. PALEOMAGNETISM

Paleomagnetism as a dating technique relies on the fact that the Earth's molten core is not static but dynamic, causing the resultant magnetic field of the earth to vary over time. When heated to a high enough temperature, usually 500–700 °C, certain minerals, such as hematite and magnitite, acquire magnetic alignments to the Earth's magnetic field, which persist upon cooling. These remnant fields can be measured and compared with the known reversals in the Earth's magnetic field. If a single date can be fixed by an independent technique, then the correspondence of the changing fields allows dating of the remaining samples. Details of magnetic analysis and its use as a dating tool are presented in PHYSICS IN ARCHAEOLOGY and in Aitken (1990).

5. AMINOSTRATIGRAPHY AND AMINOCHRONOLOGY

Amino acids are the building blocks of proteins. Once an organism dies, chemical changes due to hydrolysis take place, which gradually break the proteins into smaller groups called peptides. Amino acids may exist as internal constituents of a peptide chain, as terminal amino acids at either end of a protein or peptide, as a cyclic structure formed from two small peptides, or as free amino acids. The specifics of the diagenetic process are dependent on protein composition and thus the organism involved. After fossilization of shell material as much as 50% of the original protein content may be lost. However, that which remains is relatively stable and can be used as a relative dating tool.

Natural amino acids (except glycine) in proteins from living organisms exist in a form which has a unique stereospecific configuration due to the spatial arrangements of their atoms. This is termed L and is distinct from another spatial configuration of the same molecule (termed D). When synthesized in a laboratory, equal amounts of L and D are formed, but nature produces only L to give a ratio of D/L of 0.0. Over time some amino acids change their configurations from L to D, such that the equilibrium ratio of D/L is equal to 1.0 in very old fossils. This process of conversion is called racemization, and the 50–50% mixture is termed racemic. When an amino acid contains two groups that have this symmetry difference, but only one changes, the term epimerization is used. This is true of leucine, which can exist

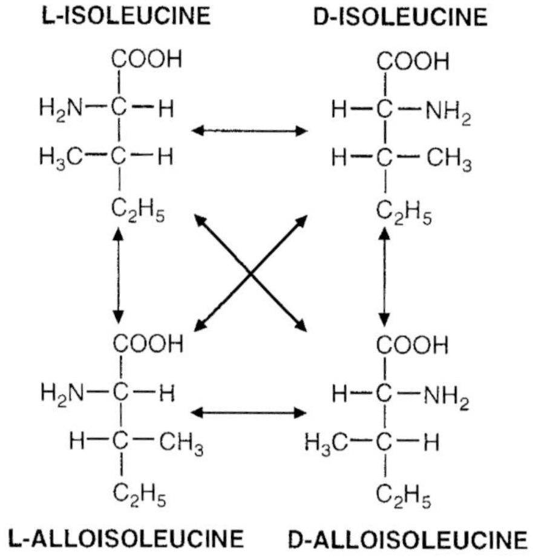

L-ISOLEUCINE D-ISOLEUCINE

L-ALLOISOLEUCINE D-ALLOISOLEUCINE

FIG. 13. Stereochemical configurations of isoleucine (after Mitterer *et al.*, 1993).

as epimers isoleucine or alloisoleucine, each of which also has L and D configurations. This difference is demonstrated in Fig. 13 (after Mitterer, 1993). The general term racemization is usually also applied to epimerization.

The rate at which racemization occurs is a very sensitive function of temperature. For this reason, it is most often used as a relative measure in aminostratigraphy, the ordering of what are termed aminozones by their D/L ratios for a specific amino acid. These zones can be dated if a specific group or zone can be correlated with a known age, as determined by an independent technique such as radiometric dating. Aminochronology is the application of these calibrated D/L ratios to predict ages for uncalibrated samples. This involves modeling the time dependence of the racemization for the appropriate temperature conditions (Wehmiller, 1993).

Racemization rates are also dependent on the specific type of organism studied. For example, under the same conditions of temperature, floraminifera from the genus *G. tumida* has a different rate of racemization from that of the genus *O. universa*. This effect is shown in Fig. 14(a), where the ratio of D-alloisoleucine to L-isoleucine is plotted as a function of age for the two genera, obtained from the same deep-sea sediment core (Müller, 1984). In some cases concordia plots for two genera can be constructed and used to determine age in the same manner as those for radiogenic nuclides (see Sec. 2.1.6). Figure 14(b) shows this for the data displayed in Fig. 14(a).

The very sensitive temperature dependence of the rate of racemization can be used in the framework of a kinetic model to predict temperature variations in Quaternary deposits during the last ice age. This technique has been applied to mollusk shells in the Mississippi Valley. Radiocarbon and thermoluminescence dating were used to obtain the time scale. Figure 15 shows the effective diagenetic temperatures (EDT) derived from modeling the rates of epimerization of isoleucine in five genera of gastropods (Oches *et al.*, 1996). The results are plotted along with the mean annual air temperature from 1951 to 1980.

Paleontological applications of aminostratigraphy and aminochronology can be found in Wehmiller (1993) and the references therein. Specific applications in archaeology are discussed in Aitken (1990).

6. CHLOROFLUOROCARBONS

An example of an anthropogenic source that can be turned into a modern dating tool is the release of compounds classified as chlorofluorocarbons (CFCs). First manufactured in the 1930s, these compounds were valued for their stability and used extensively as refrigerants and propellants until they became implicated in the destruction of stratospheric ozone. They are highly problematic because they have been accumulating rapidly in the atmosphere and have residence times of 120 y for CCl_3F (trade name F-11) and 60 y for CCl_2F_2 (trade name F-12) (Derra, 1990). It is just this stability which makes them excellent tracers of modern processes.

The detection of CFCs in groundwaters indicates post-1945 input, and studies have correlated CFC concentrations with tritium dating. Its use as an actual dating technique requires accurate profiles of the change in atmospheric CFC concentrations and the temperature of the recharge zone over time. Sorption onto soils is also a consideration. Details of the use of CFCs and the analytical techniques required can be found in Busenberg and Plummer (1992).

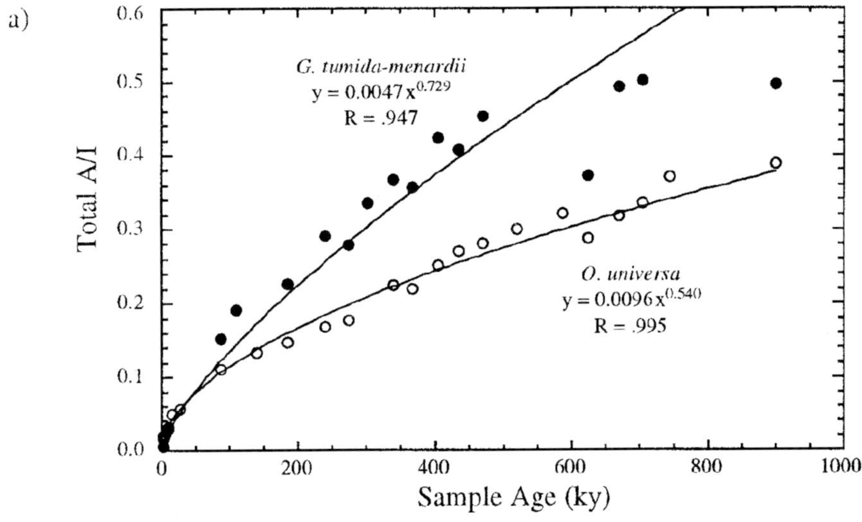

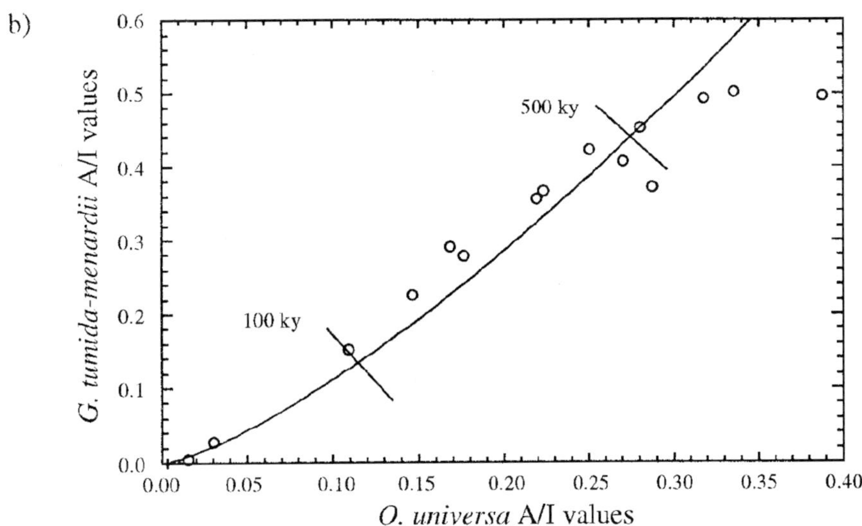

FIG. 14. (a)Ratio of D-alloisoleucine to L-isoleucine (A/I) as a function of time for floraminifera *G. tumida-menardii* (solid circles) and *O. universa* (open circles), measured in the same deep-sea sediment. (b) Concordance in the (A/I) ratios for *O. universa versus G. tumida-menardii*, plotted from the data displayed in part (a). Data from Müller (1984).

7. CONCLUSION

The application of techniques that can provide either absolute or relative chronologies is central to a wide range of research in the fields of geology, hydrology, oceanography, archaeology, paleontology, cosmology, and the environmental sciences. This article has provided an overview of techniques used today and some of the possible applications. Each has its strengths and weaknesses, requiring different sets of assumptions for validity. Hence, the study of a specific process or system is best served when a range of dating tools is

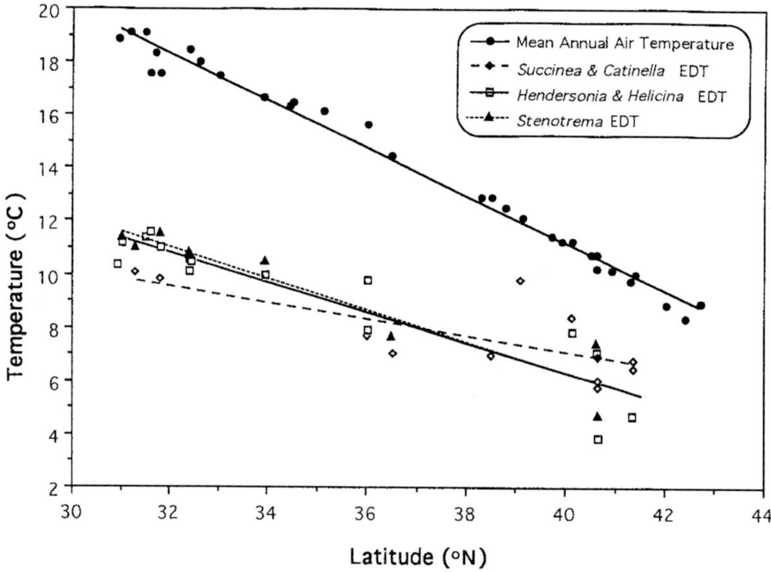

FIG. 15. Temperature profile for ice-age gastropods as a function of latitude in Mississippi Valley loess, as derived from modeling the epimerization rates of leucine. The mean annual air temperature for the period 1951–1980 is shown in the upper curve for comparison (from Oches *et al.*, 1996).

drawn upon to corroborate results. As in the cases of AMS and TIMS, analytical advances are constantly making these tools accessible to new areas of research.

GLOSSARY

Alpha Particle (α): Nucleus of a helium atom, consisting of two protons and two neutrons.

Anthropogenic: Influenced by the impact of man.

Apatite: A group of hexagonal minerals consisting of calcium phosphate together with fluorine, chlorine, hydroxyl, or carbonate.

Basalt: A dark-colored igneous rock, commonly extrusive, composed primarily of calcic plagioclase and pyroxene. Apatite and magnetite are almost always present.

Beta particle (β): A positive or negative electron that is emitted from the nucleus during radioactive decay.

Calcite: A common rock-forming mineral, $CaCO_3$, a chief constituent of limestone and most marble.

Connate: Water trapped in the interstices of sedimentary rock at the time the rock was deposited.

Dose: A measure of energy deposited in a material by ionizing radiation (α, β, γ).

Feldspar: A group of abundant rock-forming minerals of the general formula $M\mathrm{Al(Al, Si)_3O_8}$, where M can be K, Na, Ca, Ba, Rb, Sr or Fe.

Floraminifera: Large single-celled animals with a calcareous shell of calcium carbonate or calcite.

Gamma ray (γ): High-energy electromagnetic radiation emitted when a nucleus converts from a higher energy state to a lower.

Gastropod: A class of mollusks that includes snails and slugs.

Glauconite: A green mineral closely related to micas and essentially a hydrous potassium iron silicate.

Hematite: A common mineral containing Fe_2O_3, found in igneous, sedimentary and metamorphic rocks.

Hydrosphere: The waters of the Earth, as distinguished from the rocks (lithosphere), living things (biosphere) and the air (atmosphere).

Igneous: A rock or mineral that has solidified from molten material, e.g., magma.

Lithosphere: The solid portion of the earth.

Loess: A blanket deposit of buff-colored calcareous silt, homogeneous, nonstratified, weakly coherent, and porous; considered to be windblown dust of the Pleistocene age.

Magnetite: A black, opaque, strongly magnetic mineral of the group $(Fe,Mg)Fe_2O_4$.

Muon: A weakly interacting particle, similar to an electron but more massive.

Nuclide: A specific type of nucleus, with a given number of protons and neutrons.

Paleo: Old or ancient.

Pleistocene: The early epoch of the Quaternary period, which ended around 8000 y ago.

Pluton: An igneous intrusion.

Quarternary: Geologic time period from present back to 1.8 My ago.

Quartz: Crystalline silica, an important rock-forming mineral, SiO_2.

Radiogenic: Formed as the result of radioactive decay.

Radionuclide: A nuclide that is unstable and decays to another nuclide.

Tectonic: Pertaining to the forces involved in or the resulting structures of tectonics.

Tectonics: Branch of geology dealing with the broad architecture of the outer part of the earth.

Zircon: A mineral, $ZrSiO_4$.

Works Cited

Aitken, M. J. (1985), *Thermoluminescence Dating*, London: Academic Press.

Aitken, M. J. (1990), *Science-Based Dating in Archaeology*, London: Longman.

Andrews, J. N., Fontes, J.-C., Michelot, J.-L., Elmore, D. (1986), *Earth Planet. Sci. Lett.* **77**, 49–58.

Andrews, J. N., Davis, S. N., Fabryka-Martin, J., Fontes, J.-C., Lehmann, B. E., Loosli, H. H., Michelot, J.-L., Moser, H., Smith, B., Wolf, M. (1989), *Geochim. Cosmochim. Acta* **53**, 1803–1815.

Bard, E., Hamelin, B., Fairbanks, R. G., Zindler, A. (1990), *Nature (London)* **345**, 405–410.

Beer, J., Blinov, A., Bonani, G., Finkel, R. C., Hofmann, H. J., Lehmann, B., Oeschger, H., Sigg, A., Schwander, J., Staffelbach, T., Stauffer, B. R., Suter, M., Wolfli, W. (1990), *Nature (London)* **347**, 164–166.

Bentley, H. W., Phillips, F. M., Davis, S. N., Gifford, S., Elmore, D., Tudds, L. E., Gove, H. E. (1982), *Nature (London)* **300**, 737–740.

Bentley, H. W., Phillips, F. M., Davis, S. N., Habermehl, M. A., Airey, P. L., Calf, G. E., Elmore, D., Gove, H. E., Torgerson, T. (1986), *Water Resour. Res.* **22**, 1991–2002.

Bierman, P., Gillespie, A., Caffee, M., Elmore, D. (1995), *Geochim. Cosmochim. Acta* **59**, 3779–3798.

Bollhöfer, A., Mangini, A., Lenhard, A., Wessels, M., Giovanoli, F., Schwarz B. (1994), *Environ. Geol.* **24**, 267–274.

Brannon, J. C., Cole, S. C., Podosek, F. A., Ragan, V. M., Coveney Jr., R. M., Wallace, M. W., Bradley, A. J. (1996), *Science* **271**, 491–493.

Busenberg, E., Plummer, N. (1992), *Water Resour. Res.* **28**, 2257–2283.

Cerling, T. E., Craig, H. (1994), *Ann. Rev. Planet. Sci.* **22**, 273–317.

Clark, D. H., Bierman, P. R., Larsen, P. (1995), *Quarter. Res.* **44**, 367–377.

Creaser, R. A., Papanastassiou, D. A., Wasserburg, G. J. (1991), *Geochim. Cosmochim. Acta* **55**, 397–401.

Derra, S. (1990), *Res. Develop.* **32**, 54–66.

Dickin, A. P. (1995), *Radiogenic Isotope Geology*, Cambridge, UK: Cambridge University Press.

Ding, G.-J., Kilius, L. R., Wilson, G. C., Zhao, X.-L., Rucklidge, J. C. (1997), *Nucl. Instrum. Methods Phys. Res.* **B123**, 424–430.

Drimmie, R. J., Aravena, R., Wassenaar, L. I., Fritz, P., Hendry, M. J., Hut, G. (1991), *Appl. Geochem.* **6**, 381–392.

Edwards, R. L., Chen, J. H., Wasserburg, G. J. (1987), *Earth Planet. Sci. Lett.* **81**, 175–192.

Edwards, R. L., Beck, J. W., Burr, G. S. (1993) *Science* **260**, 962–967.

Edwards, R. L., Cheng, H., Murrell, M. T., Goldstein, S. J. (1997), *Science* **276**, 782–786.

Elmore, D., Phillips, F. M. (1987), *Science* **236**, 543–550.

Esperança, S., Carlson, R. W., Shirey, S. B., Smith, D. (1997), *Geology* **25**, 651–654.

Fabryka-Martin, J. T., Davis, S., Elmore, D. (1987), *Nucl. Instrum. Methods Phys. Res.* **B29**, 361–371.

Fabryka-Martin, J. T., Davis, S. N., Elmore, D., Kubik, P. W. (1989), *Geochim. Cosmochim. Acta* **53**, 1817–1823.

Faure, G. (1986), *Principles of Isotope Geology*, 2nd ed., New York: John Wiley & Sons Inc.

Freydier, C., Ruiz, J., Chesley, J., McCandless, T., Munizaga, F. (1997), *Geology* **25**, 775–778.

Fröhlich, K., Ivanovich, M., Andrews, J. N., Davis, S. N., Drimmie, R. J., Fabryka-Martin, J., Florkowski, T., Fritz, P., Lehmann, B., Loosli, H. H., Nolte, E. (1991), *Appl. Geochem.* **6**, 465–472.

Gillespie, A. R., Bierman, P. R. (1995), *J. Geophys. Res.* **100**, 24 637–24 649.

Graham, I. J., Ditchburn, R. G., Sparks, R. J., Whitehead, N. E. (1997), *Nucl. Instrum. Methods Phys. Res.* **B123**, 301–318.

Gu, Z. Y., Lal, D., Liu, T. S., Southon, J., Caffee, M. W., Guo, Z. T., Chen, M. Y. (1996), *Earth Planet. Sci. Lett.* **144**, 273–287.

Innocent, C., Parron, C., Hamelin, B. (1997), *Geochim. Cosmochim. Acta* **61**, 3753–3761.

Johnson, C. (1997), *Geology* **25**, 623–626.

Knoll, G. F. (1979), *Radiation Detection and Measurement*, New York: John Wiley & Sons, Inc.

Kretschmer, W., Anton, G., Bergmann, M., Finckh, E., Kowalzik, B., Klein, M., Leigart, M., Merz, S., Morgenroth, G., Piringer, I., Küster, H., Low, R. D., Nakamura, T. (1997), *Nucl. Instrum. Methods Phys. Res.* **B123**, 455–459.

Lal, D., Peters, B. (1967), in: K. Sitte (Ed.), *Handbuch der Physik*, Vol. 46/2, Berlin: Springer, pp. 551–612.

Lal, D. (1991), *Earth Planet. Sci. Lett.* **104**, 424–439.

Laslett, G. M., Green, P. F., Duddy, I. R., Gleadow, A. J. W. (1987), *Chem. Geol.* **65**, 1–13.

Lehmann, B. E., Davis, S. N., Fabryka-Martin, J. T. (1993), *Water Resour. Res.* **29**, 2027–2040.

Libby, W. F. (1955), *Radiocarbon Dating*, Chicago: Univ. of Chicago Press.

Lui, B., Phillips, F. M., Fabryka-Martin, J. T., Fowler, M. M., Stone, W. D. (1994), *Water Resour. Res.* **30**, 3115–3125.

Mazaud, A., Laj, C., Bender, M. (1994), *Geophys. Res. Lett.* **21**, 337–340.

Michlovich, E. S., Vogt, S., Masarik, J., Reedy, R. C., Elmore, D., Lipschutz, M. E. (1994), *J. Geophys. Res.* **99**, 187–194.

Milton, G. M., Andrews, H. R., Causey, S. E., Chant, L. A., Cornett, R. J., Davies, W. G., Greiner, B. F., Koslowsky, V. T., Imahori, Y., Kramer, S. J., McKay, J. W., Milton, J. C. D. (1994), *Nucl. Instrum. Methods Phys. Res.* **B92**, 376–379.

Mitterer, R. M. (1993), in: M. H. Engel, S. A. Macko, (Eds.), *Organic Geochemistry: Principles and Applications*, New York: Plenum, Chap. 35.

Morgenstern, U., Gellerman, R., Hebert, D., Börner, I., Stolz, W., Vaikmäe, R., Rajamäe, R., Putnik, H. (1995), *Chem. Geol.* **120**, 127–134.

Morgenstern, U., Taylor, C. B., Parrat, Y., Gäggeler, H. W., Eichler, B. (1996), *Earth Planet. Sci. Lett.* **144**, 289–296.

Müller, P. J. (1984), *"Meteor" Forsch.-Ergebnisse, Reihe C.* **38**, 25–47.

Murphy, E. M., Davis, S. N., Long, A., Donahue, D., Jull, A. J. T. (1989), *Water Resour. Res.* **25**, 1893–1905.

NADP/NTN (1990), National Atmospheric Deposition Program, *Annual Data Summary: Precipitation Chemistry in the United States*, Natural Resource Ecology Laboratory, Colorado State University, Fort Collins, CO, U.S.A.

Nishiizumi, K., Kohl, C. P., Shoemaker, J. R., Arnold, J. R., Klein, J., Fink, D., Middleton, R. (1991), *Geochim. Cosmochim. Acta* **55**, 2699–2703.

Noble, S. R., Aspden, J. A., Jemielita, R. (1997), *GSA Bull.* **109**, 789–798.

Nolte, E., Krauthan, P., Korschinek, G., Maloszewski, P., Fritz, P., Wolf, M. (1991), *Appl. Geochem.* **6**, 435–445.

Oches, E. A., McCoy, W. D., Clark, P. U. (1996), *GSA Bull.* **108**, 892–903.

Ortega-Rivera, A., Farrar, E., Hanes, J. A., Archibald, D. A., Gastil, R. G., Kimbrough, D. L., Zentilli, M., Lopez-Martinez, M., Feraud, G., Ruffet, G. (1997), *GSA Bull.* **109**, 728–745.

Phillips, F., Zreda, M. G., Smith, S. S., Elmore, D., Kubik, P. W., Dorn, R. I., Roddy, D. R. (1991), *Geochim. Cosmochim. Acta* **55**, 2695–2698.

Purdy, C. B., Helz, G. R., Mignerey, A. C., Kubik, P. W., Elmore, D., Sharma, P., Hemmick, T. (1996), *Water Resour. Res.* **32**, 1163–1171.

Raisbeck, G. M., Yiou, F., Bourles, D., Kent, D. V. (1985), *Nature (London)* **315**, 315–317.

Raisbeck, G. M., Zhou, Z. Q., Kilius, L. R. (1994), *Nucl. Instrum. Methods Phys. Res.* **B92**, 436–439.

Ravenhurst, C. E., Willet, S. D., Donelick, R. A., Beaumont, C. (1994), *J. Geophys. Res.* **99**, 20023–20041.

Santschi, P. H., Schlink, D. R., Oktay-Marshall, S., Corapcioglu, O., Fehn, U., Sharma, P. (1996), *Deep Sea Res.* **43**, 259–265.

Schlosser, P., Stute, M., Dörr, H., Sonntag, C., Münnich, K. O. (1988), *Earth Planet. Sci. Lett.* **89**, 352–363.

Schlosser, P., Kromer, B., Ekwurzel, B., Bönisch, G., McNichol, A., Schneider, R., von Reden, K., Ostlund, H. G., Swift, J. H. (1997) *Nucl. Instrum. Methods Phys. Res.* **B123**, 431–437.

Smith, P. E., Farquhar, R. M. (1989), *Nature (London)* **341**, 518–521.

Smoliar, M. I., Walker, R. J., Morgan, J. W. (1996), *Science* **271**, 1099–1102.

Solomon, D. K., Poreda, R. J., Cook, P. G., Hunt, A. (1995), *Ground Water* **33**, 988–996.

Somayajulu, B. L. K., Rengarajan, R., Lal, D., Weiss, R. F., Craig, H. (1987), *Earth Planet. Sci. Lett.* **85**, 329–342.

Steig, E. J., Polissar, P. J., Stuiver, M., Grootes, P. M., Finkel, R. C. (1996), *Geophys. Res. Lett.* **23**, 523–526.

Stuiver, M., Reimer, D. J. (1993), *Radiocarbon* **35**, 215–230.

Sutton, S. R. (1985), *Geophys. Res.* **90**, 3690–3700.

Synal, H.-A., Beer, J., Bonani, G., Suter, M., Wölfli, W. (1990) *Nucl. Instrum. Methods Phys. Res.* **B52**, 483–488.

Taylor, R. E. (1987), *Anal. Chem.* **59**, 317A–331A.

Titayeva, N. A. (1994), *Nuclear Geochemistry*, Moscow: Mir Publishers.

Turner, G. M., Kamp, P. J. J. (1990), *Earth Planet. Sci. Lett.* **100**, 42–50.

von Reden, K. F., McNichol, A. P., Peden, J. C., Elder, K. L., Gagnon, A. R., Schneider, R. J. (1997), *Nucl. Instrum. Methods Phys. Res.* **123**, 438–442.

Wang, L., Ku, T. L., Luo, S., Southon, J. R., Kusakabe, M. (1996), *Geochim. Cosmochim. Acta* **60**, 109–119.

Wang, Y., Amundson, R., Trumbore, S. (1996), *Quater. Res.* **45**, 282–288.

Wehmiller, J. F. (1993), in: M. H. Engel, S. A. Macko, (Eds.), *Organic Geochemistry: Principles and Applications*, New York: Plenum, Chap. 36.

Wolf, R. A., Farley, K. A., Silver, L. T. (1996), *Geochim. Cosmochim. Acta* **60**, 4231–4240.

Wolf, R. A., Farley, K. A., Silver, L. T. (1997), *Geology* **25**, 65–68.

Further Reading

Aitken, M. J. (1985), *Thermoluminescence Dating*, London: Academic Press.

Aitken, M. J. (1990), *Science-Based Dating in Archaeology*, London: Longman.

Choppin, G., Rydberg, J., Liljenzin, J. O. (1995), *Radiochemistry and Nuclear Chemistry*, 2nd ed., Oxford: Butterworth-Heinemann Ltd.

Dickin, A. P. (1995), *Radiogenic Isotope Geology*, Cambridge, UK: Cambridge University Press.

Faure, G. (1986), *Principles of Isotope Geology*, 2nd ed., New York: John Wiley & Sons Inc.

Knoll, G. F. (1979), *Radiation Detection and Measurement*, New York: John Wiley & Sons, Inc.

Tuniz, C., Bird, J. R., Fink, D., Herzog, G. F. (1998), *Accelerator Mass Spectrometry, Ultra Sensitive Analysis for Global Science*, Boca Raton, FL: CRC Press.

INTERFEROMETRIC TECHNIQUES

JOANNA SCHMIT, *Veeco Instruments Incorporated, Tucson, Arizona, U.S.A.*

KATHERINE CREATH, *Creath Optineering Services, Tucson, Arizona, U.S.A.*

INTRODUCTION

This article focuses on interferometric techniques, which are various methods for obtaining and analyzing an interferogram (fringe pattern) in an interferometer (see INTERFEROMETERS AND INTERFEROMETRY). An interferogram is the recorded interference signal between two beams of light exiting from the same radiation source. An interferogram carries information about the phase (the difference between the two wave fronts of the beams), the spectral content of the source, and its spatial distribution. These three phenom-

ISBN 3-527-29308-6

ena, and the major interferometric techniques used to collect these data, provide the general topics discussed in this article. First is a discussion of phase measurement (see Secs. 1 and 4); next is a brief examination of the spectral distribution of the source (see Sec. 6), followed by a look at the spatial distribution of the source (see Sec. 7). By combining techniques it is possible to measure more than one of these phenomena at a time (see Sec. 6.3).

In the discussion of phase measurement a distinction between techniques that detect phase at a single point and those that detect across the plane of an interferogram is made. Techniques that measure phase are used to image test-object characteristics such as surface topography, optical-system quality, homogeneity of media, relative position of targets, and existence of gravity waves. Measuring the spectral distribution of radiation is one of the oldest applications of interferometry; a description of two basic interference spectrometers is provided in this article. The techniques that astronomy has developed for testing the spatial distribution of source radiation in the measurement of stellar diameters are also briefly described. Finally, because phase can represent so many characteristics, a number of techniques adapted from those based on phase measurement are mentioned. Experimental mechanics tests for object displacements; refractometry and ellipsometry measure the material refractive index and thin-film thickness; and optical-fiber interferometers sense temperature and rotation. All these are applications based on interferometry. This article covers neither interference phenomena nor basic types of interferometers, as these topics were discussed in INTERFEROMETERS AND INTERFEROMETRY and OPTICS, LINEAR.

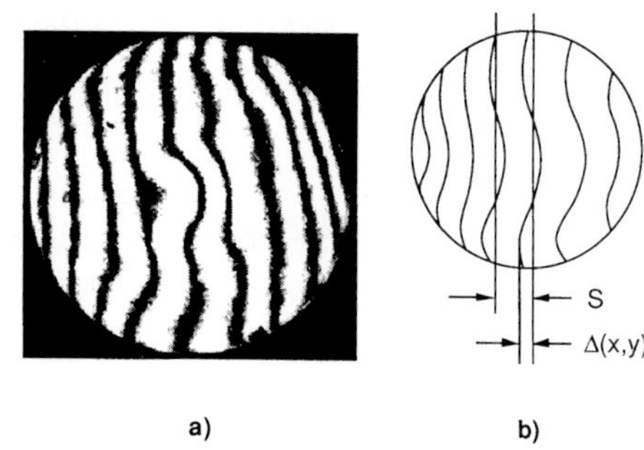

FIG. 1. (a) Typical interferogram for surface contouring. (b) Surface height deviation from a plane $= (\lambda/2)[\Delta(x, y)/S]$.

1. PHASE ACROSS THE PLANE

The techniques that measure the phase across the plane employ a two-beam interferometer to record an interference signal in the form of an interferogram (fringe pattern). These interferograms, consisting of fringes of equal thickness called Fizeau fringes (see INTERFEROMETERS AND INTERFERENCE), carry information about the phase, which corresponds to the relative difference between the test and the reference wave front. Phase variation across the plane can represent surface shape, material homogeneity or thickness, or optical-component quality. A simple way to approximate the maximum deviation of phase from the plane involves introducing tilt between the test and reference wave fronts. In this way almost straight and parallel fringes are introduced into the interferogram, Fig. 1(a). Phase deviation from the plane is seen as a departure from straight lines in the fringe pattern, Fig. 1(b). In the two interferometers most commonly used for surface contouring, the Fizeau and Twyman–Green (see INTERFERENCE AND INTERFEROMETRY), the spacing between two neighboring fringes corresponds to a surface height difference of 1/2 of the source wavelength used for testing (1 fringe $= \lambda/2$). The maximum surface height deviation from the plane, Fig. 1(b), can be found by measuring the maximum fringe deviation from a straight line $[\Delta(x, y)]$ and the average spacing between two neighboring fringes (S). However, the relationship between fringe spacing and surface height depends not only on the wavelength used for measurement but also the interferometric configuration and the indices of refraction of the media the beam passes through. Interferometric techniques used to measure the phase across the plane typically use a coherent light source such as a laser so as to achieve good-contrast fringes; these techniques may involve single or multiple interferogram analysis to retrieve the phase.

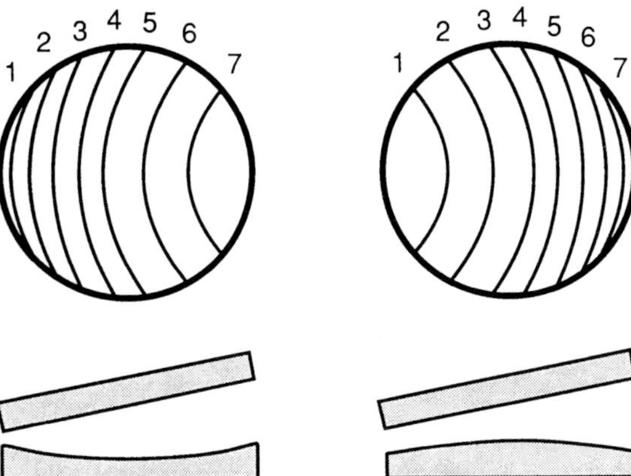

FIG. 2. Interferograms for (a) concave and (b) convex test surfaces relative to a flat reference surface.

1.1 Fringe Tracking

Until the late 1970s, fringe tracking (Yatagai, 1993) was the main way to analyze interferograms quantitatively, and although this method required only a single interferogram, it was relatively imprecise; the accuracy obtained was only one tenth of the fringe spacing. Fringe tracking works by recording the positions of the fringe intensity maxima and minima; the rest of the data are interpolated in order to create a phase map across the plane. Fringe tracking requires that the fringes be ordered so as to reflect changes in the phase between fringes, a process simplified by introducing tilt between the reference and test wave fronts. However, the direction of the tilt must be known beforehand in order to determine the shape of the tested surface correctly, *i.e.*, concave or convex (Fig. 2). If closed fringes are present in the interferogram, then fringe ordering also requires some knowledge about the tested element.

1.2 Interferogram Plus Intensities

The phase across the plane can be determined at each point without interpolation if the intensity distribution of each beam is recorded separately and one interferogram is taken. After these three intensity distributions have been determined, the phase variation can be calculated from the following equation that describes the intensity distribution I in an interferogram:

$$I = I_1 + I_2 + 2(I_1 I_2)^{1/2} \cos \varphi(x, y), \qquad (1)$$

where I_1 and I_2 are the intensities of the two beams, $2(I_1 I_2)^{1/2}$ is the interference visibility (modulation) and $\varphi(x, y)$ is the measured phase. Michelson or Mach–Zehnder interferometers (see INTERFERENCE AND INTERFEROMETRY) can use this method since the beams in the reference and test arms can be blocked separately. However, blocking each beam separately is not always possible, as when using a Fizeau interferometer, and alternative methods for retrieving the phase must be used.

1.3 Temporal Phase Measurement—Temporal Heterodyning

Temporal phase-measurement techniques (also known as temporal heterodyning, phase stepping, or phase shifting) introduce multiple times a known increment in the relative difference between the test and reference beams, called a phase shift. At least three interferograms are necessary to determine the phase encoded in the intensity distribution, Eq. (1), that can be rewritten as

$$I = I_1 + I_2 + 2(I_1 I_2)^{1/2} \cos[\varphi(x, y) + m\tau] \qquad (2)$$

where τ is the phase shift induced m times. The

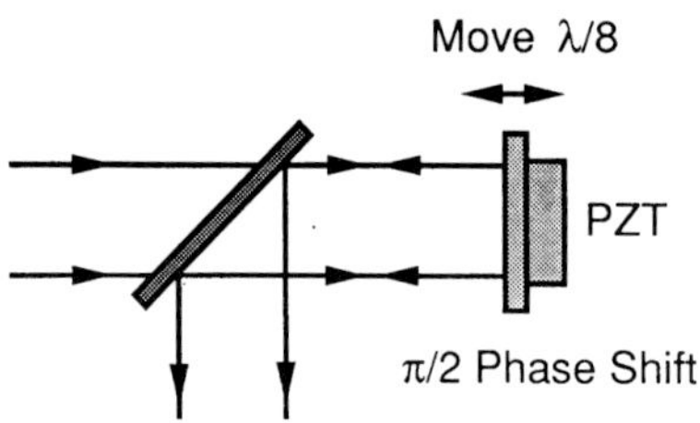

FIG. 3. Phase shift generated using a mirror moved by a piezo-electric transducer (PZT).

most common way to accomplish this phase shift is by changing the optical path difference between the interfering beams through a shift of the reference mirror along the optical axis, Fig. 3, although other ways exist—for example, by tilting a glass plate, moving a grating, rotating a half-wave plate or analyzer, or using an acousto-optic modulator (Creath, 1993). While some of these ways of shifting the phase alter the optical path between the interfering beams, others change the optical phase of one of the beams; nevertheless, both methods lead to an incremental difference (phase shift) between beams. Figure 4 shows three typical interferograms registered in the temporal phase measurement technique.

The measurement precision can be improved using more samples. An example of a common algorithm employing five intensity samples $I_1, \ldots, I_5$ with a relative $\pi/2$ phase shift to retrieve the phase $\varphi(x, y)$ is (Creath, 1993)

$$\varphi(x, y) = \arctan\left(\frac{I_1 + I_5 - 2I_3}{2I_2 + 2I_4}\right) \tag{3}$$

Because an arctangent function is not continuous and gives multiple solutions, an unwrapping procedure is employed to remove the discontinuities in the retrieved phase and essentially automatically orders the fringes (Robinson, 1993). Unwrapping procedures assume that the maximum phase discontinuity equals the difference between two neighboring solutions (2π, which is a property of the arctangent function). The values of the phase, which are measured in radians, are then converted into the values that the phase represents—for example, surface height or thickness variations. Temporal phase measurement is 10–100 times more precise than tracking the position of the fringe maxima and minima, since the repeatability of temporal phase measurement is 1/100–1/1000 of a wavelength.

Temporal phase-measurement techniques generally require that each interference fringe be sampled at least twice per fringe spacing; this constraint limits the slope of the phase that can be measured. However, a few techniques work around this limitation (Greivenkamp and Bruning, 1992). An often-used solution, white-light interferometry (see Sec. 1.5), is suitable when measuring rough surfaces or surfaces with discontinuities where sampling twice per fringe is not possible (Larkin, 1996).

Finally, although temporal phase-measurement techniques usually capture interferograms sequentially in time, and only phase not varying in time can be analyzed, a few techniques modify the configuration of the interferometer so the that *time-varying phase* can be measured, *e.g.*, gas flow. These techniques record interferograms simultaneously using multiple cameras. However, accomplishing

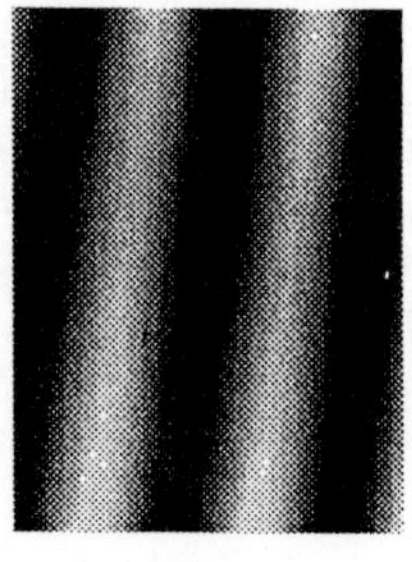

no phase shift (0°)

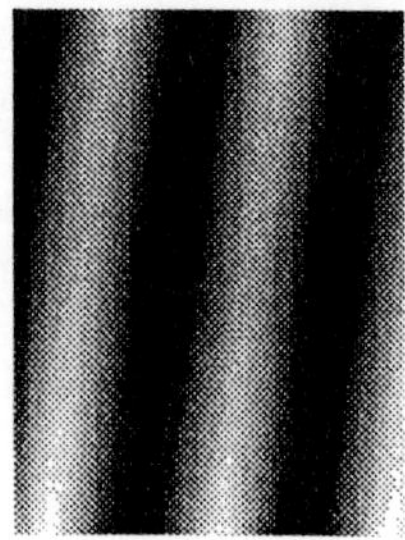

π/2 phase shift (90°)

π phase shift (180°)

FIG. 4. Interferograms showing sequence of phase shifts.

this end requires complicated hardware and alignment is critical (Kujawinska, 1993).

1.4 Single Interferogram Analysis—Spatial Heterodyning

Measuring time-varying phase can also be accomplished through the analysis of a single interferogram. Two common techniques discussed below require that a large tilt be induced between the reference and test wave fronts and that the direction of the tilt or knowledge about the tested object be known beforehand. In addition, like the temporal phase-measurement techniques described above, these techniques also require a phase unwrapping procedure in order to obtain a continuous phase map, and each fringe must be sampled at least twice per fringe spacing.

1.4.1 Fourier-Transform Technique In the Fourier-transform technique (Takeda, 1990) the discrete Fourier transform of a single interferogram is calculated revealing three distinct spectral orders, typical for a consinusoidal function, which also represents fringes (see FOURIER AND OTHER MATHEMATICAL TRANSFORMS; Ramirez, 1985). The number of almost parallel fringes from the induced tilt between the reference and test wave fronts must be large enough to separate the spectral orders to enable filtering one of these orders at the spatial frequency of the fringes. The inverse Fourier transform is then performed and the phase encoded in the interferogram is retrieved from the arctangent function of the real and imaginary parts of the inverse Fourier transform.

1.4.2 Synchronous Detection Synchronous detection (Womack, 1984) calculates the phase by sequentially applying a temporal phase-measurement technique algorithm to a few consecutive intensity samples across the plane of a single interferogram. A known phase shift is induced between adjacent sampling points, rather than between interferograms as discussed above, by tilting one wave front with respect to the other. This phase shift manifests itself as a set of straight parallel fringes. Fringe spacing must be chosen carefully so that the phase difference between adjacent sampling points is close to the phase shift required by the algorithm employed. Uniform sample reflectivity and intensity of both beams are assumed in this technique.

1.5 White-light Interferogram Analysis

Interference fringes can also be observed by use of a light source with a low degree of coherence such as a halogen lamp. Interferograms obtained with these "white light" sources show fringes of good modulation only when the optical paths of the two beams match closely. This modulation characteristic becomes important when one is trying to determine large discontinuities in phase where temporal phase-shifting interferometry is unable to assign the correct order to the fringes. *White-light interferometry* (also known as *low-coherence interferometry*) allows for the easy identification of the fringe with the best visibility because the contrast falls off so quickly. For example, Figs. 5(a) and 5(b) show interferograms of a step object; the interferogram in Fig. 5(a) is achieved with a laser as a source, and in Fig. 5(b) a white light source is used. In Fig. 5(a) there are multiple solutions to the ordering of the fringes. Thus determining the height of the step (size of the phase discontinuity) is impossible without additional information. In Fig. 5(b) the fringes can be ordered fairly simple by tracking the highest visible fringe, and the step height can then be easily determined. White-light interferometry is generally used for measuring rough surfaces or any surface with discontinuities where sampling twice per fringe is not possible.

To measure a surface using white-light interferometry, the surface is scanned in height relative to the interferometer while frames of data are acquired. As the separation between the surface and the interferometer changes, the interference fringes will intersect heights on the test surface where the optical paths are equal. A complete scan conceptually includes a three-dimensional volume of data corresponding to the location of the white-light fringes in space. The highest-contrast fringe or the peak of the fringe modulation can be used to determine the height at a single point on the test surface. This simple approach has spurred a number of new techniques for determining surface structure (Larkin, 1996). Some techniques follow

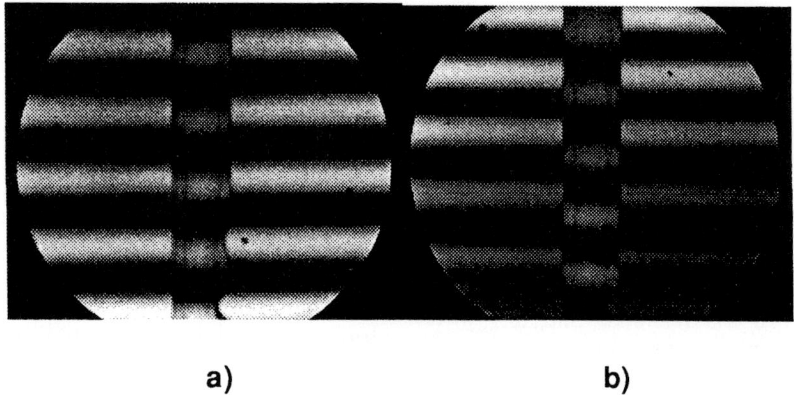

FIG. 5. Interferogram of a line feature with (a) a laser source and (b) a white-light source.

the highest-contrast fringe and others look at the peak of the fringe modulation. The relative position of these two quantities depends upon the phase change upon reflection of the light from the test surface.

1.6 Summary

The various interferogram techniques described in this section complement one other. The temporal technique and the single-interferogram techniques are used for high-precision measurement of smooth surfaces, while the fringe-modulation peak-sensing technique can examine very rough and discontinuous surfaces. Moreover, interferometric techniques analyze not only static but also time-varying phase. Techniques such as synchronous detection, Fourier transform, and temporal-phase measurement employing a spatial separation of interferograms can measure time-varying phase.

2. MEASUREMENT SENSITIVITY AND LIMITATIONS

2.1 Measurement Sensitivity

The sensitivity of an interferometric measurement can be changed by altering the test wavelength, the angle of incidence of the test beam illumination, and the number of passes through or reflections from the tested element. Varying the sensitivity of the interferometer allows for the measurement of rougher or smoother test objects. For example, a source that emits a shorter wavelength will produce a greater number of fringes in an interferogram; this choice of source wavelength increases the sensitivity of the interferometer and allows for the testing of smoother objects with high precision. A large angle of incidence of test beam illumination decreases sensitivity by producing fewer fringes in the interferogram and allows for rougher test-object measurement.

2.2 Removal of Systematic Errors

The reference wavefront in an interferometric setup is not ideal, for it carries information about imperfections in both the reference mirror and the optical system of the interferometer. If an ideal surface were available, all systematic errors could be measured and then subtracted. However, because such a surface does not exist, a set of multiple measurements is necessary in order to subtract these systematic errors (Schultz and Schwider, 1976). In addition, because various interferogram analysis techniques contain systematic sources of error, ways have been developed to decrease this error. For example, phase-shifting interferometry is vulnerable to phase-shifter miscalibration. Methods of calibrating the phase shifter precisely and the use of error-reducing algorithms help reduce any systematic errors (Creath, 1993).

2.3 Reduction of Nonsystematic Errors

High-precision interferometric measurements require special laboratory conditions. Care needs to be taken to ensure vibration and acoustic isolation as well as temperature stability and controlled air flow within the exposed optical path. If the laboratory conditions are not well controlled, nonsystematic errors may dominate a measurement. However, as long as these errors are random, statistically independent measurements can be averaged to reduce their effects and improve measurement precision and accuracy.

3. FORM MEASUREMENT

Testing form requires the measurement of phase across the plane. Form is defined as object shape or optical-system quality, in order to differentiate it from other measurable quantities such as object displacement and vibration. This section divides the examination of form measurement into optical-component testing and engineering-surface testing. The optical-component testing section is broken down into the types of components tested because similar components are tested in similar interferometric configurations. The engineering-surface testing section is organized around measurement techniques because the wide range of engineering surfaces requires a variety of different testing methods. The interferometric technique used to measure form depends on both the features of the tested object, such as its size, shape, or roughness, and the type of data that needs to be obtained.

3.1 Optical-System and -Component Testing

Interferometric techniques have proven extremely useful in testing optical systems and components, especially with the strict tolerances imposed on high-precision components and the smooth wave fronts generated by optical systems. The techniques detailed below for testing optical elements are applicable to most types of interferometers but are mainly used with the Fizeau or Twyman–Green interferometers (see Interferometers and Interferometry, Secs. 2.1 and 2.4). The two interferometers have different strengths. The Fizeau interferometer is less sensitive to vibration than the Twyman–Green because the two optical paths of the Fizeau interferometer are virtually common. However, because the test and reference optical paths differ in length, the Fizeau requires a source with higher coherence than a Twyman–Green does.

3.1.1 Flats Optical flats are typically measured against a reference flat in a collimated beam. Often the flatness is specified as the maximum surface irregularity over a given surface area. Commercially available instruments can routinely measure surface irregularities as small as 1/100 of the source wavelength. If more accurate measurements are needed, flat surfaces can be tested relative to a liquid surface (Bünnagel, 1956); the surface of the liquid, which must be clear and highly viscous, works as a reference flat because its radius equals that of the Earth's and for a diameter of 0.5 m the maximum error is $\lambda/100$ for $\lambda = 500\,\mathrm{nm}$. However, practicalities such as the influence of vibration and dust settling on the surface are problems. If a good flat is not available to test against, three flats can be tested in triple combinations in order to obtain absolute line profiles of each flat (Dew, 1966; Fritz, 1984).

3.1.2 Spherical and Aspheric Surfaces Spherical surfaces may be tested for their departure from spherical shape by comparing them with an ideal spherical reference wave front. When a spherical surface is tested in a Fizeau interferometer, Fig. 6, a transmission reference sphere is placed in the common path of the beams and the radius of the beam incident on the tested spherical surface must fill and match the radius of that surface.

Aspheric surfaces are difficult to test against a spherical wave front because they may depart too much from a spherical shape and generate too many fringes to resolve the interferogram. Null lenses and computer-generated holograms or real holograms (Offner and Malacara, 1992) are used to produce a wave front matching a perfect aspheric wave front, thereby reducing the number of fringes in the interferogram. Shearing interferometry reduces the number of fringes by interfering the tested wave front with its own replica, which can be

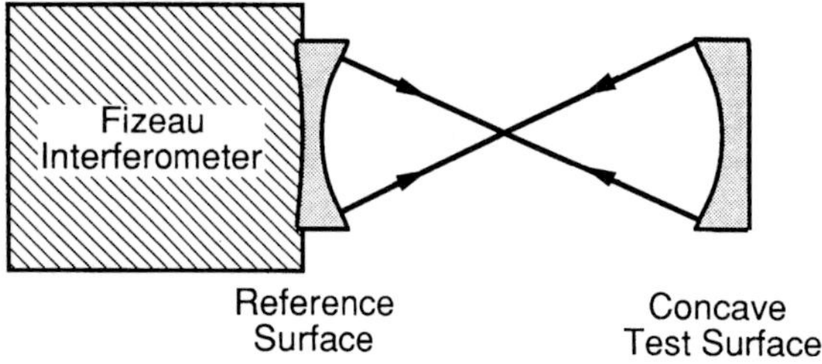

FIG. 6. Configuration for testing a concave spherical surface in a Fizeau interferometer.

displaced laterally, rotated, radially expanded or contracted. The number of fringes varies with the amount of shear, and fringes do not represent the shape of the aspheric surface directly (Malacara, 1992; Mantravadi, 1992).

3.1.3 Prisms Although prism angles may be measured with an autocollimator and a goniometer, interferometers are the most accurate method for measuring angle deviations. The procedure simply involves inserting the test prism and return flat into the test arm of the interferometer. Prisms with roof angles of 90° may be put into the test arm with (Fig. 7) or without the return mirror, as the beam can be retroreflected back to the system. The maximum deviation from 90° that a phase measuring interferometer can measure is about 1 min of arc.

3.1.4 Light Sources The quality of an optical wave front exiting from a light source may be tested using a point-diffraction interferometer or a special configuration of a Mach–Zehnder interferometer (Greivenkamp and Bruning, 1992). Both setups induce a spherical reference wave front through the use of a pinhole in the reference arm.

3.1.5 Homogeneity Homogeneity can be measured by preparing a plane-parallel sample and placing it in the test arm of an interferometer. A sample does not have to be specially prepared if it is submerged in a refractive-index–matching oil. Systematic errors can be subtracted by making a measurement before placing the sample in the test arm. Surface-independent measurements (Schwider, 1990) require multiple takes to remove variations due to the test sample's surface; this is done to isolate the refractive-index variations in the sample.

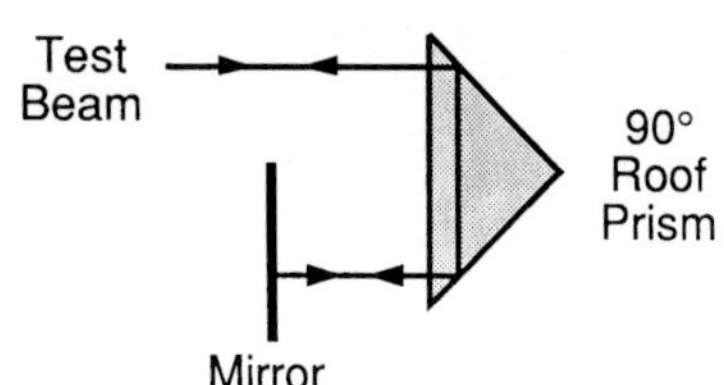

FIG. 7. Configuration for testing a 90° roof prism.

3.2 Testing Engineering Surfaces

Interferometry can also be used to measure the shape and properties of surfaces not used as optical components, such as machined metal or molded plastics. Because testing engineering surfaces is rather a specialized application of interferometric techniques, this section focuses on the methods developed to test these surfaces. The types of surfaces to be tested vary from very large ones that require overall testing of shape to very small ones that require a high resolving power to measure their microroughness. Speckle interferometry or holography may also be used for testing engineering surfaces, but since these methods can also be used for determining object displacement and vibration, two parameters important in experimental mechanics, they are discussed in the nondestructive testing section.

3.2.1 Fringe Projection Many engineering surfaces are too complex to be measured with standard interferometric techniques. Rowe and Welford (1967) devised a simple approach for measuring surface contours by projecting interference fringes or a grating onto an object and then viewing the fringes from another angle. Fringe projection is related to optical triangulation, which uses a single point of light and light sectioning; a single line is projected onto an object and then viewed from a different direction.

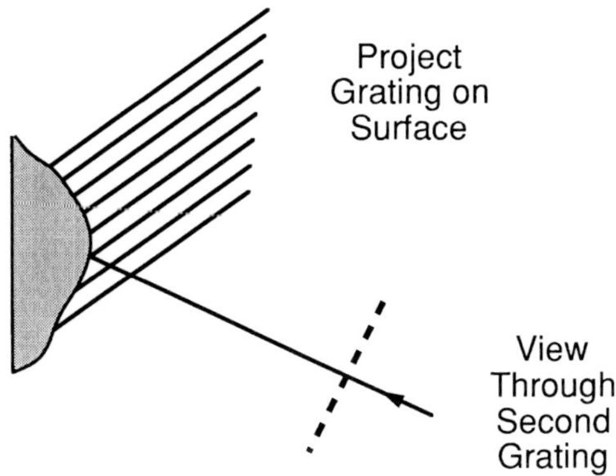

FIG. 8. Moiré technique for testing engineering surfaces.

3.2.2 Moiré In optics the term moiré (Rayleigh, 1874) refers to a pattern arising when two gratings of approximately equal spacing are superimposed. Interferometry using moiré (Patorski, 1993) can be implemented in a number of ways. A common approach is to project interference fringes (or a grating) onto an object and view them through a second grating placed in front of the viewer; see Fig. 8.

These techniques produce fringes corresponding to contours of equal height on the object. The contour interval is determined by the fringe (grating line) spacing and the angle between the illumination and viewing directions. Phase shifting can be introduced by shifting one grating or the projected fringes. Moiré and fringe-projection interferometry complement conventional interferometry, as they can contour objects with an effective contour interval of 10 μm and larger.

3.2.3 Interference Microscopy Interference microscopy combines an interferometer and a microscope into one instrument and is used for measuring engineering surfaces that demand testing with high resolving power. Interference microscopy uses three types of interferometric objectives: Michelson, Mirau, and Linnik (all are modifications of the Michelson interferometer). In a Michelson interferometer objective, which is used for small magnifications, a beam-splitter cube and reference mirror are inserted between the objective and the tested surface. At magnifications above five times, the working distance, which refers to the space between the last surface of the lens and the tested surface, becomes too small to squeeze in a beam splitter cube; instead a Mirau interferometric objective (described in OPTICAL INSTRUMENTATION, Sec. 3.2.1) is employed. For magnifications of one hundred times and above, a Linnik interferometric objective (Linnik, 1933) is used since the working distance for such high magnification is even smaller and the Mirau design is impossible to implement. In a Linnik configuration a beam-splitter cube placed before the light reaches the objective directs the beam onto two objectives; one beam is directed to the reference mirror and the other is directed to the test surface. Any type of phase-across-the-plane analysis technique discussed in this article may be used with these microscopes. Other interference microscopes are reviewed by Steel (1983).

3.2.4 Fringes of Equal Chromatic Order Interferometry Fringes of equal chromatic order (FECO) interferometry (Fig. 9) is based on multiple-beam interference between two highly reflective and nearly parallel plates and results in very sharp fringes (Tolansky, 1970). Multiple-beam interference magnifies the extent of the deformation, thereby making it easier to detect small imperfections on the tested surface. FECO uses a white-light source and is combined with a spectrometer to test thickness variation and surface roughness, which are determined from the width of the fringes. Small sections of the test surface are observed through a spectrometer that displays fringes of equal chromatic order. A FECO interferometer can achieve a precision level of $\lambda/500$ and is capable of distinguishing hills from valleys by use of the direction of the bending of the colored fringes. However, the test surface must be overlaid with a highly reflective coating and the reference surface rms (root mean square) roughness should be better then $\lambda/20$.

4. PHASE AT ONE POINT

Phase-at-one-point interferogram detection is commonly used to measure length, the spectral content of a source (see Sec. 6), or its polarization (see Sec. 8.4). For length measurement these interferometric techniques

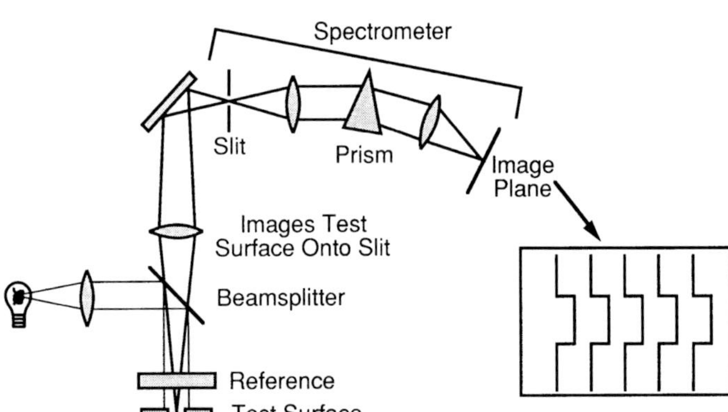

FIG. 9. Interferometer for measuring fringes of equal chromatic order (FECO).

(Hariharan, 1985; Steel, 1983) generally utilize a single detector to measure the phase at a single point of the interferogram where the fringe spacing is much larger than the size of the detector. The interferometric techniques that test the phase at one point measure length with very high precision, up to $1/10^4$ of a fringe, and are extensively used to measure relative and absolute distance. A few common single-point–phase detection techniques are described below. While measuring length is an important application of these techniques, the phase at a single point may represent other parameters, such as changes in pressure, temperature, or rotation.

4.1 Heterodyne Interferometry

Heterodyne interferometry observes interference between two beams with slightly different optical frequencies. This frequency difference is usually achieved by introducing a frequency shift into one of the beams. This frequency shift may be a result of a linear change in the phase difference and is achieved by rotating a radial grating, combining stationary and rotating quarter-wave plates, or moving a mirror in one of the arms of the interferometer. When the two beams interfere, a photodetector records a traveling interference signal at the beat frequency, which is equal to the difference in the two optical frequencies of the beams.

Heterodyne interferometry is often used to detect small changes in length. This detection is accomplished by evaluating changes in the beat frequency between the stable laser and the test (slave) laser frequency. Heterodyne interferometry measures changes in length by measuring changes in frequency.

4.2 Fringe Counting

These techniques essentially count the changes in intensity as they scan across a point detector while increasing the optical path difference between two beams. If the signal is lost for any reason, the measurement must be repeated. Two basic approaches to phase detection using fringe counting are often used. The first is based on observing the intensity changes in two interferograms. Into one interferogram an additional $\pi/2$ phase shift is introduced between interfering beams in order to determine the direction of the phase-difference changes (signals are in quadrature). The $\pi/2$ phase shift is induced by use of a special beam splitter or through polarization effects. The phase change can be detected to within a fraction of the fringe by observing the circular pattern formed by the two intensity signals mapped onto the horizontal and vertical axes on an oscilloscope. A full circle is traced each time one full fringe is scanned. The drawback to this method is that any changes in the intensity signal (such as source intensity variations) could be interpreted as a fringe.

The second approach avoids this problem; it is based on a heterodyne system that detects the interference of beams of different

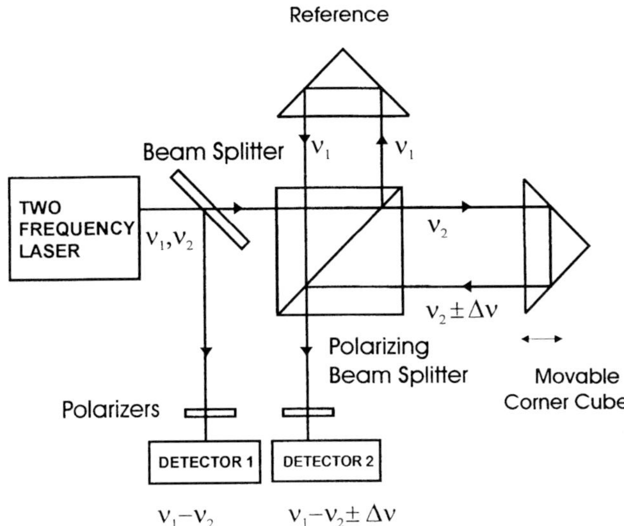

FIG. 10. Distance-measuring interferometer based on heterodyne system.

optical frequencies. Figure 10 shows a distance-measuring interferometer based on a heterodyne system that uses fringe counting. The differing frequencies can be obtained by using either a laser with two output frequencies or a single-frequency laser with two acousto-optic modulators. Either way the beams need to be orthogonally polarized. The first detector records the traveling interference reference signal at the beat frequency, which is equal to the difference in the two optical frequencies of the source. A polarizing beam splitter directs the beam of ν_1 into the reference arm and ν_2 into the test arm of the interferometer. Corner cubes return these beams to the second detector where the polarizer allows them to interfere. When the test corner cube is stationary, an intensity signal at the reference beat frequency is observed; when the test corner cube is in motion, it introduces a Doppler frequency shift in the returning beam and a different beat frequency is observed at the second detector. Signals from both detectors are forwarded to a differential counter and the change in the length of the test arm is determined.

4.3 Phase Locking

Phase-locking interferometry measures small changes in phase by locking the phase difference between the beams while detecting changes in the intensity signal (Moore, 1979). Locking the phase difference is accomplished by offsetting the phase of one of the beams and then subjecting that phase to sinusoidal modulation introduced by an oscillating mirror. This procedure induces intensity modulation that can be expressed as

$$I = I_1 + I_2 + 2(I_1 I_2)^{1/2}\cos(\varphi + \delta + a\sin\omega t). \tag{4}$$

The amplitude value a of the phase oscillation $\sin\omega t$ is much smaller than the source wavelength. The offset phase δ is set so that the average phase difference satisfies

$$\varphi + \delta = m\pi, \tag{5}$$

where m is an integer. Under this condition the intensity signal goes to zero and there is no intensity modulation due to mirror oscillation. When the average phase difference changes because of changes in tested phase φ, the intensity signal starts to modulate with its magnitude being proportional to the changes in the phase. This signal is then filtered and used to produce a feedback signal that tells the phase modulator how to alter the phase offset δ so that the condition $\varphi + \delta = m\pi$ is satisfied and the intensity modulation due to mirror oscillation is eliminated.

5. APPLICATIONS OF PHASE AT ONE POINT

5.1 Length Measurement

Phase-at-one-point detection in interferometry is very often applied to the precise measurement of length (Steel, 1983). In fact, interferometric techniques really established their value when they were applied to the measurement of material standards such as length, wavelength, line, and gauging.

5.2 Gravity-Wave Detection

Single-point phase detection has recently been employed to detect gravity waves. Much as accelerating charges produce electromagnetic waves, accelerating masses emit gravity waves. Gravity waves produced by collapsing supernovas, binary systems of neutron stars, and black holes affect the dimensions and spacing of all material objects. The movements that signal the existence of gravity waves can be detected by observing the small changes in phase at a single point in an interferometer. To register these changes requires an interferometer of very high sensitivity. A simple Michelson interferometer would need unrealistically long arms or high laser power to achieve the required accuracy. A number of new interferometers have been devised that should have sufficient sensitivity to detect gravity waves (Saulson, 1994). One of these has been coined the Laser Interferometric Gravitational-wave Observatory (LIGO). Its interferometric setup (Fig. 11) places identical Fabry–Pérot resonant optical cavities into each arm of a Michelson interferometer. The Fabry–Pérot arm cavities effectively increase the optical path length of each arm on account of the multiple reflection of light inside the cavities. The Fabry–Pérot mirrors act as inertial test masses suspended like pendulums from seismically isolated platforms. Passing gravity waves change the relative length of the resonant Fabry–Pérot cavities; this change alters the phase of the light and the phase change is picked up by the detector, thus revealing the existence of passing gravity waves. In order to detect the phase, the frequency of the laser is locked so that it is in resonance with the length of one cavity and then the length of the other cavity is adjusted to achieve the same resonance. The system detects the dark interference fringe, which allows most of the light to be reflected back to the source and back out again into the system with a recycling mirror. This additional mirror and the first surfaces of the test masses create the cavities where the light builds up, which effectively increases the optical path length and the illuminating power of the laser. The LIGO interferometer allows the detection of movements as slight as 1/1000 the diameter of a proton (10^{-18} m) and enables the identification of the presence of matter at distances of more than 70×10^6 light years from Earth. The distance separating the mirrors, more than 4 km, makes the LIGO interferometer the world's largest precision optical instrument and is pushing the limits of technology.

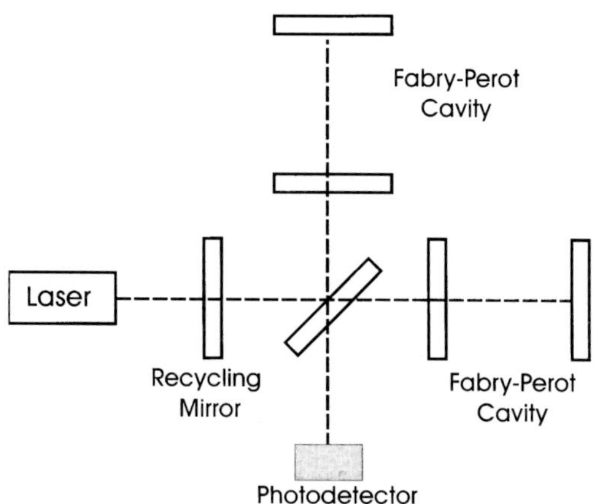

FIG. 11. Gravity-wave detection interferometer.

5.3 Other

Interferometry is used for monitoring processes such as plasma etching, film etching rate, uniformity of etching, and substrate temperature (see PLASMA ETCHING). Phase-at-one-point detection can be also used for measuring refractive index and electron density of a plasma. Temperature, pressure, strain, and rotation can be detected by optical-fiber sensors (see Sec. 8.2). Phase detection has also been used to measure the speed of light (Steel, 1983).

6. INTERFERENCE SPECTROSCOPY

Interferometric spectrometers analyze the spectral distribution of a radiating source rather than examining the phase difference as the techniques discussed previously do. Generally, spectrometers measure not just the spectrum of radiation but rather the combination of the radiating source, the detector sensitivity, and the transmittance of the interferometer. In addition, a perfectly monochromatic spectral line will result in different spectral energy distributions when detected by different types of spectrometers. This characteristic of spectrometers is called the response of the system, and it is dependent on the spectrometer's instrumental function, which determines when two spectral lines of frequencies ν_1, ν_2 (or λ_1, λ_2 in terms of wavelength) can be still resolved by the instrument. The resolving power of the spectrometer is then determined by

$$\Re = \nu/\Delta\nu = \lambda/\Delta\lambda \tag{6}$$

An important characteristic of a spectrometer, especially when faint spectra are studied, is its "light-gathering power", which is defined by the invariant known as *étendue*. For example, dispersive spectrometers, like grating or prism spectrometers, generally use a narrow slit; these systems have smaller étendue than those spectrometers that do not use this type of slit. Two types of interferometric spectrometers and their techniques for recovering the spectral content of a radiating source are discussed below, a Fabry–Pérot etalon and a version of a Michelson interferometer.

6.1 Fabry–Pérot Spectrometer

The Fabry–Pérot interferometric spectrometer (Fabry and Pérot, 1899; Vaughan, 1989) employs multiple interference between two high-quality and highly reflective plates, which can be either fixed (etalon) or variably separated. Fixed plates yield a set of circular (spatial) sharp fringes for different wavelengths of the source. This setup is often used to measure the separation of two planes in etalons (standards), the thickness of a thin film, or the wavelength of a source (Hariharan, 1985).

The instrumental function of Fabry–Pérot plates separated by distance d is the *Airy function*

$$A(\nu) = T^2/[(1-R)^2 + 4R\sin^2(\psi/2)], \tag{7}$$

where R and T are the reflectance and transmittance of the plates and $\psi = (4\pi\nu nd/c)\cos\theta$, where n is index of refraction of the medium between the plates and θ is the angle of incidence. The Airy function has a set of narrow peaks that are separated by a frequency difference equal to $c/2dn$; this frequency difference describes the *free spectral range* where the fringes will not overlap. The half-width of the peaks $\Delta\nu$ is described by

$$\Delta\nu = c/2ndF, \tag{8}$$

where F is the *finesse* of the system,

$$F = \pi\sqrt{R}/(1-R). \tag{9}$$

The limited spectral range of the etalon can be overcome by separating the ranges in order to determine the frequency they represent without any ambiguity; however, this procedure is accomplished by imaging fringes onto a slit of the spectrometer, which then limits the étendue of the system.

The scanning Fabry–Pérot interferometric spectrometer has significantly higher étendue and is often used as a spectrometer for testing faint light sources. In this system the optical separation between the two plates changes linearly and the transmitted intensity signal (section of spectrum) is recorded by a point detector placed in the focal plane of the lens, which is placed behind the interferometer. Many modifications can be introduced to increase the performance of this spectrometer. For example, spherical mirrors may replace plates and thus increase light usage efficiency.

6.2 Fourier-Transform Spectroscopy

Interferometry may also be used to obtain the spectral energy distribution of a source through a technique called Fourier-transform spectroscopy (Michelson, 1902; Vanasse and Sakai, 1967; Bell, 1972). Fourier spectroscopy records the variation in intensity at a single point while varying the optical path difference between interfering wavefronts in a two-beam interferometer, which is typically a Michelson with the effective source at infinity. The optical path difference is realized by moving one of the mirrors, as in phase-measurement techniques. The detected intensity of the interference signal is the sum of independent interference signals for each spectral component and differs with the spectral content of the tested source. This intensity can be represented by

$$I(\tau) = \int G(\nu)[1 + \cos(\tau 2\pi\nu)]\, \mathrm{d}\nu \tag{10}$$

where τ is the introduced optical path difference and $G(\nu)$ is the spectral distribution of the source. Since the recorded intensity signal is not the direct spectrum signal as in the Fabry–Pérot interferometer but rather the cosine Fourier transform of the spectrum, the Fourier transformation of the recorded signal must be calculated to achieve the spectrum of the source. The intensity of the interference signal must be sampled at least twice per fringe (as in phase-detection techniques). Because the fringe spacing in an interference signal corresponds to the wavelengths of the source, the minimum wavelength that can be detected is limited (as the slope is limited in phase-across-the-plane detection). Unequal sampling intervals and finite changes in optical path difference will introduce errors into the calculated spectrum. A process called *apodization*, where a measured interference signal is multiplied by a symmetrical signal with gradually decreasing values, like a triangle function, reduces the magnitude of errors in the calculated spectrum.

Fourier spectroscopy may be relatively difficult to implement; but when compared with prism or grating spectroscopy, Fourier spectroscopy has a very high étendue and better resolution than either of the other two methods. Another advantage of Fourier spectroscopy is its *multiplex advantage*: the signal-to-noise ratio is significantly improved as a result of simultaneous registration of all spectral components of radiation. Major applications of this technique include detection of emission spectra and absorption spectra of aqueous solutions, gases, and transient species. Because of its efficient use of light, Fourier spectroscopy is used often in astronomy to record spectra (usually infrared—see SPECTROMETERS, INFRARED) from very faint sources.

Fourier transformation is a powerful tool in interferometry; it can be used with either a coherent or an incoherent spectral source to analyze the phase across the plane. In addition, Fourier techniques are used for analyzing both the spatial and spectral distribution of a radiating source.

6.3 Spectral Imaging

The interferometric instruments described so far have been able to acquire either phase or spectral information about an object. However, many disciplines such as astronomy, robotic vision, or remote sensing are interested in obtaining both types of information at once. Recent progress in optics and electronics has allowed researchers to register multiple spectral images, thus spurring the development of new spectral imaging techniques (Itoh, 1996). Many of these techniques combine interferometric methods or couple interferometry with other optical testing procedures. For example, a double Fourier spatio-spectral interferometer merges a Michelson interferometer for Fourier spectrometry with a Michelson stellar interferometer for star diameter determination.

7. INTERFERENCE IMAGERY

Interference imagery (Steel, 1983) is an interferometric technique for testing the spatial distribution of a radiating source and is often used in astronomy to measure stellar diameters that are nearly impossible to resolve with conventional telescopes. In 1921 Michelson first developed a method for testing stellar diameters; his stellar interferometer could resolve distant objects with angular diameters on the order of 10^{-2} arc s (Michelson, 1921; see also INTERFEROMETERS AND INTERFERENCE, Sec. 1.3.2). Since that time, interferometric techniques have become more precise, using visible light as well as radio waves to measure spatial distributions. The resolution of stellar interferometers depends on both the distance between the mirrors (antennas) and the number of mirrors (antennas) used. Stellar interferometers are sometimes referred to as long-baseline or very-long-baseline (VLBI) interferometers or aperture-synthesis arrays.

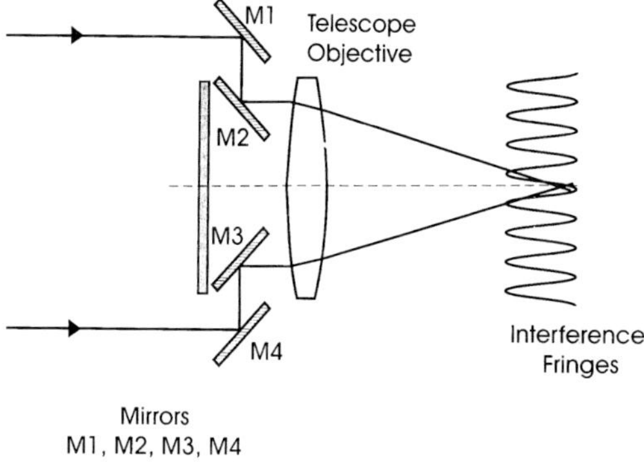

FIG. 12. Michelson stellar interferometer for measuring stellar dimensions.

Stellar interferometers today test many parameters in order to achieve maximum precision, including the phase across the plane for fringe tracking and heterodyne techniques for delay.

7.1 Michelson Stellar Interferometry

A Michelson stellar interferometer is a two-beam interferometer that uses wave-front division rather than amplitude division as the interferometers and techniques described thus far do. A Michelson stellar interferometer (Fig. 12) consists of two widely spaced mirrors (whose separation can be changed) pointed at the same stars. Light from the two mirrors is combined to form interference fringes on the image plane of a telescope. If the interferometer is pointing at two stars, two point sources, then two slightly shifted interference patterns of mean wavelength λ, one from each star, are formed. Because the source radiation from the stars is spatially incoherent, the sum of the interference patterns' intensities observed on the detector forms a new fringe pattern. The visibility of these new fringes depends on the shift between the interference patterns from each star. This shift is dependent on both the angular separation θ of the stars, which is fixed, and the separation d between the mirrors, which is referred to as the baseline and varies. The visibility of the fringes versus the mirror separation is plotted in Fig. 13(a). This visibility of the fringes is an envelope function of the Fourier transform of the spatial distribution of the source. Thus, if the mirror separation d is found for which the visibility of fringes equals zero, then the distance between the stars can be calculated.

When one is measuring the diameter of a star, the star is treated not as a point source but as a set of incoherent point sources that fills the star's diameter. The visibility of the fringes, Fig. 13(b), changes in a different fashion than for two point sources. By adjusting the separation of the two input mirrors until the fringe visibility reaches zero, which occurs when the spacing between the two mirrors is approximately $1.22\lambda\theta$, where θ is the stellar diameter, the diameter of the star can be determined but only in the direction of the mirror separation. Another procedure records the fringe pattern and then calculates the zero fringe visibility in order to determine the stellar diameter. In practice stellar interferometers are difficult to build, since the optical paths between the arms of the interferometer need to match as stars emit incoherent light. Stabilizing a system such as this requires very stable mechanical construction of the interferometer

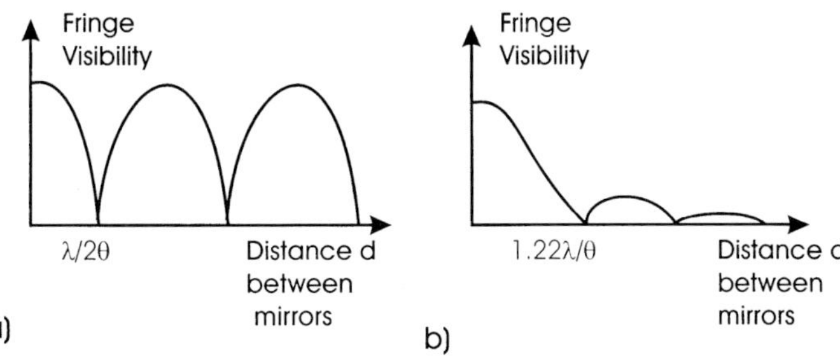

FIG. 13. Fringe visibility in a Michelson stellar interferometer for (a) two point sources and (b) a disk source.

and correction for atmospheric turbulence. In addition, if a star is not radially symmetrical, the mirrors need to be rotated or an additional set of mirrors implemented in order to determine the diameter of the star in a different direction. Finally, if radiation from the stellar source is not uniform, measurements that employ multiple separations of the mirrors need to be taken.

7.2 Stellar Intensity Interferometry

A major step in the development of stellar interferometers was the invention of the intensity interferometer by Hanbury Brown and Twiss (1954). The intensity interferometer is essentially a Michelson stellar interferometer that observes the intensity variation coming from each mirror separately; these intensity variations are then electronically correlated. The detectors in the intensity interferometer can be separated by thousands of kilometers, thereby significantly increasing the resolution of the system.

7.3 Radio Interferometry

Radio astronomy uses radio waves with wavelengths ranging from millimeters to hundreds of meters for observations in a different spectral range. Similarly, as stellar interferometry increases the resolution of visible light telescopes, radio interferometry, which is based on stellar interferometry (Steel, 1983), improves the resolution of radio telescopes (see Radio Telescopes). Radio interferometry may use antennas for radiation collection instead of lenses and mirrors, but its principle of operation is still based on interference phenomena. Radio interferometry has many uses, ranging from listening for signs of earthquakes and volcanic activity to measuring the thickness of ice sheets and monitoring sea levels (see Geodesy).

7.4 Stellar Speckle Interferometry

Stellar diameters can be determined using a single telescope and a technique developed by Labeyrie (Labeyrie, 1976; Dainty, 1984). This technique allows a telescope to measure angular distances at its diffraction limit regardless of the light conditions. Individual speckles formed at the focus of the telescope have angular sizes determined by the diffraction limit of the telescope. When short exposures are made, the speckle pattern can be frozen in time. By taking the square modulus of the Fourier transform of a speckle pattern and averaging a number of these, a fringe pattern will be seen whose spacing is proportional to the stellar diameter or separation.

8. OTHER APPLICATIONS

In addition to the applications discussed so far in this article, there are many other applications that do not fit into the previous categories. These include the nondestructive testing of mechanical displacement and strain, the use of optical fibers in interferometers, the measurement of polarization and material properties, and interference between particles rather than waves. Since it is not possible to describe every application, this section highlights a few of these other applications.

8.1 Nondestructive Testing with Interferometry

Nondestructive evaluation of the mechanical properties of technical components is increasingly important for process control during manufacturing. Information about components' deformation due to an applied force or a change in temperature can be delivered by different interferometric techniques. As these techniques are applied to experimental mechanics, changes in phase can represent the displacement both in and out of the plane of the object.

8.1.1 Holographic Interferometry Holographic interferometry is an often-used technique because both amplitude and phase and in- and out-of-plane displacement information can be obtained about a surface as a known force is applied to it. These results can be compared with simulations generated by finite-element analysis to see whether mechanical parts behave as they are modeled. For example, holographic interferometry can provide information about whether a part in an airplane motor will perform as expected when it is stressed and loaded while in operation. The mathematical theory and basics of holographic (hologram) interferometry are outlined in OPTICAL HOLOGRAPHY, Sec. 4.3.

Recent developments in nondestructive testing using holographic interferometry have enabled holograms to be made in media that are erasable and reusable (Hariharan, 1994). Phase-measurement techniques can be applied to holographic interferometers to make quantitative analysis possible. Quantitative data enable a better comparison with simulation data for structural analysis. Applications include the testing of airplane tires, audio speakers, turbine blades, satellite fuel tanks, musical instruments, and infrastructure such as pipes and bridges.

8.1.2 Speckle Interferometry Speckle interferometry is similar in many ways to holographic testing; the difference is that information about object displacement is derived from the speckle patterns. Correlating speckle patterns before and after a force is applied yields information about the surface displacements. When speckle techniques were first developed, they relied on film as the recording medium. Sensor arrays have replaced film as the recording media, and advances in signal processing have allowed for fast quantitative analysis of object displacements (Jones and Wykes, 1983).

8.1.3 TV Holography By combining holographic and speckle interferometry, it is possible to record holograms on a CCD (charge-coupled device) sensor array and do image processing essentially to create digital holograms (Creath and Wyant, 1992; Kreis, 1996; Rastogi, 1997). Because the hologram must be resolvable with a TV camera, the frequency information content has to be within the Nyquist sampling limit of the sensor. Another consideration is that image processing must remove the self-interference terms from the hologram so that the cross-interference term is not corrupted. Many different approaches have been developed to perform this processing. Displacement measurements are easily obtained by computationally subtracting two holograms. Vibration analysis can be carried out stroboscopically or by adding a known vibration to the system and looking at the difference in frequency. Instruments using these techniques can be made to be quite insensitive to vibration. These instruments are often portable so that measurements are not limited to laboratory settings.

8.1.4 Grating Interferometry High-sensitivity grating interferometry (Patorski, 1993; Post *et al.*, 1994), has become an important method in experimental mechanics to measure, with submicrometer sensitivity, the in-plane displacements of nearly flat objects under load. In conventional high-sensitivity grating interferometry, the object under test with a reflection-type grating adhered to it is illuminated by two coherent beams symmetrical to the grating normal. The interference between the two diffraction orders reflected from the grating is recorded, resulting in a map of the in-plane displacements. Polarization techniques can be used in this system for phase shifting.

8.2 Optical-fiber Interferometry

Optical-fiber interferometers do not differ in principle from conventional two-beam interferometers; the difference is that in optical-fiber interferometers the beams travel through single-mode optical fibers, allowing for compact construction of an interferometer (Hariharan, 1985; also see SENSORS, OPTICAL, Sec. 2.8).

Optical-fiber interferometers are often used as optical sensors to detect changes in pressure or temperature, as these effects alter the optical path length of the fiber. The sensitivity of the interferometric system increases as longer fibers are used; thus optical-fiber interferometers find their advantage in combining compactness with sensitivity, since long optical paths can be contained in compact areas.

The optical-fiber interferometer can use a Sagnac-type (Vali and Shorthill, 1976) interferometric configuration to sense rotation (see INTERFEROMETERS AND INTERFEROMETRY). This is done by replacing the ring cavity of a conventional Sagnac interferometer with a multi-turn loop made of single-mode optical fiber. The phase of the counterpropagating beams can then be detected using a phase-sensitive detector (Ezekiel, 1984).

8.3 Refractometry

Refractometers use interferometric techniques to measure changes in the refractive index of solids, liquids, or gases more precisely than any other method (see REFRACTOMETERS, OPTICAL). The first interferometer to measure a refractive index was devised by Michelson. Later Rayleigh and Jamin interferometers were developed specifically for this purpose (INTERFEROMETERS AND INTERFEROMETRY).

8.4 Ellipsometry

Ellipsometry (Azzam and Bashara, 1987; ELLIPSOMETERS) employs light polarization to characterize surfaces, interfaces, and thin films. In an ellipsometer linearly polarized incident light reflects from the test surface as elliptically polarized light. By superimposing two orthogonal components of an electric field, the elliptically polarized light can then be measured. Ellipsometry is a variation of common-path polarization interferometry where one of two copropagating orthogonally polarized waves acts as the reference for the other. Some ellipsometers use interferometric arrangements (mainly Michelson) to measure the index of refraction or film thickness over an entire surface, not just at a single point.

8.5 Atom Interferometry

Waves are characterized by phase and amplitude while particles are characterized by position and momentum. Both atoms and light can be treated as waves and particles, and both can be described by similar sets of equations. However, atoms have mass; therefore atomic waves obey a different dispersion relationship than does light. As the interference of wave fronts is an inherent feature of interferometry, atom interferometry (Adams *et al.*, 1994; Berman, 1996; see also INTERFERENCE AND INTERFEROMETRY) observes atom interference patterns. (Similarly, electron and neutron interference patterns are observed in electron and neutron interferometry.) Using methods similar to classical interferometry, atomic interference patterns are analyzed for modulation and fringe spacing and position. With much better precision than other interferometric methods, atom interferometry is able to detect and test gravitational and inertial effects, atomic properties, and some effects in quantum mechanics. Atom interferometry replaces the optical elements from conventional interferometers with microfabricated elements that diffract or split thermal atom beams, and tunable lasers that manipulate the trajectories of neutral atoms.

GLOSSARY

Fringe Order: The number assigned to a fringe in a single interferogram; it represents the direction of the changes in the recorded phase wavefront.

Optical Path: The length of the path that light takes through the medium multiplied by the index of refraction of the medium.

Optical Phase: Optical path of the light modified by phase changes on reflection of the light from the medium, where the phase changes depend on angle of incidence of the light and on the medium's complex index of refraction.

Phase Shift: The controlled change of the optical phase between interfering wavefronts.

Phase: Optical phase difference between test and reference wavefronts. The phase retrieved from an interferogram may represent the shape of the tested surface, its slope, or its displacement, among other things.

Works Cited

Adams, C. S., Sigel, M., Mlynek, J. (1994), *Phys. Rep.* **240**, 143–210.

Azzam, R. M. A., Bashara, N. M. (1987), *Ellipsometry and Polarized Light*, Amsterdam: Elsevier Science.

Bell, R. J. (1972), *Introductory Fourier Transform Spectroscopy*, New York: Academic Press.

Berman, P. R. (Ed.) (1996). *Atom Interferometry*, New York: Academic Press.

Bünnagel, R. (1956), *Z. Angew. Phys.* **8**, 342–350.

Creath, K., Wyant, J. C. (1992), in: D. Malacara (Ed.), *Optical Shop Testing*, New York: John Wiley and Sons, Chap. 15.

Creath, K. (1993), in: D. W. Robinson, G. T. Reid (Eds.), *Interferogram Analysis: Digital Fringe Pattern Measurement Technique*, Bristol: IOP Publishing, Chap. 4.

Dainty, J. C. (Ed.) (1984), *Laser Speckle and Related Phenomena*, Topics in Applied Physics Vol. 9, 2nd ed., Berlin: Springer.

Dew, G. D. (1966), *J. Sci. Instrum.* **43**, 409–415.

Ezekiel, S. (1984), in: S. F. Jackobs (Ed.), *Physics of Optical Ring Gyros: 7–10 January 1987, Snowbird, Utah*, SPIE Proceedings No. 487, Bellingham, WA: SPIE, pp. 13–20.

Fabry, C., Pérot, A. (1899), *Ann. Chim. Phys.* **7**, 115.

Fritz, B. S. (1984), *Opt. Eng.* **23**, 379–383.

Greivenkamp, J. E., Bruning, J. H. (1992), in: D. Malacara, (Ed.), *Optical Shop Testing*, New York: John Wiley and Sons, Chap. 14.

Hanbury Brown, R., Twiss, R. Q. (1954), *Philos. Mag.* **45**, 663.

Hariharan, P. (1985), *Optical Interferometry*, Sydney: Academic Press.

Hariharan, P. (1994), in: P. K. Rastogi (Ed.), *Holographic Interferometry*, Berlin: Springer-Verlag, Chap. 2.

Itoh, K. (1996), in: E. Wolf, (Ed.), *Progress in Optics*, Vol. 35, Amsterdam: North-Holland, pp. 147–196.

Jones, R., Wykes, C. (1983), *Holographic and Speckle Interferometry*, Cambridge: Cambridge University Press.

Kreis, T. (1996), *Holographic Interferometry: Principles and Methods*, Berlin: Academie Verlag.

Kujawinska, M. (1993), in: D. W. Robinson, G. T. Reid (Eds.), *Interferogram Analysis: Digital Fringe Pattern Measurement Technique*, Bristol: IOP Publishing, Chap. 5.

Labeyrie, A. E. (1976), in: E. Wolf (Ed.), *Progress in Optics*, Vol. **14**, Amsterdam: North-Holland, pp. 47–87.

Larkin, K. G. (1996). *J. Opt. Soc. Am. A* **13**, 832–843.

Linnik, W. (1933), *C. R. Acad. Sci. URSS* **1**, 18.

Malacara, D. (1992), in: D. Malacara (Ed.), *Optical Shop Testing*, New York: John Wiley and Sons, Chap. 5.

Mantravadi, M. V. (1992), in: D. Malacara (Ed.), *Optical Shop Testing*, New York: John Wiley and Sons, Chap. 4.

Michelson, A. A. (1902, 1961), *Light Waves and their Uses*, Chicago: Chicago University Press.

Michelson, A. A. (1921), *Astrophys. J.* **53**, 249.

Moore, D. T. (1979), *Appl. Opt.* **18**, 91.

Offner, A., Malacara, D. (1992), in: D. Malacara (Ed.), *Optical Shop Testing*, New York: John Wiley and Sons, Chap. 12.

Patorski, K. (1993), *Handbook of the Moiré Fringe Technique*, Amsterdam: Elsevier.

Post, D., Han D., Ifju, P. (1994), *High Sensitivity Moiré*, New York: Springer.

Ramirez, R. W. (1985), *The FFT: Fundamentals and Concepts*, Englewood Cliffs, NJ: Prentice-Hall, Inc.

Rastogi, P. K. (1997), *Optical Measurement Techniques and Applications*, Cambridge, MA: Artech House.

Rayleigh, L. (1874), *Philos. Mag.* **47**, 81–93 and 193–205.

Robinson, D. W. (1993), in: D. W. Robinson, G. T. Reid (Eds.), *Interferogram Analysis: Digital Fringe Pattern Measurement Technique*, Bristol: IOP Publishing, Chap. 6.

Rowe, S. H., Welford, W. T. (1967), *Nature* **216**, 786–787.

Saulson, P. R. (1994), *Fundamentals of Interferometric Gravitional Wave Detectors*, Singapore: World Scientific.

Schultz, G., Schwider, J. (1976), in: E. Wolf (Ed.), *Progress in Optics*, Vol. 13, Amsterdam: North-Holland, pp. 93–167.

Schwider, J. (1990), in: E. Wolf (Ed.), *Progress in Optics*, Vol. 29, Amsterdam: North-Holland, pp. 271–359.

Steel, W. H. (1983), *Interferometry*, 2nd ed., Cambridge: Cambridge University Press.

Tolansky, S. (1970), *Multiple Beam Interferometry of Surfaces and Thin Films*, Oxford: Oxford University Press.

Vali, V., Shorthill, R. W. (1976), *Appl. Opt.* **15**, 1099–1100.

Vanasse, G. A., Sakai, H. (1967), in: E. Wolf (Ed.), *Progress in Optics*, Vol. 6, Amsterdam: North-Holland, pp. 261–330.

Vaughan, J. M. (1989), *The Fabry–Pérot Interferometer*, Bristol: Adam Hilger.

Takeda, M. (1990), *Industrial Metrology* **1**, 79–99.

Womack, K. H. (1984), *Opt. Eng.* **23**, 391–395.

Yatagai, T. (1993), in: D. W. Robinson, G. T. Reid (Eds.), *Interferogram Analysis: Digital Fringe Pattern Measurement Technique*, Bristol: IOP Publishing, Chap. 3.

Further Reading

Bass, M. (Ed.) (1978), *Handbook of Optics*, New York: McGraw-Hill, Inc.

Born, M. Wolf. E. (1970), *Principles of Optics*, Oxford: Pergamon.

Françon, M., Mallick, S. (1971), *Polarization Interferometers*, New York: Wiley Interscience.

Gasvik, K. J. (1987), *Optical Metrology*, New York: Wiley.

Hariharan, P. (1985), *Optical Interferometry*, Sydney: Academic Press.

Hariharan, P., Malacara, D. (Eds.) (1995), *Selected Papers on Interference, Interferometry, and Interferometric Metrology*, SPIE Milestone Series No. MS 110, Bellingham, WA: SPIE.

Malacara, D. (Ed.) (1978, 1992), *Optical Shop Testing*, New York: John Wiley and Sons.

Steel, W. H. (1983), *Interferometry*, 2nd ed., Cambridge: Cambridge University Press.

Udd, E. (Ed.) (1991), *Fiber Optic Sensors: An Introduction for Engineers and Scientists*, New York: John Wiley and Sons.

Udd, E., Tatam, R. P. (Ed.) (1994), *Interferometry '94: Interferometric Fiber Sensing: 16–20 May 1994, Warsaw, Poland*, SPIE Proceedings No. 2341, Bellingham, WA: SPIE.

NEGATIVE HYDROGEN-ION SOURCES

J. H. WHEALTON, *Oak Ridge National Laboratory, Oak Ridge, Tennessee, U.S.A.*

INTRODUCTION

A number of low-energy collision processes can result in electron attachment to a hydrogen atom, producing a negatively charged ion. Being loosely bound, the electron can easily be removed, making a neutral atom once again. An important objective of negative-ion sources is how to attach this extra electron onto a hydrogen atom and keep it there throughout the extraction process. Early attempts to extract negative ions yielded currents in the microampere range. Two significant discoveries allowed the extracted current to increase by many orders of magnitude. These are covered in Secs. 1.1, 1.2, and 1.3.

The negative-ion source development, at the Budker Institute of Nuclear Physics, was motivated by the invention of charge-exchange injection into proton storage rings (Budker *et al.*, 1967). However, one inconvenience of the charge-exchange method is that it required negative-ion beams to start with. Ironically, one of the preferred embodiments for negative-ion sources also used a charge-exchange principle. Another negative-ion source develop-

ISBN 3-527-29308-6

ment at Ecole Polytechnique found an increase in beam current thousands of times more than expected through discovery of new atomic processes, brought about by a series of insightful experiments.

Originally needed for high-energy physics, and then for the Strategic Defense Initiative, application spread to high-energy neutral-beam heating and current drive for fusion plasma-confinement experiments. These activities led to a massive effort to develop and optimize these sources. Results of this development enabled applications, such as nuclear-physics materials studies, using the spallation neutron source and the next generation of semiconductor chip manufacture.

All these negative-ion sources involve the generation and maintenance of a plasma, consisting of positive ions, negative ions, and electrons, while optimizing the actual negative-ion generation and extraction. In each case there are several significant parameters which must be jointly optimized. We discuss many of the issues and some of the laboratory data and the theory necessary to guide the development of these negative-ion sources.

1. PHYSICS OF HYDROGEN NEGATIVE-ION PRODUCTION

In this section several types of hydrogen negative-ion production mechanisms that have found use in ion sources are discussed. Embodiments that illustrate the principles are presented along with some degree of comparison. Surface production from cesiated surfaces (Dudnikov and Derevyankin 1994; Belchenko *et al.*, 1973, 1997) is considered in some detail in Sec. 1.1. Volume sources are of two forms:

1. dissociative attachment of vibrationally excited hydrogen molecules, considered in Sec. 1.3, and
2. charge exchange, considered in Sec. 1.2.

1.1 Cesiated-Surface Production (Dudnikov, Belchenko, Dimov)

Positive ions hitting a molybdenum surface convert into negative ions with a significant probability (Greer and Seidl, 1984; Seidl *et al.*, 1980, 1982; Melnychuk *et al.*, 1991). As many as a third of the positive ions hitting a suitably cesium-coated surface will result in a negative ion being produced on or near this surface (Seidl *et al.*, 1984). Since positive ions are generally easier to create than negative ions, this surface process represents a significant advantage. Moreover, since there is a force at the surface due to the plasma sheath fields, the negative ions once generated are accelerated approximately normal to the surface, which means that the shape of the cesiated surface itself may have a significant influence on the ion optics. This allows for the possibility of focusing the negative-ion beam and in some cases amplifying the beam current. Many surface sources make use of this feature. If it were not for the fact that the negative-ion "temperature" or transverse energy can be fairly high (7 eV for a 150-V plasma-sheath potential drop) these sources would probably lead the other concepts of negative-ion sources, since the current density can be as high as 1 A cm^{-2} of H$^-$, and early results by Belchenko *et al.* (1994), of the Institute of Nuclear Physics (INP), yielded 11 A of total current.

1.1.1 Cold-Cathode Penning (Dudnikov Source) While this is apparently the first source using surface production from cesiated surfaces, the actual beam formation occurs as a result of charge exchange and is therefore considered in Sec. 1.2.2.

1.1.2 Semiplanotron (Belchenko Source) This form was developed by Belchenko, Dimov, and Dudnikov at the Institute of Nuclear Physics at Novosibirsk (Belchenko *et al.*, 1973), and two views of an embodiment (Belchenko and Kupriyanov, 1990) can be seen in Fig. 1.

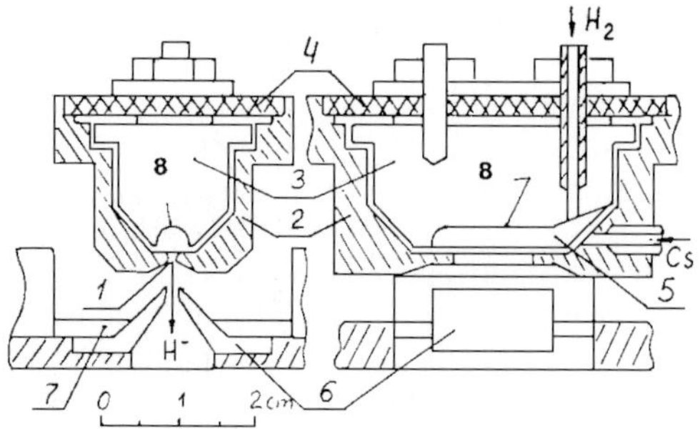

FIG. 1. Two views of semiplanotron surface source showing 1, emission slit (0.5 × 10 cm^2); 2, anode; 3, cathode; 4, insulator; 5, cathode groove; 6, extraction electrode; 7, mild steel inserts; and 8, converter.

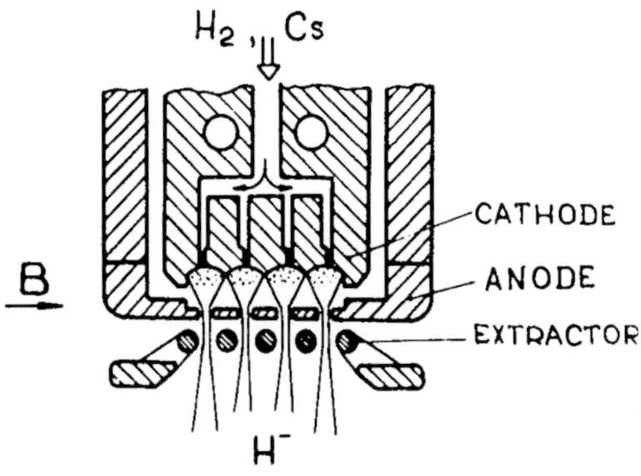

FIG. 2. Multibeamlet variation of semiplanotron.

A Penning discharge is established in region 5; this discharge consists of H^+ ions and electrons formed adjacent to a cesiated plate, 3, biased negatively by 100–300 V. If the cesium coating on the converter 8 is very thin (Seidl *et al.*, 1982), a significant amount of negative ions will result from double electron attachment to the incident positive ions or from electron attachment to sputtered hydrogen atoms. These H^- ions will accelerate through the 100–300-V sheath, go through the Penning discharge, 5, and be ready for extraction at 1. Since the H^- ions have about 100–300 V of energy, the inevitable plasma sheaths will have less of an effect than if their energy were smaller. This is one of the advantages of this mode of operation; another is that an amplification of beam current density occurs at 1 because of the concave shape of 8. However, the H^- ions, upon formation at the surface, acquire a transverse energy of a few electronvolts (Ehlers and Leung, 1980; York *et al.*, 1983), which results in a high beam emittance.

This configuration was used in multiple-beamlet configurations as illustrated in Fig. 2. 11 A of H^- were extracted from a 600-beamlet configuration with geometric focusing (Belchenko *et al.*, 1984, 1993). This multi-beamlet configuration could be useful in certain applications, such as neutral injection into magnetic-confinement fusion experiments where high beam currents are needed without a concomitant increase in brightness.

1.1.3 Hot-Cathode Semiplanotron (Stirling, Soloshenko Source) This source was developed at ORNL (Dagenhart *et al.*, 1980, 1983, 1984; Stirling *et al.*, 1984), at Kiev (Soloshenko 1997; Goretsky 1996), and at BINP (Bashkeev *et al.*, 1990). It is illustrated in Figs. 3 (end view), 4 (top view) and 5 (oblique view). Referring to the end view, electrons are emitted from a filament and ionize hydrogen gas in a Penning discharge, which is confined in a strong magnetic field (see regions labeled "positive ion arc column" in Fig. 4, or the region labeled "discharge" in Fig. 3). The discharge starts at an anode aperture and passes

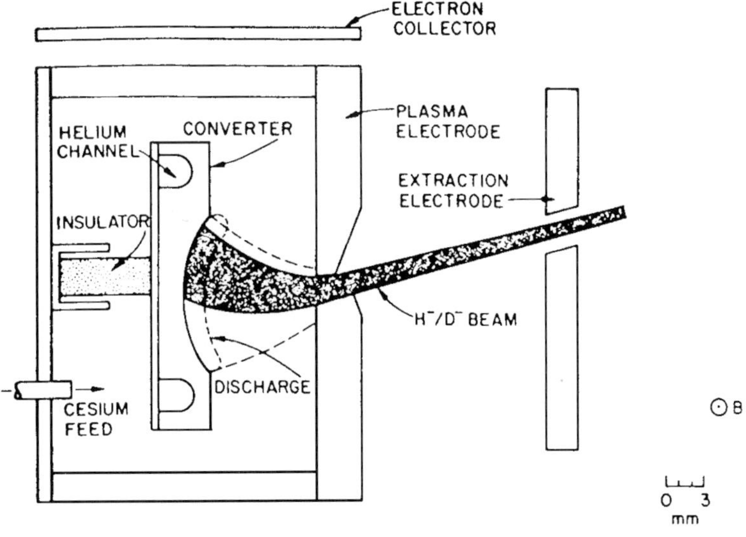

FIG. 3. Hot-cathode semiplanotron—SITEX end view.

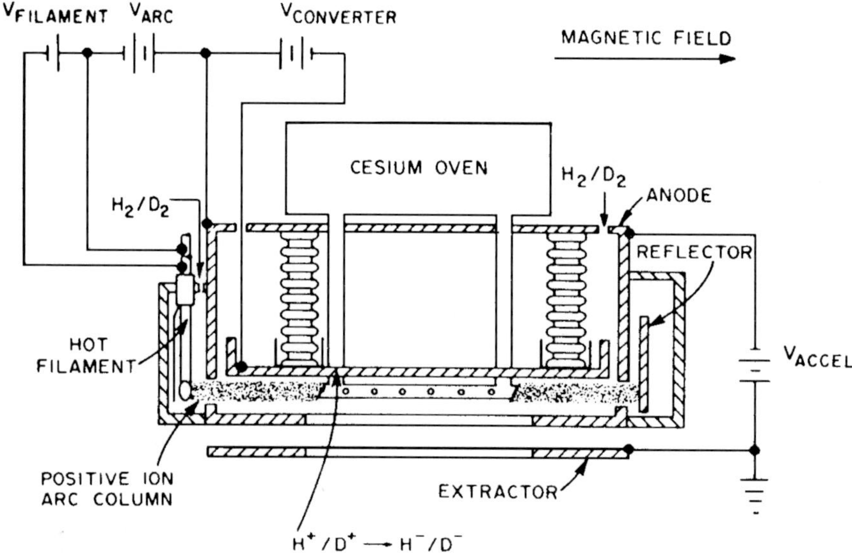

FIG. 4. Hot-cathode semiplanotron—SITEX top view.

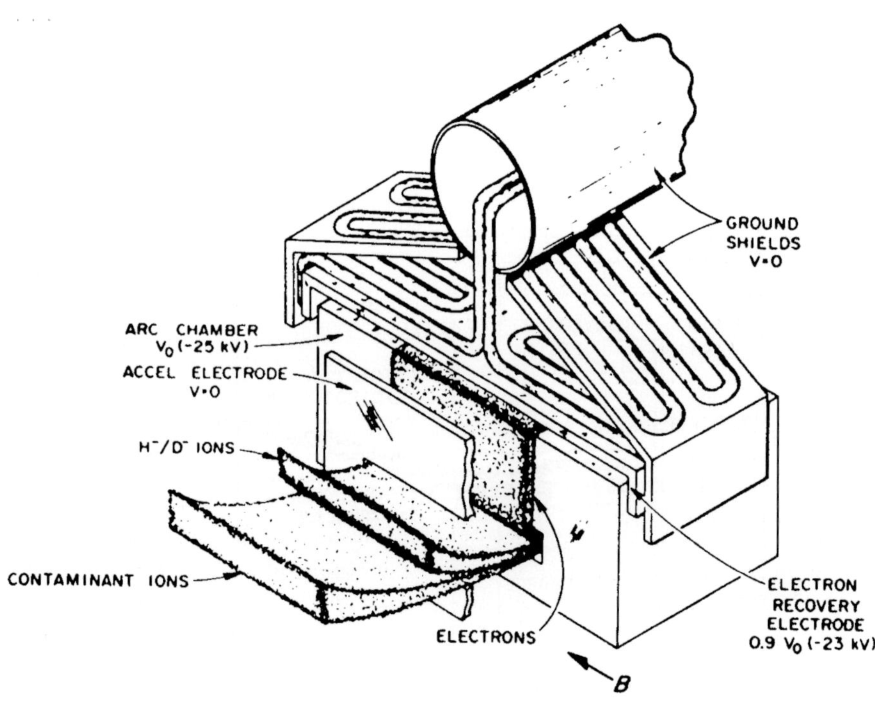

FIG. 5. SITEX—oblique view showing especially the dispatch of the electrons.

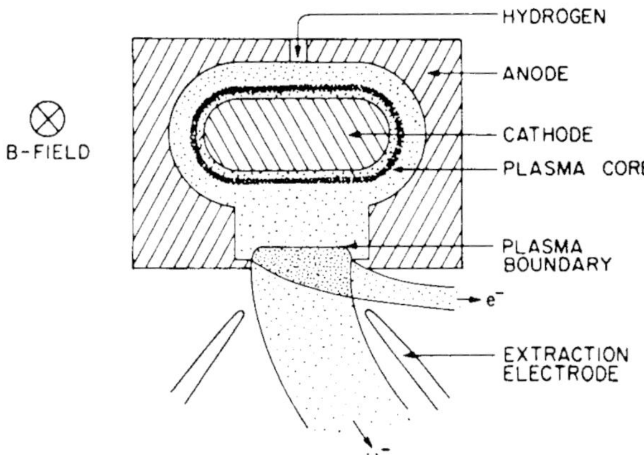

FIG. 6. Magnetron negative-ion surface source.

next to the converter surface shown most clearly in Fig. 3. This anode aperture is intended to reflect the electrons back into the main part of the column in front of the converter where they can continue to ionize the gaseous hydrogen. Since the plasma is so dense (electron density up to $10^{14}\,\mathrm{cm}^{-3}$), it requires a thick anode to provide field penetration into the column. This device has produced 650 mA of H^- for 10-s-long pulses (Dagenhart *et al.*, 1983). An oblique view is shown in Fig. 5, which shows how the electrons are captured at only a small loss. As in common with the Belchenko source, Sec. 1.1.2, positive H^+ from the plasma is attracted to the converter, which is biased relatively negative. The positive ions are converted into H^- at the surface, and these H^- are accelerated back through the sheath ($\approx$100–300 V) and continue on through the extraction aperture. An advantage of this source over that of the cold-cathode semiplanotron is in long-pulse, or cw, applications. A disadvantage is lower current densities.

1.1.4 Planotron/Magnetron A Planotron, whose principle of operation is depicted in Fig. 6 (Sluyters, 1979; Prelec, 1977a,b, 1980), differs from the semiplanotron in that the electrons in the primary discharge can go around the cathode in an $\boldsymbol{E} \times \boldsymbol{B}$ cycloidal drift determined by the applied magnetic field and the electric field that develops between the plasma and the cathode. The extra path made by the electrons, compared with the semiplanotron, in many cases provides less loss than the reflections made on a negatively biased anode Penning configuration (shown in Fig. 4). Other than providing an $\boldsymbol{E} \times \boldsymbol{B}$ recirculating electron stream, the source operates similar to the semiplanotron. An improvement made by Alessi and Sluyters (1980) is to increase the space in the back of the magnetron channel. This is depicted in Fig. 7 and has resulted in a 50% increase of H^- ion current to 600 mA. The increased negative-ion extracted current is depicted in Fig. 8. Also shown is the amplification ($\times$5) of the beam current effected by the focusing due to the groove. This source was also developed at the Fermi National Laboratory (Schmidt *et al.*, 1977, 1979, 1980).

The magnetron geometry is also scalable to multiple apertures for high current (250 mA for several hours) (Prelec 1984). An example is shown in Fig. 9.

1.1.5 Field-Free Surface Source Developed at Lawrence Berkeley Laboratory (LBL) (Ehlers and Leung, 1980; Leung and Ehlers, 1980), Los Alamos Scientific Laboratory (LASL) (York *et al.*, 1983; York and Stevens, 1984), and the National Laboratory for High Energy Physics, Japan (KEK) (Takagi *et al.*, 1984; Mori 1987), the concept is shown in Fig. 10. Ions are generated on a spherical cesium-coated converter surface as shown and, as a result of the presence of a plasma, immediately accelerate down the 100–300-V sheath normal to the surface. These ions "self-extract" by moving through the intervening plasma at high speed to an awaiting accelerator, shown at the right of the figure. The electrons are contained in the plasma by cusp-field confinement pro-

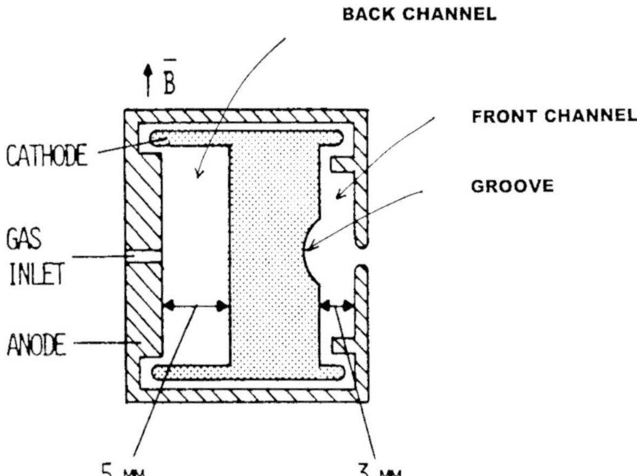

FIG. 7. The asymmetric magnetron configuration.

vided by many small magnets on the ion source chamber. 20 mA of H^- were extracted at 10% duty cycle (York and Stevens, 1984; Takagi *et al.*, 1984); 400 mA of H^- were extracted from the LBL source (Ehlers and Leung, 1980). A variation (Kwan *et al.*, 1993) produced 35 mA of D^- current.

This source is called field-free because the magnetic fields in the region of negative-ion production are small. An example of the field penetration (Lietzke *et al.*, 1984) is shown in Fig. 11. The cusp field enables confinement of plasma electrons because electrons approaching them start spiraling towards the cusp and reach a point where their longitudinal energy gets totally converted into transverse energy, and so they are reflected. Without such cusp fields the wall of the source would have to be biased highly negative ($3kT_e$), repelling the electrons electrostatically. The trouble with this arrangement is that all the positive ions are attracted to the wall and are lost (albeit at a smaller rate than the electrons would be lost if the walls were not biased). However, when the electrons are trapped by the cusp field the wall can be biased positively, or at least less negatively, and thus electrostatically confine the positive ions as well (Stirling *et al.*, 1979) for volume sources discussed in the next section.

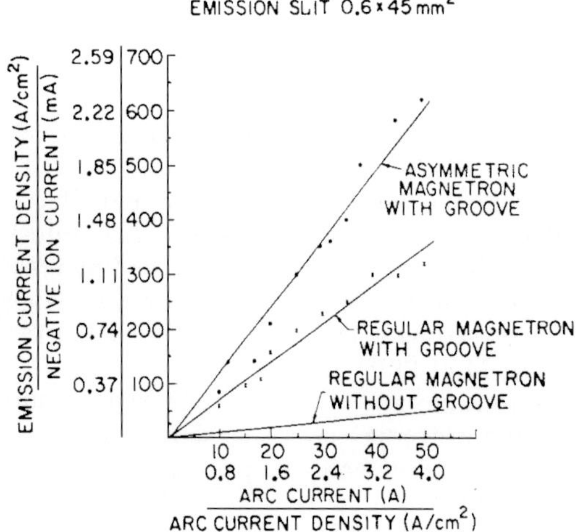

FIG. 8. H^- current versus arc current for the normal magnetron with flat cathode, normal magnetron with grooved cathode, and asymmetric magnetron with grooved cathode.

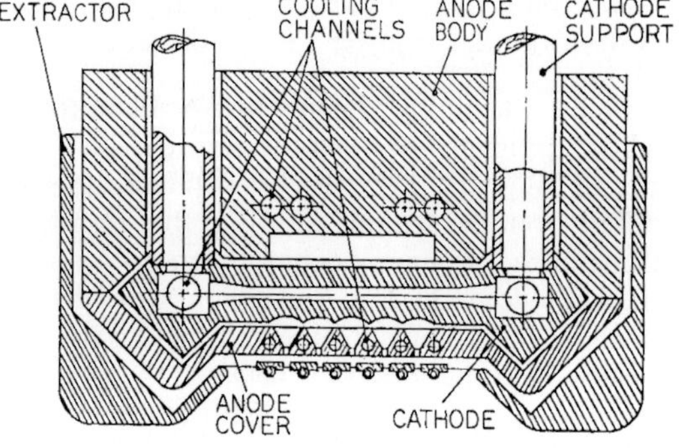

FIG. 9. Cross section of a steady-state multi-aperture magnetron source.

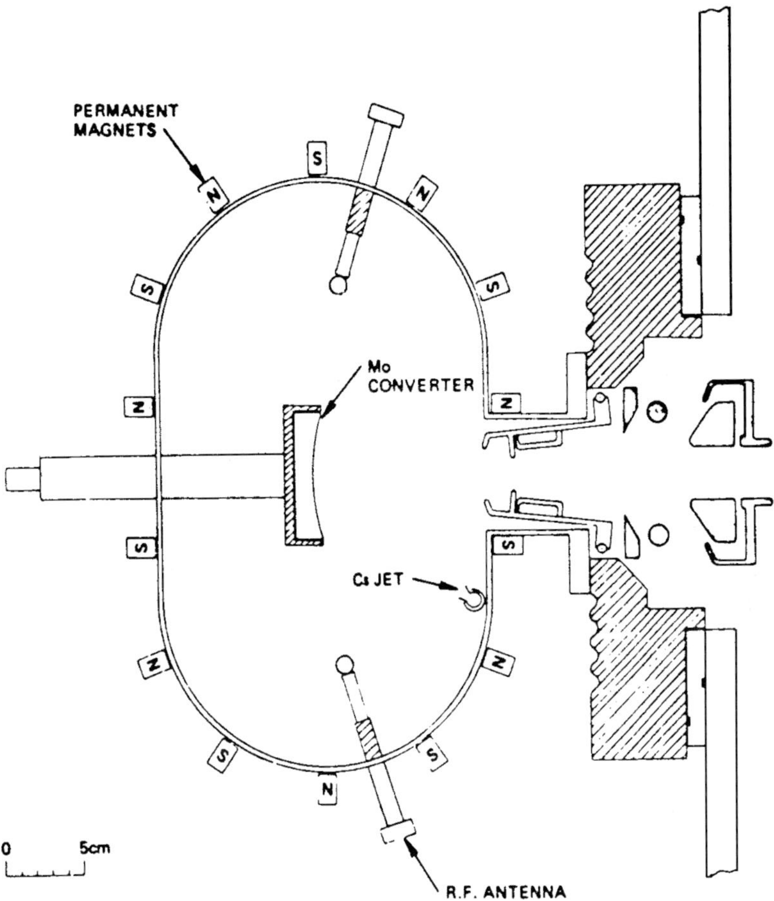

FIG. 10. Schematic diagram of the LBL self-extraction negative-ion source.

1.2 Charge-Exchange (Non-Liouville) Beam Production

One of the earliest negative-ion sources consisted of generating first a positive-ion beam and accelerating it by at least a few kilovolts and then, by running it through a gas cell, causing the positive-ion beam to undergo double charge exchange, turning it into a negative-ion beam (Semashko *et al.*, 1987). This process is inefficient compared with today's ion sources and resulted in a rather poor-quality beam. What we discuss here are the modern varieties, starting with the Penning source of Dudnikov, in which the negative ions are produced directly on a surface within the source and then in turn undergo charge exchange before extraction. This effectively lowers the ion temperature by an order of magnitude from that of the originating surface-produced negative ions.

Unfortunately, an unlucky accident occurred, which inhibited development of the low-energy, charge-exchange negative ion source—especially in the West. An inadvertent error in what is perhaps the most widely used data compilation (Barnett *et al.*, 1977) mistakenly tabulated a charge-exchange cross-section for the process

$$H^0 + H^- \rightarrow H^- + H^0,$$

that was too low by as much as a factor of eight at energies around 100 eV (Fig. 12). This was corrected in the next edition of the work (Barnett *et al.*, 1990). Anyone using these

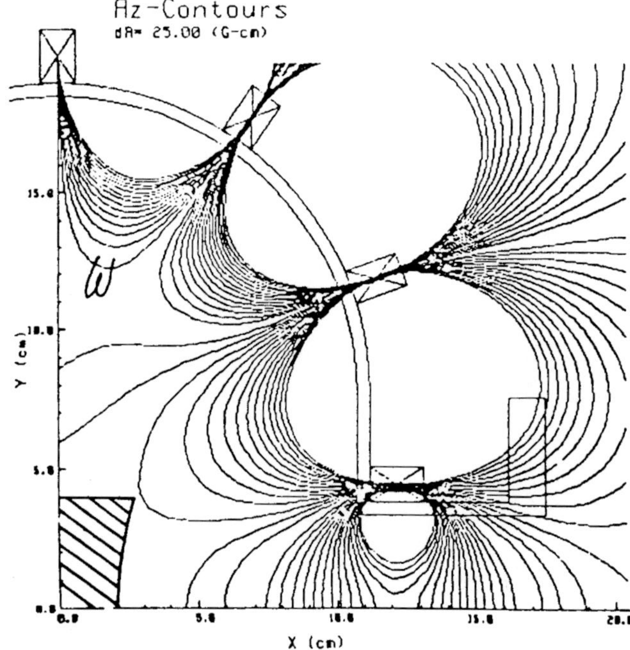

FIG. 11. Calculated magnetic field structure in one quadrant of the plasma chamber showing the filament filter and exit filter in relation to the filaments and the converter.

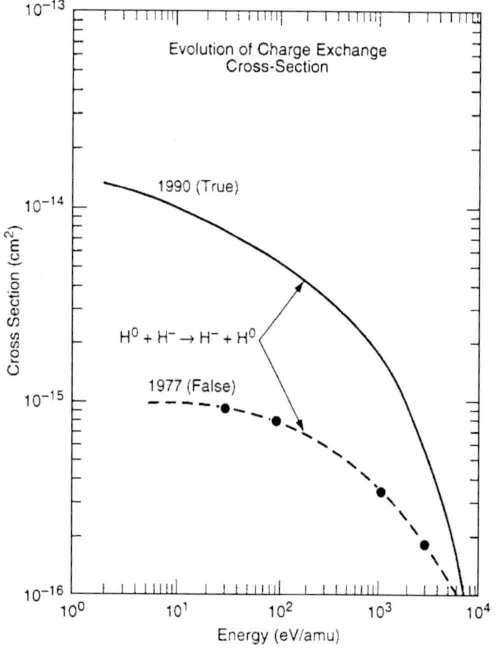

FIG. 12. Charge-exchange cross section, as published in 1977 and 1990, showing a large discrepancy especially at low energy.

erroneous data would underestimate the charge-exchange probability at these energies, resulting in the possible rejection of viable concepts. An illustration of the confusion that can arise is recorded in the comments published after the paper of Alessi and Sluyters (1980). Indeed, the discovery of the low-energy charge-exchange source was made on the other side of the world (Siberia), where the Western database might have received less attention.

1.2.1 Dudnikov Source

1.2.1.1 Cold-Cathode Penning (Dudnikov Source) The earliest negative-ion source using cesium, this configuration (see Figs. 13 and 14) is capable of prodigious output on account of a very interesting combination of brilliant inventions (Dudnikov, 1972; Belchenko et al., 1973; Dudnikov, 1980; Derevyankin and Dudnikov, 1984). This embodiment starts with a cesiated surface converter labeled as shown in Fig. 14 for initial production, but there the similarity ends; a significant variation allows a large non-Liouvillian increase in beam brightness. The negative ions once created undergo charge exchange (see Fig. 15) and may be extracted (sideways). The advantage of this

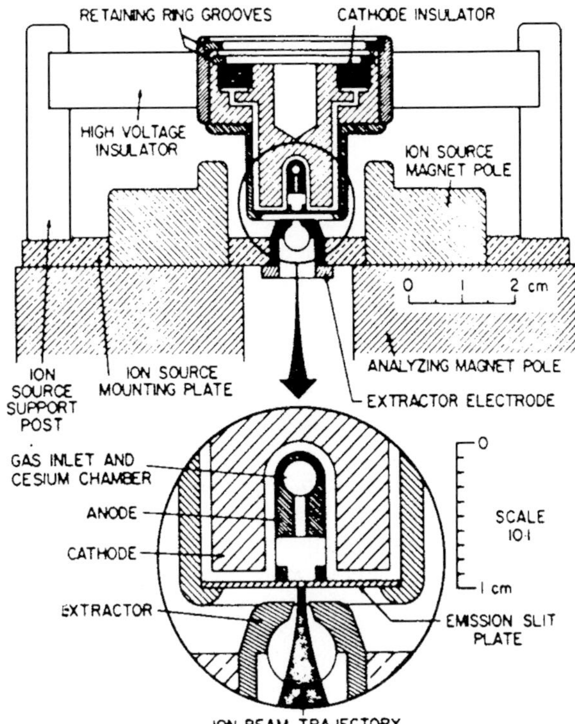

FIG. 13. Cross section of LASL version of Dudnikov source.

source over the direct surface extraction described in the previous section is that the charge-exchanged ions have a "temperature" that is likely to be an order of magnitude lower than that of the surface-produced ions.

An embodiment of one of the early versions of this source is illustrated in Fig. 13 showing the placement of magnets, hydrogen feed, cesium chamber, anode, and cathode. The region bounded by the anode and cathode is depicted in Fig. 14 showing the anode, cathode, converter, direction of magnetic field, and special anode chamber between the central discharge region and the accelerator. Some of the ion

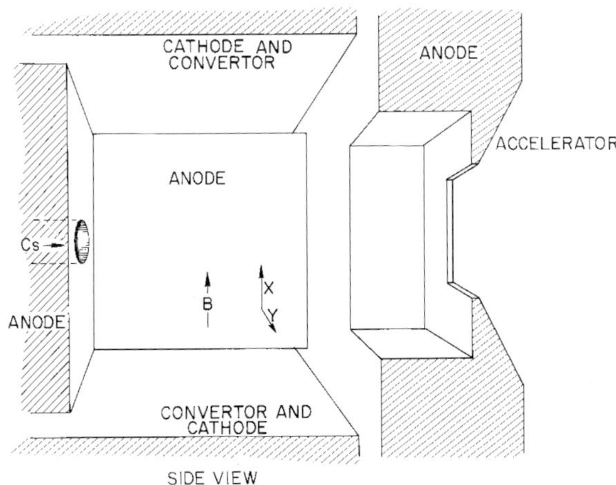

FIG. 14. Blowup of plasma-formation area in Dudnikov source.

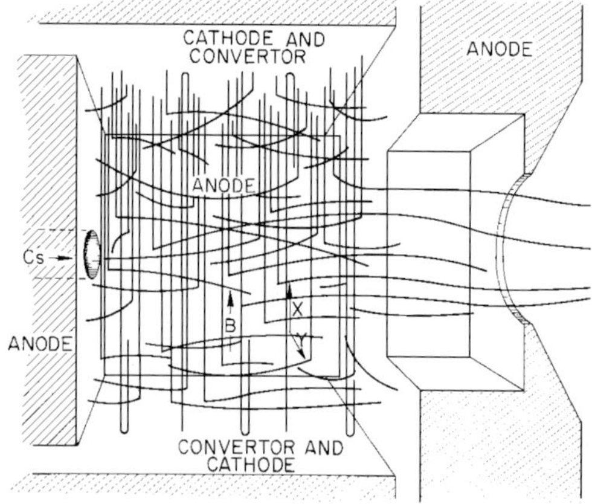

FIG. 15. Ion motion due to surface-produced ions and charge-exchanged ions within the Dudnikov source.

motion ensuing in this device is depicted in Fig. 15. Within this region, first positive ions (mostly H^+) and electrons are generated from a striker arc forming a dense, almost neutralized plasma (up to $10^{14}\,cm^{-3}$) with a plasma potential positive with respect to the cathode. In the absence of a magnetic field, the plasma potential would also be positive compared with the anode (which in turn is biased positive with respect to the cathode). However, the strong magnetic field inhibits the flux of electrons to the anode by means of the Lorentz force ($\boldsymbol{F} = q\boldsymbol{v} \times \boldsymbol{B}$), which is perpendicular to the direction of motion. The positive plasma ions are attracted to the cathode, and they intercept at an energy on the order of a hundred volts. Then, through surface production processes described above, a negative ion may be emitted [as many as one for every three incident positive ions (Seidl *et al.*, 1982)] and accelerate back down the plasma sheath, acquiring an energy of about a hundred electron volts. Because of the magnetic field the ions move largely perpendicular to the cathode (in helices). The hole in the anode shown for acceleration of these ions (*e.g.*, right side of Fig. 14) is not generally reachable by these negative ions, since they are largely confined along magnetic field lines to the region between the cathodes. These surface-produced ions would not be especially desirable for beam formation, since they have several electronvolts of random transverse energy upon leaving the converter surface (cathode) (Seidl, 1994). What is done instead is to let the surface-generated H^- ions undergo charge exchange with the gas—largely atomic H^0 under these conditions:

$$H^- + H^0 \rightarrow H^0 + H^-.$$

Some of these charge-exchanged ions do get out into the accelerator region via the anode hole depicted in Fig. 15. The virtue of this process is that the charge-exchanged ions are expected to have a much lower ion temperature than the surface-produced ones, since they originated from the much slower moving neutral atomic hydrogen gas.

Another aspect to this revolutionary source is the anode web, now known as a collar, shown in Figs. 14 and 16. Some of the effects of this web are still being explored, but a preliminary study by Allison *et al.* (1980) showed that there was an optimum web thickness whereby the extracted beam could be increased by over a factor of two (see Fig. 16). Additional meaning will be discussed in Sec. 2.2. At LANL a large effort was undertaken to vali-

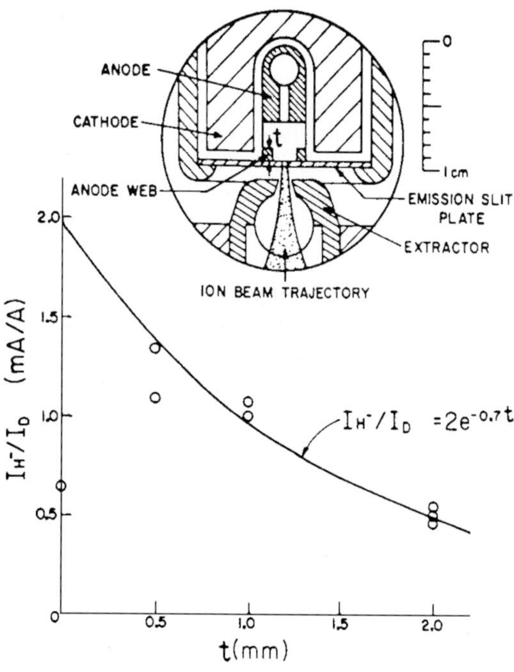

FIG. 16. H^- production efficiency (I_{H^-}/I_D) versus anode web thickness t. A curve with exponential dependence on t has been fitted to the data.

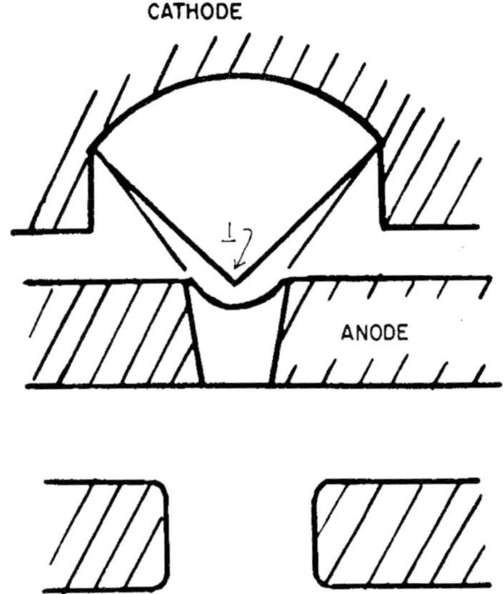

FIG. 17. The rotating ion source.

date, scale up, and deploy this Dudnikov source (Allison, 1977; Allison *et al.*, 1980) for various purposes; the 4X (Smith *et al.*, 1984, 1985), 8X (Smith *et al.*, 1993), and 16X sources and the ORNL-produced "VITEX" (Dagenhart *et al.*, 1987) are examples of such scaleups.

Additional attempts to get to a higher duty factor operation include the rotating Dudnikov source (Smith *et al.*, 1980) (Fig. 17); 160 mA of H^- were reported in the 1X source.

1.2.2 Semiplanotron and SITEX Variation By focusing the ions ahead of the extraction aperture (Dudnikov, 1980; Fig. 18), the charge-exchange reaction can be made to dominate. This was possibly achieved inadvertently in some SITEX experiments (Dagenhart *et al.*, 1984; Stirling *et al.*, 1984), because the beam-profile measurements correspond to a significantly lower ion temperature at the source as deduced from Liouville's theorem—although the charge-exchange mode

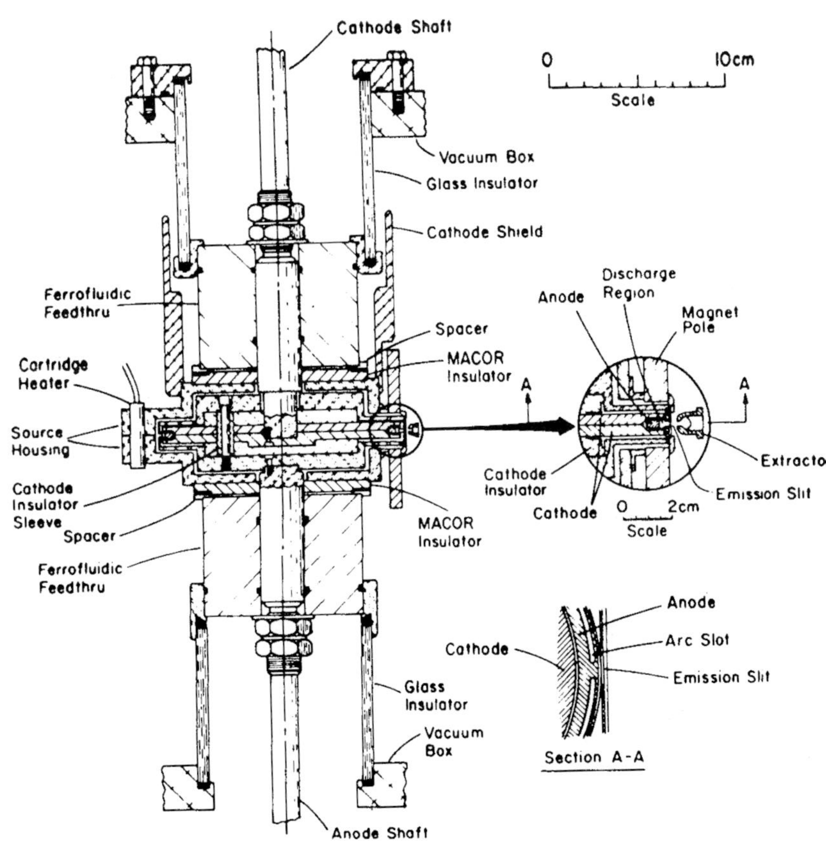

FIG. 18. Charge-exchange variation of semiplanotron.

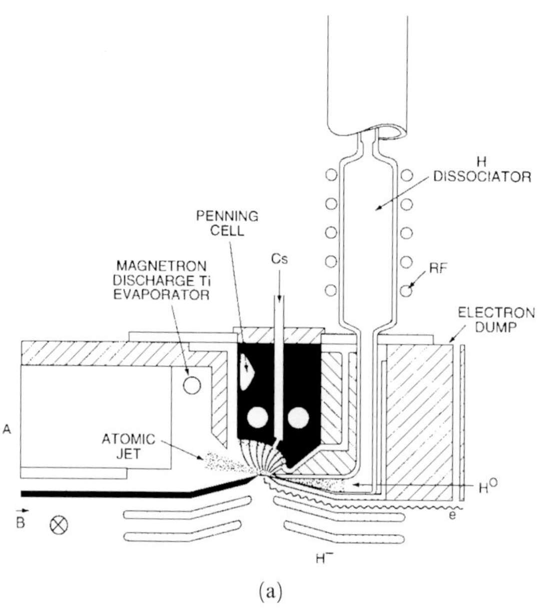

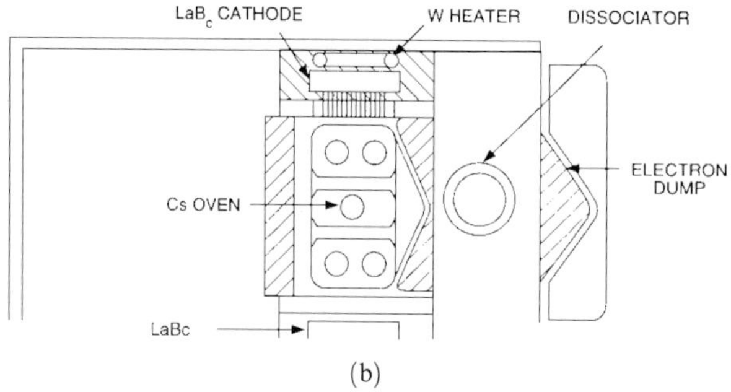

FIG. 19. Optimized version of SPS of surface plasma source. (a) Elevation view; (b) end view; (c) front view; (d) blowup of the charge-exchange region showing the configuration of the surface-generated negative-ion beam, the supersonic gas jet, and the charge-exchange region. Other features shown are the pumping, in (a), the gas dissociator for the atomic jet, in (a), the Penning discharge, in (a–c) and the cesium feed, in (a).

was not anticipated by the authors. An optimized configuration is shown in Fig. 19 (Dudnikov *et al.*, 1997). The benefit of this variation is that the advantage of high–current-density surface production can be combined with the current amplification due to geometrical focusing to produce high-density, low-temperature, charge-exchanged volume production. The focused surface-produced ions are focused to a region just before the extraction aperture where they undergo charge exchange with a gas (a supersonic jet may enhance the charge-exchange reaction rate). This is shown in Fig. 19. This source has the potential to deliver an H^- beam with a virtual source size of 10 μm, an emission current density of $3\,A\,cm^{-2}$, an ion temperature of $\frac{1}{2}$eV, and a normalized brightness of $10^{13}\,A\,mrad^{-2}$, which is very high (Dudnikov, 1993; Gudharay *et al.*, 1994).

1.2.3 Elizarov Source The Elizarov ion source (Antipov *et al.*, 1990) is taken from the positive-ion sources used in the Russian ion-thruster rocket-engine space program. A current of 1 A of H^- was extracted (Elizarov, 1972). The source (Fig. 20) consists of a hollow cathode (1) with an abundance of Cs. A cesium plasma is formed in region 9. Holes drilled in the outside of this region's cylindrical hollow cathode (2) allow plasma to escape initially into region 3, in which a radial magnetic field $\boldsymbol{B}$ is present. A series of Penning-discharge spokes is obtained. The H^- ions

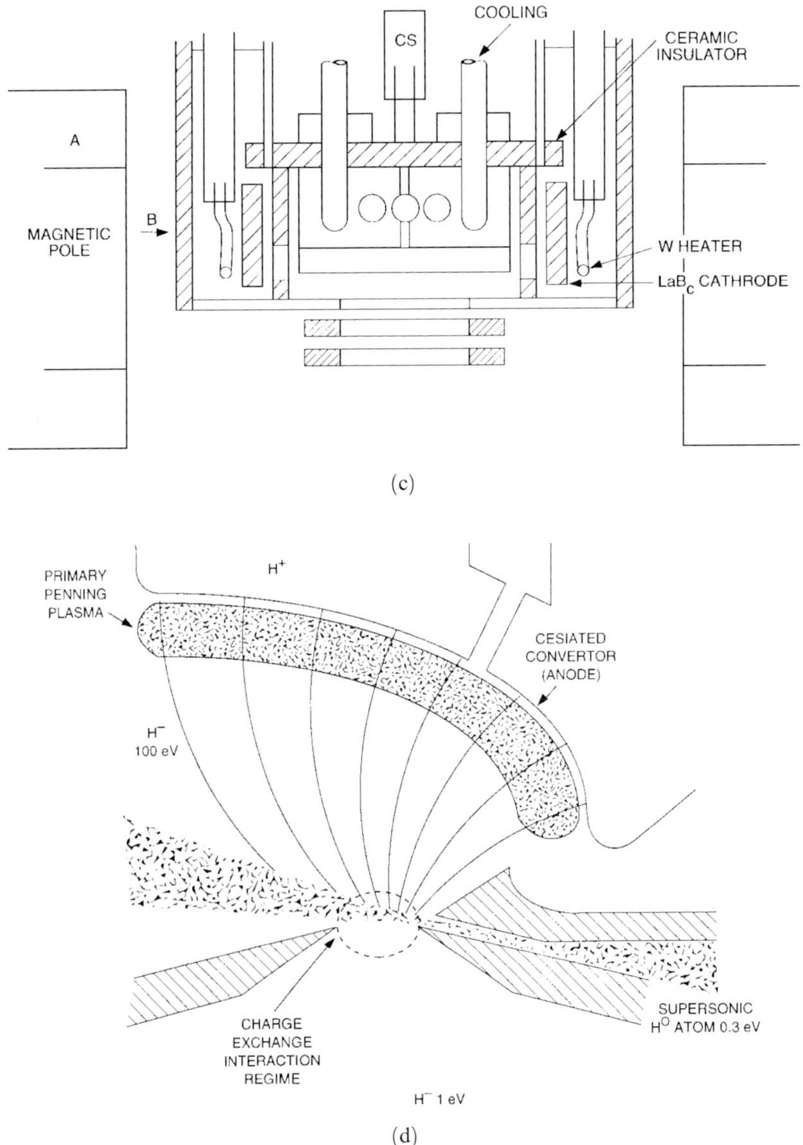

FIG. 19. (continued)

drift through charge exchange toward the extraction electrodes toward region 8, with anode 1 in combination with the externally applied magnetic field providing an intraplasma field. A source built as in Fig. 20 produced 1000 mA. The Elizarov source may also be configured to produce enhanced current density in a non-Liouvillian manner. This is achieved by forming a conical channel with resistive walls or separate wall electrodes to produce the intraplasma electric fields (in combination with the transverse magnetic field). An embodiment is shown in Fig. 21.

1.3 Excited-State Volume Production (Bacal)

Researchers at Ecole Polytechnique (Bacal *et al.*, 1977) invented a pure hydrogen hybrid multicusp volume negative-ion source. It con-

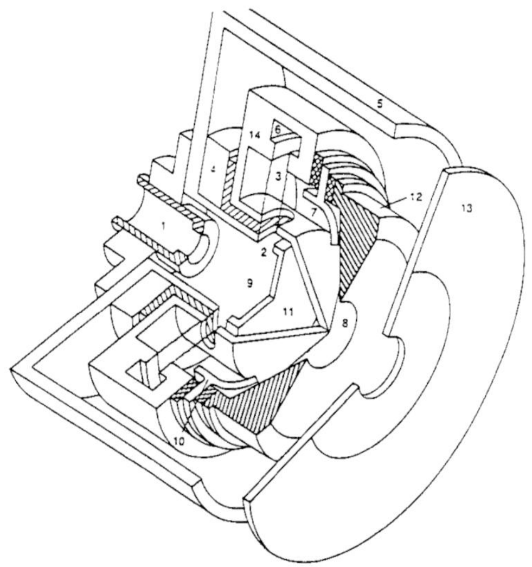

FIG. 20. Compression variation of Elizarov ion source. Shown is the cesium feed, 1; the cesium–hydrogen plasma, 9; the extraction, 2, of the cesium hydrogen plasma into the weak radial Penning discharge, 3, generated by cathode, 6; magnetic pole pieces 5, 11, 12 abide 7; funnelling electrodes 11, 12; beam extraction 8; and accelerator electrode 13.

sisted of two distinct plasmas. One, called a driver plasma (in most configurations confined by cusp field magnets; see Sec. 1.1.4) consists of a conventional discharge where hot electrons are created, causing ionization of the gaseous hydrogen and generation of vibrationally excited hydrogen molecules:

$$H_2 + e^- \rightarrow H_2^* + e^-. \quad (1)$$

These excited hydrogen molecules drift into a second plasma region, called the target plasma, where the electron energy is arranged to be much lower ($\approx 1\,eV$). Then the reaction

$$e^- + H_2^* \rightarrow H^- + H^0 \quad (2)$$

has a high probability.

For the 7th vibrational level of H_2^*, for example, the reaction is 10^5 times greater than the corresponding reaction for ground-state H_2. Absence of high-energy electrons in the target plasma region reduces the rate of H^- destruction processes. An illustration of this separation in regions for energetic electrons ($>3\,eV$) and cold electrons ($\approx 1\,eV$) is shown in Fig. 22. The isolation of the target plasma, where cold electrons dominate, from the driver plasma, with its dominantly energetic electrons, occurs because of the presence of filter fields. Electrons going from the driver plasma to the target plasma need to undergo hundreds of collisions (some of which are inelastic) with neutral H_2, which reduces their energy so that when (and if) they finally arrive at the target plasma they are cooled. The filter fields provide a path length for the electrons that is several orders of magnitude higher than what would obtain without it, and thus increase the probability of energy-losing collisions.

1.3.1 Early Versions Two early embodiments are shown in Figs. 23 and 24 (Bacal *et al.*, 1984, 1986, 1987, 1988). In the first concept, the hot electrons are confined to the larger radii by direct insertion of multiple magnets between the filament region (driver plasma) and the target plasma. The extraction hole is shown in the bottom. In this configuration there is an opportunity for geometrical amplification of excited molecules into this excited region. The magnetic filter between the driver plasma and the target plasma is furnished by the external cusp fields, with the filaments shown being in the fringe of these cusp fields. In this case the target plasma is actually divided in turn by the fringe fields of the confining cusp fields, which serve to decrease the electron temperature in the post-extractor region, shown explicitly in Fig. 24 (Bruneteau 1986; Bacal *et al.*, 1988) and is described in the extraction section (Sec. 2). In any event, the fields further retard transport of high-energy electrons, which would otherwise quickly destroy the newly formed negative ions. The main driving force for this flurry of activity in volume sources was the requirements of the neutral particle-beam (NPB) "Star Wars" program of the 1980s where low beam emittance was critical.

1.3.2 Variations More recent embodiments of the volume source use cusp-field confinement of the driver plasma and are typified by the devices produced in the UK (Holmes and Green, 1984, 1987; McAdams *et al.*, 1984, 1987), France (Bacal and Leung, 1984), Japan (Okumura *et al.*, 1987), and the United States (Leung and Ehlers, 1984). In

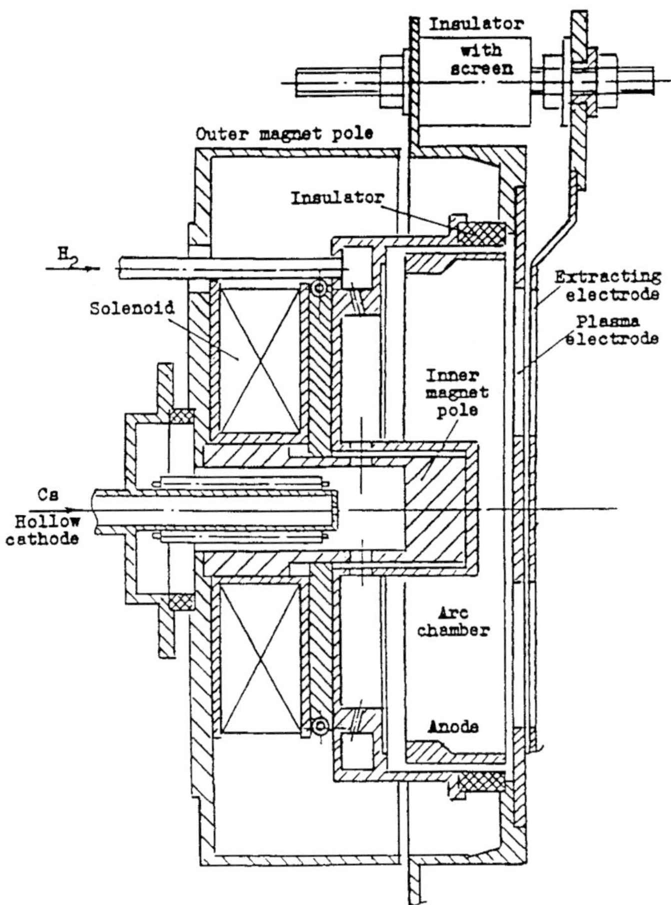

FIG. 21. Schematic diagram of Elizarov ion source.

these sources the filaments are kept totally inside the cusp field for maximum efficiency of gas ionization. The cusp fields surrounding the driver plasma enable superior confinement, since all but the lowest-energy electrons can be confined by mirror trapping resulting in their being bounced back into the plasma. The need for electrostatic electron confinement is thus reduced or even eliminated. The magnetic cusp fields also allow some reflection of the positive ions back into the driver plasma. This effect can be enhanced by biasing the containment wall less negative (or even positive) relative to the plasma.

Several more recent embodiments of this source have been vigorously pursued. One source pursued at Culham (McAdams *et al.*, 1987) (Fig. 25) employs cusp-field confinement on all but one side, as shown. The magnetic filter in this approach is provided by the two rows of linear cusp-field magnets that provide a transverse magnetic field throughout the device. A plasma with hot electrons (driver plasma) is situated on the filament side of these filter fields, while the negative-ion generation (target) plasma is shown on the extraction side.

Large-scale, multiaperture sources of this general type are being designed and built at the Japan Atomic Energy Research Institute (JAERI) (Okumura *et al.*, 1986). The electrons have to be collected locally in these sources, since the apertures are very close together. In addition, the proximity of apertures precludes the use of collars and filter-field tailoring, both of which may be useful, as discussed in Sec. 2. These sources have produced 3 A of H^- (by 1991). On a $\frac{1}{5}$-scale cesiated hydrogen source, 3.9 A of H^- at 73 keV were extracted

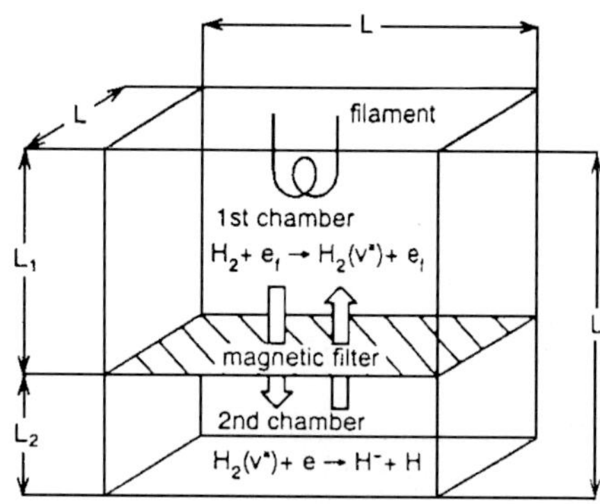

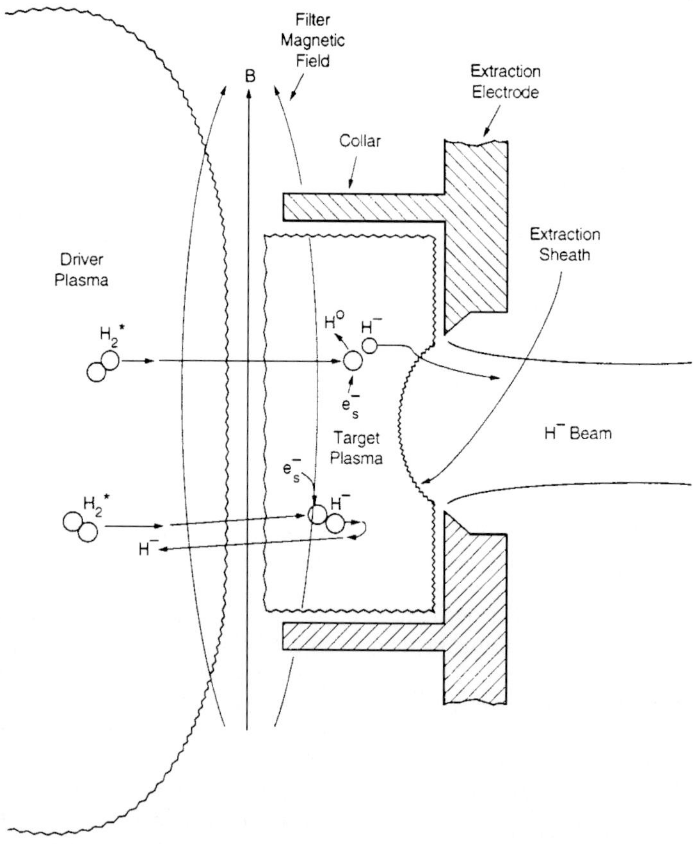

FIG. 22. Top: Illustration of principle of dissociative-attachment ion source. Bottom: Actual embodiment illustrating the role of collar and magnetic filter.

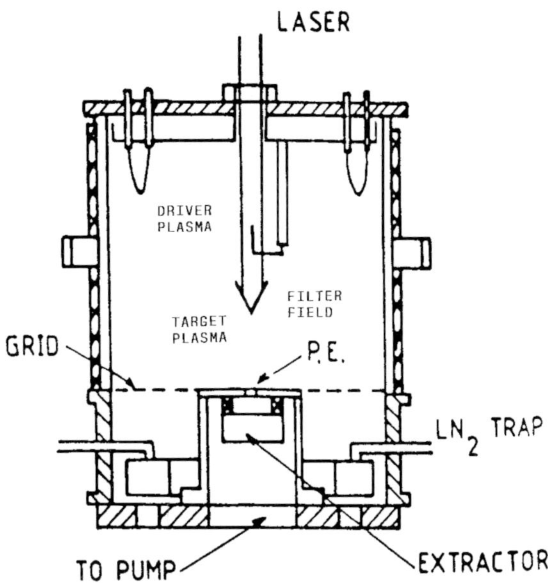

FIG. 23. Early Bacal source.

at NIFS (Kuroda *et al.*, 1996) and 14 A of H^- were extracted and accelerated to 125 keV in a cesium-seeded source. At JAERI, 10 A of H^- were produced (Inoue 1993); 13.6 A of H^- were extracted without cesium and accelerated to 125 keV (Takeiri *et al.*, 1996); and 5.9 A of D^- were extracted to 400 keV, with cesium (1.7 A of D^- without cesium) (Okumura *et al.*, 1996).

Multibeamlet ion sources of this type are also being actively developed at JAERI as shown in Figs. 26 and 27. The former shows the arrangement of edge magnets, which provide the plasma configuration and the magnetic filter. Also shown is the electrostatic accelerator including a thick extraction grid. This thick electrode is intended to catch the tangentially extracted electrons, assuming that no secondary electrons are produced from this surface being bombarded with 10-keV electrons, tangentially. These volume sources have been found to work better with the injection of Cs, as shown, which usually results in a significant (frequently a factor of 3) increase in the negative-ion output accompanied with a decrease in extracted-electron output (sometimes a factor of 1/100). Possible explanations will be discussed in Sec. 2.2. Another version is named the Kamaboko source (shown in Fig. 27) because the plasma containment vessel is said to be shaped like a fish. This source has been developed for thermonuclear fusion experiments and has delivered 18.4 amps of H^- current at 350 keV energy. The current density at the source is about 10 mA cm^{-2} (Ohara, 1998).

Development of simple beamlet sources, now mostly for accelerator applications for high-energy physics and materials science (spallation neutron sources), has proceeded along several lines. One of those is the spallation-neutron source design initiated at Berkeley (Leung, 1998).

In 1993, a variant of the Bacal source employing cusp fields, a magnetic filter, and a

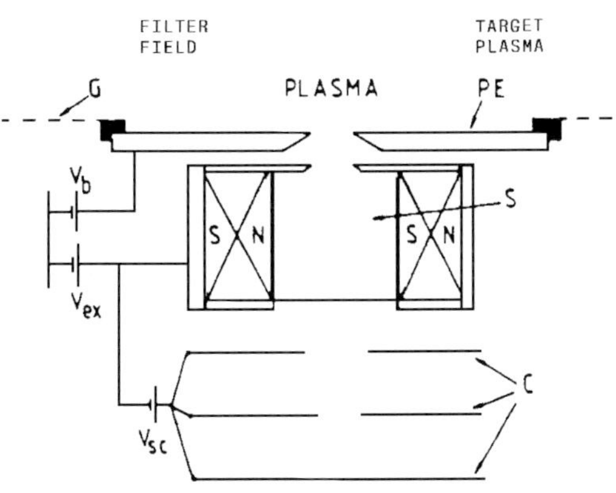

FIG. 24. Blowup of extractor region.

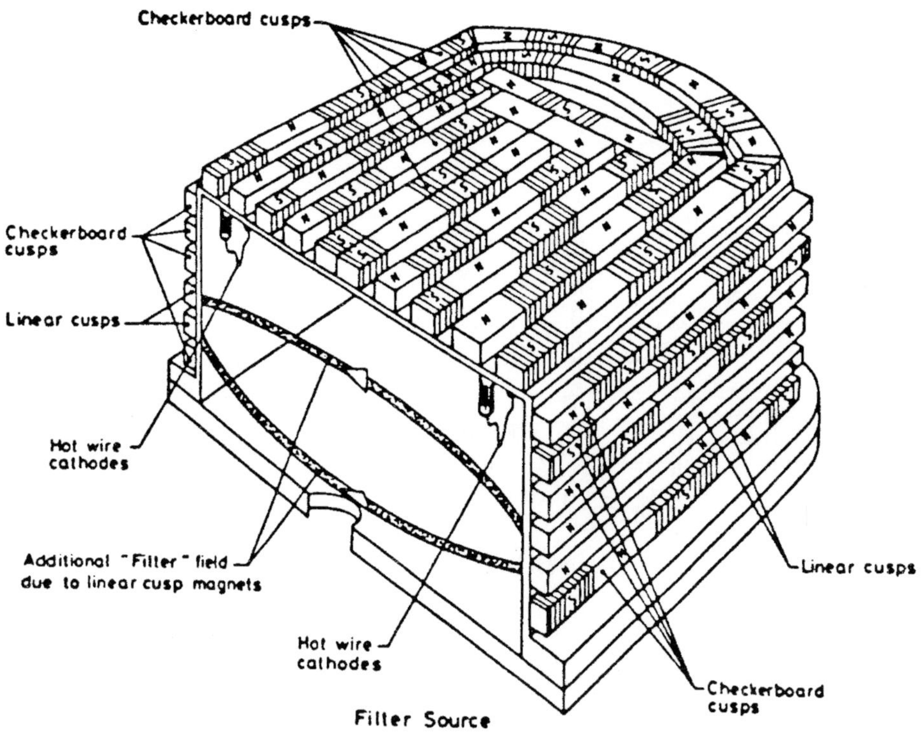

FIG. 25. Culham volume source showing a variation in the magnetic-filter configuration and cusp-field confinement on the end wall as well.

collar electrode (introduced earlier in the Dudnikov source) was produced at LBL (Leung 1994). Current densities of 250 mA cm^{-2} have been extracted for small apertures with gas pressures of 50 m Torr and $e^-/H^- \geq 100$. Ion currents of 30 mA have been extracted with a 5.4-mm-diam aperture (Leung, 1991).

An illustration of a collar arrangement is shown in Fig. 28. Notice that the target and driver plasmas are isolated from each other by the "magnetic" insulation provided by the filter fields. This insulation derives from the lower mobility of the electrons across the magnetic field, enabling the buildup of potential differences. This also means that the target plasma may be biased somewhat independently with respect to the driver plasma. The driver-plasma potential is usually determined largely by the outer walls of the ion source, which supply the arc current. The electrons drift into the target plasma via collisions. Each electron undergoes on the order of 100 collisions between the electrons and the H_2 within the filter fields and 200 collisions within the collar region. Each collision serves to cool the 8-eV electrons with the cold H_2. In addition, as depicted by the bottom half of Fig. 29, the electrons bounce off the collar electrode many hundreds of times. The probability of being captured by the collar electrode is very small per approach, but is magnified by as much as two orders of magnitude during the electron transport from the driver region to the extraction sheath. This probability is very sensitive to the plasma self bias Φ_0, so that any reduction in Φ_0 would increase the absorption per pass exponentially. A conceptual diagram is shown in Fig. 28 and a "how it goes together" diagram is shown in Fig. 30. This source uses a collar electrode, which is known to increase H^- output, especially with the addition of cesium. Discussion of this effect appears in Sec. 2.2.6.

1.3.3 rf Plasma Generation In this type of source, rf energy is injected and used to generate and maintain the plasma, instead of a filament, hollow cathode, or cold cathode.

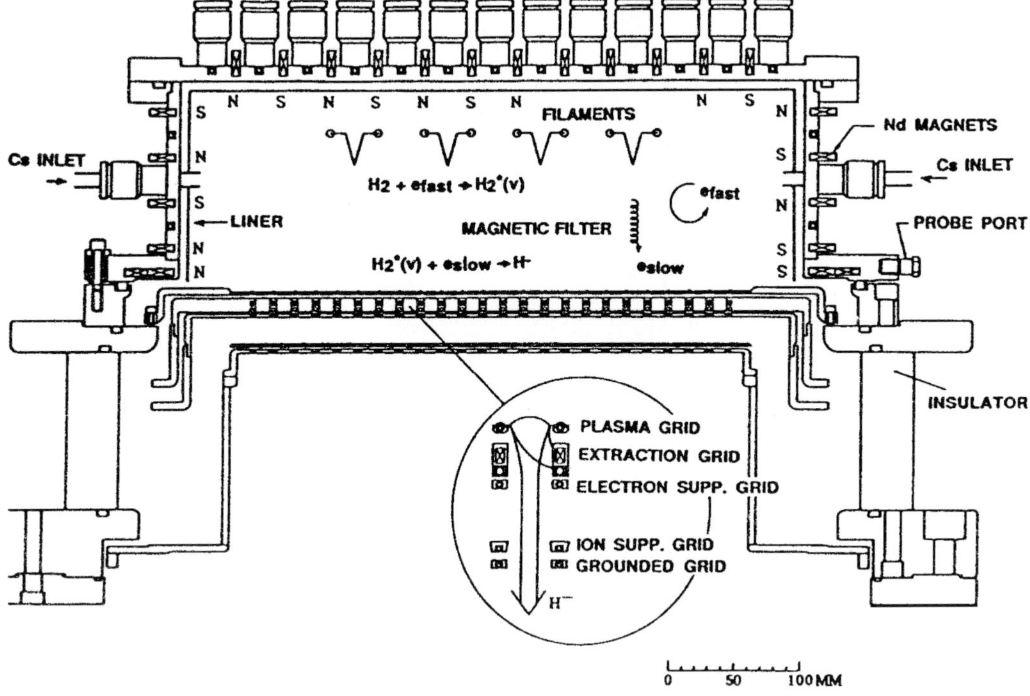

FIG. 26. A schematic diagram of JAERI multiampere H^- ion source. Cesium oven is attached in the present experiment.

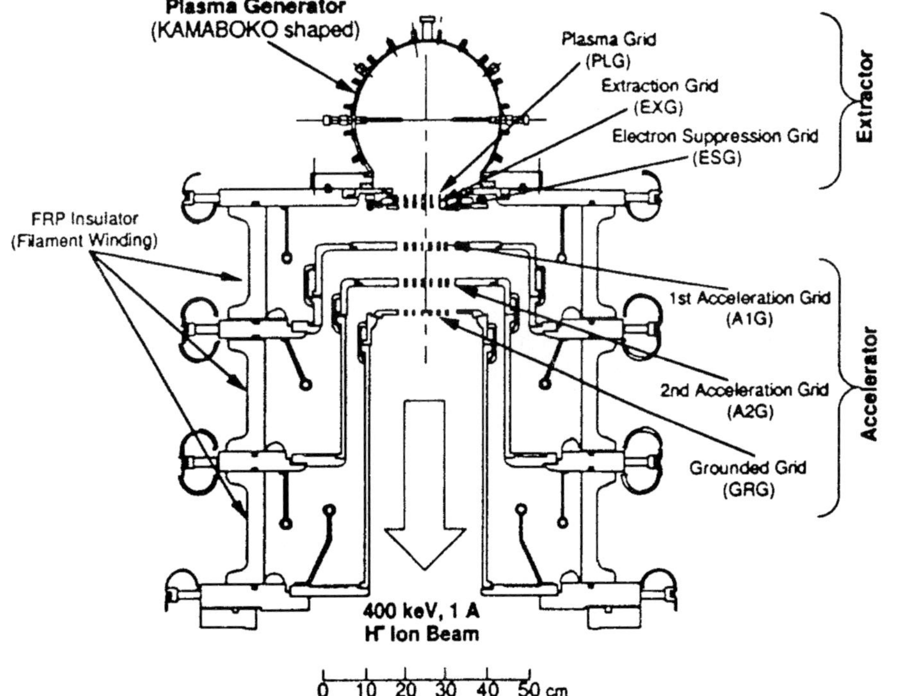

FIG. 27. A cross-sectional view of the 400-keV negative-ion source.

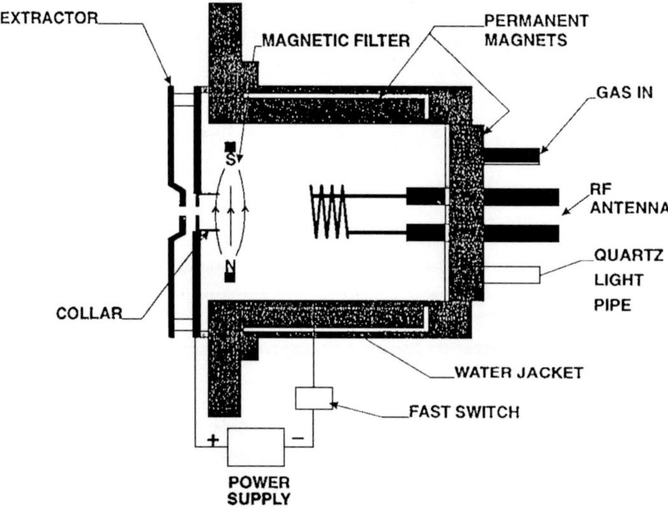

FIG. 28. Preliminary version of Spallation Neutron Source H$^-$-ion source.

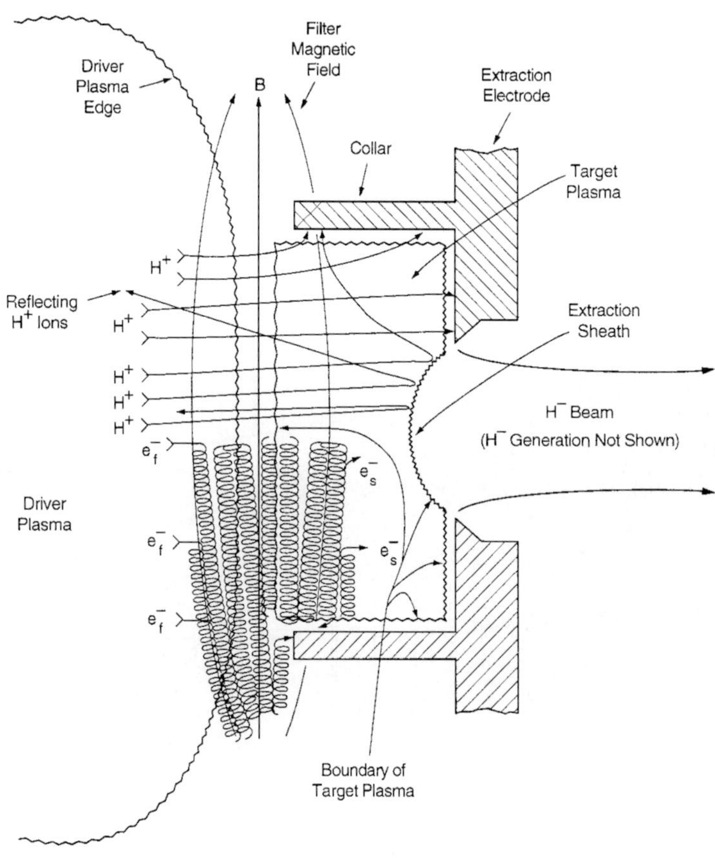

FIG. 29. Charged-particle flow in volume negative-ion source.

Antenna lifetime has been studied by Peters (1998), and so far in a particular configuration the mean lifetime is somewhat under a week with continuous operation. The magnetic filter shown in Figs. 28 and 30 allows tailoring of the magnetic field to the collar geometry, which could be important. An rf-driver volume ion source was pursued at LBL (Leung 1991, 1994; Kwan *et al.*, 1993). Development in antenna design is continuing to achieve long-pulse operation at high power (Lee *et al.*, 1998). For long-term reliable operation without injection of impurities, filaments might be replaced by rf injection by antennas or horns. Several sources of this variety have been built (Fig. 28). Recent data are questioning the reliability of present development (Peters, 1998).

1.3.4 Penning Source A Penning configuration for the driver plasma is possible with a natural insulation of the driver and target plasmas taking place through the configuration of the end walls (see Fig. 31). This configuration was explored at ORNL and was named the VITEX, for Volume Ion Transverse EXtraction (Dagenhart *et al.*, 1987).

1.3.5 Toroidal Discharge One embodiment (Alessi and Sluyters, 1990), which combines the geometry of the driver plasma indicated in Sec. 1.3.1 with magnetic cusp confinement, is shown in Fig. 32. It produces 35 mA with an $e^-/H^- = 5$ and 20 mA with $e^-/H^- = 3$ (Alessi and Prelec, 1991). The normalized rms emittance of this source is measured to be 0.32π mm-mrad for a 13-mA beam, which corresponds to an ion temperature of 0.6 eV, an unusually low value (Alessi and Sluyters, 1990). This source generates a plasma in region 1 by electron-impact ionization from the electrons generated by the filament and accelerated through the plasma sheath surrounding the filament. This driver plasma is confined by the cusp fields on the edge of the container as shown. By drifting through the magnetic fields within the plasma, low-energy electrons arrive in the target region 2. The device was subsequently embellished (Alessi *et al.*, 1994) by providing a stronger conical intraplasma magnetic field (Fig. 33). This increased strength provides fewer unwanted high-energy electrons in region 2. The conical shape ensures that the maximum pos-

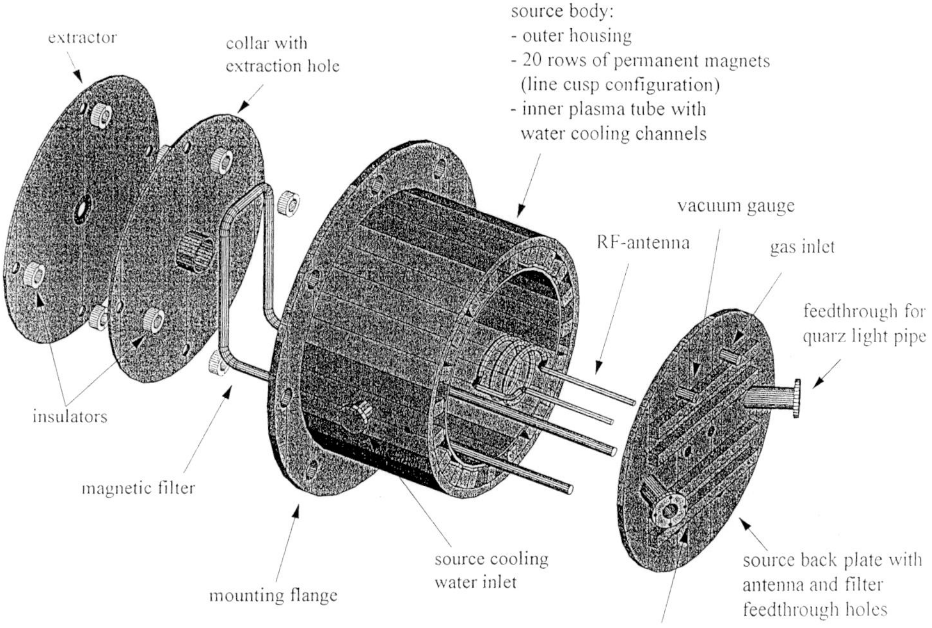

FIG. 30. Details of a preliminary version of the SNS ion source.

sible flux of low-energy electrons appears in the dissociative attachment regions. Also, because of the diverging magnetic fields near the extraction aperture 3, electrons are discouraged from being extracted. This is very important as indicated in Sec. 2.2.4. This concept has been subsequently explored at LANL (York *et al.*, 1993) with some variations in the filter-field configuration. H^- ion currents of 20 mA cm^{-2} (without cesium) and 40 mA cm^{-2} (with cesium) have been achieved.

1.3.6 Adding Cesium The effect of adding cesium to a volume source is generally that the negative-ion current increases by as much as a factor of 3. Also, and no less important, the extracted electron current decreases by a factor of as much as a hundred. This is shown in Fig. 34. Adding cesium in the compact multicusp source described in Sec. 1.3.2 increased the current from 7 to 34 mA. The current density with cesium was as much as 1.2 A cm^{-2} (Leung *et al.*, 1989). The addition of cesium in the toroidal filter source being explored at Brookhaven National Laboratory (BNL) (York *et al.*, 1993) provided a doubling of the negative-ion current and a reduction of e^-/H^- extracted from 5:1 to 2:1.

2. PHYSICS OF HYDROGEN NEGATIVE-ION EXTRACTION

It is one matter to produce negative ions as discussed in Sec. 1 but quite another to extract them into a beam with low emittance. In this section we discuss two classes of extraction: direct extraction from surface sources and extraction from volume sources. Each has its peculiarities, and the physics and computational issues of each are discussed.

2.1 Ion-Extraction Physics for Surface Sources

Surface sources are divided into two categories depending on whether the magnetic field is large or small.

2.1.1 Low-Field Surface Sources For direct-extraction sources, the extraction physics may seem relatively straightforward, since the negative ions travel through the intervening plasma at a high energy of 100 eV. However, for the cases where there is a large distance between the converter and the extraction region, finite-T_e effects may be important and need to be treated with care.

For volume extraction, all situations need special treatment since the analysis used for positive-ion extraction is not appropriate. Lawrence Berkeley Laboratory proposed a direct-extraction negative-ion source (Ehlers and Leung, 1980), which is shown in Fig. 35. Measured beam widths are shown in Fig. 36. The common interpretation of the large beam width at the focal point is that it is due to the high ion temperature (≈ 7 eV) coming off the converter. While this is a significant cause for this beam divergence, another one is the high space-charge forces within an imperfect ($T_e > 0$) plasma in which they are transported. An illustration of this latter effect is provided in Fig. 37, which shows the ion trajectories for three different plasma electron temperatures. At higher electron temperatures, deviations from perfect plasma neutrality occur, causing the beam optics to deviate from the idealistic case of Fig. 37 (top) with $T_e = 0.25$ eV. The solution of the Vlasov–Poisson equations for this type of source is usually done with the assumption (Whealton *et al.*, 1984, 1987, 1988) that the sum of the positive ions and electrons is a Boltzmann distribution. The temperature, denoted as the electron temperature, is actually that of the Boltzmann distribution. This type of analysis is expected to give a lower bound to various optical aberrations, since it neglects any Bohm ion-acoustic instabilities at the beam edge and also neglects finite negative-ion temperature (as much as 6 eV) coming off the converter. Finite ion temperatures in the limit of zero plasma temperature were considered separately (Anderson, 1984). For finite plasma temperature, $T_e = 4$ eV, which is the range commonly expected, noticeable but perhaps not drastic deviations occur; for $T_e = 16$ eV, which is higher than expected, the crossover essentially disappears, and the optics are seriously affected for this geometry. Other geometries could produce better options for these high–electron-temperature situations. The fraction of beam transmitted and beam diver-

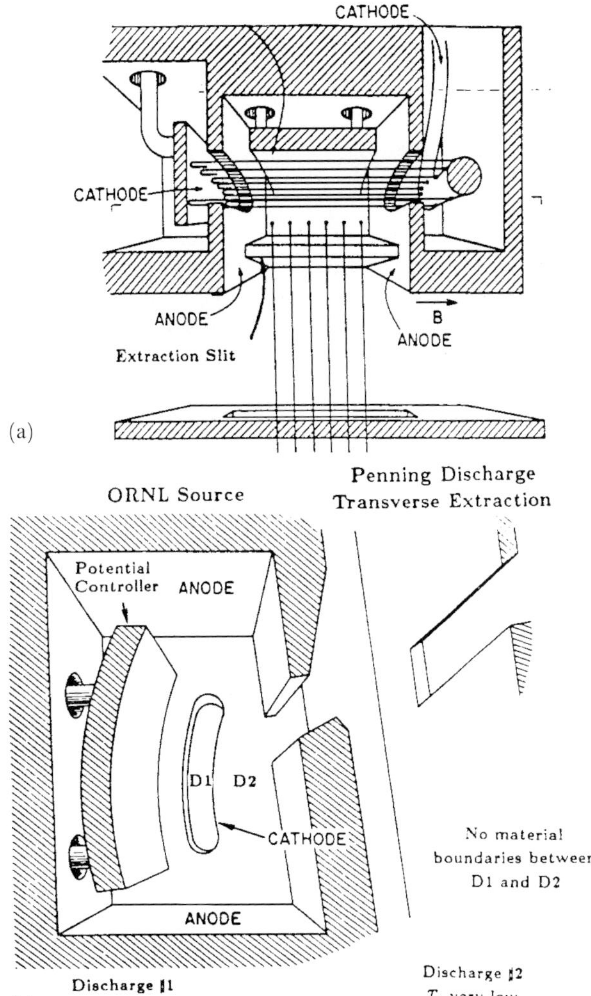

FIG. 31. (a) Top view and (b) elevation of VITEX source. Two plasmas are defined by the cathode slot, discharge DI being the Driver plasma and DZ being the target plasma.

gence as functions of electron temperature in eV, derived by the accelerator potential, are given in Fig. 38, showing the regions of single vignetting and multiple vignetting.

A more extreme situation of imperfect transport is shown for the Los Alamos Meson Pion Factory (LAMPF) injector (Whealton *et al.*, 1984), on account of a much more elongated region between the converter and extraction. The idealistic case of 0.0009 eV shows the optics as intended (York, 1983). However, even for $T_e = 0.1$ eV, which is far lower than expected, deviations from a perfect crossover point are expected. For the more realistic electron temperature of 10 eV, as shown by Whealton *et al.* (1984), the optics are significantly affected.

2.1.2 High-Field Surface Sources The long transport region that was at least conceivable in the field-free sources discussed in the previous section is not possible in the high (magnetic)-field sources where the transport region must be short compared with an ion gyroradius. For the SITEX source, Fig. 39, the negative-ion trajectories from the converter in two dimensions (2D) are shown in Fig. 40 (Whealton *et al.*, 1984), and a scheme to use a finite ellipsoidal segment converter to obtain an approximately round beam is shown by the

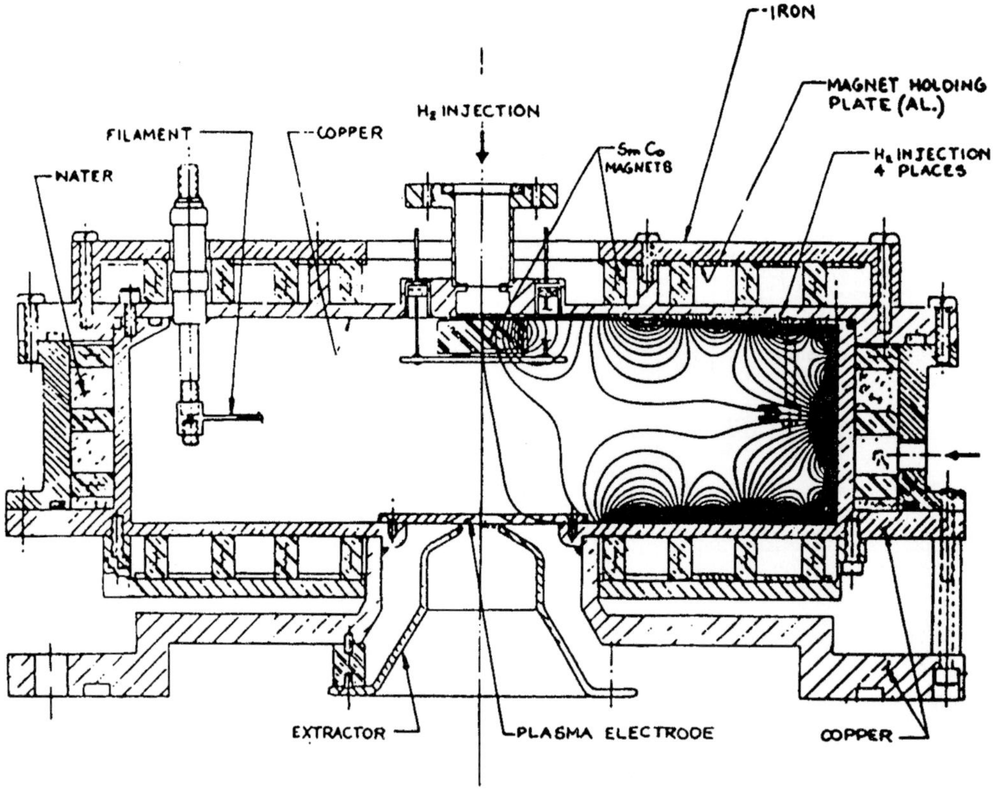

FIG. 32. BNL volume negative-ion source.

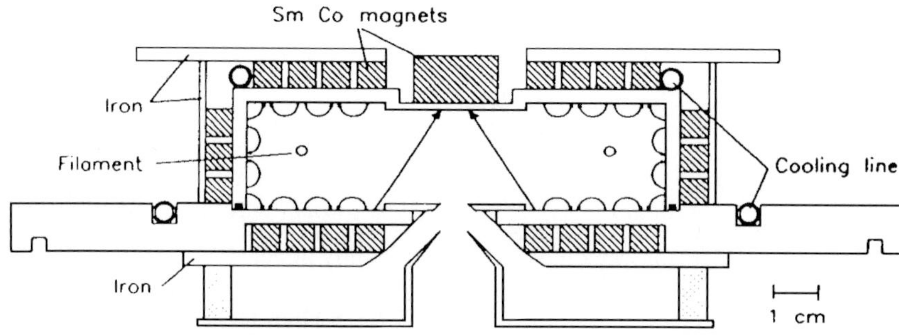

FIG. 33. Improvement on toroidal volume ion source.

3D calculation of Fig. 41 (Whealton *et al.*, 1985). The ellipsoidal connector ion source was designed such that the magnetic field was parallel to the surface and a Penning discharge would feed positive ions to the converter. A better scheme is to use the magnetron architecture described previously and also to use the charge-exchange focusing technique. An optimized version of all this is discussed in Sec. 1.2.2.

2.1.3 Dudnikov Source The Dudnikov source is considered to be a volume-production source since the actual ions extracted are

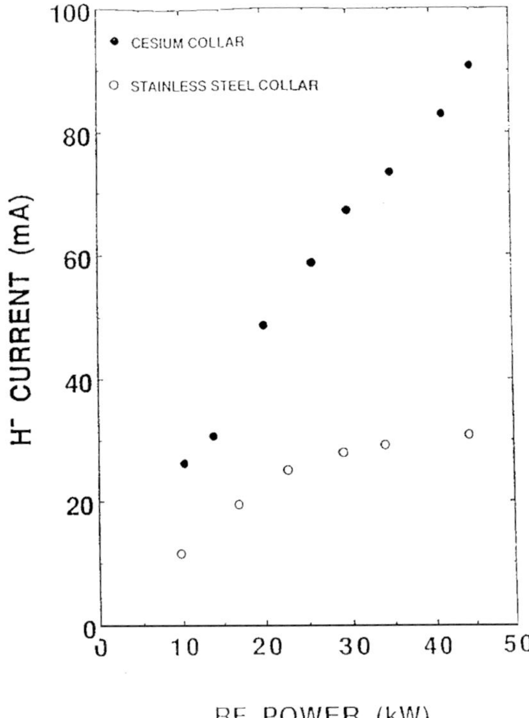

FIG. 34. Effect of cesium on negative-ion production.

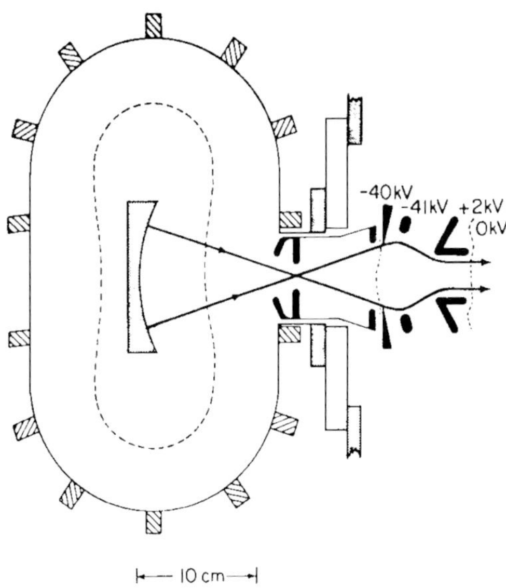

FIG. 35. Field-free converter source considered at LBL showing converter, ideal ion envelope, various vignetting and accelerating electrodes and cusp-field plasma containment.

produced through charge exchange from surface-produced H^-. Initial attempts to model this source assumed perfect symmetry with positive-ion extraction, showing something like Fig. 42 (Smith *et al.*, 1984) in 2D, and Fig. 43 (Bell *et al.*, 1985) in 3D. This approach generally was at significant variance with experiments that showed a variation of ion current with extraction voltage. The oblique sheath produced by the $\boldsymbol{E} \times \boldsymbol{B}$ drifting electrons, which also must share the transverse space-charge limit, is important but was not considered at the time. The existence of a drift region just before the extraction plane was suggested by Dudnikov and has an optimum thickness that is the balance between negative-ion depletion and extracted-electron depletion. The electron current decreases much faster with thickness than the ion depletion (Allison *et al.*, 1980).

2.2 Ion-Extraction Physics for Volume Sources

Negative-ion extraction from volume sources is less robust than extraction from

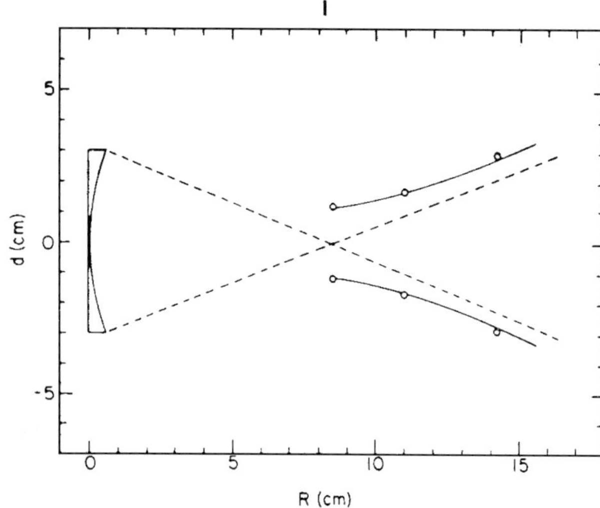

FIG. 36. Measured beam width (FWHM) compared with the idealistic zero-ion-temperature envelope trajectory.

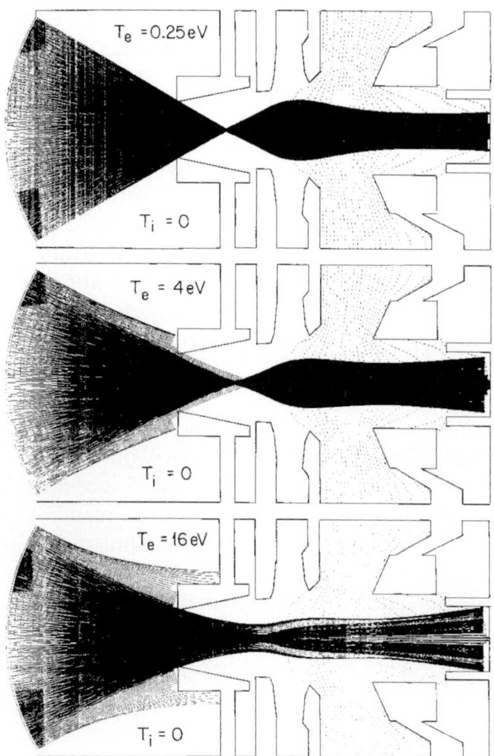

FIG. 37. Ion trajectories, still assuming zero ion temperature but considering a finite electron temperature.

surface sources, where the ions are rocketing toward the extraction aperture at energies of 100 eV or more. In fact, volume negative-ion extraction is less robust than volume positive-ion extraction for a number of reasons. For volume sources, creation of initial energies on the order of only 1 eV are expected. If the intraplasma electric fields accelerated the negative ions toward the extraction aperture, as is largely the case with positive ions in positive-ion sources, then the low initial kinetic energy of the negative ions would not be so important. The ionic flow is illustrated in Fig. 44(a) for a steady-state positive-ion source. For cases where ion-acoustic plasma-sheath instabilities develop, the situation is typified by Fig. 44(b). These positive ions, whether stable or otherwise, are electrostatically attracted toward the extraction region and either are extracted or intercept the plasma electrode. Unfortunately, these very intraplasma electric fields that tend to accelerate the positive ions toward the extraction region accelerate most of the newly formed negative ions in the opposite direction, toward the source plasma (Whealton *et al.*, 1987).

2.2.1 Asymmetries Between Positive- and Negative-Ion Extraction For positive-ion extraction, the analysis is well developed (Whealton, 1981) using the following form of the Vlasov–Poisson equations:

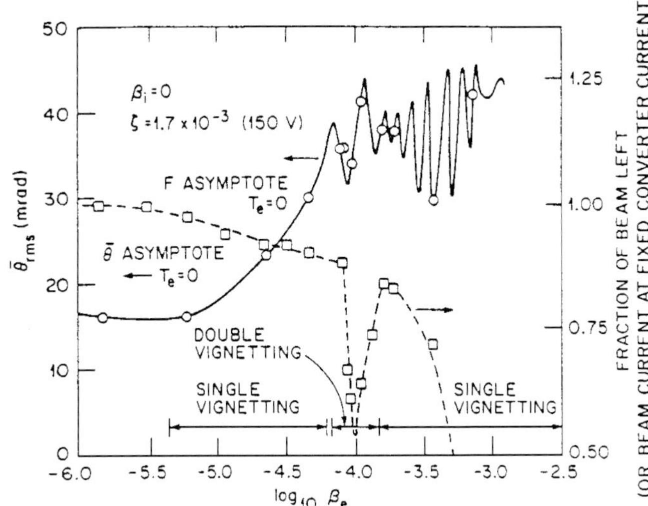

FIG. 38. Beam divergence and transmission as functions of electron temperature assuming an ion temperature of zero.

$$\nabla^2 \varphi(\boldsymbol{r}, t) = -\int f(\boldsymbol{r}, \boldsymbol{v}, t)\, \mathrm{d}\boldsymbol{v} + \exp[-\varphi(\boldsymbol{r}, t)], \quad (3)$$

$$\frac{\partial f(\boldsymbol{r}, \boldsymbol{v}, t)}{\partial t} + \boldsymbol{v} \cdot \nabla f(\boldsymbol{r}, \boldsymbol{v}, t) + \{\boldsymbol{v} \times \boldsymbol{B} - \nabla \varphi\} \cdot \nabla_v f(\boldsymbol{r}, \boldsymbol{v}, t) = 0, \quad (4)$$

where φ is the electrostatic potential; f is the ion distribution function, which is a function of position $\boldsymbol{r}$, velocity $\boldsymbol{v}$, and time t; $\boldsymbol{B}$ is the magnetic field.

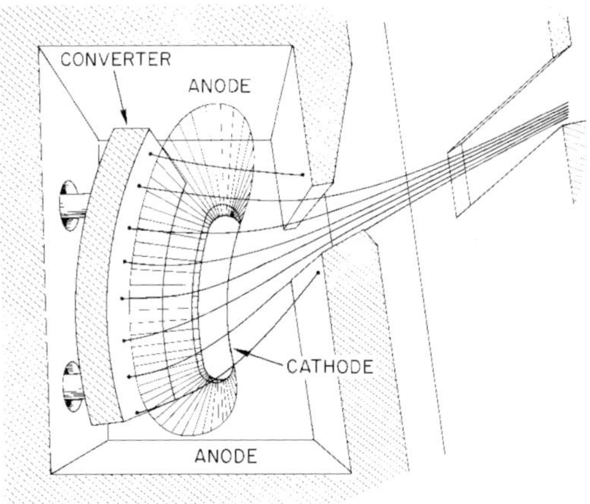

FIG. 39. Elevation view of the ORNL SITEX surface negative-ion source showing the converter, surface negative-ion trajectories, Penning discharge as projected from anode slot and extractor slot.

The first convergent solution to these equations in 2D was obtained by Cooper *et al.* (1974), for a partial sheath, and by Whealton *et al.* (1978) and Whitson *et al.* (1978) for a full sheath in 2D. The three-dimensional solution came later (Wooten *et al.*, 1979, 1981; Whealton *et al.*, 1986) and ion-acoustic time scales yet later (Whealton 1985).

The first asymmetry between negative-ion sources and a reversed-potential positive-ion source is that the electrons cancel charge imbalance excursions in positive-ion sources much better than positive ions do in negative-ion sources. This observation led to the heuristic model of Whealton *et al.* (1989), which appears to be in agreement with the laboratory data of Bacal and Stevens; see Figure 45. An example of the actual trajectory computations, at points A, B, C, and D in Fig. 45(c), is shown in Fig. 46. The plasma sheath condition in these cases varies from extremely under-dense (A) to extremely over-dense (D), all falling within the laboratory observations at Ecole Polytechnique. Similar results were obtained for the LASL data [Fig. 45(a)]. It is common to model negative-ion extraction from volume sources using the above model (Tanaka *et al.*, 1998; Watanabe *et al.*, 1996; McAdams, 1994; Cowan, 1994; Bell *et al.*, 1985; McGaffey *et al.*, 1984; Smith *et al.*, 1984; Gudharay *et al.*, 1994; Chan, 1993, 1994; Maeser *et al.*, 1996; Whealton *et al.*, 1986, 1988), an example of which is shown in Fig. 47. How-

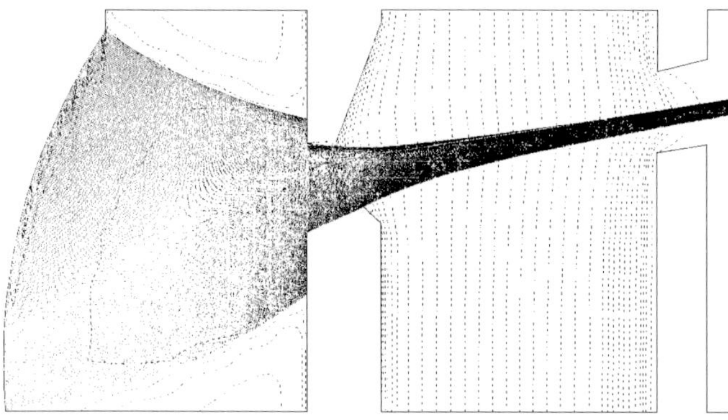

FIG. 40. Computer simulation (2D) of trajectories of negative ions from converter and extracted into slot beam.

ever, this procedure has little meaning, since it assumes that the positive ions are in equilibrium (which is not true) and neglects any explicit consideration of plasma electrons, which could be very important. Furthermore, the model assumes that the negative ions enjoy the same downhill run from plasma to extraction that positive ions have in positive-ion sources.

2.2.2 Saddle-Point Extraction The second asymmetry between negative-ion sources and reversed-potential positive-ion sources arises from the observation that the plasma electrode is biased negative to the local plasma potential for negative-ion sources just like in positive-ion sources and for the same reason—to contain the plasma electrons electrostatically. Therefore, in a positive-ion source, there is a continuous, monotonic downhill run for the positive ions generated deep in the plasma from formation through extraction. For negative ions, there is a saddle point formed within the plasma, since the plasma potential lies intermediate between those of the plasma electrode and the acceleration electrode. This saddle point is shown in Fig. 48, which for illustrative purposes is calculated in the limit of zero space charge (vacuum fields). For the positive ions [Fig. 48(a)], there is acceleration toward extraction in the region where $A_z < 0$ until they reach the saddle range ($A_z = 0$), after which they decelerate, turn around, and either return to the plasma or intercept the plasma electrode. The created negative ions, shown in Fig. 48(b), are electrostatically repelled from the extractor in regions labeled $A_z < 0$ and are electrostatically attracted toward the extractor in the region $A_z > 0$. A large volume of negative ions created is repelled from the extraction region. There is no extraction sheath in this case, since the plasma density is zero. This observation of a potential hill was made ten years ago (Whealton *et al.*, 1987) but has attracted scant attention. Recent notice has resulted in a heuristic model of a virtual cathode (Becker *et al.*, 1998), which may eventually provide a useful approach. Another approach in order to avoid the complex nonlinear theoretical problem is to use steady-state intraplasma fields derived from experiments and then plot trajectories

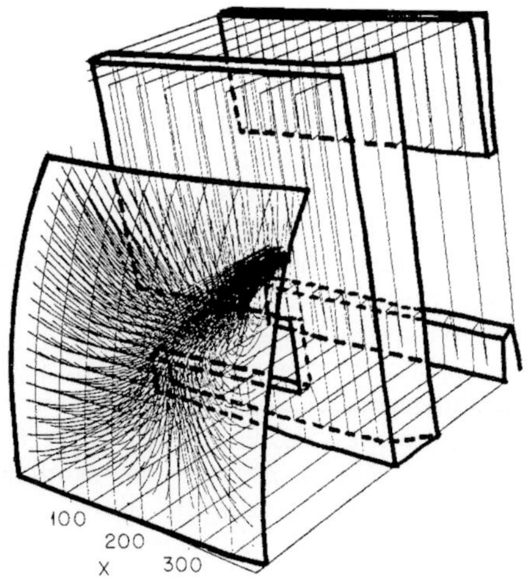

FIG. 41. Computer simulation (3D) of trajectories of negative ions from connecter and extracted into finite slot beam. The slot ends have been tailored as shown to minimize aberrations from slot end effects.

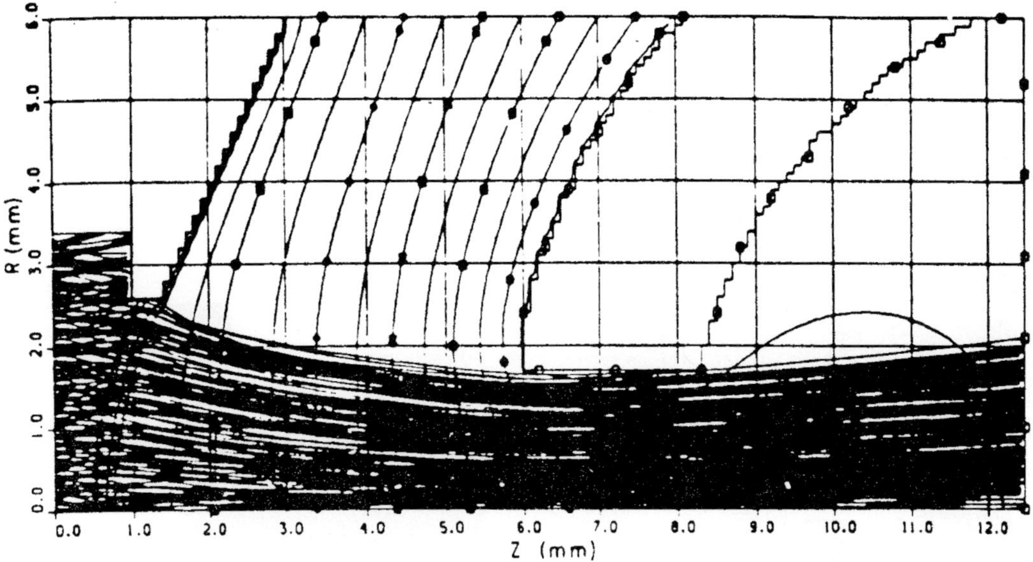

(a)

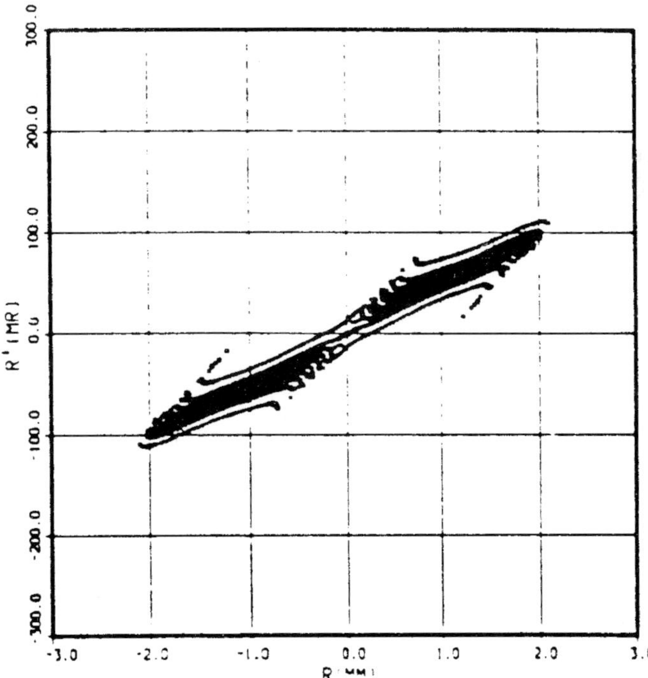

(b)

FIG. 42. (a) Snow-code calculation of the H^- ion trajectories for the 4X (Dudnikov) source extraction optics; (b) phase-space diagram at the maximum axial position shown in (a).

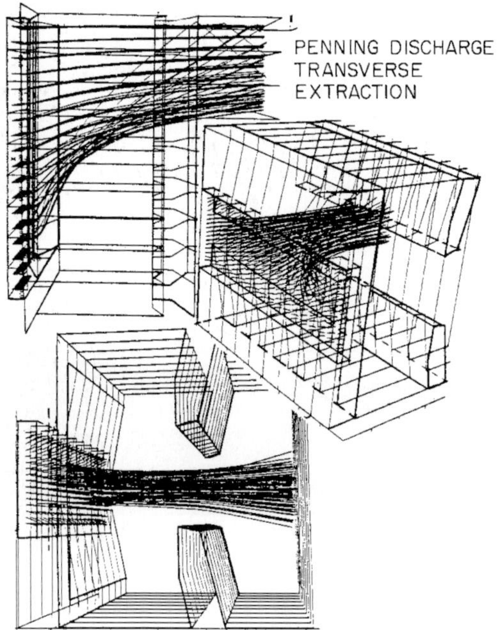

FIG. 43. A 3D calculation of the H^- ion trajectories for the Dudnikov negative-ion source, in three perspective views, showing slot end effects.

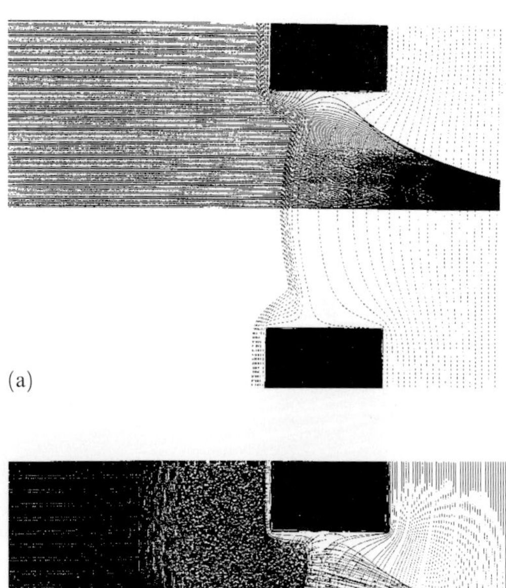

FIG. 44. (a) Plot of ion trajectories (solid lines) and equipotentials (dotted lines) in the extraction region bounded by an electrode (solid) with an aperture for ions. Beam current due to accelerator electrode interception. (b) Same as (a), except that an ion time-scale solution to the Vlasov–Poisson equations is shown for positive-ion extraction (electrons are still Boltzmann) in a region where the Bohm sheath stability criteria is not satisfied, resulting in ion-acoustic waves in the presheath and concomitant rms emittance growth.

numerically through these fields (Riz and Pamela, 1996). This makes the theoretical problem linear and straightforward. Unfortunately, the experimental measurements of plasma potential have, so far, not been resolved to provide reliable fields. Furthermore, ion-acoustic instabilities may further complicate the measurement so that a steady-state treatment of ion kinetics is questionable. Negative ions formed on one side of the ridge go toward the source plasma instead of the extraction aperture. In these cases, the location and control of such ridges could be important.

A third asymmetry results in electrons being extracted along with the negative ions, unlike the case for positive-ion sources. A fourth asymmetry is that the Bohm sheath-stability criterion is expected to be violated more extensively within certain regions for negative-ion sources than is the case for positive ion sources, at least in the absence of cesium. An illustration of the consequences of the violation of the Bohm stability criterion is shown in Fig. 44(b) [although this example is for a single Vlasov positive-ion model (Whealton *et al.*, 1987)]. An example of an actual situation where ion-acoustic instabilities may occur is illustrated in Fig. 49, where the ion temperature appears to be increasing with plasma density. This instability occurs when any process that increases the negative space charge in the pre-extraction region causes the curvature of the potential to become positive (as evidenced from the structure of the Poisson equation). Evidence for the existence of ion-acoustic instabilities in the extraction sheath is present in the disparity between ion temperatures measured in the beam, normally 1–7 eV, and within the plasma (Bacal *et al.*, 1990) of

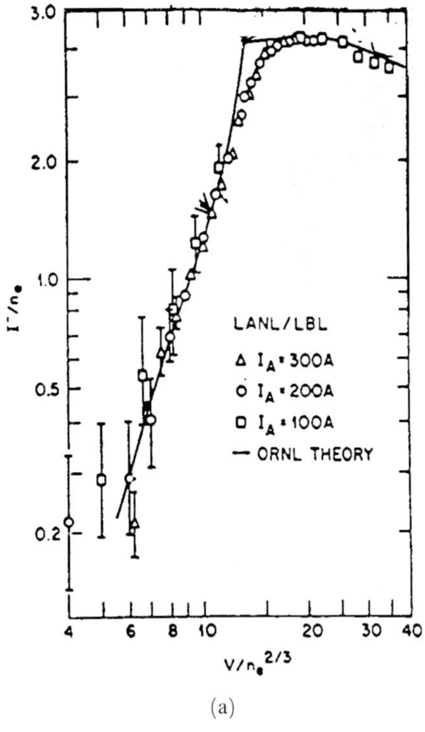

(a)

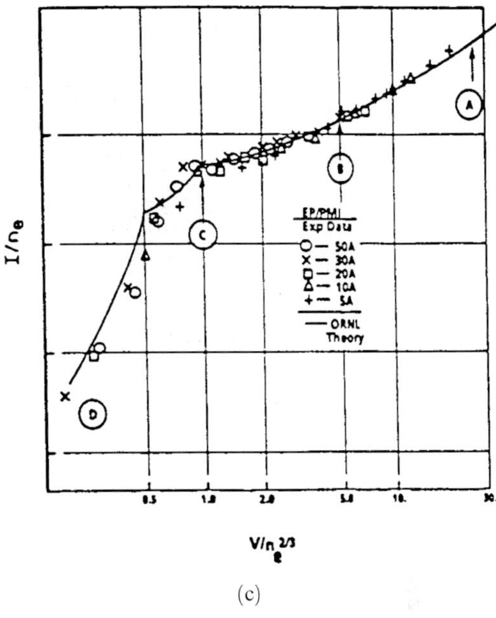

(c)

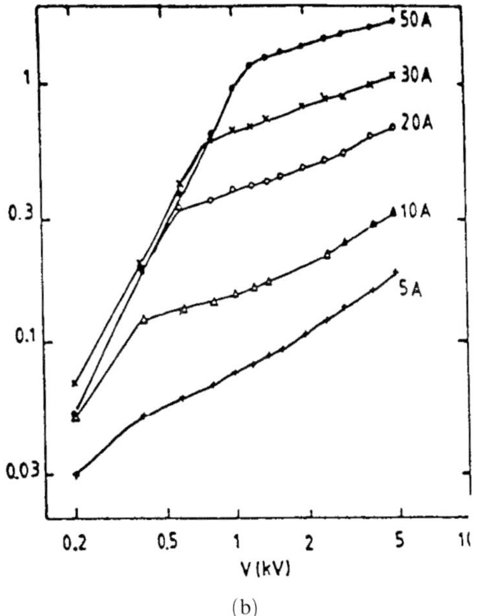

(b)

FIG. 45. (a) Data from the source described by Stevens *et al.*, showing negative-ion current as a function of accelerating voltage for several different arc currents. In order to produce all of the data on one curve the data are scaled as shown: dividing the ordinate by the plasma density and the abscissa by the two-thirds power of the plasma density. The solid line is the theoretical result. (b) Data from Ecole Polytechnique showing negative-ion current as a function of accelerating potential for several different arc currents. (c) The family of curves shown in (b) scaled as shown: dividing the ordinate by the plasma density and the abscissa by the two-thirds power of the plasma density. The solid line is the theoretical result.

0.1–0.7 eV. Analysis capable of considering sheath-produced ion-acoustic waves, as a result of the Bohm instability, will be a necessary future step in the accurate modeling of negative-ion sources. These instabilities may actually increase the availability of negative ions for extraction. Still another process, known as stochastic resonance, can occur whereby some negative ions are pushed over the hill by the fluctuating fields.

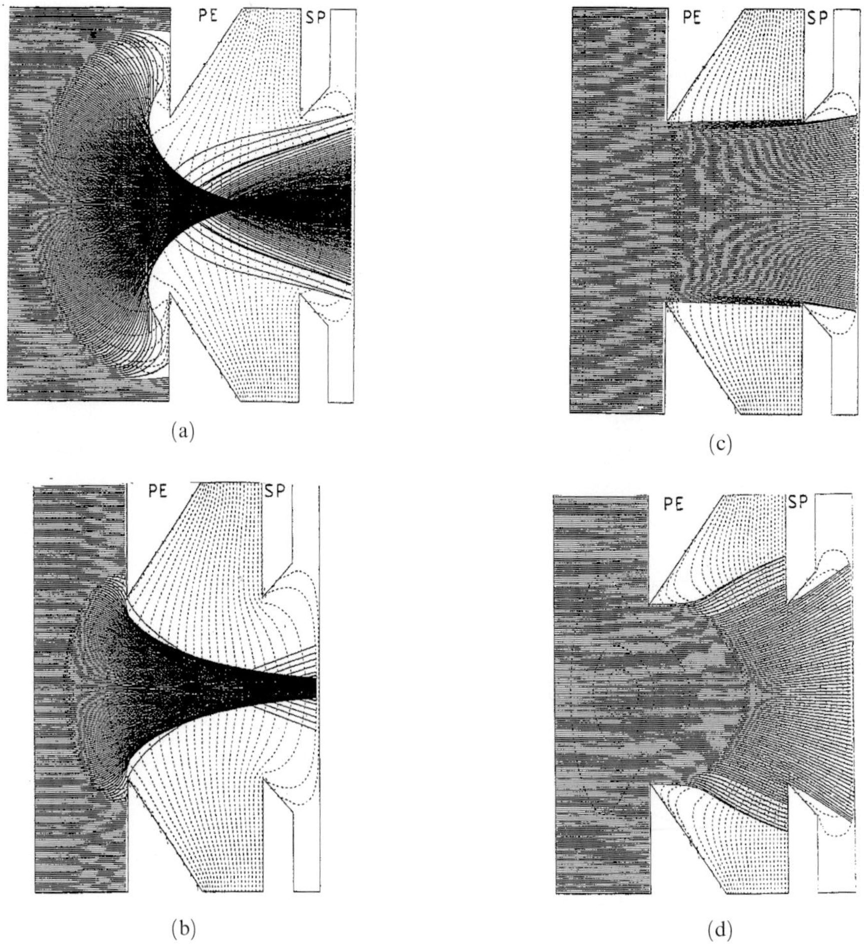

(a) (c)

(b) (d)

FIG. 46. (a)–(d) Calculated potentials and H^- ion orbits for the perveances marked A–D, respectively, in Fig. 45 (c).

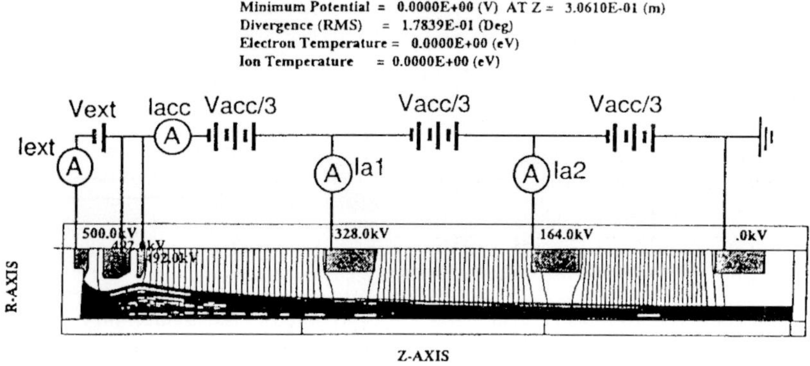

FIG. 47. The 500-keV accelerator and beam orbit simulation.

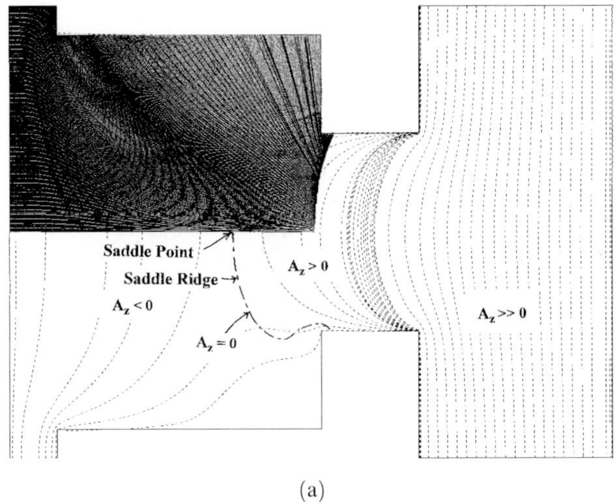

(a)

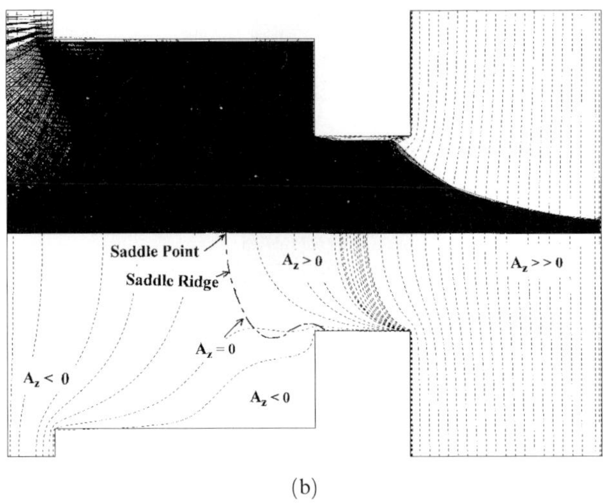

(b)

FIG. 48. (a) Positive-ion trajectories shown for negative-ion source with geometry similar to that shown in Fig. 28 or 30. Equipotentials are shown by dashed lines. In this very low-density case (the length of the figure is one Debye length), a saddle point in the potential is clearly shown; the region where volume-produced negative ions are kicked in the direction of the extraction are denoted by $A_z > 0$; regions denoted by $A_z > 0$ indicate that created negataive ions are given a kick back into the plasma. The positive-ion trajectories shown all end up hitting the electrode since they are repulsed by the beam accelerating fields. (b) Volume-created negative ion trajectories for the same case as discussed as above.

2.2.3 Double-Vlasov Model For negative ions it is useful to consider the plasma positive ions using a Vlasov description. It is necessary to consider these plasma positive ions kinetically, Eq. (5), because they are far from equilibrium, executing at most two passes through the region. It is still appropriate to consider the electron density in the presheath to be a Boltzmann distribution since the electrons are collision dominated and generally have to transverse a magnetic field in the pre-extraction region. A description is as follows:

$$\frac{\partial f_+}{\partial t} + \boldsymbol{v} \cdot \nabla f_+ + [\boldsymbol{v} \times \boldsymbol{B} - \nabla\varphi] \cdot \nabla_v f_+ = \delta(S_1), \tag{5}$$

$$\frac{\partial f_-}{\partial t} + \boldsymbol{v} \cdot \nabla f_- - [\boldsymbol{v} \times \boldsymbol{B} - \nabla\varphi] \cdot \nabla_v f_- = G(\boldsymbol{r}), \tag{6}$$

$$\frac{\partial f_e}{\partial t} + \boldsymbol{v} \cdot \nabla f_e - \frac{M}{m_e}[\boldsymbol{v} \times \boldsymbol{B} - \nabla\varphi] \cdot \nabla_v f_e = \delta(S_2), \tag{7}$$

$$\nabla^2\varphi = \int [f_+(\boldsymbol{v}) - f_-(\boldsymbol{v}) - f_e(\boldsymbol{v})]\, d^3v - e^{-\varphi}. \tag{8}$$

A case with significant plasma density (which corresponds to 1000 Debye lengths λ_D for the length of the device) is shown in Fig. 50. At these densities, positive ions falling downhill from the center of the plasma, as shown in Fig. 50, are accelerated until they reach the saddle

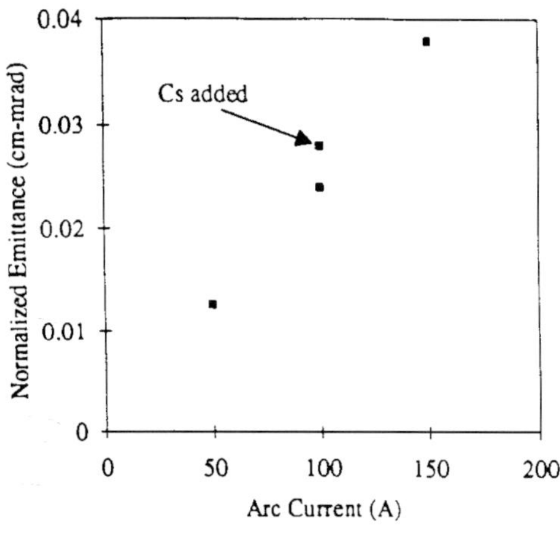

Normalized emittance vs. arc current.

FIG. 49. Normalized emittance as a function of arc current. This indicates that the ion "temperature" depends on plasma density. Increasing ion-acoustic waves in the presheath is a likely explanation for this effect.

ridge. Then the positive ions are repelled by the accelerator fields and are attracted into the plasma electrode as shown. This is in contrast to the case for positive ions, shown in Fig. 44(a), some of which can get extracted. For volume-produced negative ions, the trajectories are shown in Fig. 50(b). Here, only a fraction of the negative ions produced are extracted, and the rest are attracted to the body of the plasma since they are repulsed by the plasma electrode.

2.2.4 Transverse Space-Charge Limits When the extracted current depends on extraction voltage [see Fig. 9a of Ando (1994), or Fig. 9 of McAdams *et al.* (1987), or Fig. 45], which is usually the case at high currents, the beam is said to be transverse space-charge limited (TSCL). This is because the extraction voltage has very little effect on the negative-ion generation rate, and so the beam is limited by interception on one or more electrodes. This interception comes about as a result of poorly matched optics because of an oversupply of space charge (like charges repel—Coulomb's law). An example of experimental results is shown in Fig. 45, which illustrates this space-charge–limited flow. This is further substantiated by noting that the separate curves

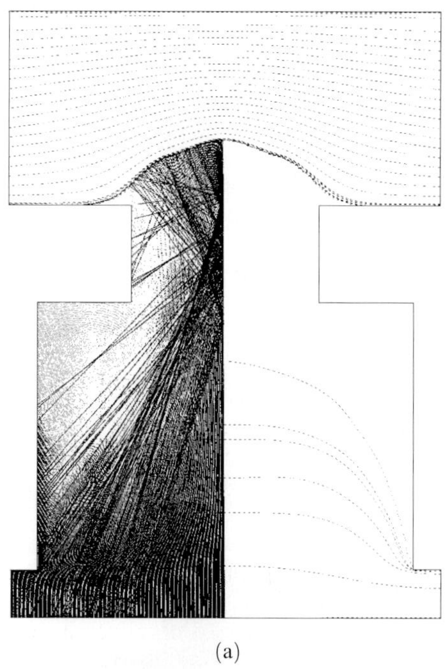

(a)

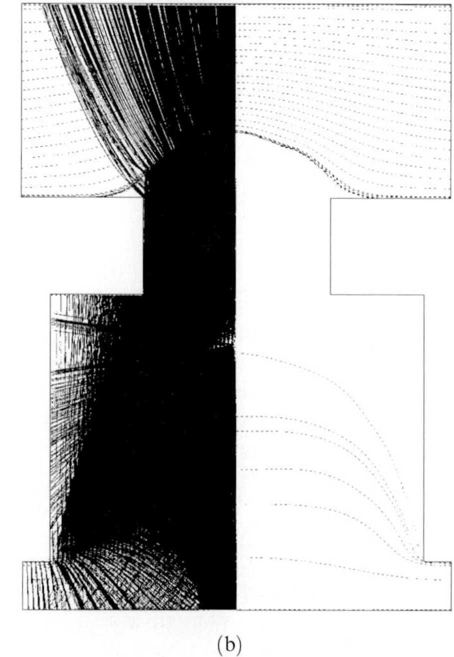

(b)

FIG. 50. (a) Same as Fig. 48(a), except that the plasma density is higher. Instabilities have been suppressed by space charge under relaxation. (b) Same as Fig. 48(b), except that the plasma density is higher.

in Fig. 45(b) actually scale by the Child–Langmuir proportionality [a consequence of dimensional analysis for positive ions (Whealton and Jaeger, 1977)], as proven by the coalescence of the curves in Fig. 45(b) into one curve, Fig. 45(c). The very nature of a transverse space-charge limit means that if the generation rate is increased (beyond any decrease in electron space charge), the beam current actually decreases. This is seen for example in Fig. 45(b), where at 200 V, the extracted H^- current for an arc current of 50 A is less than that for 30 A. The reason for this decrease is that the excess negative-ion beam generation not only gets intercepted by the electrodes but causes some of the formerly transmitted ions to be intercepted by the accelerator structure.

2.2.5 Effect of Cesium In volume sources the addition of cesium can cause an increase in the negative-ion current by up to a factor of 3 and a decrease in electrons extracted by a factor of as much as 100. One reason for the increase in negative-ion current extracted is that cesium converts positive ions into negative ions at a suitably coated surface. However, since the source typically operates at the space-charge limit, addition of more negative ions by surface production does not account for the increase in negative-ion beam current. Alternatively, the concomitant decrease in electrons extracted (frequently a factor of 100) may increase the partition of the space-charge limit occupied by the negative ions. If, for example, the electrons extracted were 100 times the ion current, not unusual for a negative-ion source operating without cesium, then the electrons could be taking up as much as two-thirds of the space-charge limit. The elimination of most of these electrons frees up two-thirds of the transverse space-charge limit, enabling the ion current to be 3 times higher, if one assumes that no emission limit exists at these higher currents.

An important question is how the presence of cesium so drastically decreases the number of extracted electrons. The effect of cesium on negative-ion output and electrons extracted varies from source to source, but the cases where the effect is greatest usually involve the application of magnetic fields and frequently such accoutrements as a collar electrode (see Sec. 1.3). What occurs is that the self-biasing

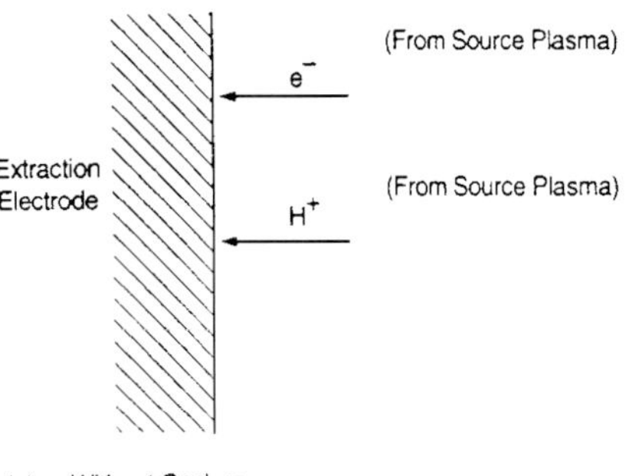

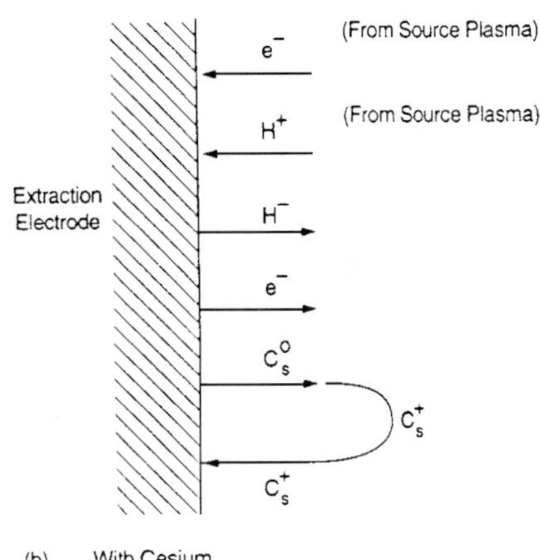

FIG. 51. Charge balance in ion source (a) without and (b) with cesium.

of the plasma with respect to the extraction electrode can change with the presence of cesium from that which it would have without it. What determines the self-bias is the flux of positive and negative charges from the electrode. For a noncesiated electrode, in the simplest case, positive ions are drawn into the electrode and are balanced by the flux of the much more mobile electrons shown simplistically in Fig. 51(a). In this simple case, the positive charge flux is balanced by retarding the source-plasma electrons electrostatically. This retardation of electrons is done by a self-biasing of the plasma positive with respect to the electrode, as shown in Fig. 52. As an example for H^+ and a Boltzmann equilibrium electron distribution, a bias V_b of $3.6kT_e/e$, as shown in Fig. 52, will effect a balance between the electron and ion fluxes.

A cesium coating, with its lower work

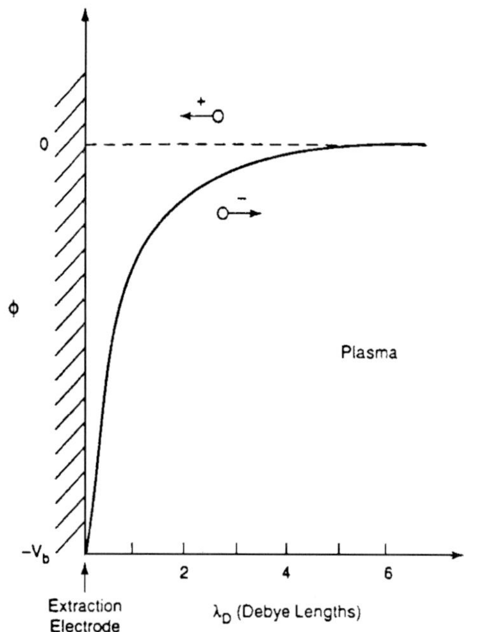

FIG. 52. Positive and negative charge flow near negatively biased electrode.

function, allows other processes to occur. One process is the conversion into negative ions of the positive ions intercepting the surface. Another process is the ejection of electrons from the surface. Finally, from neutral sputtered cesium (due to positive-ion bombardment) come positive ions, Cs^+, due to ionization by electrons and H^+; these ions return to the electrode. The Cs^+ has been observed to be a dominant species of positive ions (Ogasawara *et al.*, 1998). All three of these new processes, due to the cesium coating, cause an additional net flux of positive ions, as shown in Fig. 51(b). All of these additional processes tend to lower the magnitude of the negative wall bias so that more plasma electrons are absorbed by the extraction electrode.

To see what a profound effect this has on the electrons extracted, we need to have the transport configuration typified by Fig. 29. This is the simplest case without cesium, with only the positive ions and electrons shown. The electrons execute multiple bounces with the extraction collar electrode. The increased probability of absorption on each pass, due to decreased magnitude of bias, will dramatically decrease the number of electrons arriving at the extraction sheath. An illustration of the various additional surface mechanisms due to cesium is shown in Fig. 53.

Therefore, the reasons why a cesium coating can reduce the electrons extracted are twofold:

1. The self bias of the target plasma is diminished, allowing increased absorption of the electrons drifting in from the driver plasma; and
2. the additional flux of surface-produced negative ions partially goes back to the driver plasma, which reduces the electron flux drifting in from the driver plasma since fewer are needed for charge neutrality.

2.2.6 Increasing the Transverse Space-Charge Limit (TSCL) Now that it has been shown that cesium allows the negative ions to occupy virtually the entire transverse space-charge–limit budget, it remains to consider how the space-charge limit itself can be increased. The transverse space-charge limit is usually thought to depend on the geometry of the electrodes, the potentials applied, the beam current density, and the mass and charge of the ions. Actually, the transverse space-charge limit also depends on the shape of the extraction sheath, which by artificial means can be further manipulated. For example, the deliberate introduction of an additional positive charge into the pre-extraction sheath region would have the effect of changing the curvature of the sheath, effecting a *higher* space-charge limit (Whealton *et al.*, 1998). (Actually, this could be done also for positive-ion sources by the introduction of negative space charge in the presheath region.) Ionized sputtered cesium is just such a source of positive ions. There is an abundance of experimental evidence indicating the increase in space-charge limit for negative ions. The Dudnikov source, Sec. 1.2.1 (see Fig. 13), is perhaps the first source to use a collar with significant improvement (Allison 1980). Therefore, a cesium addition could increase the beam current beyond the transverse space-charge limit.

The Cs^+ space charge in the region just before extraction not only serves to increase the transverse space-charge limit by causing the extraction sheath to be more concave (Fig. 54 compared with 50), but can also reduce Bohm instabilities, through a reduction of the regions

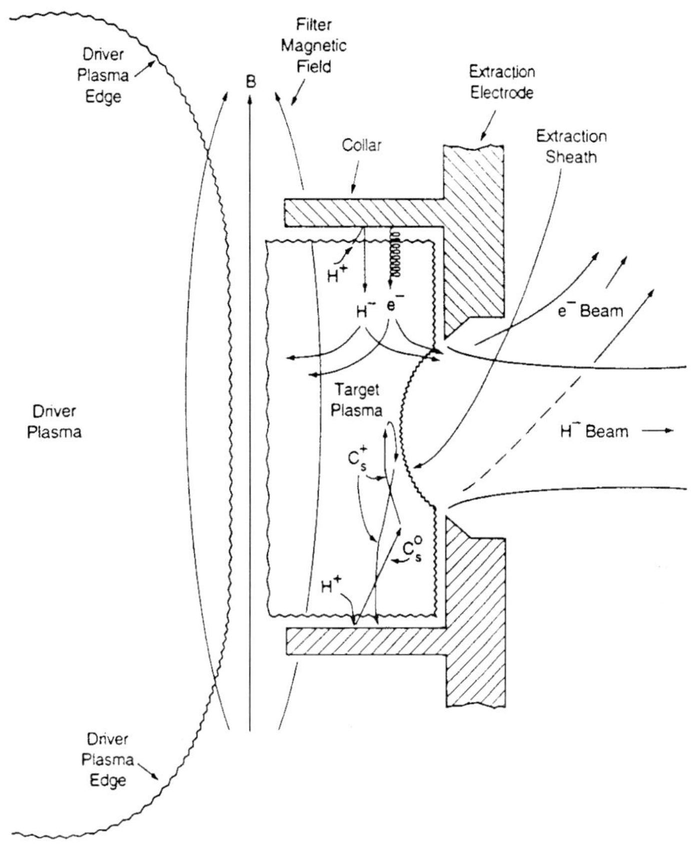

FIG. 53. Surface production of negative ions and production of Cs^+.

where there is an excess of negative space charge in the pre-extraction volume.

3. APPLICATIONS

Several applications require beams of H^-. They are in the areas of material sciences, semiconductor processing, high-energy physics, nuclear physics, energy technology, and weapons.

3.1 Beam-Storage Rings—Pulsed Spallation Neutron Sources

Pulsed spallation neutron sources require proton energies of 1 GeV or so and as high a current as possible. This is usually done by storing GeV protons in an accumulator ring. To reduce the emittance of the ring current, negative-ion injections are used to avoid the consequences of Liouville's theorem. The negative ions are produced in a single-aperture source and accelerated with an rf LINAC. For a review, see Alonso (1996).

3.2 High-Energy Neutral Beam for Fusion-Plasma Heating

Neutral beams (300 keV) of hydrogen, made by stripping an H^- beam, that pass through ambient magnetic fields become ionized (positive), which puts a positive potential in the end-plug plasmas. Positive ions escaping from the mirror device are repelled by these intraplasma fields, thus enhancing plasma confinement in the mirror field.

Neutral beams of high energy, 0.5–2 MeV, are required to heat, fuel, and provide a current driver for Tokamak fusion experiments. These beams need to be neutral to get past the in-

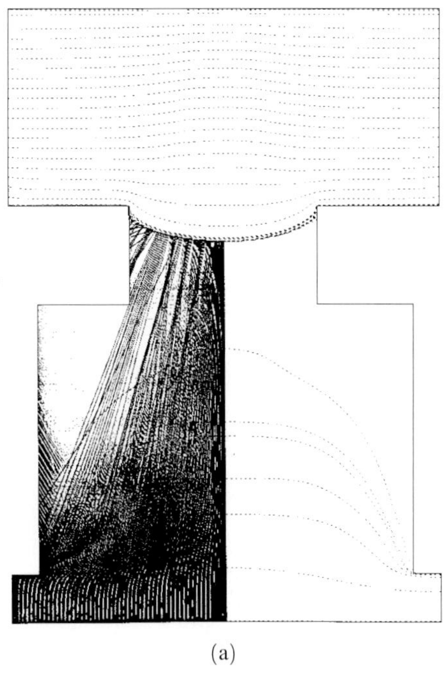

(a)

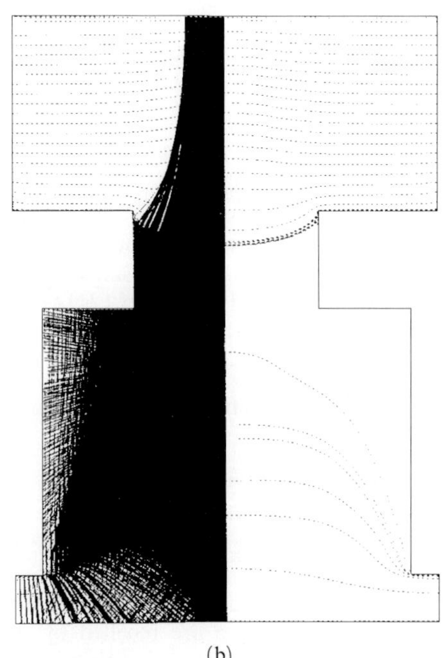

(b)

FIG. 54. (a) Same as Fig. 50(a), but positive space charge added in the presheath. (b) Same as Fig. 50(b), but positive space charge added in the presheath region.

tense magnetic fields at the tokamak's edge. The best way to produce significant H^0 currents is to start them as negative ions and accelerate them to the required energy, then pass the beam through a gas (or plasma or perhaps a laser beam), stripping off the excess electron and making them neutral. It is far easier to remove the electron than to add an electron to a positive ion at such energies.

3.3 High-Energy Neutral Beam for the Strategic Defense Initiative

In order to fly through space unaffected by the Earth's magnetic field, neutral beams are desired. These neutral beams need to be on the order of 10% of the speed of light, and so the best way they can be produced is by generation from H^- (or D^-). These negative ions are accelerated, and then the excess electron is stripped off, leaving a neutral beam. Of course, these neutral beams need to be steered in a specific direction without significant aberrations.

3.4 Ion-Projection Lithography

Advantages of using negative ions for ion projection lithography are that the loosely bound electrons fly off as secondary-emission electrons, preventing charging (see Fig. 55), a major limitation in reliability for small feature size (Gudharay *et al.*, 1994, 1996; Ishikawa, 1994, 1996).

4. CONCLUSIONS

Negative-ion sources during the last quarter century have advanced from microampere to ampere current production. This advance has been associated with many measurements; however, the improvement has been largely empirical. A deep understanding of the beam formation mechanism has yet to be demonstrated. The half dozen or so nonlinear processes simultaneously occurring have prevented a self-consistent synthesis.

The material presented and interpreted here summarizes the major components of our understanding. Using negative-ion current as a marker, six orders of magnitude improvement

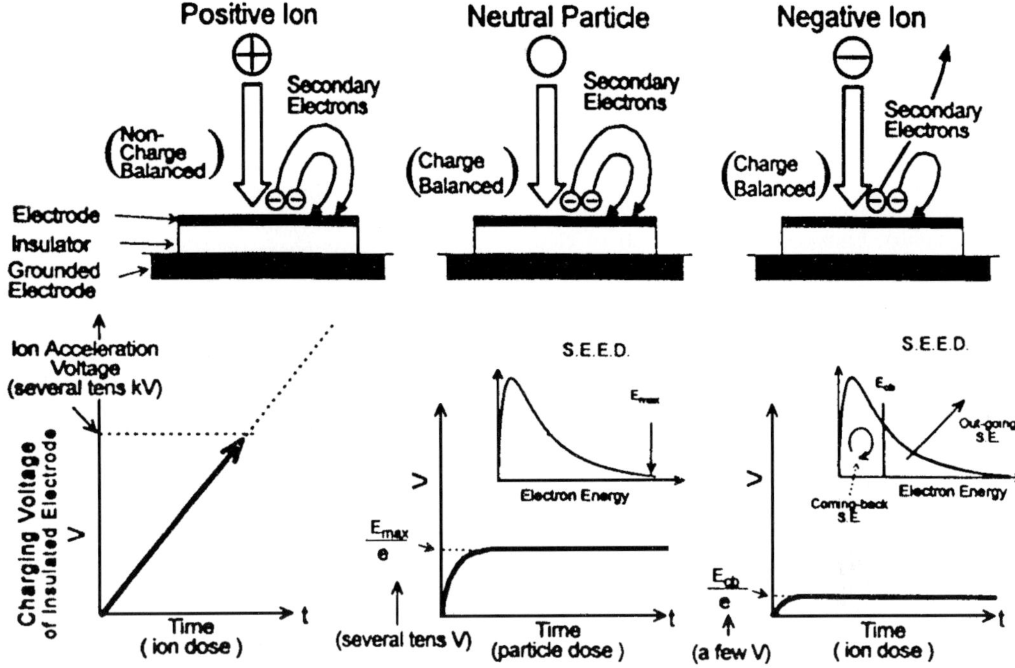

FIG. 55. Schematic representation of particle trajectories and charging potentials for the case of positive-ion, neutral-particle, and negative-ion implantation.

has taken place, opening up applications otherwise unfeasible. We can expect that additional applications will keep pace with further improvements in ion sources.

ACKNOWLEDGMENTS

The authors would like to thank Jeanie Shover and Al Akerman for their help in preparing this article.

GLOSSARY

Brightness: The specific current of particles emitted by unit surface into unit solid angle.

Calutron: A powerful plasma ion source with hot cathodes, gas discharge with electron oscillation in a strong magnetic field, and transverse extraction of the ion beam, developed by E. O. Lawrence for isotope separation.

Cesium Catalysis: The significant enhancement of negative-ion formation in a gas discharge by the addition of a small admixture of cesium or other substances with low ionization potential, decreasing the electrode work function.

Child–Langmuir Law: The equation for space-charge–limited current density in a vacuum diode.

Electron Affinity: The binding energy of the additional electron in a negative ion.

Electron Cyclotron Resonance (ECR): Resonance absorption of electromagnetic radiation by electrons in a static magnetic field when the electromagnetic frequency is equal to the cyclotron frequency of the electrons. It is a most powerful method of heating electrons to a high temperature.

Emittance: A measure for the volume in phase space occupied by a beam, indicative of the reciprocal of the quality of the beam.

Ionization Potential: The smallest potential difference necessary for electron acceleration to produce atom or molecule ionization; equal to the binding energy of an electron in the atom or molecule.

Penning Discharge: A low-pressure gas discharge with oscillation of electrons along the anode tube in a magnetic field between two cathodes.

Perveance: Beam current divided by square root of the acceleration potential; used as a measure of the ratio of current to the transverse space-charge–limited current.

Secondary Emission: The emission of electrons or ions initiated by bombardment with electrons, ions, excited or fast atoms, *etc.*

Surface-Plasma Sources (SPS): Ion sources with negative-ion formation by the interaction of plasma particles with electrode surfaces in gas discharges.

Thermionic Emission: The emission of electrons or ions from a solid heated to a high temperature, taking place by the "evaporation" of the particles.

Thermionic-Emission Converter: A power supply for conversion of heat into electricity by emission of electrons from a cathode with a higher work function and collection of them by an anode with a lower work function.

Work Function: The energy difference between the vacuum energy level of an electron and the Fermi level in a solid: the evaporation energy of an electron.

Works Cited

Alessi, J. G. (1990), in: A. Hershcovitch (Ed.), *Production and Neutralization of Negative Ions and Beams*, AIP Conference Proceedings No. 210, New York: American Institute of Physics, pp. 526–533.

Alessi, J. G., Prelec, K. (1991), in: *Proceedings of the 1991 Particle Accelerator Conference, Washington D. C.*, pp. 1913–1915.

Alessi, J. G., Prelec, K., McCafferty, D. (1994), in: J. G. Alessi (Ed.), *Production and Neutralization of Negative Ions and Beams*, AIP Conference Proceedings No. 287, New York: American Institute of Physics, pp. 387–396.

Alessi, J. G., Sluyters, T. (1980), *Rev. Sci. Instrum.* **51**, 1630–1635.

Allison, P. W. (1977), in: K. Prelec (Ed.), *Proceedings of the Symposium on the Production and Neutralization of Negative Hydrogen Ions and Beams*, Brookhaven National Laboratory Report No. BNL-50727, 119–122.

Allison, P. W., Smith, H. V., Sherman, J. D. (1980), in: T. Sluyters (Ed.), *Proceedings of the Second International Symposium on the Production and Neutralization of Negative Hydrogen Ions and Beams*, Brookhaven National Laboratory Report No. BNL-51304, pp. 171–177.

Alonso, J. R. (1996), *Rev. Sci. Instrum.* **67**, 1308–1313.

Anderson, O. A. (1984), in: K. Prelec (Ed.), *Production and Neutralization of Negative Ions and Beams*, AIP Conference Proceedings No. 111, New York: American Institute of Physics, pp. 473–488.

Ando, A. (1994), in: J. G. Alessi (Ed.), *Production and Neutralization of Negative Ions and Beams*, AIP Conference Proceedings No. 287, New York: American Institute of Physics, pp. 339–352.

Antipov, S. P., Elizarov, L. I., Martynov, M. I., Chesnokov, V. M. (1990), in: A. Hershcovitch (Ed.), *Production and Neutralization of Negative Ions and Beams*, AIP Conference Proceedings No. 210, New York: American Institute of Physics, pp. 184–197.

Bacal, M., Leung, K-N. (1984), in: K. Prelec (Ed.), *Production and Neutralization of Negative Ions and Beams*, AIP Conference Proceedings No. 111, New York: American Institute of Physics, pp. 105–112.

Bacal, M., Nicolopoulou, E., Doucet, H. J. (1977), in: K. Prelec (Ed.), *Proceedings of the Symposium on the Production and Neutralization of Negative Hydrogen Ions and Beams*, Brookhaven National Laboratory Report No. BNL-50727, pp. 26–34.

Bacal, M., Bruneteau, A. M., Doucet, H. J., Graham, W. G., Hamilton, G. W. (1980), in: T. Sluyters (Ed.), *Proceedings of the Second International Symposium on the Production and Neutralization of Negative Hydrogen Ions and Beams*, Brookhaven National Laboratory Report No. BNL-51304, pp. 95–105.

Bacal, M., Hillion, F., Nachman, M., Steckelmacher, W. (1984), in: K. Prelec (Ed.), *Production and Neutralization of Negative Ions and Beams*, AIP Conference Proceedings No. 111, New York: American Institute of Physics, pp. 418–428.

Bacal, M., Bruneteau, A. M., Bruneteau, J., Devynck, P., Hillion, F. (1986), in: *Proceedings of the 2nd European Negative Ion Source Conference, Ecole Polytechnique*, pp. 17–24.

Bacal, M., Bruneteau, J, Devynda, P., Hillion, F. (1987), in: J. G. Alessi (Ed.), *Production and Neutralization of Negative Ions and Beams*, AIP Conference Proceedings No. 158, New York: American Institute of Physics, pp. 246–258.

Bacal, M., Bruneteau, J., Devynck P. (1988), *Rev. Sci. Instrum.* **59**, 2152.

Bacal, M., Berlemont, P., Bruneteau, J., Devynck, P., Konieczny, C., Leroy, R., Stern, R. A. (1990), in: A. Hershcovitch (Ed.), *Production and Neutralization of Negative Ions and Beams*, AIP Conference Proceedings No. 210, New York: American Institute of Physics, pp. 489–503.

Barnett, C. F., Ray, J. A., Ricci, E., Wilker, M. I., McDaniel, E. W., Thomas, E. W., Gilbudy, H. B.

(1977), Oak Ridge National Laboratory Report No. ORNL-5206 (Vol. 1) p. A.4.3.

Barnett, C. F., Hunter, H. T., Kirkpatrick, M. I., Alvarez, I., Cisneros, C., Phaneuf, R. A. (1990), Oak Ridge National Laboratory Report No. ORNL-6086 (Vol. 1) pp. A-21.

Bashkeev, A. A., Dudnikov, V. G. (1990), in: A. Hershcovitch (Ed.), *Production and Neutralization of Negative Ions and Beams*, AIP Conference Proceedings No. 210, New York: American Institute of Physics, pp. 329–339.

Becker, R., Leung, K. N., Kunkel, W. (1998), *Rev. Sci. Instrum.* **69**, 1107–1109.

Belchenko, Y. I., Dimov, G. I. (1984), in: K. Prelec (Ed.), *Production and Neutralization of Negative Ions and Beams*, AIP Conference Proceedings No. 111, New York: American Institute of Physics, pp. 363–375.

Belchenko, Y. I., Kupriyanov, A. S. (1990), in: A. Hershcovitch (Ed.), *Production and Neutralization of Negative Ions and Beams*, AIP Conference Proceedings No. 210, New York: American Institute of Physics, pp. 198–213.

Belchenko, Y. I., Kupriyanov, A. S. (1994), in: J. G. Alessi (Ed.), *Production and Neutralization of Negative Ions and Beams*, AIP Conference Proceedings No. 287, New York: American Institute of Physics, pp. 255–270.

Belchenko, Y. I., Dimov, G. I., Dudnikov, V. G., Ivanov, A. A. (1973), *Dokl. Akad. Nauk. SSSR* **213**, 1283–1786.

Belchenko, Y. I., Dimov, G. I., Dudnikov, V. G. (1977), in: K. Prelec (Ed.), *Proceedings of the Symposium on the Production and Neutralization of Negative Hydrogen Ions and Beams*, Brookhaven National Laboratory Report No. BNL-50727, pp. 79–96.

Bell, M. A., Whealton, J. H., Raridon, R. J., McGaffey, R. W., Wooten, D. E. (1985), *Bull. Am. Phys. Soc.* **30**, 1609.

Bruneteau, J. (1986), in: *Proceedings of the 2nd European Negative Ion Source Conference, Ecole Polytechnique*, pp. 81–86.

Budker, G. I., Dimov, G. I., Dudnikov., V. G. (1967), *Sov. At. Energ.* **22**, 441–449.

Chan, C. F., Leung, K.-N. (1993a), in: *Proceedings of the 1993 Particle Accelerator Conference, Washington, D. C.*, pp. 3160–3162.

Chan, C. F., Leung, K. N. (1994), in: J. G. Alessi (Ed.), *Production and Neutralization of Negative Ions and Beams*, AIP Conference Proceedings No. 287, New York: American Institute of Physics, pp. 762–769.

Cooper, W. S., Halbach K. (1974), in: C. Pezzoti (Ed.), *Proceedings of the 2nd Symposium on Ion Sources and Formation of Ion Beams*, Lawrence Berkeley Laboratory Report No. LBL-3399. II-1-1.

Cowan, R. G., Niemel, J., Campbell, R. L., Raridon, R. J., Whealton, J. H., Michaut, C., Bacal, M. (1994), in: J. G. Alessi (Ed.), *Production and Neutralization of Negative Ions and Beams*, AIP Conference Proceedings No. 287, New York: American Institute of Physics, pp. 621–633.

Dagenhart, W. K., Stirling, W. L., Haselton, H. H., Kelley, G. G., Kim, J., Tsai, C. C., Whealton, J. H. (1980), in: T. Sluyters (Ed.), *Proceedings of the Second International Symposium on the Production and Neutralization of Negative Hydrogen Ions and Beams*, Brookhaven National Laboratory Report No. BNL-51304, pp. 217–224.

Dagenhart, W. K., Gardner, W. L., Stirling, W. L., Whealton, J. H. (1983), *Nucl. Technol./Fusion* **4**, pp. 1430–1442.

Dagenhart, W. K., Stirling, W. L., Barric, G. M., Barber, G. C., Ponte, N. S., Whealton, J. H. (1984), in: K. Prelec (Ed.), *Production and Neutralization of Negative Ions and Beams*, AIP Conference Proceedings No. 111, New York: American Institute of Physics, pp. 353–362.

Dagenhart, W. K., Tsai, C. C., Stirling, W. L., Ryan, P. M., Schechter, D. E., Whealton, J. H., Donaghy, J. J. (1987), in: J. G. Alessi (Ed.), *Production and Neutralization of Negative Ions and Beams*, AIP Conference Proceedings No. 158, New York: American Institute of Physics, pp. 366–377.

Derevyankin, G. E., Dudnikov, V. G. (1984), in: K. Prelec (Ed.), *Production and Neutralization of Negative Ions and Beams*, AIP Conference Proceedings No. 111, New York: American Institute of Physics, pp. 376–397.

Dudnikov, V. G. (1972), SV Autor Certificat., No. 411542.

Dudnikov, V. G. (1980), in: T. Sluyters (Ed.), *Proceedings of the Second International Symposium on the Production and Neutralization of Negative Hydrogen Ions and Beams*, Brookhaven National Laboratory Report No. BNL-51304, pp. 137–144.

Dudnikov, V. G. (1993), *Rev. Sci. Instrum.* **63**, pp. 2660–2662.

Dudnikov, V. G., Derevyankin, G. E. (1994), in: J. G. Alessi (Ed.), *Production and Neutralization of Negative Ions and Beams*, AIP Conference Proceedings No. 287, New York: American Institute of Physics, pp. 239–254.

Dudnikov, V. G., Stirling, W. L., Whealton, J. H. (1997), in: *Proceedings of the 1997 Particle Accelerator Conference, Vancouver, B.C.*, pp. 2714–2716.

Ehlers, K. W., Leung, K. N. (1980), *Rev. Sci. Instrum.* **51**, 721–730.

Ehlers, K. W., Leung., K. N. (1981), *Appl. Phys. Lett.* **38**, 287–315.

Elizarov, L. I., Chesnokov, V. M. (1994), in: J. G. Alessi (Ed.), *Production and Neutralization of Negative Ions and Beams*, AIP Conference Proceedings No. 287, New York: American Institute of Physics, pp. 305–315.

Goretsky, V. P., Ryabtsev, A. V., Soloshenko, I. A., Tarasenko, A. F., Schedrin, A. I. (1996), in: K. Prelec (Ed.), *Production and Neutralization of Negative Ions and Beams*, AIP Conference Proceedings No. 384, New York: American Institute of Physics, pp. 142–152.

Greer, J. A., Seidl, M. V. (1984), in: K. Prelec (Ed.), *Production and Neutralization of Negative Ions and Beams*, AIP Conference Proceedings No. 111, New York: American Institute of Physics, pp. 220–238.

Gudharay, S. K., *et al.* (1994), *Rev. Sci. Instrum.* **65**, 1745–1748.

Gudharay, S. K., *et al.* (1996), *J. Vac. Sci. Technol.* **B14**, 3907–3716.

Gudharay, S. K., Allen, C. K., Reiser, M., Saadatmaud, K., Chang, C. R. (1994), in: J. G. Alessi (Ed.), *Production and Neutralization of Negative Ions and Beams*, AIP Conference Proceedings No. 287, New York: American Institute of Physics, pp. 672–685.

Holmes, A. J. T., Green, T. S. (1984), in: K. Prelec (Ed.), *Production and Neutralization of Negative Ions and Beams*, AIP Conference Proceedings No. 111, New York: American Institute of Physics, pp. 429–437.

Inoue, T., Hanada, M., Mizuno, M., Ohara, Y., Okumura, Y., Suzuki, Y., Tanaka, M., Watenabe, K. (1994), in: J. G. Alessi (Ed.), *Production and Neutralization of Negative Ions and Beams*, AIP Conference Proceedings No. 287, New York: American Institute of Physics, pp. 316–325.

Isenberg, J. D., Kwon, H. J., Seidl, M. (1994), in: J. G. Alessi (Ed.), *Production and Neutralization of Negative Ions and Beams*, AIP Conference Proceedings No. 287, New York: American Institute of Physics, pp. 38–47.

Ishikawa, J. (1994), *Rev. Sci. Instrum.* **65**, 1290–1292.

Ishikawa, J. (1996), *Rev. Sci. Instrum.* **67**, 1410–1415.

Kuroda, T., *et al.* (1996), in: K. Prelec (Ed.), *Production and Neutralization of Negative Ions and Beams*, AIP Conference Proceedings No. 380, New York: American Institute of Physics, pp. 201–213.

Kwan, J. W., *et al.* (1993), in: *Proceedings of the 1993 Particle Accelerator Conference, Washington, D. C.*, pp. 3169–3171.

Kwan, J. W., Gough, R. A., Leung, K. N., Perkins, L. T., Pickard, D. S., Vujic, J., Wu, L. K., Olivo, M., Einenkel, H. (1994), in: J. G. Alessi (Ed.), *Production and Neutralization of Negative Ions and Beams*, AIP Conference Proceedings No. 287, New York: American Institute of Physics, pp. 295–304.

Lee, Y., Gough, R. A., Leung, K. N., Perkins, L. T., Pickard, D. S., Vujic, J., Wu, L. K., Olivim, M., Einenkel, H. (1998), *Rev. Sci. Instrum.* **69**, 1023–1025.

Leung, K.-N. (1998), *Rev. Sci. Instrum.* **69**, 998–1002.

Leung, K.-N., Ehlers, K. W. (1980), in: T. Sluyters (Ed.), *Proceedings of the Second International Symposium on the Production and Neutralization of Negative Hydrogen Ions and Beams*, Brookhaven National Laboratory Report No, BNL-51304, pp. 65–73.

Leung, K.-N., Ehlers, K. W. (1984), in: K. Prelec (Ed.), *Production and Neutralization of Negative Ions and Beams*, AIP Conference Proceedings No. 111, New York: American Institute of Physics, pp. 67–81.

Leung, K.-N., Ehlers, K. W., Bacal, M. (1983), *Rev. Sci. Instrum.* **54**, 56–59.

Leung, K.-N., Hauck, C. A., Kunkel, W. B., Walther, S. W. (1989), *Rev. Sci. Instrum.* **60**, 531–537.

Leung, K.-N., DeVries, G. J., Divergilio, W. F., Hamm, R. W., Hauck, C. W., Kunkel, W. B., McDonald, D. S., Williams, M. D. (1991), *Rev. Sci. Instrum.* **62**, 100–108.

Leung, K.-N., Bachman, D. A., McDonald, D. S. (1994), in: J. G. Alessi (Ed.), *Production and Neutralization of Negative Ions and Beams*, AIP Conference Proceedings No. 287, New York: American Institute of Physics, pp. 368–374.

Lietzke, A. F. K., Ehlers, W., Leung, K. N. (1984), in: K. Prelec (Ed.), *Production and Neutralization of Negative Ions and Beams*, AIP Conference Proceedings No. 111, New York: American Institute of Physics, pp. 344–352.

McAdams, R., Holmes, A. T. J., Nightingale, M. P. S., Lee, L. M., Hinton, M. D., Neumann, A. F., Green, T. S. (1987), in: J. G. Alessi (Ed.), *Production and Neutralization of Negative Ions and Beams*, AIP Conference Proceedings No. 158, New York: American Institute of Physics, pp. 298–308.

McAdams, R., King, R. F., Proudfoot, G., Holmes, A. T. J. (1994), in: J. G. Alessi (Ed.), *Production and Neutralization of Negative Ions and Beams*, AIP Conference Proceedings No. 287, New York: American Institute of Physics, pp. 353–367.

McGaffey, R. W., Meszaros, P. S., Whealton, J. H., Raridon, R. J., McCollough, D. H. (1984), in: K. Prelec (Ed.), *Production and Neutralization of Negative Ions and Beams*, AIP Conference Proceedings No. 111, New York: American Institute of Physics, pp. 533–543.

Maesser, A., Volk, K., Weber, M., Klein, H. (1996), *Rev. Sci. Instrum.*, **67**, 1054–1056.

Melmychuk, S. T., *et al.* (1991), *J. Vac. Sci. Technol.* **A9**, 1650–1655.

Mori, Y., Takagi, A., Ikegami, K., Fukumoto, S. (1987), in: J. G. Alessi (Ed.), *Production and Neutralization of Negative Ions and Beams*, AIP Conference Proceedings No. 158, New York: American Institute of Physics, pp. 378–383.

Ogasawara, M., Marishita, T., Miura, M., Shibayama, N., Hatayama, A. (1998), *Rev. Sci. Instrum.* **69**, 1132–1137.

Ohara, Y. (1998), *Rev. Sci. Instrum.* **69**, 908–913.

Okumura, Y., Horiike, H., Inone, T., Kurashima, T., Matsuda, S., Ohara, Y., Tanaka, S. (1987), in: J. G. Alessi (Ed.), *Production and Neutralization of Negative Ions and Beams*, AIP Conference Proceedings No. 158, New York: American Institute of Physics, pp. 309–318.

Okumura, Y., *et al.* (1996), *Rev. Sci. Instrum.* **67**, 1018–1020.

Peters, J. (1998), *Rev. Sci. Instrum.* **69**, 992–994.

Prelec, K. (1977a), in: K. Prelec (Ed.), *Proceedings of the Symposium on the Production and Neutralization of Negative Hydrogen Ions and Beams*, Brookhaven National Laboratory Report No. BNL-50727, p. 111.

Prelec, K. (1977b), *Nucl. Instrum. Meth.* **144**, 413–422.

Prelec, K. (1980), in: T. Sluyters (Ed.), *Proceedings of the Second International Symposium on the Production and Neutralization of Negative Hydrogen Ions and Beams*, Brookhaven National Laboratory Report No. BNL-51304, pp. 145–152.

Prelec, K. (1984), in: K. Prelec (Ed.), *Production and Neutralization of Negative Ions and Beams*, AIP Conference Proceedings No. 111, New York: American Institute of Physics, pp. 333–343.

Riz, D., Pamela, J. (1996), in: K. Prelec (Ed.), *Production and Neutralization of Negative Ions and Beams*, AIP Conference Proceedings No. 380, New York: American Institute of Physics, pp. 315–323.

Riz, D., Pamela, J. (1998), *Rev. Sci. Instrum.* **69**, 914–919.

Schmidt, C. W. (1980), in: T. Sluyters (Ed.), *Proceedings of the Second International Symposium on the Production and Neutralization of Negative Hydrogen Ions and Beams*, Brookhaven National Laboratory Report No. BNL-51304, pp. 189–196.

Schmidt, C. W., Curtis, C. D. (1977), in: K. Prelec (Ed.), *Proceedings of the Symposium on the Production and Neutralization of Negative Hydrogen Ions and Beams*, Brookhaven National Laboratory Report No. BNL-50727, pp. 123–128.

Schmidt, C. W., *et al.* (1979), *IEEE Trans. Nucl. Sci.*, **NS-26**, 4120.

Seidl, M., Pargellis, A. N., Greer, J. (1980), in: T. Sluyters (Ed.), *Proceedings of the Second International Symposium on the Production and Neutralization of Negative Hydrogen Ions and Beams*, Brookhaven National Laboratory Report No. BNL-51304, pp. 111–118.

Seidl, M., Ciu, H. L., Isenberg, J. D., Kwan, H. J., Lee, B. S. (1997), in: J. G. Alessi (Ed.), *Production and Neutralization of Negative Ions and Beams*, AIP Conference Proceedings No. 287, New York: American Institute of Physics, pp. 25–37.

Seidl, M., *et al.* (1982), *Phys. Rev. B*, **26**, 1.

Semashko, N. N., Kuzltsov, V. V., Krylov, A. I., Firsov, P. S. (1987), in: J. G. Alessi (Ed.), *Production and Neutralization of Negative Ions and Beams*, AIP Conference Proceedings No. 158, New York: American Institute of Physics, pp. 334–345.

Simonin, A., *et al.* (1996), *Rev. Sci. Instrum.* **67**, 1102–1107.

Sluyters, T. (1974), Lawrence Berkeley Laboratory Report No. LBL-3399, p. II-1-1.

Smith, H. V., Allison, P., Sherman, J. D. (1980), in: T. Sluyters (Ed.), *Proceedings of the Second International Symposium on the Production and Neutralization of Negative Hydrogen Ions and Beams*, Brookhaven National Laboratory Report No. BNL-51304, pp. 178–183.

Smith, H. V., Allison, P., Sherman, J. D. (1984), in: K. Prelec (Ed.), *Production and Neutralization of Negative Ions and Beams*, AIP Conference Proceedings No. 111, New York: American Institute of Physics, pp. 458–462.

Smith, H. V., Allison, P., Sherman, J. D. (1985), *IEEE Trans. Nucl. Sci.* **NS32**, 1797–1800.

Smith, H. V., Allison, P., Geisik, C., Schmitt, D. R. (1993), in: *Proceedings of the 1993 Particle Accelerator Conference, Washington, D. C.*, pp. 3172–3174.

Smith, H. V., Allison, P., Geisik, C., Orbesen, S. D., Schmitt, R. D., Schneider, J. D., Stelzer, J. E. (1994), in: J. G. Alessi (Ed.), *Production and Neutralization of Negative Ions and Beams*, AIP Conference Proceedings No. 287, New York: American Institute of Physics, pp. 282–294.

Soloshenko, I. A., Shchedrin, A. I., Ryabtev, A. V. (1998), in: C. Jacquot (Ed.), *Production and Neutralization of Negative Ions and Beams*, AIP Conference Proceedings No. 439, New York: American Institute of Physics, pp. 139–157.

Stevens, R., private communication, 1987.

Stirling, W. L., *et al.* (1979), *Rev. Sci. Instrum.* **50**, 523.

Stirling, W. L., Dagenhart, W. K., Whealton, J. H., Donaghy, J. J. (1984), in: K. Prelec (Ed.), *Production and Neutralization of Negative Ions and Beams*, AIP Conference Proceedings No. 111, New York: American Institute of Physics, pp. 450–457.

Takagi, A., Mori, Y., Igarashi, Z., Ikegami, K., Kubota, C., Fukumoto, S. (1984), in: K. Prelec (Ed.), *Production and Neutralization of Negative Ions and Beams*, AIP Conference Proceedings No. 111, New York: American Institute of Physics, pp. 520–523.

Takeiri, Y., *et al.* (1996), *Rev. Sci. Instrum.* **67**, 1021–1023.

Tanaka, M., Ose, Y., Koizumi, M., Yamashita, Y., Kawakami, H., Takeiri, K., Kaneko, Oka, Y.,

Tsumori, K., Osakabe, M. (1998), *Rev. Sci. Instrum.* **69**, 1144–1146.

Watanabe, K., *et al.* (1996), in: K. Prelec (Ed.), *Production and Neutralization of Negative Ions and Beams*, AIP Conference Proceedings No. 380, New York: American Institute of Physics, pp. 309–319.

Whealton, J. H. (1981), *Nucl. Instrum. Methods.* **189**, 55–77.

Whealton, J. H. (1985), *Bull. Am. Phys. Soc.* **30**, 1513.

Whealton, J. H. (1990), in: A. Hershcovitch (Ed.), *Production and Neutralization of Negative Ions and Beams*, AIP Conference Proceedings No. 210, New York: American Institute of Physics, pp. 539–556.

Whealton, J. H. (1987), in: J. G. Alessi (Ed.), *Production and Neutralization of Negative Ions and Beams*, AIP Conference Proceedings No. 158, New York: American Institute of Physics, pp. 482–506.

Whealton, J. H., Jaeger, E. F. (1977), *Rev. Sci. Instrum.* **48**, 829.

Whealton, J. H., Jaeger, E. F., Whitson, J. C. (1978), *J. Comput. Phys.* **27**, 32.

Whealton, J. H., Raridon, R. J., McGaffey, R. W., McCollough, D. H., Stirling, W. L., Dagenhart, W. K. (1984), in: K. Prelec (Ed.), *Production and Neutralization of Negative Ions and Beams*, AIP Conference Proceedings No. 111, New York: American Institute of Physics, pp. 524–532.

Whealton, J. H., *et al.* (1985), *IAEA Negative Ion Conference, Grenoble.*

Whealton, J. H., McGaffey R. W., Meszaros, P. S. (1986), *J. Comput. Phys.* **63**, 20–33.

Whealton, J. H., Bell, M. A., Raridon, R. J., Rothe, K. E., Ryan, P. M. (1988), *J. Appl. Phys.* **64**, 6210–6224.

Whealton, J. H., Meszaros, P. S., Raridon, R. J., Rothe, K. E., Bacal, M., Bruneteau, J., Devynda, P. (1989), *Rev. Phys. Appl.* **24**, 945–951.

Whealton, J. H., Olsen, D. K., Raridon, R. J. (1998), *Rev. Sci. Instrum.* **69**, 1103–1105.

Whitson, J. C., Smith, J., Whealton, J. H. (1978), *J. Comput. Phys.* **28**, 408–415.

Wooten, J. W., *et al.* (1979), Oak Ridge National Laboratory Report No. ORNL/TM-6740.

Wooten, J. W., Whealton J. H., McCollough, D. H. (1981), *J. Comput. Phys.* **43**, 95.

York, R. L., *et al.* (1983), *IEEE Trans. Nucl. Sci.*, **NS30**, 2705–2707.

York, R. L., Stevens, R. (1984), in: K. Prelec (Ed.), *Production and Neutralization of Negative Ions and Beams*, AIP Conference Proceedings No. 111, New York: American Institute of Physics, pp. 410–417.

York, R. L., Tupa, D., Swenson, D. R., Damjanovich, R. (1993), in: *Proceedings of the 1993 Particle Accelerator Conference, Washington, D. C.*, pp. 3175–3177.

Further Reading

Alessi, J. G. (Ed.) (1987), *Production and Neutralization of Negative Ions and Beams (1986)*, AIP Conference Proceedings No. 158, New York: American Institute of Physics.

Alessi, J. G. (Ed.) (1993), *Production and Neutralization of Negative Ions and Beams (1992)*, AIP Conference Proceedings No. 287, New York: American Institute of Physics.

Bacal, M., *et al.* (1996), *Rev. Sci. Instrum.* **67**, 1138–1143.

Belchenko, Y. I. (1996), *Rev. Sci. Instrum.* **67**, 1108–1110.

Forrester, A. T. (1988), *Large Ion Beams*, New York: Wiley, Chap. 10.

Hershcovitch, A. (Ed.) (1990), *Production and Neutralization of Negative Ions and Beams (1989)*, AIP Conference Proceedings No. 210, New York: American Institute of Physics.

Jacquot, C. (Ed.) (1998), *Production and Neutralization of Negative Ions and Beams (1997)*, AIP Conference Proceedings No. 430, New York: American Institute of Physics.

Pezzoti, C. (Ed.) (1974), *Proceedings of the 2nd Symposium on Ion Sources and Formation of Ion Beams*, Lawrence Berkeley Laboratory Report No. LBL-3399.

Prelec, K. (Ed.) (1977), *Proceedings of the Symposium on the Production and Neutralization of Negative Ions and Beams*, Brookhaven National Laboratory Report No. BNL-50727.

Prelec, K. (Ed.) (1984), *Production and Neutralization of Negative Ions and Beams (1983)*, AIP Conference Proceedings No. 111, New York: American Institute of Physics.

Prelec, K. (Ed.) (1996), *Production and Neutralization of Negative Ions and Beams (1995)*, AIP Conference Proceedings No. 380, New York: American Institute of Physics.

Sluyters, T. (Ed.) (1971), *Proceedings of the Symposium on Ion Sources and Formation of Ion Beams*, Brookhaven National Laboratory Report No. BNL-50310.

Sluyters, T. (Ed.) (1980), *Proceedings of the Second International Symposium on the Production and Neutralization of Negative Hydrogen Ions and Beams*, Brookhaven National Laboratory Report No. BNL-51304.

Whealton, J. H., Raridon, R. J., Bell, M. A., Rothe, K. E., Wooten, D. W., McGaffey, Wooten, J. W., Whitson, J. C. (1988), *J. Appl. Phys.* **64**, 6210–6221.

PHOTOGRAPHIC IMAGING: SPECIAL TECHNIQUES

Gunther Kürbitz, *Zeiss Optronik GmbH, Oberkochen, Germany*

INTRODUCTION

Why arrange to see in nonvisible spectral regions and at very low light levels? As a matter of fact some animals do have the capability to see in "nonvisible" (for the human eye) spectral regions or at very low light levels. Bees (*e.g.*, *Apis mellifera*), for example, have some sensitivity in the UV region of the electromagnetic spectrum. On the other hand some vipers (*viz.* the family of pit vipers, *Crotalidae*, with the rattlesnake as prominent member) obviously have some good ability to "see" heat radiation (this is done by a special organ named the pit organ). Owls are known to be active under low light levels at night. This leads to the idea that radiation outside the "visible" spectrum (and at low light levels) may contain useful information at least for special purposes of benefitting by "seeing"

ISBN 3-527-29308-6

even for human purposes. These ideas of benefitting under special constraints are the justification for this article.

In this context the benefits may reach from medical via environmental to military applications. They are discussed in the corresponding sections (see Secs. 1.3, 2.3, 3.1.3).

Imaging as discussed in this article covers imaging under terrestrial conditions; that means under consideration of atmospheric effects on the propagation of radiation. (Space astronomers are free from that limitation. If their equipment has luckily reached its orbit/position/flight path it may sense all the radiation it is designed for.) Otherwise imaging is understood as "real-time imaging" or a least as "near real-time imaging", allowing the observation of natural scenes containing moving objects. This excludes the detailed discussion of photographic films. Standard photographic films have some sensitivity in the near-UV region (*i.e.*, below 380 nm: Suits, 1978), which leads to the use of UV-blocking filters to avoid color shifting when photographing in the mountains. Special films are available with some response in the near-IR region (*i.e.*, beyond 780 nm). Their features are presented in Vol. 16 of this Encyclopedia under RECORDING, PHOTOGRAPHIC, as well as in this supplement under PHOTOGRAPHY: PHYSICS AND TECHNOLOGY.

The term "TV resolution" is occasionally used in this article. By this term should be understood an imagery on a TV screen that can resolve pixel numbers in the order of 480 × 640 pixels (US TV standard) or 576 × 768 pixels (European TV standard).

Normally—when seeing over distances of more than a few meters—the transmission characteristic of the atmosphere becomes of importance. The atmospheric propagation of radiation (visible as well as nonvisible) is completely different in the aforementioned spectral regions. Therefore every section contains a concise discussion of the influence of atmospheric transmission on system performance.

Technical imaging in nonvisible spectral regions and at very low light levels requires the use of sensors that convert nonvisible or very low-light-level radiation into electrical—or optical—signals that can be amplified and processed to generate a visible representation of the observed scenery. It covers the short-waveband extension (UV) of the visible region as well as the infrared (IR) regions.

In fact for the "nonvisible" spectral regions appropriate detectors are required as well as appropriate optics. Because these topics are essential, they are given special account in every corresponding section. It may happen that the optics for a special waveband are not significantly different from those for the visual spectral region; in this case there is no coverage of the topic of optics (this applies especially to Sec. 2, Imaging in the Near and Short-Wave Infrared and at Very Low Light Levels).

Although this article covers a rather broad spectral region besides the visible one, it seems appropriate to have a common way to proceed. Normally it is a good idea to follow the path of the radiation starting from the source (the illuminated or emitting scenery) through

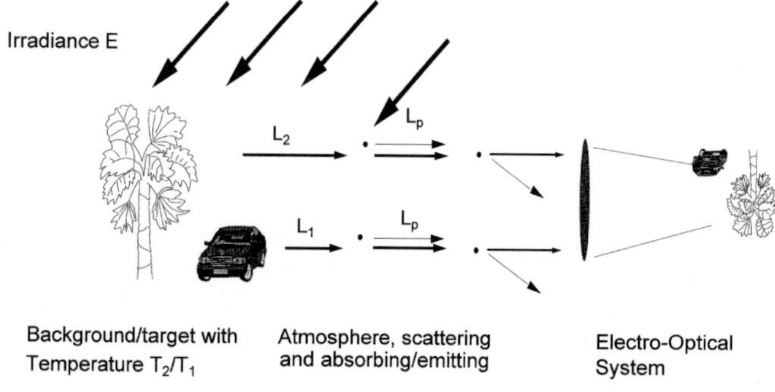

FIG. 1. Transmission chain for radiation illuminating a scene and emanating from that scene through the atmosphere to the electro-optical system. T_1 and T_2 denote the temperatures, L_1 and L_2 the radiances of target and background, respectively. L_p denotes the path radiance caused by scattering, absorption, and emission by aerosols of the atmosphere. The thicknesses of the arrows give an indication of the relative intensities of the radiation. The arrow length is arbitrary.

the atmosphere to the system (see Fig. 1); then follow the path through the system via the optics and a scanner (if necessary) to the detector. Special means to cool the detector (if necessary) have to be dealt with. System considerations concerning the performance of the respective electro-optical imaging systems are discussed where appropriate. Application information is included as far as available.

1. UV IMAGING

It is a historical fact that the photoelectric effect was originally detected in the UV region. At the end of the 19th century H. Hertz, P. Lenard, and others detected that UV radiation can release charged particles from metals. Though all of the imaging sensors working in the UV or in the visible or IR region of the electromagnetic waveband are based on that effect, it took indeed nearly a hundred years to develop UV sensors that are able to create two-dimensional images with TV resolution. These sensors are now capable of producing imagery in the waveband region up to 400 nm (the beginning of the visual spectral region). The following discussion concentrates on the waveband from 200 to 400 nm. The reason to start at about 200 nm is the fact that the atmosphere is nearly totally opaque below 200 nm.

1.1 Physical Properties of the UV Spectral Region

To see in the UV spectral region from 200 to 400 nm requires sources that emit or remit radiation in that waveband. A natural source illuminating a scene is, of course, the sun. Its radiation has to cross the Earth's atmosphere to be able to illuminate a scene. So only that portion of the solar radiation spectrum that can penetrate the atmosphere is capable of illuminating natural scenes. On the other hand, there may be artificial hot sources emitting radiation (*i.e.*, as a result of Planck's law) in the UV spectral band that only has to penetrate the atmosphere from the source to the observer. Consequently, it is necessary to investigate the atmospheric transmission from the sun to natural Earth-bound targets on the one hand and from targets to observers via the lower Earth's atmosphere on the other hand. Nowadays a typical means to carry out such an investigation is a computer code called LOWTRAN, which is a standard method of performing the required atmospheric transmission calculations for electro-optical systems. Figure 2 shows the atmospheric transmission characteristics in the 200–400-nm waveband

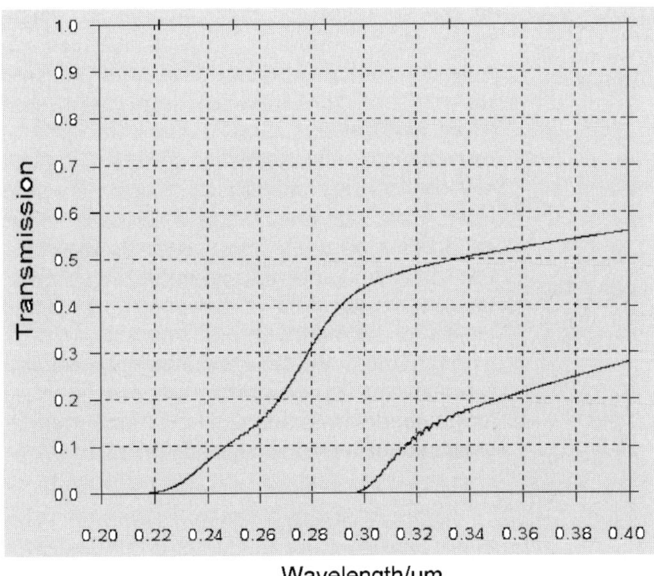

FIG. 2. Atmospheric transmission in the UV waveband below 0.4 μm. The upper curve shows the atmospheric transmission for a near-ground path of length 1000 m. The lower curve shows the transmission of the Earth's atmosphere via a vertical path from 25 km (stratosphere) to the ground. The difference in wavelength between the points where these two curves touch the zero-transmission line is called the solar-blind region.

for two cases. The upper curve shows the atmospheric transmission for a near-ground path and a path length of 1 km at a summer situation in middle Europe. In clearly shows that atmospheric transmission at the surface of the earth starts at about 220 nm and that it increases up to 400 nm. The second (lower) curve shows the transmission of the Earth's atmosphere via a vertical path from 25 km height down to the ground under the same conditions. It is obvious that below a wavelength of 300 nm the transmission for this radiation is essentially negligible. This curve shows that no solar radiation below the wavelength of 300 nm is able to illuminate any target on the surface of the Earth. The reason for that fact is the existence of the ozone layer in the outer atmosphere (the stratosphere), which absorbs UV radiation below the wavelength of 300 nm almost completely. On the other hand near the ground there is indeed less ozone and therefore some atmospheric transmission is found for radiation between 220 and 300 nm. Consequently, the waveband between 220 and 300 nm has been called the *solar-blind region*. (Some workers prefer to limit the solar-blind region to 230–290 nm; Hacker *et al.*, 1996). The existence of that solar-blind region opens the possibility of observing artificial UV sources without disturbance by the solar background or clutter. The quantitative investigation of the curves of Fig. 2 indeed shows that already for a path length of 1 km there is less than about 20% of average transmission. For longer path lengths the transmission is lowered exponentially. So evidently the observation of artificial targets in the solar-blind region is limited to a few kilometers.

1.2 Two-dimensional UV Cameras

To build a UV camera requires a UV optical lens system and a UV sensor. Today the UV sensor is a two-dimensional sensor array producing a TV-type signal that is serially (line after line) organized (similar to the signals produced by CCD-camera chips working in the visual spectral region—*e.g.*, in camcorders). Because the readout circuit used in the sensor is state of the art it is not covered by this article (see CHARGE-COUPLED DEVICES). The same applies for the video electronics that produce an electrical TV signal driving a TV monitor that allows the visualization of the UV image. Instead special attention is given to the problems associated with the UV optics and the UV sensors.

1.2.1 UV Optics The development of a well-corrected UV optical lens system requires the possibility of choosing between a number of different available optical materials. The glass catalogs of the leading manufacturers of optical glass list more than a hundred different optical glasses for the visual spectral region. The situation is completely different for the UV region. The number of optical materials suitable at all for that region is about a dozen. (Incidentally, the same situation applies for the infrared region of the spectrum that is dealt with in Sec. 3). The list of UV optical materials contains, for example, LiF, MgF_2, CaF_2, SrF_2, BaF_2, KF, quartz, NaCl, KCl, CsCl, sapphire, and diamond. Many of these materials (mainly the halides) are more or less soluble in water (LiF, NaF, SrF_2, BaF_2, KF, NaCl, KCl, CsCl); others (quartz, sapphire, diamond) are extremely hard and difficult to machine; and some (MgF_2, quartz) are birefringent. These facts complicate the design of well-corrected optical lens systems for the UV considerably. The refractive indices of most of the aforementioned materials vary in the region of about 1.4–1.8. This has the consequence that well-corrected UV lens systems require a relatively high number of single lenses. So a thorough tradeoff between required geometrical resolution and expense in view of the intended application has to be performed. Figure 3 shows the spectral transmission characteristics of two typical UV-transmitting materials: calcium fluoride (CaF_2) and sapphire. Figure 4 shows the optical scheme of a UV lens system designed for the solar-blind spectral region of 230–300 nm. It covers a 90° field of view and has a focal length of 13 mm with an *f*-number of 1.5. Though the resolution requirements are not too high, the lens system consists of nine single lenses made from fused silica (fused quartz, which is not birefringent) and CaF_2.

1.2.2 UV Sensors There are four main different techniques for building UV-sensitive

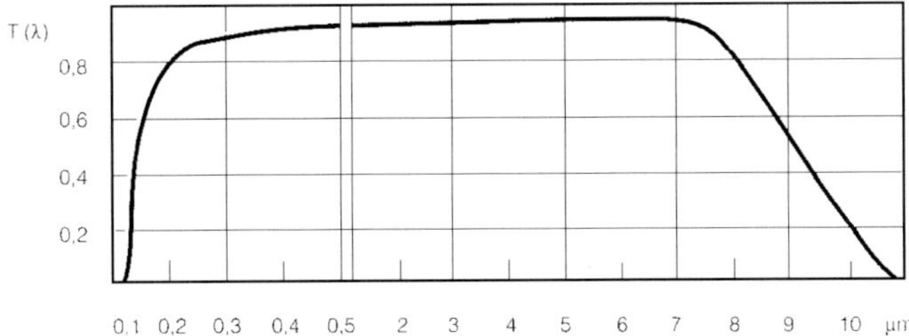

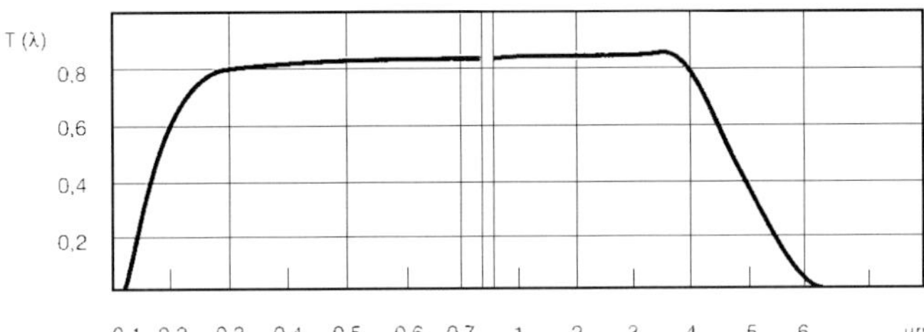

FIG. 3. Transmission characteristics of two typical UV-transmitting materials: calcium fluoride (CaF_2) and sapphire (Al_2O_3). Both materials have good transmission in the solar-blind region between 220 and 300 nm.

two-dimensional sensors. These are characterized by the terms

1. UV spectral converter intensifier with fiber-optical coupled CCD-sensors ("UV-i-CCD"),
2. UV-to-VIS converters,
3. back-illuminated CCDs, and
4. UV-enhanced photodiode arrays.

The different methods for creating TV-type signals from the aforementioned techniques are discussed in the following.

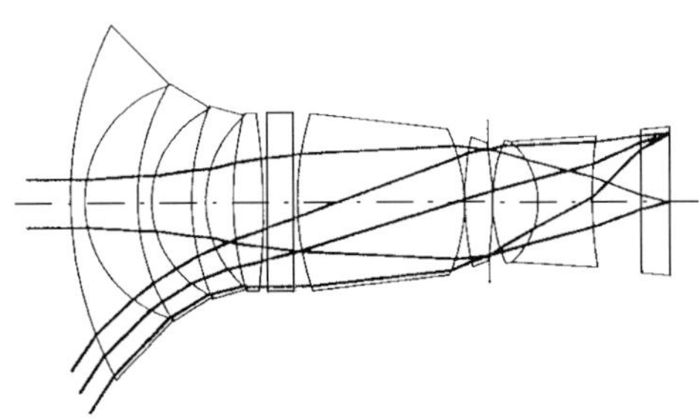

FIG. 4. Optical scheme of a UV lens system with a 90° field of view designed for the solar-blind spectral region of 220–300 nm. The system consists of nine single lenses.

1.2.2.1 The "UV-i-CCD" Method. This solution for converting UV radiation into electrical signals is based on vacuum-tube technology. A diode-type image-intensifier tube (this image intensifier tube type is described in more detail in Sec. 2.2.2) has a UV-sensitive RbTe (rubidium telluride) cathode (Fig. 5). The electrons released from the cath-

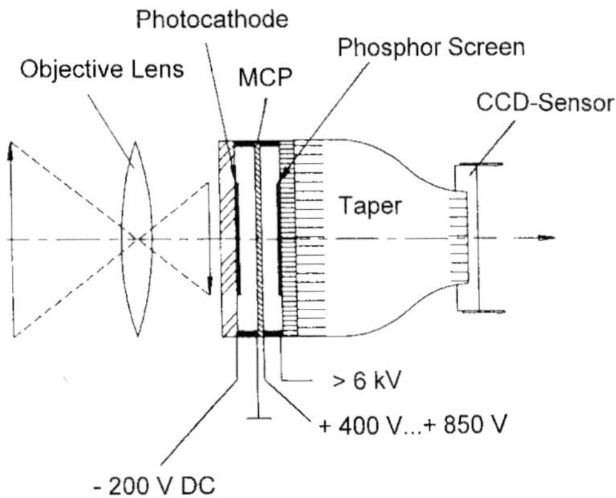

FIG. 5. UV-i-CCD sensor. This sensor is composed of a few main constituents: a vacuum-tube type image converter with a UV-sensitive cathode, an internal MCP (multichannel plate), a phosphor screen, a fiber-optic taper, and a CCD sensor.

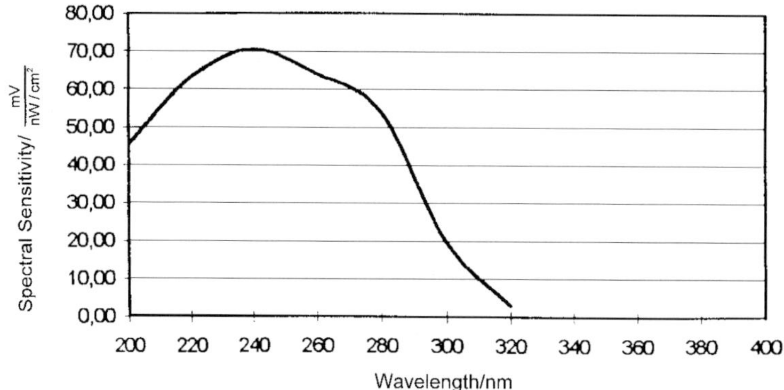

FIG. 6. Spectral sensitivity curve of a typical "solar-blind" (RbTe) UV spectral-converter intensifier cathode.

ode by the UV radiation are multiplied by a so-called multichannel plate (MCP), are accelerated in the electrical field of the vacuum tube, and produce visible photons when hitting the phosphor screen. This UV image converted by the UV-image intensifier into a visible representation hits a fiber-optical taper (an image transferring fiber-optic system that uses orderly organized optical fibers), which adjusts the geometry of the image intensifier to the geometry of the following visual CCD chip. The CCD chip produces electrical signals in a TV-standard manner.

Figure 6 shows a typical spectral sensitivity curve for this type of UV sensor. State-of-the-art sensors achieve more than 20% quantum efficiency in the UV waveband.

1.2.2.2 UV-to-VIS Converters. It is possible to achieve a UV response by coating the surface of conventional (visual) CCDs with a very thin layer of converter material that absorbs UV radiation and re-emits radiation in the visible region. A standard coating for UV applications is a material called Lumogen^{TM}. Lumogen is an organic chemical that converts light with wavelengths less than 480 nm into visible light with wavelengths between 510 and 580 nm, which CCDs detect with good sensitivity. Lumogen coating has to be performed using very accurate vapor deposition. An annealing process (mostly company proprietary) is necessary to generate a highly uniform coating capable of sustaining continuous thermal cycling. Lumogen, like any other organic UV coating, degrades over time through photodecomposition (bleaching). Under normal illumination this process is slow, but care must be taken to reduce excessive exposure to UV radiation. Some companies delivering CCD sensors with Lumogen coating offer in-house recoating to their customers.

1.2.2.3 Back-Illuminated CCDs. Standard CCD sensors for the visual spectral range show considerable loss of response in the UV due to absorption and by reflection from the polysilicon gate electrodes on the front surface of the CCD. Therefore special CCD arrays that can detect radiation through the rear surface are used. These CCDs are usually thinned to a thickness of about 10 μm and light is brought to the array from the rear side.

The quantum efficiency of back-illuminated CCD arrays is limited by reflection from the back side. As the index of refraction of silicon is as high as 2.5, this reflection can be very significant (see Sec. 3.2.1); so antireflection coatings (AR coatings) are used to reduce this reflection. Unfortunately, the chemistry of materials for normal UV AR coatings interferes with the surface of the silicon. As new materials are developed for the AR coating, increased quantum efficiency can be expected to reach values of more than 40%. In the meantime back-illuminated CCDs can be coated additionally with a scintillator material like Lumogen. Such CCD arrays achieve quantum efficiencies of nearly 30% in the 200–400 nm waveband.

1.2.2.4 UV-Enhanced Photodiode Arrays. Unlike the polysilicon gates of conventional

CCD arrays, the silicon dioxide passivation layer on top of typical silicon photodiodes is UV transparent. Silicon photodiodes are therefore sensitive from 200 nm (and below) up to 1100 nm. This fact is used by some manufacturers to build photodiode arrays (with up to 256 × 256 pixels) with a self-scanning matrix-type readout structure that is controlled by external CMOS address decoders. These decoders select the pixels through their outputs that are connected to the diodes' multiplexing gates. When the p–n junction of a photodiode is reverse biased and then open circuited, the charge stored on the depletion-layer capacitance decays at a rate proportional to the incident illumination. Each time the multiplexing gate is activated, the photodiode is reverse biased again and the internal capacitor is recharged. The amount of charge flowing on the video line is sensed as a measure for the integrated intensity. The UV enhancement is achieved by Lumogen and/or AR coating. Quantum efficiencies of more than 30% are achievable.

1.3 Applications

In the following three types of typical applications for UV imaging systems are described. All of them take advantage of the solar-blind effect described in Sec. 1.1. To repeat: this effect has the consequences that the natural scenery is completely dark. Only artificial sources or illuminated scenes are recognizable. The three, however, use different properties of that part of the UV region.

1.3.1 Missile Warning The solar-blind spectral region is of particular interest for military missile warning because in that region natural background radiation plays only a negligible role (Dirscherl, 1996). Detection systems with sensitivity in the solar-blind region of the UV spectrum are effective because of the *de facto* absence of natural background radiation. On the other hand, the extinction (absorption and scattering) within the atmosphere is usually higher in the UV than in the IR spectral range. It is necessary to use optical band-pass filters with steep edges to avoid unwanted radiation reaching the UV sensor.

Missile flares are the typical artificial UV light sources that must be detected for the purpose of missile warning. The flares of missiles threatening airborne or ground-based targets can adequately be described as black bodies of temperatures in the 2000–2500 K range. These sources emit some portion of their radiation in the solar-blind region. Because of the absence of solar illumination, the signal-to-noise ratio is mainly governed by the residual atmospheric transmission and the noise produced by the UV sensor. Well-designed UV systems can perform missile warning over distances of a few kilometers.

1.3.2 Covert Operations In a waveband region near 220 nm there exists a certain spectral region where there is only a small amount of near-surface atmospheric transmission (see Fig. 2). This waveband may for example extend from 220 to 230 nm. It opens the possibility for covert operations by using an artificial and band-limited UV source at night and at very small ranges. Examples for these applications are the repair of damaged equipment at night in good brightness without the danger of being detected from large distances.

1.3.3 Penetration of Fog, Heavy Rain and Snow Visible moisture is known to degrade the performance of infrared and, to a lesser extent, passive millimeter-wave sensors. UV performance is less affected by reason of its wavelength of 0.2–0.3 μm, which is much smaller than fog particles. Water-particle size generally varies with the type of fog, growing as the fog thickens. Larger particles and higher concentrations are found in advection fog, which often develops in coastal regions. Radiation fog, as a rule, develops inland and has smaller particles. UV radiation, with its relatively short wavelength, interacting with fog particles produces mainly forward scattering with minimum attenuation. Forward scattering is more pronounced with advection fog, and analytical models predict that it may be 100 times greater than from radiation fog. Possible applications of these facts include the high-intensity strobes used for runway thresholds, runway edges and taxiway marking at airports. Preliminary tests showed that with a

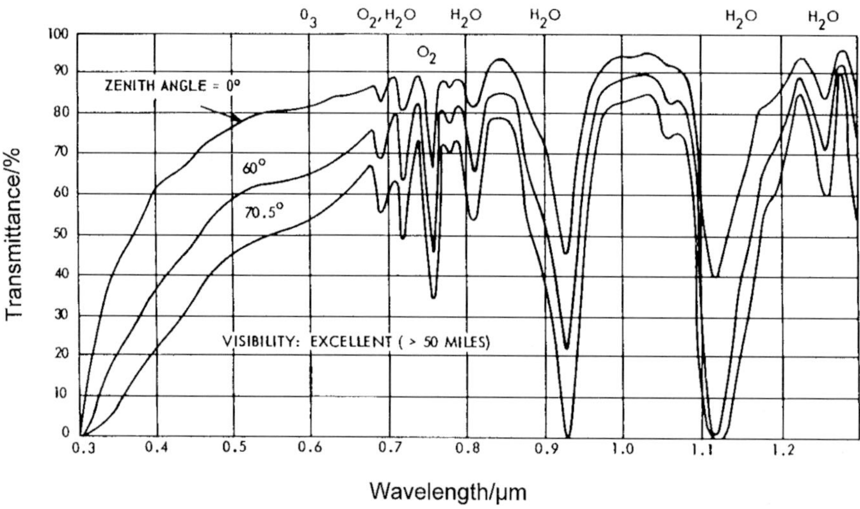

FIG. 7. The transmission of the atmosphere between 0.3 and 1.3 μm is shown for a set of standard conditions. Path length is 1 km at ground level, humidity is 60%, temperature is 25 °C. The zenith angle is the parameter, indicated for 0°, 60°, and 75°.

visibility of 250 m the UV systems had a range of up to 990 m (Nordwall, 1997).

2. IMAGING IN THE NEAR AND SHORT-WAVE INFRARED AND AT VERY LOW LIGHT LEVELS

The spectral range of this imaging region extends from about 400 nm up to about 2.0 μm. It is perhaps helpful to define what the terms "near infrared", "short-wave infrared" and "very low light level" mean. *Near infrared* describes the spectral region from about 700 to 1100 nm, where the sensitivity of silicon sensors ends. *Short-wave infrared* characterizes the waveband from about 1100 to 2000 nm (2.0 μm). *Very low light level* means the spectral region between 400 and 1100 nm (eventually up to 2000 nm), but at illumination levels on the order of less than 1 lx. All of this imaging is based on external illumination. In this context "external" means the sun, the moon and the stars, scattered night light, airglow, or artificial sources. The intensity of such illumination varies over many orders of magnitude. Technical means for seeing in the spectral regions from 400 to 700 nm are required only if the illumination by the sun is insufficient—for example, at dawn or dusk, in twilight, or at night. In the spectral region above 780 nm technical devices are always necessary to fulfil any observation requirements.

2.1 Physical Properties of the Visible and the Near and Short-Wave Infrared Regions

It is important to know the atmospheric transmission characteristics in a particular waveband when developing and designing an electro-optical device for that waveband. Figure 7 shows the transmission of the Earth's atmosphere in the spectral region from 0.3 to 1.3 μm at various zenith angles. That transmission is valid for a 1-km path length at ground level and at a humidity of 60% and a temperature of 25 °C. It must be emphasized that in this waveband there does not exist an "ozone-like" absorption effect corresponding to the one described in Sec. 1.1. The transmission characteristic of the Earth's atmosphere is quite isotropic in the vertical and horizontal directions. The absorbing molecules are included in Fig. 7.

In the visible spectral region from 400 to 700 nm there is almost no degradation due to atmospheric extinction. This may be the reason that the human eye is sensitive just in this region.

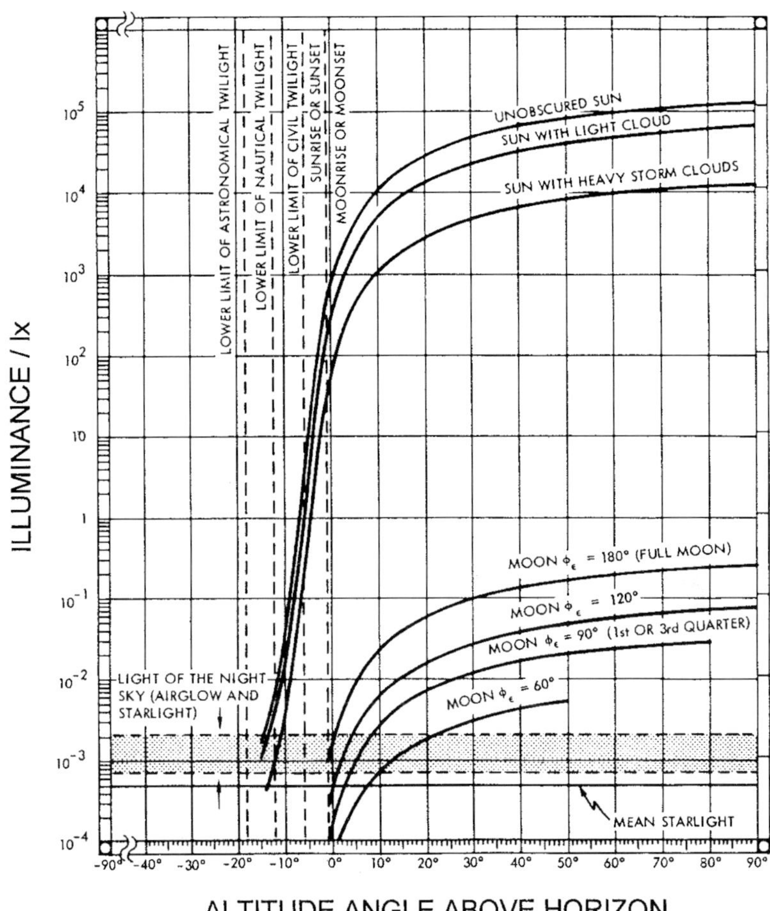

FIG. 8. Illuminance levels on the surface of the Earth due to the sun, the moon, and the sky as a function of the altitude angle of the illuminating source above the horizon.

To recognize the situation for seeing under very low light-level conditions one must look at these conditions. Figure 8 shows the illuminance in the visible (and near-IR) region of the spectrum for different illumination situations (sun, sun with clouds, full moon, *etc.*) as a function of the altitude angle of the illuminating source above the horizon, which otherwise is dependent on time of day. It is clearly recognizable that viewing at very low light levels requires the possibility of creating images at illuminance levels in the 10^{-3} lx region. This situation is the domain of image-intensifier or I^2 techniques.

There seems to be a region where night vision capabilities can he extended to longer wavelengths (up to 1.7–2.0 μm, the short-wave infrared region). This is due to the so-called *night-airglow* phenomenon. Airglow may be defined as the nonthermal radiation emitted by the Earth's atmosphere. Night airglow emissions in the short-wave infrared (or SWIR) are caused by transitions between vibrational states of the OH radical. The exact mechanism of excitation is still unclear, but the effect is to release energy from solar radiation stored during the daytime. Airglow occurs at all latitudes. There is some evidence that the distribution of night airglow is related to that of ozone. Figure 9 shows the typical night-airglow spectrum, which has its maximum intensity in the 1.4–1.9 μm region.

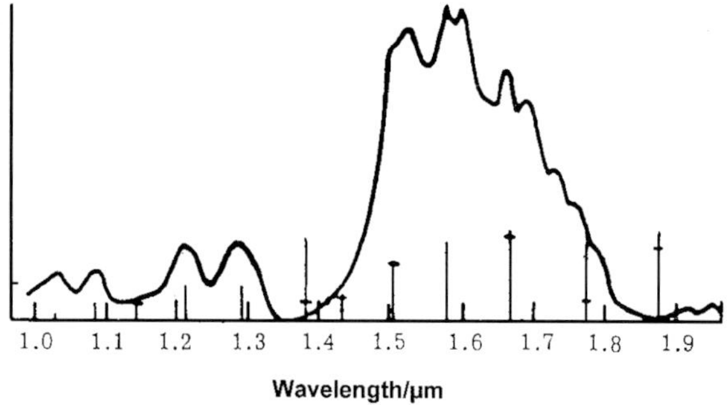

FIG. 9. Night-airglow spectrum in the wavelength region from 1.0 to 2.0 μm. The vertical lines indicate expected intensities of OH bands.

2.2 Imaging Techniques

The technical devices that convert low light level or near or short-wave infrared radiation into visible imagery are—with the exception of the SWIR region from 1.4 to 2.0 μm—all based on vacuum-tube technology. This technology is therefore described in some detail. Concerning the CCD technology a reference is made to the article CHARGE-COUPLED DEVICES. So only the special SWIR features of modern CCD sensors are treated.

2.2.1 Active Infrared Imaging and Gated Viewing Historically the technique of active infrared imaging was the first one to be applied for covert observations. That technique requires an illumination source that emits radiation outside the visible spectrum, *i.e.*, an infrared lamp or a near-IR laser, to illuminate the scene. On the observer side an image converter is required that transforms that radiation into visible light. The radiation source may be a tungsten-filament lamp or a high-pressure gas-discharge lamp equipped with an optical high-pass filter with a cuton wavelength of about 800 nm, or a near-IR laser, to prevent visual radiation from being emitted. To achieve sufficient illumination levels at the scene, the radiation of these sources has to be bundled appropriately. This is done by lenses or reflecting mirrors in the case of lamps or by means of a telescope in the case of lasers. On the receiver side a near-IR–sensitive imaging device is required. This device converts the IR radiation into a visible image. The general situation is depicted in Fig. 10. It must be emphasized that all these active IR illumination techniques are subject to detection by possible foes. Their use is therefore limited to more or less "friendly" applications.

2.2.1.1 Active Infrared Imaging The first electro-optical devices to convert IR radiation into a visible representation were the so-called image-converter tubes. Figure 11 shows a schematic diagram of a typical image-converter tube. An electrostatic field directs the electrons released by an IR-sensitive photocathode through an anode cone and focuses an inverted image onto the phosphor screen. The photocathodes for IR image-converter tubes are of the so-called S-1 type. Their sensitivity is shown in Fig. 12. Figure 12 also shows the quantum efficiency of these types of vacuum

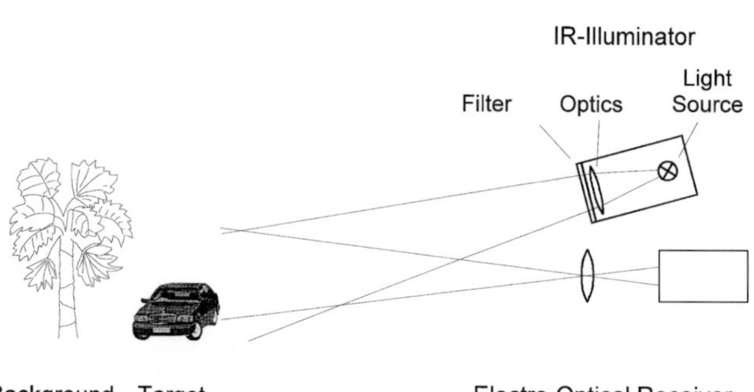

FIG. 10. General situation for an active IR-imaging system. It is composed of an IR illuminator and an EO receiving system.

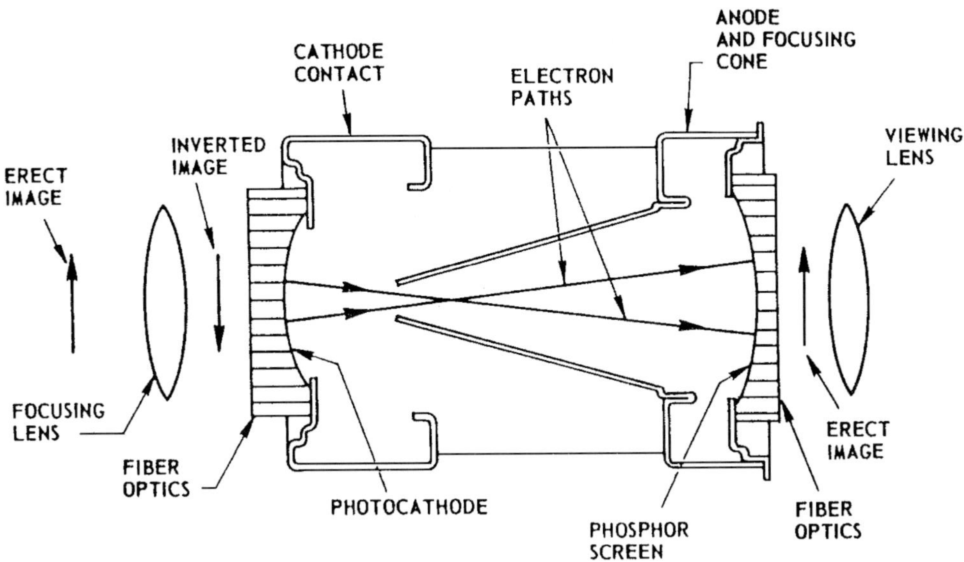

FIG. 11. Schematic diagram of an electrostatic-type image converter tube. The IR image is projected onto the input of a fiber optics, which transfers the image to the photocathode at the far end of the fiber optics. Electrons released by the photocathode are directed by an electrostatic field through an anode cone and focused on a phosphor screen.

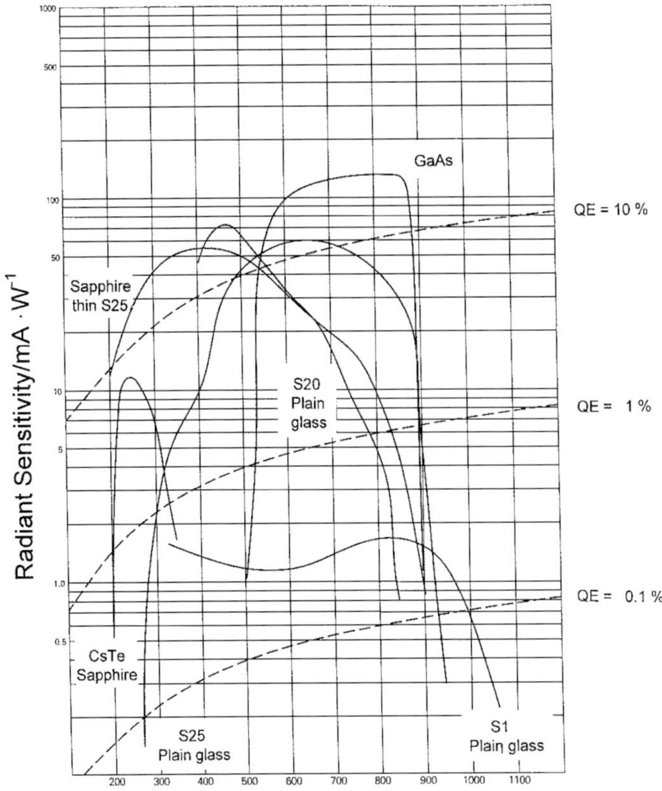

FIG. 12. Responsivities of various photocathodes used for image-converter and image-intensifier tubes as functions of wavelength.

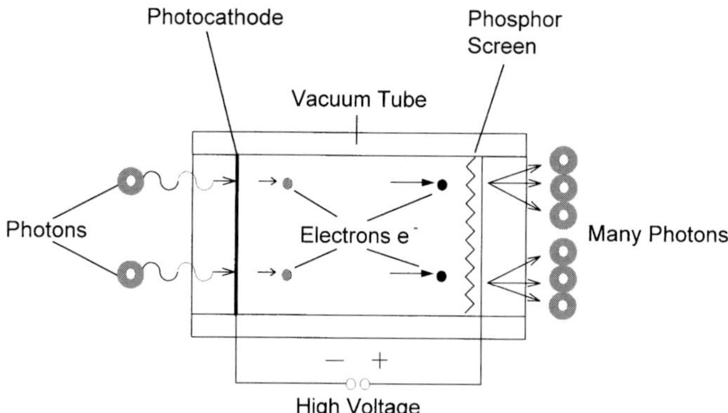

FIG. 13. An image-converter or image-intensifier tube is a vacuum-type tube that has a photocathode as the input. Incoming photons release (by the photoelectric effect) electrons from the photocathode. These electrons are accelerated by means of the electrostatic field of the applied high voltage between cathode and anode. They gain energy during this acceleration process. When they hit the phosphor screen (the anode) their energy is converted into visible light (many photons).

tubes, which is on the order of $<1\%$ for the spectral region shown. Consequently rather high illumination levels are required to obtain good range performance for these systems. Because of the active nature of these systems their use for military applications is nowadays very limited.

2.2.1.2 Gated Viewing Gated viewing is an active observation means using a pulsed illumination source and a gated receiver. The idea is to illuminate the scene in depth by a short pulse for only a small amount of time (for example, a pulse duration of 100 ns corresponds to a momentarily illuminated scene depth of 30 m) and to vary the observed scene distance by the delay time between the transmitted illuminating pulse and the receiver-gate opening time. Since the illuminating light has to cross the atmosphere between the source and the target and back to the receiver twice, the time delay constant is $150\,\mathrm{m}\,\mu\mathrm{s}^{-1}$. Nowadays the pulsed source is conveniently built up by a semiconductor laser array whereas the receiver uses a gated vacuum-tube image converter or image intensifier (see Sec. 2.2.2). The general schematic given in Fig. 10 also applies for a gated-viewing system.

2.2.2 I^2 Tubes Image-intensifier devices (I^2 devices) all follow a relatively simple operational principle:

1. The observed scenery is imaged by means of an objective lens onto the photocathode of an I^2 tube.
2. The photocathode then emits electrical particles (electrons), the number of which corresponds to the intensity of the incoming light, (many incoming light particles—photons—release many electrons).
3. The electrons are accelerated by a high electrical field and gain energy in this field.
4. At the end of the acceleration path there is a phosphor screen that transforms the energy of the impacting electrons into visible light. In effect a single incoming photon at the photocathode results in the emission of very many (typically hundreds of) photons from the phosphor screen.
5. The phosphor screen is observed through an ocular lens.

The process described results in an amplification of the incoming low–light level radiation through the chain 1 photon $\rightarrow$ 1 electron $\rightarrow$ acceleration $\rightarrow$ many photons (see Fig. 13). There is a variety of realizations for this process. These are divided into so-called "generations" of I^2 tubes.

Image-intensifier or I^2 tubes of the first generation are directly based on the aforementioned process. Their photocathode is of a so-called multi-alkali S-20 or S-25 type and has a quantum efficiency of about 10% (see Fig. 12). An electron-optical system assures, through a conical shaped anode, that the

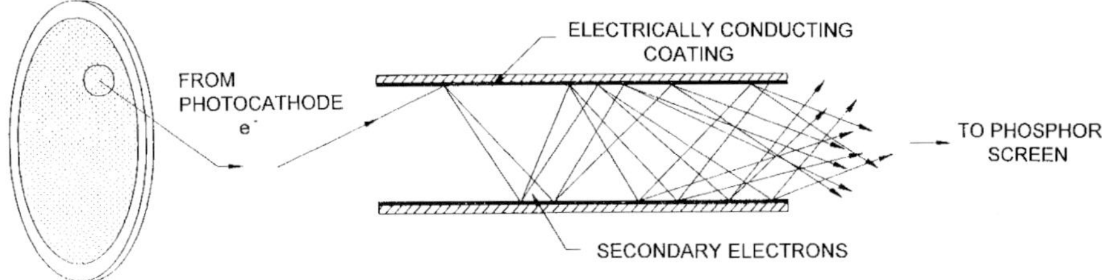

FIG. 14. Principle of the microchannel plate (MCP). The MCP multiplies the number of incoming electrons according to the principle of multiplying secondary (emitted) electrons.

charge distribution released by the incoming photons is imaged onto the phosphor screen (see Fig. 11). An I^2 tube of this type is able to achieve a light amplification factor of about 100. Normally three coupled tubes of this type are necessary—because of unavoidable coupling losses—to achieve the required amplification factors on the order of 10 000–100 000. This results in relatively lengthy image-intensifier devices of the first generation.

I^2 tubes of the second generation are based on vacuum tubes that contain a so-called microchannel plate (MCP) as a central component (see Fig. 14). This MCP has the property of multiplying the number of incoming electrons immensely (for example, by a factor of between 10^5 and 10^6). The MCP consists of about 1 million single channels each of which works like a photomultiplier tube with continuously distributed electrodes. This MCP has a high multiplication ratio for electrons without requiring extremely high voltages. An additional benefit of this type of I^2 tube is that as a result of the discrete channel nature of the MCP, blooming effects caused by high-intensity light spots in the observed scenery are limited. The gain effect caused by the MCP in second-generation I^2 tubes is sufficient to achieve the required light amplification in a single-stage tube.

I^2 tubes of the third generation are very similar to the second-generation tubes. The main difference is the type of cathode material. Whereas (first- and) second-generation I^2 tubes have so-called multi-alkali cathodes, third-generation tubes contain GaAs cathodes. The sensitivity of these cathodes is especially well adapted to the spectrum of night illumination. In the wavelength region between 600 and 900 nm the sensitivity is about 2–3 times greater than that of the second-generation tubes (see Fig. 12). Because of the complexity of their manufacturing process third-generation I^2 tubes are inherently more expensive than second-generation I^2 tubes. Thus a trade-off always has to be made as to whether to use second- or third-generation I^2 tubes.

2.2.3 Imaging in the SWIR Region Imaging devices for the SWIR region are in fact different from the devices used in the near-IR waveband region. The near-IR waveband region is governed by the use of vacuum tubes. In contrast, the SWIR waveband region is based on InGaAs technology. As already discussed in Sec. 2, the term SWIR covers the waveband from about 1100 to 2000 nm (2.0 μm). In the last few years sensitive matrix detectors have been developed with up to 128 × 128 pixels based on InGaAs CCD technology with an extension to 256 × 256 pixels and even 512 × 512 pixels foreseeable within a few years. These devices are available in uncooled and thermoelectrically cooled versions. They show good quantum efficiency together with high geometric resolution (pixel count). For details concerning the CCD technology, again reference is made to the corresponding article.

2.3 Applications

The range of applications of radiation in the near and short-wave infrared regions and at very low light levels covers a rather broad spectrum. This spectrum extends from passive

Table 1. The Johnson criterion.

Task	Minimal number of resolved line pairs on target
Detection	1
Recognition	3
Identification	6

night vision, through fog penetration, to very specific applications.

There is a set of standard conditions that is normally used to describe the range performance of night-vision devices. According to a large number of field trials, human observers are known to be able to perceive typical targets as a function of the number of resolved line pairs per target dimension as shown in Table 1. The content of this table is very often referred to as the so-called "Johnson criterion" (Johnson, 1966).

2.3.1 Passive Night Vision Passive night vision is based on the residual light levels at night. These may be very different in the various wavebands (*i.e.*, 0.4–1.0 µm and 1.0–1.7 µm). There is, however, a general relationship describing the range performance of a passive night-vision system and the corresponding illumination level. Figure 15 shows a typical range performance curve as a function of scene illumination. This curve can be used as a reference for a well-designed I^2 imaging system working with I^2 tubes of second or third generation with a typical field of view of 6°. Since up to now there are no experimental data available for the SWIR region, an equivalent graph taking into account airglow data can unfortunately not be shown. The general relationship between range and illumination level will be the same as in Fig. 15. The background for this figure is that at very low light levels the range performance of low–light-level devices (sometimes called L^3 devices) is limited by the signal-to-noise ratio of the imaging device. At higher light levels the range performance is governed by the limiting geometrical resolution. (In the case of second- and third-

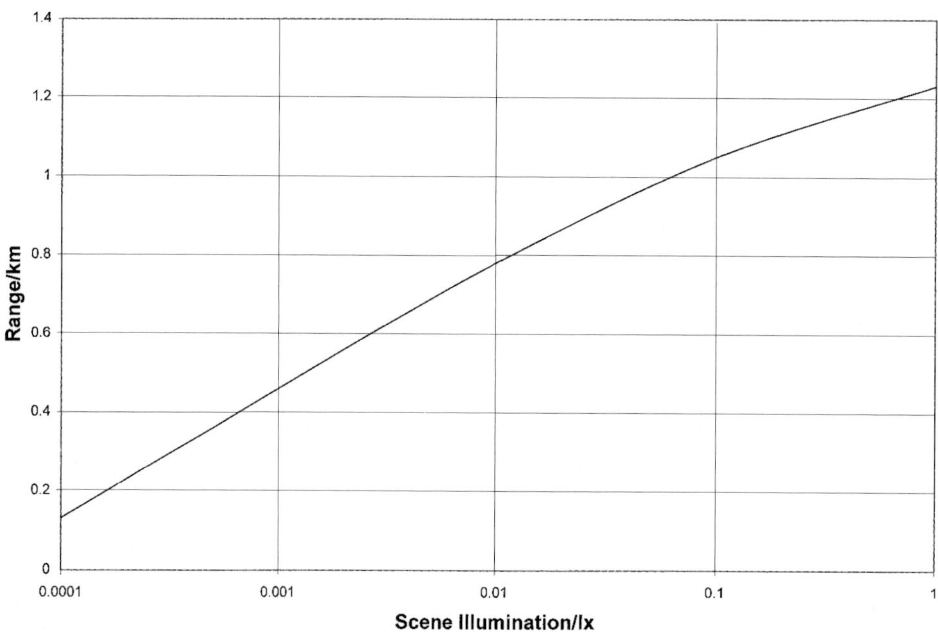

FIG. 15. Typical range performance curve for an I^2-tube system. The curve shows the range performance as a function of scene illumination. The curve is valid for the identification criterion. The field of view of the I^2 system is 6°.

generation I^2 tubes this is the modulation transfer function MTF of the MCP plate.)

2.3.2 Fog Penetration "Fog penetration" in this section refers not only to the different atmospheric conditions of penetration in the different (VIS, SWIR, MWIR, LWIR) wavebands but also mainly to active means of achieving improved penetration performance. There is one prominent means of achieving "fog penetration". It is called "gated viewing". The principle has already been described in Sec. 2.2.1.2. Nowadays "active gated viewing" is out of the perspective of military applications. There may, however, be a field of civil applications.

It is a well-known fact that traffic on motorways is sometimes troubled by fog and mist. Nevertheless the speed of some car drivers is not particularly well adapted to these situations. Very often this behavior of motorists results in a series of accidents at dangerous traffic areas. It may be worthwhile to install at these critical traffic areas gated-viewing systems that are controlled by fog density and car velocity to take images of vehicles endangering the traffic by excessive speed. The gated-viewing system will be able to identify the number plates of these cars.

2.3.3 A Very Special SWIR Application: Glucose Concentration in Blood Millions of diabetics determine their insulin dosages by drawing blood samples several times a day. The result is an inconvenient and painful procedure costing hundreds of dollars per year. However, it has been demonstrated that the glucose concentration of blood near the surface of the skin can be measured by the reflection of light at wavelengths in the vicinity of 2 µm (the SWIR region). A number of research institutions and medical instrumentation companies are exploring ways to reduce the technique to a practical, compact instrument suitable for use in hospitals and, eventually, in homes. The goal of rapid measurement by an instrument that needs no alignment and has no moving parts will require an array detector. The result will be analogous to blood oxymeters, which are based on the transmission of light in the vicinity of 0.7 µm (Cohen and Olsen, 1994).

3. VISION IN THE THERMAL INFRARED

In contrast to the situation described in the foregoing sections, vision in the thermal infrared is governed by the radiation emitted from the observed targets. The illumination caused by external sources is almost negligible. So one has to consider the radiation characteristics of typical targets as well as the atmospheric transmission characteristics in the corresponding wavebands. The optics, detectors, and system characteristics differ rather considerably from the system parameters described in Secs. 1 and 2. It is necessary to discuss these topics in detail. The following description considers the general scheme shown in Fig. 1 of this article.

3.1 Planck's Law and Atmospheric Transmission

In the thermal infrared, the typical scenery behaves nearly perfectly as a so-called blackbody. The radiation characteristics of a blackbody are described by Planck's law. Ideal blackbodies have emission and absorption coefficients of exactly 1. The typical scenery at a temperature level of about 288 K (15 °C) has an emission coefficient of nearly 0.99 in the 3–5 µm waveband as well as in the 7–12 µm waveband. The treatment of natural scenes as blackbodies is therefore a good approximation to reality.

The atmosphere creates the existence of the two infrared wavebands: 3–5 µm, called medium-wave infrared (MWIR), and 7–12 µm, called long-wave infrared (LWIR). The atmospheric transmission characteristics are discussed later.

3.1.1 Implications of Planck's Law on the Design of Electro-optical Sensors Planck's law describes the amount of radiation emitted by a blackbody in a small waveband as a function of temperature and wavelength. Figure 16 shows the graphic representation of Planck's law for blackbodies of temperature 300 K. This graph shows clearly that the maximum amount of radiation emitted by blackbodies around ambient temperature (*i.e.*, 288 K) is located at a wavelength of about

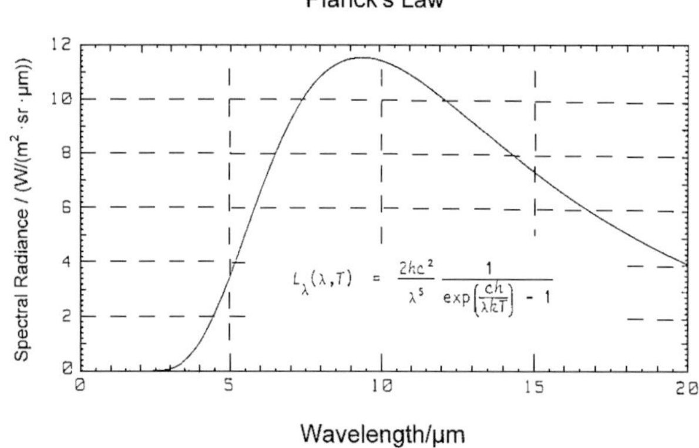

FIG. 16. Graphic presentation of Planck's law for a blackbody of about 300 K. The illustration shows also the formula for Planck's law.

10 µm. This fact is especially important for instruments designed for radiometric (for example, medical) measurements in the whole field of view (FOV). IR systems designed for creating good images of natural scenes must achieve the goal of displaying these scenes with high contrast for the observer. So it is necessary to investigate the behavior of the Planck function with respect to its sensitivity to small temperature variations. Mathematically speaking, this is shown by the thermal derivative of Planck's function. This function is often called *spectral radiance contrast*. Figure 17 shows the spectral radiance contrast as a function of wavelength. The maximum of this function is located at about 8 µm, which again is contained within the 7.0–11.5-µm waveband. Taking into account the results of Figs. 16 and 17, it clearly proves advantageous to work in the 7–12-µm waveband when the scene temperature is at a level around 300 K.

3.1.2 Atmospheric Transmission The transmission of the atmosphere in the IR region is shown in Fig. 18. This figure clearly shows that there are so-called atmospheric "windows". Outside the visible and short-wave infrared region there are two wavebands clearly recognizable: the medium-wave infrared (MWIR) in the 3–5 µm waveband and the long-wave infrared (LWIR) in the 7–12 µm region. If we compare the emission characteristics of typical blackbodies in the temperature interval from 283 to 313 K with the transmission windows there is an important serendipity, insofar as the transmission characteristics of blackbodies around 300 K coincide with the atmospheric window in the 7–12-µm region. As a consequence it has to be stated that there exist two IR wavebands allowing the thermal observation of natural scenes. The 3–5-µm waveband is optimal for targets at a temperature level of about 600 K (equivalent to about

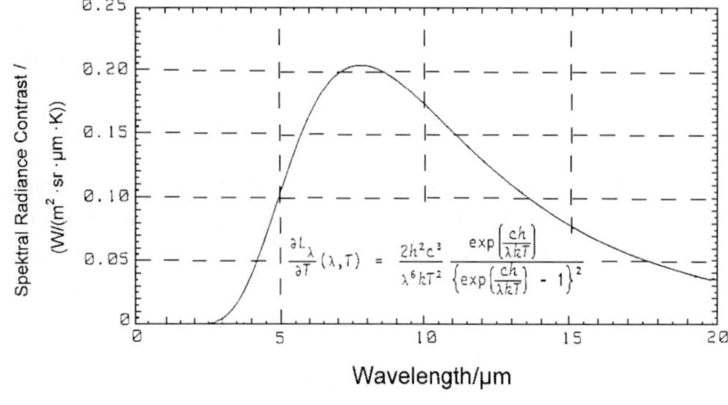

FIG. 17. Graphic representation of the thermal derivative of Planck's function for a blackbody of about 300 K. The formula for the spectral radiation contrast is also shown.

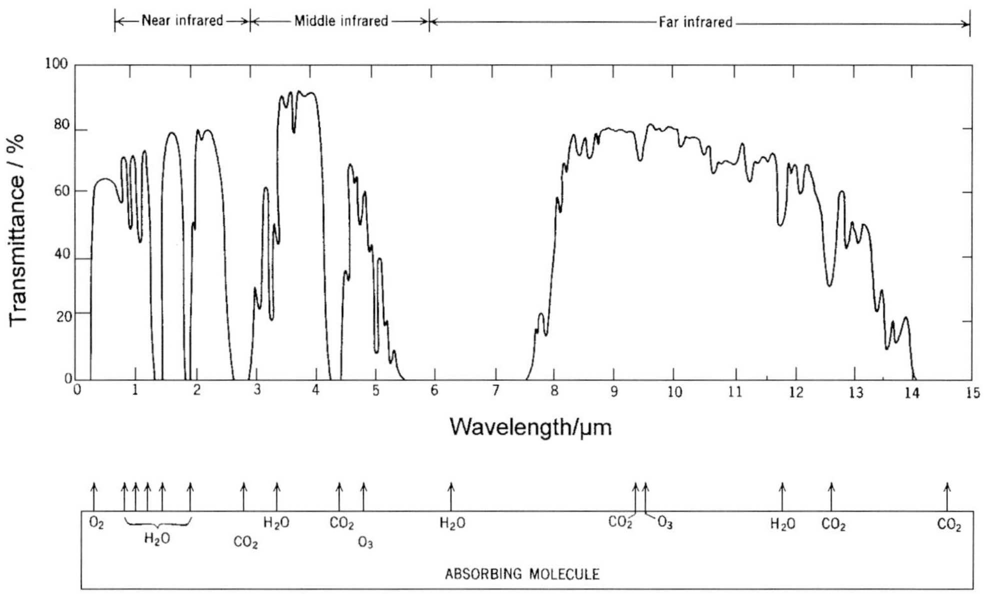

FIG. 18. Transmission of the atmosphere as a function of wavelength in the waveband from 0 to 15 μm. The parameters for the transmission conditions and the typical absorbing molecules are also shown.

330 °C) whereas the 7–12-μm waveband is ideal for the observation of blackbodies at ambient (*i.e.*, 273–310 K) temperatures.

3.1.3 Applications Since vision in the thermal infrared is based completely on the radiation emitted by the observed scenery it does not need any illumination. Therefore it is clear that thermal imaging is almost ideally suited for vision in complete darkness. The atmospheric transmission under fog and mist conditions is nearly always in favor of the MWIR and LWIR waveband as compared with the VIS waveband. The military were the first to detect and use the benefit of thermal imaging for their purposes, such as observation, detection, recognition, and identification and even driving and flying under bad weather conditions and by night. Nearly all tanks of the main western armies are equipped with thermal imagers (sometimes referred to as FLIR, Forward-Looking Infrared device). Thermal imagers are used by the fire brigade for looking through smoke and haze, *etc.* Even submarine periscopes are equipped with thermal imagers to fulfil their observation tasks.

The relatively high price of thermal imaging systems—which reflects the high complexity of these devices—has up to now prevented a broader spreading into civilian use. This situation may change if the technology of uncooled detectors ripens further (see Sec. 3.6).

3.2 Infrared Optics

As already mentioned in Sec. 1.2.1, there is a rather limited choice of infrared-optical materials available. Looking at the list of materials from well-established manufacturers one finds the materials sapphire, Ge, Si, GaAs, ZnSe, ZnS, CdTe, various IR glasses, and diamond. Some of these materials are suited for the 3–5 μm waveband and others are for the 7–12 μm waveband. A limited number of materials are suited for both of the above-mentioned wavebands 3–5 μm and 7–12 μm. These are Ge, GaAs, ZnSe, ZnS, and CdTe, as well as certain IR glasses. Figure 19 shows a typical transmission curve for one of the most often used IR materials, Ge.

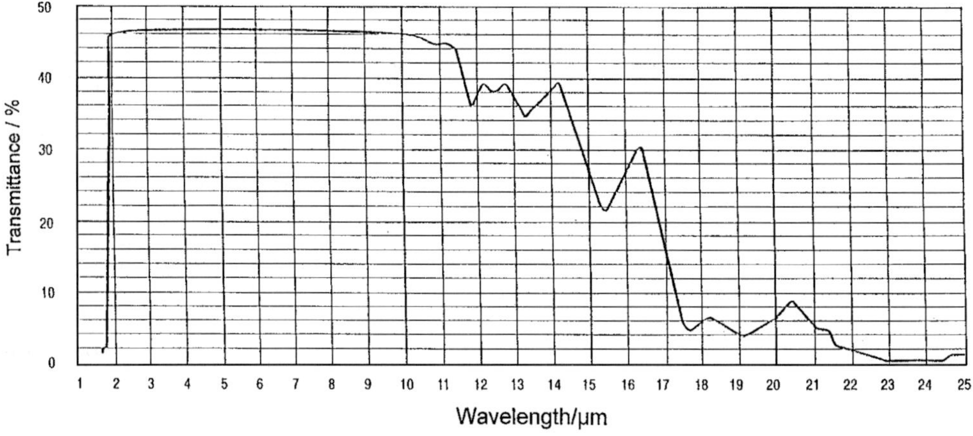

FIG. 19. Transmission of a 1-mm Ge plate as a function of wavelength between 1 and 25 µm. The curve shows clearly that the transmission of an uncoated plane-parallel Ge plate is only about 47%.

3.2.1 Antireflection Coating Most of the aforementioned optical materials for the IR region have rather high indices of refraction. An extreme example is Ge with a refractive index of 4. Beer's law (which is the special case for normal incidence of the more general Fresnel-formula system) describes the reflection characteristics of optical materials at normal incidence as a function of index of refraction for a single surface:

$$\delta = \left(\frac{n-1}{n+1}\right)^2$$

Applying the formula to the refractive index of Ge ($n = 4$) results in a 36% reflection (loss) for a single Ge surface. Taking into account that from the 64% of light energy entering a Ge plate another 36% will be reflected back and bounce back and forth between the two surfaces of a Ge plate—losing 36% at every bounce—one finds that a plane-parallel Ge plate transmits only 47% of the incoming energy. Taking into account that a realistic IR-optical system may be composed of six or more single lenses easily leads to the conclusion that some action has to be taken. Fortunately there is a means on hand, called "antireflection coating".

An antireflection coating is a thin film, ideally having a refractive index that is the square root of the refractive index of the substrate and has an optical path length of one quarter of the wavelength for which the substrate is to be antireflection-coated. This coating is a typical single-wavelength antireflection coating. Today very sophisticated multilayer coatings exist that can achieve broadband antireflection properties, very little remaining reflection, notch filter characteristics, *etc*. The know-how of companies doing IR coating is almost always treated as "company proprietary".

3.2.2 Infrared Optics Design Though the number of different optical materials suitable for use in the infrared is rather limited when environmental, thermal, and mechanical characteristics and cost are considered, the selection is adequate to design and fabricate the needed optical systems. Optical-system configurations in the infrared are often simpler than their visible counterparts, by virtue of the inherently higher refractive index of the materials and the inherently low dispersion of many IR materials (Fischer, 1991). An important factor in the evaluation of infrared-optical systems is the development of computer-based lens-design programs. As the memory capacity and computational speed of computers has increased drastically in the past, the capability of optical-design software has rapidly increased as well. This allows the design of sophisticated infrared-optical systems that are nearly diffraction limited (at least) on axis.

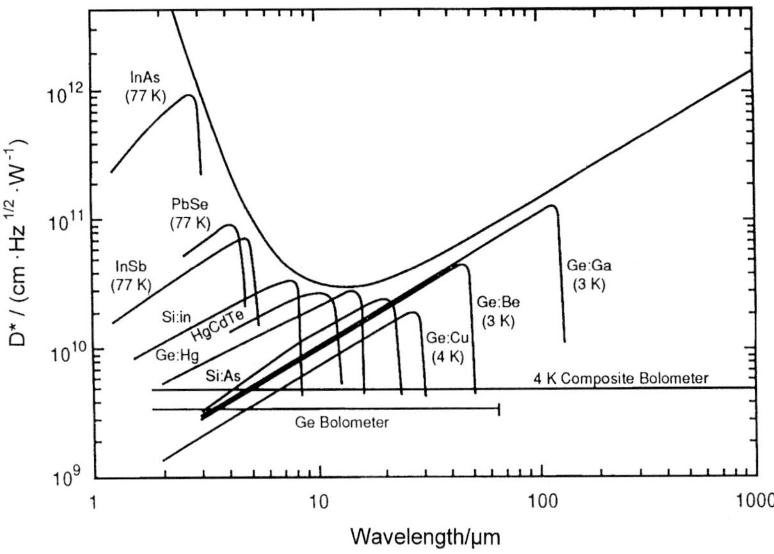

FIG. 20. Typical D^* curves for some selected IR-detector materials. The enveloping curve for the D^* peak values of the various detector curves represents the theoretical limit for background-limited detection at 300 K and 180° field of view, and a photovoltaic detector.

3.3 Infrared Detectors

The sensitivity range—in terms of wavelength—of detectors is based on the detector material. All semiconductor or quantum detector devices exhibit a so-called cutoff wavelength. This cutoff wavelength characterizes a very small waveband region where the sensitivity of the detector decreases from a maximum to nearly zero. For silicon that cutoff wavelength is at about 1 μm. InGaAs devices as described in Sec. 2 may have a cutoff wavelength dependent on the relative concentration of the constituents within 1.7–2.8 μm. There is one prominent material used for infrared detectors: cadmium mercury telluride (CMT). This ternary semiconductor material can be tuned to a wide range of cutoff wavelengths by the relationship of the cadmium to the mercury content. The tuning range for the cutoff wavelength extends from about 2 μm to nearly 40 μm. Of course other detector materials are available: PtSi, InSb, PbSe, GaAs for the 3–5-μm region; and PbSnTe, GaAs for the 7–12-μm region. Some of them, like PtSi, suffer from a very low quantum efficiency ($<1\%$), require very low cryogenic temperatures of less than 80 K (PtSi, InSb, GaAs) even though they are working in the 3–5 μm waveband, or are very limited in sensitivity (PbSe) or availability (PbSnTe). Various new exotic materials—for example, PbSnSe—have been discussed at conferences (see Zogg *et al.*, 1994) but have had no impact on reality up to now. For the foreseeable future CMT remains the material of choice for the 3–5-μm as well as for the 7–12-μm waveband.

A measure for the performance of detector materials is the detectivity D^*. The detectivity is the output signal-to-noise ratio of the input power to the detector intrinsic noise, normalized to detector area and electrical bandwidth. The units for D^* are cm $Hz^{1/2}$ W^{-1}. Figure 20 shows detectivity curves for some of the aforementioned detector materials (see RADIATION DETECTORS, INFRARED).

3.4 Cooling

It has already been mentioned in Sec. 3.3 that almost all of the detectors suitable for the IR wavebands 3–5 μm or 7–12 μm have to be cooled to cryogenic temperatures in order to achieve acceptable sensitivity. There are three different methods of cooling detectors down

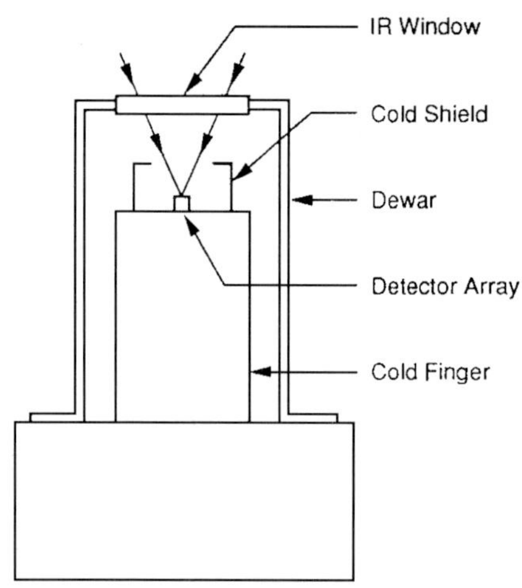

FIG. 21. Typical detector–dewar interface. The IR radiation reaches the detector via an IR window, which seals the vacuum of the dewar against the environment, and a cold shield. The detector array is mounted on top of the cold finger. The cold finger is in contact with the low-temperature end of an appropriate cooling system.

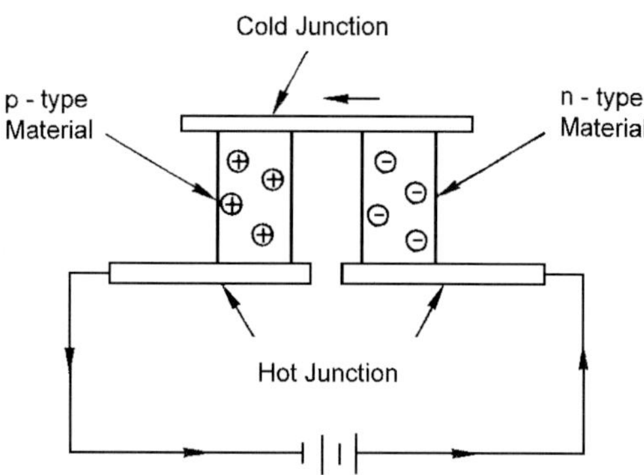

FIG. 22. Peltier thermoelectric couple. Two dissimilar materials are connected in series with a source of electric current. In such a case the cold junction will be cooled and the hot junction will be heated; it has to transport the thermal energy away.

to their working temperature: thermoelectric cooling, Joule–Thomson cooling, and Stirling-engine cooling. An interface problem exists between the IR detector to be cooled and the cooling system. The solution to this interface problem is called a *dewar*. The dewar contains the detector in a vacuum-flask type of vessel. The vacuum is necessary to avoid icing of the detector, which is cooled down far below the freezing point of water. A typical dewar is shown in Fig. 21.

3.4.1 Thermoelectric Cooling Thermoelectric coolers use the Peltier cooling effect. When two dissimilar metals are connected in series with a source of electrical current, one junction will be cooled while the other will be heated. The generation or absorption of heat depends on the direction of current, while the rate of heat pumping is a function of the current and the material properties. This basic phenomenon was first observed by Jean C. A. Peltier in 1834 (Egli, 1958). The outstanding features of thermoelectric coolers are simplicity and reliability. There are no moving parts. The primary limitation of thermoelectric coolers is the maximum temperature difference attainable: about 150 K based on available materials. This limits the use of thermoelectric coolers to systems working in the MWIR waveband. Figure 22 shows a typical single-stage thermoelectric cooler that consists of a p- and an n-type semiconductor connected to each other by a metallic conductor. Thermoelectric couples can be arranged in parallel and in series (cascaded) to pump more heat and to reach larger temperature differences. Manufactured units are generally available with up to four cascaded stages in which temperature differences of 120 K are achieved.

3.4.2 Joule–Thomson Cooling In open-cycle Joule–Thomson (J–T) cooling systems, high-pressure gas (200–400 kPa) combined with a J–T cooler expansion valve produces the necessary cooling (Fig. 23). The J–T effect involves the ratio of temperature change to pressure change of an actual gas in the process of expansion (during a constant-enthalpy process) without doing work. The change in internal energy during the expansion process results in cooling of the gas. The cooled, expanded gas is passed back over the incoming gas to cool it. This results in regenerative cooling. The process continues until liquid begins to form at the orifice to produce a bath of liquid at the boiling temperature of the gas. For nitrogen (N_2) this boiling temperature is at 77 K. Of course, high-pressure gas bottles become empty after some time. Considerable logistics are needed to supply a frequent user with filled high-pressure bottles.

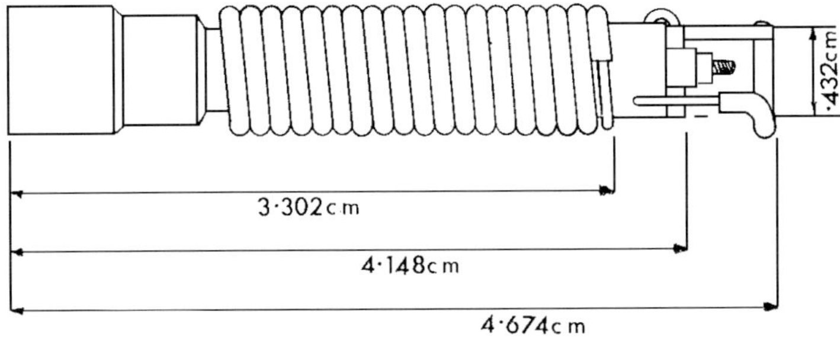

FIG. 23. Open-cycle Joule–Thomson cooler. The high-pressure gas enters the cooler from the left side. The orifice is on the right side, where the gas expands.

3.4.3 Stirling-engine Cooling Stirling cooling engines are mechanical refrigerator systems (see also the article STIRLING ENGINES). A mechanical refrigerator is a device that absorbs heat at a low temperature and rejects it at some higher temperature. To perform this task, expenditure of work is required. Advanced Stirling coolers consist of a compressor and a cold head. The two constituents can be separated (split) by a single (more or less) flexible line. The compressor is of a so-called linear-driven type, which comprises two pistons electromagnetically driven by coils in a push-pull manner. Figure 24 shows a typical linear-driven split Stirling-engine cooling system.

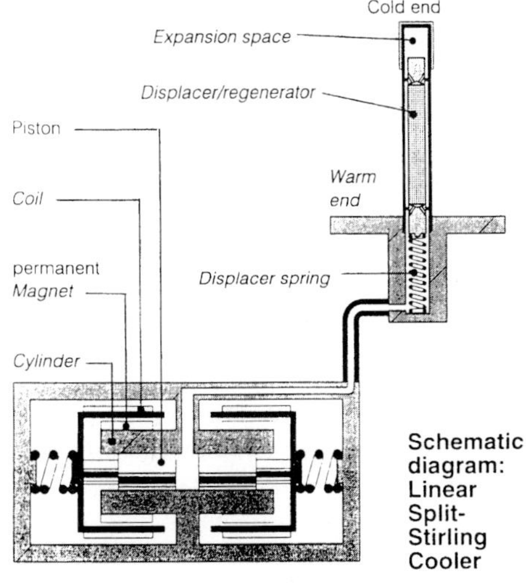

FIG. 24. Linear-driven split-Stirling cooler.

3.5 Infrared Systems

The design of a complete infrared-imaging system depends to a high degree on the availability of the necessary components. Ideally one would like to follow the method as for systems working in the visual spectral region: using a two-dimensional detector array that has an integrated readout circuit producing TV-type signals. Unfortunately some physical and technical constraints make it difficult to achieve that goal for IR detectors. This fact sometimes leads to the need for scanning. Scanning in this context means that the image of the scene is moved—by optomechanical means—over the detector/detector array in a timely consecutive manner (see also OPTICAL SCANNERS).

Nowadays it is possible to build detector arrays without the need for a scanning mechanism. Mainly these detectors work in the 3–5-μm waveband. For PtSi and InSb detectors in particular, very high degrees of integration density have been achieved. The relative benefits of scanned and staring systems are discussed in the following sections.

3.5.1 Scanning Systems Scanning is always necessary if the number of desired pixels in the displayed image is considerably larger than the number of available detector elements. If, for example, there is a detector available with about 1000 detector elements and an image is to be displayed with about 500 000 pixels, then scanning is the method to achieve that goal. Figure 25 shows some typical scanning schematics. Modern scanning

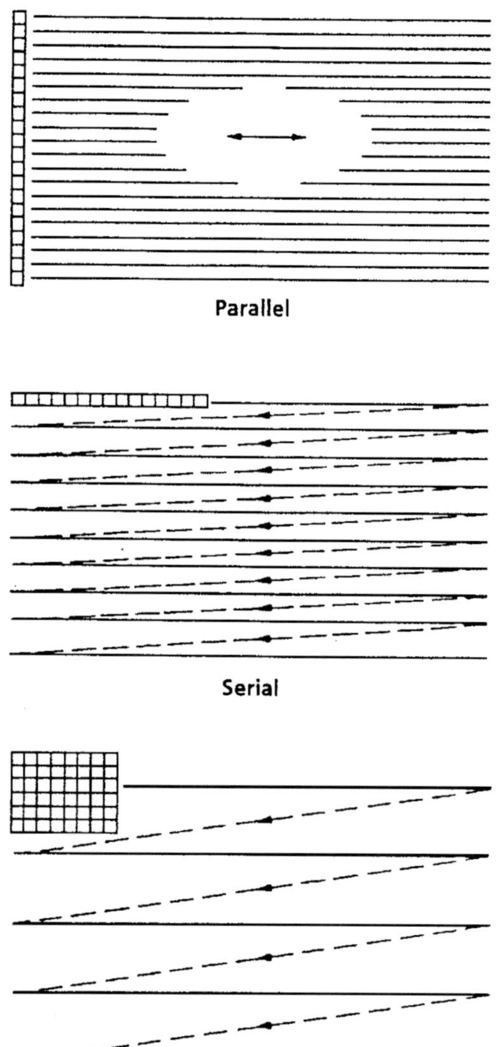

FIG. 25. Typical scan schematics for thermal imagers. The upper drawing shows a parallel scan scheme. The scene is scanned over a vertically oriented detector array in the horizontal direction for all detector elements in parallel. The central drawing depicts a serial scanning pattern. All scene spots are seen by each element of the horizontal oriented detector array in a sequential (serial) manner. The signals of the single detector elements are delayed and summed up synchronously. To achieve this the delay time constant and the scan velocity must be matched, of course. This procedure is called time delay and integration (TDI). The lower drawing shows a scan pattern that is a combination of the upper two.

thermal imagers use the so-called parallel scanning scheme almost exclusively. One typical realization of that principle is now discussed. It is assumed that there is a detector with 288 (240) single detector elements available; 288 (240) is the number of lines in a standard European (US) TV field. A whole TV frame has 576 (480) TV lines in two fields. Therefore an optomechanical mechanism is necessary to create a thermal TV image with 576 (480) lines and, for example, 768 (640) pixels in each line. The scanning mechanism must be able to do the main sweep in 1/50 (1/60) s and to perform a vertical interlace also in 1/50 (1/60) s. A typical thermal imager designed to achieve these requirements is depicted in Fig. 26. In the odd field the scanner scans the scene through the lines $1, 3, 5, \ldots, 575$, in the even field through lines $2, 4, 6, \ldots, 576$. Therefore the displayed image has 576×768 (480×640) TV pixels.

This method of operation is typical for thermal imaging systems working in the 7–12-μm waveband. On account of the high photon flux in that waveband (see Fig. 16) typical silicon circuits designed to process the detector signals are only able to integrate the charge from the CMT detector diodes for about 20–100 μs. In this respect the scanning principle does not necessarily degrade the performance of a thermal imager working in the 7–12-μm waveband.

3.5.2 Staring Systems The optimal realization of a thermal imager is a staring system working like a camcorder in the visible range. As already mentioned, TV systems with good resolution have a pixel count of 576×768 (European standard) or 480×640 (US standard). An IR-detector array with 576×768 (480×640) pixels seems to be necessary to generate such a TV-type signal. However there is a means at hand that can reduce the number of IR pixels by a factor of 2 or 4.

That means is called *Micro-Scan* (or *μ-scan*). In the case of the so-called 4:1 μ-scan the detector is illuminated for one fourth of the frame time, then the μ-scan mechanism moves the scene by a fraction of the detector pitch in a horizontal direction, then down in a vertical direction, back in a horizontal direction and finally up to the starting position (Fig. 27).

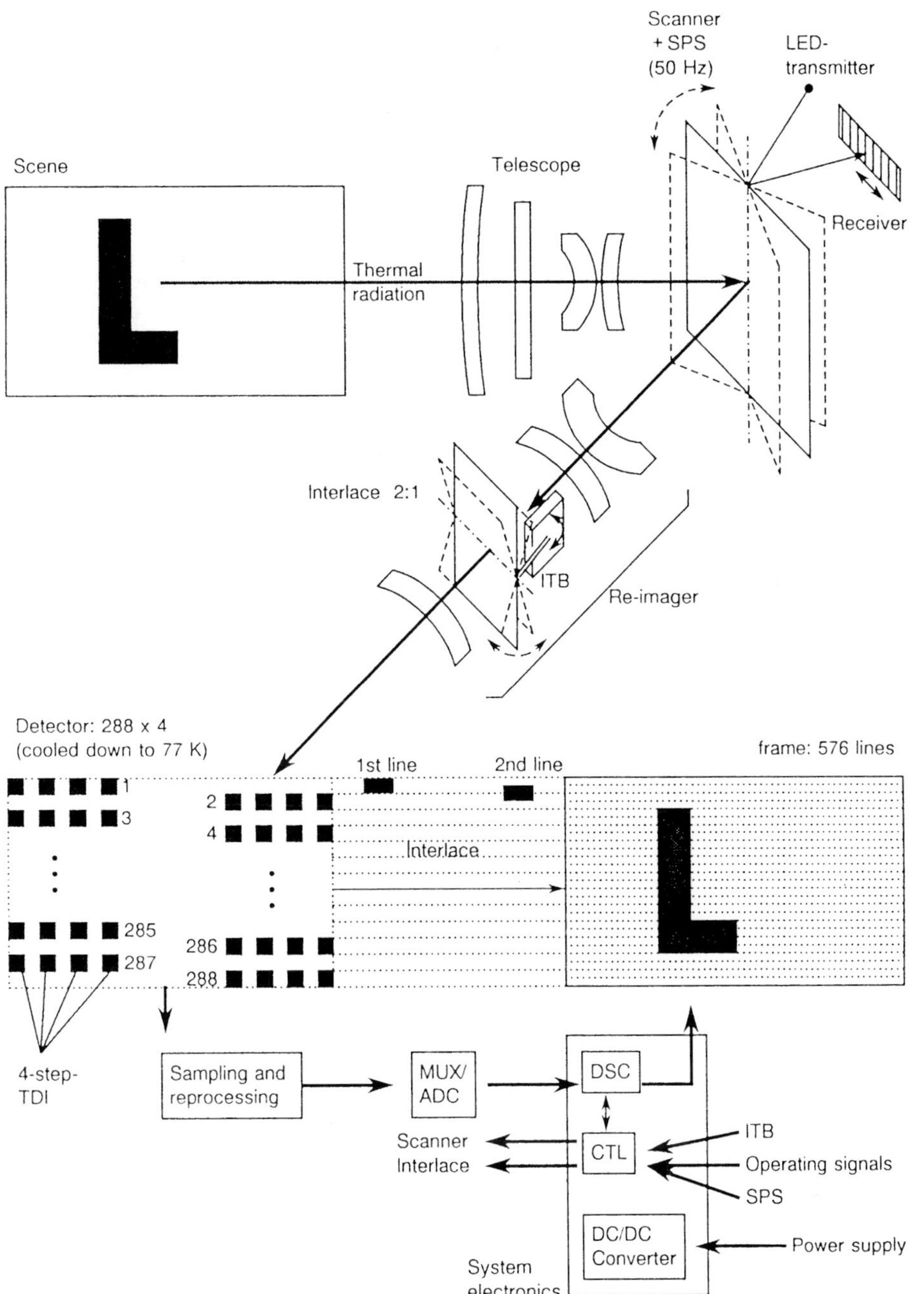

FIG. 26. The working principle of a typical parallel-scan thermal imager. The SPS (scan-position sensor) measures the actual position of the scan mirror. The ITB (interlace transducer board) determines the position of the interlace plate that creates the interlace between the even and odd fields of a TV signal.

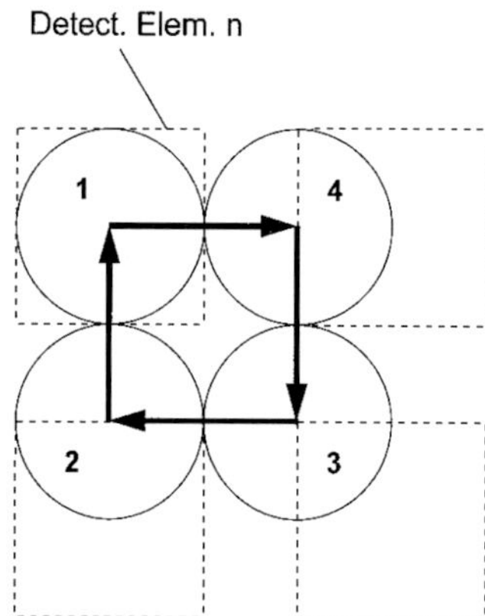

FIG. 27. Principle of a fourfold μ-scan. The dashed squares show the detector elements of a staring detector array. The circles represent the blur figures of four adjacent scene spots, which are scanned by the μ-scanner over detector element n in the sequence indicated by the arrows.

Such IR cameras are presently available mainly for the MWIR band with PtSi detectors with 480 × 640 pixels (no μ-scan necessary) and as prototypes with CMT detectors with 288 × 384 pixels (with or without μ-scan).

3.6 Uncooled Systems

All of the above-mentioned IR systems are based on the photoeffect. This effect means that an incoming photon separates charge carriers in the photosensitive material with a rather high degree of probability. The working principle of "uncooled systems" is based on a thermal effect. The incoming IR radiation heats up a detector element. The temperature change is used to detect a change of an electrical parameter to generate, finally, an electrical signal. As a consequence of the principle of heating up a detector element corresponding to the incoming radiation, it is always necessary to insulate thermally the individual elements of an uncooled-detector system from their neighbors as well as possible.

This thermal insulation is a severe problem for uncooled thermal detectors (the following is adapted from Hopper, 1993). The other challenge is to choose an effective way to detect the temperature change of the detector material. The possibilities are almost limitless. The question is not which parameters change with temperature, but which do not. Volume, pressure, reflectivity, polarization, resistance, and many other parameters have been used to make a variety of IR detectors since the discovery of electromagnetic energy beyond the visible spectrum by Sir William Herschel in 1800. Today two methods, each for solving the tasks of signal generation and insulation, are used. For signal generation in uncooled thermal imagers usually one of the effects, the bolometric effect or the pyroelectric effect, is applied.

The bolometric effect is a change in the electrical resistance of the responsive element due to temperature changes produced by absorbed incident radiation. The pyroelectric effect is based on a temperature-change–induced variation of the value of a capacitor built from thin pyroelectric material as a dielectric, with metallized opposing surfaces. Because only temperature changes produce a signal, pyroelectric detectors need a chopper to create imagery from nonmoving scenes.

For the purposes of thermal insulation the silicon-bridge and reticulation techniques are in use.

The silicon-bridge technique is characterized by very small ($< 50\,\mu m$) silicon bridges produced by anisotropic etching of Si, which is the carrier material for the thermal detector. These silicon bridges thermally isolate the detectors from the substrate to a certain degree and simultaneously provide the electrical contacts to the silicon readout circuit. Reticulation means the detector elements (pixels) are separated out of the detector material, which is deposited on a conducting substrate by means of laser radiation.

Historically the bolometric detectors come combined with silicon-bridge insulation; the pyroelectric detectors, with reticulation (Fig. 28). In the future it may happen that pyroelectric detectors—which because of their high impedance are better matched to the MOS readout circuits than bolometers—will be combined with the Si-bridge insulation technique, which provides better thermal insulation. Nowadays thermal cameras with staring uncooled detectors are available that have pixel numbers of up to 240 × 320. TV resolution is foreseen for the next generation of uncooled thermal imagers.

	Signal Generation	
	pyroelectric	Bolometer
Si-Bridges		**Honeywell**
Reticulation	**Raytheon TI**	

FIG. 28. Principles of signal generation and insulation for uncooled thermal detectors.

3.7 System Performance

Performance prediction models for thermal imaging systems are highly elaborate. Therefore this section deals with systems performance of thermal imaging systems in some detail.

There are three main performance parameters characterizing the efficiency of a thermal imaging system:

1. NETD (noise-equivalent temperature difference);
2. MRTD (minimum resolvable temperature difference); and
3. range.

While the parameter NETD is a single number, the parameter MRTD represents a function, *i.e.*, a set of temperature differences depending on spatial frequency (target distance). The range again is a single number for a well-defined observation task.

3.7.1 Noise-equivalent Temperature Difference, NETD There is a relatively simple definition of NETD (see Lloyd, 1975): NETD is the blackbody target-to-background temperature difference for a large test pattern that produces a peak-signal to rms-noise ratio (SNR) of 1 at the output of the system when the system views the test pattern. The theoretical derivation of NETD numbers is often cumbersome and sometimes produces erroneous results. The definition of NETD, however, allows the measurement of that number with good precision. Table 2 contains NETD numbers for typical thermal imagers of first, second and third generations. The table clearly shows that the NETD is typically divided by a factor of 2 for each successive generation of thermal imagers.

3.7.2 Minimum Resolvable Temperature Difference, MRTD It has been found that four factors heavily influence subjective image quality: sharpness, graininess, contrast rendition, and interference by artifacts. It is desirable to have a summary measure that includes all four elements of image quality, that is, one which is a unified system–observer performance criterion. This function is now generally called the minimum resolvable temperature difference (MRTD). The definition of the MRTD uses some assumptions about the perception capabilities of the observer's eye–brain system. For example, it is assumed that the observer is equipped with an integration time of 0.2 s. Another assumption is that a signal-to-noise ratio of 2.25 is an adequate threshold for the human eye–brain system for a 50% perception capability. In summary, the MRTD function describes the target tempera-

Table 2. Typical NETD numbers for three generations of thermal imagers.

Generation of thermal imager	Number of detector elements	Wavelength/μm	NETD/mK
First	120	7.5–11.5	<200
Second	288×4	7.5–10.5	<100
Third	288×384	3–5	<50

ture difference necessary to obtain a specific observation task. The MRTD is dependent on target size and target distance. A combined measure for target size at a certain distance is the so-called spatial frequency. It is analogous to the familiar temporal or electrical frequency. While the temporal frequencies have the unit Hertz (Hz), spatial frequencies have the unit of cycles per radian (or milliradian). The meaning of spatial frequency should be clear from Fig. 29, which represents a one-dimensional sinusoidally varying light source with a spatial period of T_x. Obviously a target of constant size represents an increasing spatial frequency if it is moved to a growing distance from the observer. On the other hand the ability of an electro-optical system to distinguish between faint object structures is also dependent on spatial frequency. It is a matter of fact that all optical and electro-optical systems suffer from poor response to high (spatial) frequencies. As a consequence thermal imagers require a higher target temperature difference for the observation of more distant targets.

3.7.3 Range Performance Range is the predominant factor governing the performance of an electro-optical system as judged by the end user (especially the military one). Therefore the method of predicting range performance is briefly described. A starting point is the definition of a standardized target. That target is the so-called NATO standard target as defined in "STANAG 4347". It has a size of 2.3 m $\times$ 2.3 m with a thermal pattern of 3 line pairs having a temperature difference of 2 K each. Different observation requirements are again governed by the Johnson criterion (Johnson, 1966), as already mentioned in Sec. 2.3 and Table 1.

It is assumed that the target temperature difference of 2 K is apparently lowered by atmospheric extinction on the way to the observer. The extinction coefficient is dependent on atmospheric conditions. For typical summer situations the extinction coefficient is $\leq 0.2\,\text{km}^{-1}$ for 50% of the weather conditions in midlatitude countries (USA, Europe, *etc.*). The resulting effective temperature difference reaching the thermal imaging system due to target distance and atmospheric extinction is the so-called "effective temperature difference" ETD.

By definition the intersection of the ETD curve with the MRTD curve marks the range for a given target and observation criterion (see Fig. 30). This means that the "range" is achieved if the effective temperature difference ETD at the location of the thermal imaging system equals the minimum resolvable temperature difference MRTD for that spatial frequency. The spatial frequency can easily be converted into a range for specified target size and observation criterion. Figure 30 shows a typical situation where the range is defined by the intersection of the ETD with the MRTD functions.

GLOSSARY

CMT: A preferred detector material for infrared applications. CMT stands for cadmium mercury telluride. By changing the relative content of cadmium to mercury in the ternary compound the cutoff wavelength of detectors can be tuned between approximately 2 and 40 µm. Today CMT is the material of choice for high-performance IR detectors.

Cryogenic: Refers to cooling of a physical device to very low temperatures, in the vicinity of the boiling temperature of nitrogen (*i.e.*, 77 K) or even helium (*i.e.*, 4 K). Typical IR detectors with high sensitivity often need to be cooled to nitrogen temperature ($\approx$ 80 K).

InSb: Indium antimonide, a material suitable for IR detectors working in the 3–5-µm waveband. It offers good sensitivity and homogeneity but has to be cooled down to about 80 K for good performance.

IR: Abbreviation for infrared. The electromagnetic spectrum with wavelengths beyond the long-end boundary of the visible spectrum (*i.e.*, at wavelengths $>$700 nm) is called the IR region. It extends to about 100 µm. The wavelength region between 700 nm and 2.0 µm is called the near and short-wave infrared (NIR and SWIR), the region from 3 to 5 µm the medium-wave infrared (MWIR), the band between 7 and 12 µm the long-wave infrared (LWIR).

LOWTRAN: The name of a computer code, with the most recent version LOW-

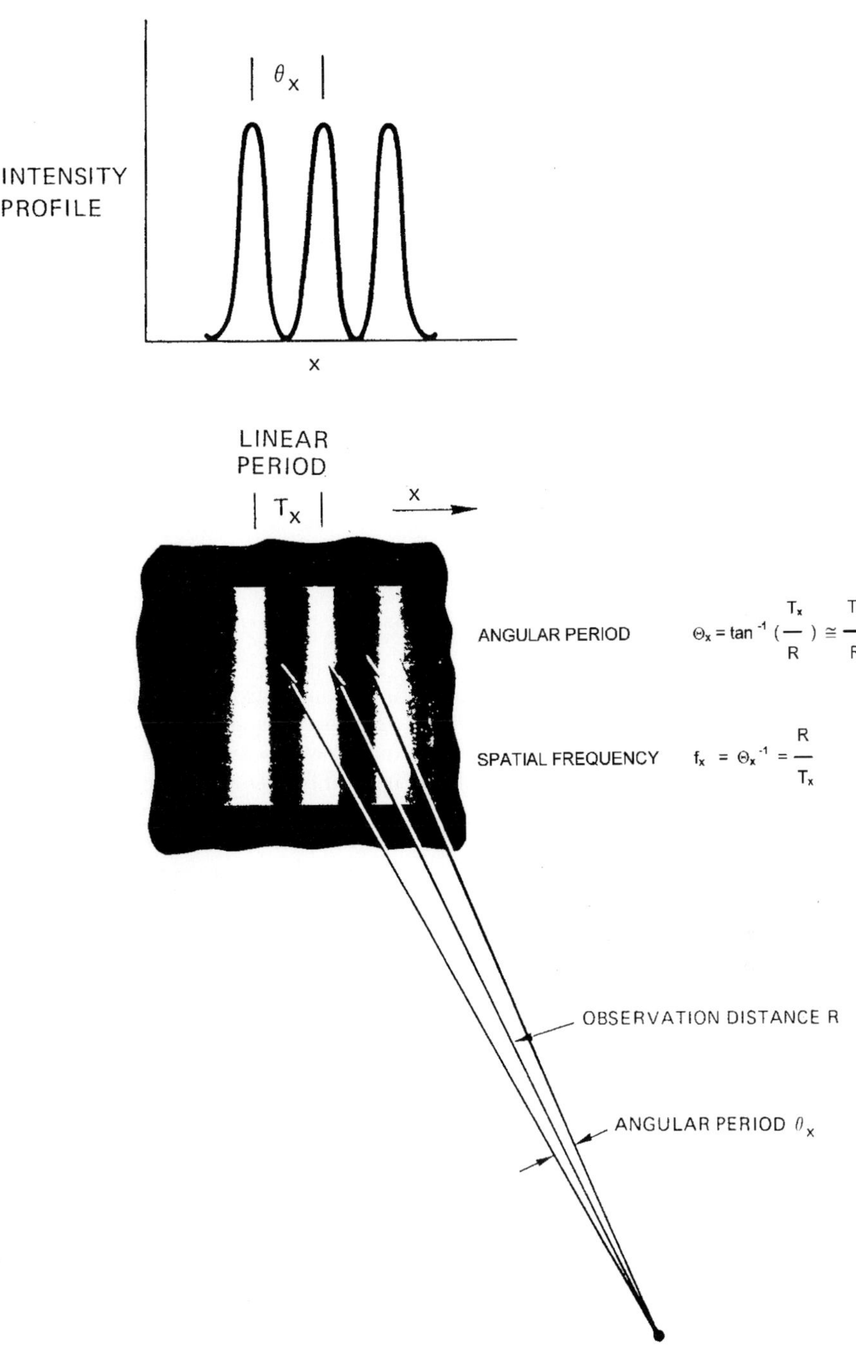

FIG. 29. Meaning of spatial frequency.

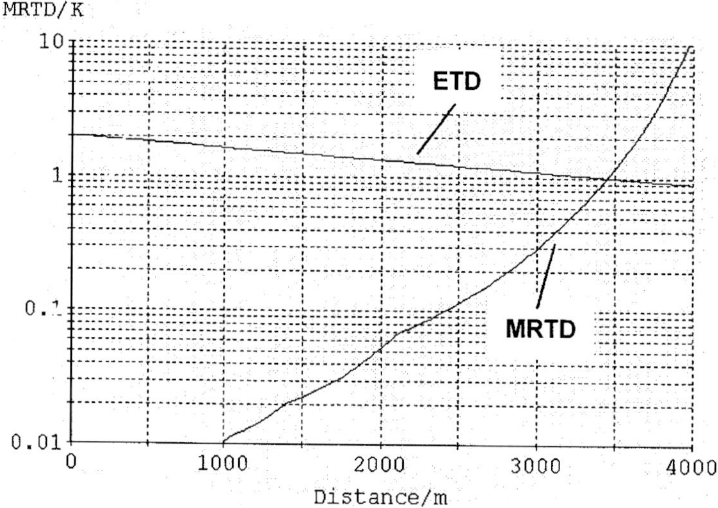

FIG. 30. "Range performance". The graph shows the intersection of the ETD with the MRTD curve. By definition the intersection point at about 3450 m defines the range performance of that thermal imager.

TRAN 7. LOWTRAN calculates the transmittance and/or radiance for a specified path through the atmosphere. It has been validated against field measurements and is widely used for many broadband system performance studies. It is supported by the U.S. Air Force Philips Laboratory.

Lumogen: A trade mark of the BASF company, denoting an organic material that converts UV radiation into visible by fluorescence.

LWIR: Wavelength region between about 7 and 12 μm.

MWIR: Wavelength region between about 3 and 5 μm.

NIR: Wavelength region between about 0.7 and 1.0 μm.

Pixel: An abbreviation for picture element. It characterizes a single sensitive detector element of a detector array.

PtSi: Platinum silicide, a detector material that allows building of detector arrays with large pixel numbers (10^5–10^6) sensitive in the MWIR region and with limited thermal sensitivity. PtSi detectors have to be cooled to about 80 K for good performance.

Scanner: An optomechanical means to cover a relatively large field of view with a limited number of detector pixels. The scanner sweeps the image of the observed scene in a time-consecutive manner over the available detector elements. Few detector elements require rapid scanning; many detector elements need only slow scanning.

Solar Blind: Term characterizing the UV waveband between 200 and 300 nm. In that waveband no solar radiation reaches the Earth because of ozone absorption in the stratosphere.

STANAG: Abbreviation for Standardisation Agreement. It describes a set of NATO-wide accepted standards for different technical tasks. STANAG 4347 and 4350 describe the calculation and measurement of range performance for thermal imagers.

SWIR: Wavelength region between about 1 and 2 μm.

Uncooled: Description of a class of thermal imagers where the detector needs no cryogenic cooling. That class of IR detectors is based on the heating of a detector element due to the incoming radiation. The temperature change of the detector element results in an electric signal.

UV: Ultraviolet region of the electromagnetic spectrum. It extends from about 300 nm to lower wavelengths as low as about < 100 nm.

VIS: The so-called visual spectral region from about 400 to 700 nm.

Works Cited

Cohen, M., Olsen, G. (1994), *Pho. Spectra* **28** (7), 132–134.

Dirscherl, R. (1996), *Proc. NATO-IRIS* **41** (1), 221–230.

Egli, P. H. (1958), *Thermoelectricity*, New York: John Wiley and Sons.

Fischer, R. E. (1991), in: R. Hartmann, W. J. Smith (Eds.), *Infrared Optical Design and Fabrication*, Critical Reviews of Optical Science and Technology No. 36, Bellingham, WA: SPIE Optical Engineering Press, pp. 19–43.

Hacker, F., Lauth, H., Schlichting, J. (1996), *Proc. NATO-IRIS* **41** (1), 263–276.

Hopper, G. S. (1993), in: S. B. Campana (Ed.), *Passive Electro-Optical Systems*, Vol. **5** of J. S. Accetta, D. L. Shumaker (Eds.), *The Infrared and Eectro-Optical Systems Handbook*, Bellingham, WA: SPIE Optical Engineering Press.

Johnson, J. (1966), in: L. Biberman (Ed.), *Proceedings of the Seminar on Direct-Viewing Electro-optical Aids to Night Vision*, Institute for Defense Analyses Study S254, Washington: U.S. GPO.

Lloyd, J. M. (1975), *Thermal Imaging Systems*, New York: Plenum Press.

Nordwall, B. D. (1997), *Aviation Week Space Technol.* **147** (6), 81–84.

Suits, G. H. (1978) in: W. W. Wolfe, G. J. Zissis (Eds.), *The Infrared Handbook*, Ann Arbor, MI: Environmental Research Institute of Michigan (ERIM).

Zogg *et al.* (1994), *Wiss. Z. Techn. Univ. Dresden* **43**, 8–12.

Further Reading

Acetta, J. S., Shumaker, D. L. (Eds.) (1993), *The Infrared and Electro-Optical Systems Handbook*, Bellingham, WA: SPIE Optical Engineering Press.

Hartmann, R., Smith, W. J. (Eds.) (1991), *Infrared Optical Design and Fabrication*, Critical Reviews of Optical Science and Technology No. 38, Bellingham, WA: SPIE Optical Engineering Press.

Hudson, R. D. (1969), *Infrared System Engineering*, New York: John Wiley & Sons.

N. N. (1992), *Electro-Optics Handbook (BURLE)*, Lancaster, PA: Burle Industries, Inc., Tube Products Division.

Wolfe, W. W., Zissis, G. J. (Eds.) (1978), *The Infrared Handbook*, Ann Arbor, MI: Environmental Research Institute of Michigan (ERIM).

PHOTOGRAPHY: PHYSICS AND TECHNOLOGY

Norbert Schuster and Jürgen Mangelsdorf, *Rollei Fototechnic, Braunschweig, Germany*

INTRODUCTION

"A picture is worth a thousand words." Man's dream of a quick picture was realized by Daguerre in 1838, when he found that mercury vapor condensed preferentially on exposed areas of silver iodide deposited on a copper plate. At this time, the projection of object scenes by lenses and mirrors was well known. In 1840 Petzval created the first high-aperture portrait lens. In 1867, the first lenses with a wide field of view were introduced by Steinheil.

The development of classical photographic material began about 1875 (Mutter, 1962). It was a new target to improve photographic lenses: the first Planar (1892) by Rudolph at Zeiss in Jena, the Taylor & Hobson Triplett (1895), and the first Tessar (1907) also by Rudolph. Pictures for everyone became a reality. Photographers with great bellows cameras opened their offices and recorded marriages and other important events on photosensitive plates.

In the "golden twenties," the photographic camera became portable. Roll and flat films reduced camera weight. New viewfinder systems, like the Rollei twin-lens from 1929 in the $6 \times 6\text{-cm}^2$ format and the single-lens reflex system from Exakta Dresden in 1930, guaranteed an accurate framing of the object. The first antireflection coating (Zeiss Jena, 1935) improved the transmission and the image quality of lenses.

After World War II, the development was directed toward the mass market: the photographic camera for everybody. Color slide films became affordable. Amateurs demanded different focal lengths for their cameras. The first multilayer antireflection coating (Minolta, 1956) and the Voigtländer Zoomar 2.8/36–82 (1958) (see Sec. 1 for notation) traced the further development of lenses.

In 1960, Asahi Pentax presented the first

ISBN 3-527-29308-6

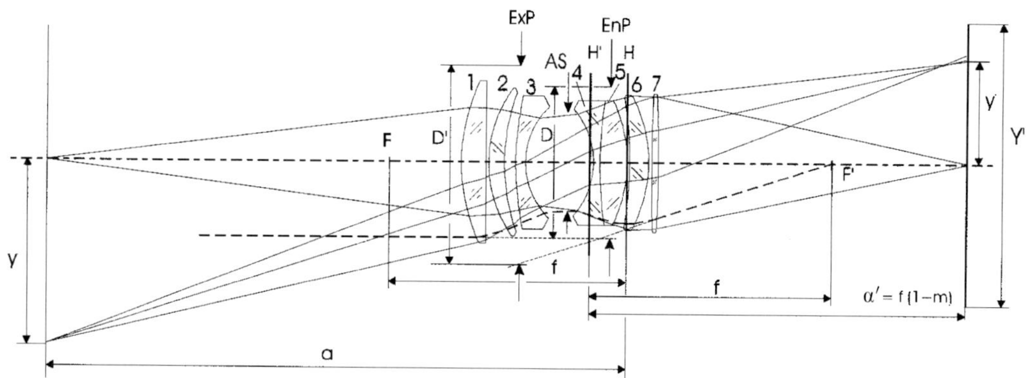

FIG. 1. Ray tracing through the standard lens 1.4/50 with the following paraxial variables: 1–7, glass lenses; AS, aperture stop; *a*, object distance with object size *y*; *a′*, image distance with image size *y′*; EnP, entrance pupil with diameter *D*; ExP exit pupil with diameter *D′*; F, *F′* foci; *f* focal length; H, H′ principal planes; *Y′* detector size.

through-the-lens (TTL) exposure measurement. In the smallest photographic camera with a film size of $24 \times 36\,\text{mm}^2$, the Rollei 35 (1966), optomechanical features still dominated. However, the implementation of electronic assemblies increased through the years. The first electronically controlled shutter (Pentax, 1971) and the automatic limited flash time by measurement in the camera (Olympus, 1973) avoided misexposures. In lens development, Canon presented in 1971 the first commercial aspherical photographic lens, 1.2/55, and in 1975 the first manual focusing 3× zoom lens.

The focusing problem was also solved by modern electronics. In 1981, Pentax launched the first AutoFocus (AF) camera, and many companies offered large sets of interchangeable AF lenses for the 24 × 36 film format. Technical achievements created a new type of compact camera: equipped with fully automatic exposure control, AutoFocus, flash, and a zoom lens.

The market of 50×10^9 consumer images worldwide per year in the 1990s, based on the silver halide technique, will be changed. The new solid state detectors offer new orientations: digital storage and picture transfer, direct print on normal paper without chemical procedure, smaller cameras.

1. FORMATION OF THE PHOTOGRAPHIC IMAGE

The photographic image is generated by the accumulation of the luminous flux, emerging from the object scene, onto the photographic detector. This accumulation must be realized with a certain amount of spatial resolution. The laws to describe this spatial resolution are the topic of this section.

1.1 Standard Photographic Lenses

The image formation by photographic lenses will be described by the paraxial approximation, which defines for each lens the foci F and F' and the principal planes H and H'. An object situated a very long distance in front of the lens is projected in the focal plane F'. The image of an object situated in the focal plane F is projected a very long image distance behind the lens. An object situated in the principal plane H is projected in the principal plane H' with a constant size. The focal length of photographic lenses is $f = \overline{H'F'} = \overline{FH}$. Figure 1 demonstrates the basic paraxial relationships on a standard lens.

The image distance and the object distance are related as follows:

$$1/a' + 1/a = 1/f. \tag{1}$$

Their proportion is the lateral magnification m, which defines the ratio of the imaged object scene:

$$m = a'/a = y'/y. \tag{2}$$

The photographic image requires a positive

Table 1. Size of typical still-camera detector formats with standard focal lengths.

Detector name	Size $X' \times Y'$ (mm^2)	Approximate standard focal length (mm)
Olympus 1 K × 0.8 K	4.8 × 3.6	7
Sony 1.3 K × 1 K	8.6 × 6.9	13
Minox film	10.5 × 7.5	14
Half 135 film	17 × 13	25
APS-film C/H/P	16.7 × 25.1/16.7 × 30.2/10.1 × 30.2	37
Kodak 3 K × 2 K	28 × 18	37
LORAL 2 K × 2 K	30.7 × 30.7	50
135 film	36 × 24	50
Medium-format film 6 × 6	56.0 × 56.0	80
DICOmed 4 K × 4 K	61.5 × 61.5	80
60 × 90 planfilm	56.0 × 82.6	100
90 × 120 planfilm	114 × 82.6	140
5″ × 7″ planfilm	121 × 170	220
180 × 245 planfilm	171 × 231	260
8″ × 10″ planfilm	194 × 245	300

focal length f and a magnification $m \geq 0$. The limit $m = 0$ is a very practical case: the object distance a is very large in relation to the focal length f, and the image plane approaches the focus F'.

The field of view of a photographic camera is determined by the size of the photographic detector (length X' is perpendicular to the drawing plane in Fig. 1 and Y' is vertical in the drawing plane) and the lateral magnification m. The field definition as an angle is also applicable for the infinite object distance:

$$2\omega = 2\arctan\frac{\sqrt{X'^2 + Y'^2}}{2f(1+m)}. \tag{3}$$

Commonly, the field of view of a photographic lens is given for the infinite object position $a \rightarrow \infty$ or $m = 0$. This field of view becomes comparable to the field of view of the human eye because the normal human eye forms an image of an object situated at infinity without accommodation effort. Its field of view is $2\omega =$ 45–50° (see Fig. 28) and serves as a measure for standard photographic lenses. Corresponding to Eq. (3), the standard focal length depends on the picture format. Table 1 lists some still-camera detector formats and their corresponding standard focal lengths. The digital detector examples are explained with the pixel number in each direction. Note that for professional large-format cameras the typical object distance is not at infinity and that the orientation of the pictures can be changed. In accordance with Table 1 the focal length corresponds to the detector size.

All detectors are sensitive to the image brightness, which is determined by the cone of light rays entering one point of the image plane. In Fig. 1 the on-axis ray bundle is limited by the aperture stop, situated between several lenses. The top view of the lens shows an image of the aperture stop, called *entrance pupil.* A view from the back side offers another stop image, the so-called exit pupil. Their different diameters D and D' and their different locations are caused by different stop projections: in Fig. 1, in the front by the lenses 1, 2, 3 and in the back by the lenses 4, 5, 6, 7. The ratio of pupil diameters is called lateral pupil magnification:

$$m_{\mathrm{p}} = D'/D. \tag{4}$$

This value in the lens example in Fig. 1 is greater than 1.

The measure for the ray bundle size is the "f-number" k, defined by

$$k = f/D. \tag{5}$$

This ratio of focal length to entrance-pupil diameter characterizes the size of the on-axis ray bundle for an object at infinity, which is represented in Fig. 1 by a dashed line. This ray

Table 2. Typical focal lengths for the most frequently used film formats.

Lens type	Field angle interval	Typical focal length (mm)	
		24×36-mm^2 film	6×6-cm^2 film
Fish-eye	$2\omega \geq 180°$	<16	30
Superwide-angle	$114° > 2\omega > 80°$	13–25	40
Wide-angle	$80° > 2\omega > 55°$	28–35	45–65
Standard	$55° > 2\omega > 40°$	50, 55	80, 100
Long focal lenses	$40° > 2\omega > 20°$	85, 100	120–180
Telephoto	$20° > 2\omega > 8°$	135–300	250–600
Supertelephoto	$2\omega < 8°$	400–1200	1000

bundle defines the maximum brightness in the image plane. A small k value indicates a great relative aperture and a great brightness.

The f-number and the focal length are the most important basic values of a photographic lens. Two presentations are usual. The lens example in Fig. 1 can be represented as 50 mm 1 : 1.4 (focal length $f \approx 50$ mm and the relative aperture $1:k$) or as 1.4/50 (f-number or stop number $k = 1.4$ and focal length $f \approx 50$ mm).

1.2 Field of View

Equation (3) describes the field of view as a function of the detector size and the focal length. With the object position at infinity ($m = 0$) the field angle defines the type of photographic lens. Photographic cameras with film formats of 24×36 mm^2 and 6×6 cm^2 are equipped with a wide range of changing lenses. For these most frequently used film formats the field angles and the corresponding focal lengths are listed in Table 2. Note that Eq. (3) is not applicable to fish-eye lenses.

Figure 2 demonstrates the changing field angle for different focal lengths. The increasing focal length decreases the registered field of view and increases the size of the object details in the picture.

The perspective is the representation of the depth of field in the picture. The change of focal length permits the perspective to be shaped by the photographer. This effect is

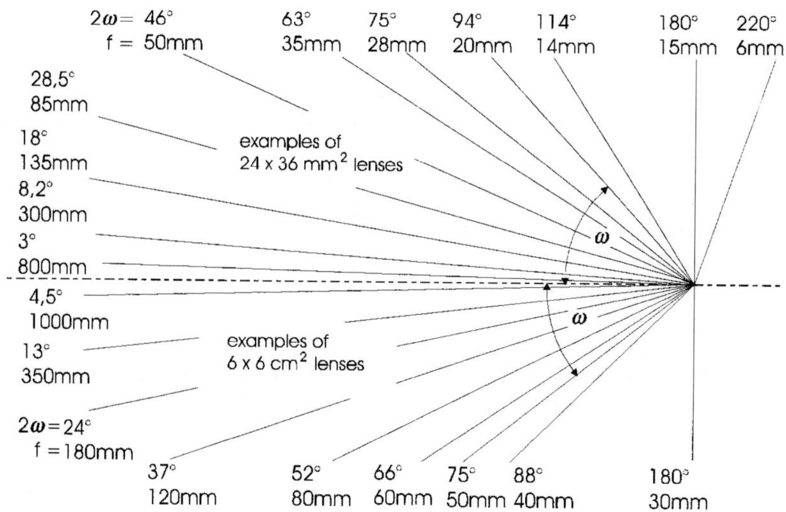

FIG. 2. The half-field angle of different lenses for the object position at infinity. The top number indicates the field of view, the bottom number the focal length.

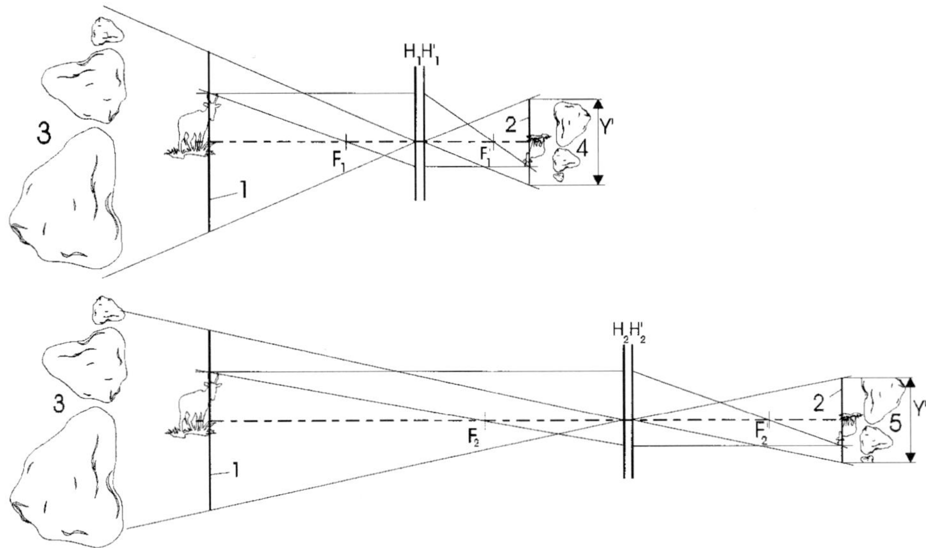

FIG. 3. Changing perspective by different focal lengths at a constant magnification of $m = 0.5$ of the main object: 1, sharply imaged object plane; 2, image plane with the format $X'Y'$; 3, object background; 4, background imaged by short focal length; and 5, background imaged by long focal length.

coupled with the change of the field of view in the picture. A clear change of perspective is arranged if the main object detail is projected with a constant magnification by lenses having different focal lengths. Figure 3 demonstrates the principle. In accordance with Eqs. (1) and (2) this condition implies that the photographer must change the object distance. The short–focal-length lens images a greater section of the background scene than the long–focal-length lens.

1.3 Spatial Resolution of Lenses

The paraxial relationships are the first approximation to the photographic image, giving a definitive relationship between the object point and the image point, and *vice versa*. In practice, the photographic lens transforms the object point into an image having a certain circle of confusion. This confusion arises from the light diffraction on the stop and lens edges (physical limit of lens resolution), from the lens aberrations, and from stray light. The compensation of aberrations defines the sophisticated lens arrangement and the price of a photographic lens.

A general representation of the lens resolution is the *modulation transfer function* (MTF) (Goodmann, 1968). According to the linear filter theory, a photographic lens maps the light distribution from the object plane into the light distribution in the image plane by the process of convolution. The theoretical basis of the MTF has been presented by Gaskill (1978).

Figure 4 explains the nature of the MTF. A sinusoidal light distribution in the object plane is projected by the photographic lens in the image plane. The light distribution in the image plane is less modulated because diffraction, aberrations, and stray light diffuse the incoming light. This effect is quantified by the contrast in the object and image plane:

$$C = \frac{L_{\max} - L_{\min}}{L_{\max} + L_{\min}} \quad \text{and} \quad C' = \frac{E_{\max} - E_{\min}}{E_{\max} + E_{\min}}.$$

It is also evident that the loss of contrast C' increases for shorter periods p'. In accordance with electronics the period is replaced by the spatial frequency $R' = 1/p'$. Finally, the MTF is the ratio between the image contrast and the object contrast for different spatial frequencies:

$$M_0(R' = R'_1) = [C'/C]_{R'=R'_1}. \tag{6}$$

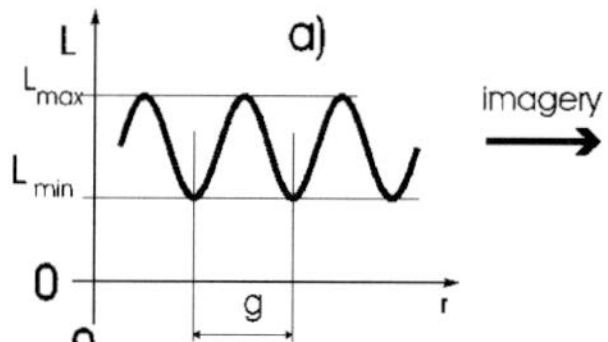

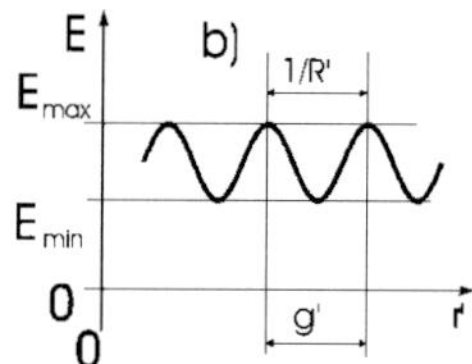

FIG. 4. The nature of the MTF. (a) Light distribution in the object plane with the period g and the distance r perpendicular to the optical axis, represented by the luminance L. (b) Light distribution in the image plane with the period g' and the distance $r' = mr$ perpendicular to the optical axis, represented by the illuminance E.

For a frequency $R' \to 0$ that corresponds to a very long period, the loss of contrast is negligible: $M_0(R' = 0) = 1$. On the other hand, over a certain short period p', diffraction, aberrations, and stray light diffuse the incoming light so that $E_{\max} = E_{\min}$: the MTF becomes $M_0(R' \geq R'_{\max}) = 0$. The behavior of the function $M_0(R')$ in the interval $0 < R' < R'_{\max}$ depends on the photographic lens. When the diffraction on the aperture stop dominates the circle of confusion, a fixed MTF relation exists. The best MTF is

$$M_{0\,\mathrm{lim}} = \frac{2}{\pi}\left(\arccos X - X\sqrt{1 - X^2}\right) \quad \text{with}$$

$$X = \lambda k R'(1 + m/m_p). \tag{7}$$

The maximum frequency limited only by diffraction is found at $M_{0\,\mathrm{lim}} = 0$ or correspondingly at $X = 1$:

$$R'_{\mathrm{lim}} = \frac{1}{\lambda k(1 + m/m_p)}. \tag{8}$$

The MTF is a measure for the circle of confusion in the photographic image. In practice, for large apertures the MTF is much lower then the diffraction-limited case (Fig. 5).

The circle of confusion changes with the image height r'. For off-axis image points, the figure of confusion loses the rotational symmetry. However, it remains symmetrical relative to the drawing plane in Fig. 1, which includes the optical axis. This plane is called the meridional or tangential plane. All rays in Fig. 1 that pass the center of the aperture stop are called chief rays. Finally, the plane that is perpendicular to the drawing plane and strikes the chief ray is called sagittal or radial plane. Consequently, the image quality in the meridional plane differs from that in the sagittal plane. Figure 6 demonstrates this effect by M_0 functions diverging with increasing image height.

The MTF representation in Fig. 6 demonstrates the change in the confusion figure over the image plane. The decreasing MTF with increasing frequencies corresponds with the preceding figure. Typically, the image quality decreases on the edges. Furthermore, Fig. 6 demonstrates a better MTF for a smaller aperture. This trend is true down to the critical stop where the aberrations and the diffraction form the smallest figure of confusion. In our example, the critical stop is at the f-number $k = 5.6$. For smaller apertures with greater f-numbers, the diffraction dominates and decreases the MTF.

The aberration theory distinguishes between several types of aberrations (Welford,

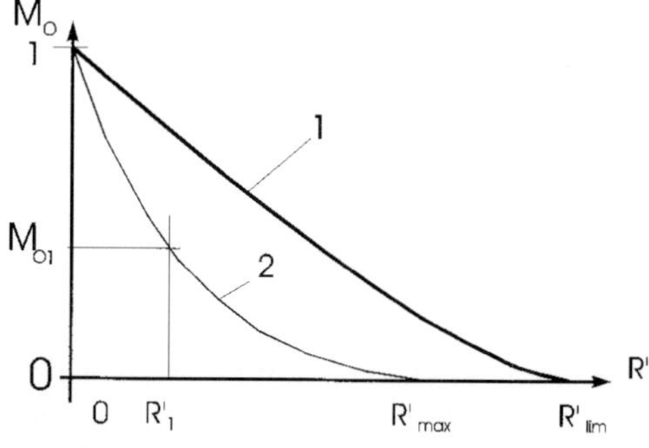

FIG. 5. MTF of the lens 1.4/50 from Fig. 1 at f-number $k = 3.4$ and the object distance infinity. Curve 1, diffraction limit with $R'_{\mathrm{lim}} = 500\,\mathrm{mm}^{-1}$; curve 2, real MTF diminished by aberrations at the on-axis object point, $R'_{\max} = 300\,\mathrm{mm}^{-1}$; $M_{01} = 0.35$ at frequency $R'_1 = 80\,\mathrm{mm}^{-1}$.

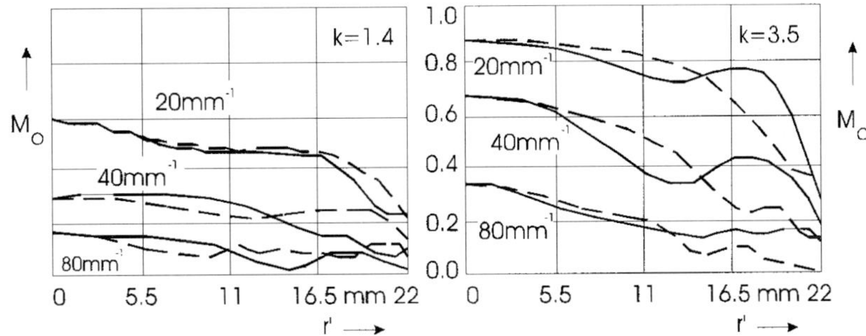

FIG. 6. MTF of the lens 1.4/50 from Fig. 1 at different f-numbers *versus* the image height r', parameter R': —— meridional, - - - - sagittal.

1974). The user of photographic lenses must distinguish between the circle of confusion, well described by the MTF, and the *distortion*, which is the displacement of the center of the confusion figure relative to the paraxial image height $r' = mr$ (Haferkorn and Richter, 1984). The distortion turns a quadratic grid into a distorted figure (see Fig. 7).

The compensation of different aberrations is a special task of the optical design (Malacara and Malacara, 1994; Laikin, 1995). On the basis of the mathematical model the optical designer uses an archive of lenses (Cox, 1964). New lens materials provide new possibilities of lens arrangement with improved image quality. The compensation of lens aberration can be arranged by variation of mechanical parameters such as lens radii, lens thicknesses, and lens distances. For this purpose, the lens aberrations at different field points are summarized in an evaluation function that must be minimized. The mathematical problem is the optimization of many variables under certain constraints. Modern optical-design computer programs use both global and local optimization procedures. The success of the lens optimization strongly depends on the starting point.

1.4 Imaging Space

The imaging space is the depth of space in which the photographic imagery is realized with a certain resolution. The realized resolution power in the image plane is determined by the lens and the detector and limits the depth of focus on the rear side of the lens. The corresponding space on the front side of the lens, called the depth of field, is gained by the application of the paraxial approximation of the imagery.

1.4.1 Interaction between Lens and Detector

The interaction between lens and detector is well described by the linear filter theory: the

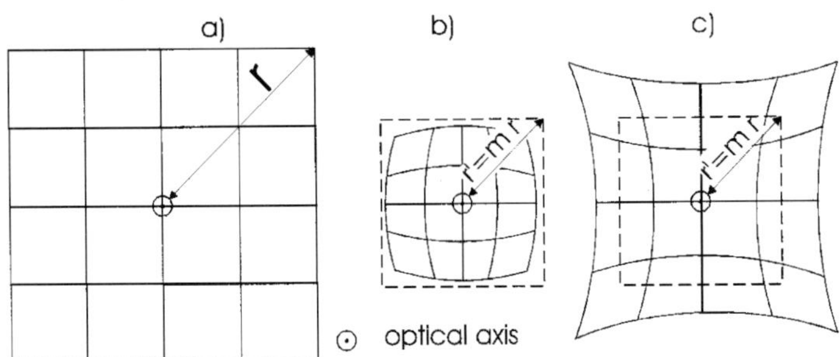

FIG. 7. Effect of distortion on a quadratic object grid. (a) Grid in the object plane with the object height r. (b) Image in the presence of barrel distortion. (c) Image in the presence of pincushion distortion.

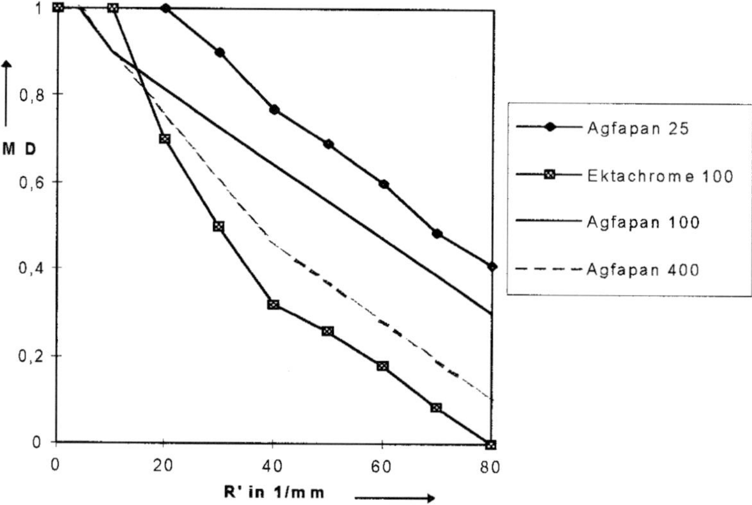

FIG. 8. MTF of classical photographic films with ASA index of sensitivity.

detector scans the light distribution in the image plane by the process of convolution. In this model, the resulting MTF after imagery by the lens and after light reception by the detector is

$$M_{sc}(R') = M_0(R')M_D(R') \tag{9}$$

with $M_D(R')$ the MTF of the detector.

The MTF of classical photographic film is determined by measurement. For this purpose, a sinusoidal grid with varying frequencies is projected onto the film. After development of the film, the blackness distribution is measured by a microdensitometer (Bumba and Pospisil, 1969). Applying the contrast definition and Eq. (6), the measured MTF of the film is found. Some results are represented in Fig. 8. The MTFs of the films start with a constant modulation $M_D \approx 1$ and continue up to a break frequency. Afterwards the MTFs decrease almost linearly. Typically, the resolution of the black-and-white Agfa films decreases with increasing sensitivity. Mostly, the resolution of color diapositive film is poorer than that of a black-and-white film with the same sensitivity.

The resolution power of digital detectors is determined by the active pixel area, which convolutes the projected light distribution. The MTF of a matrix with quadratic pixels having a size x_D yields after the Fourier transform

$$M_D(R') = \frac{\sin(\pi x_D R')}{\pi x_D R'}. \tag{10}$$

The MTF of digital detectors decreases continuously to zero (see Fig. 9). In practice, frequencies $R' > 1/x_D$ are ineffective by the reduced lens resolution. A supplementary problem of digital imaging arises at frequencies above the Nyquist frequency $R'_N = 1/2p'$ with the pixel pitch p': aliasing effects by superposition of the regular object structure with the regular matrix structure. The random distribution of silver halide crystals in classical photographic emulsion avoids this problem.

The optimized tuning of lens resolution and matrix resolution depends on the object scene: point objects are recognized from aliasing-free, out-of-focus images by subpixel interpolation; classical photographic scenes permit a certain aliasing effect (Jahn and Scheele, 1997).

A very often used parameter of resolution power is the diameter of the circle of confusion D_r. It results in the interaction of lens and detector. Based on the linear filter theory, the following relationship was found (Volosov, 1971):

$$D_r \approx \frac{2}{\pi R'_1}\sqrt{2.5[1 - M_{sc}(R'_1)]} \quad \text{valid at}$$

$$M_{sc}(R'_l) > 0.5. \tag{11}$$

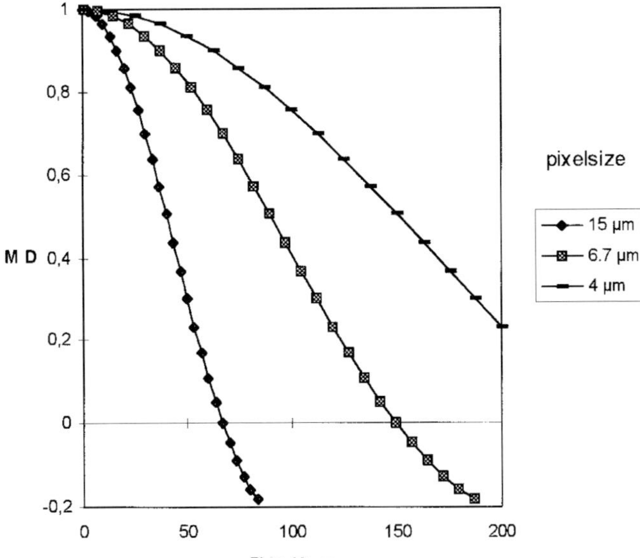

FIG. 9. MTF of digital matrix detectors with different active pixel sizes x_D.

Equation (11) describes the circle of confusion as a function of several variables: lens field and lens aperture corresponding to Fig. 7, and detector resolution corresponding to Fig. 8 or Fig. 9.

For projecting purposes it is necessary to define an admissible circle of confusion. Franke (1964) has defined it for classical photographic films: starting from the viewing distance for classical photographs and the resolution of the human eye, the circle of confusion depends on the detector format. These values are listed in Table 3.

For digital detectors we find a comparable value by application of Eq. (11) when a real lens MTF is supposed. The diffraction-limited MTF, defined by Eq. (7), at small apertures ($k \geq 11$) fixes a practical lens resolution. The resulting circles of confusion are listed in Table 3. With the increasing detector size the circle of confusion D_r also increases. The calculated values for digital detectors correspond with the results for classical film. The new capability of digital photography demonstrates the DICOmed-matrix in Table 3. Realizing a detector size of the 60 × 60 format, the circle of confusion is the same as the circle of confusion of the classical 24 × 36 film.

1.4.2 Depth of Focus and Depth of Field

The circle of confusion is a reasonable param-

Table 3. Circle of confusion of different photographic detectors.

Detector name	Pixel pitch/size	D_r (µm)	Conditions
Olympus 1 K × 0.8 K	4.8 µm	12	Eq. (11) at $k = 11, R'_1 = 40\,\text{mm}^{-1}$
Sony 1.3 K × 1 K	6.7 µm	15	Eq. (11) at $k = 11, R'_1 = 40\,\text{mm}^{-1}$
Minox film	10.5 × 7.5 mm^2	15	Franke, 1964
Half 135 film	17 × 13 mm^2	25	Franke, 1964
Kodak 3 K × 2 K	9 µm	20	Eq. (11) at $k = 11, R'_1 = 25\,\text{mm}^{-1}$
LORAL 2 K × 2 K	15 µm	28	Eq. (11) at $k = 16, R'_1 = 20\,\text{mm}^{-1}$
135 film	36 × 24 mm^2	33	Franke, 1964
Medium-format film 6 × 6	56 × 56 mm^2	60	Franke, 1964
DICOmed 4 K × 4 K	15 µm	33	Eq. (11) at $k = 16, R'_1 = 15\,\text{mm}^{-1}$
60 × 90 planfilm	56 × 83 mm^2	75	Franke, 1964
90 × 120 planfilm	114 × 83 mm^2	100	Franke, 1964

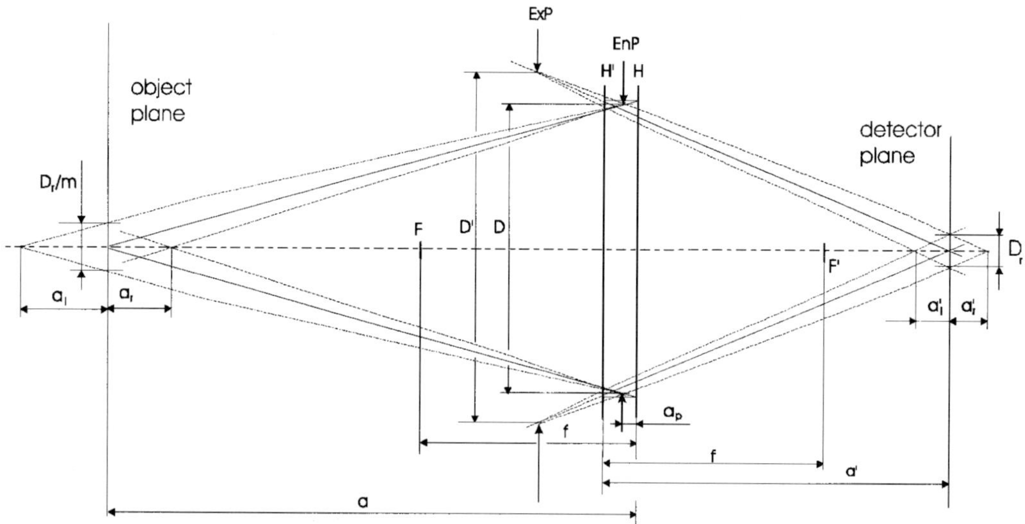

FIG. 10. Paraxial scheme of the lens 1.4/50 from Fig. 1: a, object distance; a', image distance; $a_l + a_r$ depth of field; $a'_l + a'_r$ depth of focus; D_r, circle of confusion; EnP, entrance pupil with the diameter D at distance a_p; ExP, exit pupil with the diameter D'; F, F', foci; f, focal length; and H, H', principal planes.

eter to define the depth of focus. In Fig. 10 the origin of the depth of focus is demonstrated at an on-axis point. The "sharp" paraxial image is realized by the axial-ray bundle sized by the entrance-pupil diameter or the exit-pupil diameter. When we accept a circle of confusion in the image plane, the dashed-line ray bundles have a confusion figure with diameter D_r in the image plane. All on-axis ray bundles with intersection points from $a' - a'_l$ to $a' + a'_r$ realize a confusion diameter smaller than D_r in the detector plane, when the pupil diameter remains constant. The dashed-line ray bundles characterize the depth of focus on the rear side of the lens and the depth of field on the front side.

The corresponding circle of confusion in the object plane D_r/m [m = magnification in accordance with Eq. (2)] fixes the relationships

$$a_r = \frac{D_r(a - a_p)}{Dm + D_r} \quad \text{and} \quad a_l = \frac{D_r(a - a_p)}{Dm - D_r}. \tag{12}$$

The depth of field depends on the object distance, the pupil diameters and the admissible circle of confusion D_r.

The object distance at $a_l \to \infty$ has a great practical significance. Objects situated beyond this distance realize a circle of confusion smaller than D_r in the image plane. If we set $Dm - D_r = 0$, the hyperfocal distance is found as

$$a_\infty = f(1 + D/D_r). \tag{13}$$

At this distance, the right-hand side depth of field is $a_\infty/2$; this means that the total depth of field becomes $a_\infty/2 \leq a < \infty$. Figure 11 demonstrates the hyperfocal distance for different lenses. Generally, the depth of field increases with the f-number k and decreases with the focal length. In traditional photographic lenses, the depth of field with varying f-number is indicated on the focusing distance scale.

1.4.3 Tilted and Shifted Image Formation

In the previous paragraphs it was supposed that the optical axis passes through the centers of the image plane and the object plane. In the paraxial approximation, this arrangement guarantees a proportional and sharp imagery. Beyond that, there are two more general arrangements that guarantee a sharp paraxial image.

In the first, the optical axis stands orthogonal to the object and image planes, and the detector plane is shifted laterally by the dis-

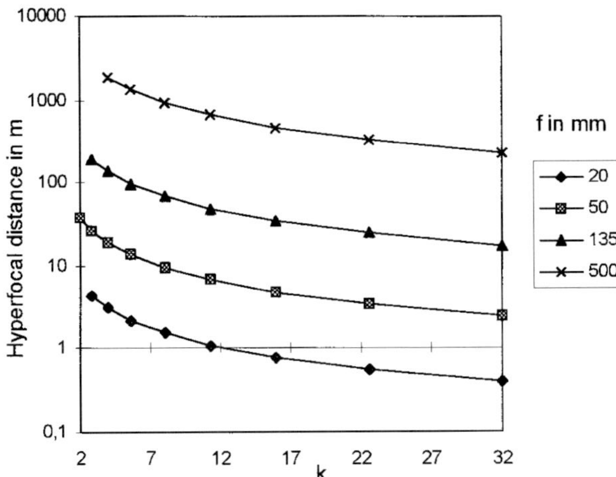

FIG. 11. Hyperfocal distance a_∞ for wide-angle, standard, telephoto, and super-telephoto lenses in the 24 × 36 film format ($D_r = 33\,\mu\text{m}$) *versus* the f-number.

tance s [see Fig. 12(a)]. Thus, the photographer avoids converging lines in the sharp paraxial image. The lens needs an enlarged field angle $2\omega_1 > 2\omega_2$ because it is used asymmetrically.

Without the shift motion, the photographer must tilt the camera to capture the whole object [Fig. 12(b)]. The object distance changes with the object height. In accordance with Eq. (2), the magnification becomes variable versus the image height and produces converging lines. The image appears "sharp" only within the borders of the depth of focus.

Secondly, a sharp paraxial image is guaranteed by the Scheimpflug condition (1898):

$$m = \frac{\tan\sigma}{\tan\sigma'}. \tag{14}$$

The angles σ, σ' mark the inclination of the object and image plane to the optical axis. The imagery can be strongly distorted (see Fig. 13). Ray 3 respects the principle of the paraxial image construction and fixes the geometrical arrangement of the Scheimpflug condition. The sharp imagery of the tilted object plane is realized when the intersection line of the object plane and detector plane coincides with the principal planes of the lens. In practice, the interval between the two principal planes i of

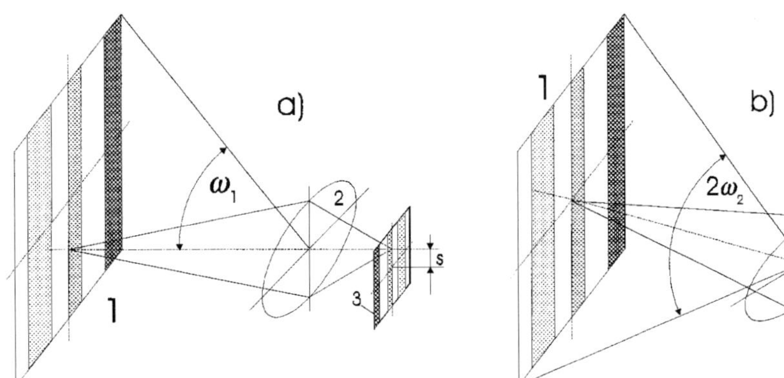

FIG. 12. Avoidance of converging lines shifting the image plane: 1, object plane; 2, lens with optical axis; and 3, detector size. (a) Proportional sharp image by means of orthogonality between the object, its image and the optical axis. (b) Converging lines in a typical camera with an optical axis fixed in the center of the image plane.

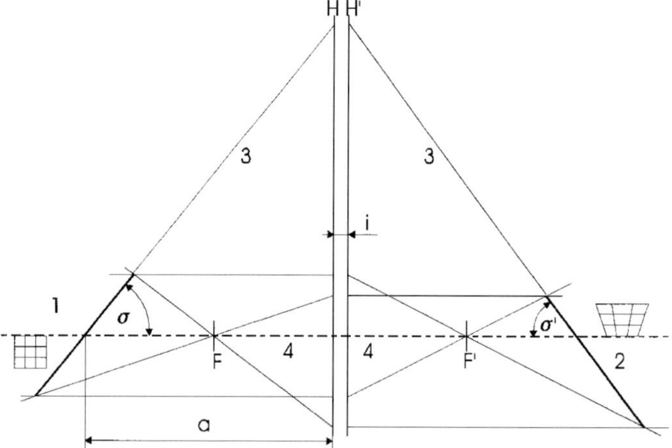

FIG. 13. Paraxial image formation and Scheimpflug condition: 1, object plane with quadratic grid perpendicular to the drawing plane; 2, image plane with strongly distorted grid picture; 3, paraxial ray; and 4, optical axis.

photographic lenses is small in comparison with the object distance a.

Around the detector plane a changing depth of field provides a new imaging space with a variable perspective. Professional photography uses this enlarged depth of field by shifting and tilting the image plane relative to the optical axis.

1.5 Examples of Photographic Lenses

The range of photographic lenses available is manifold. The interchangeable lenses of single-lens reflex (SLR) cameras provide the possibility of changing the field of view within a wide range. The fixed-detector format requires a changing focal length in accordance with Table 2.

Lenses with a constant focal length attain the highest image quality. Figure 14 represents some examples from the fish-eye to the supertelephoto lens including traced rays to the on-axis point and to the extreme field points.

The special requirement of the SLR lenses is the distance between the last lens surface and the detector plane, called back focal length. This spacing is reserved to the hinged mirror (see Sec. 4). The lens arrangements of different lenses vary strongly. The lens scale in Fig. 14 is the same except for the 5.6/1000. The standard lens 1.8/50 resembles the lens in Fig. 1. Typically, a nearly symmetrical lens arrangement around the aperture stop permits a good image quality at high relative apertures.

The specialty of the superwide-angle 2.8/20 is the small effective focal length f in relation to the back focal length. This result requires a divergent front lens group and a convergent lens group behind the aperture stop. These types of lenses get a corrected distortion over the full detector format.

Fish-eye lenses produce an unnatural picture that can be circularly bordered and heavily distorted. The ray tracing in example 3.5/14 demonstrates the reasons. The bundle with the 90° incident field angle does not reach the detector edge, and the projection of the full half space onto the detector plane produces a strong barrel distortion.

The telelens 3.5/200 demonstrates the inverse spacing problem: the effective focal length f is much longer than the back focal length. A divergent back lens group reduces the construction length of the complete lens in relation to the focal length. Nevertheless, refracting supertelephoto lenses are long and weighty.

Short construction lengths at long focal distances become possible by use of reflecting surfaces. Although the example of the 5.6/1000 is represented in a reduced scale, the focal length is not marked. Based on the Cassegrain reflecting telescope, the catadioptric photo-

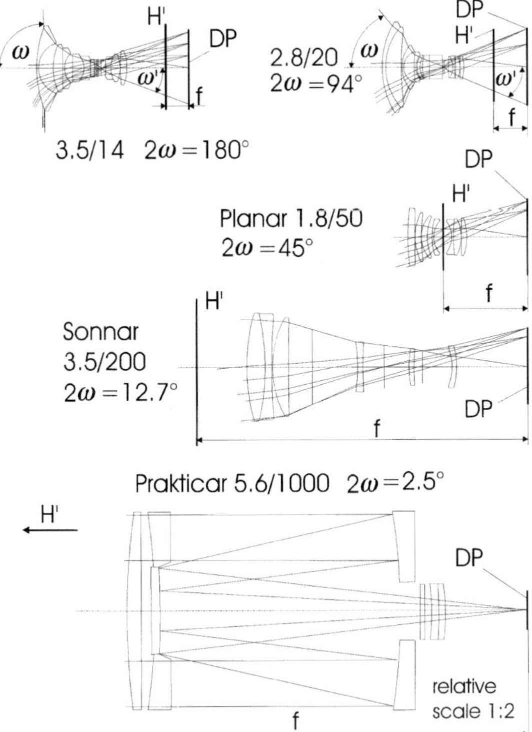

FIG. 14. Examples of interchangeable SLR-camera lenses in the format 24 × 36 with constant focal length and infinite object distance. DP, detector plane, *f*, focal length, H′, image principal plane.

graphic supertelephoto lenses use a large collecting mirror with a hole and a diverging secondary mirror that reflects the light into the detector plane. These spherical mirrors yield the main refractive power without color dispersion of light. The supplementary lens groups improve the image quality.

Zoom lenses offer the fascinating feature of photography with a continual change of the field of view. The focal length is changed by the displacement of any lens group having a constant focal length. During the zoom movement, the image plane is fixed at the photographic detector by means of a coupled nonlinear displacement of the lens groups. This principle is called mechanical compensation (Mann, 1993). Zoom lenses with a constant maximal relative aperture are realized by a four-group lens design (see Fig. 15). The front group focuses at different object distances. The second group, called the variator, changes the focal lens by a linear axial movement. The third group, called the compensator, is mechanically coupled with the variator. It equalizes the displacement of the image plane by means of a nonlinear axial movement. The back lens group is the basic focusing lens and includes the aperture stop. This arrangement guarantees a constant axial-ray bundle cone to the on-axis image point at all focal lengths. Consequently, the f-number is not changed. Other zoom-lens arrangements omit the constant f-number, increase the focal length, and realize smaller lens diameters. Table 4 lists some examples of modern zoom lenses including their particular correcting elements.

The surface shape of aspherical lenses (AL) diverges from the classical sphere. This supplemental degree of freedom improves the spatial resolution of lenses. The production of aspherical glass lenses, however, is very expensive. The floating element (FE) principle (Kämmerer, 1979) avoids increasing aberrations of photographic lenses at shorter object distances by changing an internal vertex distance of the lenses during the focusing displacement. Internal focusing (IF) reduces the focusing movement to a displacement of only an internal group of lenses. The front group of the lens is fixed and the center of gravity is not

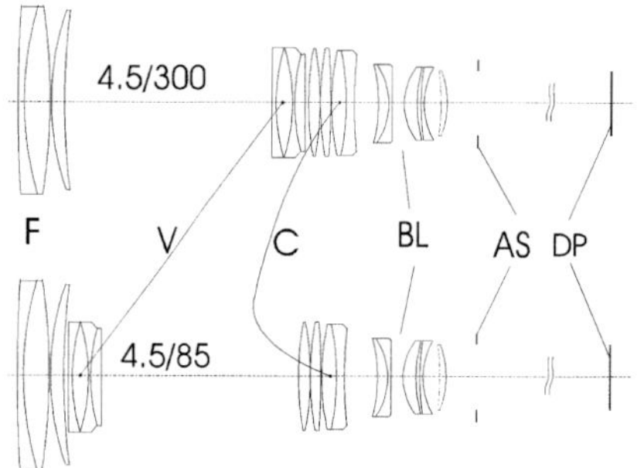

FIG. 15. Zoom-lens design for 24 × 36 SLR cameras with constant f-number: AS, aperture stop; BL, basic lens; C, compensator; DP, detector plane; F, focusing group; and V, variator.

moved. Lenses with low or extreme dispersion (LD) are used as converging lenses. In combination with traditional diverging lenses, LD lenses yield a better suppression of the color light dispersion.

2. ENERGY BALANCE IN THE PHOTOGRAPHIC IMAGE

The incoming luminous flux onto the image plane is accumulated by the detector. The adaptation of detector capacity to the object-plane brightness is the topic of this section.

2.1 Illuminance in the Image Plane

The light and radiation transfer from the object plane to the detector is described by analogous formulae. All photometric quantities respect the variation of photopic human eye response with the wavelength, denoted by the normalized function $V(\lambda)$ (see Fig. 16). Outside of the wavelength range $380\,\mathrm{nm} \leq \lambda \leq 780\,\mathrm{nm}$ all photometric values are zero. The adaptation of detector capacity to the object plane brightness needs radiometric relations.

The illuminance or irradiance near the optical axis is

$$E_0 = \frac{L'\pi\Omega_0}{4k^2(1 + m/m_{\mathrm{p}})^2}. \tag{15}$$

The quadratic dependence on k explains the fundamental role of the f-number in photography. The magnifications m and m_{p} present the influence of the image formation. The quantities describing the radiation transfer are listed in Table 5.

Table 4. Examples of modern AF zoom lenses for 24 × 36-cm^2 cameras with special correcting elements: AL, aspherical lenses; LD, low or extreme dispersion lenses; FE, floating element; IF, internal focusing.

Zoom type	Lens examples	Lenses/ groups	Construction length (mm)	Weight (g)	Special correcting elements
Wide range	Sigma 3.8–5.6/28–200	14/13	77.5	450	2 AL
	Tamron 3.8–5.6/28–200	16/14	82	465	2 AL, 2 LD, IF
	Canon 4–5.6/35–350	21/15	167	1385	2 LD, IF
Wide angle	Canon 2.8/17–35	15/10	96	545	2 AL, IF
	Pentax 3.5–4.5/17–28	9/7	61	260	
	Tokina 2.8/20–35	15/11	86		1 AL, IF
	Tamron 2.7–3.5/20–40	15/12	81	525	2 AL, IF
Standard	Tamron 3.3–5.6/24–70	8/7	60	270	2 AL
	Canon 3.5–4.5/24–85	15/12	70	380	1 AL, IF
	Minolta 3.5–4.5/24–85	14/12	73	415	2 AL, IF
	Sigma 2.8–4/28–70	11/8	63.5	285	FE
	Canon 2.5–4.5/28–105	15/12	75	370	IF
	Nikkor 3.5–4.5/35–135	15/12	109	685	IF
Telephoto	Canon 2.8/70–200	18/15	194	1310	4 LD, IF
	Tokina 2.8/80–200	17/11	184	1350	1 LD, FE
	Tokina 4.5–5.6/80–400	16/10	136	960	1 LD
	Minolta 4.5–6.7/100–400	14/11	149	840	2 LD
	Sigma 4.5–5.6/135–400	13/11	181	1210	2 AL, 1 LD, IF
	Tamron 5.6/200–400	13/11	180	1320	2 LD, IF
	Pentax 5.6/250–600	18/16	442	5050	4 LD, IF

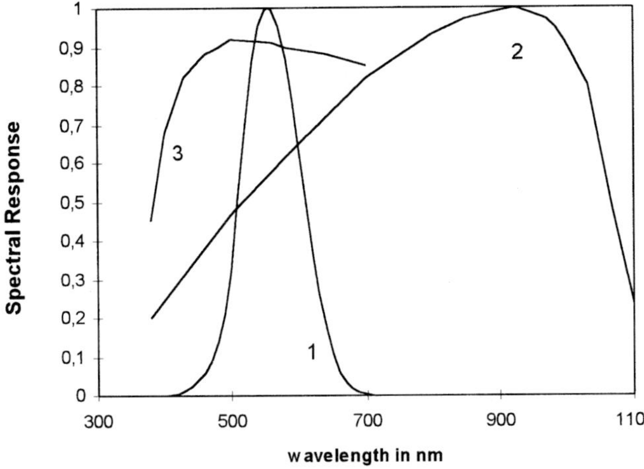

FIG. 16. Spectral response *versus* wavelength. Curve 1, relative spectral response of photopic human eye $V(\lambda)$; curve 2, relative spectral response of typical silicon detector $R(\lambda)$; and curve 3, transmission of the 6 × 6 lens Xenotar 2.0/80.

The quantity L' in Eq. (15) represents the equivalent luminance or equivalent radiance that acts on the detector. It considers all spectral dependences:

$$L' = L^* \bar{\tau}_0 = \int_{\lambda_1}^{\lambda_2} L_\lambda \tau_0(\lambda) R(\lambda)\, \mathrm{d}\lambda, \qquad (16)$$

where L_λ represents the spectral brightness in the object plane. Light sources have a characteristic spectral radiance; illuminated object scenes reflect and scatter the incoming radiation with a certain spectral and spatial distribution. The resulting object brightness is described by L^*. $\tau_0(\lambda)$ represents the transmission of the lens, $\bar{\tau}_0$ its mean value in the range λ_1–λ_2. The spectral transmission interval of typical optical glasses is 380–2500 nm. In this region the main transmission losses are produced by reflection on the glass–air surfaces. Antireflection (AR) coatings minimize the natural reflecting power of each glass–air surface from over 4% to less than 0.5% by internal interference in the AR layer. The resulting transmission of the lens changes slowly with the wavelength (see Fig. 16).

The quantity $R(\lambda)$ represents the relative spectral detector response. The response of the human eye $V(\lambda)$ demonstrates the limits of our visual perception. The typical silicon response covers the visible region. When a Si detector measures the radiation perceived by the human eye, it must be covered by a special conversion filter having a typical blue–green color.

In the case of a constant luminance L', the brightness in the image plane decreases versus the image height. The change of the solid angle formed by the active exit pupil area $A'_\mathrm{p}(\omega)$ and the distance between the exit pupil center and the image point on the detector explains this phenomenon. Typically, the solid angle to the detector edge is smaller than to the on-axis detector point. The decreasing illuminance E with the increasing field angle in several types

Table 5. Photometric and radiometric quantities and units.

Symbol	Photometric		Radiometric	
	Quantity	Unit	Quantity	Unit
E	Illuminance	lx	Irradiance	$\mathrm{W\,m^{-2}}$
L	Luminance	$\mathrm{cd\,m^{-2}}$	Radiance	$\mathrm{W\,sr^{-1}\,m^{-2}}$
Ω_0	Solid angle	sr	Solid angle	sr
L_λ	Spectral luminance	$\mathrm{cd\,m^{-2}\,nm^{-1}}$	Spectral radiance	$\mathrm{W\,sr^{-1}\,m^{-2}\,nm^{-1}}$
H	Illumination	lx s	Irradiation	$\mathrm{W\,s\,m^{-2}}$
I	Luminous intensity	cd	Radiant intensity	$\mathrm{W\,sr^{-1}}$

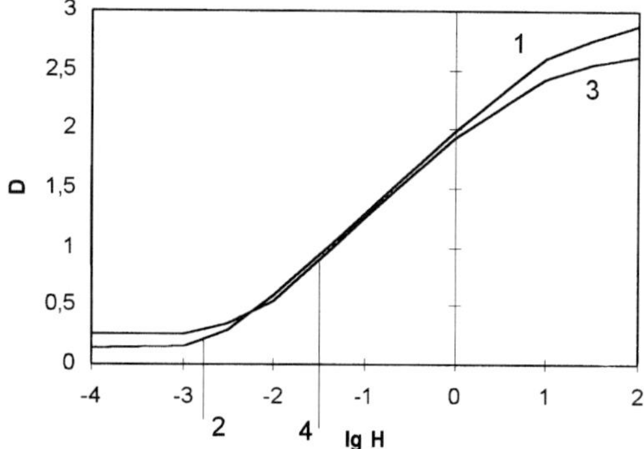

FIG. 17. Sensitometric curves of black-and-white films: 1, Agfapan 400 (ISO 400/27°) with minimum illumination H_{min} at 2; and 3, Agfapan APX 25 (ISO 25/15°) with minimum illumination H_{min} at 4.

of projection including fish-eye lenses is analyzed by Rusinov (1979). The general relationship for plane detector surfaces is

$$\frac{E}{E_0} = \frac{A'_p(\omega)}{A'_p(\omega = 0)} \cos^4\omega', \quad (17)$$

where $A'_p(\omega)$ is the active exit-pupil area at the object-side field angle ω; $\omega = 0$ means the active exit pupil area on the axis; ω' is the image-side field angle. The difference between ω and ω' is demonstrated in Fig. 14 by the fish-eye lens and the superwide-angle lens, where ω is strongly reduced to ω'. Generally, Eq. (17) recommends a small image-side field angle to minimize the illuminance loss by the $\cos^4$ relationship.

Figure 17 shows the typical brightness distribution in the photographic detector plane. The analyzed standard lens 2.8/80 has a constant field angle of 2ω. The illuminance in the center of the picture is described by Eq. (15). The maximum aperture $k = 2.8$ produces double the amount of illuminance from the next f-number, $k = 4$. However, the next f-number, $k = 8$, yields a fourth part of the illuminance value from $k = 4$.

2.2 Photographic Speed

Photographic detectors are sensitive to the incoming light energy. Illuminance and irradiance represent the light power per detector surface. The detector reacts to their product with the time t_e of exposure: $H = Et_e$. The relationship between the exposure time and the brightness in the object plane as a function of the f-number k and the film sensibility S is the topic of this section.

The photographic speed is deduced from the circumstances of traditional photographic films. The film development depends on the illumination H and is represented by sensitometric curves (see Fig. 17). Usually, the development is described by the optical density $D = -\log \tau_{film}$ with τ_{film} being the measured film transmission after development under defined conditions. The double logarithmic diagram corresponds to the logarithmic brightness sensitivity of the human eye. The film sensitivity is defined by the minimum illumination H_{min}. Two presentations are used:

$$S(\text{ASA}) = 0.8/H_{min}$$

and

$$S(\text{DIN}) = 10° \log H_{min} \quad (18)$$

A good picture of an object scene with varying spectral luminance produces an optical density in the linear part of the sensitometric curve. For this purpose, the illumination H in the picture plane must be adapted to the varying brightness in the object plane by means of the exposure time t_e and the choice of f-number k. Referring to Eq. (15), these quantities relate to H as

$$2^B = k^2/t_e; \quad (19)$$

B is the "exposure value". Equation (19) demonstrates the origin of the graduation at the photographic camera. The exposure time is geometrically stepped by the value 2: $t_e =$ 16 s, 8 s, 4 s, 2 s, 1 s, 1/2 s, 1/4 s, 1/8 s, 1/15 s, 1/30 s, 1/60 s, 1/125 s, 1/250 s, ..., 1/8000 s. The f-numbers of lenses are stepped by the value $\sqrt{2}$: $k = 1.0$, 1.4, 2.0, 2.8, 4.0, 5.6, 8.0, 11, 16, 22, 32.... The change of one f-number step guarantees the same change of illumination H as one exposure time step, as displayed in Table 6 for a constant exposure value.

2.3 Flash Photography

When the object brightness is not sufficient to obtain a desired exposure value B, the object

Table 6. Combination of f-numbers and exposure times for a constant exposure value $B = 14$.

k	t_e (s)	k	t_e (s)
1.4	1/8000	8.0	1/250
2.0	1/4000	11	1/125
2.8	1/2000	16	1/60
4.0	1/1000	22	1/30
5.6	1/500	32	1/15

scene must be illuminated by supplementary light sources. Modern electronic flash systems send a light impulse during a few microseconds with a time-changing luminous intensity $I(t)$ (see Fig. 18). The illuminance by the flashlamp in the object scene is $E^* = [I(t)/d^2]\Omega_0$ with d the distance between the lamp and the object scene. The reflected luminance at a diffuse reflecting surface is $L^* = \rho^* E^*/\pi\Omega_0$, so that the illumination in the detector plane is determined, by using Eq. (15), as

$$H = \frac{\rho^* \bar{\tau}_0 \int I(t)\,\mathrm{d}t}{(kd)^2 4(1 + m/m_p)^2}\Omega_0. \tag{20}$$

The integral over the time considers the changing flash luminous intensity.

The performance of a photographic flashlamp is described by the guide number as the product of f-number k and the distance d in the denominator of Eq. (20): $Z = kd$. At several conditions such as $H = H_{\min}$, $S = \text{ISO } 100/21°$, and $m = 0$, Eq. (20) yields an expression for the guide number. Considering all test conditions according to DIN 19011, the guide number is described as

$$Z = (0.713\ \text{m cd}^{-1/2}\text{s}^{-1/2})\left(\int I\,\mathrm{d}t\right)^{1/2}. \tag{21}$$

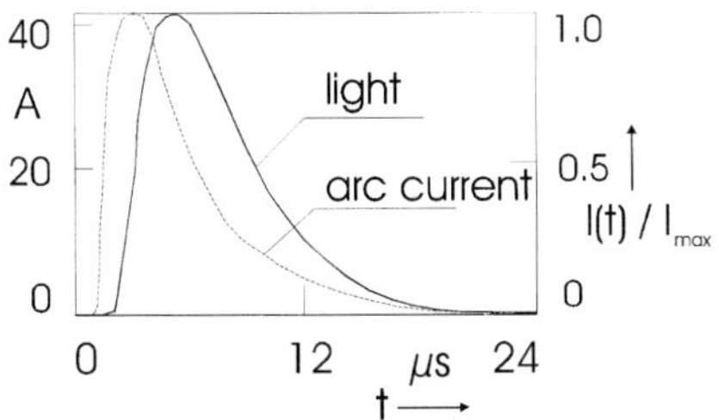

FIG. 18. Time characteristic of a typical flashlamp.

Therefore, the flashlamp performance depends on the area below the discharging curve expressed in cd s and the light distribution in the solid angle.

Modern electronic flash installations make different working principles possible.

1. The complete flash duration in accordance with the guide number is used without consideration of the object scene.
2. On the basis of measurement of reflected light in the object scene, the flashlamp is switched off when the illumination H is sufficient for the sensitivity S and the f-number k (computerflash).
3. The flash time is automatically limited by the exposure control of the photographic camera. For this very reliable solution, the incoming flash light is measured in the camera. The selected f-number, exposure time, film sensitivity, object reflectance, and the various object distances are considered.

3. TECHNOLOGY OF THE PHOTOGRAPHIC CAMERA

A photographic camera must realize the following functions: show the chosen object field by a viewfinder, open the shutter during the exposure time, measure the incoming light, and focus the chosen object.

3.1 Viewfinders

There are two types of viewfinders, parallax and through-the-lens viewfinders. Parallax viewfinders are mounted parallel to the camera lens. Through-the-lens finders are the core part of SLR cameras, with the advantage of always having the right image frame without aiming and parallax error.

The SLR viewfinder permits precise controlling of focus and image boundaries. The total system of the SLR camera constitutes a telescope with magnifying power in the range of unity, depending on the focal length of the objective lens used (see Fig. 19). The image orientation is indicated by coordinates x, y. The lens inverts the object orientation. In the finder screen the side-orientation is reverted, and the height appears true. A roof pentaprism

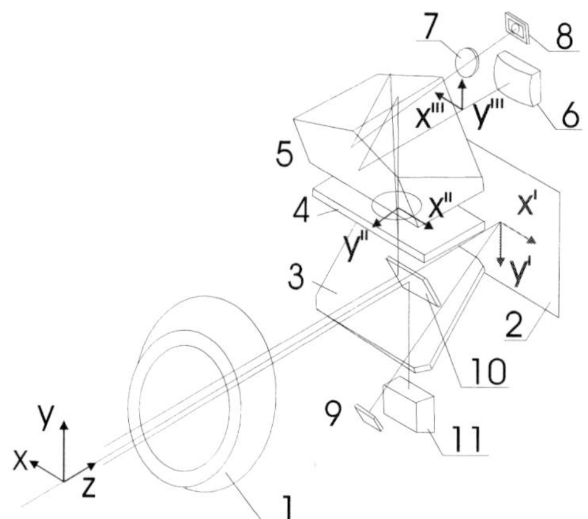

FIG. 19. Main elements of modern single-lens reflex camera: 1, photographic lens; 2, image plane; 3, main mirror; 4, finder-screen ground glass; 5, roof pentaprism; 6, focusing-plane magnifier; 7, light meter lens; 8, light-meter device; 9, flash-light exposure meter; 10, Autofocus mirror; and 11, autofocus module.

or roof mirror is used as an image inverter. The roof is needed for the left–right inversion and the pentaprism in order to get an upright image.

The traditional finder screen has a fine-ground front surface and a Fresnel lens at the rear surface. The Fresnel lens guarantees the light transport from the object plane to the photographer's eye. To control the focus manually, a crossed-wedge focusing device is implemented in the center of the ground surface (see Fig. 20). In the out-of-focus position, vertical object lines are separated by the cross-wedge and the ground-surface roughness yields a nonsharp image outside of the wedges. In focus position, the ground-surface roughness becomes ineffective and vertical object lines appear without breaks.

The arrangement of the Albada viewfinder is derived from the reversed Galilean telescope, which consists of a negative front lens and a positive lens on the eye side (Fig. 21). The image of the object is magnified by a power of less than unity. In order to get a viewfinder mask, the rear face of the negative lens is made partially reflecting, combining the light reflected from the mask with the transmitted light from the object. Both images are formed in the same virtual image plane 7. After

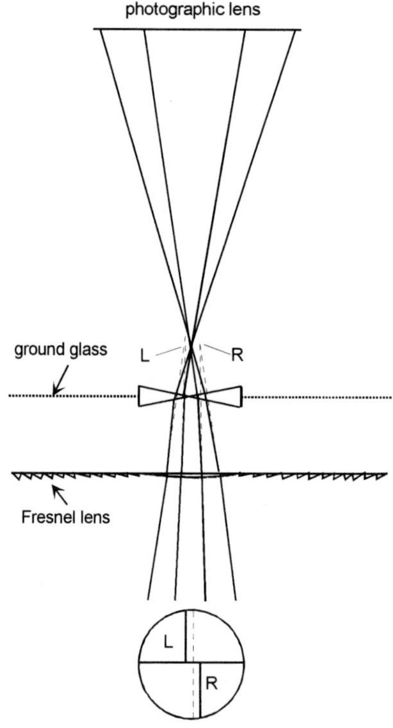

FIG. 20. Crossed-wedge focusing device for SLR cameras.

projection by lens 4, the eye sees the object image usually with the metallic mask.

Especially for viewfinder cameras with zoom lenses, real-image viewfinders are applied. In principle, this type consists of its own simply

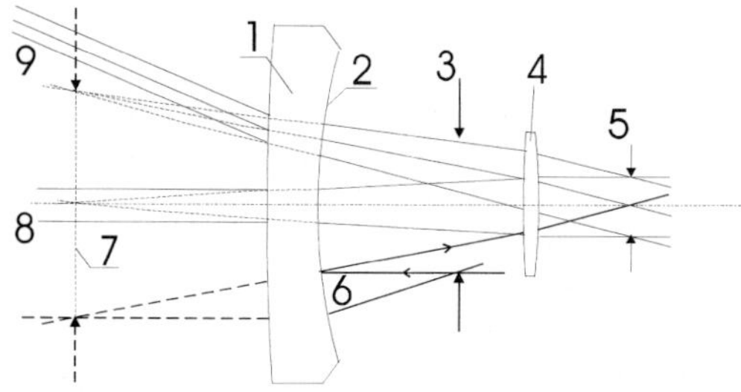

FIG. 21. Modern Albada viewfinder: 1, diverging front lens with partially reflecting rear surface 2; 3, field of view limiting metallic mask; 4, converging lens; 5, iris of human eye; 6, reflected light path from mask; 7, virtual image plane of object and mask; 8, on-axis ray bundle; and 9, maximum-field ray bundle.

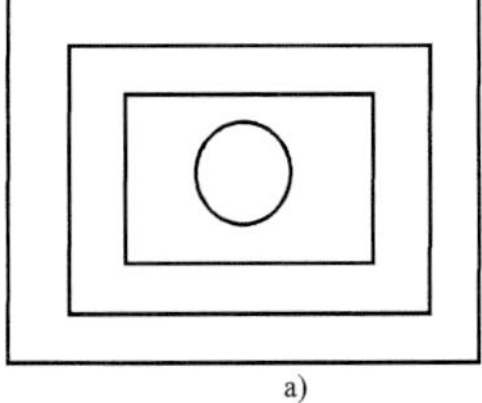

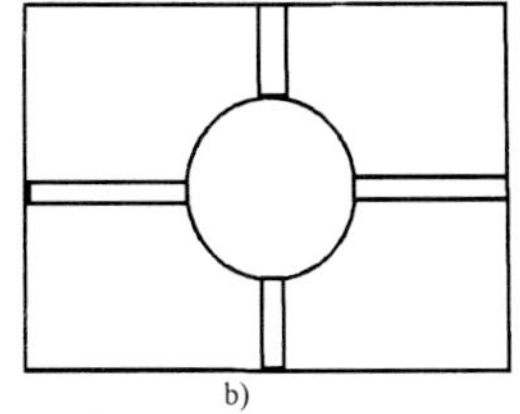

FIG. 22. Types of light-meter structure. (a) Sensor for measuring at different fields of view and (b) sensor for area-weighted measurement in SLR cameras.

designed zoom lens, an image inverter and a magnifier lens. By zooming the main lens system, the zoom of the finder is positioned synchronously.

3.2 Exposure Measurement

To measure the exposure, cameras are equipped with two different types of light meter, through the lens and parallel to the lens. Besides these, external exposure meters are also used, which give the photographer a measure of the scene brightness to use for calculating the exposure time.

The outside-the-lens type of light metering is very common with automatic-viewfinder cameras. The light meter device, consisting of sensor and lens, is mounted near the objective lens, to eliminate parallax error. To reduce the high IR sensitivity, photodiodes are equipped with a blue filter. The photocurrent will be intensified by a logarithmic amplifier. To ensure the right scene brightness, the light sensor is structured as multi-area type [see Fig. 22(a)]. The active sensor area corresponds to the actual field of view of the camera zoom lens.

Modern SLR cameras are equipped with *through-the-lens* (TTL) light metering, which is the best solution among the various metering types. The arrangement is presented in Fig. 19. The image on the integrated finder screen is projected by lens 7 into the light-meter device 8, placed nearby the magnifier lens. A multi-area sensor [see Fig. 22(b)] allows the measuring of separate areas of the image. Center-weighted, integrating, selective or spot metering are customary.

In some cameras the light reflected by the surface of the film is measured (9 in Fig. 19). This method can be used to turn off a flash automatically, controlling the quantity of light.

3.3 Types of Shutters

There are two common types of shutters used for controlling the exposure time. According to their position, they are called between-the-lens (or simply lens) shutter or focal-plane shutter, respectively. Focal-plane shutters are a fixed part of the camera. Therefore, they allow the lens to be changed without further protection of the film. Professional studio and medium-size cameras are mostly equipped with lens shutters because comparable focal-plane shutters would be too large or too slow.

The focal-plane shutter consists of two opaque curtains of staggered lamina blades (see Fig. 23). Instead of blades, classical types were made out of flexible cloth. These curtains pass the film window one after another. For fast exposure times, the second curtain is released shortly after the first one so that a slit is formed between the curtains. A slit width smaller than the film window results in an effective exposure time that is less than the time to cover the whole film frame. The individual image zones are exposed one after another. The use of a flash light restricts the synchronization time to 1/250 s or more. If the object moves very rapidly, the proportions will

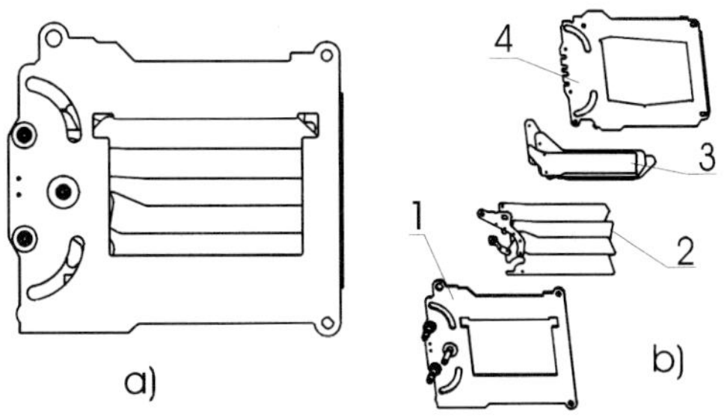

FIG. 23. Focal-plane shutter: (a) top side and (b) exploded view. 1, Top plate; 2, curtain with shutter blades; 3, folded curtain; 4, bottom plate.

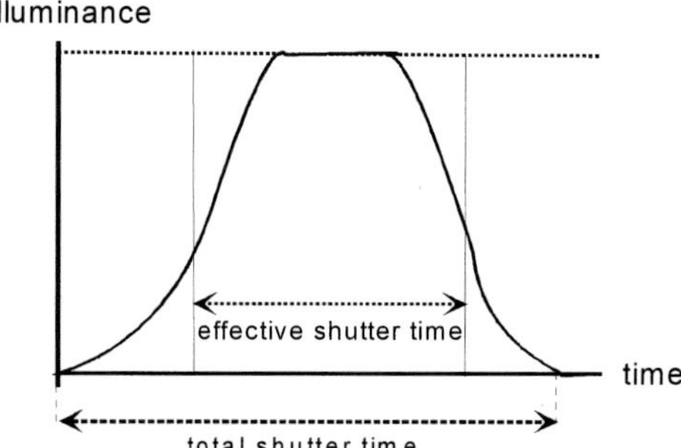

FIG. 24. Time characteristic of between-lens shutter.

be distorted. In modern focal-plane shutters, the curtains move perpendicularly along the short side of the film window.

As the speed of the curtain edges varies, the effective time is kept constant by varying the slit width. For mechanical reasons, the curtain is placed at a certain distance from the film. The cone of rays from the lens has a finite diameter in the plane of the shutter. Therefore, a small slit width results in horizontal vignetting. Focal-plane shutters are typically used in 35-mm SLR cameras. Modern lamina shutters allow exposure times of as little as 1/8000 s to be used.

Between-the-lens shutters are situated near the plane of the aperture stop of the photographic lens, the position at which the cross section of the light beam reaches its minimum. Several blades (normally five) cover the aperture. The ends of those blades are seated in a sector ring. The blades are levered by control cams that are arranged on the cam disk. A diagram of luminous flux versus time is shown in Fig. 24. The maximum of illuminance is given by full aperture, the lateral gradient by the speed of opening and closing action.

In order to calculate the effective shutter time, the shutter curve is transfigured into a rectangle with the same height and area. This is the time a perfect shutter would require in order to provide the same exposure. The total opening time of the shutter is the time from the beginning of the opening action of the curve up to the end of the closing action. The ratio between the effective time and the total opening time is known as the efficiency of the shutter. Today, fast between-the-lens shutters reach effective shutter speeds of up to 1 ms with a diaphragm of 24 mm diameter.

Some cameras use a diaphragm shutter situated next to the rear lens in order to enable the use of interchangeable lenses. In order to avoid vignetting, the diameter of the shutter has to be very large, which, however, reduces the shutter's speed efficiency. If the functions of f-stop and shutter are combined, the device is called an iris shutter. In this case, the shutter only opens up to the corresponding f-stop area. Therefore, the shutter time is linked to the f-stop and cannot be selected separately. Iris shutters are common in modern compact cameras as a fixed part of the integrated lens. The fastest speed that can be achieved with an aperture of $f/22$ or more is 1/400 s.

3.4 Automatic Exposure Programs

For controlling the lens aperture, an iris diaphragm consisting of several diaphragm blades is applied (see Fig. 25). The diaphragm is placed in the plane of the aperture stop of the lens. Several blades cover the aperture. Similar to the design of lens shutters, the ends of the segments are seated in a sector ring and are levered by diaphragm cams. According to the angle of turning, the aperture closes to a determined f-number.

Modern automatic cameras are equipped with automatic exposure devices that control shutter speed and lens aperture. The user has three options to choose from:

1. shutter priority: the user is able to set the shutter time, and the automatic camera system sets the diaphragm;
2. aperture priority: the aperture is set by the user, and the camera sets the shutter time;
3. programmed automatic exposure. In this mode, both shutter speed and aperture are set by the CPU of the camera.

In some cases, different program modes are available: for fast moving objects, preferably a high shutter speed or alternatively a program for high depth of field is used.

3.5 Focusing Devices

In order to make SLR photographers certain of accurate manual focusing, a finder

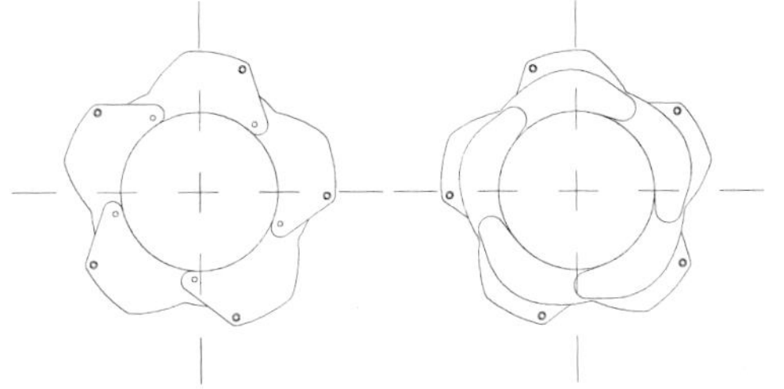

FIG. 25. Top and back view of an iris diaphragm of the same diameter.

screen with a central crossed-wedge device is applied (Fig. 20). If the image is above the crossing point of the prism, perpendicular lines will be apparently divided, one half to the right and one half to the left. If the image is in plane, no displacement will be recognized.

An *active autofocus* device measures distances by emitting the necessary (IR) light pulses itself. The system consists of an IR-emitting source and a receiver—usually a CCD line sensor (see CHARGE-COUPLED DEVICES)—each combined with lenses [Fig. 26(b)]. The reflected signal is received by a row in the detector corresponding to the distance of the object. The distance is given by $d = fb/p(N - N_{\text{inf}})$ with b the basic distance between the lenses, f the focal length, N the row number on the CCD line sensor, and p the CCD pitch. Each arrangement must be calibrated individually by measuring the row number for infinity and a defined, close distance. The calculated set of parameters is stored in the CPU of the camera in order to control the focus motor of the lens system.

The *passive autofocus* device consists of two parallel receiver systems, which are combined to one double lens and two CCD-line sensors fixed on the IC (integrated circuit). Both right and left sensor arrays receive a signal of the image data shifted in relation to infinity. Using, for example, a line sensor with 256 pixels per line, N is calculated as $N = (a + b)/256$, where a and b are the interpolated fractional zone numbers of the right and left sensor arrays.

There are disadvantages of passive autofocus: the system fails under twilight condition and darkness, and the system cannot measure objects with large monotone areas or horizontal structure. The accuracy of the device depends strongly on its base width and is insufficient for the use of telephoto lenses with high apertures.

The arrangement of passive AF systems in modern SLR cameras is presented in Fig. 19 by the elements 10 and 11. The structure of the complete AF module 11 is demonstrated in

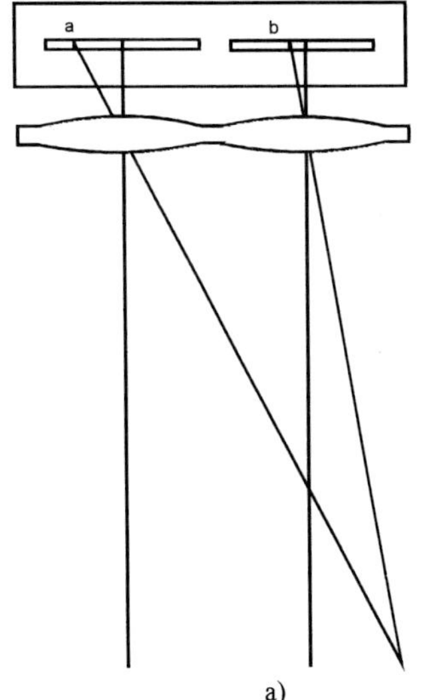

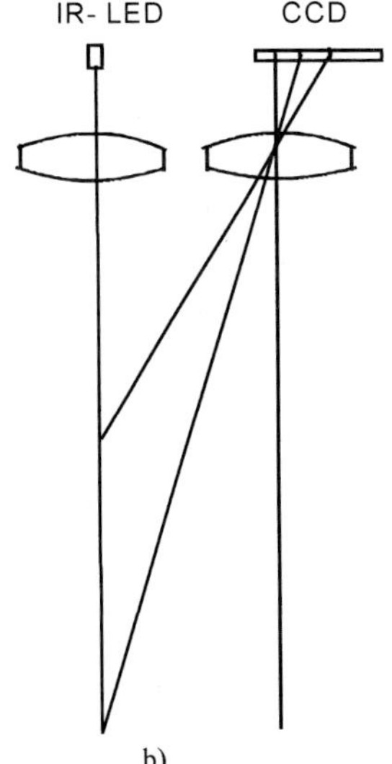

FIG. 26. Autofocus systems of compact cameras: (a) passive; and (b) active.

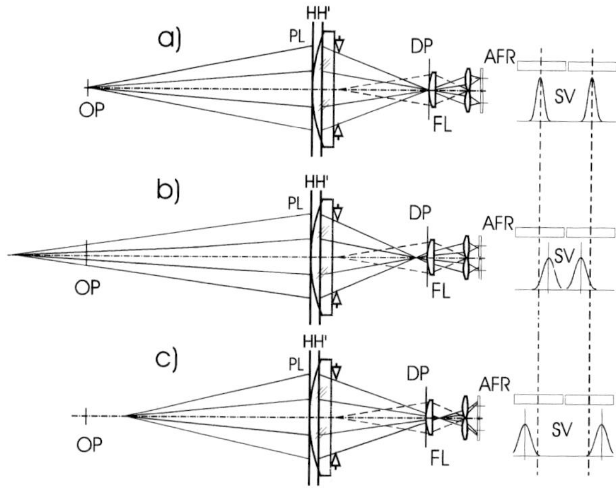

FIG. 27. Passive autofocus arrangement in the SLR camera: (a) sharp image; (b) object distance to long; (c) object distance to short. AFR, Autofocus detector rows; DP, corresponding detector plane; FL, field lens; OP, object plane; PL, photographic lens; and SV, signal voltage *versus* detector rows.

Fig. 27. The AF module consists of the field lens and pairs of relay lenses with line detectors. The field lens projects the exit pupil of the photographic lens in the relay lens plane. The brightness distribution on both rows of the pair analyses the displacement of focus. In the in-focus position both rows obtain the same light distribution [Fig. 27(a)]. The hyperfocal position [Fig. 27(b)] yields a contrast-reduced light distribution, displaced to the common center of the rows. On the other hand, in the confocal position [Fig. 27(c)], the contrast-reduced light distribution is displaced away from the common center.

4. TYPES OF MODERN PHOTOGRAPHIC CAMERAS

Three cameras will be selected from the most popular ones: the view-finder camera, the single-lens reflex camera (SLR), and the bellows camera. Typical technical items are presented in Table 7.

4.1 Viewfinder Cameras

The most popular commercial cameras are the viewfinder cameras, especially for 35-mm film. This type of camera is very compact and includes features like integrated autoexposure and autofocus, automatic film winding, integrated flash, and, in most models, a zoom lens. In order to make the camera very compact, the lenses move into the rest position, whereas the rear lens is positioned directly in front of the film surface. These cameras are equipped with Albada viewfinders or with real-image viewfinders, especially for the zoom types. The main parts of the camera are lens with integrated motor (*e.g.* zoom lens, 38–145 mm, or single-focus, 35 mm), autofocus system (active and passive types are used), viewfinder, integrated flash, light meter, and LCD control panel.

4.2 Single-lens Reflex (SLR) Cameras

For professional or semiprofessional purposes, SLR (single-lens reflex) cameras are used. These cameras were very popular, especially in the years between 1960 to 1985. Their popularity declined with the introduction of fully automatic compact viewfinder cameras.

Most system cameras, *i.e.*, cameras with interchangeable lenses and various accessories, are designed as SLR cameras. This design has the advantage of controlling the focus and image boundaries accurately without any parallaxes. The exposure measurement is taken through the lenses (TTL).

Both SLR 35-mm and SLR medium-sized cameras are suitable for professional mobile use. Their general design is shown in Fig. 19. The lenses form the image of the object on the finder screen, the position of which is equivalent to the plane of the film. A 45° tilted mirror is positioned in the path of the principal

Table 7. Typical technical items of modern photographic cameras.

Item	Type of camera			
	Viewfinder compact	35 mm SLR	Medium-format SLR	Professional bellows
Typical image size	$24 \times 36\,mm^2$, APS	$24 \times 36\,mm^2$	$4.5 \times 6\,cm^2$–$6 \times 7\,cm^2$	$6 \times 9\,cm^2$–$18 \times 24\,cm^2$
Typical weight with standard lens	350 g	1 kg	2–5 kg	5–20 kg
Interchangeable magazines	None	None	Yes	Yes
Shutter type	Between lens	Focal plane	Between lens 1/100–1/1000 s,	Between lens, rear lens
Shortest exposure time	1/500 s	1/4000 s	Focal plane 1/30–1/90 s	1/30–1/500 s
Auto exposure	Yes	Yes	Rare	None
Integrated flash	Yes	Often	None	None
Integrated driver for film winding	Yes	Yes	Rare	None
Viewfinder	With parallax by telescope	Parallax-free by projected image plane	Parallax-free by projected image plane	Parallax-free directly in image plane
Change of imagined field	Only zoom lenses	Zoom lenses, interchangeable lenses	Rare zoom, interchangeable lenses	Only inter-changeable widefield lenses
Focal length range	$f = 28$–$120\,mm$	$f = 6$–$1200\,mm$	$f = 30$–$1000\,mm$	$f = 65$–$500\,mm$
Shift and tilt motion	None	Only by special lenses	Only by special lenses	Yes
Manual focusing	Rare	Yes	Yes	Yes
Autofocus	Yes	Yes	Rare	None

rays, causing a 90° deflection to the screen. Immediately before the picture is taken, the mirror is turned up. The shutter is positioned directly in front of the film. Sharpness and image frame can be viewed either through the light shaft (which is realized in medium-sized cameras) or through a magnifier combined with an inverting prism. For light measurement, a light-sensor device is installed near the magnifier. Modern light-meter devices use a multiple-area sensor that allows separate measurements of the light to be taken for each zone. According to size and weighting, the measurement is called either integral, selective, center-weighted, or spot measurement.

For valuation of focusing, an AF module is installed at the bottom of the camera. Therefore the main mirror is made partially reflecting. Part of the light beam from the object is directed to the AF sensor passing it and another reverting mirror. The use of specifically located sensors allows the operator to take measurements of focus at different locations of the image. Behind the plane of proper sharpness, the image is divided by two auxiliary lenses and formed on two separate line sensors. The focal deviation is calculated by a CPU recording the difference in the phases of the two histograms and is immediately used for the adjustment of the focus motors of the objective lenses.

4.3 The Professional Bellows Camera

Studio cameras are designed for professional purposes and allow the photographer to use large-sized films, many-pixel solid state detectors (2 K × 3 K and more), or magazines with roll film, and to use interchangeable lenses with different focal lengths. The sharpness has to be adjusted on a ground glass, which is positioned alternatively in the plane of the film. The exposure is measured by using a separate light meter.

The concept of a studio camera is based on modular design systems. Both lenses and film magazines are interchangeable. A flexible bellows connects the frames of the lenses and magazines. Axial shifting of the frames adjusts the local sharpness for different focal lengths or ratios of magnification. Optimal image inclination is achieved by tilting the frames according to the Scheimpflug law. The frames can be shifted perpendicularly and in the vertical direction in order to obtain satisfactory perspectives. The focus problem of an arbitrary object scene is three-dimensional, which can be resolved by special algorithms (Mangelsdorf and Schuster 1996). The extensive adjustments require the use of a tripod.

5. FINAL PICTURE GENERATION

The target of photography is to preserve pictures for the human eye. Therefore, the nature of perception determines the requirements of the generation of the final picture.

5.1 The Human Eye as the Receptor of the Final Picture

The nature of human vision is referred to in the article Physiological Optics. Various aspects are specific to the photography problem.

When one looks at a photograph, the field of view is determined by the sharp perception using both eyes (Hering, 1931). The mean range of sharp perception with both eyes for different persons is $2\omega = 45$–$55°$, corresponding to the field of view of standard photographic lenses. The other parts of the visual field shown in Fig. 28 include zones of nonsharp vision (Schober, 1950), perceived outside of the fovea with reduced spatial resolution. The physiological limit of visual perception, $\delta\omega_E \approx 0.3\,\text{mrad} = 1'$, relates to recognition of two object points with equal brightness. These

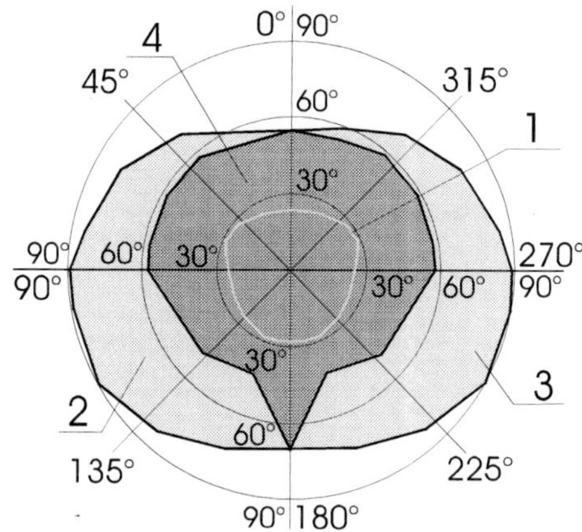

FIG. 28. Field of human eyes keeping head and eyes still: 1, sharp-perception field with both eyes; 2, only right eye visual field; 3, only left eye visual field; and 4, both eyes visual field.

points are easily visible at an angular distance from $\delta\omega_E \approx 1\,\text{mrad} = 3'$. For approximate purposes, we use $\delta\omega_E = 0.5\,\text{mrad}$.

5.2 Classical Photographs

In 1996, classical photographs on paper constituted the majority of the consumer market (83% of prints). Color photographs are present in every family. Black-and-white photography has become more and more the domain of professional photographers.

Different paper formats are looked at from different viewing distances because the viewer wishes at first to look at the complete picture, corresponding to the sharp visual field of view $2\omega \approx 50°$. Then, for more details, the viewer can shorten the viewing distance until the maximum accommodation is reached but losing the survey of the complete picture. The normal minimum viewing distance is 25 cm; young and myopic people are also capable of seeing sharp images as shorter distances. The viewing distances listed in Table 8 are calculated by means of Eq. (3) after replacing the denominator. The minimum resolvable distance on paper is calculated at each viewing distance for the eye resolution limit $\delta\omega_E = 0.5\,\text{mrad}$.

The printing of photographs is realized by projection of the negative film onto the paper plane. The enlarger exposure system checks

Table 8. Typical photographic paper formats.

Format (cm^2)	Typical viewing distance at $2\omega = 50°$ (cm)	Minimum resolvable distance on paper at $\delta\omega_E = 0.5$ mrad (μm)
8.9 × 12.7	17	83
10.5 × 14.8	19	97
12.7 × 17.8	23	117
16.5 × 21.6	29	146
20.3 × 25.4	35	174
24.0 × 30.5	42	208
40.6 × 50.8	70	349
50.8 × 61.0	85	426

the luminance in the plane of the negative and controls the exposure time of the paper.

5.3 Slide Projection

The fascination of slide projection is due to the excellent picture quality offered simultaneously to a large public. The principle of the slide projector is demonstrated in Fig. 29. The projector lens forms the image of the illuminated slide on a reflecting screen. The public perceives the reflected light distribution. Disturbing light sources are also reflected, so that a dark projection room is desirable. The main projection problem is the realization of a homogenous light distribution on the screen. For this purpose, the condenser guides the rays from each lamp filament point to all slide points. Then, all rays pass the projector lens. Ray bundles starting in the slide plane converge in the screen plane.

The screen is situated at a long distance in relation to the focal length of the projector lens. Typical relationships are represented in Fig. 30. This representation covers the typical consumer applications with slides, having a holder format 5 × 5 and 7 × 7 cm^2. Their picture formats on the film are listed in Table 9.

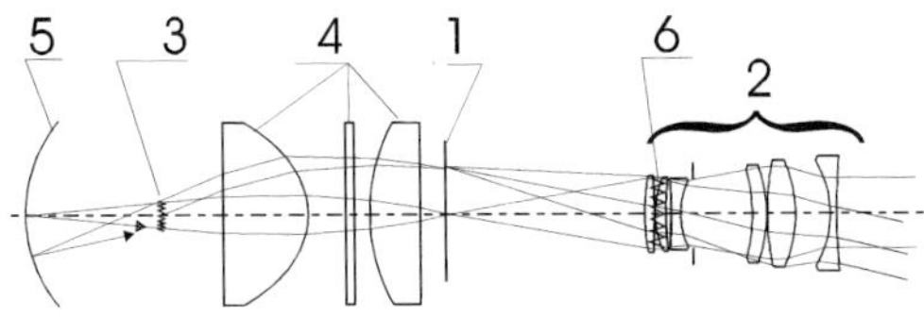

FIG. 29. Optical arrangement of 6 × 6 slide projector Rollei Vision 66; 1, slide plane; 2, projection lens Apogon 2.8/90; 3, filament of halogen lamp; 4, condenser system with heat filter; 5, spherical reflector; and 6, filament image.

5.4 Printed Digital Pictures

Modern printers and cheaper solid state detectors determine the success of digital photography. The preservation of pictures without any silver halide–based technology will replace the classical photographic film step by step. Different techniques are referred to in the article PHOTOGRAPHY, DIGITAL.

The acceptance of a printing technology depends on the resolution reached on the hard copy. It is defined by the print pattern measured in lines per inch (lpi) (see Table 10). The print pattern is formed by a quadratic number of printed dots. This number of dots (4 × 4–16 × 16) defines the gray-level steps (16–256) printable in one pattern. The density of spots is expressed as dots per inch (dpi). Typical values for a PC monitor are 72–120 dpi.

The gray-level number $GL = (DPI/LPI)^2$ for each color and the resolution LPI define the print quality. An advantageous combination depends on the printer. For example, a thermosublimation printer needs a smaller gray-level step number because the printing procedure melts the pattern.

An important criterion is the comparison with classical photographs. If we assume a print resolution of 200 lpi, which corresponds to the eye resolution limit at a 25-cm viewing distance, and that each detector pixel has one print pattern, the printed pictures reach the formats listed in Table 11. A classical photograph observed at a viewing distance of 25 cm has a format of about 13 × 18 cm^2 (see Table 8). The 1.3 K × 1 K detector fulfills this condi-

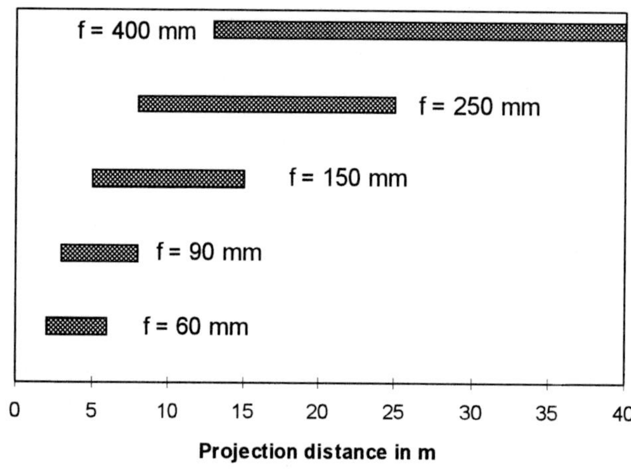

FIG. 30. Typical slide projection distances with different focal lengths of the projector lens.

Table 9. Typical consumer slide formats.

Specification	Picture size (mm^2)	Projected field (mm^2)
35 mm slide,	18 × 24	16.5 × 23
5 × 5 cm^2 holder size	24 × 24	23 × 23
	24 × 36	23 × 35
Medium-format slide,	45 × 60	39 × 55
7 × 7 cm^2 holder size	60 × 60	55 × 55

Table 10. Typical specifications for printed photographs.

	Resolution (lpi)	Dot density of different printers (dpi)	
Newspaper print	50–80	Thermosublimation	300
High-quality prints	150–200	wax	300–600
Eye limit at 25 cm		laser	600–1200
viewing distance	200	ink jet	600–1440

Table 11. Printable formats of digital pictures having a resolution of $\delta\omega_E = 0.5$ mrad at a viewing distance of 25 cm.

Pixel number of solid state detector	Printed picture format (cm^2) with 125-μm print pattern spacing ≈ 200 lpi
0.64 K × 0.48 K	8 × 6
1 K × 0.8 K	12.5 × 10
1.3 K × 1 K	16.3 × 12.5
3 K × 2 K	37.5 × 25
4 K × 4 K	50 × 50

tion. The 1 K × 0.8 K detector is also acceptable; lower detector pixel numbers offer resolution deficits in the printed 9 × 13-cm^2 format. For professional and technical purposes, the higher pixel numbers are useful.

GLOSSARY

Aperture Stop: Iris stop limiting the light cone to the image.

Autofocus (AF): Motorized focusing on modern cameras.

Chief Ray: Ray striking the optical axis in the center of the aperture stop.

Close-Up Focusing at Infinity: Shortest object distance with a depth of field to infinity.

Critical Stop: f-number with the best resolution of a photographic lens.

Diffraction Limit: Physical limit of resolution as a result of light diffraction on the aperture stop.

Distortion: Displacement of the center of confusion figure in the image plane in relation to the paraxial image height.

f-Number: Ratio of focal length f to the entrance pupil diameter D, after f the most important quantity of photographic lenses.

Focal Length (f)**:** Most important quantity of a lens describing in first approximation the ray deflection.

Field of View: Size of object space in angular measure, covered by the detector.

Floating Element (FE): Axial displacing lens group of a photographic lens to improve the image quality at different magnifications.

Magnification: Ratio of image height y' to object height y measured perpendicularly to optical axis; for object planes at infinity, $m = 0$.

Manual Focus (MF): Traditional focusing with visual control.

Meridional: Plane in rotationally symmetric lenses that contains the optical axis and the chief ray.

Modulation Transfer Function (MTF): Complete measure of the spatial resolution for lenses, detectors and their interaction.

Paraxial Approximation: Linear approximation of image formation by means of principal planes and foci; each point of the object plane corresponds to one point of the image plane and *vice versa*.

Perspective: Presentation of depth of field in a photographic picture.

Pupil: Image of the aperture stop. After projection in the top side of the lens it is called entrance pupil, having the diameter D; after projection in the back side of the lens it is called exit pupil, having the diameter D'.

Sagittal: Plane in rotationally symmetric lenses, perpendicular to the meridional plane, that contains the chief ray.

Scheimpflug Condition: Relation for a paraxial sharp image formation with shifted and tilted object and image planes.

Through-the-Lens (TTL): Light metering in the camera behind the lens.

Zoom Lens: Lenses with changing focal length.

Works Cited

Bumba, V., Pospisil, J. (1969), *Optik* **30 H 1**, 84–92.

Cox, A. (1964), *A System of Optical Design*, London: Focal Press.

Franke, G. (1964), *Photographische Optik*, Frankfurt am Main: Akademische Verlagsgesellschaft.

Gaskill, J. D. (1978), *Linear Systems, Fourier Transforms, and Optics*, New York: Wiley.

Goodman, J. W. (1968), *Introduction to Fourier Optics*, New York: McGraw-Hill.

Haferkorn, H., Richter, W. (1984), *Synthese Optischer Systeme*, Berlin: VEB Deutscher Verlag der Wissenschaften.

Hering, E. (1931), *84 wissenschaftliche Abhandlungen*, Leipzig: Verlag G. Thieme.

Jahn, H., Scheele, M. (1997), "Zur optimalen Anpassung von Objektiven an CCD-Sensoren," *Photogram. Fernerk. Geoinf.* **1** (2), 103–115.

Kämmerer, J. (1979), *Wann sind Qualitätssteigerungen bei photographischen Objektiven sinnvoll?*, Oberkochen: Carl Zeiss.

Kingslake, R. (1992), *Optics in Photography*, Bellingham, WA: SPIE Press.

Laikin, M. (1995), *Lens Design*, New York: Marcel Dekker.

Malacara, D., Malacara, Z. (1994), *Handbook of Lens Design*, New York: Marcel Dekker.

Mangelsdorf, J., Schuster, N. (1996), *Feinwerktechnik & Meßtechnik* **104**, 728–734.

Mann, A. (1993), *Zoom Lenses*, Milestone Series Vol. MS 85, Bellingham, WA: SPIE Optical Engineering Press.

Mutter, E. (1962), *Kompendium der Photographie*, Berlin: Verlag für Radio-Foto-Kinotechnik.

Rusinov, M. M. (1979), *Tekhnicheskaya Optika*, Leningrad: Maschinostroenie.

Scheimpflug, T. (1898), *Photogr. Korresp.* **35**, 114–121.

Schober, H. (1950), *Das Sehen*, Vol. 1, Darmstadt: Markewitz-Verlag.

Schröder, G. (1981), *Technische Fotografie*, Würzburg: Vogel-Verlag.

Volosov, D. S. (1971), *Fotograficheskaya Optika*, Moskva: Iskusstvo.

Welford, W. T. (1974), *Aberrrations of the Symmetrical Optical System*, London: Academic Press.

Further Reading

Born, M., Wolf, E. (1975), *Principles of Optics*, 5th ed., Oxford: Pergamon Press.

Haferkorn, H. (1986), *Bewertung optischer Systeme*, Berlin: VEB Deutscher Verlag der Wissenschaften.

Kowalski, P. (1980), *Théorie Photographique Appliquée*, Paris: Masson.

Marchesi, J. (1990), *Photokollegium*, Schaffhausen, Switzerland: Verlag Photographie.

Papoulis, A. (1968), *Systems and Transforms with Applications in Optics*, New York: McGraw-Hill.

Ray, S. F. (1988), *Applied Photographic Optics*, London: Focal Press.

SAFETY AND HEALTH IN THE PHYSICAL SCIENCE LABORATORY*

DANIEL T. CRANE, *Salt Lake Technical Center, Occupational Safety and Health Administration, United States Department of Labor, Salt Lake City, Utah, U.S.A.*

INTRODUCTION

Most accidents can be foreseen and avoided. Accidents usually occur either because a known hazard went unabated or because a new, unforseen hazard is encountered. Understanding the nature of common hazards

*The opinions and statements of the author are completely his own and do not establish policy of the United States of America or necessarily establish interpretations of federal regulations. Any mention of a specific product name does not imply endorsement or preferred use.

ISBN 3-527-29308-6

in the physical science laboratory is key to managing safety and health. The purpose of this article is to examine the role of the safety and health professional in activities and processes that may exist or occur in physical science laboratories. These professionals are known variously as industrial hygienists, safety officers, or chemical hygiene officers, or by other suitable titles. They may be certified by a professional organization.

1. RESPONSIBILITY FOR SAFETY

Safety is the responsibility of every person in a facility, beginning with top management and extending even to visitors. It is therefore important that potential hazards be communicated to all involved individuals at a facility. The control of the processes within physical science laboratories lies with the management of those facilities. While the laboratory workers perform the work directly, management ultimately has the basic responsibility to ensure that the workers have a workplace free of recognized hazards [OSHA, 1970, Section 5(a)].

1.1 Management Commitment

No program of safety and health can be effective without active support and impetus from the highest management officials. This support must be sincere and adequate. Safety and health is not the sole responsibility of the safety and health professional. All levels of management must be committed to the safety and health of all the employees whether they are researchers, graduate students, or custodial staff. The reason for this is clear. The response of employees in the work environment will follow immediately from what supervisory and management staff emphasize. Top laboratory management must actively pursue safety and health in a visible manner such that middle and first-line managers and supervisors understand that worker safety cannot be sacrificed for production or research. No influence from independent researchers must interfere with the safe operation of the facility. The safety and health professional is usually charged with the direct responsibility to provide information to laboratory personnel so that they may be aware of potential hazards and workable remediations.

1.1.1 Occupational Safety and Health Act Safety and health in the workplace is an international concern. Many nations have instituted particular regulations to limit exposure to health and physical hazards. Efforts in the United States of America are illustrative, in general, of international activities. Although the specific regulations will vary by nation, the concepts used by safety and health professionals are not tied to specific regulations. In 1970, in the United States, the Congress passed the Occupational Safety and Health Act, which states that every employer has the responsibility to provide a workplace free of recognized health or safety hazard [OSHA, 1970, Section 5(b)]. Other countries have established similar requirements (Kortsha, 1991). This article uses the United States regulations to illustrate particular concepts used by safety and health professionals.

1.1.2 Responsibility to Inform It is an important function of management, through the safety and health professional, to ensure that employees understand the hazards of the processes and materials that are present in the laboratory. The Occupational Safety and Health Administration (OSHA) has established guidelines for employee right to know (OSHA, 1997, Subpart Z, Sections 1200 and 1450). Specifically, this standard requires a number of things in the laboratory. Labels on containers of hazardous chemicals may not be removed or defaced. Material Safety Data Sheets (MSDS) received with incoming shipments of hazardous chemicals must be maintained by the laboratory and be readily accessible to laboratory personnel. Laboratory personnel must be appraised of the hazards of the chemicals and processes in the workplace through information and training at the time of their initial work assignments and whenever a new hazard is introduced into the work area. Employees must be made aware of and trained about the measures they can take to protect themselves from these safety and health hazards. This would include such measures as appropriate work practices, emergency procedures, and personal protective equipment. The safety and health officer is generally charged with making certain that such training occurs.

1.1.3 Management Responsibility Laboratory management must have an active involvement with safety. An effective means to establish and maintain a working safety and health program is to provide an empowered safety office with safety and health professionals who understand the particular issues of the physical science laboratory. The most effective safety programs are administered by joint labor and management personnel. Small operations may have a single employee serving part-time in this capacity, while in large operations the safety office or officer may have literally hundreds of employees devoted to maintaining safety in the extensive operations of the laboratory system, such as within the United States Department of Energy (Lawrence Livermore National Laboratory, 1997).

1.1.4 Empowerment of Safety and Health Professional The safety and health professional must be empowered to make administrative decisions and take immediate action regarding safety. This is why management participation is essential. He or she has the direct responsibility (with management support) to determine the content and provide the training required in each work area. Employee input is essential for determining the hazards that are present in the various work areas. Top management may be unaware of some of the hazards and must therefore be responsive to employee suggestions.

1.1.5 Safety and Health Professional Responsibilities The safety and health professional has the responsibility to provide initial training for new employees and for transferred employees before they are exposed to potential hazards. The safety office or officer has the responsibility to maintain records of all training offered and to whom it was given. The safety office or officer has the responsibility, with management, to conduct inspections to identify and correct hazards or see that the inspections are done. Effective internal safety inspections are made with a team composed of both management and labor, including employees from within the area being inspected and others who have no direct involvement so that they can take an unbiased view of the operations being inspected.

1.2 Responsibility of Employees

While management has the primary responsibility to provide a workplace free of recognized hazard, it remains with the employees of the laboratory to follow safe work practices and to work with management to abate unsafe conditions. It is not appropriate for employees purposely to put themselves at unreasonable risk to meet institutional deadlines or demands. In the course of work, employees at the work-detail level frequently are the first to become aware of new hazards. They are certainly the ones who are most often injured by not following accepted work practices. Employees must take a personal interest in using existing and appropriate personal protective equipment and engineering controls and following established safety guidelines and regulations. Employees must follow guidelines such as the wearing of safety glasses, protective clothing, or other personal protective equipment including respirators, hearing protection, face shields, *etc.*

2. PLANNING FOR SAFETY AND HEALTH

One of the best ways to avoid accident and injury is to plan ahead for foreseeable hazards. Management, along with the safety and health professional, has the first responsibility to assess the presence of potential hazards.

2.1 Hazard and Risk

To plan effectively for safety, it is necessary first to understand the difference between hazard and risk.

2.1.1 Hazard A hazard exists if a particular substance, process or situation has the potential to cause harmful effects. Given such a broad definition, there are few materials or processes which do not have some potential to cause harm. It is necessary to establish a system for determining the likelihood that a particular hazard might cause a harmful effect—that is, to determine risk (Newsom, 1996).

2.1.2 Risk Risk is defined as the probability that a substance or situation will produce harm under specified conditions. Risk is a combination of two factors: the probability that an adverse event will occur (such as a specific disease or type of injury) and the consequences of the adverse event. Risk encompasses impacts on public health and on the environment, and arises from exposure and hazard. Risk does not exist if exposure to a harmful substance or situation does not or will not occur. Hazard is determined by whether a particular substance or situation has the potential to cause harmful effects.

2.1.3 The Presidential/Congressional Commission on Risk Assessment and Risk Management This commission has published effective guidelines for determining the extent of risk. It is beyond the scope of this article to explore fully the intricacies of risk management. However, the Commission has established a six-stage framework for risk management (Newsom, 1996):

1. Define the problem and put it in context.
2. Analyze the risks associated with the problem in context.
3. Examine options for addressing the risks.
4. Make decisions about which options to implement.
5. Take actions to implement the decisions.
6. Conduct an evaluation of the actions results.

2.2 Written Safety and Health Plan

Safety and health professionals aid employers by providing written safety and health plans that are required if there is a potential for release of hazardous materials (OSHA, 1997, Subpart Z, Section 120). Such a plan assures that there is a coherent and controlled approach to the safety and health in the laboratory. A written plan provides a framework against which to build effective prevention and response.

2.2.1 Chemical Hygiene Plan The safety and health professional is instrumental in writing and maintaining the chemical hygiene plan, which is a comprehensive document that covers all aspects of safe laboratory operation. It sets forth procedures, equipment, laboratory equipment, personal protective equipment, and work practices that are capable of protecting employees from the safety and health hazards presented by hazardous chemicals and processes used in that particular workplace (OSHA, 1997, Subpart Z, Section 1450). It prescribes the proper storage and handling of chemicals, radionuclides, and physical agents. It prescribes the appropriate handling of lasers, magnetic fields, and the specific processes and hazards present in the laboratory. It is beyond the scope of this article to prescribe specific plans. However, the plan must be capable of protecting employees from safety and health hazards associated with hazardous chemicals and processes in that laboratory and capable of keeping exposures below the limits specified by OSHA or other governing national standards organization. The Chemical Hygiene Plan must also be readily available to employees, employee representatives, and, upon request, government regulators. The Chemical Hygiene Plan includes specific measures that the employer must take to ensure laboratory employee protection. These include written standard operating procedures, special engineering controls, personal protective equipment and its use, provisions for employee information and training, required medical monitoring, and emergency response procedures as well as specific instructions on how to revise the plan as needed. The safety and health professional is very familiar with all aspects of the Chemical Hygiene Plan and keeps it current and effective.

2.2.2 Emergency Response Plan The Emergency Response Plan establishes the procedures to follow in case of emergency. The plan outlines the steps to be taken following an accident. It will include who is to be notified in the laboratory organization, the local and state governments, and the national government along with required notification deadlines. It will include any specific guidelines that must be given to company or local emergency personnel in order for them to carry out their emergency response. This might include certain areas where not to fight a fire because of explosion hazard, or areas in which radiation is a concern. These instructions must be shared with the expected emergency providers before an emergency occurs to avoid any unnecessary confusion. Knowing ahead of time what hazards are present can allow professional emergency personnel to develop appropriate action plans and procedures.

2.3 Monitoring Safety and Health

It is necessary to know when and if laboratory personnel or others are affected by operations in the laboratory. For this reason, laboratory management must monitor potential exposures to hazards and then maintain for each employee an accurate record of any measurements taken to monitor employee exposures and any medical consultation and examinations, including tests or written opinions.

2.3.1 Monitoring and Sampling Generally, the safety and health official conducts or oversees monitoring or sampling to determine whether or not an employee is exposed to hazards. There are various devices for this purpose, ranging from filters to sorbent tubes to passive monitors. In some cases, there are direct-reading instruments to give an estimate of the hazard. Filters, tubes, and passive devices usually need to be analyzed by an industrial hygiene laboratory, while direct-reading instruments give an immediate indication of the exposure. Other devices are used to monitor exposures to radiation or noise. Again, some devices give an immediate reading while others such as film badges must be sent to a laboratory for interpretation. It is the responsibility to the safety and health officer to understand the proper use of these devices and techniques.

2.3.2 Medical Examinations for Routine Exposures Whenever an employee develops signs or symptoms associated with a hazardous chemical or process to which the employee may have been exposed in the laboratory, the employee must be provided an opportunity to receive an appropriate medical examination. If exposure monitoring reveals an exposure level routinely above the action level [or, in the absence of an action level, the permissible exposure limit (PEL)] for an OSHA-regulated substance for which there are exposure monitoring and medical surveillance requirements, medical surveillance must be provided for the affected employee.

2.3.3 Medical Examinations After an Accident Whenever an event takes place in the work area such as a spill, leak, explosion, or other occurrence resulting in the likelihood of a hazardous exposure, affected employees must be provided an opportunity for medical consultation.

2.4 Novel Hazards

The nature of a laboratory is generally to investigate processes that are not routine, and may be new, novel, and undocumented outside the immediate work environment. Such situations must be examined for the potential to cause harm to any exposed personnel. The safety and health professional helps researchers control hazards through the systematic use of industrial hygiene controls. Engineering controls are the first line of protection followed by good work practices, administrative controls, and, where no other mode of protection is feasible, personal protective devices. The principles of engineering control are substitution, isolation, and ventilation (Soule, 1991).

2.4.1 Substitution If a health or safety problem is presented by the use of a specific material, wavelength, or process, substitution of a more benign material, wavelength, or process should be investigated as a remediation of a hazard. For example, enclosures might be made of plastic or safety glass rather than regular glass, or cleaning chemicals might be chosen that are safer. An externally vented hood might be substituted for a stand-alone unit. Care must be taken when substituting one chemical or process for another so as not to cause further hazard.

2.4.2 Isolation The second principle of engineering control is isolation. This is frequently some barrier such as a blast shield, or may consist of remotely conducting the operation in a separate room. Also, simply putting distance between the operator and the operation can be considered isolation. In addition, compartmentalization such as is done for tank farms is a form of isolation that limits the involvement of an entire facility in the event of an accident. It may be possible to separate different procedures, isolating the most dangerous from other, more benign ones.

2.4.3 Ventilation The third principle of engineering control is ventilation. Ventilation provides control of the environment by varying the airflow. The airflow can be used to control the temperature of the space, remove a contaminant, or dilute the concentration of the contaminant to below permissible exposure levels. Control of ventilation requires a detailed knowledge of air flow and the materials or processes to be controlled. Such information is beyond the scope of this article. Information geared to industrial hygiene can be found in standard industrial hygiene references (Soule, 1991).

2.4.4 Personal Protective Equipment Personal protective equipment includes safety glasses, safety shoes, special clothing, respiratory protection, and any other device or system that is designed to be worn or used by an individual. Personal protective equipment should never be considered an acceptable alternative to a well-designed engineering approach to safety except where such is shown to be either unfeasible or more unsafe than the use of personal protective equipment. Certain pieces of personal protective equipment such as safety glasses should be used regardless of the presence of engineering controls. Where respirators are used, a respiratory protection program is required to be in place. Certain operations require specialized personal protective equipment. For example, when working with lasers, special eye protection is required. The safety and health professional must be knowledgeable as to the proper use of this type of personal protective equipment.

3. RECOGNITION AND REMEDIATION OF SAFETY AND HEALTH HAZARDS IN THE PHYSICAL SCIENCE LABORATORY

Awareness of recognized hazards is the primary tool used in recognition of the potential hazards. Management along with the safety and health professional, as well as involved workers, must review the processes used or expected to be used in the laboratory. Forethought and planning to perform the work safely knowing recognized hazards will prevent many accidents.

3.1 Planning for Safety

Looking for hazards in the physical science laboratory begins with the planning stages of a project or process by looking for recognized hazards in the processes to be used. For example, if gas cylinders are to be used, the cylinders must be properly secured and vented. If a process involves ionizing radiation, the proper safeguards, such as shielding, must be established. Laboratory personnel must continue to look for hazards after the project begins and throughout the project to remediate any significant hazard which may appear. Looking for hazards must not be limited to items which are recognized, codified into regulation, or found on a list.

3.1.1 Exposure Routes The routes of exposure to chemical and physical hazards are dermal, respiration, and ingestion.

Dermal includes any skin or eye exposure which would result in

1. direct absorption of a chemical into the skin or eye;
2. radiation exposure of the skin; or
3. temperature extreme.

Exposure through respiration is by

1. inspiration of a chemical or radiation hazard;
2. extremely hot or cold air;
3. inhalation of toxic gases;
4. exposure to atmospheres deficient in oxygen;
5. in hyperbaric conditions involving too much oxygen; or
6. the absence of air or exposure to vacuum.

Ingestion hazards are from anything that might be taken by mouth intentionally or unintentionally, by transmission from hand to mouth, or otherwise transmitted to the mouth.

3.1.2 Looking for Hazards Safety personnel, along with representatives from the work area being inspected, should look for any process, employee action, chemical storage, electrical wiring, virtually anything which might cause an accident. This inspection should not be limited to recognized hazards. Rather, the safety and health professional should look for any process or situation which *may* present a hazard. Immediately dangerous situations should be remediated as rapidly as possible. Once the team has identified potential hazards, those for which there are specific rules or regulations should be remediated. New or novel potential hazards should be evaluated for risk and remediated as appropriate.

3.2 Common Hazards Found in the Physical Science Laboratory

There are a number of common hazards to be found in the physical science laboratory. Most accidents occur as a result of slips, trips, and falls (OSHA, 1997, Subpart E, Section 38). Many other hazards are a result of environmental exposures such as extreme heat or cold (Mutchler, 1991; OSHA, 1997, Subpart D, Section 22). Confined spaces may present a hazard in breathing air quality. These spaces may be too small to contain sufficient air for the duration of the task, or a concentration of explosible gases may easily be reached when welding gases or other generated gases are being used. Certain spaces will have insufficient oxygen for breathing, concentrated toxic gases, or vacuum. (OSHA, 1997, Subpart J, Section 146).

3.2.1 Ergonomic Hazards Ergonomic hazards are generally the result of three factors: force, frequency, and posture. In the physical science laboratory, there are many tasks involving one or more of these recognized stressors that may result in repetitive motion injury, back injury, or other upper extremity cumulative trauma disorder. Back injuries can usually be limited by implementing training programs that teach employees the proper techniques for lifting and by limiting the amount of weight that individual employees are required to lift (Waters, 1993). Often, the weight involved is very little, but the frequency of repetition may reach very high levels, as in manual data input on a keyboard. Examples of this sort of stress may be seen in keyboard entry personnel and other employees in related tasks. Posture is a strong factor in cumulative trauma disorder especially when developing new tasks. Employees must not be allowed to let their dedication to the project overshadow good posture, or other good work habits. Thought must be given to service access when designing test enclosures. Proper access allows the employee to address the task with good posture. Repetitive tasks should be evaluated for good ergonomic practice, and employees must be trained in the principles of biomechanics and how good work practice will prevent ergonomic injury.

3.2.2 Chemical Exposure Much of the health hazard in the physical science laboratory comes from exposure to chemicals. Common problems arise from improper use and storage of chemicals. Incompatible chemicals should never be stored in the same cabinets. Before using any chemical in the laboratory, the employee should review the material safety data sheet, which must be available for all employees. Particular attention should be paid to the physical properties of the chemical, to ensure that the chemical will be used within safe environmental limits and stored safely and that the proper safety and spill clean-up materials are on hand. Employees handling chemicals must wear appropriate personal protective gear as necessary. This may include single or double gloves, safety glasses, respirators, or special clothing prescribed by regulation or by the producer of such chemicals.

3.2.3 Ignition Sources Generally, as part of the chemical safety requirements, the laboratory must assure that equipment used in the presence of chemicals that may form an explosible atmosphere are certified and maintained as explosion resistant (OSHA, 1997, Subpart H, Section 106 and Subpart S, Sections 304, 305 and 307).

3.2.4 Cryogens Cryogenic liquids or solids at very low temperatures pose several hazards. The first and most obvious is the low temperature itself, which is capable of freezing flesh. Appropriate personal protective equipment must be used when working with cryogens. Transfer pipes, hoses, and vessels often provide a location for condensation of air or other gas in contact with the vessel surface.

This condensate may be enriched with oxygen. As a result, such pipes and vessels should not be in contact with or suspended above asphalt surfaces or other combustible surface materials (OSHA, 1997, Subpart Z, Section 1051).

3.2.5 Physical Hazards Physical hazards may include loose objects on high ledges or shelves, improperly guarded machines, or even loose tools near large magnets. Physical hazards may also include unguarded pulleys, swinging arms, low ceilings, or stacks of unused equipment or boxes that might fall on a worker. Physical hazards might also include loose items that may fall or be thrown during an earthquake (in areas prone to earthquakes). Physical hazards might also include objects that are ejected from experimental devices, or a whole variety of electrical hazards. Some of these may include ungrounded wiring, loose cables on the floor, unshielded cables in wet areas, and too many plugs for a given circuit rating.

3.2.6 Inhalation When looking for inhalation hazards, it must be remembered that many substances when inhaled are toxic but may not necessarily present a hazard otherwise. Any dust that can be inhaled must be controlled to acceptable levels, preferably by the use of engineering controls such as fume hoods, local exhaust, barriers, or remote operation (OSHA, 1997, Subpart Z, Section 1000). Protection from exposure to gases to prevent toxic exposure or suffocation must be provided. Again, engineering controls are preferred to personal protective devices such as respirators. Survey the materials being used and check for any applicable standards such as OSHA standards, HSE standards, ISO standards, EPA standards where appropriate, DOE standards, DOT regulations where appropriate, or special industry standards. In certain emergency situations, as specified in OSHA standards, respirators may be used when

1. they meet use criteria;
2. there is a respirator program in place to assure safe use of the respirator; and
3. the safe capacity of the particular respirator meets or exceeds the expected concentration of contaminant.

Other criteria may apply and the laboratory should consult current OSHA requirements on this issue (OSHA, 1997, Subpart I, Sections 132–136).

3.2.7 Radiation and Radioactive Materials Most generally, radiation is divided into two classes, ionizing and nonionizing. Ionizing radiation includes particulate, beta, alpha, and gamma radiation. Nonionizing radiation generally includes ultraviolet (UV), infrared (IR), radio frequency (rf), extremely low-frequency (ELF), and sonic energy. The distinction between the two classes is generally made by considering whether or not a particular type of radiation can cause ionization in human skin. This energy level is often taken to be 10 eV. Further specific information on appropriate levels of exposure will be found in several of the references given in the Further Reading section. See also Wilkening (1991); OSHA (1997), Subpart G, Section 97.

3.2.8 Noise Noise exposures may come from a variety of sources including equipment, tools, radios, and explosions. Care must be taken to assure that exposures to noise are not excessive. Ear damage due to noise exposure is generally permanent. OSHA has established permissible exposure levels based upon the A-weighted scale of a standard noise meter (OSHA, 1997, Subpart G, Section 95).

3.2.9 Biological and Nontraditional Hazards Physical science laboratories are becoming more interdisciplinary in nature, with equipment being developed for medical diagnostics and other nontraditional markets. It is therefore necessary to be aware of and follow requirements for safe handling of biological materials. It is possible that biological samples such as microorganisms, including viruses, may be present in the physical laboratory. Laboratory animals or pathogenic tissue samples may be found in some laboratories. OSHA regulates the handling of biological materials primarily through the bloodborne pathogens regulations, which are aimed at HIV and HIB (OSHA, 1997, Subpart Z, Section 1030).

3.2.10 Heat Stress or Exposure to Temperature Extremes Precautions must be taken to ensure that employees do not suffer either hyperthermia (high temperature) or hypothermia (low temperature). Hyperthermia is generally considered to exist when the body core temperature exceeds 38 °C and there is no underlying physical condition to cause the high temperature. Hypothermia is generally considered to exist when the core body temperature falls below about 32 °C.

4. WHAT TO DO WHEN AN ACCIDENT HAPPENS

The definition of an accident includes incidents involving spills of hazardous materials, bomb threats, civil disturbance, explosion, fire, natural disaster such as an earthquake, terrorism, or any other act or happenstance that could put laboratory personnel or the community at risk. Specific requirements vary by locality and nation. Many community governments and most national governments can provide contacts and specific information necessary to meet legal requirements in their respective venues.

4.1 Emergency Response

It is beyond the scope of this article to describe procedures for every possible accident. A laboratory chemical response plan such as that required by OSHA will establish the levels of response based upon the relative severity of the accident. It includes whom to call for particular events, what to do, and whom to inform. All employees in a facility must be trained in the contents of this plan and how to respond if an accident occurs.

4.1.1 Responsibility of the Safety and Health Professional Depending upon the size of the facility, the safety and health professional and one or more employees may be specially trained to respond to medical emergencies, chemical spills, radiation releases, or other hazards. When such employees or teams are used, they must be appropriately trained according to applicable local and federal regulations.

4.1.2 Injuries When an injury occurs, immediate notification of in-laboratory first aid responders or local emergency medical providers is imperative. For other emergencies such as fire or other large incidents, an automatic call system may be used to contact emergency responders. Such procedures will be set out clearly in the written safety and health plan for the laboratory. Generally, a good relationship should be established between the laboratory and the local emergency responders. Where there are security areas or special hazards, it is important for the local emergency responders and laboratory management to work out agreeable procedures for emergency response. The safety and health official is usually the point of contact with local service providers in setting up such procedures.

4.2 Notification Requirements

Every incident will be reported to immediate supervisors, while more severe incidents will be required to be reported to local, state, and national agencies.

4.2.1 National Government Notification In the United States, OSHA requires employers to maintain a log of injury and illness that is occupationally related. This is kept on the OSHA 200 log (OSHA, 1997, Subpart F, Section 67). Moreover, OSHA must be notified within 8 h of any fatal accident or any work-related incident requiring the hospitalization of three or more employees (OSHA, 1997, Subpart Z, Section 8). The Environmental Protection Agency requires notification of spills through the Resource Conservation and Recovery Act (RCRA) (EPA, 1976a), Toxic Substances Control Act (TSCA) (EPA, 1976b), Comprehensive Environmental Response Compensation and Liability Act (CERCLA) (EPA, 1977), and Superfund Amendments and Recovery Act (SARA) (EPA, 1986). This response is required within 24 h of the incident to the closest EPA office. Other nations may require differing levels of reporting. The safety and health professional will be familiar with all requirements for reporting.

4.2.2 Local Government Notification Many local governments have regulations requiring notification in the event of an accident. As it is a good practice to involve local service providers in safety planning, these same local providers should prove to be valuable sources for specific local reporting requirements. Along with these providers the local air-quality board or its equivalent, the waste-management authority, the EPA, and the OSHA office or their equivalents are excellent sources of specific and current information.

4.3 Evaluation and Remediation

Once an incident has occurred, it is important to evaluate the causes and ramifications of the incident to avert further incidents. Management, along with the safety team and the principal investigators, should evaluate all pertinent information to establish appropriate remediations in whatever form necessary to assure safe working conditions for laboratory personnel, visitors, and the community in which the laboratory is located.

5. CONCLUSION

Safety and health in the physical science laboratory is the responsibility of everyone in the facility. The safety and health professional acts as an integral part of the laboratory to help each employee to understand his or her role in managing safety in the laboratory. There may be specific requirements to be followed by the laboratory imposed by national or local government as well as by contracting authorities. No element of a sound safety plan can be abrogated by the pressure of project completion deadlines or cavalier attitudes of investigators. It is the responsibility of management to assure that laboratory employees and invited investigators adhere to established procedures and that where new processes are being developed, appropriate planning and care is taken to evaluate potential hazards and risk for those processes. Because of the intricacy of the processes and governmental requirements, it is suggested that competent professional safety and health consultants be retained either on an as-needed basis or on staff. In addition, all the regulatory agencies provide help for specific processes and can be very helpful in setting up and maintaining effective safety and health programs.

GLOSSARY

Accident: Any unexpected happening that interrupts the work sequence or process and that may result in injury, illness, or property damage to the extent that it causes loss.

Ergonomics: The science of fitting the job to the worker in a manner so as to lessen the risk of injury from operations involving repetitious, forceful or awkward postures.

Hazard: A situation or process that has the potential to cause harm.

MSDS: Material safety data sheet. A document that communicates the physical and chemical hazards of chemicals to the employee using or handling the chemical. The MSDS contains detailed information about handling, storage and use of a chemical and how to control unexpected events.

OSHA: Occupational Safety and Health Administration.

PEL: Permissible exposure limit. An exposure level to a chemical or other hazard, established by OSHA, above which employees may not be exposed.

PPE: Personal protective equipment. Items such as safety glasses, gloves, respirators, or similar apparel that are used by individuals as protection from hazard.

Risk: The likelihood that a particular hazard will result in an accident.

Works Cited

EPA (Environmental Protection Agency) (1976a), *Resource Conservation and Recovery Act* (RCRA), 42 USC 321 *et seq.*, Washington: GPO.

EPA (Environmental Protection Agency) (1976b), *Toxic Substances Control Act* (TSCA), 15 USC 2601 *et seq.*, Washington: GPO.

EPA (Environmental Protection Agency) (1977), *Comprehensive Environmental Response, Compensation, and Liability Act* (CERCLA), 42 USC 9601 *et seq.*, Washington: GPO.

EPA (Environmental Protection Agency) (1986), *Superfund Amendments and Reauthorization Act* (SARA), 42 USC 9601 *et seq.*, Washington: GPO.

Kortsha, G. X. (1991) in: G. D. Clayton, F. E. Clayton (Eds.), *Patty's Industrial Hygiene and Toxicology, General Principles*, Vol. 1, Part A, New York: John Wiley and Son, Chap. 3.

Lawrence Livermore National Laboratory (1997), *The Environment, Safety, and Health Program at the Lawrence Livermore National Laboratory*, Livermore, CA: www.llnl.es_and_h/esh_prog.

Mutchler, J. E. (1991) in: G. D. Clayton, F. E. Clayton (Eds.), *Patty's Industrial Hygiene and Toxicology*, Vol. 1, Part A, New York: Wiley and Sons, p. 763 *et seq.*

Newsom, S. (Ed.) (1996), *Risk Assessment and Risk Management in Regulatory Decision Making*, The Presidential/Congressional Commission on Risk Assessment and Risk Management, Vols. 1, 2, Washington: GPO.

OSHA (Occupational Safety and Health Administration) (1970), *OSHA Act of 1970*, amended on November 5, 1990, Washington: GPO.

OSHA (Occupational Safety and Health Administration) (1997), *Code of Federal Regulations*, Title 29, Washington: Office of the Federal Register, National Archives and Records Administration.

Soule, R. D. (1991) in: G. D. Clayton, F. E. Clayton (Eds.), *Patty's Industrial Hygiene and Toxicology*, Vol. 1, Part B, New York: Wiley and Sons, p. 137 *et seq.*

Waters, T., Putz-Anderson, V., Garg, A., Fine, L. J. (1993), *The Revised National Institute for Occupational Safety and Health Lifting Equation*, London: Taylor and Francis.

Wilkening, G. M. (1991) in: G. D. Clayton, F. E. Clayton (Eds.), *Patty's Industrial Hygiene and Toxicology*, Vol. 1, Part B, New York: Wiley and Sons, p. 599 *et seq.*

Further Reading

American Conference of Governmental Industrial Hygienists (ACGIH) (1998), *Threshold Limit Values for Chemical Substances and Physical Agents, Biological Exposure Indices*, Cincinnati, OH: ACGIH.

Clayton, G. D., Clayton, F. E. (Eds.), 1991, *Patty's Industrial Hygiene and Toxicology*, New York: Wiley and Sons.

Earley, M. W. (Ed.) (1996), *National Electrical Code Handbook*, Quincy, MA: National Fire Protection Association.

Furr, A. K. (Ed.) (1990), *Handbook of Laboratory Safety*, 2nd ed., Boca Raton: Chemical Rubber Company.

Newsom, S. (Ed.) (1996), *Risk Assessment and Risk Management in Regulatory Decision Making*, The Presidential/Congressional Commission on Risk Assessment and Risk Management, Vols. 1 and 2, Washington: GPO.

Occupational Safety and Health Administration, *OSHA Computerized Information System*, CD-ROM, 729-013-00000-5, Washington: GPO.

Occupational Safety and Health Administration, *OSHA Computerized Information System*, Internet Web site, Washington, D.C. *http://www.osha.gov.*

Waters, T. R., *et al.* (1993), *The Revised NIOSH Lifting Equation*, London: Taylor and Francis.

SILICIDES

JEFFREY P. GAMBINO, *IBM Microelectronics, Hopewell Junction, New York, U.S.A.*

EVAN G. COLGAN, *IBM Research, Yorktown Heights, New York, U.S.A.*

INTRODUCTION

Silicon forms a wide variety of compounds with metals; these compounds are called silicides. Silicides have a number of useful properties including high thermal stability, high mechanical hardness, and low resistivity. Silicides are used in a wide variety of applications. In microelectronics, silicides are used as Ohmic contacts, Schottky contacts, and interconnects. In these applications, the high thermal stability of the silicides and the low resistivity provide an advantage compared with other conductors, such as Al, which has much lower thermal stability, or polycrystalline silicon, which has much higher resistivity. Silicides are often used as furnace elements, on account of their thermal stability in oxidizing ambients. Because of their high strength, silicides are also being considered for use as high-temperature structural materials, in aircraft engines for example. For these applications, silicides have an advantage of higher thermal stability and lower density compared with Ni-based super-

ISBN 3-527-29308-6

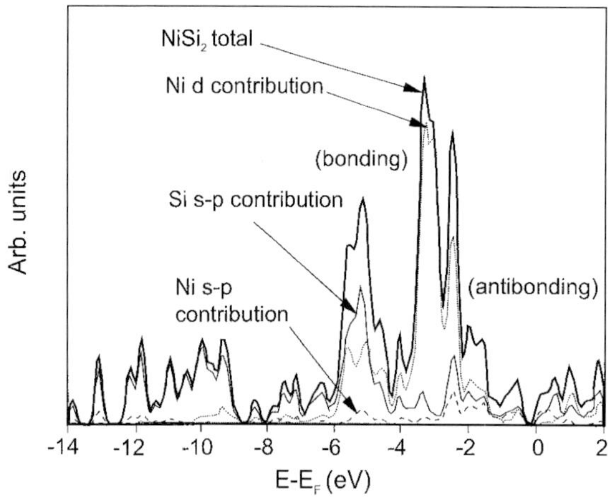

FIG. 1. Theoretical density of states and partial density of states for $NiSi_2$ (Bisi and Calandra, 1981).

alloys, and higher ductility and lower cost compared with Si-based ceramics. Other applications where silicides are being considered include thermoelectric devices, magnets and superconductors.

This article surveys the properties, processing, and applications of transition-metal silicides. The first section focuses on the properties of silicides, including the electronic and atomic structure, reactions and stability, and the mechanical, optical, electrical, and magnetic properties of these materials. The second section describes processes for forming silicides, such as the salicide process, the polycide process, physical vapor deposition, chemical vapor deposition, ion implantation, melt growth, sintering, and plasma-spray deposition. The final section discusses applications of silicides, including microelectronics, structural materials, coatings, thermoelectric materials, magnetic materials, and optical materials. Silicides formed from alkali metals, alkaline-earth metals, and rare-earth metals (Samsanov and Vinitskii, 1980) are not discussed in detail here, because there are very few potential applications for these materials at present.

1. PROPERTIES

1.1 Structure

1.1.1 Electronic Structure

1.1.1.1 Bulk Electronic Structure Many of the useful properties of silicides, such as high melting point and high mechanical hardness, arise from the formation of strong covalent bonds. Despite these strong covalent bonds, in most silicides there is a continuous density of states across the Fermi level, which provides high conductivity. Knowledge of the electronic structure of silicides is also useful for understanding properties such as Schottky-barrier heights (see Sec. 1.1.1.2) and chemical bonds and compound stability (see Sec. 1.1.2.1).

The main factor in determining the electronic structure of transition-metal silicides is the formation of hybrid orbitals between d states in the transition metal and 3p states in Si, resulting in strongly covalent metal d–Si p bonds (Lange, 1996; Bisi and Calandra, 1981; Speier *et al.*, 1989; Mattheiss, 1992; Reader *et al.*, 1992). These bonds produce bonding and antibonding states above and below the Fermi level (Fig. 1). In addition, there are some nonbonding d states with energies close to those of the pure metals, but of reduced width as a result of the greater separation between metal atoms in the silicides compared with the pure metal. The Si s states are only weakly involved in chemical bonding and form a relatively broad band several electronvolts below the Fermi level. The sequence of these levels with increasing energy is as follows: Si s, Si p/metal d bonding, metal d nonbonding, and Si p/metal d antibonding. The Fermi level is generally between the nonbonding and the antibonding states; hence the nonbonding states are usually occupied while the antibonding states are usually empty (Derrien, 1995).

Some trends in the band structure of silicides are observed with increasing atomic

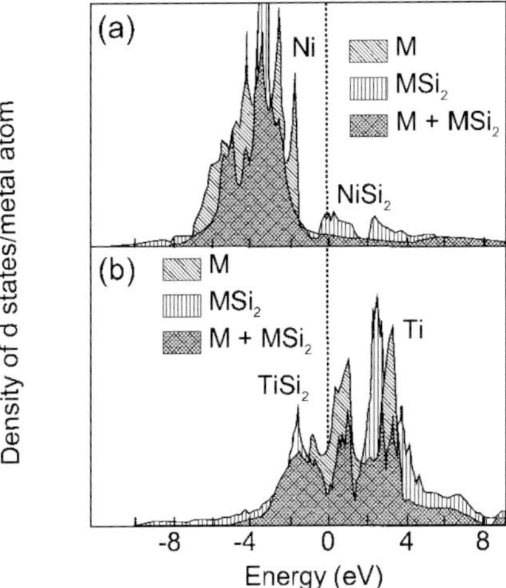

FIG. 2. Comparison of the calculated metal d-state densities in the pure metal and the disilicides for (a) Ni and (b) Ti. The pure metal bands have been shifted so that the centroids of the d band match those of the disilicides. (Reprinted from Weijs *et al.*, 1990, with permission from Springer-Verlag. Copyright 1990.)

number, moving through a series in the periodic table (Derrien, 1995; Reader *et al.*, 1992). First, the metal d states move through the Fermi level with increasing atomic number. In addition, the shape of the d states varies within a series, starting as a broad band of mostly empty states and finishing as a narrow band of mostly filled states. These trends are apparent when one compares the band structures of $NiSi_2$ and $TiSi_2$ (Weijs *et al.*, 1990; Reader *et al.*, 1992) (Fig. 2). The reduction in band width of the nonbonding metal d states compared with the elemental metal is greater for $NiSi_2$ than for $TiSi_2$, because of a stronger interaction with the Si p states in the case of $TiSi_2$. In addition, compared with $TiSi_2$, the Ni d character in the band structure of $NiSi_2$ is relatively weak in the bonding–antibonding regions above and below the metal d bands. The difference is due to the smaller radial extent of the Ni 3d wave function compared with that in Ti, which results in less mixing of the Si p and metal d states.

In most of the transition-metal silicides, there is a continuous density of states across the Fermi level, resulting in a metallic conduction. However, some silicides are semiconductors (see Sec. 1.3.3.1), with the minimum in the density of states at the Fermi level opening up into an energy gap. For a given metal, the density of states at the Fermi level increases as one goes from the Si-rich phases to the metal-rich phases, as a result of an increasingly strong interaction between metal d states (Speier *et al.*, 1989).

1.1.1.2 Interfacial Electronic Structure

One of the major applications for silicides is as contacts to semiconductors (see Sec. 3.1). The electronic structure of interfaces is also of interest for determining the adhesion strength of metal–silicide interfaces in structural materials (Hong *et al.*, 1993).

The electrical properties of the contacts are determined by the electrical barrier that forms at the metal–silicon interface. The height and width of the barrier determine the ease of electron transport across the interface, and hence determine whether the contact is Ohmic or rectifying, and whether it has high or low resistance. The barrier height depends on the work-function difference between the metal and semiconductor, the interface state density, the atomic structure at the interface, and the doping level in the semiconductor.

The mechanism of Schottky-barrier formation in covalent semiconductors is still not well understood, despite years of study. The barrier height ϕ_b is determined by the requirement that in equilibrium, the Fermi levels of the metal and semiconductor must be equal. This is achieved by creation of an electric dipole at the interface. In the absence of interface states, the charge at the interface is related to the difference in Fermi levels between the metal and the semiconductor, and is given by

$$\Phi_b = \varphi_m - \chi_s, \qquad (1)$$

where φ_m is the metal work function and χ_s is the electron affinity of the semiconductor. This model works fairly well for ionic semiconductors, where the interface state densities are low, and the barrier height does in fact show a strong dependence on the electronegativity of the metal (which is related to the work function of the metal; see Louie *et al.*, 1977; Schluter, 1977; Kurtin *et al.*, 1969). However, for covalent semiconductors, there is much less dependence of the barrier height on the electronegativity of the metal, as a result of the presence of interface states that pin the

Fermi level (Cowley and Sze, 1965; Heine, 1965). For covalent semiconductors, an extra term is added to Eq. (1), corresponding to the electrostatic interface dipole, that is determined by the electron distribution at the interface (Flores and Ortega, 1992). In fact, interface states associated with the silicide have been postulated from recent band-structure calculations of the silicide–Si interface (Fujitani and Asano, 1994). Some of these interface states have energies within the Si band gap and spatially extend a few atomic layers into the Si.

Recently it has been shown that Schottky-barrier heights depend on the local atomic structure of the interface (Tung, 1992). For example, the barrier height of epitaxial $NiSi_2$ on *n*-type Si can range from 0.40 eV [on (100) Si] to 0.79 eV [for type B $NiSi_2$ on (111) Si] (Tung, 1993). (For the type A interface, the silicide has the same orientation as the substrate, whereas for the type B interface, the silicide also has a ⟨111⟩ orientation, but is rotated by 180° about the ⟨111⟩ surface normal with respect to the substrate.) Fujitani and Asano (1994) have proposed that the atomic-structure dependence of the barrier height is due to bond bending at the interface, though more studies are required to determine if this hypothesis is correct.

Most of the Schottky-barrier heights on Si reported in the literature have been measured on polycrystalline silicides (Table 1). However, the different orientations in the polycrystalline silicide are likely to have different barrier heights, so that the contacts are electrically nonhomogeneous. The Schottky-barrier heights measured from polycrystalline silicides will be determined by the silicide grains with the lowest barrier heights (Tung, 1992).

The Schottky-barrier heights of silicides have been characterized on a number of compound semiconductors (Liu *et al.*, 1997; Shih *et al.*, 1988, Yokoyama *et al.*, 1986; Porter *et al.*, 1995a,b,c) including SiC, GaAs and GaN (Table 2). GaN and SiC have a greater degree of ionic bonding than Si or GaAs; hence, the barrier heights are expected to show a greater dependence on the type of metal or silicide.

Table 1. Schottky-barrier heights for various silicides on *n*-type Si (Tung, 1993; Sullivan *et al.*, 1993).

Silicide	Silicon substrate	Schottky barrier (eV)
IrSi (polycrystalline)	(100)	0.90
PtSi (polycrystalline)	(100)	0.86
$OsSi_2$ (polycrystalline)	(100)	0.85
Pt_2Si (polycrystalline)	(100)	0.77–0.85
MnSi (polycrystalline)	(100)	0.76
Pd_2Si (polycrystalline)	(100)	0.72
$ReSi_2$ (polycrystalline)	(100)	0.68
NiSi (polycrystalline)	(100)	0.65
$NiSi_2$ (type B, epitaxial)	(111)	0.79
$NiSi_2$ (type A, epitaxial)	(111)	0.65
$NiSi_2$ (epitaxial)	(100)	0.40
$CoSi_2$ (polycrystalline)	(100)	0.65
$CoSi_2$ (type B, epitaxial)	(111)	0.69
WSi_2 (polycrystalline)	(100)	0.65
$MoSi_2$ (polycrystalline)	(100)	0.64
RhSi (polycrystalline)	(100)	0.63
VSi_2 (polycrystalline)	(100)	0.63
FeSi (polycrystalline)	(100)	0.62
$TiSi_2$ (polycrystalline)	(100)	0.60
$CrSi_2$ (polycrystalline)	(100)	0.60
$TaSi_2$ (polycrystalline)	(100)	0.59
$NbSi_2$ (polycrystalline)	(100)	0.59
$ZrSi_2$ (polycrystalline)	(100)	0.55
HfSi (polycrystalline)	(100)	0.53
YSi_2 (polycrystalline)	(100)	0.39
$GdSi_2$ (polycrystalline)	(100)	0.37

1.1.2 Atomic Structure

1.1.2.1 Crystal Structure The crystallographic structures of transition-metal silicides are varied and complex (Fig. 3). Some structures form over significant regions; an MSi compound with the FeSi prototype is formed for nine different transition metals. Typically, an identical or related structure is found for silicides of the same composition and group (Murarka, 1983), though this is not without

Table 2. Schottky-barrier heights for various silicides on n-type compound semiconductors. In all cases, the silicide is polycrystalline.

Silicide	Substrate	Schottky barrier (eV)	Reference
Ti_5Si_3	α-SiC (1000)	0.90	Porter *et al.*, 1995b
CoSi	α-SiC (1000)	1.05	Porter *et al.*, 1995a
Pt_2Si	α-SiC (1000)	1.26	Porter *et al.*, 1995c
$WSi_{0.6}$	GaAs	0.70	Shih *et al.*, 1988
$MoSi_{0.7}$	GaAs	0.70	Yokoyama *et al.*, 1986
PtSi	GaN	0.87	Liu *et al.*, 1997

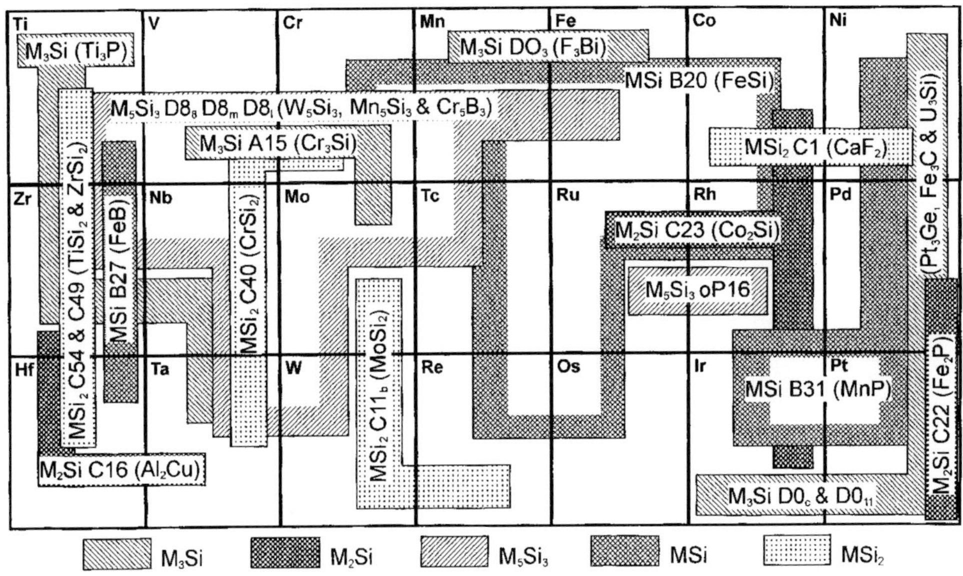

FIG. 3. Fields of structures for transition-metal silicides (from Goldschmidt, 1967; Massalski, 1990).

exception (for example, V_3Si and $CrSi_2$; see Fig. 3 and Table 3). Further details on crystal structure are available in standard references (Pearson, 1972).

The $MoSi_2/CrSi_2/TiSi_2$ prototypes are closely related and are just variations in stacking sequences of close-packed triangular layers (Fig. 4). The tetragonal $MoSi_2$ structure results from a twofold repeat, the hexagonal $CrSi_2$ structure results from a threefold repeat, and the orthorhombic $TiSi_2$ results from a fourfold repeat. The $ZrSi_2$ structure is also related to these structures. The structural differences correlate with d-band filling, with C40 being more stable than C54, and $C11_b$ being more stable than C40, as the number of d electrons in the transition metal increases. This suggests that the structural energies of transition metal disilicides are mainly determined by electronic band effects rather than atomic size effects (Carlsson and Meschter, 1991).

1.1.2.2 Interfacial Atomic Structure: Epitaxial Silicide–Silicon Interface In Sec. 1.1.1.2, it was noted that the Schottky-barrier height depends on the atomic structure of the silicide–silicon interface. In addition, novel device structures are made possible by growing silicides epitaxially on Si (see Sec. 3.1.4). Hence, there has been much interest in the atomic structure of epitaxial silicide–Si interfaces.

Factors that affect epitaxial growth are the lattice mismatch between the two materials and the chemical bonding at the interface (Chen and Tu, 1991; van den Hoek *et al.*, 1988; Zur *et al.*, 1985). Epitaxial growth is often observed in systems where the two lattices have identical or quasi-identical forms and have similar lattice spacings. The interfacial chemistry also plays an important role in epitaxial growth; epitaxial growth can occur if the interfacial energy is sufficiently low, even if the lattice mismatch is relatively high (an example is C54-$TiSi_2$ on Si).

Recently, the differences in interface structure between epitaxial $CoSi_2$ and epitaxial $NiSi_2$ on (111) Si have been explained in terms of the interfacial bonds (van den Hoek *et al.*, 1988). For $NiSi_2$, the metal atoms closest to the interface are seven-fold coordinated (seven nearest-neighbor Si atoms), and the Si atoms of the silicide form bonds to the Si substrate, whereas for $CoSi_2$, the metal atoms closest to the interface are eight-fold coordinated (eight nearest neighbor Si atoms) and form bonds to the Si substrate. The most energetically favorable structure depends on the number of electrons in bonding orbitals. In $CoSi_2$/(111), only

Table 3. Crystal structures of some transition-metal silicides (from Pearson, 1972).

Composition	Prototype	Strukturbericht designation	Pearson symbol	Space group
M_3Si	Ti_3P	—	tP32	$P4_2/n$
	Cr_3Si	A15	cP8	$Pm\bar{3}n$
	Pt_3Ge	—	mC16	$C2/m$
	Fe_3C	DO_{11}	oP16	$Pnma$
	U_3Si	DO_c	tI16	$I4/mcm$
	Fe_3Bi	DO_3	cF16	$Fm\bar{3}m$
M_2Si	Co_2Si	C23	oP12	$Pnma$
	Al_2Cu	C16	tI12	$I4/mcm$
	Fe_2P	C22	hP9	$P\bar{6}2m$
M_5Si_3	W_5Si_3	$D8_m$	tI38	$I4/mcm$
	Mn_5Si_3	$D8_8$	hP16	$P6_3/mcm$
	Cr_5B_3	$D8_l$	tI32	$I4/mcm$
	Ru_5Si_3, Rh_5Si_3[a]	—	oP16	$Pbam$
MSi	FeSi	B20	cP8	$P2_13$
	MnP	B31	oP8	$Pnma$
	FeB	B27	oP8	$Pnma$
MSi_2	$CrSi_2$	C40	hP9	$P6_222$
	$MoSi_2$	$C11_b$	tI6	$I4/mmm$
	$ZrSi_2$	C49	oC12	$Cmcm$
	$TiSi_2$	C54	oF24	$Fddd$
	CaF_2	C1	cF12	$Fm\bar{3}m$

[a] Prototype not known for Ru_5Si_3 and Rh_5Si_3.

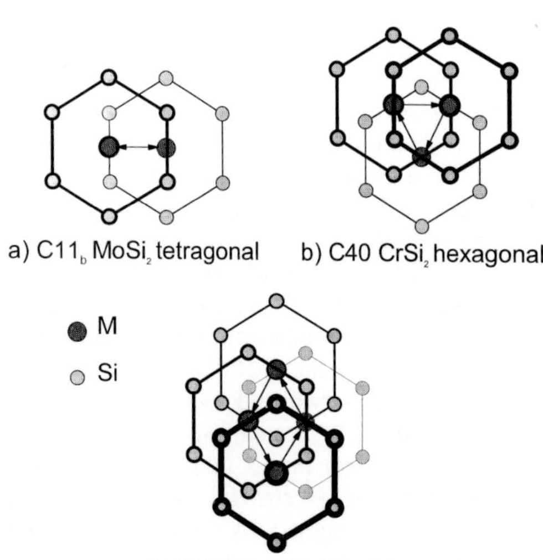

FIG. 4. Stacking relationships between $MoSi_2$, $CrSi_2$, and $TiSi_2$ structures.

the metal–Si bonding orbitals are filled with electrons; the Si–Si bonding orbitals have no electrons, and hence are energetically unfavorable. For $NiSi_2$/(111), the extra d electron from Ni occupies the Si–Si bonding orbital; this is energetically more favorable than metal–Si bonding orbitals with two electrons, because of the large overlap of the Si–Si bonds (Fig. 5).

1.1.2.3 Phonons There have been relatively few studies on phonon spectra (atomic vibrations) in silicides. Most of the studies have focused on semiconducting silicides, where the phonon spectra can be used to analyze phases and crystalline quality (Fenske *et al.*, 1996). A number of studies have also been made of specific heat and low temperature resistivity of silicides, which are related to phonons (Nava *et al.*, 1993). Infrared spectroscopy or Raman scattering has been used to study phonons in PtSi, $TiSi_2$, WSi_2, $CoSi_2$, $NiSi_2$, and β-$FeSi_2$ (Miglio and Meregalli, 1996; Guizetti and Marabelli, 1995).

The phonon spectra of semiconducting silicides show high complexity (Fig. 6), due to the

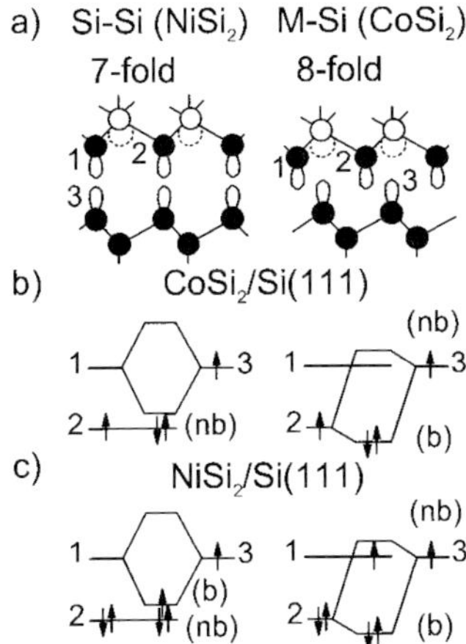

FIG. 5. Energy-level diagrams of interface bonds for (a) seven-fold (left) and eight-fold (right) structures, showing a comparison between (b) $CoSi_2$/Si (111) and (c) $NiSi_2$/Si (111). (Reprinted with permission from van den Hoek *et al.*, 1998. Copyright 1988 by the American Physical Society.)

large number of atoms per unit cell (Miglio and Meregalli, 1996). The phonon spectra for these silicides show a high absorption coefficient, which is a typical feature of ionic bonding, despite the fact the these silicides have strong covalent bonds. A possible explanation is that dynamical charges arise when the atoms vibrate on account of their large polarizability.

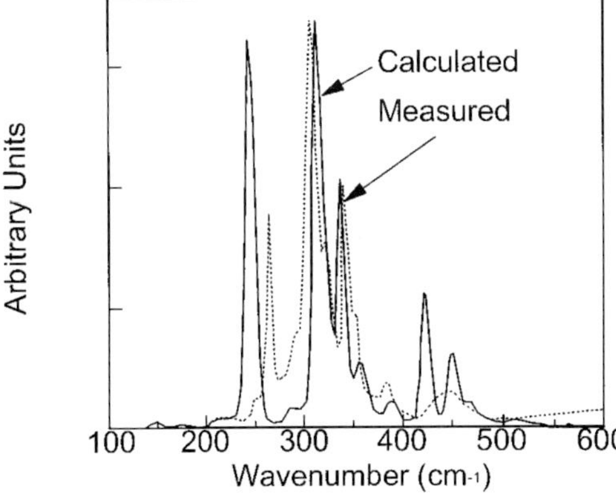

FIG. 6. Comparison between simulation (solid line) and experiment (dotted line) for infrared spectra of β-$FeSi_2$. The peaks in the spectra are probably related to lattice vibrations in the orthorhombic CaF_2 structure. (Reprinted with permission from Miglio and Meregalli, 1996.)

In addition, the phonon spectra for the semiconducting silicides are in nearly the same frequency interval, despite the large mass differences between the metals (for example, Fe and Ir). This suggests that the spectra are mainly determined by the bonding properties, which are similar in these silicides (Fenske *et al.*, 1996; Lange, 1996). Finally, phonons may be responsible for the low mobilities observed in semiconducting silicides (Christensen, 1990).

For silicides with cubic CaF_2 structures, such as $NiSi_2$ and $CoSi_2$, two phonon modes are expected, one Raman-active and the other (at higher frequency) IR-active (Bocelli *et al.*, 1995). The Raman-active mode consists of counterphase motion of the Si atoms along the $\langle 111 \rangle$ direction, while the metal atom is at rest (*i.e.*, only the Si atoms move), whereas for the IR-active mode, there is counterphase motion of the Si atoms with respect to the metal along the $\langle 111 \rangle$ direction (*i.e.*, both Si and metal atoms move) (Malegori and Miglio, 1993).

1.2 Reactions and Stability

1.2.1 Metal–Si Reactions One of the most common methods to form silicides is by reacting a thin film of metal with Si. This approach is used in microelectronics manufacturing to form silicides on polycrystalline Si gates or single crystal Si contacts (see Sec. 3.1). Reactions between metals and Si are also used in diffusion bonding of Si-based ceramics to metals (see Sec. 3.3). In addition, metal–Si reactions are of scientific interest and have revealed much information on mechanisms of thin-film reactions.

1.2.1.1 Silicide Phases In general, when Si comes in contact with a metal it is not thermodynamically stable and will react to form silicides, with the reaction depending on the annealing temperature, the metal–Si phase diagram, and kinetics (Gas and d'Heurle, 1993; Zhang and d'Heurle, 1994). For transition metals, a series of intermediate silicide phases will form. In principle, all the stable intermediate phases in the phase diagram at the annealing temperature should form. However, often for thin-film reactions some stable phases are not observed on account of kinetics factors (see below).

A number of general observations can be made about the binary phase diagrams of

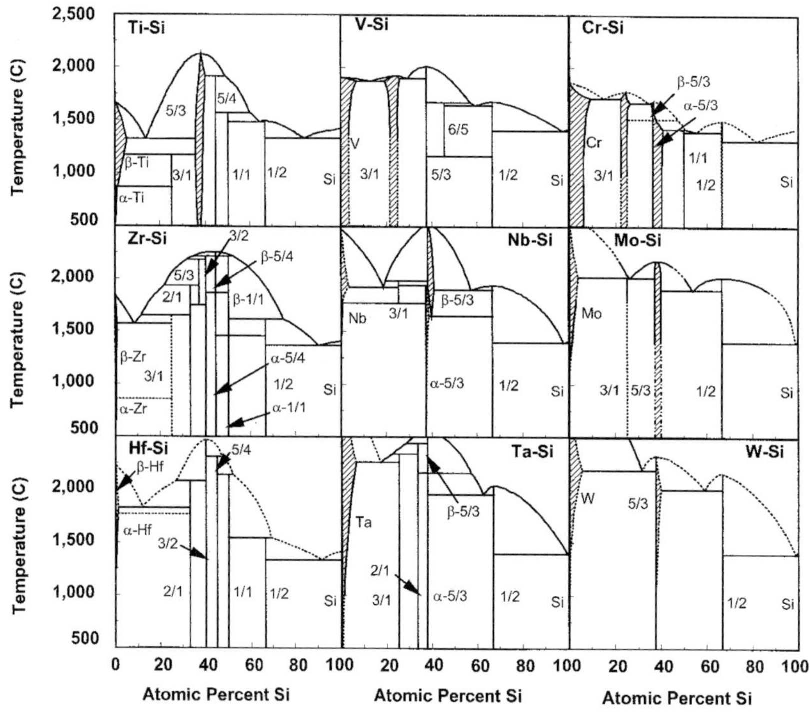

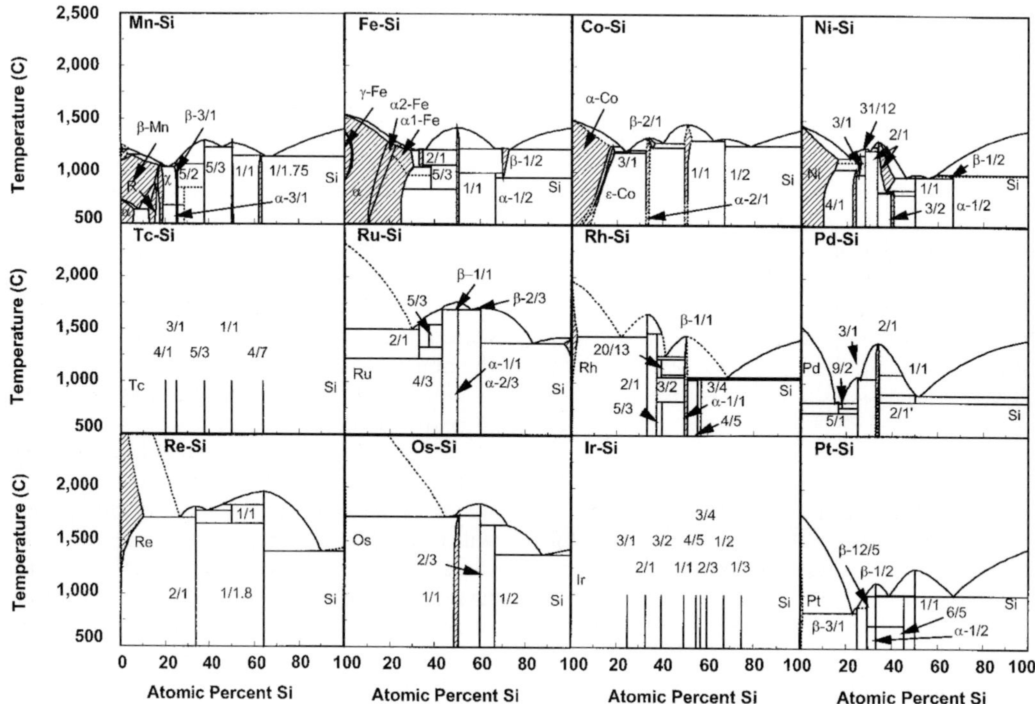

FIG. 7. Phase diagrams of transition-metal silicides (from Massalski, 1990).

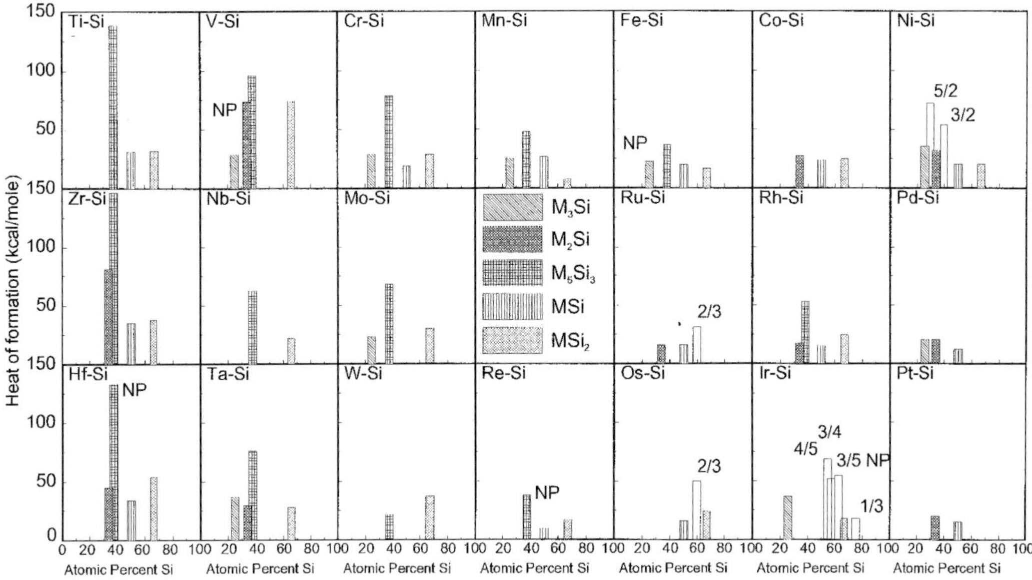

FIG. 8. Heats of formation for transition metal silicides. The M/Si composition is indicated by shading for common stoichiometries or by text. The NP designation is for compositions not present in the equilibrium phase diagram (from Mann and Clevenger, 1995).

transition-metal silicides (Massalski, 1990) (Fig. 7). As noted above, the Si-rich phase is typically MSi_2 except in the case of Ir where a more Si-rich phase, $IrSi_3$, is formed. The most metal-rich phase is generally M_4Si or M_3Si with a number of exceptions. Typically, about five silicide phases are formed for each element, with a minimum of two for W–Si or a maximum of almost ten for Ni, Rh, or Ir. There is significant Si solubility for the first row of transition metals (Ti to Ni), about 3–20 at.% Si, and for Nb, Mo, Ta, W, and Re in the second and third rows, about 3–10 at.% Si. In general, silicides have very limited composition ranges. Significant deviations from stoichiometry are found in Ti_5Si_3, Nb_5Si_3, Cr_5Si_3, Mo_5Si_3, W_5Si_3 (W_5Si_3 and Mn_5Si_3 are prototype structures; see Fig. 3), V_3Si, Cr_3Si (Cr_3Si prototype), FeSi, OsSi, CoSi (FeSi prototype), and a number of other silicides.

Typical values of the transition-metal silicide heats of formation are arranged according to the periodic table (Fig. 8) (Mann and Clevenger, 1995). The heat of formation is related to the d-band occupancy of the silicide. In refractory-metal silicides, only the bonding orbitals are occupied, so that the heat of formation directly reflects the bonding energy. However, in near-noble–metal silicides, nonbonding and antibonding orbitals are occupied, which reduces the bonding energy. Hence, the heat of formation is lower for near-noble–metal silicides compared with refractory-metal silicides (Hara and Ohdomari, 1988).

1.2.1.2 Kinetics Silicides formed by reacting a metal with Si require nucleation and growth of the new phase. If the growth rate is faster than the nucleation rate, then the silicide formation is nucleation controlled. If the nucleation rate is faster than the growth rate, then the silicide formation is either diffusion limited or interface reaction-rate limited (Mayer and Lau, 1989).

Diffusion-limited reactions generally exhibit laterally uniform growth with well-defined kinetics, whereas nucleation-limited reactions exhibit laterally nonuniform growth and a critical temperature dependence (no reaction below the critical temperature and extremely rapid reaction above the critical temperature). In most applications (and especially for integrated circuits), a uniform reaction is required; hence diffusion-limited kinetics are preferred.

Nucleation effects are often observed during formation of new phases in thin films (d'Heurle, 1988). Nucleation-limited reactions are expected when a new phase is forming at an interface and the free-energy change for the transformation is small with respect to the increase in surface energy. Nucleation is generally not rate limiting in reactions between transition metals and Si, because the free-energy change is usually large. Nucleation-limited reactions are often observed during the formation of the Si-rich silicide phases or when the transition results in only a small change in composition or structure (for example, the transformation of C49-$TiSi_2$ to C54-$TiSi_2$). The films produced from nucleation-limited reactions are often rough; the film nucleates in isolated regions, then grows rapidly both laterally and vertically.

After nucleation, growth of the new phase is limited either by the interface reaction or by diffusion (Gas and d'Heurle, 1993). As the thickness increases, the growth of the new phase becomes limited by the flux of atoms through the new phase to the growing interface(s). The growth of the new phase will be diffusion limited and will increase with the square root of time,

$$L(t)^2 = K_d t \tag{2}$$

where K_d is related to the diffusivity of the dominant moving species through the new phase. The analysis is much more complicated if two or more phases grow simultaneously (Zhang and d'Heurle, 1994).

For thin-film reactions such as those used in CMOS processes (*i.e.*, formation of NiSi from Ni_2Si and Si), it is generally observed that the silicide phases form sequentially rather than simultaneously. The new phase (NiSi) only begins to grow after the most metal-rich phase (Ni) is completely consumed. In contrast, in bulk diffusion couples the Ni_2Si phase can reach the critical thickness such that Ni_2Si and NiSi (as well as other phases) can exist simultaneously. Of course, if the diffusion of reactants is altered (by impurities in the films, for example) then multiple silicide phases may be observed simultaneously even in the thin-film case.

1.2.1.3 Results on Metal–Si Reactions Diffusion-limited kinetics have been observed for both bulk and thin-film reactions between metals and Si (Lien *et al.*, 1986; d'Heurle and Gas, 1986; Mayer and Lau, 1989; Gas and d'Heurle, 1997) (Table 4). The reaction kinetics are similar for bulk and thin-film couples. However, the diffusivities are higher in the thin-film case as a result of the contribution from grain-boundary diffusion (Barge *et al.*, 1995). The kinetics can be analyzed by plotting the square of thickness versus time for different annealing temperatures. For diffusion-limited kinetics and assuming that only one phase forms at a time, a series of straight lines is ob-

Table 4. Silicide phase sequence for thin metal films on single-crystal Si. The silicides are listed in the order of their appearance from the top down. The diffusing species during silicide formation is italicized. Where known, the reaction kinetics, t or $\sqrt{t}$, are indicated. (Reprinted from Colgan *et al.*, 1996, copyright 1996, with permission from Elsevier Science.)

Ti[a] Ti_5Si_3 (?) Ti*Si* TiSi_2 t, $\sqrt{t}$(?)	**V** VSi_2 t, $\sqrt{t}$	**Cr** CrSi_2 t	**Co** Co_2Si $\sqrt{t}$ Co*Si* $\sqrt{t}$ *Co*Si_2 t[b]	**Ni** Ni_2Si $\sqrt{t}$ *Ni*Si $\sqrt{t}, t$ *Ni*Si_2 t[b]
Zr $ZrSi_2$	**Nb** NbSi_2	**Mo** MoSi_2 t	**Rh** Rh*Si* $\sqrt{t}$ Rh_4Si_5[b] Rh_3Si_4[b]	**Pd**[c] Pd_2Si $\sqrt{t}$ PdSi[b]
Hf Hf*Si* $\sqrt{t}$ $HfSi_2$[b]	**Ta** TaSi_2 t	**W** WSi_2 t, $\sqrt{t}$	**Ir** Ir*Si* $\sqrt{t}$ $IrSi_{1.75}$[b]	**Pt** Pt_2Si $\sqrt{t}$ *Pt*Si $\sqrt{t}$

[a] Whether Ti_5Si_3 and TiSi form before $TiSi_2$ in a clean system is not clear. The growth kinetics of $TiSi_2$ are linear on single crystal Si and $\sqrt{t}$ on α-Si.

[b] Silicide formation occurs in a laterally nonuniform manner.

[c] The diffusing species during Pd_2*Si* formation is probably both Pd and Si.

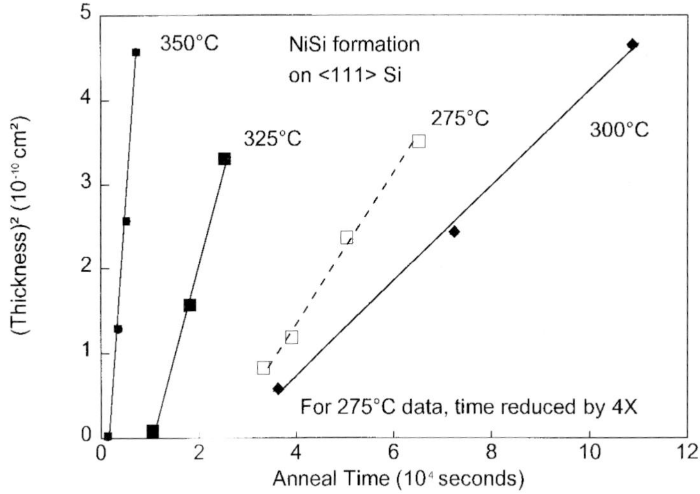

FIG. 9. The thickness of NiSi squared as a function of time, showing diffusion-limited kinetics. (Reprinted with permission from d' Heurle *et al.*, 1984. Copyright 1984 American Institute of Physics.)

tained (d'Heurle *et al.*, 1984) (Fig. 9). The slopes of the lines are related to the diffusivity of the dominant diffusing species, while an Arrenhius plot of the slopes is related to the activation energy for diffusion.

In most cases (Co in Co_2Si, Ni in Ni_2Si, Si in $TiSi_2$), the majority atoms in the silicide are the most mobile (d'Heurle and Gas, 1986). In general, these atoms are next-nearest neighbors to one another, so that a vacancy can propagate along the network of majority atoms with only small perturbations of the lattice. In contrast, the motion of minority atoms requires the coexistence of either two vacancies (minority and majority) or local destruction of the lattice, resulting in a high activation energy.

Nucleation-limited reactions have been observed in a number of silicides where the free energy of formation is small (d'Heurle, 1988). These nucleation effects are eliminated when these phases are formed on amorphous Si, because of the additional free energy in the system associated with crystallization of amorphous Si. Similarly, there is evidence that grain boundaries and grain-boundary triple points are preferred sites for the nucleation of both $CoSi_2$ (Applebaum *et al.*, 1985) and C54-$TiSi_2$ (Ma and Allen, 1994), presumably because of the additional gain in free energy produced by eliminating grain boundaries. The nucleation of C54-$TiSi_2$ is an important problem in microelectronics processing (see Sec. 3.1).

Metal–Si reactions can also be affected by impurities, which generally inhibit silicide formation. The impurities can be films at the metal–Si interface, such as native oxides or surface contamination, contaminants in the metal film from either the metal deposition or the anneal ambient, or dopants in the Si substrate. The most common contaminants during processing of silicides for integrated circuits are oxygen and nitrogen, as well as dopants such as As, P, and B (Gambino and Colgan, 1998).

1.2.2 Results on Metal Reactions with Si-Based Compounds Metals can react with a number of Si-based compounds to form silicides, including SiO_2, Si_3N_4, SiC, and SiGe. For Si integrated-circuit processing, SiO_2 and Si_3N_4 are used as insulators and often react with transition metals used for contacts and interconnects, such as Ti or Co. These reactions can be beneficial, for example by providing good adhesion between the metal and the insulator, or they can be detrimental, for example by leaving conducting filaments on insulating regions. For structural materials, SiO_2, Si_3N_4, or SiC is used either as reinforcing fibers or as the matrix in composites, and often reacts with transition metals that are in the composites. These reactions are usually beneficial, providing strong adhesion between the different components of the composites. Reactions between metals and ceramics are also used to join structural metal alloys to structural ceramics, by diffusion bonding. Finally, SiC or SiGe is used in novel microelectronic devices, and reacts with metals in contacts to form Ohmic or Schottky contacts.

1.2.2.1 Metal–SiGe Reactions There are many studies of reactions of metals with SiGe alloys, including Ti, Zr, W, Co, Ni, Fe, Cr, Mn, Pd, and Pt (Aldrich *et al.*, 1996; Aldrich *et al.*, 1995; Hong and Mayer, 1990; Luo *et al.*, 1997; Carmondy *et al.*, 1996; Boutarek and Madar, 1993). Si and Ge are very similar chemically; hence many of the silicide phases observed in metal–Si reactions are also observed in metal reactions with SiGe alloys. However, there are a number of notable differences, including formation of metal–Ge compounds or metal–SiGe solid solutions, and segregation of Ge.

A number of general trends are observed during reactions of metals with SiGe alloys. Silicide formation is favored over germanide formation, because of the higher heat of formation for silicides (Aldrich *et al.*, 1996). For the same reason, the silicide formation temperatures are higher on SiGe alloys compared with pure Si (Aldrich *et al.*, 1995; Luo *et al.*, 1997). The phase sequence is silicide-like for Si-rich SiGe alloys and germanide-like for Ge-rich SiGe alloys (Aldrich *et al.*, 1995). The dominant diffusing species is usually the same as during silicide formation (*i.e.*, generally metal for near-noble metals and Si/Ge for refractory metals).

The phases that form depend on the miscibility of the silicides and germanides of a given metal (Boutarek and Madar, 1993). For Ti (and Zr), the C54 phases of $TiSi_2$ and $TiGe_2$ have the same structure and very similar unit cell dimensions, and therefore are completely miscible, forming a $Ti(Si_{1-y}Ge_y)_2$ solid solution. Metals where the silicides and germanides are partially miscible, such as V, form solid solutions over a limited range of SiGe alloy compositions. Finally, metals for which the silicides and germanides are immiscible, such as Fe, Cr, and Mn, form separate silicide and germanide phases (Fig. 10). For some metals both types of behavior are observed, depending on the phase that forms; CoSi and CoGe form solid solutions whereas $CoSi_2$ and $CoGe_2$ do not.

1.2.2.2 Metal–SiO₂ Reactions The transition metals exhibit a wide range of reactivities with SiO_2. Titanium and Ta will reduce SiO_2, forming metal oxide and silicide compounds, whereas Co, Mo, and W do not react significantly with SiO_2. These differences can be

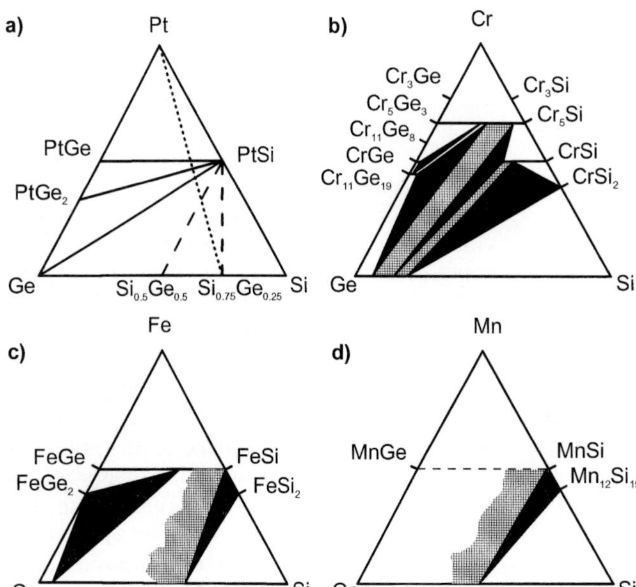

FIG. 10. Phase diagrams for ternary metal–Si–Ge systems; (a) Pt–Si–Ge (Hong and Mayer, 1990), (b) Cr–Si–Ge, (c) Fe–Si–Ge, and (d) Mn–Si–Ge (reprinted from Boutarek and Madar, 1993, copyright 1993, with permission from Elsevier Science).

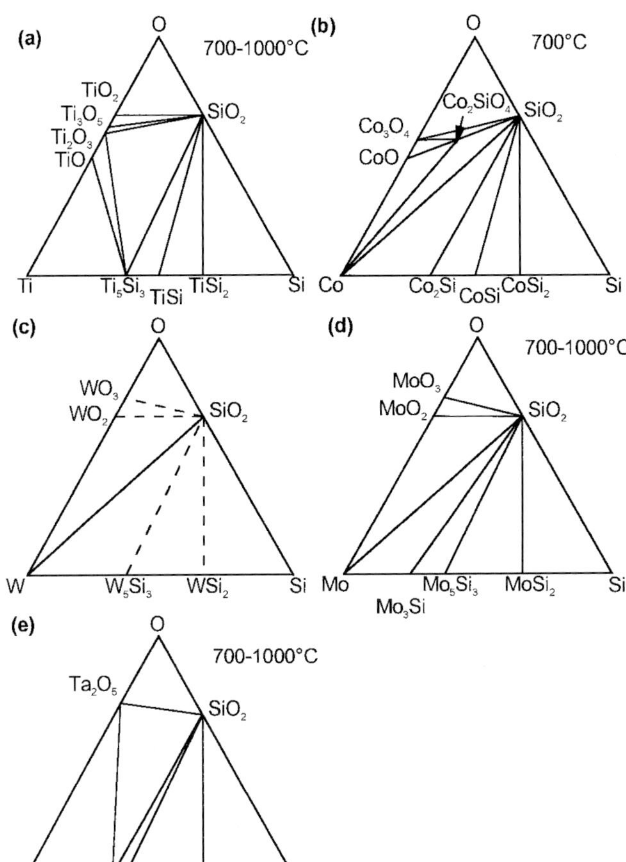

FIG. 11. Ternary phase diagrams for (a) Ti–Si–O, (b) Co–Si–O, (c) W–Si–O, (d) Mo–Si–O and (e) Ta–Si–O. [Parts (a) and (c)–(e) reprinted with permission from Bayers, 1984. Copyright 1984 American Institute of Physics. Part (b) from Maex, 1993, with permission from Elsevier Science.]

understood in terms of the M–Si–O phase diagrams (Fig. 11) (Beyers, 1984; Maex 1993); no reaction will occur between the metal and SiO_2 if there is a tie line between the metal and SiO_2.

Titanium is a good example of a metal that reacts with SiO_2. Reactions between Ti thin films and SiO_2 result in the formation of Ti_5Si_3 at the interface and TiO at the surface (Wang and Mayer, 1990). Oxygen from the consumed SiO_2 is pushed into the unreacted Ti and eventually forms TiO on top of the Ti_5Si_3. A metal-rich silicide (Ti_5Si_3) forms rather than a Si-rich silicide ($TiSi_2$) because the supply of Si is limited (since it is bonded to SiO_2) whereas the supply of Ti is relatively large. Similarly, a Ti-rich oxide is formed (TiO) rather than an oxygen-rich oxide (TiO_2), because the supply of oxygen is limited. Cobalt is a good example of a metal that is relatively unreactive with SiO_2. In fact, no reaction is observed at temperatures below 700 °C (Nguyen *et al.*, 1996).

1.2.2.3 Metal–Si_3N_4 Reactions There have been relatively few studies of metal reactions with Si_3N_4. However, the results are expected to be qualitatively similar to metal reactions with SiO_2: metals that react with SiO_2 will generally also react with Si_3N_4. Reactions between Ti and Si_3N_4 have been studied in detail. The ternary phase diagram (Fig. 12) shows that reactions are expected between Ti and Si_3N_4 (Sambasivan and Petuskey, 1994). At low temperatures (500 °C), the reaction between Ti and Si_3N_4 results in Ti_5Si_3 at the interface, with N dissolving into the unreacted Ti, analogous to the case of Ti on SiO_2. However, at higher temperatures (800 °C), $TiSi_2$ forms at the interface and TiN

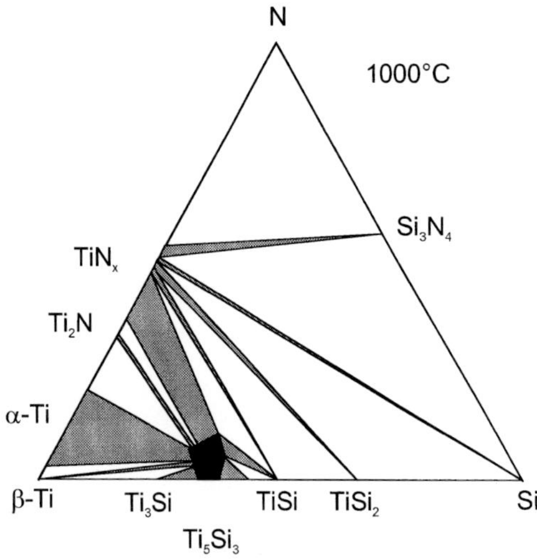

FIG. 12. Ternary phase diagrams for Ti–Si–N (reprinted with permission from Sambasivan and Petuskey, 1994, Fig. 2).

forms on the surface (Barbour *et al.*, 1987), presumably because the reaction is no longer limited by the supply of Si.

1.2.2.4 Metal–SiC Reactions There are many studies of reactions of metals with SiC, including Ti, Hf, Mo, Co, Ni, and Pt. At low temperatures (<750 °C) the results are qualitatively similar to studies on metal reactions with SiO_2 or Si_3N_4. The Ti–Si–C phase diagram (Fig. 13) shows that reactions are expected between Ti and SiC (Taubenblatt and Helms, 1986). Experiments show the Ti_5Si_3 forms at the interface and TiC forms at the surface (Porter *et al.*, 1995a,b). No reactions are observed between Pt, Ni, or Co and SiC at low temperatures. However, at high temperatures (>900 °C), extensive reactions are observed associated with interfacial melting between the metal and SiC. The most likely mechanism is formation of a thin molten silicide layer at the interface (*i.e.*, if the anneal temperature is above the eutectic temperature of the silicide), followed by decomposition of the SiC, with the decomposition of SiC expected to be the rate-limiting step. Because of the limited supply of Si, the reacted layer generally consists of a metal-rich silicide, such as Pt_2Si, Co_2Si, or Ni_3Si (Chou *et al.*, 1991; Porter *et al.*, 1995a,b), though Si-rich silicides are sometimes observed at the highest reaction temperatures. Carbon is also observed, rather than metal carbides, probably in consequence of the low heats of formation for these metal carbides.

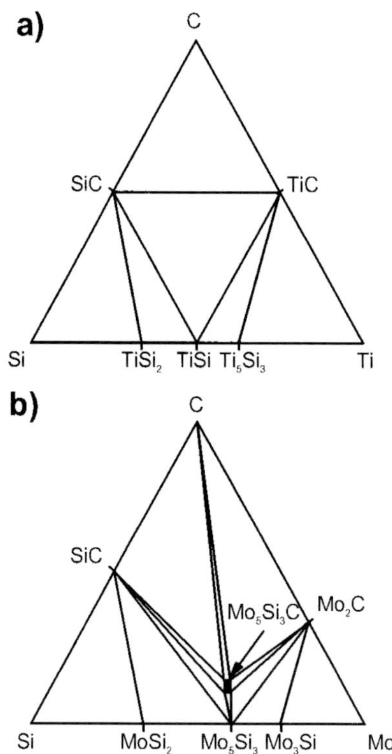

FIG. 13. Ternary phase diagrams for (a) Ti–Si–C (reprinted with permission from Taubenblatt and Helms, 1986. Copyright 1986 American Institute of Physics) and (b) Mo–Si–C (from van Loo *et al.*, 1982, Fig. 2, reproduced with permission).

1.2.3 Thermal Stability The thermal stability of silicides is one of the main reasons that they are used in so many different applications. In microelectronics, the silicide must be stable with respect to the underlying semiconductor during annealing steps used in integrated-circuit manufacturing (such as glass reflow). In structural materials, the silicide must be stable in high-temperature environments, such as the turbines in jet engines.

1.2.3.1 Melting Point Using the traditional definition of "refractory" (stable at >1538 °C, the melting point of Fe), the majority of transition-metal silicides are refractory compounds (Fig. 7). The melting points of the transition metals range from 1246 °C for Mn to 3422 °C for W (Massalski, 1990) and

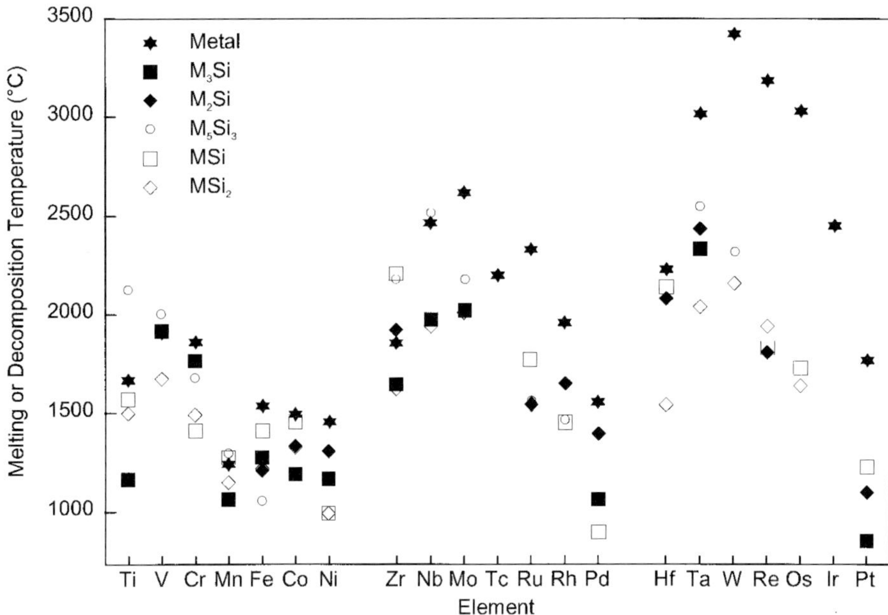

FIG. 14. Melting or decomposition temperature for transition metals and some transition metal silicides by period and group in the periodic table (from Massalski, 1990).

the decomposition temperature of the most stable transition-metal silicides range from 1229 °C for PtSi to 2550 °C for Ta_5Si_3. For comparison, the melting temperature of Si is 1414 °C. The melting or decomposition temperatures increase with the period number (Fig. 14), reaching their maximum values for groups V or VI (V, Cr, Nb, Mo, Ta and W), and are at their minimum values for the near-noble metals (Ni, Pd and Pd). Interestingly, there are silicide phases that have higher melting points than that of the parent metal for Ti, Zr, Hf, V, Nb, and Mn (Figs. 7 and 14). The thermal stability of the near-noble–metal silicide thin films on silicon is also limited by the presence of low-temperature (<1000 °C) Si-rich eutectics (Fig. 7). The metal melting point and the melting point of the most stable silicide are somewhat correlated (Fig. 15).

1.2.3.2 Stability of Thin Films The stability of silicide thin films is of interest in microelectronics applications. In Si integrated circuits, silicides must be stable on both single-crystal Si and polycrystalline Si. High-temperature annealing can cause agglomeration of silicides on either single-crystal or polycrystalline Si, and can cause silicide-enhanced grain growth in polycrystalline Si (Colgan *et al.*, 1996a).

Agglomeration of silicides on single-crystal Si is often observed after high-temperature annealing, especially for near-noble–metal silicides. There have been many studies of the agglomeration of silicide thin films on Si, including $TiSi_2$, $CoSi_2$, and $NiSi_2$ (Colgan *et al.*, 1996a,b). Agglomeration starts with grain-boundary grooving in the silicide, followed by grain separation and finally the formation of silicide islands. The driving force for grain-boundary grooving is the reduction of grain-boundary energy in the system (Nolan *et al.*, 1992; Mullins, 1957; see also GRAIN BOUNDARIES).

Agglomeration depends on a number of factors, including film thickness, grain size, grain-boundary energy, surface energy (of the film and the substrate), and interface energy (of the film–substrate interface). Agglomeration will increase as the groove depth increases; hence, agglomeration increases with decreasing film thickness, increasing grain size, and increasing grain-boundary energy. Agglomeration is probably limited by the slowest-diffusing species, because there must be transport of

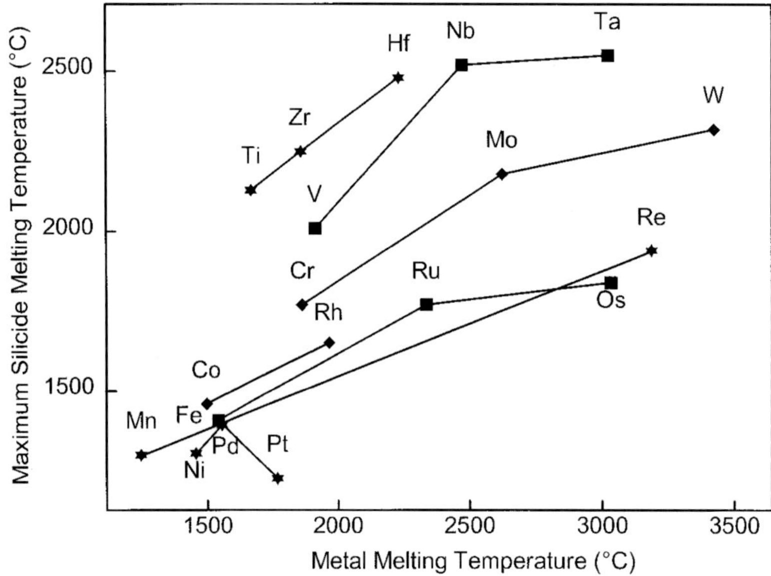

FIG. 15. Maximum silicide melting temperature *versus* parent metal melting temperature. Elements in the same group are connected by lines (Massalski, 1990).

both Si and the metal. Hence, it is expected that the agglomeration temperature for a given silicide will scale with the melting point of the silicide.

Silicide thin films are less stable on polycrystalline Si than on single-crystal Si, because there is an additional driving force for morphological changes: the grain-boundary energy of the polycrystalline Si. Grain growth in the polycrystalline Si is enhanced by the silicide, which provides a fast diffusion path for transport of material. The grain growth in the polycrystalline Si can cause severe roughening of the silicide–polycrystalline Si interface, and is undesirable in integrated circuits (Colgan *et al.*, 1996a; Gambino *et al.*, 1998). On the other hand, similar silicide-enhanced grain growth has been used to form large-grain Si films, which are desirable for thin-film transistors (Lee *et al.*, 1996). The temperature at which silicide-enhanced grain growth occurs for a given silicide scales with the deformation temperature of the silicide (Fig. 16). The growing Si grains protrude into the silicide; the growth

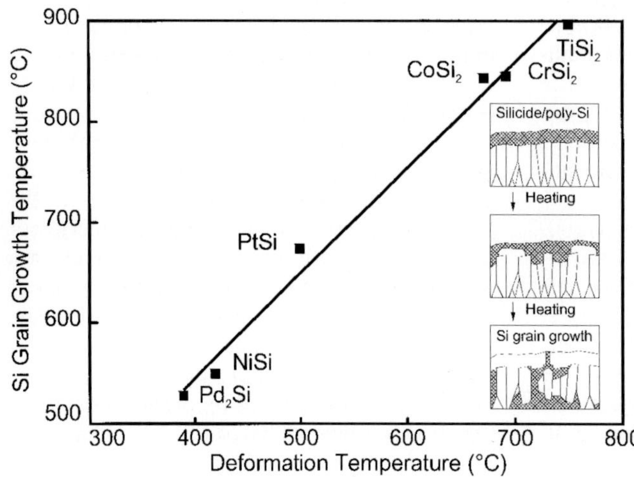

FIG. 16. Silicide degradation temperature as a function of silicide deformation temperature for various silicides (from Hong *et al.*, 1994).

of the grains will be constrained if the silicide is difficult to deform (Hong *et al.*, 1994).

1.2.4 Chemical Stability The chemical stability of silicides is important in a number of applications. In structural materials, it is desirable that the silicides withstand corrosive environments such as high-temperature oxidation. The oxidation resistance of silicides is also of interest in microelectronics applications, where silicides are exposed to high-temperature oxidations used to fabricate devices. However, during integrated-circuit fabrication it is also necessary to pattern silicides; in this case etching of the silicides is necessary and desirable.

1.2.4.1 Oxidation The oxidation of silicides has been widely studied for both microelectronics applications and structural materials. Oxidation can result in either SiO_2 formation or metal oxides, depending on the availability of Si at the oxide–silicide interface. In general, it is preferred to form SiO_2 rather than metal oxides, because SiO_2 is a stable oxide that grows relatively slowly without affecting the integrity of the underlying silicide. In contrast, many metal oxides (such as W oxides and Mo oxides) are volatile and can produce cracks or blisters in the oxide film.

Metal-free oxides can be formed on most of the silicides (including WSi_2, $MoSi_2$, $CoSi_2$, $TiSi_2$, $NiSi_2$ and PtSi) if there is an adequate supply of Si. This can be seen from the ternary phase diagrams (Fig. 11); metal oxides will not form if the system stays in the MSi_2–SiO_2–Si triangle (Beyers *et al.*, 1984). The supply of Si at the oxide–silicide interface can be analyzed in terms of the activity of Si at the interface, which is related to the chemical potential (Zhang and d'Heurle, 1992; Lee, 1992). If the silicide is saturated with Si, the activity of Si at the oxide–silicide interface approaches unity and SiO_2 is formed. However, if the supply of Si is limited, the activity at the oxide–silicide interface will decrease, and may become low enough to allow oxidation of metals to occur.

For refractory-metal silicides (such as $TiSi_2$, $MoSi_2$, and WSi_2), Si diffusion through the silicide provides the excess Si required for SiO_2 formation. The excess Si must be supplied from the underlying Si (in the case of a thin silicide film) or from excess Si in the bulk of the silicide. In either case, the diffusion of Si to the oxide–silicide interface must be unhindered, to ensure that there is an adequate supply for SiO_2 formation. Because the activation energy for Si diffusion throught the silicide is higher than that for the growth of SiO_2, high oxidation temperatures (800 °C or above) are required to form SiO_2. At low temperatures, the supply of Si will be limited and metal oxides will form (Zhang and d'Heurle, 1992).

For near-noble–metal silicides (such as $NiSi_2$, $CoSi_2$, and PtSi), the metal diffusivity is high and oxidation occurs by a different mechanism. During oxidation, the silicide decomposes at the oxide–silicide interface, providing Si for oxidation. The free metal then diffuses to the underlying Si, forming more silicide. During oxidation of $NiSi_2$ and $CoSi_2$, there is also diffusion of excess Si away from the oxide–silicide interface, due to the volume increase (and hence compressive stress) associated with SiO_2 formation (d'Heurle *et al.*, 1987). In either case, the diffusivity of the metal must be high enough to ensure an adequate supply of Si at the oxide–silicide interface.

The formation of SiO_2 on silicides and on Si occur by similar mechanisms (Zhang and d'Heurle, 1992). During the initial stages of the reaction, oxidation is limited by the interfacial reaction between Si and oxygen, and oxide growth will be linear with time. As the oxide grows, the reaction will eventually be limited by diffusion of oxidizing species through the SiO_2, and the oxide thickness will increase with the square root of time. For dry oxidations, the diffusing species is molecular oxygen, whereas for wet oxidations (*i.e.*, steam), the diffusing species is H_2O (Irene, 1988). Although the kinetic processes are similar, the oxidation rates of silicides are generally higher than that of Si, with near-noble–metal silicides usually having higher oxidation rates than refractory metal silicides (Frampton *et al.*, 1987) (Fig. 17).

The oxidation of WSi_2 and $MoSi_2$ can result in the formation of volatile metal oxides (such as WO_3 and MoO_3) if there is insufficient Si at the surface. The formation of these volatile metal oxides causes volume expansion and blistering of the protective SiO_2 films, resulting in futher oxidation of the silicide. The formation of metal oxides can cause peeling of silicide thin films (Yoo *et al.*, 1990) and can cause disintegration of bulk silicides; this is

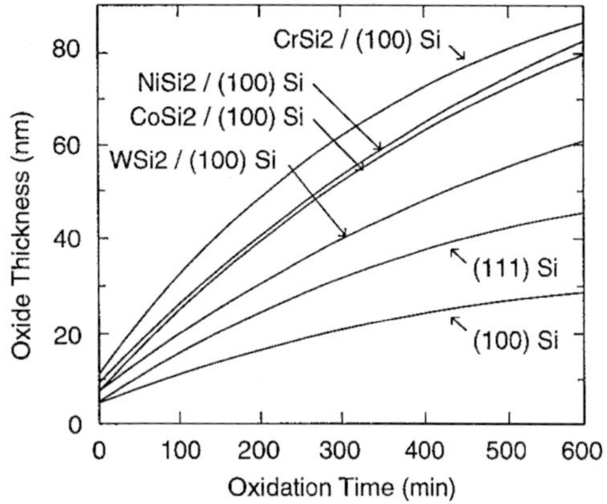

FIG. 17. The kinetics of SiO_2 formation for several silicides at 800°C in nominally dry O_2 (reprinted with permission from d'Heurle *et al.*, 1987, and Taylor & Francis, Ltd.).

often referred to as a "pest" reaction (Fig. 18) (Grabke and Meier, 1995; Chou and Nieh, 1993).

A number of methods have been investigated for preventing metal oxidation. For

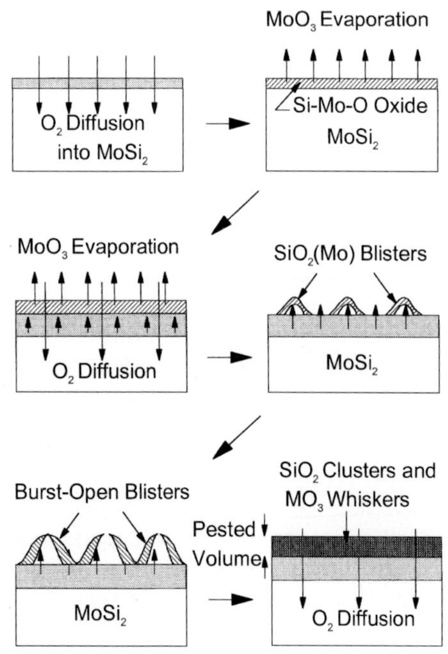

FIG. 18. Schematic of pest reaction in single crystal $MoSi_2$ (reprinted with permission from Chou and Nieh, 1993, Fig. 17).

thin films used in microelectronics, a Si-rich silicide is used to ensure that there is an adequate supply of Si at the surface during oxidation. For bulk $MoSi_2$ structural materials, one approach that has been used is to alloy the silicide with Ge. Germanium forms a glass (GeO_2) at lower temperatures than SiO_2; the GeO_2 glass protects the silicide at low temperatures. The GeO_2 also lowers the viscosity of SiO_2 at high temperatures, making it easier to seal cracks if they form. In addition, GeO_2 increases the coefficient of thermal expansion, which reduces the expansion mismatch between the $MoSi_2$ and the protective SiO_2 layer (Mueller *et al.*, 1992).

1.2.4.2 Silicide Reactions with Dopants (As, P, B) In microelectronics applications, silicides are often in contact with heavily doped Si (*i.e.*, Si doped with As, P, or B). For these applications, it is desirable to maximize the active dopant concentration in the Si. Hence, the thermodynamic and kinetic stability of the silicide on doped Si is important. Diffusion of dopants into the silicide can lower the dopant concentration in the Si, causing a higher contact resistance for contacts to diffusions and a threshold-voltage shift for contacts to gates. At the same time, dopant diffusion from the silicide can be beneficial, making it possible to form shallow junctions. Interactions between dopants and silicides are also relevant for structural materials; B doping in Ni_3Si improves the ductility of the alloy (Takasugi and Yoshida, 1991). Hence, an understanding of the interactions between silicides and dopants is required, including compound formation, solubility, diffusivity, and segregation (Maex *et al.*, 1989; Lien and Nicolet, 1984; Murarka and Williams, 1987).

There have been relatively few studies on the solubility of dopants in silicides (Shone *et al.*, 1986; Jahnel *et al.*, 1982); most of the data on solubilities are from studies on dopant diffusion in polycrystalline silicides. In most cases, the solubilities were estimated on the basis of the average dopant concentration in the silicide. It is expected that dopants segregate to grain boundaries in many cases, and so the actual solubility (*i.e.*, in the grains) will be less than that estimated from measurements on polycrystalline thin films.

Silicides can react with dopants to form metal–dopant compounds. Most silicides are

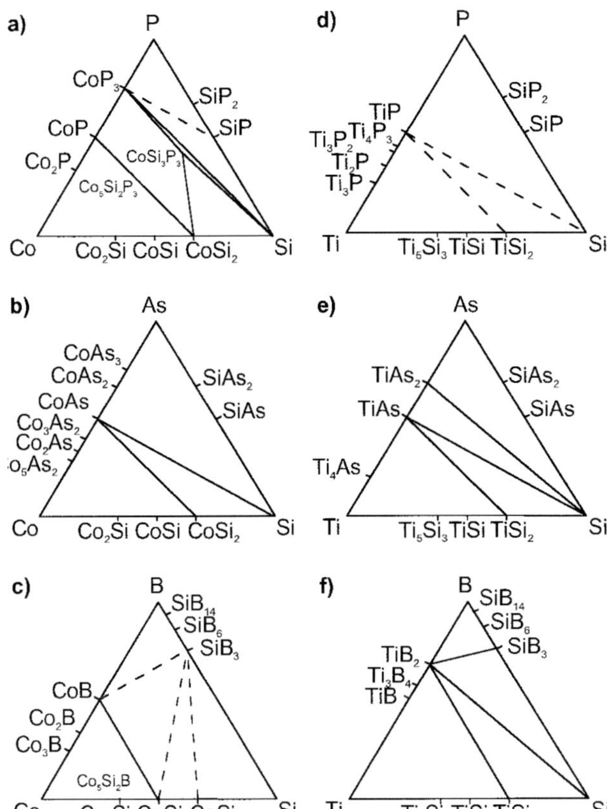

FIG. 19. Ternary phase diagrams for metal–Si–dopant phases: (a) Co–Si–P, (b) Co–Si–As, (c) Co–Si–B, (d) Ti–Si–P, (e) Ti–Si–As and (f) Ti–Si–B (reprinted with permission from Madar, 1995).

expected to form binary or ternary compounds with dopants at sufficiently high dopant concentrations (Fig. 19, from Madar, 1995). However, these compounds are often not observed in microelectronics applications (Maex, 1993), perhaps because dopant concentrations are below the solubility limit in the silicide. In contrast, binary and ternary dopant compounds are often observed in structural silicides (Briant and Taub, 1988).

There have been many studies on dopant diffusion in polycrystalline silicides (Gas *et al.*, 1995). With the exception of a few studies where silicides with large grain sizes were used, diffusivities measured in most studies are probably dominated by grain-boundary diffusion. This is directly relevant for microelectronics applications, where the silicides are polycrystalline thin films. A comparison of dopant diffusion studies (Fig. 20) indicates that grain-boundary diffusion greatly enhances dopant transport in silicides with small grains

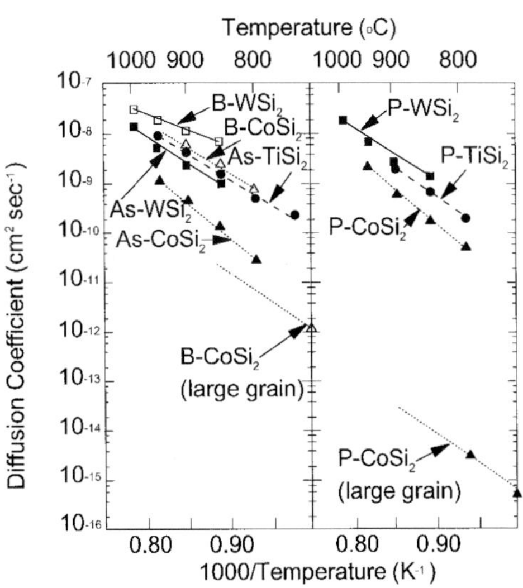

FIG. 20. Diffusion coefficients of B, As and P in various silicides (reprinted from Gambino and Colgan, 1998, with permission from Elsevier Science).

(Chu *et al.*, 1992) compared with silicides with large grains (Thomas *et al.*, 1988b; Zaring *et al.*, 1990). In most cases, the diffusion coefficients of the dopants in silicides are higher than those in Si. The activation energies for volume diffusion are typically 2–3 eV, and are similar to those for silicide self-diffusion, suggesting a substitutional diffusion mechanism (Gas *et al.*, 1995).

1.2.4.3 Chemical Reactivity The chemical reactivity of silicides is important in microelectronics applications. When silicides are patterned by dry etching methods a high reactivity is required. However, during subsequent wet or dry etch steps, a low reactivity is required. Low reactivity is also required for silicides used for coatings in corrosive environments.

Vijh (1994) has studied the hydrogen evolution reaction ($H_2O \rightarrow H_2 + \frac{1}{2}O_2$) in a number of silicides; the hydrogen-evolution reaction is of relevance for corrosion and electrolysis reactions. In general, silicides show high chemical and electrochemical stability in electrolyte solutions, especially at cathodic potentials. At anodic potentials, some oxide growth and dissolution of the silicide can occur. In addition, a variety of electrode reactions, including the hydrogen-evolution reaction, can be sustained at high rates on silicide electrode materials. Silicides formed from a catalytic metal (such as PtSi) show higher rates for the hydrogen-evolution reaction in acidic solutions compared with alkaline solutions; the reverse is true for silicides formed from noncatalytic metals (such as Zr).

In microelectronics applications, silicides must withstand a number of wet etches used for cleaning the substrates or for removing excess metal (see Sec. 2.1), including HF, $H_2O_2 + H_2SO_4$, $H_2O_2 + NH_4OH + H_2O$ (SC-1 solution), and $H_2O_2 + HCl + H_2O$ (SC-2 solution). Most commonly used silicides etch slowly if at all in $H_2O_2 + H_2SO_4$ cleans and SC-1 or SC-2 cleans. One exception is $TiSi_2$, which can be etched in $H_2O_2 + NH_4OH + H_2O$ solutions. Another exception is As-doped $CoSi_2$, which can be etched and oxidized in $H_2O_2 + HCl + H_2O$ solutions (Gambino *et al.*, 1990). Most silicides etch to some degree in HF. A common method for patterning silicide thin films formed by metal-Si reactions is to use a wet etch that attacks the nonsilicided metal over the insulator regions, but does not attack the silicide that forms in the contacts (*i.e.*, salicide process; see Sec. 2.1). A number of etches have been developed for $TiSi_2$, $CoSi_2$, NiSi, and PtSi (Gambino and Colgan, 1998).

Reactive ion etching (RIE) is commonly used for patterning codeposited silicide thin films for microelectronics applications (*i.e.*, polycide process, see Sec. 2.2) (Flamm *et al.*, 1984). Anisotropic etching is achieved by using ion bombardment to accelerate the etch reactions. Reactive ion etching can occur if the material to be etched forms volatile compounds with the etch gases. Tungsten, Ti, and Si all form relatively volatile fluorides, chlorides, and bromides, and therefore can be etched with either F-, Cl-, or Br-based chemistries (Chow *et al.*, 1984). In contrast, Ni, Pt, and Co fluorides, chlorides and bromides have high boiling points; hence $NiSi_2$, $CoSi_2$, and PtSi are difficult to pattern by RIE.

1.3 Properties

1.3.1 Mechanical Properties The mechanical properties of silicides, such as creep rate, hardness, yield strength, and brittle–ductile transition temperature, are obviously relevant for structural materials. Because of this, there have been many studies on the mechanical properties of bulk silicides (Kumar, 1994). Mechanical properties are also important for microelectronics applications, where high stresses can be detrimental to underlying devices.

A high value of hardness is required for applications such as coatings, and is also a measure of strength. In general, hardness increases with increasing melting point of the material (Fig. 21), and so most silicides have high values of hardness (Fleischer, 1994; Ostling and Zaring, 1995). The hardness and fracture toughness of silicides can often be increased by alloying (Fig. 22) (Maloy *et al.*, 1991).

The yield strength of a material is the minimum stress that must be applied to induce plastic deformation of the material, and is usually proportional to the elastic modulus (Young's modulus) of the material (Barrett *et al.*, 1973). The yield strengths and elastic moduli for silicides are generally high compared with other materials (Fig. 23) (Shah *et al.*, 1992), Ostling and Zaring, 1995).

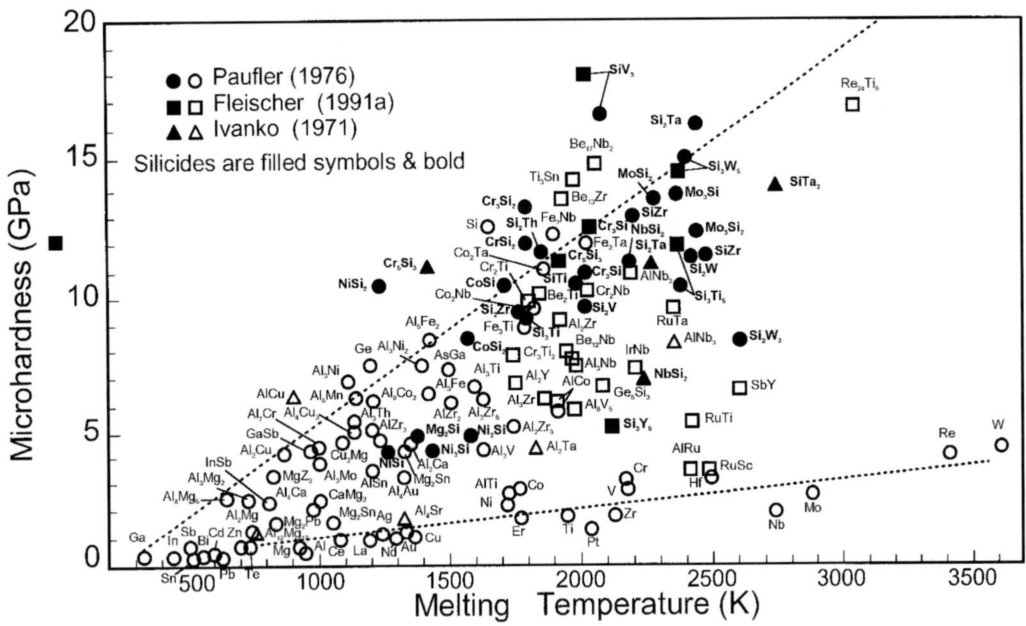

FIG. 21. Hardness at room temperature of metals and intermetallic compounds. (From Fleischer, 1994. Copyright John Wiley & Sons Limited. Reproduced with permission.)

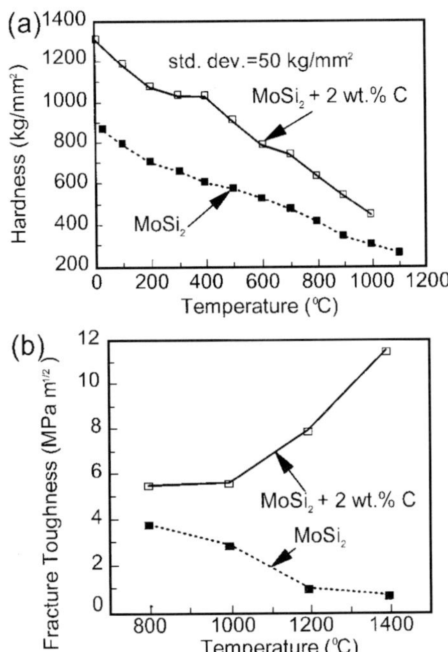

FIG. 22. (a) Microhardness and (b) fracture toughness versus temperature for $MoSi_2$ and $MoSi_2 + 2$wt% C. (From Maloy *et al.*, 1991, reprinted with permission of the American Ceramic Society, P.O. Box 6136, Westerville, OH 43086-6136. Copyright 1991 by The American Ceramic Society. All rights reserved.)

The yield strength and ductility of a material are related to the slip systems that are available in the structure for dislocation motion. Slip is difficult in most silicides, because of complicated crystal structures and covalent bonding. The mobility of dislocations increases as the ratio of the lattice spacing to the Burgers-vector length (*i.e.*, the width of the dislocation) decreases. In silicides, the lattice spacing for the slip planes is often large, on account of the complicated crystal structures. In addition, the Burgers vectors are often small, as a result of dissociation of dislocations caused by the highly directional covalent bonds. Hence, dislocations in silicides generally have low mobility. As a result, silicides have high yield strength but low ductility. Silicides are generally brittle at low temperatures, which is a disadvantage for structural materials, because they are weak against fracture. Slip becomes easier as temperature increases, because dislocation motion, which requires an atom at the core of the dislocation to jump a distance on the order of the Burgers vector length, increases with temperature (Hertzberg, 1976).

The temperature of the brittle-to-ductile transition generally increases with increasing

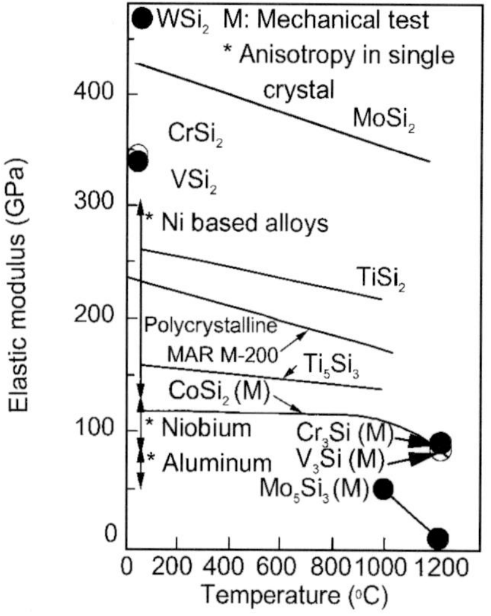

FIG. 23. Comparison of average elastic modulus as a function of temperature (reprinted from Shah *et al.*, 1992, copyright 1992, with permission from Elsevier Science).

melting point of the material (Fig. 24) (Fleishcher, 1994). Because of their low temperature brittleness, silicides by themselves cannot be used as structural materials. However, this problem can be overcome by forming a composite, adding a second phase to the silicide that improves the low temperature ductility. The fracture toughness is a measure of the work expended at a crack tip during crack propagation, and gives a measure of brittleness (*i.e.*, brittle materials have low fracture toughness). The fracture toughness of silicide composites is much greater than that of the silicide alone (Vasudevan and Petrovic, 1992).

The high-temperature (>900 °C) yield strength of $MoSi_2$ is anisotropic (Fig. 25) and is much higher at orientations near [001] compared with other orientations (Umakoshi *et al.*, 1990; Kimura *et al.*, 1990). The dominant high-temperature slip planes in $MoSi_2$ (and also WSi_2, which has the same structure) are the {110} and {013} planes, with the {110}

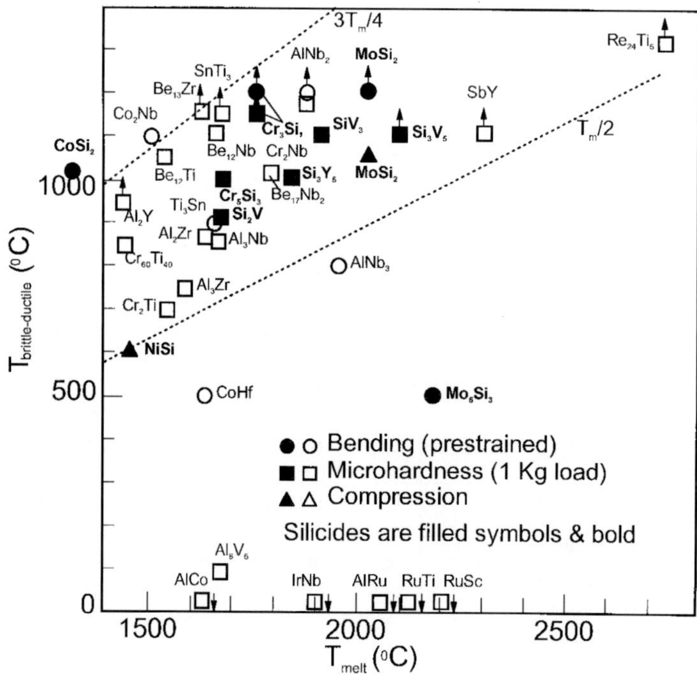

FIG. 24. Brittle-to-ductile transistion temperatures as a function of melting points of the material. Data are from bending, microhardness, and compression experiments. (From Fleischer, 1994. Copyright John Wiley & Sons Limited. Reproduced with permission.)

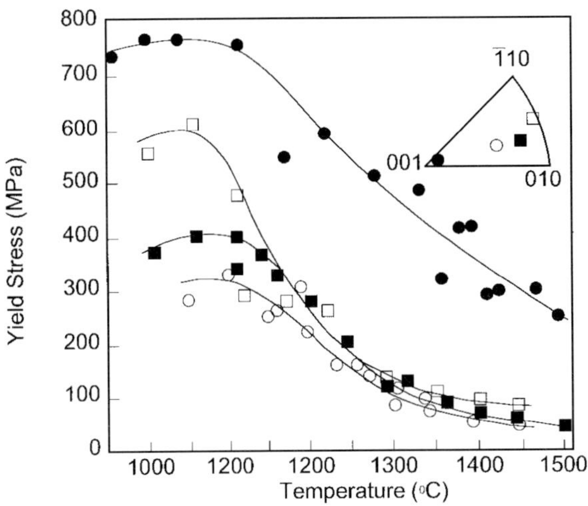

FIG. 25. Temperature dependence of yield strength for several orientations of $MoSi_2$ single crystals (from Umakoshi *et al.*, 1990, copyright 1990, with permission from Elsevier Science).

planes being dominant; the slip direction is ⟨331⟩ in both cases. For $MoSi_2$, the critical resolved shear stress for {013} planes is higher than that for {110} planes, perhaps because the cores of dislocations on {013} planes are nonplanar and spread onto several planes (the {110} planes have a simple ABAB stacking sequence, whereas the {013} planes have a long-period stacking sequence). Orientations near [001] will deform on the [013] slip system, and hence will have a higher yield strength. In contrast, for WSi_2, the critical resolved shear stress for {110} planes is higher than that for {013} planes, perhaps due to a large number of stacking faults on {110} planes.

$CoSi_2$ is more ductile than $MoSi_2$, on account of a more active slip system. $CoSi_2$ has the fluorite structure, and has an fcc crystal symmetry (compared with the more complicated tetragonal $C11_b$ structure of $MoSi_2$). Possible slip systems include {110}, {111}, and {001} planes, all in the ⟨110⟩ direction, as well as {001} planes in the ⟨100⟩ direction. The {111} planes are probably the dominant slip planes at high temperatures (>800 °C), as in fcc metals (Takeuchi *et al.*, 1992). However, at room temperature the {001} planes are the dominant slip planes, in the ⟨100⟩ direction (Ito *et al.*, 1992). Orientations near [100] will deform on the [110] slip system, and hence will have a higher yield strength.

The yield strength of $MoSi_2$ decreases abruptly at high temperatures (between 1100 and 1200 °C, see Fig. 25), indicating an increase in ductility (Umakoshi *et al.*, 1990). The improvement in ductility is due to activation of the {110} and {013} slip systems. At these temperatures, most dislocations have ⟨331⟩ Burgers vectors. At higher tempertures (above 1300 °C), ⟨100⟩ and ⟨110⟩ dislocations are activated, and the ductility is greatly improved.

At high temperatures (greater than half the melting point), the plastic deformation of a material under an applied stress varies with time, independent of whether the applied stress is greater than or less than the yield strength. This deformation is called creep. Many silicides have low creep rates, which are desirable in structural materials (Fig. 26) (Shah *et al.*, 1992). The creep rates of $MoSi_2$ can be further improved by forming composites (Bose, 1992). The activation energy for creep in $MoSi_2$ is high in comparison with lattice diffusion (4.48 eV compared with 2.4 eV for Si diffusion and 3.6 eV for Mo diffusion in $MoSi_2$). Possible creep mechanisms include grain-boundary sliding and dislocation climb at higher applied stresses, and grain-boundary sliding at lower applied stresses (Vasudevan and Petrovic, 1992; Sadananda *et al.*, 1992). Dislocation climb is expected to be limited by diffusion of Mo, which is the slower diffusing species.

1.3.2 Optical Properties The optical reflectivity of metallic silicides can be used to examine the electronic structure such as the

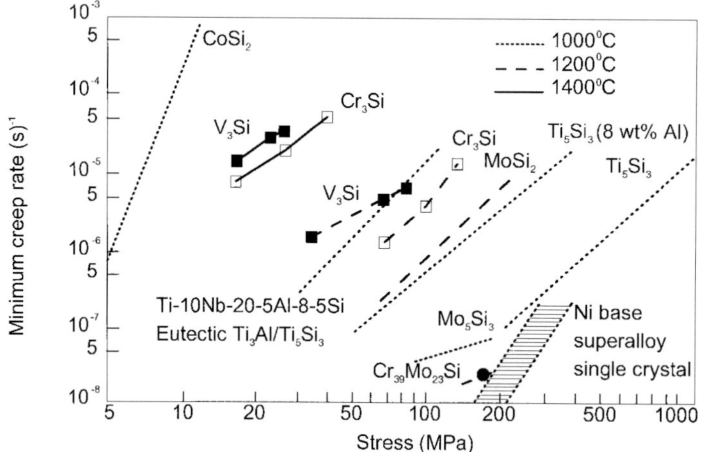

FIG. 26. Comparison of minimum creep rates in compression *versus* stress for several silicides and alloys (reprinted from Shah *et al.*, 1992, copyright 1992, with permission from Elsevier Science).

density of states and character of the bands through the complex dielectric function. The complex dielectric function, $e(\omega) = \varepsilon_1(\omega) + i\varepsilon_2(\omega)$, is related to the complex refractive index, $n = n + ik$ so that $\varepsilon_1 = n^2 + k^2$ and $\varepsilon_2 = 2nk$ and can be calculated from reflectance or other optical data. A recent review (Guizetti and Marabelli, 1995) summarizes information on the complex refractive index for silicides. Examples of the interpretation of the optical response of metallic silicides in terms of the electronic structure can be found in the work of Nava *et al.* (1993). Following Nava *et al.*, for the free-electron region at low photon energies, the classical Drude model (Ashcroft and Mermin, 1976) can be used to determine the plasma frequency (ω_p) and the relaxation time (τ) from the complex refractive index:

$$\varepsilon_1(\omega) = 1 - \omega_p{}^2/[\omega^2 - (1/\tau)^2] \tag{3}$$

$$\varepsilon_2(\omega) = 1 - (\omega_p{}^2/\tau)/\omega[\omega^2 - (1/\tau)^2] \tag{4}$$

The optical resistance at zero frequency, $\rho_{opt} = 4\pi/\omega_p{}^2\tau$, was also calculated and compared well with the dc resistance measurements. The optical reflectivity and dielectric functions for VSi_2, $NbSi_2$, and $TaSi_2$ thin films are shown in Fig. 27 (Nava *et al.*, 1993). At low photon energy, the reflectivity is very high and rapidly decreases for energies around 1 eV. The value of ε_2 drops from a large free-elec-

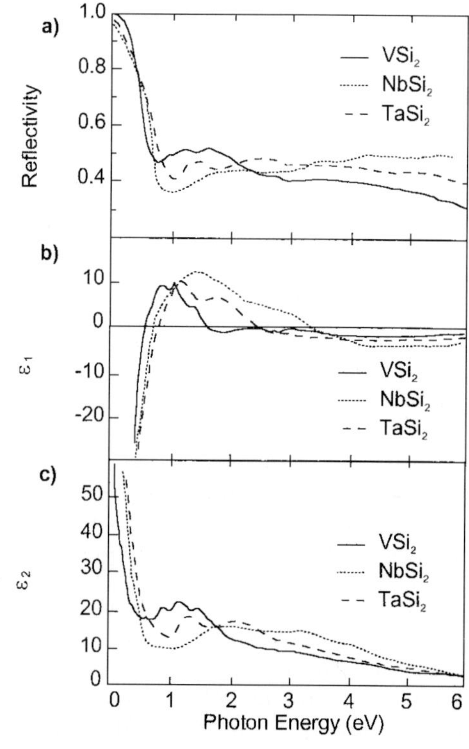

FIG. 27. (a) Optical reflectivity, (b) real part of the dielectric constant, and (c) imaginary part of the dielectric constants for VSi_2, $NbSi_2$ and $TaSi_2$ (reprinted from Nava *et al.*, 1993, copyright 1993, with permission from Elsevier Science).

tron value at low photon energy and passes through a minimum around 1 eV.

The optical properties of semiconducting silicides can be used to determine the band gap and whether the transition is direct or indirect. The absorption coeffient, $\alpha = 4\pi k/\lambda$, is calculated from k, the extinction coefficient, where λ is the wavelength. For an indirect transition,

$$\alpha = 0 \quad \text{for } h\nu < E_g - E_p, \tag{5}$$

$$\alpha = A(h\nu - E_g + E_p)^2 \quad \text{for} \quad E_g + E_p > h\nu > E_g - E_p, \tag{6}$$

$$\alpha = A(h\nu - E_g + E_p)^2 + B(h\nu - E_g - E_p)^2 \quad \text{for } h\nu > E_g + E_p, \tag{7}$$

where E_g is the energy gap, E_p is the phonon energy, h is Planck's constant, ν is the frequency of the absorbed light, and A and B are constants. For a direct transition,

$$\alpha = 0 \quad \text{for } h\nu < E_d, \tag{8}$$

$$\alpha = C(h\nu - E_d)^{1/2} \quad \text{for } h\nu > E_d, \tag{9}$$

where E_d is the energy of the direct transition and C is a constant. A good example of a semiconducting silicide with a direct energy gap is β-$FeSi_2$, with a direct band-gap energy of 0.89 eV (Fig. 28) (Bost and Mahan, 1988; Guizetti and Marbelli, 1995).

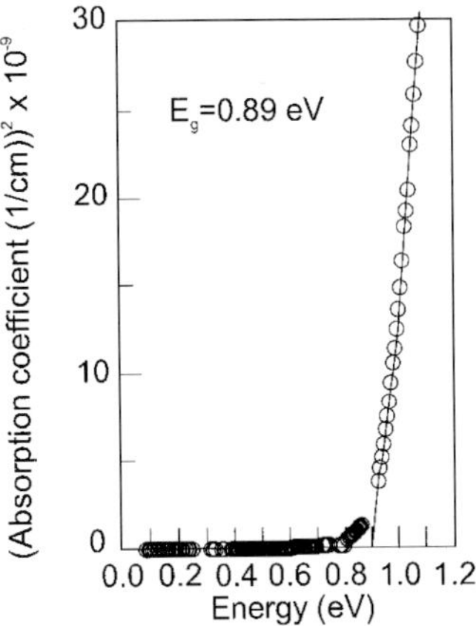

FIG. 28. Plot of the β-$FeSi_2$ absorption coefficient squared versus energy. The solid line is a model based on direct transitions of 0.89 eV and 1.1 eV. (Reprinted with permission from Bost and Mahan, 1988. Copyright 1988 American Institute of Physics.)

1.3.3 Electrical Properties The electrical properities of silicides have been studied extensively, since the resistivity is of critical importance for many applications (Nava *et al.*, 1993; Gottlieb *et al.*, 1995). The electrical transport properties of silicide–semiconductor interfaces have also been widely studied.

1.3.3.1 Resistivity In general, silicides are metallic conductors with room-temperature resistivities on the order of 10^{-6}–$10^{-3}\,\Omega$ cm, but a number of silicides are superconducting or semiconducting (Fig. 29). For the metallic silicides, the resistivity increases with temperature and can be approximated by Matthiessen's rule (Kittel, 1976), where the temperature-dependent resistivity is equal to an electron-impurity scattering component that is temperature independent plus temperature-dependent terms that include electron–phonon scattering, electron–electron scattering, and other temperature-dependent effects. The temperature dependence of the resistivity for a number of silicide single crystals has been modeled by Nava *et al.* (1993), and it was found that in several cases there were well defined temperature ranges in which a specific electron–phonon scattering mechanism dominated.

The resistivity of noncubic single crystals exhibits anisotropy due to the shape of the Fermi surface (Fig. 30 and Table 5) (Nava *et al.*, 1993). The room-temperature anisotropy is nearly 2 for the hexagonal silicides (VSi_2, $NbSi_2$, $TaSi_2$ and Pd_2Si) and almost undetectable for the higher-symmetry orthorhombic $TiSi_2$.

The semiconducting silicides (Fig. 29) are generally formed from the middle groups of transition metals. As mentioned in Sec. 1.1.1.1, for the semiconducting silicides, a minimum in the density of states at the Fermi level opens into a gap. In moving from $NiSi_2$ to $TiSi_2$, the density of states due to the metal d electrons moves to lower energy, passing through the Fermi level, with increasing atomic number (see Fig. 2). The shift of the minimum in the density of states relative to the Fermi level changes the conduction in the disilicide. The conduction in $NiSi_2$ and $CoSi_2$ is metallic, but the majority carriers are holes (Nicolet and Lau, 1983). The next silicides, $CrSi_2$, $MnSi_{1.7}$,

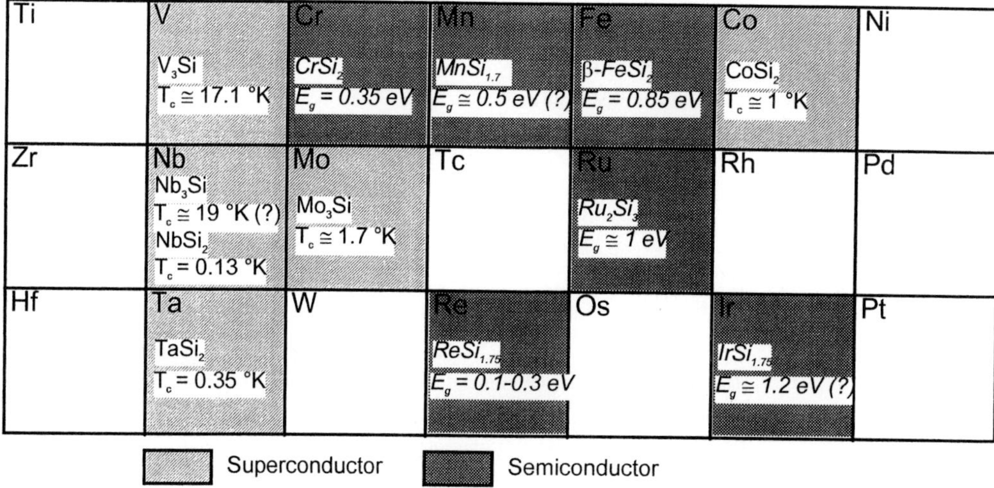

FIG. 29. Superconducting and semiconducting transition-metal silicides (from Gottlieb *et al.*, 1995; Mantl, 1994).

and b-$FeSi_2$, are semiconductors. Finally, $TiSi_2$ is a metallic conductor with electrons as the majority carriers (the carrier type for VSi_2 has not been reported). "Undoped" semiconducting silicides (*i.e.*, no intentional dopants) usually have holes as the majority carriers. They can be made *n*-type by the addition of transition metals with a larger number of d electrons (Co, Ni, Pd, or Pt has been used with b-$FeSi_2$), or *p*-type by the addition of transition metals with a smaller number of d electrons (Mn, Cr, V, or Ti has been used with b-$FeSi_2$), than the metal atom they replace (Lange, 1996). The activation energies for most dopants vary between about 0.08 and 0.14 eV.

There have been relatively few studies on the superconducting properties of silicides, probably because the critical temperature (T_c) is low for most superconducting silicides (see Fig. 29). One exception is V_3Si, which has a relatively high critical temperature (≈ 17 K) compared with other intermetallic superconductors. However, all of the intermetallic superconductors have much lower critical temperatures than oxide superconductors (Bednorz and Müller, 1986), and therefore have been less studied in recent years.

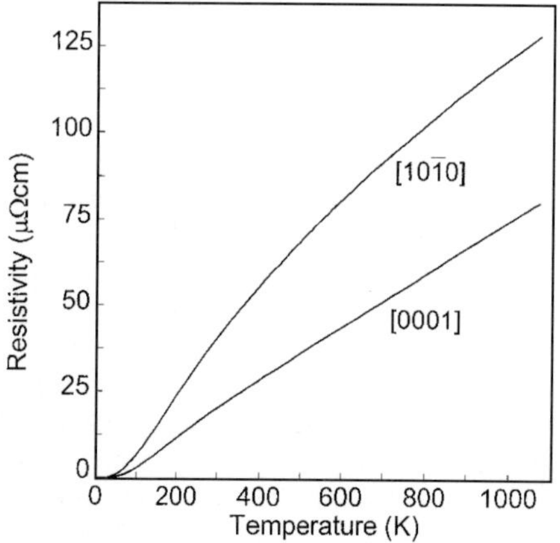

FIG. 30. Temperature dependence of single-crystal $TaSi_2$ resistivity with the current parallel to the [0001] and [10$\bar{1}$0] crystallographic directions (reprinted from Nava *et al.*, 1993, copyright 1993, with permission from Elsevier Science).

V_3Si, like other intermetallic superconductors with high T_c, has the A15 crystal structure. The Si atoms are arranged on a body-centered cubic (bcc) lattice, and the V atoms are in pairs that straddle the center of a face of the cube. This results in orthogonal chains of closely spaced V atoms running through the crystal structure. It is believed that these chains lead to the good superconducting properties of these materials (Stekly and Gregory, 1994).

1.3.3.2 Ohmic, Rectifying, and Gate Contacts Metal contacts on lightly doped Si form rectifying contacts (*i.e.*, nonlinear current–voltage

Table 5. Electrical properties of some transition-metal silicides (from Nava *et al.*, 1993).

Silicide	Structure/prototype	Orientation	Resistivity (RT) ($\mu\Omega$ cm)
$TiSi_2$	Orthorhombic/$TiSi_2$	[100]	11.0
		[010]	10.0
		[001]	12.0
VSi_2	Hexagonal/$CrSi_2$	[0001]	34.1
		[$10\bar{1}0$]	59.6
$NbSi_2$	Hexagonal/$CrSi_2$	[0001]	22.0
		[$10\bar{1}0$]	39.2
$TaSi_2$	Hexagonal/$CrSi_2$	[0001]	20.2
		[$10\bar{1}0$]	40.1
$MoSi_2$	tetragonal/$MoSi_2$	[001]	12.6
		[110]	16.9
WSi_2	tetragonal/$MoSi_2$	[001]	12.9
		[110]	11.9
Pd_2Si	Hexagonal/Fe_2P	I $\parallel$ to c-axis	13.7
		I $\perp$ to c-axis	18.9

characteristics, with high current in one polarity and low current for the opposite polarity), which are used to make Schottky diodes. For Schottky diodes, the most important property is the barrier height, which depends on the properties of the metal and the semiconductor surface. Metal contacts on heavily doped Si can be Ohmic (*i.e.*, linear current–voltage characteristics) and are used to make contacts to diffusions. The most relevant parameter for Ohmic contacts is the contact resistance, which depends on the metal and the doping in the Si. For gate contacts to metal–oxide–semiconductor (MOS) transistors, the most relevant parameter is the threshold voltage, which determines when the device "turns on" (*i.e.*, when the MOS transistor conducts current). Silicide or metal contacts on top of polycrystalline Si in polycide structures (see Sec. 2.2) do not directly affect the threshold voltage of the device. However, the silicide can indirectly affect the threshold voltage by depleting dopant from the underlying polycrystalline Si.

The current–voltage characteristics of Schottky diodes are determined by the barrier height at the metal–semiconductor interface. For lightly doped Si at room temperature, the current transport across the interface is dominated by thermionic emission of majority carriers over the barrier (Sze, 1981).

For heavily doped Si, the potential barrier at the metal–Si interface becomes very narrow, and tunneling of electrons from the conduction band of the Si into the metal is the dominant transport mechanism. A critical parameter for characterizing Ohmic contacts is the specific contact resistance, which depends strongly on the both the barrier height and the active dopant concentration (Sze, 1981).

Most Ohmic contacts in microelectronics applications are made with Ti or $TiSi_2$. The barrier height of Ti on Si is between 0.5 and 0.6 eV, depending on the annealing temperature after metal deposition (Aboelfotoh and Tu, 1986). Hence, the resistance for these contacts is mainly dependent on the doping level. In order to minimize contact resistance, the contacts are often doped to above the solubility limit. A lower contact resistance is expected for *n*-type contacts than *p*-type contacts, because the *n*-type dopants such as As and P have a higher solubility than *p*-type dopants such as B (Nobili, 1990). In fact, data on $TiSi_2$ contacts (Hui *et al.*, 1985) show a 2× higher contact resistance for *p*-type contacts compared with *n*-type contacts (Fig. 31).

1.3.4 Magnetic Properties The magnetic properties of some silicides are of interest for permanent magnets (Buschow, 1991). There are very few reports on the magnetic properties of silicides. All the studies on the magnetic properties of silicides involve ferromagnetic compounds, where Si is combined with Fe, Co, or rare earths. One type of ferromagnetic material consists of ternary Fe–Al–Si alloys, which exhibit high magnetic permeability (Kumar, 1994). More recent studies have focused on the magnetic properties of ternary

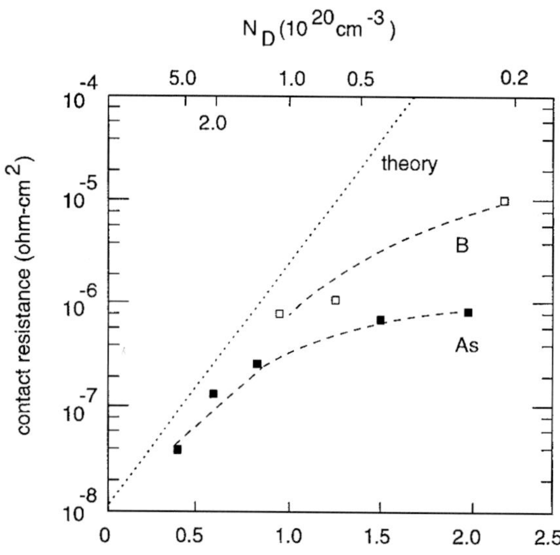

FIG. 31. Specific contact resistance *versus* dopant concentration for $TiSi_2$ on B-doped or As-doped Si (from Hui *et al.*, 1985). Theoretical curve is from Sze (1981), assuming a barrier height of 0.6 eV. (Reprinted from Gambino and Colgan, 1998, copyright 1998, with permission from Elsevier Science.)

silicides with the tetragonal $ThMn_{12}$- or $ThCr_2Si$-type structures (Buschow, 1991). The tetragonal structures provide magnetic anisotropy, which is desirable for permanent magnets. The $ThMn_{12}$ structures also possess high Curie temperatures.

The $ThMn_{12}$ compounds have a composition $RFe_{12-x}Si_x$, where R is a rare-earth element, such as Y, Nd, or Sm (note that other elements can be used instead of Si, including Al, Ti, V, Cr, Mo, and W). The role of the Si is to stabilize the structure, while the rare earths and the Fe provide the permanent magnetic moment. The magnetic properties can be described by combining the $Fe_{12-x}Si_x$ sublattice magnetization with the R sublattice magnetization. For light rare-earth elements, there is a ferromagnetic coupling between the two sublattices, whereas for heavy rare-earth elements there is an antiferromagnetic coupling (Buschow, 1991; Matar *et al.*, 1994).

2. FABRICATION

2.1 Salicide Process

Silicides in Si integrated ciruits are often formed by a solid-state reaction between the metal and Si (Fig. 32). The metal is deposited by physical vapor deposition (see Sec. 2.3) over a patterned wafer, where the pattern consists of insulators (used for isolating de-

a) Gate, Source & Drain fabricated

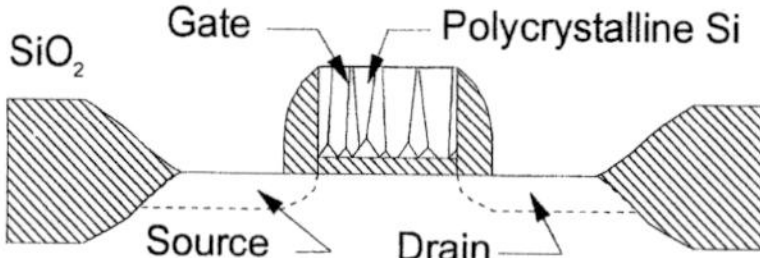

b) Metal Deposition

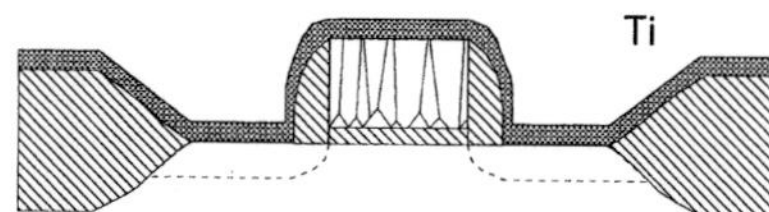

c) First anneal in Nitrogen

TiN/*C49* $TiSi_2$

TiN/Ti

d) Selective etch & second anneal

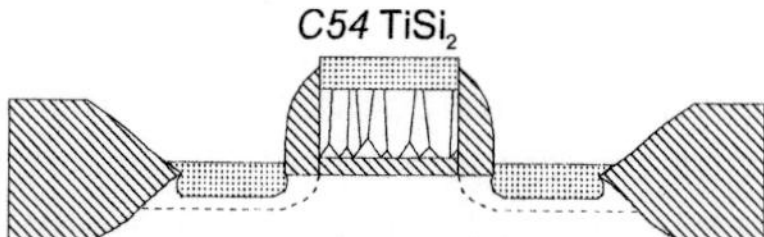

FIG. 32. Schematic of salicide process on an MOS transistor to form $TiSi_2$: (a) starting structure, (b) Ti deposition, (c) first anneal in N_2 to form C49 $TiSi_2$ (and TiN), (d) selective wet etch to remove Ti/TiN and second anneal to form C54-$TiSi_2$. (Reprinted from Gambino and Colgan, 1998, copyright 1998, with permission from Elsevier Science.)

vices) and exposed Si regions (where junctions or gates are to be formed). The wafers are then annealed, with the metal only forming silicide in regions where the Si or polycrystalline Si is exposed. For the appropriate anneal conditions (*i.e.*, sufficiently low temperatures) there is no significant silicide formed in regions where there are insulators, such as Si_3N_4 or SiO_2, since metals react much more slowly with these materials compared with Si. Hence, the silicide is self-aligned to the contacts and gates of the device. The nonsilicided metal is then removed by a wet etch, which etches the metal much faster than the silicide (see Sec. 1.2.4.3). This process is often called a self-aligned silicide or salicide process (Colgan *et al.*, 1996a).

There are a number of advantages to using the salicide process compared with other metallizations. Silicides formed by solid-state reactions often consume contamination and defects at the original metal–Si interface, resulting in low contact resistance for Ohmic contacts and reproducible rectifying behavior for Schottky diodes. Relatively planar metal–Si reactions allow metallization of shallow junctions without degradation of the junction electrical properties (*i.e*, without metal penetrating the junctions and causing high leakage currents). Finally, the self-aligned formation allows a smaller spacing between conductors or contacts than can be achieved with lithography, resulting in higher circuit densities.

There are also problems associated with the salicide process. Lateral silicide formation can result in shorting of contacts (Fig. 33); hence, the anneal temperature and time must be minimized to limit lateral silicide formation. Excessive surface contamination or doping can inhibit silicide formation, resulting in nonuniform silicide thickness and high sheet resistance (see Sec. 1.2.1.3). Nonplanar silicide formation or silicide agglomeration (see Sec. 1.2.3.2) during high-temperature annealing can result in silicide protrusions, junction leakage or high sheet resistance. For silicides on polycrystalline Si, silicide-enhanced grain growth can also occur during high-temperature annealing (see Sec. 1.2.1.3), resulting in threshold-voltage shifts or gate oxide leakage (Fig. 33). Finally, phase transformations (such as C54-$TiSi_2$) can be inhibited on narrow lines, resulting in high sheet resistance (Fig. 34).

A number of techniques have been used to avoid the problems described above, including integrated processing, capping layers, preamorphization, ion-beam mixing, epitaxial growth and nucleation enhancement. Integrated processing combines a number of steps, such as the preclean, metal deposition, and silicide formation anneal, in one cluster tool (Nulman, 1990; Aoki *et al.*, 1994). This minimizes the exposure of the Si surface and metal films to air, thereby improving silicide formation. Another approach to prevent oxygen contamination of the metal used in the salicide process (*i.e.*, during exposure to air or the anneal ambient) is to use capping layers such as TiN or Si on top of the metal film (Kaplan *et al.*, 1992; Berti and Bolkhovsky, 1992). For nucleation-limited reactions, such as $TiSi_2$ or $CoSi_2$ formation, amorphization of the substrate can enhance the nucleation of the new phase (Mogami *et al.*, 1994; Kuwano *et al.*, 1990). The free energy of the system is increased, resulting in an increased driving force for the reaction. Amorphization is usually achieved by implantation of Si, Ge, or As prior to metal deposition. Another approach to improve nucleation is ion-beam mixing (Mayer *et al.*, 1981; Maex *et al.*, 1986). In this

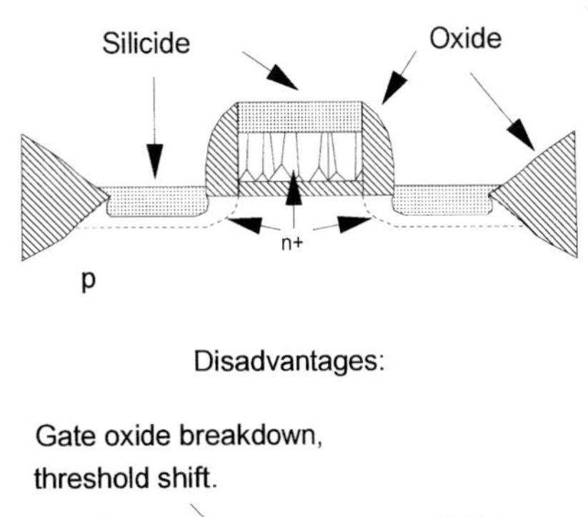

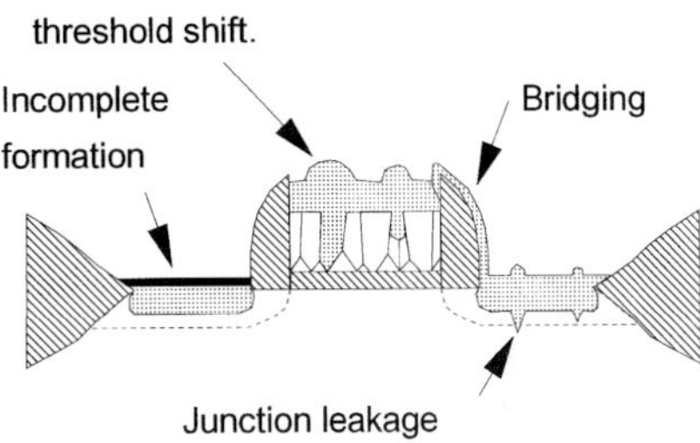

FIG. 33. Schematic showing advantages and disadvantages of salicide structure. (Reprinted from Gambino and Colgan, 1998, copyright 1998, with permission from Elsevier Science.)

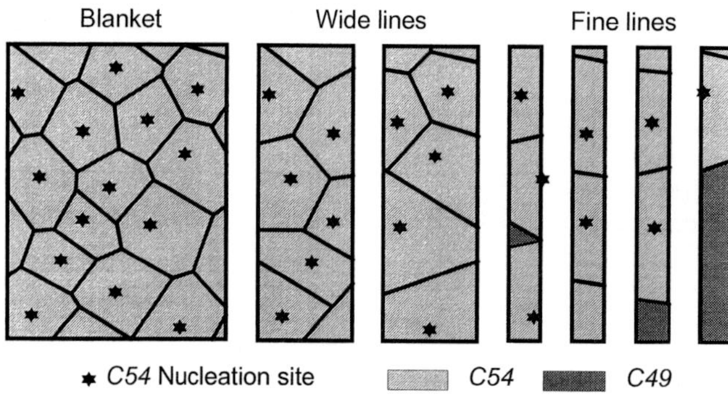

FIG. 34. Schematic of nucleation of C54-$TiSi_2$ as a function of linewidth. For narrow lines, the average distance between nuclei will increase compared with that for blanket films as a result of geometric effects. (Reprinted from Gambino and Colgan, 1998, copyright 1998, with permission from Elsevier Science.)

case, ion implantation is performed after metal deposition, to intermix the metal and Si atoms at the interface. A third approach to improve nucleation is to add impurities that aid in the nucleation of the new phase. This method has been used to improve the nucleation of C54-$TiSi_2$ (Clevenger *et al.*, 1996). The formation of epitaxial silicides has been studied as a method to improve the uniformity of the silicide–Si interface (Hsia *et al.*, 1991; Ogawa *et al.*, 1994; Tung, 1997). Epitaxial growth can be achieved in a salicide process by depositing a thin diffusion barrier (SiO_2 or a refractory metal, such as Ti, Ta, Zr, W) between the metal to be reacted (such as Co) and the Si substrate. If the diffusion barrier is sufficiently thin, the overlying metal can diffuse through it and react with the substrate to form a silicide. The reaction is slowed down, allowing the silicide to grow epitaxially.

2.2 Polycide Process

Silicides in integrated circuits are often formed by codeposition of the silicide, followed by reactive ion etching (RIE) to pattern it (Fig. 35). This process is often used to fabricate MOS transistors (see Sec. 3.1.1) and is referred to as a polycide process. In the polycide process, the polycrystalline Si for the gate is deposited and doped; doping can occur *in situ* during the deposition of the polycrystalline Si or can be done after the deposition by ion implantation (Colgan et al., 1996a). The silicide is then deposited by PVD or CVD (see Secs. 2.3 and 2.4), followed by deposition of an insulator such as SiO_2 or Si_3N_4. The gate stack

a) Deposit gate stack & pattern.

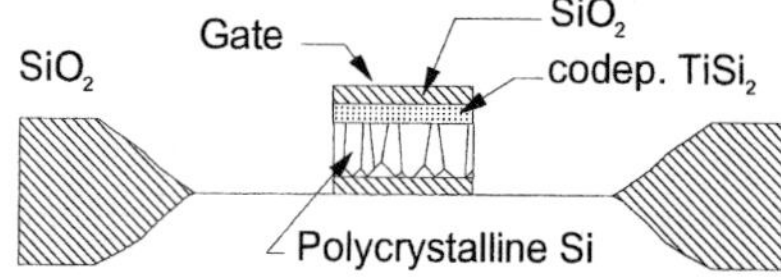

b) Thermal oxidation, fabricate sidewall spacer.

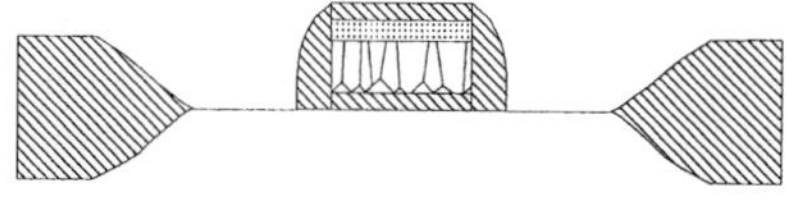

c) Ion implantation & annealing.

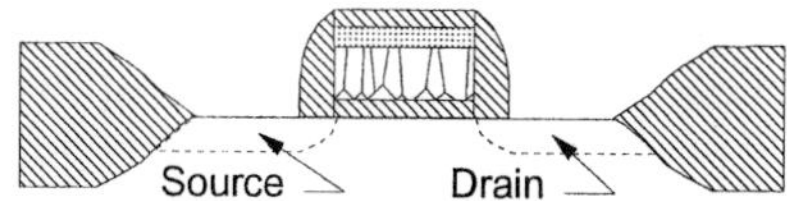

d) Salicide process for Source & Drain.

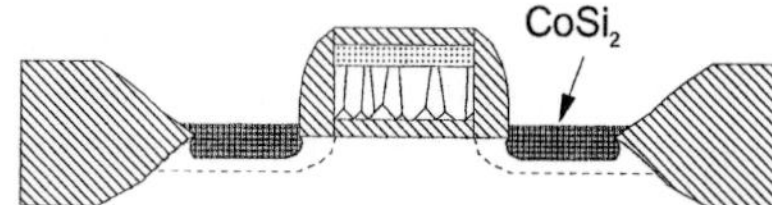

FIG. 35. Schematic of polycide process on an MOS transistor to form $TiSi_2$ on gates: (a) deposit polycrystalline Si, $TiSi_2$, and SiO_2, then pattern using lithography and reactive ion etching, (b) thermally oxidize sidewalls and form sidewall spacers using conformal deposition of an insulator and RIE, (c) form junctions by ion implantation and annealing, and (d) form $CoSi_2$ on the junctions using a salicide process. (Reprinted from Gambino and Colgan, 1998, copyright 1998, with permission from Elsevier Science.)

(*i.e.*, insulator, silicide, polycrystalline Si, and gate oxide) is patterned by lithography and RIE. Next, the structure is exposed to a brief thermal oxidation to remove etch damage at the perimeter of the gate oxide. Insulating sidewalls are formed by conformally depositing an insulator (by CVD), then removing the insulator on horizontal surfaces using RIE. Finally, junctions are formed by ion implantation and annealing. The silicide patterning is a self-aligned process because the silicide is aligned to the polycrystalline Si using no additional masks (Colgan *et al.*, 1996a).

There are a number of advantages to using codeposited silcides compared with other metalizations. Silicon-rich silicides are thermodynamically stable on Si at high temperatures (800–1000 °C) (see Sec. 1.2.1.1), and therefore are compatible with high-temperature processes such as junction anneals and glass-reflow anneals. Silicides are also generally stable during thermal oxidation (see Sec. 1.2.4.1), which is useful during the sidewall oxidation of gates in MOS transistors. The silicide–Si interface is more planar than that resulting from solid state metal–Si reactions, and no Si is consumed. Hence, the formation of silicide protrusions into the polycrystalline Si is reduced.

There are also problems associated with the polycide process. Patterning of some silicides (such as $CoSi_2$, $NiSi_2$, and PtSi) by RIE is difficult because of the relatively low volatility of the metal chlorides and bromides, resulting in incomplete removal of the silicide (see Sec. 1.2.4.3). In addition, the silicide must be defined by lithography; this is not a problem for gate conductors, but is a problem for junctions, where the silicide must be as closely aligned to the Si as possible to avoid problems with leakage or high series resistance. High-temperature annealing can lead to enhanced grain growth in the polycrystalline Si (see Sec. 1.2.3.2), resulting in threshold-voltage shifts or gate-oxide leakage. High-temperature annealing can also lead to dopant diffusion into the silicide and, hence, dopant depletion from the underlying Si, resulting in high contact resistance or threshold-voltage shifts. Finally, oxidation of the silicide can lead to localized consumption and thinning of the polycrystalline Si or oxidation of the metal in the silicide (see Sec. 1.2.4.1).

2.3 Physical Vapor Deposition

Physical vapor deposition (PVD) techniques include evaporation and sputtering, where material is removed from a target by vaporization or ion bombardment and deposited on a substrate. Sputtering (*q.v.*) is often used in integrated-circuit fabrication for depositing elemental metals for silicide reactions or for codepositing the metal and Si, because it is a simple and relatively low-cost method (Table 6). For integrated circuits, sputtering is preferred over evaporation because it is easier to obtain uniform films on large (200-mm diameter) wafers and because the deposition is more conformal over topography on the wafers. The one application where evaporation is used (at least for research) is the epitaxial growth of silicides on Si by molecular-beam epitaxy (MBE). For MBE of silicides (Table 6), metal and Si atoms are evaporated under ultrahigh vacuum on a clean Si surface at high temperatures (von Känel, 1992). PVD methods are also used to make coatings in structural materials (Mashayekhi *et al.*, 1992; Arps, 1996).

In advanced integrated circuits (sub-0.25-μm feature sizes), the metals are often deposited into features with high aspect ratios, such as contact holes. The sputtering process

Table 6. Physical vapor deposition of silicides (poly = polycrystalline silicide, epi = epitaxial silicide).

Silicide	Deposition method	Temperature (°C)	Reference
WSi_2 (poly)	Magnetron sputtering	20, 450	Sundahl *et al.*, 1995
$TiSi_2$ (poly)	Magnetron sputtering	20	Broadbent *et al.*, 1987
$TaSi_2$ (poly)	Magnetron sputtering	20	Goldsmith *et al.*, 1984
$NiSi_2$ (epi)	Molecular-beam epitaxy	250–600	von Känel, 1992
$CoSi_2$ (epi)	Molecular-beam epitaxy	250–600	von Känel, 1992
β-$FeSi_2$ (epi)	Molecular-beam epitaxy	650–700	von Känel *et al.*, 1994

Table 7. Chemical vapor deposition of silicides.

Silicide	Deposition method	Temperature (°C)	Reference
WSi_2 (blanket)	$WF_6 + SiH_4$	300–450	Saraswat *et al.*, 1983
WSi_2 (blanket)	$WF_6 + Si_2H_6$	225–425	Shioya *et al.*, 1987
WSi_2 (blanket)	$WF_6 + SiH_2Cl_2$	510–600	Telford *et al.*, 1993
$TiSi_2$ (selective)	$TiCl_4 + SiH_2Cl_2$	720	Maury *et al.*, 1997
$TiSi_2$ (selective)	$TiCl_4 + SiH_4$	650–950	Southwell & Seebauer., 1997
$TiSi_2$ (blanket)	$TiCl_4 + SiCl_4$	1000–1100	Singheiser *et al.*, 1982
$TaSi_2$ (selective)	$TaCl_5 + SiH_2Cl_2$	670	Carlsson *et al.*, 1990
$MoSi_2$ (blanket)	$MoCl_5 + SiH_4$	550–670	Bobet *et al.*, 1995
$MoSi_2$ (blanket)	$MoF_6 + SiH_4$	140–550	West + Beeson, 1987
β-$FeSi_2$ (blanket)	$Fe_2Cl_6 + SiH_4$	900	Morand *et al.*, 1994
$ReSi_2$ (blanket)	$ReCl_5 + SiH_4$	650–900	Dutron *et al.*, 1995

must be modified to ensure that there is adequate step coverage at the bottom of the hole. One approach is to use a collimator in between the target and the wafers (*i.e.*, collimated sputtering), so that only sputtered atoms with nearly normal incidence reach the substrate. Another approach is to ionize the sputtered atoms (*i.e.*, ionized PVD), so that highly directional deposition of the metal ions can be achieved (Rossnagel, 1995).

2.4 Chemical Vapor Deposition

Chemical vapor deposition (CVD) (*q.v.*) is often used to form WSi_2 in integrated circuits, providing improved step coverage and lower resistivity compared to PVD films (Domenicucci *et al.*, 1997). Recently, there has also been interest in selective CVD of $TiSi_2$ for contacts to Si devices (Maury *et al.*, 1997). Silicides formed by the selective CVD process potentially have lower resistivity and lower Si consumption compared with silicides formed by metal–Si reactions. Another advantage of CVD processes is that the as-deposited silicide is polycrystalline with a relatively large grain size, as opposed to codeposited PVD films which are an amorphous mixture of the metal and Si. CVD has also been used to form silicide coatings on structural materials, by depositing Si or silicides on metal alloys at high temperatures (Singheiser *et al.*, 1982; Nicoll *et al.*, 1979; Southwell *et al.*, 1987).

For CVD of silicides, a metal precursor (WF_6, $TaCl_5$, $MoCl_4$, or $TiCl_4$) is reacted with a Si precursor (SiH_4, Si_2H_6, or SiH_2Cl_2) at elevated temperatures (Table 7). There have been few detailed studies on the kinetics of silicide formation by CVD. Kinetics studies on CVD of WSi_2 and $TiSi_2$ suggest that there are at least two competing reactions that are depositing the metals and/or silicon (Srinivas *et al.*, 1990; Southwell and Seebauer, 1997). In both cases, Si deposition occurs by dissociation of an Si precursor, such as SiH_4, on the Si surface. During CVD of $TiSi_2$, Ti is deposited by adsorption and dissociation of $TiCl_4$. During CVD of WSi_2, W is probably deposited by adsorption and dissociation of WF_6 on the surface of the substrate, analogous to the case of W CVD (Kobayashi *et al.*, 1991; Colgan and Chapple-Sokol, 1992; Hsieh, 1993). In the case of $TiSi_2$ CVD, the rate-limiting step is believed to be the arrival of $TiCl_4$ at the surface, whereas for WSi_2 CVD, the rate-limiting step may be related to the dissociation of SiH_4 at the surface. The byproducts of these reactions are F- and Cl-based species that can etch Si. These byproducts should be minimized to avoid damaging underlying Si devices.

2.5 Ion Implantation

Ion implantation is the introduction of ionized atoms into a substrate with enough energy to penetrate beyond the surface (Seidel, 1983). The technique is widely used for doping semiconductors, but can also be used to form buried polycrystalline or epitaxial silicide layers in Si (Mantl, 1992). Formation of silicides in Si requires implanting a high dose of metal ($>10^{17}\ cm^{-3}$) followed by a high-temperature anneal ($>600\,^{\circ}C$). Ion implantation has been used to form a variety of thin silicide films, including $CoSi_2$, $TiSi_2$, $NiSi_2$, $CaSi_2$, $CrSi_2$, and $FeSi_2$ (Mantl, 1992; Audet *et al.*, 1992; Omura *et al.*, 1991). Although the technique has been widely studied, it has not been used in manufacturing on account of high cost and residual implant damage in the substrate.

Some unique mechanisms occur when silicides are formed by reacting Si with implanted metals, rather than deposited metals (Mantl, 1992). Silicide precipitates form in the Si during ion implantation. During the early stages of thermal annealing, there is competive growth of the precipitates, with larger precipitates growing at the expense of smaller precipitates (*i.e.*, Ostwald ripening). In order for a continuous silicide layer to form, there must be an interconnected network of precipitates; otherwise, the Ostwald ripening will result in large isolated grains. The coalescence of adjacent silicide precipitates is driven by a reduction of interface energy, and results in planarization of the silicide–Si interface (Mantl, 1992).

2.6 Melt Growth

Growth by solidification of a molten alloy is often used to form structural materials, including silicides. Growth from a melt is also used to form single-crystal silicides for experimental studies, using Czochralski or float zone techniques.

Bulk polycrystalline silicides can be formed by traditional metallurgy techniques such as arc melting and casting (Shah *et al.*, 1992). Novel composites can be fabricated by solidification of melt near the eutectic composition. This technique has been used to fabricate Nb/Nb_5Si_3 composites (Nb_5Si_3 matrix) and SiC/$MoSi_2$ composites ($MoSi_2$ matrix) (Mendiratta *et al.*, 1991; Tilly *et al.*, 1992).

Single crystals of silicides have been grown by either Czochralski or floating-zone techniques (Nava *et al.*, 1993). In both techniques, single crystals are grown on seed crystals from a melt. In the Czochralski method, a single crystal is pulled from a melt of the appropriate material, whereas in the float-zone method, a small region in a rod of polycrystalline material is melted, then resolidified (Ghandi, 1983). A problem with applying these methods to refractory-metal silicides is that Si has a high vapor pressure at the melting point of the silicide, resulting in loss of Si and the formation of metal-rich silicide phases (Lograsso, 1992). Crystal growth is performed in an inert ambient to minimize the concentration of oxygen or nitrogen impurities in the melt. A problem with applying the Czochralski technique to silicides is that there is no crucible material that can contain liquid metal silicides without contaminating them. This problem can be overcome by using an rf-levitated melt, and has been used to grow single crystals of $TiSi_2$, VSi_2, $CrSi_2$, $NbSi_2$, $MoSi_2$, $TaSi_2$, WSi_2, and Pd_2Si (Nava *et al.*, 1993). The floating-zone technique produces purer crystals than the Czochralski technique because no crucible is required, and has been used to grow single crystals of of $TiSi_2$, VSi_2, $CrSi_2$, $CoSi_2$, $MoSi_2$, α-$FeSi_2$, and V_3Si (Nava *et al.*, 1993).

2.7 Sintering

Sintering is often used to form high-temperature structural materials, including silicides and silicide composites (Table 8). Sintering is generally used when melt growth is not practical, because of the high melting point of the materials, or for composites, where a two- (or more-) phase structure is desired. Sintering is a process where particles are pressed into a die that has the shape of the desired part. The compact is then heated at a high enough temperature so that the particles bond together, with the driving force being the reduction in surface energy of the particles (Barrett *et al.*, 1973; Kingery *et al.*, 1976). Although most of the porosity of the original powder is eliminated, some porosity often remains. Some special types of sintering that have been applied to silicides include hot pressing or hot isostatic pressing, where the compact is sintered under high pressure and high temperature, and reaction sintering, where reactive powders (such as a metal and SiC) are mixed and then annealed.

Oxygen contamination must be minimized during the sintering of $MoSi_2$, because the presence of an SiO_2 phase at grain boundaries in the silicide reduces the high-temperature strength (see Sec. 1.3.1). Hence, sintering must

Table 8. Formation of structural silicides. The table includes the powders used, the method (HIP = hot isostatic pressing), the maximum annealing temperature, the density (% of theoretical density), and the reference.

Silicide	Powders	Fabriction method	Temperature (°C)	Density (%)	Reference
$MoSi_2$	Mo + Si	Hot pressing	1500	96.6	Schwarz *et al.*, 1992
$MoSi_2$	$MoSi_2$	Plasma spray	—	97	Castro *et al.*, 1992
$MoSi_2$–Mo_5Si_3	Mo + Si	Hot pressing	1500	97.7	Schwarz *et al.*, 1992
$MoSi_2$–WSi_3	Mo + W + Si	Hot pressing	1500	96.8	Schwarz *et al.*, 1992
$MoSi_2$–SiC	$MoSi_2$ + SiC	Hot pressing	1900	—	Sadananda *et al.*, 1992
$MoSi_2$–SiC	Mo_2C + Si	Hot pressing	1700	—	Henager *et al.*, 1992
$MoSi_2$–SiC	$MoSi_2$ + C	Hot pressing	1325	—	Maloy *et al.*, 1991
$MoSi_2$–SiC	$MoSi_2$ + SiC	Plasma spray	—	—	Tiwari *et al.*, 1992
$MoSi_2$–Al_2O_3	$MoSi_2$ + Al_2O_3	HIP	1500	—	Alman *et al.*, 1992
$MoSi_2$–Al_2O_3	$MoSi_2$ + Al_2O_3	Plasma spray	—	—	Alman *et al.*, 1992
$MoSi_2$–ZrO_2	$MoSi_2$ + ZrO_2	Hot pressing	1700	95–96	Petrovic *et al.*, 1992
$MoSi_2$–CaO	$MoSi_2$ + CaO	Hot pressing	1700	95	Gibala *et al.*, 1992
$MoSi_2$–TiC	$MoSi_2$ + TiC	Hot pressing	1700	95	Gibala *et al.*, 1992
$MoSi_2$–TiB_2	$MoSi_2$ + TiB_2	Plasma spray	—	—	Tiwari *et al.*, 1992
$MoSi_2$–Nb	$MoSi_2$ + Nb	HIP	1350	—	Alman *et al.*, 1992
$MoSi_2$–Ta	$MoSi_2$ + Ta	Plasma spray	—	—	Castro *et al.*, 1992

take place in an oxygen-free ambient, such as Ar. In addition, the starting powders must be free of oxygen contamination. This can be achieved by mixing powders in an oxygen-free environment (Schwarz *et al.*, 1992).

Reaction sintering (also referred to as solid-state displacement reactions) uses diffusional phase transformations to react two or more elements or compounds to form new compounds that are more thermodynamically stable (Hanager *et al.*, 1992). Interwoven or dispersed microstructures can be fabricated with this method, making it especially useful for making composites.

2.8 Plasma Spray Deposition

Plasma spray deposition can be used to deposit large thicknesses of materials (>30 μm) on a substrate, and has been used to form silicides and silicide composites for structural materials (Table 8). Plasma spray deposition uses a thermal plasma (plasma flame) formed by an electric arc. A powder of the material to be deposited is fed into the flame in a carrier gas (typically Ar). The injected powder accelerates, melts, and is carried to the substrate, where it solidifies rapidly, forming a thick coating (Herman, 1988). Some of the advantages of using plasma spraying for silicides include dense, oxide-free deposits, small grain size, and the possibility of making laminated composites (Tiwari *et al.*, 1992; Castro *et al.*, 1992; Alman *et al.*, 1992).

3. APPLICATIONS

3.1 Microelectronics

3.1.1 MOS Technology Most Si integrated circuits are fabricated by use of complementary metal–oxide–semiconductor (CMOS) technology. The advantages of CMOS technology compared with bipolar technology are lower power consumption, higher circuit densities, and lower manufacturing costs (Ning, 1992; Masaki, 1992).

Silicides are often used on the gates and diffusions of MOS transistors to reduce the series resistance of the device, and hence increase the circuit speed. In memory chips such as dynamic random-access memories (DRAMs), silicides are mainly used on the word lines (gates) and bit lines (first interconnect level) (Fig. 36) (Winnerl, 1994). For these devices, there is little performance gained by using silicides on diffusions since the series resistance of the device is usually dominated by the contact to the storage node. In addition, silicides on the diffusions increase cost and potentially

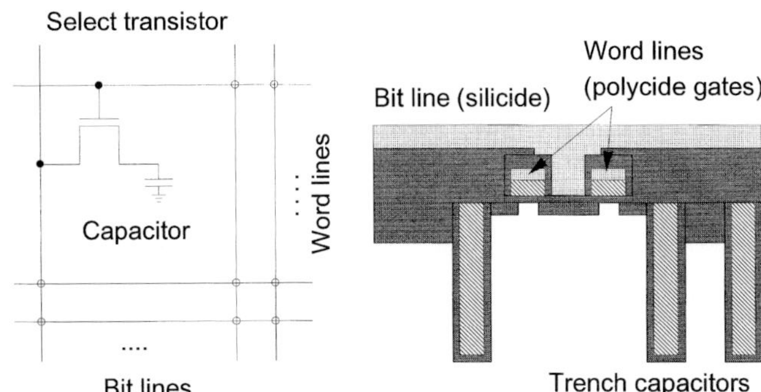

FIG. 36. Schematic of DRAM with silicide on word lines and bit lines (reprinted from Gambino and Colgan, 1998, copyright 1998, with permission from Elsevier Science.)

could decrease yield (through junction leakage associated with metal contamination). Polycides are generally used on the gates because there is a large gain in performance (gates run along the length of the array and so low resistance is required to minimize the *RC* delay) and there is less extra processing compared with a salicide process. In addition, polycide structures can be easily capped with an insulator such as Si_3N_4, which is useful for fabricating borderless contacts (a 30–40% increase in density can be achieved by using contacts that are borderless to gates and isolation; see Gambino *et al.*, 1995). Polycides are often used for bit lines as well, with the polycrystalline Si underlayer serving to act as a contact to diffusions. If lower resistance is required, a W bit line could be used instead of a polycide, with either polycrystalline Si or W studs as contacts to gates and diffusions. The most commonly used silicide in polycide structures is WSi_2, because it has good high temperature stability, is relatively easy to pattern by RIE, and can be deposited by either PVD or CVD.

In logic chips, silicides are often used on source/drain junctions as well as gates and local interconnects (Fig. 37). The improvement in performance is great enough to justify the extra cost and yield loss. In addition, contacts that are borderless to gates are seldom used in logic devices, and so insulator-capped polycide structures are not required. Silicided local interconnects can be fabricated at the same time as the gate and source/drain silicides. Alternatively, W local interconnects can be fabricated along with W studs for contacts. For logic applications, the silicide is usually formed by a salicide process using $TiSi_2$. $TiSi_2$ has a number of advantages compared with other silicides, including low resistivity, relatively high thermal stability, and relatively low sensitivity to interfacial contamination (because Ti reduces native oxides).

As with DRAMs, the switching speed of the device is increased by lowering the resistance of the gates (Fig. 38). The silicide on the source/drain junctions increases the drive current of the MOS transistor (Fig. 39). The silicide reduces the voltage drop at the source end of the device, which increases the effective applied gate potential. The reduced voltage drop is mainly the result of a reduction in the effective contact resistance, due to an increase in the contact area (Ng and Lynch, 1987). The low sheet resistance of the silicide is important for devices that have long lengths of diffusions and for wide devices that have a low channel resistance (Osburn *et al.*, 1991). Silicides can

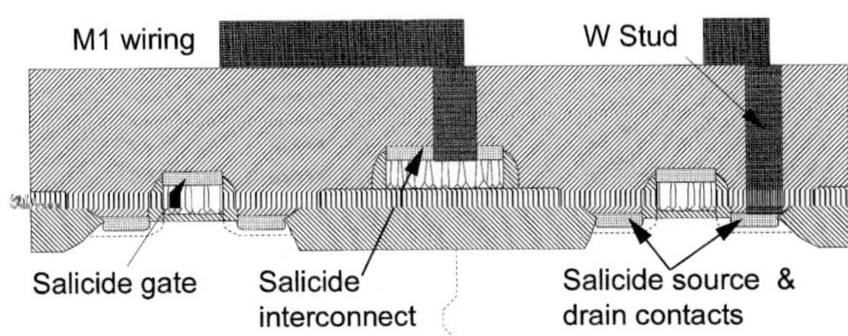

FIG. 37. Schematic of logic device showing silicides on gates and diffusions (reprinted from Gambino and Colgan, 1998, copyright 1998, with permission from Elsevier Science.)

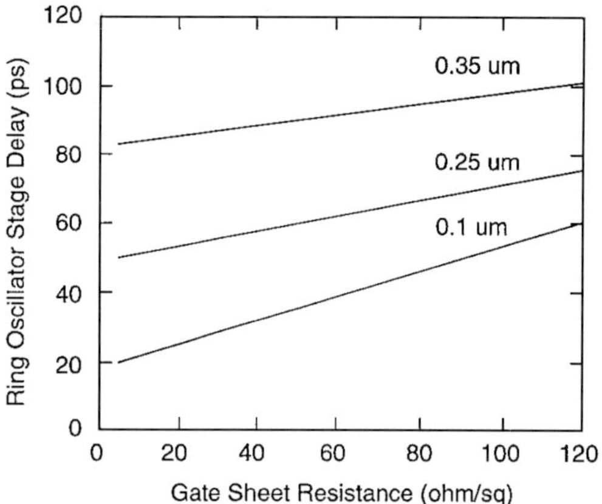

FIG. 38. Performance of ring oscillator as a function of gate resistance (reprinted with permission from Mann *et al.*, 1995).

also be used to increase the packing density of devices. For example, butting n^+ and p^+ diffusions can be connected by a silicide rather than the first metal level. The area of the diffusions can be reduced, since an extra contact hole can be eliminated.

Although silicides can enhance device performance, they can also degrade yield. Junction leakage can be increased (Osburn, 1990; Maex, 1993; Nauka *et al.*, 1986), and gate-oxide breakdown voltage decreased (Koburger *et al.*, 1982; Gambino *et al.*, 1992; Tanielian *et al.*, 1985), if too much Si or polycrystalline Si is consumed in the junctions or the gates, respectively. Interface state densities can be increased by rapid thermal anneals (Hsieh *et al.*, 1993; Liu *et al.*, 1995) used to form silicides or due to gettering of hydrogen (Chang and Chiu, 1989; Hirade *et al.*, 1995) by Ti during post-metallization anneals. Dopant depletion or counterdoping can occur during high-temperature anneals after silicide formation, resulting in high contact resistance (Scott *et al.*, 1987; Chittipeddi *et al.*, 1993) or threshold-voltage shifts (Matsuoka *et al.*, 1994; Hillenius *et al.*, 1986). Finally, excessive lateral silicide formation in a salicide process can cause leakage between junctions and gates.

3.1.2 Bipolar Technology The main use of silicides in bipolar transistors is as Schottky diodes. There have also been proposals to use silicides on base contacts (Fig. 40) or collector contacts, but these processes are not currently used in manufacturing (Shiba *et al.*, 1996; Yallup *et al.*, 1995).

Schottky diodes are used as clamping diodes between the base and collector to prevent excess minority-carrier charge storage that would limit switching speed (Sze, 1981; Tada and Laraya, 1967). PtSi is the most commonly used silicide for bipolar transistors (Li *et al.*, 1987), on account of a high barrier height and ease of patterning by a salicide process. Some problems associated with PtSi formation on bipolar transistors are incomplete silicide formation (Moy *et al.*, 1988), because Pt does not reduce native oxides, and nonuniform reactions between Pt and polycrystalline Si, which can cause junction leakage (Gambino *et al.*, 1989).

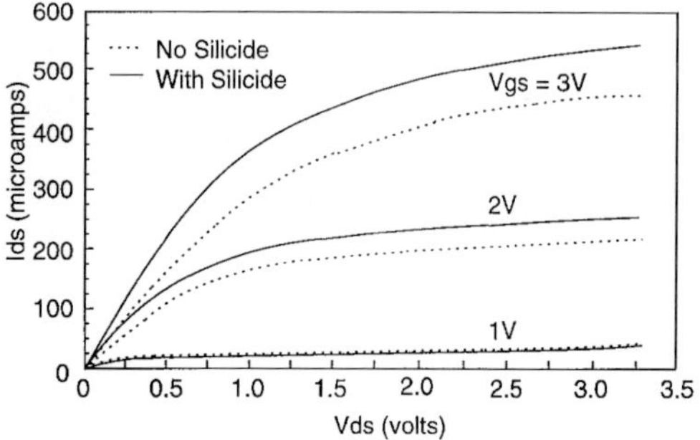

FIG. 39. Device characteristics with and without silicide on source/drain junctions (reprinted from Osburn *et al.*, 1991, copyright 1991, with permission from Elsevier Science).

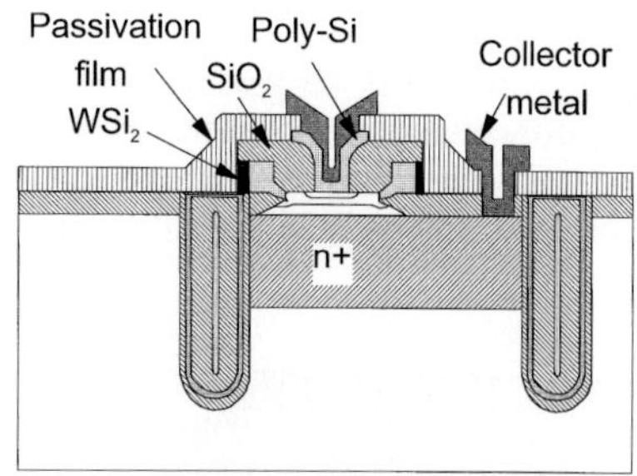

FIG. 40. Schematic of bipolar transistor with silicide on base contact to reduce series resistance (reprinted with permission from Shiba *et al.*, 1996, © 1996 IEEE).

3.1.3 Infrared Detectors Another application for silicide Schottky diodes is as infrared detectors (Kimata and Tubouchi, 1995). Schottky diodes have a number of advantages compared with other infrared detectors, including monolithic construction using standard Si processing and high uniformity. The silicides for Schottky diodes are formed by a salicide process.

The operation of Schottky-diode photodetectors is based on internal photoemission (Kimata and Toubouchi, 1995; Jimenez, 1996). In the first step (Fig. 41), incident photons are absorbed in the silicide and generate electron–hole pairs. Some of the photoexcited holes are then emitted over the Schottky barrier, becoming detected carriers. Hence, the Schottky barrier height defines a minimum energy or cutoff wavelength for detection. Because the energy of infrared radiation in the atmospheric window is in the range of 0.1–0.4 eV, a low Schottky-barrier height is required; this can be most easily achieved on p-type Si.

The quantum efficiency of Schottky-diode infrared detectors depends on the photon energy, because of the strong dependence of photoemission on the energy of the excited holes (Kimata and Tubouchi, 1995). The quantum efficiency η is given by

$$\eta = C_1(h\nu - \phi_B)^2/h\nu \qquad (10)$$

Table 9. Properties of silicides used in infrared detectors (from Kimata and Tubouchi, 1995).

Silicide	Silicon	Barrier height (eV)	Cutoff wavelength (μm)
Pd_2Si	*n*-type	0.6	1–2
$CoSi_2$	*p*-type	0.44	2.8
NiSi	*p*-type	0.40	3.1
Pd_2Si	*p*-type	0.34	3.5
PtSi	*p*-type	0.22	6
IrSi	*p*-type	0.152	8.2

where C_1 is the quantum efficiency coefficient, h is Planck's constant, ν is the wavelength of the incident radiation, and ϕ_B is the barrier height. Hence, the Schottky-barrier detector has no response to photons with energy lower than the barrier height.

There are five silicides used for Schottky-barrier infrared detectors: Pd_2Si, PtSi, IrSi, Co_2Si, and NiSi (Table 9 and Fig. 42). PtSi on *p*-type Si is most commonly used, on account of a spectral response to radiation in the range 3–5 μm. IrSi on *p*-type Si is expected to have the lowest barrier height, and therefore should be able to detect radiation at the longest wavelengths (8.2 μm).

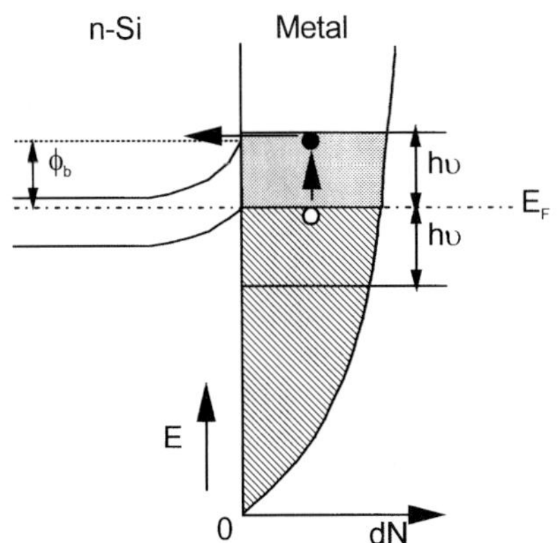

FIG. 41. Schematic of internal photoemission in a Schottky-barrier infrared detector (used with permission from Kimata and Tubouchi, 1995).

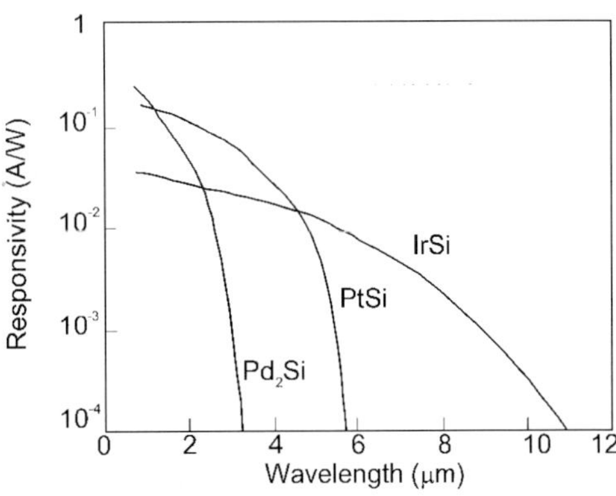

FIG. 42. Plot of $(\eta h\nu)^{1/2}$ *versus* $h\nu$ for a Schottky-barrier infrared detector. The gradient of the plot gives the square root of the quantum efficiency while the intersection with the *x* axis gives the barrier height (used with permission from Kimata and Tubouchi, 1995).

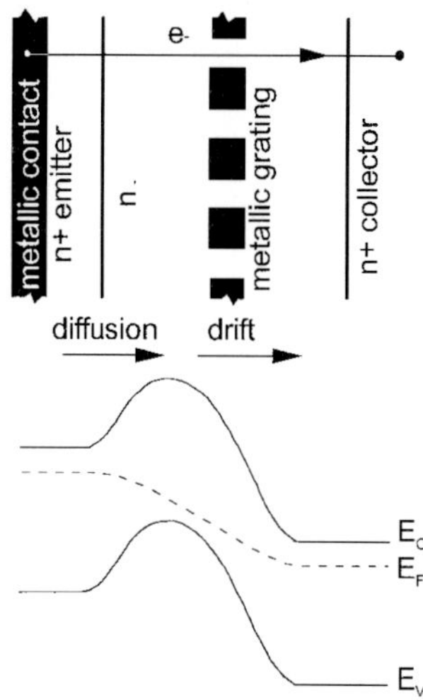

FIG. 43. Schematic cross section of permeable-base transistor with corresponding energy diagram (from Reader *et al.*, 1992).

3.1.4 Permeable-Base Transistors In permeable-base transistors (Reader *et al.*, 1992), a metal grid is used to modulate current flow from an emitter to a collector (Fig. 43). The metal grid is embedded in an n region, resulting in a Schottky diode and a depletion layer around the metal. The extension of the Schottky diode is controlled by the voltage on the metal grid, which in turn controls the flow of carriers through the grid.

There are a number of advantages of permeable-base transistors compared with bipolar or MOS transistors. The main advantage is that the width of the control electrode is determined by the thickness of the metal layer, rather than by lithography (as is the case in MOS transistors), so that the transit time for charge carriers can be very fast, resulting in high switching speeds. The main problem with this device is the difficulty of fabricating buried silicide layers in defect-free Si. Defects in the Si associated with the buried silicide can cause high leakage currents. Because of these problems, permeable-base transistors are still in the development stage and are not being manufactured.

3.2 Structural Materials

Silicides are being considered for a number of applications as structural materials in the aerospace and automotive industries. Aerospace applications include aircraft-engine turbine components, such as blades, vanes, combustors, nozzles, and seals. Automotive applications include turbocharger rotors, valves, glow plugs, and advanced turbine engine parts (Petrovic, 1993).

Structural materials can be broadly classified into two groups of applications, depending on the use temperature. For applications in the range of 800–1000 °C, nickel-based superalloys are being used. Above 1000 °C, silicon-based ceramics are being considered, because of their excellent oxidation resistance and low density. However, these materials are brittle over the entire temperature range, and the cost of fabrication is high. Hence alternative materials are being investigated for high-temperature structural materials, including silicides (Vasudevan and Petrovic, 1992). The use of high-temperature materials can provide a number of advantages, including minimized air cooling, fewer engine parts, and higher operating temperatures (which can improve fuel efficiency).

Silicides have a number of advantages compared with other structural materials, including higher melting points and lower densities compared with other intermetallics, and lower costs and greater ductility (at high tempertures) compared with Si-based ceramics. In addition, silicides can be alloyed with other materials such as SiC or other silicides, to improve their properties. This is important because the mechanical properties of the silicides by themselves are inadequate for these applications; improvements are required in the high-temperature strength and creep resistance, and in the low-temperature fracture toughness. Composite reinforcement of a silicide matrix can significantly improve these mechanical properties (see Figs. 19–21). These composites can be fabricated using a wide variety of techniques (Table 8), including melt casting and sintering.

Composites are classified into the following categories based on the morphologies, orientations, and distributions of the strengthening or toughening components in the intermetallic (silicide) matrix: discontinuous reinforcements, continuous reinforcements, and hybrids (Miracle and Mendiratta, 1994). In discontinuous composites, an intermetallic matrix contains particulates, whiskers, or chopped fibers, such as SiC whiskers, in an $MoSi_2$ matrix. At high temperatures, where there is significant plasticity of the silicide matrix, the morphology and aspect ratio of the discontinuous reinforcements strongly influences the strength and creep resistance of the composite; needle-like reinforcements with a high aspect ratio provide the greatest strengthening and toughening. Continuous composites contain continuous fibers or filaments, typically distributed in a uniaxial direction in a silicide matrix. Fibers with high strength and stiffness enhance the strength and creep resistance of the silicide. Since the mechanical properties are highly anisotropic, applications as structural materials are limited to components where transverse properties are not critical (*e.g.*, turbine blades, ring components, and structural rods). Both refractory-metal and ceramic fibers have been used to reinforce $MoSi_2$ (Maloney and Hecht, 1992). The refractory-metal fibers are ductile, and therefore increase the damage tolerance of the brittle silicide matrix. Although the ceramic fibers are brittle, the composite can have improved fracture toughness if there is a weak fiber–matrix interface; a weak interface when present in the path of a crack will debond locally and blunt the crack (Shaw *et al.*, 1995). Hybrid composites incorporate multiple types of reinforcement, designed to enhance specific mechanical properties. For example, continuous uniaxial fibers may primarily enhance high-temperature strength and creep resistance, whereas a dispersion of whiskers in the matrix may enhance toughness and transverse mechanical properties.

There are a number of other considerations in the fabrication of composites (Maloney and Hecht, 1992). The thermal expansion coefficient must be closely matched between the fiber and the matrix, in order to minimize thermal stress. For $MoSi_2$, sapphire fibers are closest in terms of thermal expansion coefficient, whereas SiC is significantly lower. This problem can be alleviated by adding a phase with a low thermal expansion coefficient to the silicide matrix (such as SiC platelets or particles). Another consideration when using refractory metal fibers is the reactivity of the metal with the silicide. Minimal reactivity is desirable, so that the refractory metal remains a metal, rather than being converted into a (brittle) silicide. For $MoSi_2$, Ta exhibits the lowest reactivity compared with other refractory metals.

3.3 Oxidation-Resistant Coatings

Silicides have been used as oxidation-resistant coatings on a number of materials, including Ni-based alloys, steels, and refractory metals and alloys (Kumar, 1994). The SiO_2 layers that grow on silicides in oxygen ambients show a low permeation rate to oxygen compared with other materials, especially above 1500 °C. The SiO_2 layers are also stable in the alkaline sulfates that are often found in gas turbines. More recently, silicides have been examined as coatings to protect Ni-based superalloys from carburization in high-temperature reactors (Singheiser *et al.*, 1982).

Silicides that are used as coatings must be "matched" to the underlying substrate, to prevent degradation by cracking, spalling, or interdiffusion. The brittleness of silicides and the thermal expansion mismatch between the coating and the substrate can lead to cracking and spalling during thermal cycling. Alloying can be used to adjust the thermal expansion mismatch of the protective SiO_2 layer, and hence reduce cracking. B_2O_3 or GeO_2 additions to the coating not only decrease the thermal expansion mismatch but also decrease the viscosity of the SiO_2 layer, making it more resistant to cracking. Interdiffusion between the substrate and the coating can alter the protective properties of the coating, by depleting Si from the silicide. Metal-rich silicides generally oxidize much faster than silicon-rich silicides, resulting in a less effective coating (Kircher and Courtright, 1992). Interdiffusion can be controlled by using an appropriate diffusion barrier between the substrate and the coating; for example, a TiC diffusion barrier can be used between a graphite substrate and an $MoSi_2$ coating (Grunling and Bauer, 1982).

Most of the applications for silicide coatings involve protecting the underlying material from oxidation. Some examples include Ni–Cr silicide coatings on Ni-based superalloys, Cr–Ti silicide coatings on Nb-based alloys, and $MoSi_2$ coatings on refractory metals. Silicide coatings have been fabricated with a wide variety of methods, including PVD, CVD, plasma spray deposition, or diffusion reactions (pack coating) (Nicholls and Stephenson, 1994).

3.4 Thermoelectric Materials

One of the earliest applications for silicides was as heating elements in high-temperature furnaces (Kumar, 1994). More recently, semiconducting silicides have been investigated as thermoelectric devices, for converting heat into electricity (Verdnikov, 1994).

Furnace elements made from silicides consist of $MoSi_2$ with about 20 vol% of a glassy phase. $MoSi_2$ has a number of advantages as a heating element compared with other materials. First, the resistivity of $MoSi_2$ increases with temperature; if a new batch of material is introduced into the furnace (causing the temperature to drop), the power output of the heating elements rises rapidly and automatically compensates. In addition, the heating elements will accept high power loads so that large amounts of power can be concentrated in a small area. Finally, a protective SiO_2 coating forms on the silicide, resulting in a long life span for the heating elements, even in oxidizing ambients.

Thermoelectric devices convert heat into electricity, and are often used in deep-space vehicles with a radioactive heat source. At present, these devices are made from SiGe alloys. However, several semiconducting silicides are promising as thermoelectric devices. The themoelectric figure of merit for $MnSi_{1.7}$ is comparable to that for SiGe alloys, and calculations suggest that the figure of merit for Ru_2Si_3 may be higher than that for SiGe alloys (Vedernikov, 1994; Vining, 1992).

3.5 Magnetic Materials

There have been two different approaches to using silicides as magnetic materials. One approach is to add Si to an Fe-based alloy to improve the wear resistance and corrosion resistance of the alloy. Ternary Fe-Al-Si alloys are an example of this approach, and have been used in magnetic heads (Kumar, 1994). The other approach is to use Si-stabilized Fe-based compounds for permanent magnets (see Sec. 1.3.4). These materials have not yet been used commercially.

3.6 Optical Materials

Silicides have been used as optical materials in attenuated phase-shift masks. Phase-shift masks are used to improve the resolution of optical lithography during fabrication of integrated circuits. An attenuated phase-shift mask allows some transmission of light through the "opaque" regions of the mask, thereby changing the phase and improving the resolution of the image (Hashimoto *et al.*, 1994). Recently, MoSi alloys have been used as the "opaque" material in attentuated phase-shift masks, instead of Cr. The MoSi alloys are easier to pattern than Cr, thus making it easier to fabricate defect-free masks (Jonckheere *et al.*, 1994).

4. SUMMARY

This article has reviewed the properties, processing, and applications of silicides. The combination of high thermal stability, high mechanical hardness, and low electrical resistivity has made the use of silicides attractive for a wide range of applications. Already, there is widespread use of silicides in the microelectronics industry. With continued studies on these materials, it is likely that silicides will also be used in other applications such as structural materials.

There are many areas of study that could be useful in terms of improving these materials. A better understanding of the atomic and electronic structures of interfaces between silicides and semiconductors could be useful for microelectronics applications. Much has been learned about metal–Si reactions, but further work is required to understand the effects of nucleation, impurities, and grain boundaries on these reactions. More work is needed to understand reactions between metals and Si compounds, for both microelectronics and structural materials applications. More details on the surface and interface energies would be useful in understanding the thermal stability of silicide thin films. Information on the effect of defects on oxidation of silicides could be used to improve the properties of silicide coatings. Detailed studies of the phase equilibria between dopants and silicides could be useful for creating new structural materials. Much information is still needed to understand the etching and passivation reactions for silicides in aqueous solutions. Additional studies are needed on the basic mechanical properties of silicides, as well as processes for improving these properties. More work is also needed to improve the electrical properties of silicide thin films used in microelectronics applications. The initial work on the optical and magnetic properties of silicides shows that there may be new opportunities to use silicides for these applications.

Because of the many potential applications, there are likely to be many new studies on silicides in the coming years.

ACKNOWLEDGEMENT

The authors thank F. M. d'Heurle and J. M. E. Harper for helpful discussions.

GLOSSARY

Agglomeration: The breaking up of a thin film on a substrate into islands, due to a reduction in surface and interface energies.

Burgers Vector: The unit slip distance for the dislocation, parallel to the direction of slip.

Composite: A polyphase material with a large fraction of a minor phase (generally at least 10%).

CVD: Chemical vapor deposition.

DRAM: Dynamic random-access memory.

Electron Affinity: The energy released when an electron is attached to an atom or molecule.

Epitaxy: Growth of a thin layer of material with the same orientation and crystal structure as the underlying single crystal substrate.

Eutectic: A solid solution of two or more components (eutectic composition), having the lowest freezing point (eutectic temperature) of all possible mixtures of the components.

Ferromagnetic: Materials that have a permanent magnetic field even in the absence of an applied field.

Fracture Toughness: Work expended at a crack tip during crack propagation.

Fermi Level: The energy at which the probability of finding an electron in an energy band in a solid is 1/2.

MBE: Molecular beam epitaxy.

MOS Transistor: Metal–oxide–semiconductor transistor.

Nucleation: The initial formation of a new phase within a material or at an interface or surface.

Phase: A homogeneous, physically distinct, and mechanically separable portion of a material with a given chemical composition and structure.

Phase Diagram: A phase equlibria diagram between two or more components (binary = two components, ternary = three components) at a constant pressure, with temperature and composition as variables.

Phonons: Lattice vibrations in a crystal.

Photoemission: Emission of electrons as a result of bombardment by photons.

Polycide: A self-aligned silicide process used in microelectronics applications, where the silicide is deposited on a polycrystalline Si gate stack and patterned by RIE.

PVD: Physical vapor deposition, such as evaporation or sputtering.

Raman Scattering: A scattering effect where the scattered light differs in wavelength from the incident light.

Refractory Materials: Materials that have a high melting point (>1538 °C, the melting point of iron).

RIE: Reactive ion etching; a directional dry etching method used in microelectronics processes.

Salicide: A self-aligned silicide process used in microelectronics applications, where the silicide is formed selectively on Si using a metal–Si reaction and a selective etch.

Schottky Barrier: Potential barrier for electronic carriers at a metal–semiconductor or metal–insulator interface.

Sintering: Compression of particles into a solid body using heat and pressure, at a temperature below the melting point of the solid.

Slip Plane: The plane on which a dislocation moves. A slip system describes all the different slip planes in a given crystal structure.

Thermoelectric Effect: The generation of electrical currents in a material due to a temperature gradient in the material. Also, the production or absorption of heat when electronic carriers are forced across a temperature gradient in a material.

Threshold Voltage: The voltage that must be applied to the gate of an MOS transistor to "turn on" the transistor.

Transition Metals: The metals between II and III in the periodic table; the subgroups starting with Sc and ending with Zn.

Work Function: The difference in energy between the Fermi level of a solid and the vacuum energy.

Yield Strength: The minimum stress that must be applied to induce plastic deformation in a material.

Works Cited

Aboelfotoh, M. O., Tu, K. N. (1986), *Phys. Rev. B.* **34**, 2311–2318.

Aldrich, D. B., Chen, Y. L., Sayers, D. E., Nemanich, R. J., Ashburn, S. P., Ozturk, M. C. (1995), *J. Mater. Res.* **10**, 2849–2863.

Aldrich, D. B., d'Heurle, F. M., Sayers, D. E., Nemanich, R. J. (1996), *Phys. Rev. B* **53**, 16279–16282.

Alman, D. E., Shaw, K. G., Stoloff, N. S., Rajan, K. (1992), *Mater. Sci. Eng.* **A155**, 85–93.

Aoki, A., Yamaguchi, H. Saito, T., Otsuka, F., Owada, N. (1994), in: *Proceedings of the Eleventh International VLSI Multilevel Interconnection Conference (VMIC)*, Santa Clara, CA: Eleventh International VLSI Multilevel Interconnection Conference.

Appelbaum, A., Knoell, R. V., Murarka, S. P. (1985), *J. Appl. Phys.* **57**, 1880–1886.

Arps, J. H., Page R. A., Dearnaley, G. (1996), *Surf. Coat. Technol.* **84**, 579–583.

Ashcroft, N. W., Mermin, N. D. (1976), *Solid State Physics*, New York: Holt, Rinehart and Winston, Chap. 1.

Audet, S. A., White, A. E., Short, K. T., Hsieh, Y.-F., Ross, F. M., Rafferty, C. S. (1992), *Appl. Phys. Lett.* **61**, 2311–2313.

Barbour, J. C., Kuiper, A. E. T., Willemsen, M. F. C., Reader, A. H. (1987), *Appl. Phys. Lett.* **50**, 953–955.

Barge, T., Gas, P., d'Heurle, F. M. (1995), *J. Mater. Res.* **10**, 1134–1145.

Barrett, C. R., Nix, W. D., Tetelman, A. S. (1973), *The Principles of Engineering Materials*, Englewood Cliffs, N. J: Prentice-Hall, Chaps. 6 and 9.

Bednorz, J. G., Muller, K. A. (1986), *Phys. Rev. B*, **64**, 189–193.

Berti, A. C., Bolkhovsky, V. (1992), in: *Proceedings of the Tenth International VLSI Multilevel Interconnection Conference (VMIC)*, Santa Clara, CA: Tenth International VLSI Multilevel Interconnection Conference.

Beyers, R. (1984), *J. Appl. Phys.* **56**, 147–152.

Beyers, R., Sinclair, R., Thomas, M. E. (1984), *J. Vac. Sci. Tech.*, **B2**, 781–784.

Bisi, O., Calandra, C. (1981), *J. Phys. C: Solid State Phys.* **14**, 5479–5494.

Bobet, J. L., Naslain, R., Bernard, C. (1995), *J. Chem. Vap. Dep.* **3**, 223–253.

Bocelli, S., Guizzetti, G., Marabelli, F., Thungstrom, G., Petersson, C. S. (1995), *Appl. Surf. Sci.* **91**, 30–33.

Bose, S. (1992), *Mater. Sci. Eng.* **A155**, 217–225.

Bost, M. C., Mahan, J. E. (1985), *J. Appl. Phys.* **58**, 2696–2703.

Bost, M. C., Mahan, J. E. (1988), *J. Appl. Phys.* **64**, 2034–2037.

Boutarek, N., Madar, R. (1993), *Appl. Surf. Sci.* **73**, 209–213.

Briant, C. L., Taub, A. I. (1988), *Acta Metall.* **36**, 2761–2770.

Broadbent, E. K., Morgan, A. E., Coulman, B., Huang, I.-W., Kuiper, A. E. T. (1987), *Thin Solid Films* **151**, 51–63.

Buschow, K. H. J. (1991), *Rep. Prog. Phys.* **54**, 1123–1213.

Carlsson, J.-O. (1990), *CRC Crit. Rev. Solid State Mater. Sci.* **16**, 161–212.

Carlsson, A. E., Meschter, P. J. (1991), *J. Mater. Res.* **6**, 1512–1517.

Carmondy, M. W., Johnson, A. S., Kvan, E. P. (1996), in: R. T. Tung, K. Maex, P. W. Pellegrini, L. H. Allen (Eds.), *Silicide Thin Films—Fabrication, Properties, and Applications*, MRS Symposium Proceedings Vol. 402, Pittsburgh: Materials Research Society, pp. 399–404.

Castro, R. G., Smith, R. W., Rollett, A. D., Stanek, P. W. (1992) *Mater. Sci. Eng.* **A155**, 101–107.

Chang, S.-T., Chiu, K. Y. (1989), *IEEE Trans. Elec. Dev.* **36**, 145–147.

Chen, L. J., Tu, K. N. (1991), *Mater. Sci. Rep.* **6**, 53–140.

Chittipeddi, S., Dziuba, C. M., Kannan, V. C., Kelly, M. J., Cochran, W. T., Rambabu, B. (1993), *J. Elec. Mat.* **22**, 785–791.

Chou, T. C., Joshi, A., Wadsworth, J. (1991), *J. Mat. Res.* **6**, 796–809.

Chou, T. C., Nieh, T. G. (1993), *J. Mat. Res.* **8**, 214–226.

Chow, T. P., Saxena, A. N., Ephrath, L. M., Bennett, R. S. (1984), in: R. A. Powell (Ed.), *Dry Etching for Microelectronics*, New York: Elsevier, Chap. 2.

Christensen, N. E. (1990), *Phys. Rev. B* **42**, 7148–7153.

Chu, C. L., Saraswat, K. C., Wong, S. S. (1992), *IEEE Trans. Elec. Dev.* **39**, 2333–2340.

Clevenger, L. A., Cabral, C., Roy, R. A., Lavoie, C., Viswanathan, R., Saenger, K. L., Jordan-Sweet, J., Morales, G., Ludwig, K. L., Stephenson, G. B. (1996), in: R. T. Tung, K. Maex, P. W. Pellegrini, L. H. Allen (Eds.), *Silicide Thin Films—Fabrication, Properties, and Applications*, MRS Symposium Proceedings Vol. 402, Pittsburgh: Materials Research Society, pp. 257–268.

Colgan, E. G., Chapple-Sokol, J. D. (1992), *J. Vac. Sci. Tech.* **B10**, 1156–1166.

Colgan, E. G., Gambino, J. P., Hong, Q. Z. (1996a), *Mater. Sci. Eng. Rep.* **R16**, 43–96.

Colgan, E. G., Gambino, J. P., Cunningham, B. (1996b), *Mater. Chem. Phys.* **46**, 209–214.

Cowley, A. M., Sze, S. M. (1965), *J. Appl. Phys.* **36**, 3212–3220.

Derrien, J. (1995), in: K. Maex, M. van Rossum (Eds.), *Properties of Metal Silicides*, London: INSPEC, p. 155.

d'Heurle, F. M., (1988), *J. Mater. Res.* **3**, 167–195.

d'Heurle, F. M., Gas, P. (1986), *J. Mater. Res.*, **1**, 205–221.

d'Heurle, F. M., Petersson, C. S., Baglin, J. E., LaPlaca, S. J., Wong, C. Y. (1984), *J. Appl. Phys.* **55**, 4208–4218.

d'Heurle, F. M., Cros, A., Frampton, R. D., Irene, E. A. (1987), *Philos. Mag.*, **B55**, 291–308.

Domenicucci, A., Dehm, C., Loh, S., Clevenger, L., Dziobkowski, C., Cabral, C., Lavoie, C., Jorden-Sweet, J. (1997), in: S. C. Moss, D. Ila, R. C. Cammarata, E. H. Chason, T. L. Einstein, E. D. Williams (Eds.), *Thin Films—Structure and Morphology*, MRS Symposium Proceedings Vol. 441, Pittsburgh: Materials Research Society, pp. 3–8.

Dutron, A.-M., Blanquet, E., Bourhila, N., Madar, R., Bernard, C. (1995), *Thin Solid. Films*, **259**, 25–31.

Fenske, F., Lange, H., Oertel, G., Reinsperger, G.-U., Schumann, J., Selle, B. (1996), *Mater. Chem. Phys.* **43**, 238–242.

Flamm, D. L., Donnelly, V. M., Ibbotson, D. E. (1984), in: N. G. Einspruch, D. M. Brown (Eds.), *Plasma Processing for VLSI*, New York: Academic Press, p. 189.

Fleischer, R. L. (1991), in: O. Izumi (Ed.), *Proceedings of the International Symposium on Intermetallic Compounds*, Sendai: The Japan Institute of Metals, p. 157.

Fleischer, R. L. (1994), in: J. H. Westbrook, R. L. Fleischer (Eds.) *Intermetallic Compounds: Vol. 2, Practice*, New York: John Wiley, Chap. 11.

Flores, F., Ortega, J. (1992), *Appl. Surf. Sci.* **56–58**, 301–310.

Frampton, R. D., Irene, E. A., d'Heurle, F. M. (1987), *J. Appl. Phys.* **62**, 2972–2980.

Frommeyer, G., Rosenkranz, R., Ludecke, C. (1990), *Z. Metallk. (Germany)* **81**, 307–313.

Fujitani, H., Asano, S. (1994), *Phys. Rev. B* **50**, 8681–8698.

Gambino, J. P., Colgan, E. G. (1998), *Mater. Chem. Phys.* **52**, 99–146.

Gambino, J. P., Monkowski, M. D., Tsang, P. J., Shepard, J. F., Ransom, C. M., Wong, C. Y. (1989), *J. Electrochem. Soc.* **136**, 2063–2067.

Gambino, J. P., Kastl, R. H., Cunningham, B., Turene, F. E. (1990), *Ext. Abstr. Electrochem. Soc.* **90–1**, Abs. No. 170.

Gambino, J. P., Cunningham, B., Buchanan, D. A. (1992), in: A. Katz, S. P. Murarka, Y. I. Nissim, J. M. E. Harper (Eds.), *Advanced Metallization and Processing for Semiconductor Devices and Circuits—II*, MRS Symposium Proceedings Vol. 260, Pittsburgh: Materials Research Society, pp. 181–186.

Gambino, J. P., Dobuzinsky, D., Armacost, M., Cunningham, B., Turene, F. E. (1995), in: *Proceedings of the Twelfth International VLSI Multilevel Interconnection Conference (VMIC)*, Santa Clara, CA: Twelfth International VLSI Multilevel Interconnection Conference.

Gambino, J. P., Colgan, E. G., Domenicucci, A. G., Cunningham, B. (1998), *J. Electrochem. Soc.* **145**, 1384–1389.

Gas, P., d'Heurle, F. M. (1993), *Appl. Surf. Sci.* **73**, 153–161.

Gas, P., d'Heurle, F. M. (1997), in: D. L. Beke (Ed.), *Diffusion in Semiconductors and Non-Metallic Solids*, New York: Springer Verlag.

Gas, P., Thomas, O., d'Heurle, F. M. (1995), in: K. Maex, M. van Rossum (Eds.), *Properties of Metal Silicides*, London: INSPEC, p. 298.

Ghandi, S. K. (1983), in *VLSI Fabrication Principles*, New York: John Wiley & Sons, p. 81.

Gibala, R., Ghosh, A. K., Van Aken, D. C., Srolovitz, D. J., Basu, A., Chang, H., Mason, D. P., Yang, W. (1992), *Mater. Sci. Eng.* **A155**, 147–158.

Goldschmidt, H. J. (1967), *Interstitial Alloys*, London: Butterworth, Chap. 7.

Goldsmith, C. C., Campbell, D. R., Mader, S. (1984), in J. E. E. Baglin, D. R. Campbell, W. K. Chu (Eds.), *Thin Films and Interfaces II*, New York: Elsevier, p. 569.

Gottlieb, U., Lasjaunias, J. C., Laborde, O., Thomas, O., Madar, R. (1993), *Appl. Surf. Sci.* **73**, 232–236.

Gottlieb, U., Nava, F., Affronte, F., Laborde, O., Madar, R. (1995), in: K. Maex, M. van Rossum (Eds.), *Properties of Metal Silicides*, London: INSPEC, p. 205.

Grabke, H. J. Meier, G. H. (1995), *Oxid. Met.* **44**, 147–176.

Grunling, H. W., Bauer, R. (1982), *Thin Solid. Films* **95**, 3–20.

Guizetti, G., Marabelli, F. (1995), in: K. Maex, M. van Rossum (Eds.), *Properties of Metal Silicides*, London: INSPEC, p. 227.

Hara, S., Ohdomari, I. (1988), *Phys. Rev.* **B38**, 7554–7557.

Hashimoto, K., Samuels, D. J., Farrell, T. R., Moy, D., Martino, R. M., Ferguson, R. A., Sato, T., Maurer, W. (1994), *Jpn. J. Appl. Phys.* **33**, 6823–6830.

Heine, V. (1965), *Phys. Rev.* **A138**, 1689–1696.

Henger, C. H., Brimhall, J. L., Hirth, J. P. (1992), *Mater. Sci. Eng.* **A155**, 109–114.

Hertzberger, R. W. (1976), *Deformation and Fracture Mechanics of Engineering Materials*, New York: John Wiley, Chap. 2.

Herman, H. (1988), *Mat. Res. Soc. Bull.* **13** (12), 60–67.

Hillenius, S. J., Liu, R., Georgiu, G. E., Field, R. L., Williams, D. S., Kornblit, A., Boulin, D. M., Johnston, R. L., Lynch, W. T. (1986), in: *IEEE International Electron Devices Meeting (IEDM), Technical Digest*, Piscataway, NJ: Institute of Electrical and Electronics Engineers.

Hirade, S., Inoue, Y., Yamaha, T., Hangasaki, O., Hotta, T. (1995), in: *Proceedings of the Twelfth International VLSI Multilevel Interconnection Conference (VMIC)*, Santa Clara, CA: Twelfth International VLSI Multilevel Interconnection Conference.

Hong, Q. Z., Hong, S. Q., d'Heurle, F. M., Harper, J. M. E. (1994), *Thin Solid. Films* **253**, 479–484.

Hong, Q. Z., Mayer, J. W. (1990), in: S. Katz, S. P. Murarka, A. Appelbaum (Eds.), *Advanced Metallization in Microelectronics*, MRS Symposium Proceedings Vol. 181, Pittsburgh: Materials Research Society, pp. 35–38.

Hong, T., Smith, J. R., Srolovitz, D. J. (1993), *Inter. Sci.* **1**, 223–235.

Hsia, S. L., Tan, T. Y., Smith, P., McGuire, G. E. (1991), *J. Appl. Phys.* **70**, 7579–7587.

Hsieh, J. J. (1993), *J. Vac. Sci. Tech.* **A11**, 3040–3046.

Hsieh, J. C., Fang, Y. K., Chen, C. W., Tsai, N. S., Lin, M. S., Tseng, F. C. (1993), *J. Appl. Phys.* **73**, 5038–5042.

Hui, J., Wong, S., Moll, J. (1985), *IEEE Trans. Elec. Dev. Lett.*, **EDL-6**, 479–481.

Irene, E. A. (1988), *CRC Crit. Rev. Solid. State Mater. Sci.* **14**, 175–223.

Ito, K., Inui, H., Hirano, T., Yamaguchi, M. (1992), *Mater. Sci. Eng.* **A152**, 153–159.

Ivanko, A. A. (1968), in: G. V. Samsonov (Ed.), *Handbook of Hardness Data*, Kiev: Academy of Science of the Ukrainian SSR [English translation (1971), Jerusalem: Israel Program of Scientific Translations].

Jahnel, F., Biersack, J., Crowder, B. L., d'Heurle, F. M., Fink, D., Isaac, R., Lucchese, C. J., Petersson, C. S. (1982), *J. Appl. Phys.* **53**, 7372–7378.

Jimenez, J. R. (1996), in: R. T. Tung, K. Maex, P. W. Pellegrini, L. H. Allen (Eds.), *Silicide Thin Films—Fabrication, Properties, and Applications*, MRS Symposium Proceedings Vol. 402, Pittsburgh: Materials Research Society, pp. 419–430.

Jonckheere, R., Ronse, K., Popa, O., Van den Hove, L. (1994), *J. Vac. Sci. Technol.* **B12**, 3765–3772.

Kaplan, W., Zhang, S.-L., Norstrom, N., Ostling, M., Lindberg, A. (1992), *Microelectron. Eng.* **19**, 661–664.

Kimata, M., Tubouchi, N. (1995), in: A. Rogalski (Ed.), *Infrared Photon Detectors*, Bellingham, WA: SPIE Optical Engineering Press, Chap. 9.

Kimura, K., Nakamura, M., Hirano, T. (1990), *J. Mater. Sci.* **25**, 2487–2492.

Kingery, W. D., Bowen, H. K., Uhlmann, D. R. (1976), *Introduction to Ceramics*, New York: John Wiley, Chap. 10.

Kircher, T. A., Courtright, E. L. (1992), *Mater. Sci. Eng.* **A155**, 67–74.

Kittel, C. (1976), *Introduction to Solid State Physics*, New York: John Wiley, Chaps. 6, 14 and 15.

Kobayashi, N., Goto, H., Suzuki, M. (1991), *J. Appl. Phys.* **69**, 1013–1019.

Koburger, C., Ishaq, M., Geipel,, H. J. (1982), *J. Electrochem. Soc.* **129**, 1307–1312.

Kumar, K. S. (1994), in: J. H. Westbrook, R. L. Fleisher (Eds.), *Intermetallic Compounds: Vol. 2, Practice*, New York: John Wiley, Chap. 10.

Kurtin, S., McGill, T. C., Mead, C. A. (1969), *Phys. Rev. Lett.* **22**, 1433–1436.

Kuwano, H., Phillips, J. R., Mayer, J. W. (1990), *Appl. Phys. Lett.* **56**, 440–442.

Lange, H. (1996), in: R. T. Tung, K. Maex, P. W. Pellegrini, L. H. Allen (Eds.), *Silicide Thin Films—Fabrication, Properties, and Applications*, MRS Symposium Proceedings Vol. 402, Pittsburgh: Materials Research Society, pp. 307–318.

Lee, H. G. (1992), *Thin Solid Films* **216**, 230–234.

Lee, S.-W., Ihn, T.-H., Joo, S.-K. (1996), *IEEE Elec. Dev. Lett.* **17**, 407–409.

Li, F., Lustig, N., Klosowki, P., Lannin, J. S. (1990), *Phys. Rev.* **B41**, 10210–10213.

Li, G. P., Ning, T. K., Chuang, C. T., Ketchen, M. B., Tang, D. D., Mauer, J. (1987), *IEEE Trans. Elec. Dev.* **ED-34**, 2246–2253.

Lien, C. D., Nicolet, M.-A., Lau, S. S. (1986), *Thin Solid Films* **143**, 63–72.

Lien, C. D., Nicolet, M.-A. (1984), *J. Vac. Sci. Technol.*, **B2**, 738–747.

Liu, Q. Z., Yu, L. S., Lau, S. S., Redwing, J. M., Perkins, N. R., Kuech, T. F. (1997), *Appl. Phys. Lett.* **70**, 1275–1277.

Liu, R., Lu, C.-Y., Sung, J. J., Pai, P.-S., Tsai, N.-S. (1995), *Solid State Electron.*, **38**, 1473–1477.

Lograsso, T. A. (1992), *Mater. Sci. Eng.* **A155**, 115–119.

Louie, S. G., Chelikowsky, J. R., Cohen, M. L. (1977), *Phys. Rev. B* **15**, 2154–2162.

Luo, J.-S., Lin, W.-T., Chang, C. Y., Tsai, W. C., Wang, S. J. (1997), *Mater. Chem. Phys.* **48**, 140–144.

Ma, Z., Allen, L. H. (1994), *Phys. Rev. B* **49**, 13501–13511.

Madar, R. (1995), in: K. Maex, M. van Rossum (Eds.), *Properties of Metal Silicides*, London: INSPEC, The Institute of Electrical Engineers, p. 109.

Maex, K. (1993), *Mat. Sci. Eng. Rev.* **R11**, 53–153.

Maex, K., Ghosh, G., Delaey, L., Probst, V., Lippens, P. Van den hove, L., DeKeersmaecker, R. F. (1989), *J. Mater. Res.* **4**, 1209–1217.

Maex, K., Van den hove, L., DeKeersmaecker, R. F. (1986), *Thin Solid Films* **140**, 149–161.

Malegori, G., Miglio, L. (1993), *Phys. Rev. B* **48**, 9223–9230.

Maloney, M. J., Hecht, R. J. (1992), *Mater. Sci. Eng.* **A155**, 19–31.

Maloy, S., Heuer, A. H., Lewandowski, J., Petrovic, J. (1991), *J. Am. Ceram. Soc.* **74**, 2704–2706.

Mann, R. W., Clevenger, L. A. (1995), in: K. Maex, M. van Rossum (Eds.), *Properties of Metal Silicides*, London: INSPEC, The Institute of Electrical Engineers, p. 55.

Mann, R. W., Clevenger, L. A., Agnello, P. D., White, F. R. (1995), *IBM J. Res. Dev.* **39**, 403–417.

Mantl, S. (1992), *Mater. Sci. Rep.* **8**, 1–95.

Mantl, S. (1994), *Nucl. Instrum. Meth. B*, **84**, 127–134.

Mashayekhi, A., Parfitt, L., Kalnas, C., Jones, J. W., Was, G. S., Hoffman, D. W. (1992), *Mater. Res. Soc.*, **273**, 275–280.

Masaki, A. (1992), *IEEE Circuits Dev.*, Nov., p. 18.

Massalski, T. B. (1990), *Binary Alloy Phase Diagrams*, Materials Park, OH: ASM International.

Matar, S. F., Chevalier, B., Etourneau, J. (1994), *J. Magn. Magn. Mater.* **137**, 293–304.

Matsuoka, F., Ishimaru, K., Gojohbori, H., Koike, H., Unno, Y., Sai, M., Kondo, T. Ichikawa, R., Kakumu, M. (1994), *IEICE Trans. Elec.* **E77-C**, 1385.

Mattheiss, L. F. (1992), *Phys. Rev. B* **45**, 3252–3259.

Maury, D., Rostoll, M. L., Gayet, P., Regolini, J. L. (1997), *J. Vac. Sci. Technol.* **B15**, 133–137.

Mayer, J. W., Lau, S. S. (1989), *Electronic Materials Science: For Integrated Circuits in Si and GaAs*, New York: Macmillan, p. 307.

Mayer, J. W., Tsaur, B. Y., Lau, S. S., Hung, L.-S. (1981), *Nucl. Instrum. Meth.* **182/183**, 1–13.

Mendiratta, M. G., Lewandowski, J. L., Dimiduk, D. M. (1991), *Metall. Trans.* **22A**, 1573–1583.

Miglio, L., Meregalli, V. (1996), in: R. T. Tung, K. Maex, P. W. Pellegrini, L. H. Allen (Eds.), *Silicide Thin Films-Fabrication, Properties and Applications*, MRS Symposium Proceedings Vol. 402, Pittsburgh: Materials Research Society, 367–372.

Miracle, D. B., Mendiratta, M. G. (1994), in: J. H. Westbrook, R. L. Fleisher (Eds.), *Intermetallic Compounds: Vol. 2, Practice*, New York: John Wiley, Chap. 13.

Mogami, T., Wakabayashi, H., Saito, Y., Matsuki, T., Tatsumi, T., Kunio, T. (1994), *IEEE Int. Elec. Dev. Meeting Proc.*, p. 687.

Morand, Y., Blanquet, E., Bourhila, N., Thomas, N., Bernard, C., Madar, R. (1994) in: R. W. Fathauer, S. Mantl, L. J. Schowalter, K. N. Tu (Eds.), *Silicides, Germanides, and Their Interfaces*, MRS Symposium Proceedings Vol. 320, Pittsburgh: Materials Research Society, pp. 91–96.

Moy, D., Basavaiah, S., Chuang, C. T., Li, G. P., Hackbarth, E., Brodsky, S. B., Polcari, M. R. (1988), *Solid State Electron*, **31**, 843–849.

Mueller, A., Wang, G., Rapp, R. A., Courtright, E. L., Kircher, T. A. (1992), *Mater. Sci. Eng.* **A155**, 199–207.

Mullins, W. W. (1957), *J. Appl. Phys.* **28**, 333–339.

Murarka, S. P. (1983), *Silicides for VLSI Applications*, New York: Academic Press, p. 50.

Muraraka, S. P., Williams, D. S. (1987), *J. Vac. Sci. Technol.* **B5**, 1674–1688.

Nauka, K., Amano, J., Scott, M. P., Weber, E. R., Turner, J. E., Tsai, R. (1986), in: M. Wittmer, J. Stimmell, M. Strathman (Eds.), *Materials Issues in Silicon Integrated Circuit Processing*, MRS Symposium Proceedings Vol. 71, Pittsburgh: Materials Research Society, pp. 319–324.

Nava, F., Tu, K. N., Thomas, O., Senateur, J. P., Madar, R., Borghesi, A., Guizzetti, G., Gottlieb, U., Laborde, O., Bisi, O. (1993), *Mater. Sci. Rep.* **9**, 141–200.

Ng, K. K., Lynch, W. T. (1996), *IEEE Trans. Elec. Dev.* **ED-33**, 965–972.

Nguyen, T., Ho, H. L., Kotecki, D. E., Nguyen, T. D. (1996), *J. Appl. Phys.* **79**, 1123–1128.

Nicholls, J. R., Stephenson, D. J. (1994), in: J. H. Westbrook, R. L. Fleisher (Eds.), *Intermetallic Compounds: Vol. 2, Practice*, New York: John Wiley, Chap. 22.

Nicolet, M.-A., Lau, S. S. (1983), in: N. G. Einspruch, G. B. Larrabee (Eds.), *VLSI Microelectronics Science*, Vol. 6, New York: Academic Press, Chap. 6.

Nicoll, A. R., Hildebrandt, U. W., Wahl, G. (1979), *Thin Solid Films* **64**, 321–326.

Ning, T. H. (1992), in: *Proceedings of the Thirteenth IEEE/CHMT International Electronics Manufacturing Technology Symposium*, Piscataway, NJ: Institute of Electrical and Electronics Engineers.

Nobili, C. (1990), in: H. R. Huff, K. G. Barraclough, J. Chikawa (Eds.), *Semiconductor Silicon 1990*, Pennington, N. J.: Electrochemical Society, p. 550.

Nolan, T. P., Sinclair, R., Beyers, R. (1992), *J. Appl. Phys.* **71**, 720–724.

Nulman, J. (1990), in: Katz, S. P. Murarka, A. Appelbaum (Eds.), *Advanced Metallization in Microelectronics*, Pittsburgh: Materials Research Society Proc. **181**, pp. 123–132.

Ogawa, S., Fair, J. A., Kouzaki, T., Sinclair, R., Jones, E. C., Cheung, N. W., Fraser, D. B. (1994), in: R. W. Fathauer, S. Mantl, L. J. Schowalter, K. N. Tu (Eds.), *Silicides, Germanides, and Their Interfaces*, MRS Symposium Proceedings Vol. 320, Pittsburgh: Materials Research Society, pp. 355–360.

Omura. T., Inokawa, H., Izumi, K. (1991), *J. Mater. Res.* **6**, 1238–1247.

Osburn, C. M. (1990), *J. Electron. Mater.* **19**, 67–88.

Osburn, C. M., Wang, Q. F., Kellam, M., Canovai, C., Smith, P. L., McGuire, G. E., Xiao, Z. G., Rozgonyi, G. F. (1991), *Appl. Surf. Sci.* **53**, 291–312.

Ostling, M., Zaring, C. (1995), in: K. Maex, M. van Rossum (Eds.), *Properties of Metal Silicides*, London: INSPEC, The Institute of Electrical Engineers, p. 15.

Paufler, P. (1976), in: *Intermetallische Phasen*, Leipzig: VEB Deutscher Verlag für Grundstoffindustrie, pp. 165–187.

Pearson, W. B. (1972), *The Crystal Chemistry and Physics of Metals and Alloys*, New York: Wiley-Interscience.

Petrovic, J. J. (1993), *Mater. Res. Soc. Bull.*, **18** (7), 35–40.

Petrovic, J. J., Bhattacharya, A. K., Honnell, R. E., Mitchell, T. E., Wade, R. K., McClellan, K. J. (1992), *Mater. Sci. Eng.* **A155**, 259–266.

Porter, L. M., Davis, R. F., Bow, J. S., Kim, M. J., Carpenter, R. W. (1995a), *J. Mater. Res.* **10**, 26–33.

Porter, L. M., Davis, R. F., Bow, J. S., Kim, M. J., Carpenter, R. W., Glass, R. C. (1995b), *J. Mater. Res.* **10**, 668–679.

Porter, L. M., Davis, R. F., Bow, J. S., Kim, M. J., Carpenter, R. W. (1995c), *J. Mater. Res.* **10**, 2336–2342.

Reader, A. H., van Ommen, A. H., Weijs, P. J. W., Wolters, R. A. M., Oostra, D. J. (1992), *Rep. Prog. Phys.* **56**, 1397–1467.

Rossnagel, S. (1995), in: D. C. Jacobson, D. E. Luzzi, T. F. Heinz, M. Iwaki (Eds.) *Beam-Solid Interactions for Materials Synthesis and Characterization*, MRS Symposium Proceedings Vol. 354, Pittsburgh: Materials Research Society, pp. 503–510.

Sadananda, K., Feng, C. R., Jones, H., Petrovic, J. (1992), *Mater. Sci. Eng.* **A155**, 227–239.

Sambasivan, S., Petuskey, W. T. (1994). *J. Mater. Res.* **9**, 2362–2369.

Samsanov, G. V., Vinitskii, I. M. (1980), *Handbook of Refractory Compounds*, New York: IFI/Plenum Data Company.

Saraswat, K. C., Brors, D. L., Fair, J. A., Monnig, K. A., Beyers, R. (1983), *IEEE Trans. Elec. Dev.* **ED-30**, 1497–1505.

Schluter, M. (1977), *Phys, Rev. B.* **17**, 5044–5047.

Schwarz, R. B., Srinivasan, S. R., Petrovic, J. J., Maggiore, C. J. (1992), *Mater. Sci. Eng.* **A155**, 75–83.

Scott, D. B., Chapman, R. A., Wei, C.-C., Mahant-Shetti, S. S., Haken, R. A., Holloway, T. C. (1987), *IEEE Trans. Electron. Dev.* **ED-34**, 562–573.

Seidel, T. E. (1983), in: S. M. Sze (Ed.), *VLSI Technology*, New York: McGraw-Hill, Chap. 6.

Shah, D. M., Berczik, D., Anton, D. L., Hecht, R. (1992), *Mater. Sci. Eng.* **A155**, 45–57.

Shaw, L., Miracle, D., Abbaschian, R. (1995), *Acta Metal.* **43**, 4267–4279.

Shiba, T., Tamaki, Y., Onai, T., Kiyota, Y., Kure, T., Nakamura. T. (1996), *IEEE Trans. Electron. Dev.* **43**, 1357–1363.

Shih, Y.-C., Callegari, A., Murakami, M., Wilkie, E. L., Hovel, H. J., Parks. C. C., Childs, K. D. (1988), *J. Appl. Phys.* **64**, 2113–2121.

Shioya, Y., Ikegami, K., Kobayashi, I., Maeda, M. (1987), *J. Electrochem. Soc.* **134**, 1220–1224.

Shone, F. C., Hansen, S. E., Kao, D. B., Saraswat, K. C., Plummer, J. D. (1986), *Proc. IEEE Inter. Elec. Dev. Mett.* p. 534.

Singheiser, L., Wahl, G., Thiele, W. (1982), *Thin Solid Films* **95**, 35–45.

Southwell, G., MacAlpine, S., Young, D. J. (1987), *Mater. Sci. Eng.* **88**, 81–87.

Southwell, R. P., Seebauer, E. G. (1997), *J. Electrochem. Soc.* **144**, 2122–2137.

Speier, W., Kumar, L., Sarma, D. D., de Groot, R. A., Fuggle, J. C. (1989), *J. Phys.: Condens. Matter* **1**, 9117–9129.

Srinivas, D., Raupp, G. B., Hillman, J. (1990), in: S. S. Wong, S. Furukawa (Eds.), *Tungsten and Other Advanced Metals for VLSI/ULSI Applications*, Pittsburgh: Materials Research Society Proceedings Vol. V-6, pp. 407–413.

Stekly, Z. J. J., Gregory, E., (1994), in: J. H. Westbrook, R. L. Fleisher (Eds.), *Intermetallic Compounds: Vol. 2, Practice*, New York: John Wiley, Chap. 16.

Sullivan, J. P., Tung, R., Eaglesham, D. J., Schrey, F., Graham, W. R. (1993), *J. Vac. Sci. Technol.* **B11**, 1564–1570.

Sundahl, M. J., Irwin, R. B., Nanda, A., Wilkins, C. W., Rambabu, B. (1995), *Phys. Status Solidi* **A147**, 477–490.

Sze, S. M. (1981), *Physics of Semiconductor Devices*, New York: John Wiley & Sons, Chap. 5.

Tada, K., Laraya, J. L. R. (1967), *Proc. IEEE* **55**, 2064–2065.

Takasugi, T., Yoshida, M. (1991), *J. Mater. Sci.* **26**, 3032–3040.

Takeuchi, S., Hashimoto, T., Shibuya, T. (1992), *J. Mater. Sci.* **27**, 1380–1384.

Tanelian, M., Lajos, R., Blackstone, S., Pramanick, D. (1985), *J. Electrochem. Soc.* **132**, 1456–1460.

Taubenblatt, M. A., Helms, C. R. (1986), *J. Appl. Phys.* **59**, 1992–1997.

Telford, S. G., Eizenberg, M., Chang, M., Sinha, A. K., Gow, T. R. (1993), *J. Electrochem. Soc.* **140**, 3689–3701.

Thomas, O., Gas, P., Charai, A., LeGoues, F. K., Michel, A., Scilla, G., d'Heurle, F. M. (1988a), *J. Appl. Phys.* **64**, 2973–2980.

Thomas, O., Gas, P., d'Heurle, F. M., LeGoues, F. K., Michel, A., Scilla, G., (1988b), *J. Vac. Sci. Technol.* **A6**, 1736–1739.

Tilly, D. J., Lofvander, J. P. A., Levi, C. G. (1992), in: D. B. Miracle, D. L. Anton, J. A. Graves (Eds.), *Intermetallic Matrix Composites II*, MRS Symposium Proceedings Vol. 273, Pittsburgh: Materials Research Society, pp. 295–300.

Tiwari, R., Herman, H., Sampath, S. (1992), *Mater. Sci. Eng.* **A155**, 95–100.

Tung, R. (1992), in: D. Wolf, S. Yip (Eds.), *Materials Interfaces: Atomic Level Structure and Properties*, New York: Chapman & Hall, p. 550.

Tung, R. (1993), *J. Vac. Sci. Technol.* **B11**, 1546–1552.

Tung, R. (1997), *Jpn. J. Appl. Phys.* **36**, 1650–1654.

Umakoshi, Y., Sakagami, T., Hirano, T., Yamane, T. (1990), *Acta Metallogr.*, **38**, 909–915.

van den Hoek, P. J., Ravenek, W., Baerends, E. J. (1988), *Phys. Rev. Lett.* **60**, 1743–1746.

van Loo, F. J. J., Smet, F. M., Rieck, G. D., Verspui, G. (1982), *High Temp. High Press.* **14**, 25–31.

Vasudevan, A. K., Petrovic, J. J. (1992), *Mater. Sci. Eng.* **A155**, 1–17.

Vedernikov, M. V. (1994), in: J. H. Westbrook, R. L. Fleisher (Eds.), *Intermetallic Compounds*, Vol. 2, *Practice*, New York; John Wiley, Chap. 20.

Vijh, A. K. (1994), in: J. H. Westbrook, R. L. Fleisher (Eds.), *Intermetallic Compounds*, Vol. 2, *Practice*, New York: John Wiley, Chap. 23.

Vining, C. B. (1992), in: C. B. Vining (Ed.), *Proceedings 9th International Conference on Thermoelectronics*, Pasadena: California Institute of Technology, p. 249.

von Känel, H. (1992), *Mater. Sci. Rep.* **8**, 193–269.

von Känel, H. Kafader, U., Sutter, P., Onda, N., Sirringhaus, H., Muller, E., Kroll, U., Schwarz, C., Goncalces Conto, S. (1994), in: R. W. Fathauer, S. Mantl, L. J. Schowalter, K. N. Tu (Eds.), *Silicides, Germanides, and Their Interfaces*, MRS Symposium Proceedings Vol. 320, Pittsburgh: Materials Research Society, pp. 73–84.

Wang, S. Q., Mayer, J. W. (1990), *J. Appl. Phys.* **67**, 2932–2938.

Weijs, P. J. W., Czyzyk, M. T., Fuggle, J. C., Speier, W., Sarma, D. D., Buschow, K. H. J. (1990), *Z. Phys.* **B78**, 423–430.

West, G. A., Beeson, K. W. (1987), in: G. W. Cullen (Ed.), *Proceedings 10th International Conference on CVD*, Pennington NJ: Electrochemical Society, p. 720.

Winnerl, J. N. (1994), in: R. W. Fathauer, S. Mantl, L. J. Schowalter, K. N. Tu (Eds.), *Silicides, Germanides, and Their Interfaces*, MRS Symposium Proceedings Vol. 320, Pittsburgh: Materials Research Society, pp. 37–46.

Yallup, K., Wilson, R., Quinn, C., McDonnell, B., Blackstone, S. (1995), *Proc. IEEE Int. SOI Conf.*, p. 137.

Yokoyama, N., Ohnishi, T., Nakamura, T., Nishi, H. (1986), in: R. J. Nemanich, P. S. Ho, S. S. Lau (Eds.), *Thin Films—Interfaces and Phenomena*, MRS Symposium Proceedings Vol. 54, Pittsburgh: Materials Research Society, pp. 401–408.

Yoo, C.-S., Lin, T.-H., Tsai, N.-S., Ijsendoom, L. J. (1990), *Jpn. J. Appl. Phys.* **29**, 2535–2540.

Zaring, C., Gas, P., Svensson, B. G., Ostling, M., Whitlow, H. F. (1990), *Thin Solid Films*, **193/194**, 244–247.

Zhang, S.-L., d'Heurle, F. M. (1992), *Philos. Mag.* **66**, 415–424.

Zhang, S.-L., d'Heurle, F. M. (1994), *Mater. Sci. Forum* **155–156**, 59–70.

Zheng, L. R., Hung, L. S., Mayer, J. W. (1985), *J. Appl. Phys.* **58**, 1505–1514.

Zur, A., McGill, T. C., Nicolet, M. A. (1985), *J. Appl. Phys.* **57**, 600–603.

Further Reading

Fathauer, R. W., Mantl, S., Schowalter, L. J., Tu, K. N. (Eds.) (1994), *Silicides, Germanides, and Their Interfaces*, MRS Symposium Proceedings Vol. 320, Pittsburgh: Materials Research Society.

Gambino, J. P. Colgan, E. G. (1998), *Mater. Chem. Phys.* **52**, 99–146.

Maex, K., van Rossum, M. (Eds.) (1995), *Properties of Metal Silicides*, London: INSPEC, The Institute of Electrical Engineers.

Murarka, S. P. (1983), in: *Silicides for VLSI Applications*, New York: Academic Press.

Nicolet, M.-A., Lau, S. S. (1983), in: N. G. Einspruch, G. B. Larrabee (Eds.), *VLSI Microelectronics Science*, Vol. 6, New York: Academic Press, Chap. 6.

Samsanov, G. V., Vinitskii, I. M. (1980), in: *Handbook of Refractory Compounds*, IFI/Plenum Data Company.

Tung, R. T., Maex, K., Pellegrini, P. W., Allen, L. H. (Eds.) (1996), *Silicide Thin Films—Fabrication, Properties, and Applications*, MRS Symposium Proceedings Vol. 402, Pittsburgh: Materials Research Society.

Westbrook, J. H., Fleisher, R. L. (Eds.) (1994), *Intermetallic Compounds*, Vol. 2, *Practice*, New York: John Wiley.

SILICON: POROUS

PHILIPPE M. FAUCHET, *Department of Electrical and Computer Engineering, The Institute of Optics, and Laboratory for Laser Energetics, University of Rochester, Rochester, New York, U.S.A.*

INTRODUCTION

Silicon is the second most abundant element in the Earth's crust. One can argue that in fact it is the second most important element in today's technological world, just after carbon, which is the essential ingredient of life. This is because microelectronics relies almost exclusively on silicon to make the chips that are the soul of every microprocessor and intelligent system. Silicon has taken such a leadership position in high technology and consumer electronics alike for two main reasons: it is cheap and abundant, and its oxide can be prepared with excellent properties.

The wafers that are used in microelectronics are made of bulk, crystalline silicon (c-Si). The properties of this form of silicon have been extensively investigated for more than 40 years and, as a result, are well known. But there are many other forms of silicon. Polycrystalline and microcrystalline silicon films are made of crystalline silicon grains with sizes ranging from $\approx 1\,\mu m$ to a few nanometers. They are routinely prepared by a variety of techniques such as chemical vapor deposition or laser recrystallization of yet another form of silicon, called amorphous silicon. The boundaries between the grains, which act as a source of traps and scattering centers, control many of the electrical properties. Amorphous silicon is a form of silicon that lacks long-range order but maintains the local tetrahedral coordination. Amorphous silicon itself is useless as an electronic material because it contains a huge density of traps, called dangling bonds. These dangling bonds can be passivated by hydrogen, which removes them from the band-gap region. Thus alloying amorphous silicon with several atomic percent of hydrogen makes it useful for photovoltaics, for example. Silicon is also present in a large variety of polymeric and organic compounds.

Porous silicon is a form of silicon with

ISBN 3-527-29308-6

unique properties, distinct from those of crystalline, microcrystalline, or amorphous silicon. It was first prepared in 1956, but its true nature was not understood until more than one decade later. The most common preparation procedure consists of etching a silicon wafer in an electrochemical cell in the presence of HF. Under the appropriate conditions, a large density of pores is produced on the surface of the wafer, which is used as the anode. These pores, which can vary in size from ≈ 1 nm to >1 μm, can run straight perpendicular to the wafer surface or form a complex mesh of interconnected pores. They can penetrate tens of micrometers below the wafer surface. The silicon material left between the pores usually retains its crystalline nature and is made of structures whose sizes can vary from ≈ 1 nm to >1 μm.

It is clear that porous silicon can be prepared with a wide range of properties, which makes it a very versatile material. All forms of porous silicon share a very large internal surface area, which can be as large as an Olympic stadium per gram of material. Thus, most properties of porous silicon are affected not only by its crystalline core but also by its surface properties. In addition, the properties of the crystalline core may differ from those of bulk c-Si when the size of the silicon structures drops below 10 nm, a size regime in which quantum mechanical effects modify its electronic states.

Although porous silicon has been studied and used for more than 40 years, it is only starting in 1990 that it attracted a lot of attention. This interest was triggered by the discovery that appropriately prepared porous silicon can emit bright visible light that can be tuned from the near infrared throughout most of the visible spectrum. Since bulk crystalline Si has an indirect band gap near 1.1 eV (or 1.1 μm), both the brightness and the color range were unexpected. For the first time, silicon light emitters appeared possible. Since then, we have learned much about the properties of this unusual form of silicon. This article reviews our understanding of many of its properties. It starts with a discussion of the preparation conditions and the post-preparation treatments that stabilize its properties. The article then moves to a review of many of its properties. Its optical properties are highlighted, since they are responsible for most of the interest in porous silicon since 1990. Finally, several applications of porous silicon are discussed, in the areas of light-emitting devices, solar cells, integrated optics, sensors, micromachining, and biomedicine.

1. PREPARATION/MATERIALS SCIENCE

1.1 Electrochemistry: Electropolishing and Porous-Silicon Formation

Porous silicon (PSi) is a form of silicon that is produced chemically or electrochemically. The electrochemical route is more flexible, in the sense that it leads to a wider range of porosities, thicknesses, *etc.* PSi is prepared by anodic etching of silicon in a solution containing HF. It was first made in 1956 (Uhlir, 1956) and later recognized as etched silicon, as opposed to a deposit on the silicon wafer. In the electrochemical cell, the anode is the silicon wafer and the cathode is a platinum wire. When a constant current is passed between the electrodes, one can distinguish three regions in the curve of current (I) *versus* polarity (V) (Fig. 1). In reverse bias, I is zero and the dissolution rate of the silicon wafer is very slow. In forward bias, at low I, the I-V curve is reminiscent of a diode; this is the region where PSi is formed. As V is increased further, a region of instability is entered, beyond which the increase of I with V becomes minimal; this is the regime of electropolishing. In this regime, the Si is dissolved layer by layer, whereas in the PSi regime, pores of different sizes, depending on several parameters, are produced (Smith and Collins, 1992). There is a

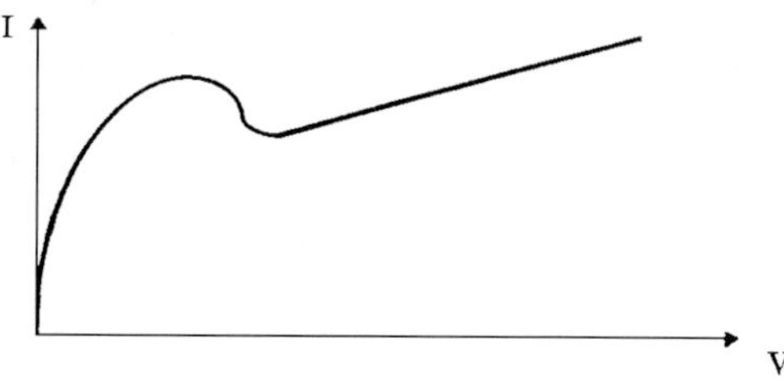

FIG. 1. Current–voltage relationship corresponding to the anodic oxidation of a p-type Si wafer in HF.

critical current density below which PSi is formed and above which electropolishing takes place. This demarcation exists because the formation of pores requires that the transport of holes from the bulk to the surface be the limiting step.

It is also possible to produce PSi starting from an *n*-type Si wafer. In this case, there are two interesting regimes in the *I-V* curve. In forward bias, *I* increases as in a diode and PSi can be formed. In reverse bias, no current flows except when the liquid/semiconductor interface is flooded by light. In this case, the *I-V* curve resembles that for *p*-type Si, except with an offset near 0 V. Note that porous silicon can also be prepared starting from polycrystalline, microcrystalline, or even heavily doped amorphous silicon. In addition, several other semiconductors have been made porous electrochemically.

The pore morphology and achievable porosities are strongly dependent on the current density, the HF concentration, the use of light during anodization, and the substrate type and doping density. Additionally, they are influenced by the *p*H of the solution, the use of alcohol in the solution, and other parameters such as the temperature. As the porosity increases,

1. the pores lose their directionality and the structure acquires a spongy character, and
2. the average size of the silicon skeleton decreases to a few nanometers.

The terminology used to described PSi has traditionally focused on the pore size (Table 1). This classification is convenient when the pores are of interest, but less useful for studies of the properties of electrons and holes in the silicon material left between the pores (Fauchet, 1997). Since Canham's work in 1990 (Canham, 1990), most of the interest in PSi has come from its strong, visible luminescence. Luminescence has been associated with the

Table 1. Traditional porous silicon classification.

Classification	Diameter of pores (nm)	Surface area ($m^2\,cm^{-3}$)
Macro	>50	10–100
Meso	2–50	100–300
Micro	≤2	300–800

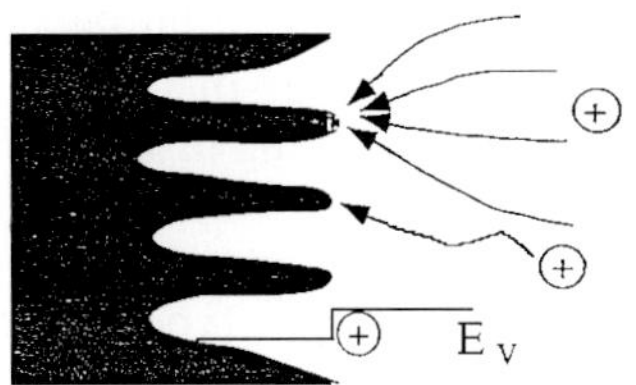

FIG. 2. Three mechanisms promote the formation of pores. From top to bottom, they are the concentration of the electric field lines at the pore's tip; random walk, which makes it easier for the holes to reach the pore's tip; and the potential barrier to hole transport from the bulk to the Si column's tip due to quantum confinement.

presence of silicon wires or nanocrystallites with typical dimensions in the nanometer size. Thus, a new term has been introduced, nanoporous silicon. Here, the prefix nano is used to describe the size of the Si objects. The porosity and the nanocrystallite size can be correlated for various types of samples. Because PSi is a very complex material, which can contain either Si wires or Si dots or even both types of structure simultaneously, the words nanocrystallites, dots, wires, and columns will be considered interchangeable. In a given section, one word will be preferred if this choice helps conveys the concept under examination.

The pore initiation and growth mechanisms are qualitatively understood. Pore growth can be explained by several models, each one of which is more relevant for a specific regime of porosity and pore size (Smith and Collins, 1992). Assume first that shortly after anodic etching has started, the silicon/electrolyte interface has become irregular (Fig. 2). The key question is whether the fluctuations of the Si/electrolyte interface will grow (PSi formation) or disappear (electropolishing). Si etching requires the injection of holes from the substrate. A depletion region is formed in the thin PSi layer itself and also in a region of the Si wafer near the PSi/c-Si interface. In forward bias, the holes can still reach the Si/electrolyte interface by one of two mechanisms. First, the electric field lines are focused at the pore's tip. Thus, holes will preferentially reach the Si/electrolyte interface deep in the pores, where etching will proceed rapidly. In contrast, no holes will reach the end of the Si rods, effectively stopping the etching there. Second, if this electrostatic effect is not strong enough, random walk of the holes toward the Si/electrolyte interface makes it more likely that they reach it at or near the pore's tip, resulting in an effect similar to the electrostatic one. Another important mechanism becomes predominant if the Si rods are narrow enough (typically less than 10 nm). In this size regime, the electronic states start to differ from those of bulk silicon. When the motion of carriers is restricted in one or more dimensions, their minimum energy can no longer be zero. As a result of this effect, which is called quantum confinement, the holes in the valence band of the Si rods are pushed to lower energy, which produces a potential barrier to hole transport from the wafer to the Si rods. The holes can no longer drift or diffuse into the Si rods and further etching stops (Lehmann *et al.*, 1992).

When the top layer of the wafer is amorphized—for example, by ion implantation at high doses—hole transport is suppressed and porous-silicon formation is inhibited. In contrast, when the wafer surface is lightly damaged—for example, by a light ion milling—a higher density of larger nanopits or nano-irregularities is produced on the surface, the electric field lines are redistributed, and porous silicon is produced at a much increased rate (Duttagupta *et al.*, 1995). When light of increasing photon energy is used during anodization, the resulting peak luminescence energy also increases. This result can be explained as follows: In the dark, the etching of the Si rods stops when their valence-band edge drops by an amount that is sufficient to present an insurmountable barrier to hole transport from the Si wafer. With light, holes continue to be generated by absorption of the incoming photons, resulting in further etching and thinning of the Si rods. The valence-band edge continues to drop and the conduction-band edge continues to increase. The band-gap energy of the Si rods, which is given by the sum of the band-gap energy of bulk Si, the drop in the valence-band edge, and the increase in the conduction-band edge, increases further. This increase in the band-gap energy should be reflected by an increase of the luminescence energy. Similar experiments performed in the electrolyte in the open-circuit configuration just after the end of anodization yield similar result and thus confirm the interpretation (Koyama and Koshida, 1993).

1.2 Other Techniques to Produce Nanocrystalline Silicon

Since the discovery that PSi can emit light efficiently in the visible range, there has been a renewed interest in the manufacture of silicon quantum dots and wires. Spark erosion (Hummel *et al.*, 1993) consists of producing an electric discharge between a silicon wafer and an electrode, either in vacuum or in the presence of a gas. The resulting Si surface is covered by pits and hills of various sizes in the micrometer to nanometer scale. Spark-processed silicon is found to be highly luminescent in the visible range. Its color depends on the processing conditions, ranging from the red to the blue.

Lithographic techniques have also been used produce nanometer-size Si pillars or wires. Optical or electron-beam lithography is first used to produce submicron features. The Si stuctures are then thinned—for example, by an anisotropic etch in KOH and/or by oxidation—until the remaining Si nanostructure is on the order of 10 nm or less. Vertical Si pillars (Nassiopoulos *et al.*, 1995) and horizontal Si wires (Zaidi *et al.*, 1995) with dimensions below 10 nm have been fabricated. In many cases, luminescence has been reported with these structures, including at least one report of electroluminescence. However, the density of the Si nanostructures is low, which severely limits the brightness.

Another interesting method to produce Si nanocrystals that emit light is to start with a microcrystalline film and then to oxidize the film such that the diameter of the Si crystallites decreases to the appropriate size (less than 5 nm) and the surface of each nanocrystallite is passivated by silicon dioxide (Tamura *et al.*, 1994). This procedure has been shown to produce films with easily detectable visible luminescence. The advantages of this technique are that it is simple and compatible with Si microelectronic technology. The highest reported efficiency to date remains more than one order of magnitude below that of a typical PSi film. In addition, the crystallite size distribution is not well controlled.

Theory predicts that it should be possible to achieve visible luminescence even if the carriers are confined in only one direction, in Si slabs. These Si slabs are very similar to the familiar quantum wells grown by molecular-beam epitaxy (MBE) or chemical vapor deposition (CVD). Si/CaF_2 films grown by MBE exhibit visible luminescence when the Si layer thickness drops below 2 nm (Vervoot *et al.*, 1995). However, the Si films are microcrystalline because the large lattice mismatch and difference in thermal expansion coefficient between Si and CaF_2 make it impossible to grow epitaxial films free of dislocations. Thus, quantum confinement probably takes place in three directions, not just one, which triples the increase in the band-gap energy. It has also been shown that amorphous silicon layers sandwiched between silicon dioxide barriers luminesce in the visible range (Lu *et al.*, 1995). Once again, there is a critical thickness of the a-Si layers above which no luminescence is detected. The maximum luminescence efficiency of these structures remains at least one order of magnitude below that of a good-quality PSi film.

Another promising technique for the production of highly luminescent Si nanostructures involves the recrystallization of an amorphous Si/SiO_2 superlattice, made of many alternating layers of amorphous Si and SiO_2 (Tsybeskov *et al.*, 1998). These recrystallized Si layers are made of identical Si crystallites whose diameter is determined by the thickness of the original Si layer. These films exhibit an external photoluminescence quantum efficiency between 0.1 and 1% and a luminescence spectrum including all the well-known phonon replicas, which coincides with that of bulk c-Si for diameters in excess of 10 nm.

Si quantum dots can be prepared by several other techniques. One example consists of implanting SiO_2 with Si and then annealing the film to form Si clusters that then recrystallize. The luminescence of these structures can be tuned throughout the visible part of the spectrum by controlling the processing steps (Petrova-Koch *et al.*, 1996). However, the control on the size of the crystallites and the crystallite packing density is limited. Another example consists of preparing clusters or crystallites in the gas phase, which are then collected on a glass plate or in a matrix (Littau *et al.*, 1993). Gas-phase synthesis of Si quantum dots, followed by gas-phase oxidation and size selection, has produced the most luminescent material to date (quantum efficiency of 50% at low temperatures). Unfortunately, the production rate of these crystallites is too low to make them technologically useful.

1.3 Post-Preparation Treatments

1.3.1 Supercritical Drying One of the major problems in the manufacture of PSi involves the removal of the etching solution from the pores after anodization. The most straightforward technique consists of letting the solution evaporate in air. This procedure can lead to the collapse of the Si rods in highly porous samples. This collapse is a consequence of the large capillary-induced tensile stresses encountered during evaporation at the vapor–liquid interface in the pores. To avoid this problem, various methods have been proposed, including the use of low–surface-tension fluids such as pentane, freeze drying, and supercritical drying, a technique well known from silica aerogels. The use of supercritical fluids as drying agents has led to the best results (Canham *et al.*, 1994). Their application requires knowledge of the phase boundary and critical point of the confined fluid in the porous material, which are usually estimated from bulk fluid properties. The phase-coexistence boundary and critical point of a confined fluid can, however, be significantly different from its bulk values.

Figure 3 shows a typical trajectory followed during supercritical drying and contrasts it to a trajectory corresponding to normal drying in air. Trajectory 1–2–3 illustrates the supercritical drying of PSi with carbon dioxide (bulk phase $T_c = 31.1\,°C$, $P_c = 73.8$ bar). The ethanol is first replaced by liquid CO_2 for 3 h at point 1. This is followed by the conversion of liquid CO_2 into a supercritical state by raising its temperature to 40 °C during the step 1–2. Finally, the CO_2 is vented at 40 °C for a period of 3 h during step 3. Similar results are obtained if another fluid is used (for example, C_2H_4). Whereas most samples prepared without supercritical drying often break for porosities greater than 80%, PSi films of porosities greater than 95% survive intact if they are prepared by supercritical drying. This is the case even for free-standing films lifted off the Si substrate (von Behren *et al.*, 1997).

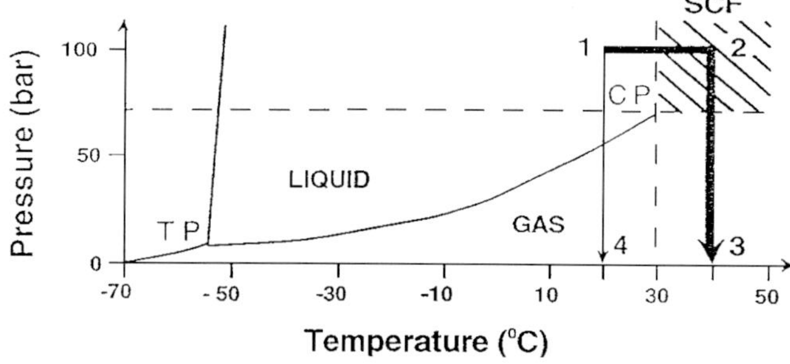

FIG. 3. Pressure–temperature diagram of bulk CO_2. Supercritical drying (trajectory 1–2–3) avoids the liquid–gas interface by converting liquid CO_2 into a supercritical fluid (SCF). Conventional drying is shown by trajectory 1–4.

1.3.2 Chemical Quenching Because of its very large internal surface, PSi is an ideal laboratory for the study of the chemistry of silicon surfaces. The surface of PSi, which is passivated by Si–H_n bonds, can be modified by chemical reactions (Sailor and Lee, 1997). The chemical compounds or molecules that become attached to the PSi surface can act as surface-state traps or sites for interfacial energy transfer and charge transfer, resulting in the quenching of the photoluminescence (PL). Reaction of PSi with iodine completely quenches the PL, on account of the production of Si–I surface traps that act as nonradiative recombination centers. This type of reaction is not fully reversible. In contrast, exposure to liquid or vapor methanol also quenches the PL, but this quenching is reversible because no net chemical reaction has taken place. Exposure to electron acceptor or donor components can also quench the PL reversibly (Rehm *et al.*, 1996), as in the case of a toluene solution of anthracene. However, if the electron acceptor (donor) state has a higher (lower) energy than the conduction- (valence-) band edge of PSi, electron (hole) transfer from PSi to the component is energetically forbidden and the PL is not quenched (*e.g.*, naphthalene). The quenching ability of a chemical species is also a function of the initial PSi surface termination. For example, reversible quenching is observed when Si–H–passivated PSi is exposed to liquid methanol. When SiO_2-passivated PSi is exposed to the same solution, no quenching is observed (Rehm *et al.*, 1995).

1.3.3 Aging and Oxidation After anodization, the internal surface of PSi is covered by silicon hydride bonds that remove any surface dangling-bond states from the band gap and thus provide excellent surface passivation. Excellent passivation is a requirement for achieving a high luminescence efficiency, since the radiative lifetime of PSi is in the microsecond to millisecond time domain because of the indirect gap of porous silicon, which is ample time for trapping in any nonradiative recombination center. Experimentally, it is observed that the PL efficiency is decreased by many factors (Collins *et al.*, 1992), including illumination with short-wavelength light, heating, exposure to various gases and liquids, and the presence of a high electric field. This PL degradation is intrinsic to the "fragile" Si–H_n bond and thus unavoidable. Hence, it is necessary to find a substitute for the Si–H bond, and the most obvious candidate is silicon dioxide. Silicon dioxide and its interface with crystalline silicon can be of very good quality, and in fact they are in large part responsible for the dominant role of silicon in microelectronics.

The stability of the luminescence efficiency of porous silicon passivated by silicon dioxide (SiO_2) has been shown to be excellent (Fauchet *et al.*, 1997). Oxide-passivated porous silicon can be prepared by different approaches: anodization at a low current density with light assistance, furnace oxidation, rapid thermal oxidation, or even aging the sample in air. A marked improvement in stability is observed when the Si–H bonds are replaced by Si–O bonds by using anodization with light assistance. After rapid thermal oxidation at 900 °C, the PL efficiency is comparable to that of Si–H–passivated PSi because the densities of dangling bonds present in as-anodized and oxidized samples are comparable. The difference is that stability of the Si/SiO_2 interface is

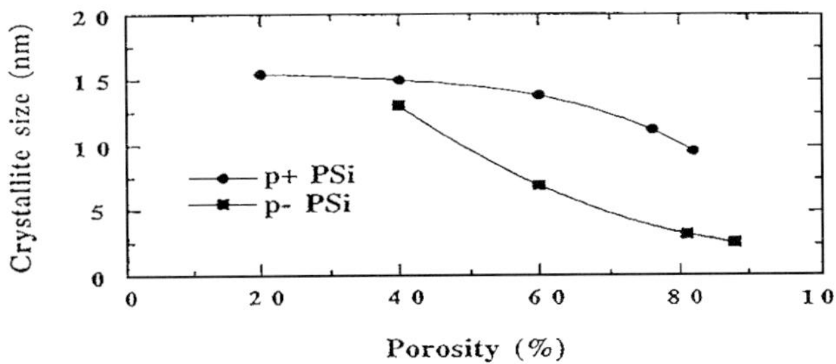

FIG. 4. The crystallite size in PSi depends on the porosity and the doping level of the starting Si wafer.

such that no additional dangling bonds are produced even under most adverse conditions. Annealing at high temperatures is necessary for forming a high-quality oxide and a high-quality Si/SiO_2 interface. To avoid the complete oxidation of the Si skeleton and the transformation of the PSi layer into a porous glass, the high-temperature oxidation is performed in an atmosphere poor in oxygen and for a short time.

2. PROPERTIES

2.1 Structural

2.1.1 X-Ray Scattering X-ray scattering from porous silicon (Lehmann *et al.*, 1993) shows that PSi simultaneously exhibits the characteristics of a monocrystalline solid (long range order) and an inhomogeneous solid (short range order) (Naudon *et al.*, 1995). The average size of the Si nanocrystallites present in PSi films can be calculated from an analysis of the x-ray lineshape by applying the Scherer equation (Cullity, 1978) to the broad peak profile. The variation of size with porosity for p^- and p^+ PSi films is plotted in Fig. 4 (Fauchet *et al.*, 1997). It should be noted that this technique is most useful when applied to samples with a single particle size. X-ray diffraction also shows that the lattice constant of PSi increases with the porosity and depends on whether the surface is covered by SiO_2 or Si–H_n. A value of the strain as large as 0.1% is possible.

2.1.2 Electron Microscopy Although techniques such as scanning-electron and atomic-force microscopies reveal some of the features of PSi, its nanostructure is best examined by transmission-electron microscopy (Berbezier, 1995; Cullis *et al.*, 1997). TEM, which is capable of a resolution down to ≈ 1 Å, has provided very detailed images of PSi. Crystallites as small as 1 nm and their lattice planes have been imaged. Scanning-electron microscopy and atomic-force microscopies do not provide the same atomic scale resolution and are better suited to the study of mesoporous and macroporous PSi.

Figure 5 shows cross-sectional TEM images of PSi prepared from a p^+ Si wafer (Beale *et al.*, 1985). In these samples, the pores are ≈ 10 nm wide and run perpendicular to the Si wafer surface (along the $\langle 100 \rangle$ direction, a general trend for all substrate orientations). For the highest porosity (79%) shown in Fig. 5, the structure becomes open, the pores are interconnected, and it is difficult to identify individual pores. In contrast, the TEM image of Fig. 6 shows that in microporous PSi formed from a p^- Si wafer, the nanostructure is coral-like or sponge-like, with characteristic dimensions on the 1–3-nm scale (Cullis *et al.*, 1992). This is the type of PSi that produces the brightest luminescence. In addition, we note that PSi can be produced with pores on two widely different length scales ($\approx$nm and sub-μm) by using light during the anodization of n^+ Si wafers.

Lattice-plane imaging shows that the crystallographic orientation of the substrate is maintained after anodization, with a trend to a small misorientation at the highest porosities. Transmission electron diffraction patterns also indicate that, in general, PSi maintains the crystalline nature of the substrate, although the presence of some disordered material has sometimes been reported.

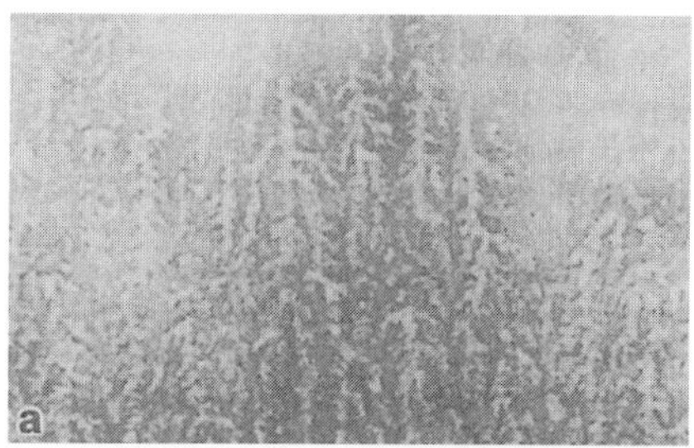

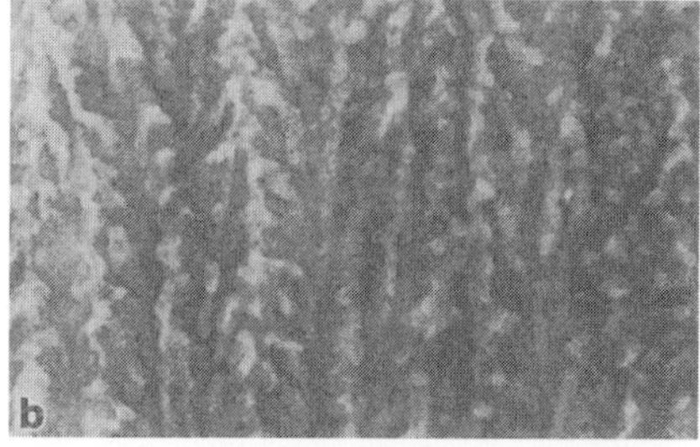

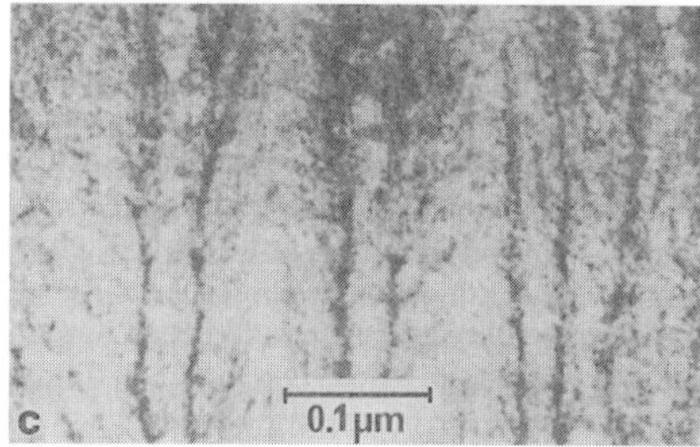

FIG. 5. Cross-sectional transmission-electron microphotographs of PSi samples prepared from a p^+ Si wafer. The porosities are 31%, 51%, and 79%, from top to bottom.

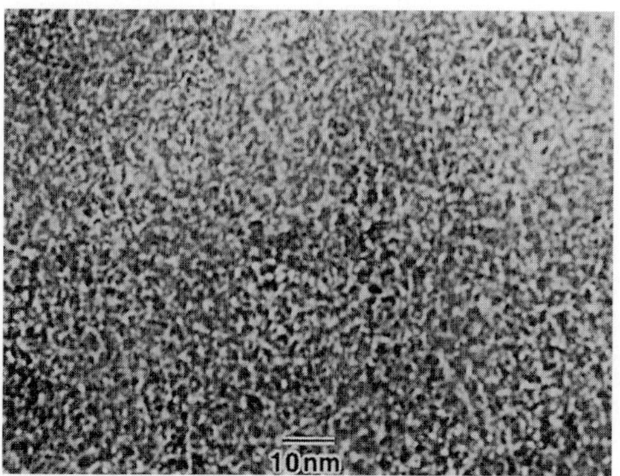

FIG. 6. Cross-sectional transmission-electron microphotographs of PSi prepared from a p^- Si wafer. The spongy structure is made of highly interconnected micropores.

2.2 Mechanical

The mechanical properties of PSi depend critically on the porosity. The open structure of high-porosity PSi makes it a very fragile material. One convenient characterization technique, which can provide useful information regarding material strength, crystallite size, and surface quality, is a hardness measurement. Hardness is defined as the resistance of a solid to plastic deformation. The method most widely used for measuring hardness is indentation. In the Vickers test, a square-based pyramidal indenter produces a diamond-shaped indent on the surface and the hardness H is related to the applied load P_L (in kg f) and the diagonal of the indent d (in mm) by

$$H = 1.8544(P_\mathrm{L} d^{-2}) \tag{1}$$

The load dependence of hardness (indentation size effect) provides useful information regarding the nature of a material (brittle, ductile). A power law relationship exists between the applied load and the indentation diagonal:

$$P_\mathrm{L} = K_\mathrm{L} d^n, \tag{2}$$

where n is called the Meyer index.

Figure 7 compares the hardness of different PSi films, as a function of porosity (Duttagupta *et al.*, 1997a). The hardness of the p^+ PSi films is comparable to that of the p^- PSi films. At low porosities, there is a moderate decrease of hardness with porosity. This is expressed by the full line in Fig. 7 as $H = H_0(1-P)^m$ $(m = 0.67)$, where H_0 is the hardness of crystalline Si (11.5 GPa). However, in the porosity range of 70–80%, the hardness drops appreciably and there is considerable deviation from the previous $(1-P)^{2/3}$ proportionality. This deviation indicates the change in the morphology of the films from isolated pores to isolated Si columns near that porosity. The trend of decreasing hardness with increasing porosity in confirmed by measurements of the Young's modules *versus* porosity in p^+ PSi films using nanoindentation (Bellet *et al.*, 1996).

The Meyer's index may be used to gauge the nature of a material. For Si, the value of n is 1.7 for c-Si (brittle, ordered), while that for

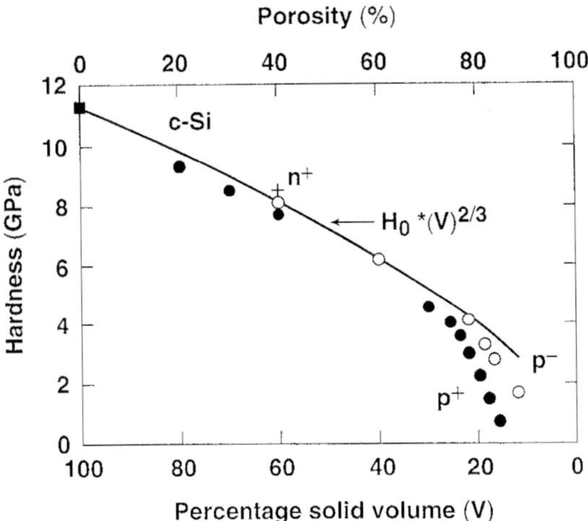

FIG. 7. The hardness of PSi depends on the porosity and the doping level of the starting Si wafer. The open circles are data from PSi prepared from p^- Si wafers, the closed circles are data from p^+ Si wafers and the cross is from n^+ Si wafers.

amorphized Si (softer, highly disordered) is 2. In PSi even at 80% porosity, $n = 1.75$ for p^- PSi and $n = 1.72$ for p^+ PSi. This is a further confirmation of the crystalline nature of porous silicon.

2.3 Thermal

The thermal conductivity of PSi depends strongly on porosity and is expected to be much below that of c-Si as the porous structure becomes less dense. At very high porosities, the preferred heat transfer could switch from conduction to radiation. The thermal conductivity of PSi has been measured by several techniques (Lang *et al.*, 1995; Duttagupta, 1997b). It depends not only on the porosity but also on the pore nanostructure (whether the material is macro-, meso-, or microporous) and on whether oxidation has been performed or not. The thermal conductivity of several PSi samples prepared from a p^+ Si wafer is shown in Fig. 8 (Duttagupta, 1997b). At high porosities, the thermal conductivity of PSi drops below that of glass.

2.4 Vibrational

The Si crystallite size in PSi films can be measured by Raman spectroscopy (Fauchet and Campbell, 1988). In unstrained bulk crystalline silicon the Si—Si peak is narrow, symmetric, and centered at $520\,\mathrm{cm}^{-1}$ with a full width at half maximum of approximately $4\,\mathrm{cm}^{-1}$. This peak broadens, becomes asymmetric, and is downshifted with increasing porosities. In addition, a broad peak often arises around $480\,\mathrm{cm}^{-1}$, which is the signature of amorphous (disordered) Si. The changes in the 520-cm^{-1} peak can be used to deduce the average crystallite size (and in some cases, its shape as well). Figure 9 shows that the crys-

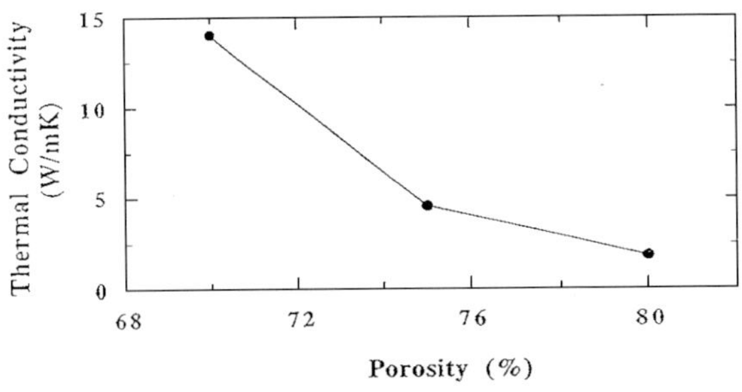

FIG. 8. The thermal conductivity of PSi decreases with increasing porosity and is much lower than that of bulk c-Si ($150\,\mathrm{W\,m^{-1}\,K^{-1}}$). The samples are prepared from a p^+ Si wafer.

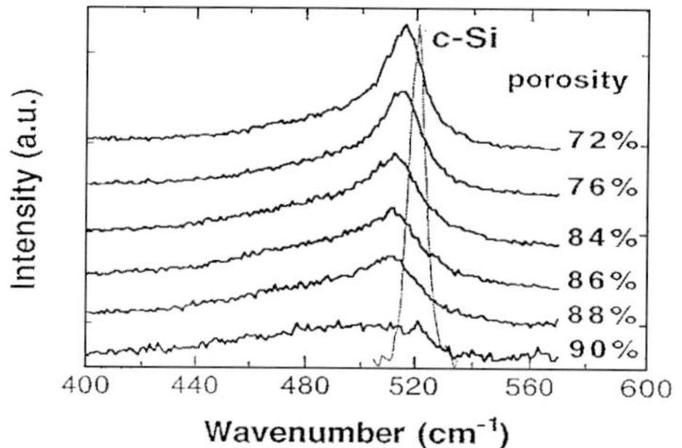

FIG. 9. Dependence of the Raman spectrum of PSi on porosity for samples prepared from a *p* Si wafer. The optic phonon Raman spectrum of bulk c-Si is shown for comparison.

tallite size decreases with increasing porosity (von Behren *et al.*, 1998). Below 25 Å, Raman spectroscopy can no longer be used as a reliable tool to measure the nanocrystal size. Additionally, stress affects the phonon line shape of nanocrystalline films, resulting in a shift of the Raman line for a uniform, homogeneous stress and a broadening of the Raman line for nonuniform, inhomogeneous stress.

2.5 Electrical

The current density (J)–voltage (V) characteristic of a simple metal/PSi/c-Si device is shown in Fig. 10 (Peng *et al.*, 1996). The J-V curve is rectifying and follows, in forward bias, a power law relationship $J = KV^m$ where $m \approx 2$. This power-law relationship is typical of a space-charge–limited current and shows that the current is dominated by carrier transport in the high-resistivity ($\approx 10^{10}\,\Omega\,\mathrm{cm}$) PSi layer. The entire device can be modeled as an insulator sandwiched between two conductors (the top metal contact and the Si substrate) with negligible band bending at the two interfaces. When the injected-carrier concentration exceeds the thermal-carrier concentration, a departure from Ohmic behavior is observed, and for double-carrier injection, the J-V relationship is given by

$$J = \varepsilon\varepsilon_0 \mu_{\mathrm{eff}} V^2 / d^3 \tag{3}$$

where $\varepsilon\varepsilon_0$ is the dielectric function of PSi (in $\mathrm{F\,m^{-1}}$), d is the thickness of the PSi layer, and μ_{eff} is the effective carrier mobility. The measured dependence of J on both V and d satisfies Eq. (3).

Two methods have been employed to measure the carrier mobility. In the first method, the response of the electroluminescence (EL) to a small alternating (ac) bias is measured as a function of frequency in a device held under a large forward bias (Peng and Fauchet, 1995). The injected carriers drift through the PSi layer where they recombine radiatively, and the transit time through the device (τ_t) is given by

$$\tau_t = d^2 / \mu_{\mathrm{eff}} V. \tag{4}$$

When the modulation frequency of the ac bias becomes large, the electric field reverses itself in a time shorter than the transit time and the magnitude of the ac EL decreases. Figure 11 plots the 3 dB frequency f_0 (the frequency

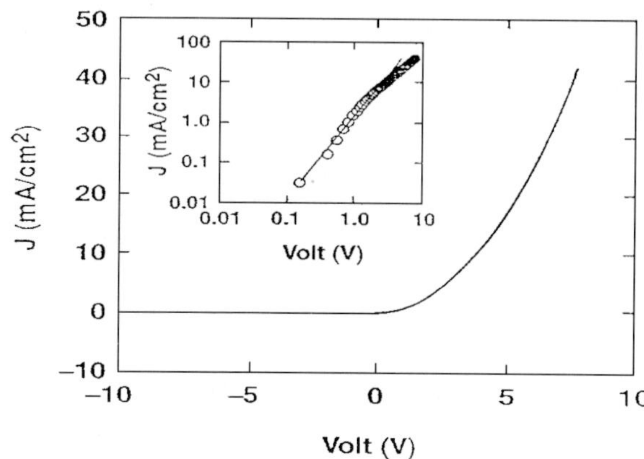

FIG. 10. Current density–voltage relation for a Au/PSi/c-Si device structure. The inset shows that the current density increases as the square of the voltage.

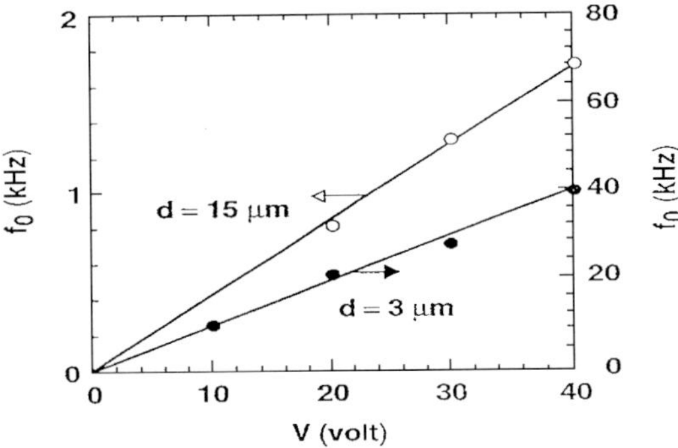

FIG. 11. The 3 dB frequency of the electroluminescence response of device structures similar to that of Fig. 10 depends on the applied voltage and the thickness of the PSi layer. The solid lines are a fit using Eq. (4) and a mobility of $10^{-4}\,cm^2\,V^{-1}\,s^{-1}$.

at which the magnitude of the ac EL drops by 50%) *versus* dc bias for two devices having PSi layers of 3 and 15 μm, respectively. Other measurements have confirmed that f_0 scales with d^{-2} and with V. Thus, f_0 is equal to the inverse of τ_t, and the effective mobility obtained in PSi of ≈75% porosity is $10^{-4}\,cm^2\,V^{-1}\,s^{-1}$. This measurement does not indicate whether the mobility is that of the electrons, the holes, or a combination of both.

Another method to measure the mobility is the time-of-flight technique. Here, the device is held in reverse bias, and a short laser pulse injects electrons and holes on one side of the PSi layer. One type of carrier is then swept through the PSi layer under the effect of the electric field, producing a current that is detected as a function of time after injection. These experiments performed on PSi layers of ≈75% porosity and passivated by ≈2 monolayers of SiO_2 have shown that the electron drift mobility is between 10^{-4} and $10^{-5}\,cm^2\,V^{-1}\,s^{-1}$ and is weakly dependent on temperature from 400 to 77 K (Rao *et al.*, 1997). Electron transport is found to be dispersive and not Gaussian. This behavior has implications for the nature of electronic transport is PSi, which have not yet been explored throroughly.

2.6 Optical: Experiments

2.6.1 Absorption/Transmission The influence of the porosity (P) on the optical properties is most easily seen in transmission experiments performed on free-standing PSi films separated from the c-Si substrate by a lift-off step. The transmission spectrum of high-quality thin films is modulated by Fabry–Pérot interference fringes in the transparency region. The transmission can reach 100%, which is indicative of high-quality films having parallel front and back surfaces and negligible light scattering. The optical transmission spectra of thicker free-standing porous silicon films of various P are shown in Fig. 12 and compared with that of c-Si (von Behren *et al.*, 1998). The absorption edge shifts by more than 1.5 eV toward the blue when the porosity increases from 0% (bulk c-Si) to 92%. Since the index of refraction is close to unity for $P \geq 80\%$ (Fauchet *et al.*, 1998), the optical thicknesses of these films are equal, and the changes in transmission with porosity are a direct proof of the opening of the band gap by quantum confinement. An additional contribution comes from the fact that as the porosity increases, a smaller fraction of the volume of film is made of Si. The absorption coefficient of Si nanocrystals of average sizes from 43 to

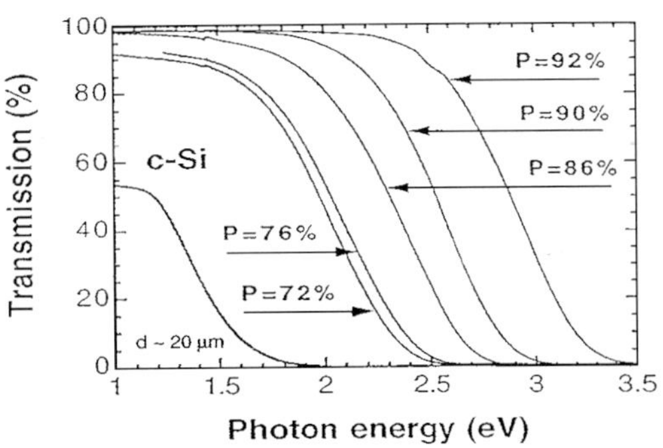

FIG. 12. The transmission spectra of 20-μm-thick PSi films depend on the porosity. The samples are prepared from a *p*-Si wafer and the spectrum of bulk c-Si is shown for comparison.

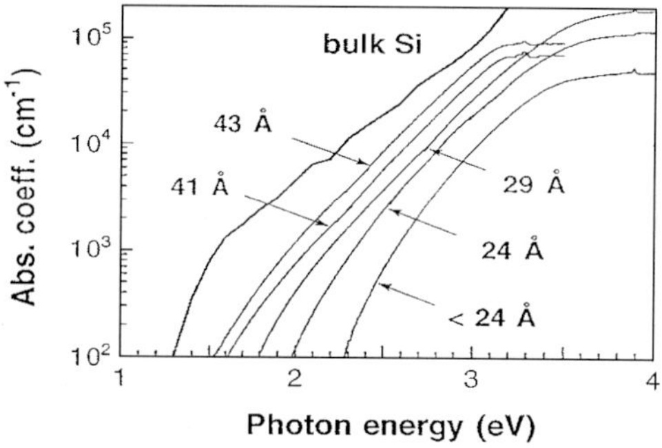

FIG. 13. The absorption spectra of the nanocrystallites present in PSi depend on the crystallite size. The samples are those of Fig. 12.

24 Å is plotted in Fig. 13 (von Behren *et al.*, 1998). The shape of the spectra shows that PSi remains an indirect-gap semiconductor even at the highest porosities/smallest crystallite sizes.

2.6.2 Dielectric Function and Index of Refraction The dielectric function ε of PSi maintains the spectral dependence of bulk crystalline Si at low porosities. With increasing porosity, the peaks in the imaginary part of ε at the E_1 and E_2 critical points broaden and decrease in amplitude. Finally, the imaginary part of ε approaches that of amorphous silicon (Koshida *et al.*, 1993), consistent with the loss of infinite long-range order. The index of refraction of PSi at or below the band gap is strongly dependent on the porosity. Figure 14 shows that the index of refraction decreases to near 1.1 at porosities of 90% and above (Fauchet *et al.*, 1998). This decrease is predicted by effective-medium theories.

2.6.3 Luminescence Radiative recombination of an electron with a hole across the band gap of a semiconductor produces luminescence at an energy equal or close to the band-gap energy. In direct-gap semiconductors such as GaAs, the electron and hole have the same momentum, the radiative recombination rate is large, and the radiative lifetime is short (a few nanoseconds). To obtain a large PL efficiency, the nonradiative recombination rate must be smaller than the radiative recombination rate. The quantum efficiency of the luminescence is defined by

$$\eta = \tau_{\text{nonrad}}/(\tau_{\text{nonrad}} + \tau_{\text{rad}}), \tag{5}$$

where η is the quantum efficiency, τ_{rad} is the radiative lifetime and τ_{nonrad} is the nonradiative lifetime. Nonradiative recombination occurs at the surface (surface recombination) and in the bulk (Auger recombination, recombination at defects).

In silicon, an indirect-gap semiconductor (see SILICON, CRYSTALLINE), the radiative recombination of an electron and a hole is possible only if another particle capable of carrying a large momentum is involved. The participation of a third particle (*e.g.*, a phonon) makes the radiative rate much lower and pushes the radiative lifetime into the millisecond regime. The quantum efficiency decreases by several orders of magnitude because the competing nonradiative processes take over, even for high-purity materials and good surface passivation. At room temperature, the quantum efficiency of bulk c-Si is usually on the order of 10^{-6}–10^{-7}. To improve this number, one can either increase the radiative rate

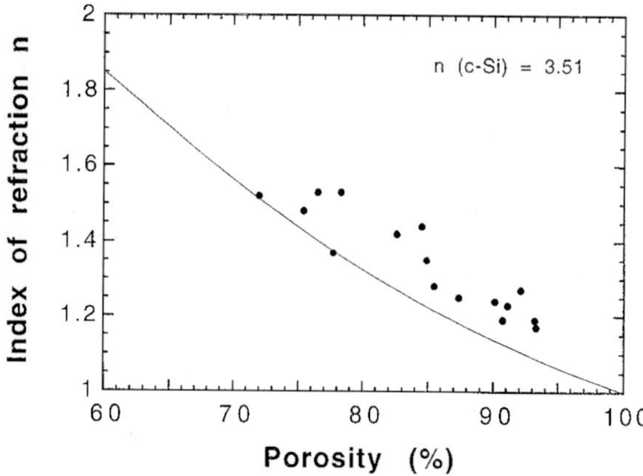

FIG. 14. The index of refraction of PSi depends on the porosity, in qualitative agreement with the prediction of an effective-medium theory, shown by the solid line.

or decrease the nonradiative rate. In PSi, both occur, although the large quantum efficiency is in large part due to the decrease in the nonradiative rate by a proper surface passivation.

The most common PL band is in the red/near-infrared part of the spectrum (Fauchet, 1996). This "red" band (or "slow" band) extends from slightly above the band gap of crystalline silicon to the blue-green part of the spectrum. The PL characteristics are wide spectrum (100–500 meV), large efficiency under UV pumping (1–10% at room temperature, greater at cryogenic temperatures), and long decay time (< 1–100 μs at room temperature, up to 10 ms at cryogenic temperatures). Figure 15 shows the evolution of a typical PSi spectrum after pulsed excitation and the temperature dependence of the PL lifetime and intensity (Vial *et al.*, 1993). In general, the shorter the emission wavelength, the shorter the PL lifetime. The PL decay does not follow a simple law and can be fitted very well using a stretched exponential function (Kanemitsu, 1993). Tuning of the red PL band can be achieved by many parameters, such as the porosity/crystallite size (by changing the HF concentration, current density, wafer doping), the use of light during anodization, or post-anodization treatments (Fauchet, 1996).

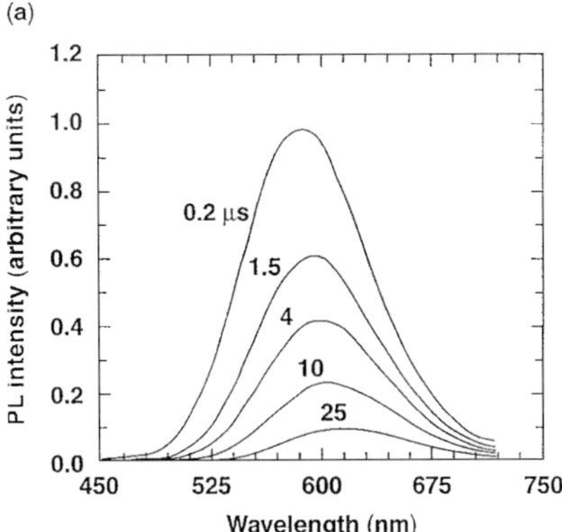

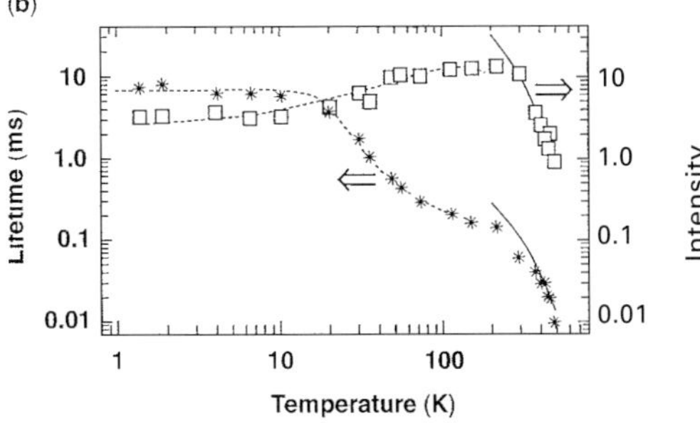

FIG. 15. (a) Temporal evolution of the PL spectrum of red-emitting PSi after pulsed UV excitation. (b) The PL intensity and the PL lifetime depend on the measurement temperature.

Although ultrahigh-porosity PSi that has not been exposed to air can luminesce weakly in the blue, strong blue PL follows high-temperature oxidation ($T_{ox} \geq 1000\,°C$) of PSi. Figure 16 shows the PL spectrum and dynamics in PSi oxidized at 1100 °C for 2 min (Tsybeskov *et al.*, 1993). The broad PL band (FWHM > 0.5 eV) with a peak near 2.6 eV and the decay on a nanosecond time scale are typical of these samples. The decay is non-exponential, as for the red band, but in contrast to the red band, no significant wavelength dependence of the "blue" (or "fast") PL decay is observed from 440 to 650 nm, the decay dynamics does not change appreciably when the measurements are performed at cryogenic temperatures, and the decay dynamics is not fitted well by a stretched exponential. The highest quantum efficiency of the blue PL can approach that of the red PL, especially after long-term exposure to water vapor and contaminants containing carbon (Loni *et al.*, 1995a).

Theory indicates that for crystallites of ≈1 nm diameter, the band gap should be in the blue/near UV (Wang and Zunger, 1996), and phononless, fast (approximately nanosecond) radiative recombination should dominate (Hybertsen, 1994). This model, however, cannot explain the reversible quenching of the blue PL in methanol after high-temperature oxidation of PSi (Rehm *et al.*, 1995), since the very small Si nanocrystallites, if they exist at all, are protected by a very thick oxide layer. Silicon dioxide itself is known to luminesce efficiently in the visible range under appropriate conditions, and a nanosecond blue PL has been reported after excitation by UV photons in high-purity, wet synthetic silica (Anedda *et al.*, 1993). However, the fact that the blue PL increases with exposure to water vapor and car-

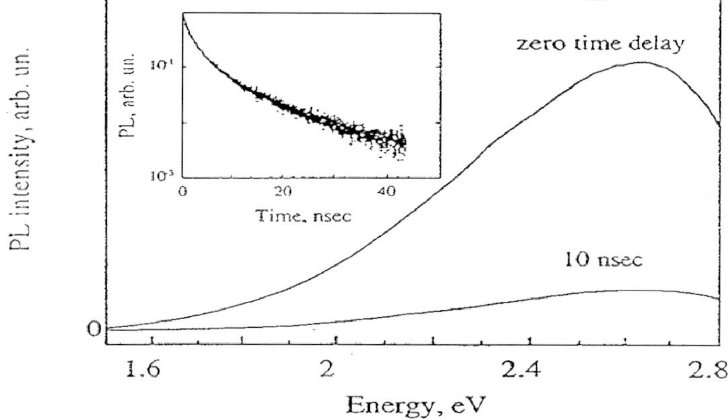

FIG. 16. Temporal evolution of the PL spectrum of blue-emitting PSi after UV excitation. The inset shows the temporal decay of the PL intensity.

bon compounds (Loni *et al.*, 1995a) strongly suggests that a chemical species adsorbed on the inner surface of the porous oxide is responsible for the blue PL. Silanol has been suggested as one candidate, but no conclusive identification has been made so far.

PSi can also emit light in the infrared. Figure 17 shows the normalized PL spectra after annealing in vacuum at different temperatures (Fauchet *et al.*, 1993). This PL band is tunable from ≈0.8 eV to past 1.2 eV, depending on the sample preparation and processing conditions, and has a nonexponential lifetime in the ≈1-μs range. Its intensity increases after annealing in ultrahigh vacuum at a temperature as high as ≈500 °C, when most of the hydrogen that passivates PSi has been desorbed and the red PL has disappeared. The tunable infrared band has been associated with dangling bonds. Figure 18 shows that the peak energies of the red and infrared bands vary linearly with each other (Petrova-Koch *et al.*, 1995). This can be explained if the infrared band corresponds to recombination of a near–band-edge electron with a hole trapped at a dangling bond whose absolute energetic position is insensitive to size, ≈0.3 eV above the bulk silicon valence-band edge.

PSi containing large crystallites also produces a PL near 1.1 eV, which coincides with

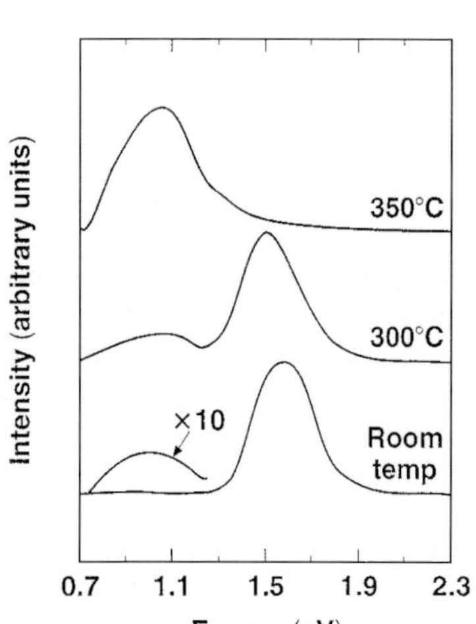

FIG. 17. Normalized PL spectra of PSi before and after annealing in vacuum at two temperatures.

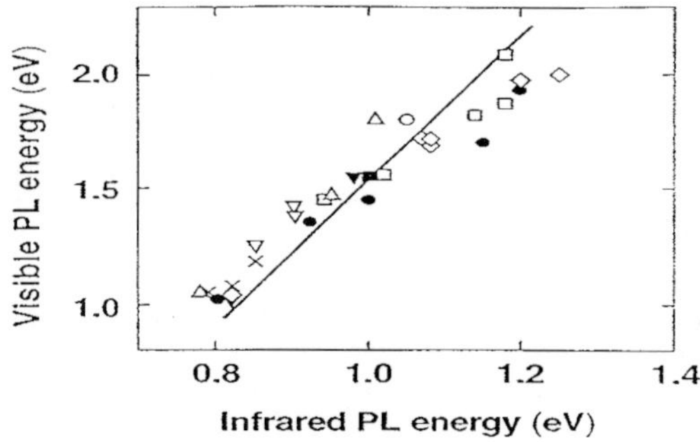

FIG. 18. Correlation between the red PL peak energy and the infrared PL peak energy over a wide range of samples prepared under different conditions. The line has a slope of 3.

the PL spectrum observed in bulk c-Si (Tsybeskov *et al.*, 1996a), including all the phonon replicas. As the temperature increases toward 300 K, the PL spectrum widens, but the PL efficiency remains unchanged near or above 0.1%. The PL is due to radiative recombination in well-passivated, large crystallites ($\gg$10 nm).

2.7 Optical: Theory

2.7.1 Band Gap The band gap of low-dimensional Si nanostructures is larger than that of bulk Si because of quantum confinement in one, two or three dimensions. In the simple particle-in-a-box model, when electrons in the conduction band and holes in the valence band are confined spatially by potential barriers, the lowest energy for optical transitions increases by

$$\Delta E = h^2/8[1/m_c + 1/m_v] \sum 1/L_i^{\,2} \qquad (6)$$

where h is Planck's constant, m_c and m_v are the effective masses in the conduction and valence bands respectively, L_i is the dimension of the confined region in direction i, and the number of confined directions may be 1, 2, or 3 corresponding to a quantum well, wire, or dot, respectively (Collins *et al.*, 1997). This expression assumes that the bands are parabolic, there is only one conduction (valence) band, and the potential barrier is infinitely large. Thus, it can only be used as a guide, since none of these assumptions is entirely valid.

Various techniques have been employed to calculate the band gap of Si quantum wells, wires, and dots. Calculations have been performed using different types of tight-binding and local-density approximations, mostly using a (semi)empirical approach (Delerue *et al.*, 1993; Wang and Zunger, 1994; Delley and Steigmeier, 1995; Ogut *et al.*, 1997). Figure 19 compares the band gap for wells, wires, and dots calculated by the effective-mass approximation and by a pseudopotential approach (Wang and Zunger, 1996). The band gap increases faster with decreasing size when the number of confined directions increases. The effective-mass approximation overestimates the band gap by a very large amount ($\approx$1 eV for $\approx$1 nm wells, $\approx$2 nm wires, and $\approx$3 nm

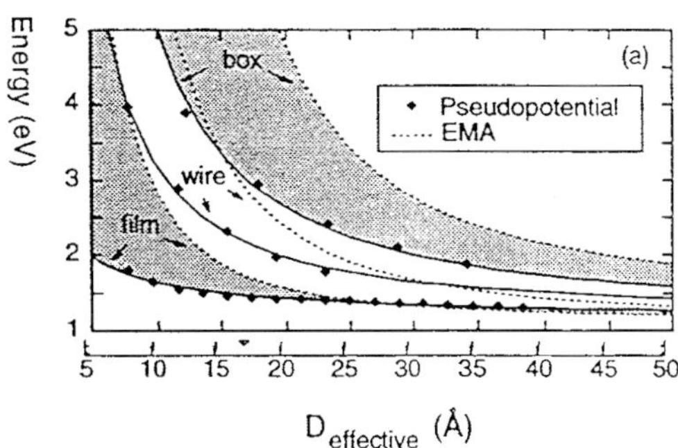

FIG. 19. Comparison of the band gap calculated by the effective-mass approximation [Eq. (6)] and a pseudo-potential approach, for Si quantum dots (box), wires, and wells (films).

dots). This overestimate occurs because theory shows that ΔE scales with size as $L^{-1.3}$ instead of L^{-2} (Delerue *et al.*, 1993). Furthermore, theory (Ren and Dow, 1992) and experiments (van Buuren *et al.*, 1993) indicate that the quantization energy in the valence band is larger than in the conduction band [$\Delta E_v \approx (1.5–2) \times \Delta E_c$], also in contrast with the effective-mass approximation (which predicts that $\Delta E_v \sim 1/m_v < \Delta E_c \sim 1/m_c$).

Most of the calculations have been performed for H-terminated Si nanostructures. It is expected that the exact numerical answers will be affected by a change in the surface termination, especially a change from H to O passivation following partial oxidation. At this time, the models are not sophisticated enough to account properly for effects such as strain that exist at such an interface.

Despite the profound effect that quantum confinement has on the magnitude of the band gap of the Si nanocrystals present in porous silicon, the nature of the optical transition remains identical to that of bulk c-Si, in the sense that phonon-assisted optical transitions continue to predominate over transitions involving no phonons. Theory suggests that the oscillator strength for phonon-assisted transitions continues to exceed that for phononless transitions down to sizes of $\approx$1 nm (Hybertsen, 1994). Thus porous silicon behaves as an indirect-gap semiconductor, as confirmed by the presence of pronounced phonon steps in resonant PL measurements performed at low temperatures (Calcott *et al.*, 1993).

2.7.2 Luminescence The models that attempt to explain the luminescence in porous silicon can be divided into two broad categories (Fauchet 1996):

1. The PL occurs in quantum-confined crystalline silicon nanostructures, and
2. the PL occurs elsewhere—for example, in a disordered tissue material, in specific bonds or molecules (such as siloxene) that would be present on or attached to the surface, or in the nonbridging oxygen hole center present in silicon dioxide.

The overwhelming experimental evidence indicates that carriers confined in the Si nanocrystals produce the strong, tunable PL. The remaining question is whether the PL energy coincides with the band-gap energy. If recombination involves carriers confined to the nanocrystal, the PL energy should equal the band gap minus the binding energy of the exciton. If, on the other hand, the PL energy is less than the band gap, one or both carriers may be trapped, either in surface states (Koch, 1993) or as self-trapped excitons (Allan *et al.*, 1996). Because the nanocrystals present in PSi do not all have the same size, the absorption edge reflects the presence of large crystallites while the PL is dominated by the smaller crystallites, which are less likely to contain a defect or to suffer from Auger recombination. Thus, the PL and absorption measurements may not sample the same fraction of the nanocrystal population, which makes a comparison difficult.

Changes in the surface chemistry and in the chemical environment can lead to modifications in the PL properties. Although this has been invoked to support recombination involving trapped carriers, changes in surface properties can also modify the electronic states in the interior of the nanocrystals. Despite the difficulties with a comparison of the band-gap and PL energies as mentioned above, this is the only type of experiment that can directly address the question.

The band gap measured from absorption measurements should be equal to the one-electron band gap minus the binding energy of the free exciton confined to the nanocrystal. The free-exciton binding energy increases with decreasing radius size (Wang and Zunger, 1996) for two major reasons: namely, because the Coulombic attraction increases when the exciton is confined to a radius smaller than the bulk-exciton Bohr radius, and because the dielectric constant decreases dramatically for sizes below 2–3 nm. Realistic calculations of the exciton binding energy indicate that the binding energy approaches 0.5 eV for sizes of ≈2 nm. The calculated excitonic band gap, defined as the one-electron band gap minus the exciton binding energy, is in agreement with measurements of the absorption edge, in the size regime below 4.5 nm (Fig. 20) (von Behren *et al.*, 1998). The measured PL peak energy is consistently lower than the band-gap energy, especially for smaller sizes. This Stokes shift between band-gap and PL energies increases with decreasing size, as shown in Fig. 21 (Fauchet, 1998). The origin of this Stokes shift is still under discussion. A possible candidate is that the exciton becomes self-trapped at the surface. Model calculations of self-trapping at the Si–H bond show that the nanocrystal size necessary for self-trapping is much too small, below 1 nm (Allan *et al.*, 1996). However, trapping of an electron, or a hole, or even an exciton, could occur for more reasonable sizes with oxide-covered nanocrystals. This would produce, for smaller nanocrystallites, an increase of the PL energy that is much smaller than the increase of the band-gap energy.

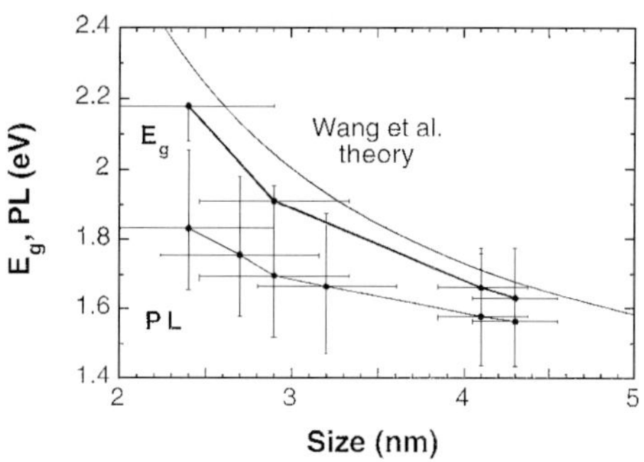

FIG. 20. Measured and calculated band gaps (E_g) for Si nanocrystallites of different sizes. The PL peak energy in samples exposed to air is also shown for comparison.

3. APPLICATIONS

3.1 Light-emitting Devices

Electroluminescence (EL) was observed in PSi shortly after Canham's report of strong

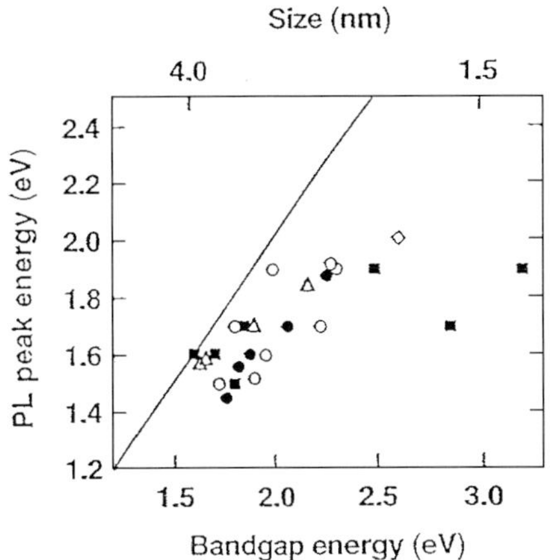

FIG. 21. Dependence of the PL peak energy on the band gap and crystallite size for the Si nanocrystals found in PSi. The data would fall on the solid line if the two energies were identical.

room-temperature PL—first in solution during anodic oxidation (Halimaoui *et al.*, 1991) and then in solid-state devices (Richter *et al.*, 1991; Kalkhoran *et al.*, 1992; Bassous *et al.*, 1992; Koshida and Koyama, 1992). Many different device structures have been demonstrated, and strong (0.1%–1% efficiency), voltage-tunable EL has even been reported under cathodic polarization of n-type PSi using liquid contacts (Bsiesy *et al.*, 1993). Research has focused on solid-state devices because of their technological importance and because liquid contacts are inherently unstable. A typical PSi light-emitting device (LED) consists of a transparent or semitransparent contact (metal, ITO, or conducting polymers) and a 1- to 10-μm-thick PSi layer on a crystalline silicon (c-Si) substrate (p or n type). Typical threshold conditions for EL are $V \approx 10\,V$ and $J \approx 10\,mA\,cm^{-2}$, although for the best devices V is now $\approx 2\,V$ and $J \ll 1\,mA\,cm^{-2}$ (Loni *et al.*, 1995b). The EL external quantum efficiency has also improved to $>0.1\%$, and the irreversible degradation that plagued the early devices has been eliminated.

Initially, the stability of the early PSi LEDs was poor because the Si–H bonds that passivate the Si nanocrystal surfaces are very fragile and can be easily broken by exposure to light, ambient air, moderate temperatures, and large electric fields. To improve the stability, the fragile Si–H bonds have been replaced by the stronger Si–O bonds and the devices have been engineered to provide better heat sinking and greater mechanical stability, leading to no degradation after weeks of operation under ambient conditions. Figure 22 shows the results of stability tests, performed under pulsed drive conditions well above EL threshold for a PSi LED (Tsybeskov *et al.*, 1996b). No degradation is seen for several weeks. A key factor in achieving stable devices is to minimize the density of states at the interfaces between the different layers present in the structure. Devices that contain too many interface states suffer from reversible degradation, produced by the charging of the interface states (Fauchet *et al.*, 1997).

The two main factors that keep the EL efficiency one to two orders of magnitude below the PL efficiency are the difficulties in injecting carriers from the contact into the PSi and the poor transport properties in PSi. By improving the contacts and keeping the PSi layer thickness below 1 μm, the power efficiency of the

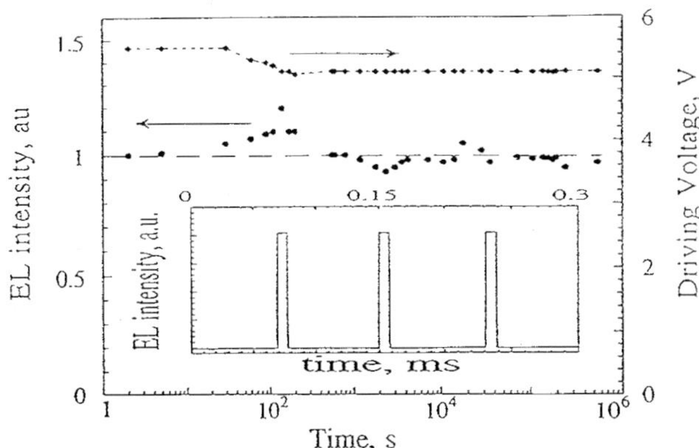

FIG. 22. Stability of the electroluminescence intensity under pulsed current excitation conditions for a PSi LED.

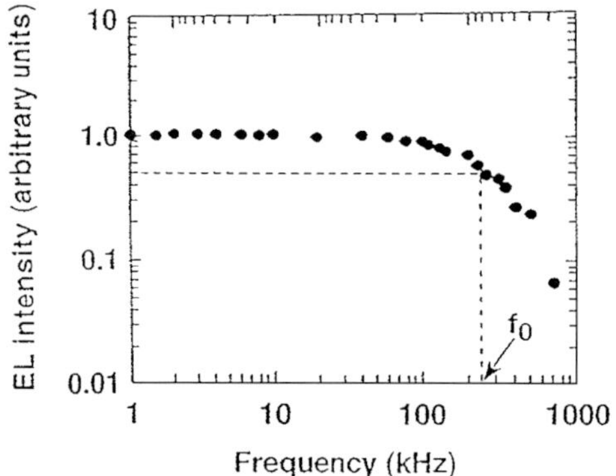

FIG. 23. The amplitude of the ac electroluminescence depends on the modulation frequency. The measurement was performed on a p–n PSi junction LED. The 3-dB frequency is indicated.

best PSi LEDs has been increased to greater than 0.1%. The internal quantum efficiency, defined as the ratio of the number of photons generated inside the PSi layer to the number of carriers injected inside the PSi layer, is estimated to be $>1\%$, which is comparable to the PL efficiency. The threshold voltage (2 V) and current density ($1\,\mu A\,cm^{-2}$) of some of these devices makes them compatible with microelectronic technology.

The response time of PSi LEDs is an important parameter that determines their suitability in applications such as optical interconnects and real-time optical signal processing. The speed of the LEDs can be obtained by measuring the frequency response in the small-signal limit. The device is held at a large forward bias, well above the threshold for detectable EL. A small ac voltage is applied, which modulates the EL. The amplitude of the ac EL is monitored as a function of frequency, as already mentioned in Sec. 2.5. Figure 23 shows that the magnitude of the ac EL signal remains constant up to a critical frequency beyond which it drops quickly (Peng and Fauchet, 1995). The 3-dB frequency, which is defined as the frequency at which the ac EL has dropped by a factor of 2, has been found in Sec. 2.5 to be the inverse of the transit time. The 3-dB frequency determines the maximum frequency response of the diode in this small-signal analog test. Thus, the response time of the EL is limited by the time it takes carriers to cross the PSi layer, and not the PL lifetime, which is usually shorter. By making the PSi layer thinner than 1 μm or by using a PSi *p-n* junction, the 3-dB frequency can approach 1 MHz, when it is limited by the luminescence lifetime. Clearly, the small-signal ac modulation speed of these PSi LEDs is not likely to approach 1 GHz, unless they can be made using blue-emitting PSi, whose PL lifetime is ≈ 1 ns.

Since the PL can span the spectrum from the infrared to the blue, it is reasonable to expect that EL can cover the same range of wavelengths. As early as 1992, LEDs with peak wavelengths ranging from the deep red to the blue were fabricated. The EL efficiency of these devices was between 0.005% and 0.01%. LEDs operating in the infrared are also attractive. For example, 1.5-μm radiation is eye-safe and compatible with fiber communication systems. In addition, it is not absorbed by the c-Si wafer, which may be an advantage for the integration of PSi LEDs in close proximity to microelectronic circuits. PSi LEDs emitting 1.1-μm light and 1.5-μm light have been demonstrated, using oxidized mesoporous Si and Er-doped oxidized PSi, respectively (Fig. 24) (Fauchet *et al.*, 1997).

Ultimately, the success or failure of PSi-based technology rests on the compatibility of PSi LEDs with standard microelectronic processing. An integrated bipolar transistor/PSi LED structure has been demonstrated (Hirschman *et al.*, 1996). The complete structure, shown in Fig. 25, was fabricated using accepted silicon microelectronic fabrication procedures. The LED could be turned on and off by applying a small current pulse to the base of the bipolar transistor. Arrays of such integrated structures can easily be fabricated. This represents a critical first step toward the integration of silicon-based LEDs into VLSI circuits.

3.2 Solar Cells

Solar cells have been made using PSi, either as the photovoltaic material or as a passive layer intended to improve the performance of the c-Si solar cell. The use of PSi in photovoltaic devices has been reported as early as 1982 (Prasad *et al.*, 1982). This involved the anodization of p^+/n c-Si solar cells to form a

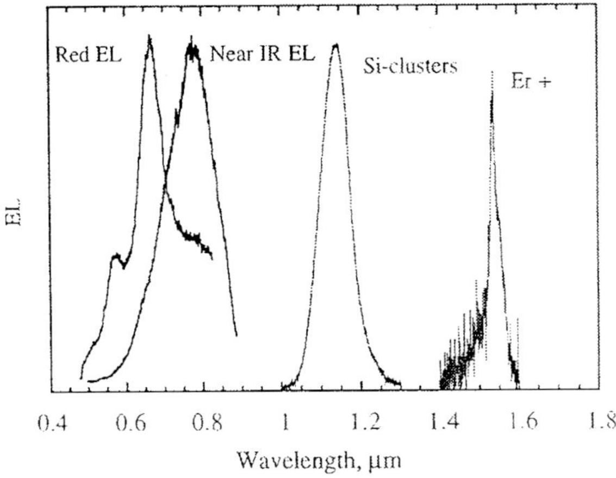

FIG. 24. Electroluminescence spectra for four PSi LEDs. The red/near-infrared LEDs were prepared with a nanoporous light-emitting layer. The two infrared LEDs were prepared with a mesoporous light-emitting layer, which was doped with erbium for the 1.5-μm device.

thin porous layer, which was oxidized at 500–600 °C. As a result, the reflectance of the solar cell was reduced from 37% to 8% in the 300- to 800-nm region, which increased the short-circuit current density (J_{sc}) by 25%. The potential advantages of PSi-based photovoltaic devices include the following:

1. Its highly textured surface increases light trapping and reduces loss due to reflection.
2. The band gap of PSi may be optimized for the absorption of sunlight (the maximum theoretical solar-cell efficiency vs. band gap curve peaks at around 1.5 eV).
3. The efficient photoluminescence of PSi might be used to down-convert higher-energy solar radiation (blue/UV) into longer-wavelength light (red/IR), which is absorbed more efficiently in bulk silicon.

Solar cells in which PSi is the photovoltaic material have been demonstrated with an efficiency in excess of 10% (Duttagupta and Fauchet, 1998). A μc-Si film deposited on *n*-type (100) c-Si substrates (1 Ω cm) at 580 °C using low-pressure chemical vapor deposition was doped *in situ* with phosphorus (SiH_4/PH_3) and boron (SiH_4/B_2H_6). The p^+-n films were anodized, and the average porosity P was varied to achieve maximum efficiency. Under illumination, an ideal solar cell is equivalent to a constant-current generator in parallel to a *p-n* junction diode. With increasing porosity, the series resistance first increases slightly and

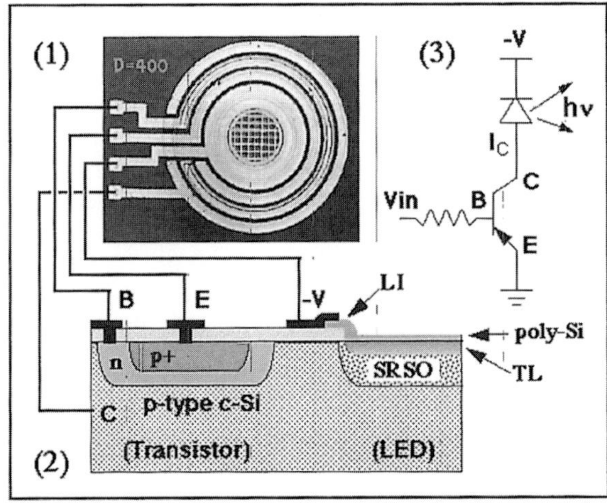

FIG. 25. Photograph, cross section, and equivalent circuit of a PSi LED integrated with a bipolar transistor. The LED surface is covered by the cross-hatched Al contacts. The transistor surrounds the LED. The cross section starts on the right at the center of the LED. (TL is a mesoporous layer and LI is the contact to the metal electrodes.)

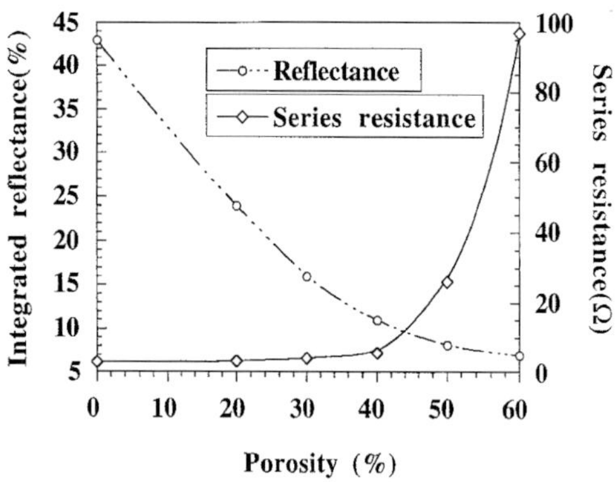

FIG. 26. Dependence of the reflectance and the series resistance on porosity for a solar-cell device made of porous microcrystalline silicon.

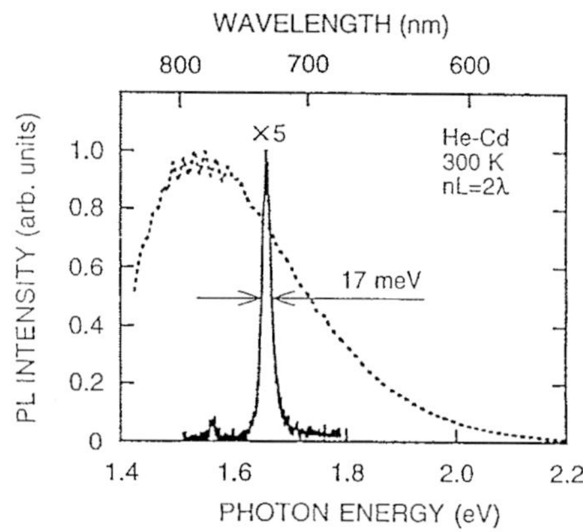

FIG. 27. Narrowing of the PL spectrum when porosity superlattices are used to shape the emission spectrum of PSi.

then, above a porosity of 50%, increases by several orders of magnitude. In contrast, the reflectance decreases monotonically with increasing porosity (Fig. 26). As a result, the short-circuit current has a maximum around 40% porosity. The best solar energy efficiency of 10.2% has been achieved for devices with a 40% average porosity, which also exhibit an enhanced blue response Note that photodetectors, which are devices closely related to solar cells, have also been made using PSi (Zheng *et al.*, 1992).

3.3 Mirrors and Photonic Band Gap

High-reflectivity mirrors made of PSi can be fabricated with reflectivities in excess of 99%. They consist of a stack of alternating high-porosity/low-porosity layers (Berger *et al.*, 1994), produced by changing the current density periodically during anodization or by anodizing under a constant current density alternating c-Si layers having high and low doping concentrations. These porosity superlattices, which can be viewed as Fabry–Pérot resonators or Bragg reflectors, have a high reflectivity in specific wavelength regions only. They have been inserted in PSi structures to narrow the spectral width of the luminescence by more than one order of magnitude, as shown by Fig. 27 (Araki *et al.*, 1996).

PSi containing periodically spaced macropores (initiated by means of a mask on the wafer) has been used as a two-dimensional photonic band gap structure (Gruning *et al.*, 1996). A photonic band gap material is to photons what a semiconductor crystal is to electrons: it prevents the propagation of light in a range of photon energies by the destructive interference produced by the alternating regions of high index of refraction (Si) and low index of refraction (pores), provided that the periodicity is on the order of the wavelength in PSi. Attenuation of electromagnetic waves by more than two orders of magnitude has been observed near 5 μm using a triangular lattice of 75-μm-deep pores spaced by 2.3 μm.

3.4 Waveguides

Optical waveguides require that a region of high index of refraction be sandwiched between regions of low index of refraction. Because the index of refraction of PSi can easily be controlled by changing the porosity, PSi waveguides can be readily fabricated. PSi waveguides can be useful in Si optoelectronic integrated circuits (OEICs), where they could manipulate optical signals. PSi waveguides with losses as small as $\leq 1\,\text{dB}\,\text{cm}^{-1}$ have been demonstrated (Bondarenko *et al.*, 1995). They have also been used to collect efficiently the electroluminescence produced by PSi LEDs.

3.5 Sensors

Because the properties of PSi are very sensitive to the state of its internal surface, it should be a useful material in detecting the presence of chemical compounds, humidity, *etc*. PSi sensors could operate on the basis of a reversible or irreversible change in optical properties (PL), electrical properties (conductivity), or even structural properties. PSi sensors have been developed for detecting humidity in integrated circuits. They have also been proposed for detecting environmentally important chemicals (smokestack gases) and alcohols (breathalyzers) (Sailor and Lee, 1997). Although PSi sensors with promising properties have been demonstrated, one common problem is the lack of specificity.

3.6 SOI Technology

In the silicon-on-insulator (SOI) technologies, the electronic devices are manufactured on a thin Si film on an insulating substrate, which minimizes problems associated with parasitic capacitances, latch up, and radiation damage. PSi was first used in the FIPOS (full isolation by porous oxidized silicon) process and then in a variety of SOI applications (Tsao, 1987). In these applications, the oxide layer was formed by the full oxidation of PSi usually formed from heavily doped layers. This approach has been abandoned in favor of alternative techniques.

3.7 Micromachining

Micromachining of Si is a technological area of rapidly growing importance because it allows the development of micro-electro-mechanical systems (MEMS) devices that can be used in a variety of applications, such as pressure sensors. PSi, as a sacrificial layer, is well suited to micromachining (Lang, 1996) because the etching rate of Si depends strongly on the doping density and the dopant type and on the current density. An etching hierarchy (from most easily etched to less easily etched) has been introduced: n^+ Si under illumination, p^+ Si under illumination, p and n Si under illumination, p Si in the dark, n Si in the dark and amorphized Si. A combination of etch stops (*e.g.*, using layers of different resistivities) and masking techniques (using photoresists and/or silicon nitride or carbide protective layers) permits the micromachining of Si in three dimensions. Examples of integrated microsensors fabricated using the PSi technology include bolometers.

3.8 Biomedical Applications

Electronic components are increasingly used inside the human body (*e.g.*, pacemaker, drug delivery systems). In these applications, the Si chip is isolated from the biological medium by a package that is more bio-compatible or has the desired bioactivity. Unlike bulk Si, which is mostly bio-inert, PSi has been shown to induce the nucleation and growth of calcium phosphate (Canham, 1995). This implies that PSi can naturally coat to bone and other soft tissues in the human body. Interestingly, the kinetics of calcification can be controlled electrically. Thus, chips containing PSi regions could be used to bind to specific sites on demand. This field is just in its early infancy, and future developments (Canham *et al.*, 1997) will tell whether the early promises transform into real applications.

3.9 Capacitors

Capacitors have been manufactured using PSi technology (Lehmann *et al.*, 1995). Macroporous PSi was formed by seeding the pores with a 3.5-μm pitch. The resulting gain in surface area over a bulk sample was 85. The PSi was then doped heavily n-type and a 30-nm-thick dielectric layer was grown inside the pores. Such devices, which showed excellent characteristics, can be integrated on the back side of Si chips, thus saving valuable real estate and avoiding parasitic effects in high-quality RC timing circuits.

GLOSSARY

Hole: An empty energy level in the valence band of a semiconductor; it behaves as a particle carrying a positive charge.

Luminescence: Emission of a photon during the recombination of an electron with a hole. If the electron and hole have been injected optically, it is called photoluminescence; if they have been injected electrically, it is called electroluminescence.

Pore: Region from which the silicon atoms have been removed by the etching process.

Works Cited

Allan, G., Delerue, C., Lannoo, M. (1996), *Phys. Rev. Lett.* **76**, 2961–2964.

Anedda, A., Bongiovanni, G., Cannas, M., Congiu, F., Mura, A., Martini. M. (1993), *J. Appl. Phys.* **74**, 6993.

Araki, M., Koyama, H., Koshida, N. (1996), *Jpn. J. Appl. Phys.* **35**, 577–580.

Bassous, E., Freeman, M., Halbout, J.-M., Iyer, S. S., Kesan, V. P., MMungia, P., Pesarcik, S. F., Williams, B. L. (1992), in: S. S. Iyer, R. T. Collins, L. T. Canham (Eds.), *Light Emission from Silicon*, MRS Symposium Proceedings Vol. 256, Pittsburgh: Materials Research Society, p. 23.

Beale, M. I. J., Benjamin, J. D., Uren, M. J., Chew, N. G., Cullis, A. G. (1985), *J. Cryst. Growth* **73**, 622.

Bellet, D., Lamagnere, P., Vincent, A., Brechet, Y. (1996), *J. Appl. Phys.* **80**, 3772.

Berbezier, I. (1995), in: J.-C. Vial, J. Derrien (Eds.), *Porous Silicon Science and Technology*, Berlin: Springer-Verlag, p. 207.

Berger, M. G., Dieker, C., Thonissen, M., Vescan, L., Luth, H., Munder, H., Thiess, W., Wernke, M., Grosse, P. (1994), *J. Phys. D: Appl. Phys.* **27**, 1333.

Bondarenko, V. P., Dorofeev, A. M., Kazuchits, N. M. (1995), *Microelectron. Eng.* **28**, 447.

Bsiesy, A., Muller, F., Ligeon, M., Gaspard, F., Herino, R., Romestain, R., Vial, J.-C. (1993), *Phys. Rev. Lett.* **71**, 637–641.

Calcott, P. D. J., Nash, K. J., Canham, L. T., Kane, M. J., Brumhead, D. (1993), *J. Phys.: Cond. Matter* **5**, L91.

Canham, L. T. (1990), *Appl. Phys. Lett.* **57**, 1046–1048.

Canham, L. T. (1995), *Adv. Mater.* **7**, 1033.

Canham, L. T., Cullis, A. G., Pickering, C., Dosser, O. D., Cox, T. I., Lynch, T. P. (1994), *Nature (London)* **368**, 133.

Canham, L. T., Reeves, C. L., Wallis, D. J., Newey, J. P., Houlton, M. R., Sapsford, G. J., Godfrey, R. E., Loni, A., Simons, A. J., Cox, T. I., Ward, M. C. L. (1997), in: R. W. Collins, P. M. Fauchet, I. Shimizu, J.-C. Vial, T. Shimada, A. P. Alivisatos, (Eds.), *Advances in Microcrystalline and Nanocrystalline Semiconductors—1996*, MRS Symposium Proceedings Vol. 452, Pittsburgh: Materials Research Society, pp. 579–590.

Collins, R. T., Tischler, M. A., Stathis, J. H. (1992), *Appl. Phys. Lett.* **61**, 1649–1651.

Collins, R. T., Fauchet, P. M., Tischler, M. A. (1997), *Phys. Today* **50**, 24–31.

Cullis, A. G., Canham, L. T., Dosser, O. D. (1992), *Mat. Res. Soc. Symp. Proc.* **258**, 7.

Cullis, A. G., Canham, L. T., Calcott, P. D. J. (1997), *J. Appl. Phys.* **82**, 909–965.

Cullity, B. D. (1978), *Elements of X-Ray Diffraction*, 2nd ed., Reading, MA: Addison-Wesley.

Delerue, C., Allan, G., Lannoo, M. (1993), *Phys. Rev. B* **48**, 11024.

Delley, B., Steigmeier, E. F. (1995), *Appl. Phys. Lett.* **67**, 2370.

Duttagupta, S. P., Chen, X. L., Jenekhe, S. A., Fauchet, P. M. (1997a), *Solid State Commun.* **101**, 33–37.

Duttagupta, S. P., Fauchet, P. M., Chen, X. L., Jenekhe, S. A. (1997b), in: R. W. Collins, P. M. Fauchet, I. Shimizu, J.-C. Vial, T. Shimada, A. P. Alivisatos, (Eds.), *Advances in Microcrystalline and Nanocrystalline Semiconductors—1996*, MRS Symposium Proceedings Vol. 452, Pittsburgh: Materials Research Society, pp. 473–478.

Duttagupta, S. P., Fauchet, P. M. (1998), unpublished.

Duttagupta, S. P., Peng, C., Fauchet, P. M., Kurinec, S. K., Blanton, T. N. (1995), *J. Vac. Sci. Technol. B* **13**, 1230.

Fauchet, P. M. (1996), *J. Luminesc.* **70**, 294–309.

Fauchet, P. M. (1997), in: *Pits and Pores: Formation, Properties and Significance for Advanced Luminescent Materials*, Pennington, NJ: The Electrochemical Society, pp. 27–60.

Fauchet, P. M. (1998), in: *Light Emission in Silicon: From Physics to Devices*, Semiconductors and Semimetals, Vol. 49, San Diego, CA: Academic Press, pp. 206–252.

Fauchet, P. M., Campbell, I. H. (1988), *Crit. Rev. Solid State Mater. Sci.* **14**, S79–S101.

Fauchet, P. M., Ettedgui, E., Raisanen, A., Brillson, L. J., Seioferth, F., Kurinec, S. K., Gao, Y., Peng, C., Tsybeskov, L. (1993), in: M. A. Tischler, R. T. Collins, M. L. W. Thewalt, G. Abstreiter (Eds.), *Silicon-Based Optoelectronic Materials*, MRS Symposium Proceedings Vol. 298, Pittsburgh: Materials Research Society, p. 271.

Fauchet, P. M., Tsybeskov, L., Duttagupta, S. P., Hirschman, K. D. (1997), *Thin Solid Films* **297**, 254–260.

Fauchet, P. M., von Behren, J., Hirschman, K. D., Tsybeskov, L., Duttagupta, S. P. (1998), *Phys. Status Sol.* **165**, 3–13.

Gruning, U., Lehmann, V., Engelhardt, C. M. (1996), *Appl. Phys. Lett.* **66**, 3254–3256.

Halimaoui, A., Oules, C., Bomchil, G., Bsiesy, A., Gaspard, F., Herino, R., Ligeon, M., Muller, F. (1991), *Appl. Phys. Lett.* **59**, 304.

Hirschman, K. D., Tsybeskov, L., Duttagupta, S. P., Fauchet, P. M. (1996), *Nature (London)* **384**, 338–340.

R. Hummel, R., Morrone, A., Ludwig, M., Chang, S. S. (1993), *Appl. Phys. Lett.* **63**, 2771.

Hybertsen, M. S. (1994), *Phys. Rev. Lett.* **72**, 1514.

Kalkhoran, N. M., Namavar, F., Maruska, H. P. (1992), in: S. S. Iyer, R. T. Collins, L. T. Canham (Eds.), *Light Emission from Silicon*, MRS Symposium Proceedings Vol. 256, Pittsburgh: Materials Research Society, p. 84.

Kanemitsu, Y. (1993), *Phys. Rev. B* **48**, 12,357.

Koch, F. (1993), in: M. A. Tischler, R. T. Collins, M. L. W. Thewalt, G. Abstreiter (Eds.), *Silicon-Based Optoelectronic Materials*, MRS Symposium Proceedings Vol. 298, Pittsburgh: Materials Research Society, p. 319.

Koshida, N., Koyama, H. (1992), *Appl. Phys. Lett.* **60**, 347.

Koshida, N., Koyama, H., Suda, Y., Yamamoto, Y., Araki, M., Saito, T., Sata, K., Sata, N., Shin, S. (1993), *Appl. Phys. Lett.* **63**, 2774–2776.

Koyama, H., Koshida, N. (1993), *J. Appl. Phys.* **74**, 6365.

Lang, W. (1996). *Mater. Sci. Eng.* **R17**, 1–55.

Lang, W., Drost, A., Steiner, P., Sandmaier, H. (1995), in: R. W. Collins, C. C. Tsai, M. Hirose, F. Koch, L. Brus (Eds.), *Microcrystalline and Nanocrystalline Semiconductors*, MRS Symposium Proceedings Vol. 358, Pittsburgh: Materials Research Society, p. 561.

Lehmann, V., Cerva, H., Gosele, U. (1992), in: S. S. Iyer, R. T. Collins, L. T. Canham (Eds.), *Light Emission from Silicon*, MRS Symposium Proceedings Vol. 256, Pittsburgh: Materials Resarch Society, p. 3.

Lehmann, V., Jobst, B., Muschik, T., Kux, A., Petrova-Koch, V. (1993), *Jpn. J. Appl. Phys.* **32**, 2095.

Lehmann, V., Honlein, W., Reisinger, H., Spitzer, A., Wendt, H., Willer, J. (1995), *Solid State Technol.*, November 1995, 99.

Littau, K. A., Szajowski, P. J., Miller, A. J., Kortan, A. R., Brus, L. E. (1993), *J. Phys. Chem.* **97**, 1224.

Loni, A., Simons, A. J., Calcott, P. D. J., Canham, L. T. (1995a), *J. Appl. Phys.* **77**, 3557.

Loni, A., Simons, A. J., Cox, T. I., Calcott, P. D. J., Canham, L. T. (1995b), *Electron. Lett.* **31**, 1288.

Lu, Z. H., Lockwood, D. J., Baribeau, J.-M. (1995), *Nature (London)* **378**, 258.

Nassiopoulos, A. G., Grigoropoulos, S., Papadimitrou, D., Gogolides, E. (1995), *Phys. Status Sol. (b)* **190**, 91.

Naudon, A., Goudeau, P., Vezin, V. (1995), in: J.-C. Vial, J. Derrien (Eds.), *Porous Silicon Science and Technology*, Berlin: Springer-Verlag, p. 255.

Ogut, S., Chelikowsky, J. R., Louie, S. G. (1997), *Phys. Rev. Lett.* **79**, 1770–1773.

Peng, C., Fauchet, P. M. (1995), *Appl. Phys. Lett.* **67**, 2515–2517.

Peng, C., Hirschman, K. D., Fauchet, P. M. (1996), *J. Appl. Phys.* **80**, 295–300.

Petrova-Koch, V., Muschik, T., Polisski, G., Kovalev, D. (1995), in: R. W. Collins, C. C. Tsai, M. Hirose, F. Koch, L. Brus (Eds.), *Microcrystalline and Nanocrystalline Semiconductors*, MRS Symposium Proceedings Vol. 358, Pittsburgh: Materials Research Society, p. 483.

Petrova-Koch, V., Fischer, T., Sheglov, K., Koch, F. (1996), in: D. J. Lockwood, P. M. Fauchet, N. Koshida, S. R. J. Brueck (Eds.), *Advanced Luminescent Materials*, Perrington, NJ: The Electrochemical Society, p. 382.

Prasad, A., Balakrishnan, S., Jain, S. K., Jain, G. C. (1982), *J. Electrochem. Soc.*, **129**, 596.

Rao, P., Schiff, E. A., Tsybeskov, L., Fauchet, P. M. (1997), in: R. W. Collins, P. M. Fauchet, I. Shimizu, J.-C. Vial, T. Shimada, A. P. Alivisatos, (Eds.), *Advances in Microcrystalline and Nanocrystalline Semiconductors—1996*, MRS Symposium Proceedings Vol. 452, Pittsburgh: Materials Research Society, pp. 613–618.

Rehm, J. M., McLendon, G. L., Tsyebskov, L., Fauchet, P. M. (1995), *Appl. Phys. Lett.* **66**, 3669–3671.

Rehm, J. M., McLendon, G. L., Fauchet, P. M. (1996), *J. Am. Chem. Soc.* **118**, 4490–4491.

Ren, S. Y., Dow, J. D. (1992), *Phys. Rev. B* **45**, 6492.

Richter, A., Steiner, P., Kozlowski, F., Lang, W. (1991), *IEEE Electron Dev. Lett.* **12**, 691.

Sailor, M. J., Lee, E. J. (1997), *Adv. Mater.* **9**, 783–793.

Smith, R. L., Collins, S. D. (1992), *J. Appl. Phys.* **71**, R1.

Tamura, H., Ruckschloss, M., Wirschem, T., Veprek, S. (1994), *Appl. Phys. Lett.* **65**, 1537.

Tsao, S. S. (1987), *IEEE Circuit Dev. Mag.* **3**, 8.

Tsybeskov, L., Vandyshev, J. V., Fauchet, P. M. (1993), *Phys. Rev. B* **49**, 7821–7824.

Tsybeskov, L., Moore, K. L., Hall, D. G., Fauchet, P. M. (1996a), *Phys. Rev. B* **54**, R8361–R8364.

Tsybeskov, L., Duttagupta, S. P., Hirschman, K. D., Fauchet, P. M. (1996b), *Appl. Phys. Lett.* **68**, 2058–2060.

Tsybeskov, L., Hirschman, K. D., Duttagupta, S. P., Fauchet, P. M., Zacharias, M., McCaffrey, J. P., Lockwood, D. J. (1998), *Phys. Status Sol. (a)* **165**, 69–77.

Uhlir, Jr., A. (1956), *Bell Syst. Tech. J.* **35**, 333.

Vervoort, L., Bassani, F., Mihalcescu, I., Vial, J.-C., Arnaud d'Avitaya, F. (1995), *Phys. Status Sol. (b)* **190**, 123.

Vial, J.-C., Bsiesy, A., Fishman, G., Gaspard, F., Herino, R., Ligeon, M., Muller, F., Romestain, R., MacFarlane, R. M. (1993), in: P. M. Fauchet, C. C. Tsai, L. T. Canham, I. Shimizu, T. Aoyagi

(Eds.), *Microcrystalline Semiconductors: Materials Science & Devices*, MRS Symposium Proceedings Vol. 283, Pittsburgh: Materials Research Society, p. 241.

van Buuren, T., Tiedje, T., Dahn, J. R., May, B. M. (1993), *Appl. Phys. Lett.* **63**, 2911–2913.

von Behren, J., Chimowitz, E. H., Fauchet, P. M. (1997), *Adv. Mater.* **9**, 921–926.

von Behren, J., van Buuren, T., Zacharias, M., Chimowitz, E. H., Fauchet, P. M. (1998), *Solid State Commun.*, **105**, 317–32.

Wang, L. W., Zunger, A. (1994), *J. Phys. Chem.* **98**, 2158.

Wang, L.-W., Zunger, A. (1996), in: P. V. Kamat, D. Meisel (Eds.), *Nanocrystalline Semiconductor Materials*, Amsterdam: Elsevier Science.

Zaidi, S. H., Chu, A.-S., Brueck, S. R. J. (1995), in: R. W. Collins, C. C. Tsai, M. Hirose, F. Koch, L. Brus (Eds.), *Microcrystalline and Nanocrystalline Semiconductors*, MRS Symposium Proceedings Vol. 358, Pittsburgh: Materials Research Society, p. 957.

Zheng, J. P., Jiao, K. L., Shen, W. P., Anderson, W. A., Kwok, H. S. (1992), *Appl. Phys. Lett.* **61**, 459–461.

Further Reading

Canham, L. T. (1997), *Properties of Porous Silicon*, London: The Institution of Electrical Engineers.

Collins, R. W., Tsai, C. C., Hirose, M., Koch, F., Brus, L. (1995), *Microcrystalline and Nanocrystalline Semiconductors*, MRS Symposium Proceedings Vol. 358, Pittsburgh: Materials Research Society.

Collins, R. W., Fauchet, P. M., Shimizu, I., Vial, J.-C., Shimada, T., Alivisatos, A. P. (1997), *Advances in Microcrystalline and Nanocrystalline Semiconductors—1996*, MRS Symposium Proceedings Vol. 452, Pittsburgh: Materials Research Society.

Fauchet, P. M., Tsai, C. C., Canham, L. T., Shimizu, I., Aoyagi, Y. (1993), *Microcrystalline Semiconductors: Materials Science & Devices*, MRS Symposium Proceedings Vol. 283, Pittsburgh: Materials Research Society.

Lockwood, D. J. (1998), *Light Emission in Silicon: From Physics to Devices*, San Diego: Academic Press.

Lockwood, D. J., Fauchet, P. M., Koshida, N., Brueck, S. R. J. (1996), *Advanced Luminescent Materials*, Pennington, NJ: The Electrochemical Society.

Schmuki, P., Lockwood, D. J., Isaacs, H., Bsiesy, A. (1997), *Pits and Pores: Formation, Properties, and Significance for Advanced Luminescent Materials*, Pennington, NJ: The Electrochemical Society.

Tsu, R., Feng, Z. C. (1994), *Porous Silicon*, Singapore: World Scientific.

Vial, J.-C., Derrien, J. (1995), *Porous Silicon Science and Technology*, Berlin: Springer-Verlag.

SUN, STRUCTURE OF

MARC H. PINSONNEAULT, *Department of Astronomy, The Ohio State University, Columbus, Ohio, U.S.A.*

INTRODUCTION

The Sun is the foundation for the study of stars. Its surface and global properties can be determined to much higher precision than for other stars; traditionally these have been combined with theoretical models to deduce the internal properties of the Sun. The study of the Sun has been revolutionized, however, by the availability of diagnostics of its internal properties. Neutrino experiments provide direct evidence for nuclear reactions in the core of the Sun; the frequencies of acoustic waves in the Sun provide a means of determining the speed of sound and the rotation rate as functions of depth. In addition, the relative abundances of elements in meteorites can be measured precisely. Meteoritic data can provide information about the composition of the material that the Sun was formed from, and a comparison of the current surface abundances with the initial abundances can be used to test for changes in the surface abundances during the lifetime of the Sun. This wealth of data has

ISBN 3-527-29308-6

Table 1. Global properties of the sun.

Mass	1.9891×10^{33} g
Radius	6.9598×10^{10} cm
Luminosity	$3.844(1 \pm 0.004) \times 10^{33}$ erg s^{-1}
Age	4.57 ± 0.02 Ga
T_{eff}	5770 K

Table 2. Changes in global solar properties as functions of time.

Time (10^9 a)	Radius ($R_\odot$)	Luminosity ($L_\odot$)	Convection-zone depth ($M_\odot$)
0.0	0.868	0.673	0.0329
0.04	0.874	0.698	0.0310
0.27	0.883	0.724	0.0300
0.5	0.889	0.736	0.0291
1.0	0.900	0.763	0.0282
1.5	0.912	0.792	0.0282
2.5	0.937	0.850	0.0268
3.5	0.965	0.916	0.0252
4.57	1.000	1.000	0.0240

provided astronomers with a stringent test of the standard models of the structure of stars, as well as evidence that allows an assessment of the importance of physical processes usually neglected in theoretical stellar models.

Section 1 of this article summarizes the global and surface properties of the Sun. Section 2 introduces the main components of the theoretical models. In Section 3, the diagnostics of the internal properties of the Sun are discussed. The inferred internal structure of the Sun is discussed in Sec. 4. The results are summarized in Sec. 5.

1. GLOBAL AND SURFACE PROPERTIES

1.1 Mass, Radius, Luminosity, Surface Temperature, and Age

Table 1 summarizes some of the important global properties of the Sun (Lang, 1991). Note that these properties, especially age, are determined far more accurately than for stars in general. For example, the solar mass is inferred from the orbital periods of the planets, and the primary uncertainty is the value of the gravitational constant *G*. It is extremely difficult to infer directly the age of a single star such as the Sun. However, the time scale for the formation of the Sun, planets, and smaller bodies of the solar system is short compared with the age of the Sun. The age of the Sun is therefore most accurately determined by radioactive age dating of meteorites, and the primary uncertainty is the possibility of a small difference in age between the meteorites and the Sun (see Bahcall *et al.*, 1995; this article also discusses updated values for the solar luminosity and other properties of the Sun).

In general the radius and luminosity (total energy output per second) of the Sun will change over both long and short time scales. The effective temperature (T_{eff}) of the Sun is the surface temperature that a blackbody emitting the same flux per unit area would have. Radius (R), luminosity (L), and the effective surface temperature are related by

$$L = 4\pi R^2 \sigma {T_{\text{eff}}}^4, \tag{1}$$

where σ is the Stefan–Boltzmann constant. In Table 2 we show the time dependence of the solar radius and luminosity inferred from theoretical models (see Sec. 2). The long-term changes in the solar energy output are of interest for understanding the time evolution of the climate of the Earth. Shorter-term changes in brightness and radius can arise from phenomena such as the solar sunspot cycle (Sec. 1.4; see also SOLAR RADIATION).

1.2 Composition—Photospheric and Meteoritic

The composition of a star exerts a strong influence on its structure and evolution. The Sun is primarily composed of hydrogen ($\approx$0.70 by mass fraction), with most of the rest of the material in the form of helium ($\approx$0.28) and only about 0.02 in the form of heavier elements. Meteorites provide the most accurate means of measuring the relative abundances of heavy elements in the protosolar nebula. They do not, however, provide a means of determining the mass fractions of abundant volatile species, such as H, He, C, N, O, and Ne. In the solar photosphere we can measure the current surface mass fractions of different species, with the notable exception of ^{4}He, relative to hy-

Table 3. Abundances in the solar photosphere ($\log N_H = 12.0$).

Element	Photosphere	Meteorites
H	12.0	—
He	[10.99±0.035]	—
Li	1.16±0.1	3.31±0.04
Be	1.15±0.1	1.42±0.04
B	2.6±0.3	2.88±0.04
C	8.55±0.05	—
N	7.97±0.07	—
O	8.87±0.06	—
Ne	[8.08±0.06]	—
Na	6.33±0.03	6.31±0.03
Mg	7.58±0.05	7.58±0.02
Al	6.47±0.07	6.48±0.02
Si	7.55±0.05	7.55±0.02
P	5.45±0.04	5.57±0.04
S	7.21±0.06	7.27±0.05
Cl	5.5±0.3	5.27±0.06
Ar	[6.52±0.10]	—
K	5.12±0.13	5.13±0.03
Ca	6.36±0.02	6.34±0.03
Ti	5.02±0.06	4.93±0.02
Cr	5.67±0.03	5.68±0.03
Mn	5.39±0.03	5.53±0.04
Fe	7.50±0.04	7.51±0.01
Ni	6.25±0.04	6.25±0.02

drogen. These measurements, however, are not as accurate as those in the meteorites.

To infer the initial composition of the Sun, data from the two sources are combined. Meteorites are used to set the relative initial abundances of the species when possible, on a logarithmic scale relative to Si, where log(Si) = 6. The photospheric abundances are measured relative to hydrogen, on a logarithmic scale where log(H) = 12. The meteoritic scale is converted into the photospheric scale by measuring the ratios of the abundances on the two different scales for well-measured species (see Anders and Grevesse, 1989). The mass fractions of abundant species in the Sun from Anders and Grevesse (1989) are given in Table 3.

The initial ^{4}He abundance of the Sun is determined indirectly. In astronomy, the mass fractions of hydrogen, helium, and all other elements combined (referred to as "metals") are denoted by X, Y, and Z, respectively, and satisfy $X + Y + Z = 1$. The surface abundances give the ratio Z/X. Because the luminosity of a star depends on its initial composition, the requirement that a solar model has the luminosity of the Sun at the age of the Sun can be used to infer the initial ^{4}He mass fraction (Y). In addition, helioseismic data can be used to infer the current surface Y (Sec. 3.2).

The light elements Li and Be are significantly more abundant in meteorites than in the solar photosphere (by factors of 100 and 2, respectively). This constitutes evidence for mild mixing in the solar envelope, as these species burn at temperatures of 2.5 and 3.5 MK, respectively (Sec. 2.4.1). The current surface Y is also lower than the initial Y, which is evidence for the gravitational settling of helium out of the surface layers of the Sun (Sec. 2.4.2). Both theoretical models and the sound speed inferred from helioseismology require the burning of helium in the core of the Sun (Sec. 2, Sec. 3.2).

1.3 Surface Rotation and Oblateness

The Sun rotates at a rate that depends on the latitude at the surface; Libbrecht and Morrow (1991) give the surface rotation frequency as a function of latitude as

$$\nu(R_\odot, \phi) = 462 - 75\sin^2\phi - 50\sin^4\phi \text{ nHz}, \quad (2)$$

which implies an equatorial rotation period of about 25.4 d and a polar period of order 34 d. Helioseismic data can be used to infer the internal rotation as a function of depth (Sec. 3.2.2).

The measured solar oblateness is

$$\varepsilon = 8.63 \pm 0.88 \times 10^{-6} \quad (3)$$

where $\varepsilon = 2(R_{eq} - R_{pole})/(R_{eq} + R_{pole})$. This provides not only a measurement of the local distortion of the surface of the Sun, but an indication of the quadrupole moment and a constraint on the internal solar rotation. The distortion from spherical symmetry induces small corrections to the equations of stellar structure. We have evidence that the Sun is experiencing angular-momentum loss (Sec. 1.5), and young solar analogs rotate more rapidly than the present-day Sun. This distortion could therefore have been larger in the past. Rotation can also induce circulation currents and mixing, which could have a more significant impact on the structure of the Sun.

1.4 The Solar Sunspot Cycle and Magnetic Fields

The existence of sunspots has been known for centuries; they are regions with high magnetic field strengths of 2–3 kG and cover a fraction of the solar surface ranging up to about 0.001. Sunspots appear dark relative to the rest of the surface because they are significantly cooler than the rest of the photosphere; they are also surrounded by active regions that are hotter than the mean surface temperature (and therefore brighter). We have extensive records that indicate that the number of such spots varies with a period of order 22 years (including a reversal in polarity; the number of sunspots varies over an 11-year cycle). The flux per unit area is different in and around sunspots than for the rest of the solar surface; as a result, these changes in spot coverage can cause changes in the energy output of the Sun. The amplitude of the modulation of the total solar irradiance during the sunspot cycle is 0.08%. In some spectral regions the variation in the amplitude is higher (see SOLAR RADIATION). For an extensive discussion of the solar activity cycle, see the chapters in Sec. IV of Cox *et al.* (1991).

The sunspot cycle is intimately linked with the generation of the solar magnetic field. The dynamo theory postulates that differential rotation in the solar interior leads to an amplification of the internal magnetic field strength. At some point the fields become strong enough to be subject to magnetic instabilities, and magnetic flux can then rise to the surface. In the outer layers ($R > 0.713R_\odot$) of the Sun, energy is transported by mass motions (convection), with an overturn time scale of the order of a month. The most likely site of the solar dynamo is therefore below the surface convection zone, where the magnetic field can grow over a longer time scale than is permitted in the convection zone.

1.5 Solar Chromosphere, Corona, and Wind

Above the solar photosphere there are three different regions. The Sun has a chromosphere hotter than the photosphere, a very hot and extended corona, and a steady stream of material escaping from the Sun referred to as the solar wind. The measured solar mass loss rate is $3 \times 10^{-14} M_\odot/\mathrm{a}$, which implies that only a small fraction of the Sun's mass has been lost over the 4.5 Ga of its lifetime.

The heating mechanism for the chromosphere and corona has been a subject of some controversy. Turbulent motions in the solar atmosphere can generate acoustic waves, which increase in velocity as the density decreases outwards until they become shock waves. Alfvén waves may also contribute to the heating of the chromosphere. The temperature in the chromosphere rises from the minimum at the top of the photosphere of 4500 K to a maximum temperature of 20 000 K. There is then a transition region above the chromosphere where the temperature rises to of order 10^6 K in the corona. The properties of the chromosphere and corona change significantly during the course of the solar sunspot cycle.

2. THEORY OF STELLAR STRUCTURE AND EVOLUTION

Much of our understanding of the interior of the Sun comes from theoretical models of its structure constrained by the surface and global properties outlined in Sec. 1. In this section the ingredients of a theoretical stellar model are described. In Sec. 4 the internal structure of a theoretical solar model is described and compared with the diagnostics of the internal properties of the Sun described in Sec. 3. We begin by outlining the equations of stellar structure and then comment on the input physics, method for constructing a solar model, and some additional physical processes not usually included in standard solar models.

2.1 Basic Equations and Assumptions

The basic equations of stellar structure are expressions of the conservation of mass, energy, and momentum, along with a prescription for the transport of energy. The continuity equation gives

$$\frac{\mathrm{d}M}{\mathrm{d}r} = 4\pi\rho r^2 \tag{4}$$

where $M(r)$ is the mass, r is the radius, and

$\rho(r)$ is the density. Hydrostatic equilibrium can be expressed as

$$\frac{dP}{dr} = -\rho g \tag{5}$$

where $P(r)$ is the pressure and $g = GM/r^2$ neglecting departures from spherical symmetry. The high mass of the Sun implies that a very large internal pressure gradient is required to balance the force of gravity. Energy generation requires

$$\frac{dL}{dr} = 4\pi\rho r^2 \varepsilon, \tag{6}$$

where $L(r)$ is the luminosity (erg s^{-1}) and $\varepsilon(r)$ is the energy generation rate (which includes nuclear energy generation, energy loss from neutrino emission, and changes in gravitational potential energy).

The temperature gradient is the minimum of that for energy transport by radiation and that for energy transport by mass motions (convection); for transport by radiation,

$$\frac{dT}{dr} = -\frac{3}{16\pi ac}\frac{\kappa L}{r^2 T^3}, \tag{7}$$

where $T(r)$ is the temperature, c is the speed of light, a is the radiation constant, and $\kappa(T, \rho, \text{composition})$ is the opacity of matter to radiation.

These equations must be supplemented by an equation of state that relates the pressure, temperature, and density; a convection theory that determines the temperature gradient in regions where energy is carried by mass motions; and a prescription for changes in abundance induced by nuclear reactions as a function of time. The standard model assumes spherical symmetry, neglects magnetic fields and mass loss, and does not include either mixing processes linked with rotation or gravitational settling of heavy elements with respect to lighter ones.

2.2 Input Physics

In Fig. 1 the energy generation rate ε and the opacity κ are shown as functions of fractional radius in a model of the present-day Sun. Energy generation is primarily through nuclear reactions in the Sun and is highly temperature sensitive. The final luminosity of the Sun, however, is set by the balance between energy generation and energy transport. In the inner part of the Sun, the opacity of matter to radiation is low and energy transport is primarily by radiation. When the opacity is too high material becomes convectively unstable; the strong increase in opacity in the outer layers of the Sun indicates that energy transport is primarily through convection above $0.713R_\odot$. The nuclear-reaction rates, equation of state and energy transport in the solar interior are reviewed below.

2.2.1 Nuclear-Reaction Rates The conversion of hydrogen into helium is the primary energy source for the Sun [see Bahcall (1989) for a discussion of nuclear-reaction rates relevant for the Sun; cross sections are summarized in Bahcall and Pinsonneault (1992) and Bahcall *et al.* (1995)]. In most cases the cross sections for the relevant reactions can be measured experimentally; because the typical energies in the Sun (≈ 1 keV at the center) are much lower than the regime which is accessible experimentally (>100 keV), the experimental results are extrapolated to the low-temperature regime. There are two principal mechanisms for hydrogen burning. In the proton–proton (pp) chain,

$$^1\text{H} + {}^1\text{H} \rightarrow {}^2\text{H}, \quad {}^1\text{H} + {}^2\text{H} \rightarrow {}^3\text{He}, \tag{8}$$

and either

$$^3\text{He} + {}^3\text{He} \rightarrow {}^4\text{He} + 2\,{}^1\text{H} \text{ (85\% of the time)} \tag{9}$$

or

$$^3\text{He} + {}^4\text{He} \rightarrow {}^7\text{Be} \text{ (15\% of the time)}. \tag{10}$$

The ^{7}Be then either experiences an electron capture or a proton capture, and is ultimately processed into two ^{4}He nuclei. Some of these reactions produce neutrinos that can be detected experimentally. This chain accounts for 98% of the energy output of the Sun.

In the carbon–nitrogen–oxygen (CNO) cycle the C, N, and O nuclei are used as cata-

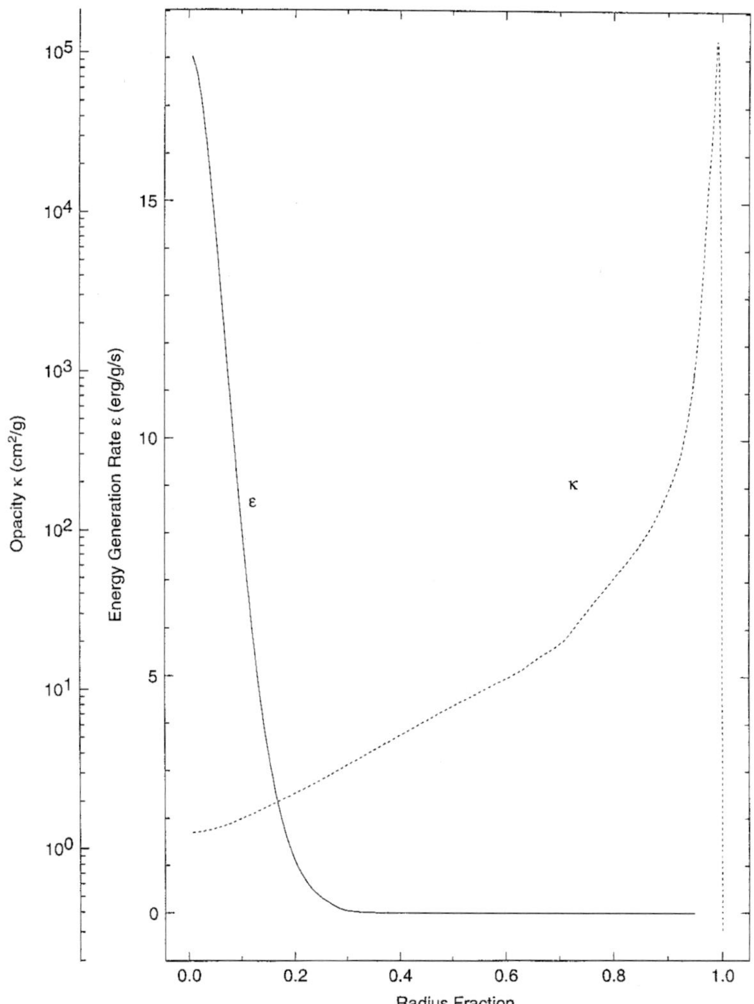

FIG. 1. The energy generation rate (solid line, first linear scale) and opacity (dashed line, second logarithmic scale) as functions of the fractional radius in a theoretical solar model.

lysts, and hydrogen is converted into helium through a series of proton captures. These reactions are more temperature sensitive than the pp chain and predominate in stars more massive than the Sun (accounting for only 2% of the solar luminosity). They produce some neutrinos that can be detected experimentally. The Sun is not hot enough to burn He and heavier elements.

2.2.2 Energy Transport The mean free path of a photon in the solar interior is of the order of 1 cm. Because the Sun is much hotter at the center (≈ 15.6 MK) than it is at the surface (≈ 5770 K), the photons produced at a given radius will on average be more energetic than those produced at a slightly larger radius; when these photons interact with matter there will be a net transport of energy outwards. As a result of the small length scale, the transport of energy by radiation can be treated as a diffusion process. Theoretical models rely on quantum mechanical calculations of the opacity of matter to radiation, such as the recent work of Iglesias and Rogers (1996), to compute the temperature gradient needed to carry the solar flux via radiation. If we denote the actual temperature gradient by $\mathrm{d}\ln T/\mathrm{d}\ln P =$

∇, then convection takes over as the dominant mode of energy transport whenever the radiative temperature gradient ∇_{rad} is greater then the adiabatic temperature gradient ∇_{ad}. The mode of energy transport switches from radiative to convective near $0.713R_{\odot}$; there is a discontinuity in the speed of sound as a function of depth at this point that can be detected through the study of solar oscillations (Sec. 3.2). The temperature gradient in the convective region of the Sun is determined using mixing-length theory. In the deeper layers of the solar convection zone, the actual temperature gradient is close to the adiabatic gradient; it can depart significantly from the adiabatic gradient in the outermost layers of the Sun.

2.2.3 Equation of State The bulk of the mass of the Sun is in the form of hydrogen and helium, which ionize at low temperatures; most of the heavier elements are in the form of C, N, and O, which are fully ionized at modest temperatures of order 1 MK. As a result, the interior of the Sun is essentially fully ionized and is also close to an ideal gas. An important quantity for the solar equation of state is the mean molecular mass μ, defined by

$$P = \rho \Re T/\mu \tag{11}$$

where $\Re$ is the gas constant.

The mean molecular mass in the solar interior changes in regions of partial ionization. For the purposes of computing the pressure, free electrons are as important as ions; for example, there will be twice as many particles per gram in an ionized hydrogen gas as in a neutral hydrogen gas. There are more particles per gram in fully ionized hydrogen (2) than there are in fully ionized helium (3/4) or heavier elements ($\approx 1/2$), and so composition changes can affect the equation of state; nuclear burning and gravitational settling in the solar interior will therefore also cause changes in μ as a function of radius and time. Modern equations of state include relativistic effects, electron degeneracy, radiation pressure, and electrostatic corrections. In the outer layers, the partial ionization of H, He, and metals needs to be computed. Partial ionization is important for the thermal structure in the outer layers because of the strong dependence of the number of free particles per gram on the ionization state of the matter. Other thermodynamic properties, such as the adiabatic temperature gradient (see Sec. 2.2.2), are also strongly affected by partial ionization. Most modern solar models use tabulated equations of state that include the above effects; examples include the OPAL equation of state (Rogers *et al.*, 1996).

2.3 Constructing a Calibrated Solar Model

Stellar evolution is an initial-value problem. A model with the mass of the Sun is evolved to the age of the Sun. A calibrated model reproduces the solar surface Z/X, radius, and luminosity at the age of the Sun. The initial Y of the Sun is adjusted to reproduce the solar luminosity, while the mixing length α (the distance in pressure scale heights that a fluid element travels before releasing its heat) in the convective portion of the Sun is adjusted to reproduce the solar radius. The inferred initial Y depends on the other input physics, but is typically in the range 0.27–0.29 by mass fraction.

2.4 Physical Processes not in the Standard Solar Model

Standard solar models do not incorporate rotation and element-separation processes. However, there is evidence that these phenomena have a measurable impact on the observed properties of the Sun; solar and stellar models that include rotation and element separation have been constructed.

2.4.1 Rotation Rotation has a direct effect on the structure of the Sun; in addition, it can induce mixing from meridional circulation and hydrodynamic instabilities triggered by differential rotation with depth. A full treatment of rotation in the Sun includes the effect of rotation on the structure of the Sun, the initial angular momentum in the protostellar phase, angular-momentum loss via the solar wind, and internal angular-momentum transport. Solar models with rotation can explain the surface light-element abundances of the Sun, but rotation has only a modest impact on the thermal structure of the solar models (Richard *et al.*, 1996). In other areas of stellar evolution rotational mixing can produce dramatic observational effects.

2.4.2 Microscopic Diffusion Heavy elements sink with respect to lighter ones in the presence of gravity; in addition there is a net downward flux of heavier elements in the presence of a temperature gradient. These diffusion effects are included in recent solar models; they result in a lowering of the surface abundances of elements heavier than hydrogen, a deepening of the surface convection zones in the models, and an increase in the inferred central temperature and the expected solar neutrino fluxes.

3. DIAGNOSTICS OF THE INTERNAL PROPERTIES OF THE SUN

The global and surface properties of the Sun can be used to infer its internal properties in conjunction with the theoretical models described in Sec. 2. However, there are also diagnostics of the internal properties of the Sun. Neutrinos are produced in nuclear reactions in the solar core; because of their weak interaction with matter, they can pass through the Sun and be detected on the Earth. Acoustic waves in the Sun also provide information about the speed of sound and rotation as a function of depth.

3.1 Solar-Neutrino Fluxes

Some of the nuclear reactions that occur in the Sun emit neutrinos. Because the cross section of neutrinos in matter is extremely low, these neutrinos escape from the Sun and can be detected in terrestrial experiments. Different reactions emit neutrinos with different energies, and there are now four experiments, which are sensitive to neutrinos with different energies. The resulting neutrino spectrum can be compared with that predicted by theoretical models. The detection of solar neutrinos constitutes evidence that nuclear reactions from hydrogen burning are the predominant solar energy source.

However, the observed fluxes are significantly less than the predicted rates. This persistent discrepancy between theory and experiment is the *solar neutrino problem*, and its solution involves either significant modifications of the theoretical solar models or new neutrino physics. In Sec. 4.6, the observed and predicted fluxes are compared. Here we briefly review the processes that produce solar neutrinos. Bahcall (1989) contains a detailed description of the different solar neutrino experiments.

The different neutrino experiments currently under way include the following:

1. The Homestake experiment is a radiochemical experiment that observes neutrinos via the reaction

$$\nu_e + \mathrm{Cl}^{37} \rightarrow e^- + \mathrm{Ar}^{37}. \qquad (12)$$

The radioactive Ar^{37} is extracted from a C_2Cl_4 fluid. The threshold energy is 0.8 MeV.
2. The Kamiokande experiment is based on neutrino–electron scattering in water. The direction of the neutrinos can be inferred in this experiment, but it has a very high threshold neutrino energy of 7.5 MeV. A scaled-up version of the Kamiokande experiment (Super-Kamiokande) is currently in progress; it will permit far better statistics and a detailed analysis of the neutrino energy spectrum.
3. The GALLEX and SAGE experiments detect the absorption of neutrinos in gallium via the reaction

$$\nu_e + \mathrm{Ga}^{71} \rightarrow e^- + \mathrm{Ge}^{71}. \qquad (13)$$

The threshold energy for this process is low (0.23 MeV), making it possible to detect the majority of the neutrinos produced in the Sun. These are radiochemical experiments, similar to the Homestake experiment.
4. The SNO experiment, currently underway, uses heavy water to observe Cherenkov light from electrons interacting with deuterium in addition to the neutrino–electron scattering to which the Kamiokande experiment is sensitive. The charged-current reaction

$$\nu_e + \mathrm{H}^2 \rightarrow e^- + p + p \qquad (14)$$

is sensitive to the electron neutrinos produced by the Sun, while the neutral-current reaction

$$\nu + H^2 \rightarrow \nu + p + n \qquad (15)$$

is sensitive to other neutrino flavors. This could provide an important test of some theories of new neutrino physics, which explain the solar neutrino deficit by postulating that some of the electron neutrinos produced in the Sun are converted into other flavors as they travel from the center of the Sun to the Earth.

The major processes that produce neutrinos in the Sun are as these:

1. The pp reaction produces a low-energy neutrino. The GALLEX and SAGE experiments are sensitive to this neutrino, as well as those from the other reactions listed below.
2. The pp reaction can be mediated with an electron (the pep reaction). All of the experiments except Kamiokande can detect pep neutrinos.
3. After ^{3}He is produced, there are two branches of the pp chain. ^{3}He + ^{3}He occurs 85% of the time and does not produce a neutrino; ^{3}He + ^{4}He occurs 15% of the time and has two possible terminations, each of which produces a neutrino (^{7}Be and ^{8}B). The ^{8}B neutrino is highly energetic and can be detected in all of the experiments; Kamiokande and SNO are sensitive only to the ^{8}B neutrinos. The other experiments can also detect some of the lower-energy ^{7}Be neutrinos.
4. The Sun can also process hydrogen into helium by using the C, N, and O nuclei as catalysts; these reactions are responsible for about 2% of the Sun's luminosity and produce neutrinos that can be detected by GALLEX, SAGE, and the Homestake experiment.

The cross section for neutrino capture is extremely small; for the Homestake experiment the measured capture rate is $2.55 \pm 0.17(\text{stat}) \pm 0.18(\text{syst})$ solar neutrino units (SNUs), which are units of 10^{-36} events per target atom per second. Detecting such rare events is technically challenging (see Bahcall, 1989, for a discussion of the experimental techniques). By comparison, the theoretical fluxes are of the order of 8–9 SNUs (Sec. 4.3.3). Different experiments are sensitive to neutrinos of different energies and the results of different experiments can therefore be combined to form a neutrino energy spectrum of the Sun. One of the most interesting developments in the solar neutrino problem has been the apparent strong dependence of the discrepancy between theory and observation on neutrino energy; this feature is difficult to explain by changing the theoretical fluxes alone. The comparison between theory and observation is discussed in Sec. 4.6.

3.2 Solar Oscillations

Solar oscillations can be observed through either intensity fluctuations or Doppler shifts on the solar surface as a function of position and time. The amplitudes are at the centimeter per second level, with characteristic periods of the order of 5 min. These fluctuations arise because internal waves in the Sun can be refracted by the strong density gradient in the outer layers of the Sun. Therefore, the Sun acts as an acoustic cavity, and there will be constructive interference for certain characteristic frequencies. The frequencies ($\approx 10^7$ modes of oscillation) with which it oscillates can be used to deduce a variety of internal properties. The restoring force for the solar oscillations detected to date is pressure, and so they are sometimes referred to as p-mode oscillations. These modes have their maximum amplitude in the outer layers, making them an ideal diagnostic of the outer layers of the Sun. Waves where the restoring force is gravity (g modes) have not yet been unambiguously identified; because they have their maximum amplitude in the core, a detection of solar g modes would provide important tests of the conditions at the center of the Sun. An extensive discussion of the observational and theoretical studies of the solar oscillations, referred to as helioseismology, can be found in Part II of Cox *et al.* (1991).

3.2.1 Oscillations as a Diagnostic of the Thermal Structure Expressed in terms of spherical harmonics (n, l, m), the depth to which a given p mode will penetrate depends strongly on l: lower-l modes can penetrate further into the Sun. Information from these modes can be combined to form a map of the speed of sound as a function of depth in the solar interior. The errors are largest in the deep interior because there are few p modes that penetrate very deeply, and the frequencies of these low-l modes have proven difficult to measure. However, high-quality data for the low-l p modes (and therefore good information on the physical conditions in the deep interior of the Sun) have recently become available (Tomczyk *et al.*, 1995). The sound speed inferred in the solar interior is compared with theoretical predictions in Sec. 4.

3.2.2 Internal Rotation In the presence of rotation, there is a splitting of the p-mode frequencies that can be used to infer the internal solar rotation as a function of latitude and depth. The rotation in the deep solar interior is currently uncertain, but there is good information on the rotational properties outside the core of the Sun.

4. INTERNAL STRUCTURE OF THE SUN

The internal properties of the Sun that are described in this section are inferred from a theoretical solar model (Sec. 2) that satisfies the global properties of the Sun as described in Sec. 1. Throughout this section theory will be compared with data on the internal properties of the Sun (Sec. 3); current theoretical models are in good agreement with solar oscillation data but conflict with solar-neutrino data. We begin with an overview of the major components of the internal structure of the Sun in Secs. 4.1–4.3; a detailed discussion of the different layers of the solar interior is then presented in Secs. 4.4–4.6.

4.1 Mass and Luminosity Distributions

The mass fraction M and luminosity L of the Sun as functions of radius are shown in Fig. 2. The Sun is highly centrally concentrated; 60% of its mass is contained within $0.3R_\odot$, 90% of the mass is contained within $0.5R_\odot$, and 98% of the mass is contained within $0.7R_\odot$. The central density of $156\,\mathrm{g\,cm^{-3}}$ is far higher than the mean density of $1.4\,\mathrm{g\,cm^{-3}}$. The energy generation of the Sun is concentrated in the deep interior; 46% of the luminosity is generated within $0.1R_\odot$, 95% of the energy is generated within $0.2R_\odot$, and effectively all of the energy (with the exception of energy from gravitational contraction) is generated within $0.4R_\odot$. Significant conversion of hydrogen into helium is also concentrated in the inner $0.2R_\odot$ (see Sec. 4.3 below). In the sections that follow, this will be referred to as the solar core.

4.2 Thermal Structure

The pressure, temperature, and density of the Sun as functions of radius are shown in Fig. 3. Note that the temperature declines much less rapidly than the density, which in turn declines much less rapidly than the pressure. The speed of sound as a function of depth inferred from seismology is shown in Fig. 4(a); the fractional difference between the sound speed of the Sun and that of the theoretical model discussed in this article is shown in Fig. 4(b). Theory agrees with observation to an accuracy of 0.2% throughout the solar interior. The deviations from the predictions of the model are interesting and will be discussed below.

The actual, adiabatic, and radiative temperature gradients are compared in Fig. 5. Energy is carried by radiation in the interior; in the outer layers energy is carried by convection (conduction is not an important mode of energy transport within the Sun). The time scale for convection is short compared with the solar age, which implies that the outer layers of the Sun are fully mixed; the interior can experience composition gradients due to nuclear burning and gravitational settling, and these gradients can be reduced by mixing. We can therefore divide the outer layers of the Sun into a surface convection zone and a radiative mantle above the nuclear burning core.

There is a change in the derivative of the speed of sound as a function of depth at the base of the surface convection zone. The base of the convection zone inferred from helioseismology is $0.713 \pm 0.003R_\odot$. Theoretical models that include the latest radiative opac-

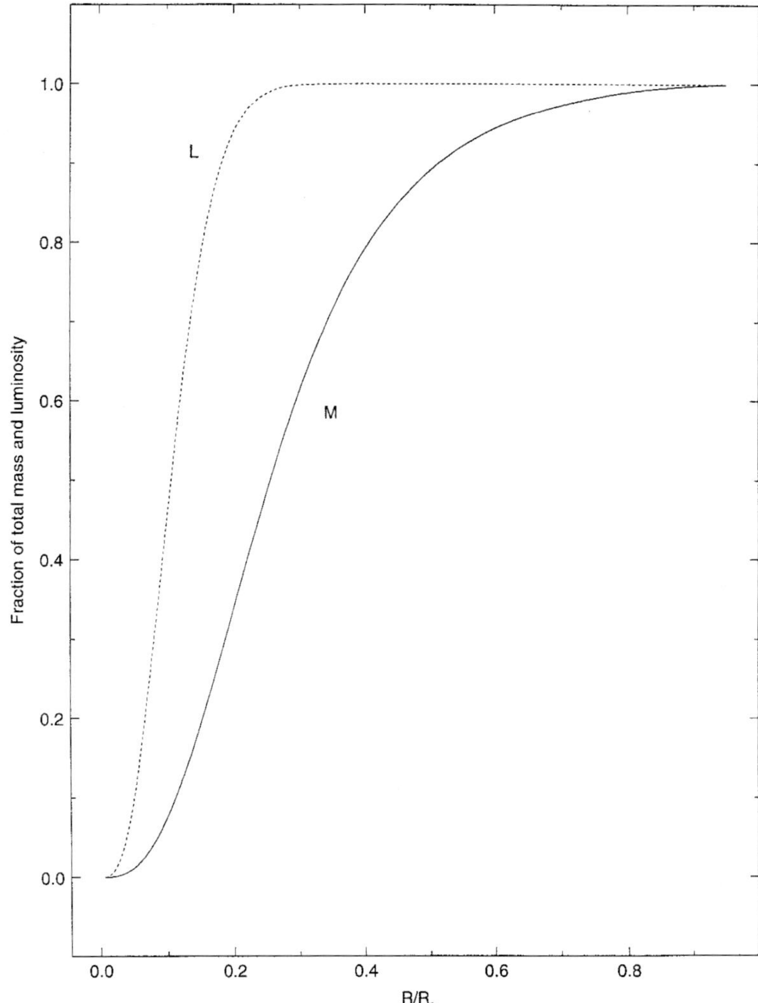

FIG. 2. The fractions of the total luminosity (dashed line) and mass (solid line) as functions of radius fraction in a theoretical solar model.

ities and the gravitational settling of helium are in good agreement with the measured depth of the surface convection zone.

4.3 Abundance Profiles

The hydrogen and metal abundances as functions of radius are shown in Fig. 6. The strong decrease in the central hydrogen abundance within $0.2R_{\odot}$ is caused by hydrogen burning; the increase in hydrogen in the surface layers is a consequence of the gravitational settling of ^{4}He. Because the Sun is not burning elements heavier than ^{4}He, the radial profile of the metal abundance shows the effect of gravitational settling of heavy elements from an initial mass fraction of 0.02. There is a 10% drop in the surface metal abundance and a 5% increase in the central metal abundance that result from settling.

Different species burn at different temperatures; the abundances as functions of radius of ^{3}He and the major isotopes involved in the CNO cycle are shown in Fig. 7. The small increases in abundance of C, N, and O below the convection zone are caused by diffusion (diffusion of ^{3}He was not included in this calcula-

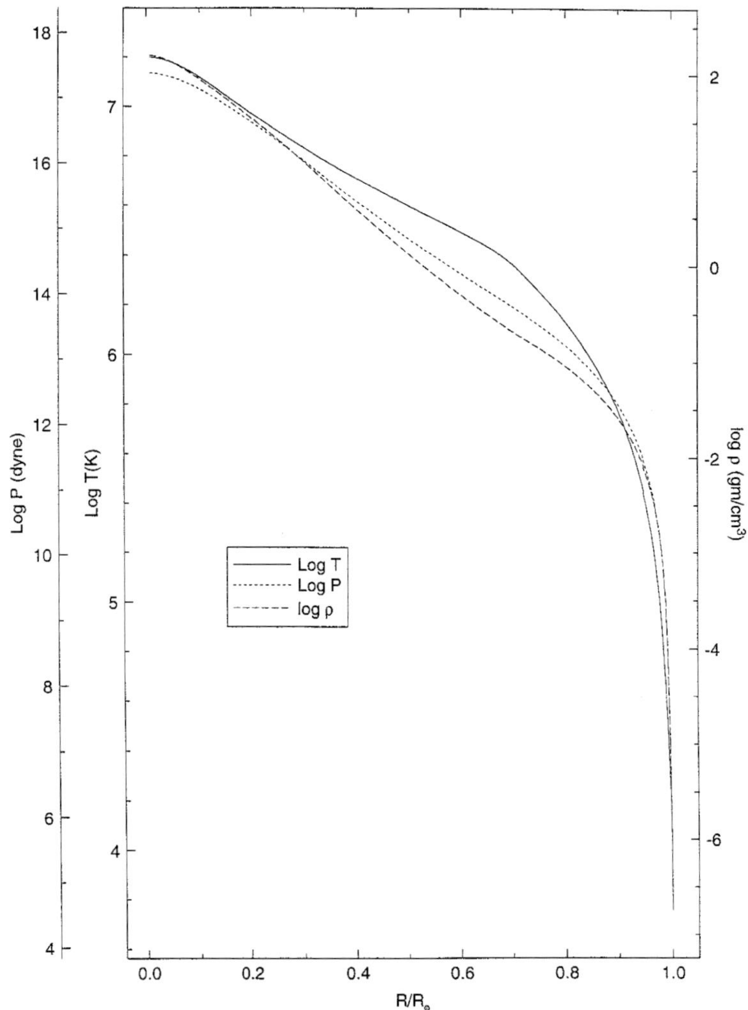

FIG. 3. Pressure, temperature, and density on a logarithmic scale as functions of radius fraction in a theoretical solar model.

tion). The peak in the abundance of ^{3}He at $0.26R_{\odot}$ and the peak in the abundance of ^{13}C at $0.18R_{\odot}$ are caused by nonequilibrium nuclear burning. ^{12}C burns into ^{13}C at a lower temperature than ^{13}C burns into ^{14}N; there will therefore be a region where ^{13}C is produced more efficiently than it is destroyed. A similar phenomenon occurs for ^{3}He. The strong increase in N and decrease in C interior to $0.15R_{\odot}$ occurs because the surface C/N ratio (3) is different from the C/N nuclear equilibrium ratio of 1/200. As the temperature rises, the rate of the reactions that destroy ^{3}He rise faster than the rate of the reaction that produces ^{3}He; as a result, the ^{3}He abundance decreases dramatically in the solar core. There is a slight depletion of O and increase in N in the very central region from partial processing of oxygen into nitrogen, which occurs only at relatively high temperatures.

We discuss the detailed properties of the three major layers of the Sun (convective envelope, radiative mantle, and nuclear burning core) separately below.

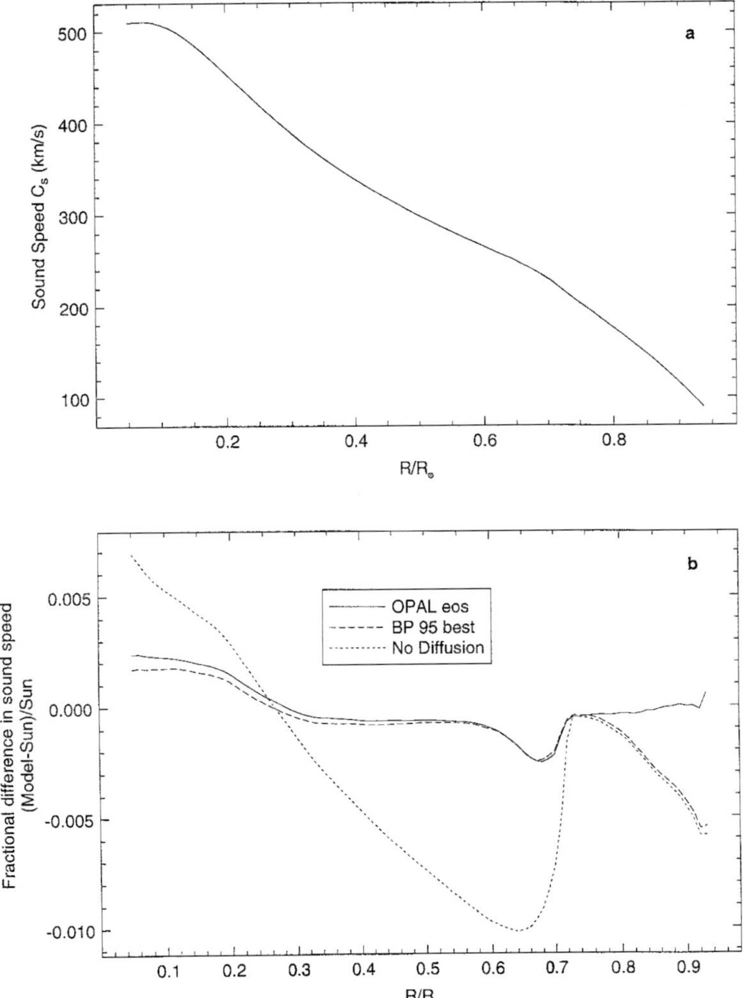

FIG. 4. (a) The sound speed as a function of radius fraction that is inferred from helioseismology; the outer layers ($R > 0.94R_\odot$) are not shown. Data are from S. Basu and J. Christensen-Dalsgaard (personal communication). (b) The difference between the real Sun and different theoretical models. The short-dashed line is the standard model of Bahcall and Pinsonneault (1992), which does not include gravitational settling; the long-dashed line is the model of Bahcall *et al.* (1995), which includes gravitational settling; the solid line is the same model with the improved OPAL equation of state (Rogers *et al.*, 1996).

4.4 Convection Zone

The opacity of matter to radiation is high at low temperatures and in regions where ionization is occurring; both conditions characterize the outer layers of the Sun. The ionization structure of the outer layers of the Sun is illustrated in Fig. 8. In this region, energy is transported through mass motions (convection). Because the time scale for energy transport via convection is short (of the order of a month) compared with the lifetime of the Sun (4.57 Ga) the convection zone is fully mixed. Typical convective velocities are of order $1\,\mathrm{km\,s^{-1}}$ and gradually decline with increased depth.

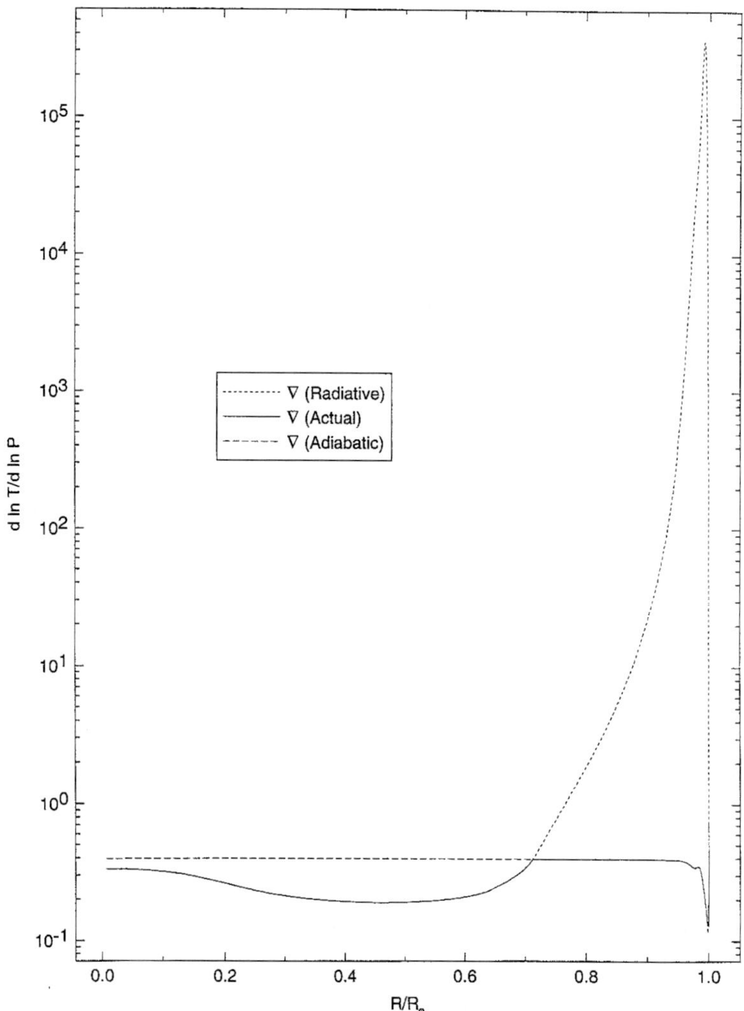

FIG. 5. Actual, radiative, and adiabatic temperature gradients as functions of radius fraction in a theoretical solar model. The actual and radiative gradients are identical for $R < 0.713R_{\odot}$; the actual gradient is very close to the adiabatic gradient for $R > 0.713R_{\odot}$ except for a thin layer near the surface (see Fig. 9).

4.4.1 Superadiabatic Layer In the outer layers of the Sun there is a transition from mostly neutral H and He at the surface to full ionization. In this region the adiabatic temperature gradient drops and the actual temperature gradient differs significantly from the adiabatic value (Fig. 9). In this superadiabatic region the model properties are sensitive to the treatment of convection. In the lower portion of the convection zone the superadiabatic temperature gradient is small.

4.4.2 Surface versus Initial Abundances The surface mass fraction of ^{4}He inferred from helioseismology is 0.242 ± 0.003, while the initial Y found from solar models ranges from 0.27 to 0.29. This is evidence for mild gravitational settling of ^{4}He; theoretical calculations predict a comparable degree of settling for heavier elements ($\approx 10\%$).

The surface abundances of both Li and Be are significantly below their meteoritic values, and the base of the convection zone is not deep

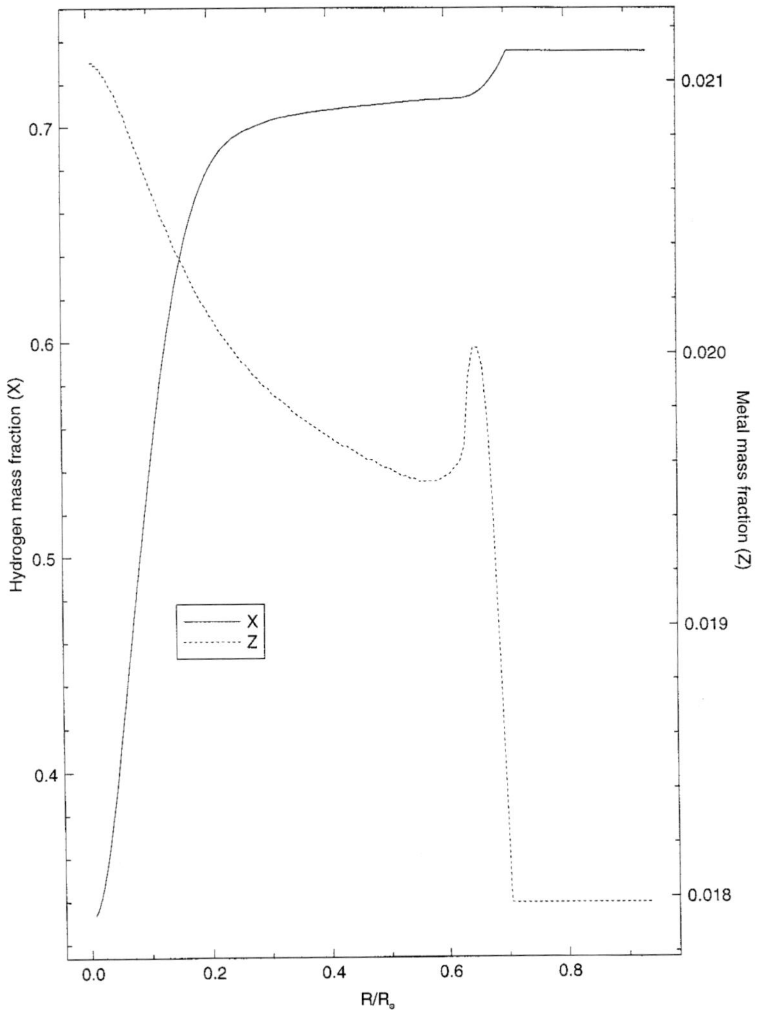

FIG. 6. Mass fractions of hydrogen (solid line, left axis) and all elements heavier than helium (dashed line, right axis) as functions of radius fraction in a theoretical solar model.

and hot enough to burn either. Although stars like the Sun do experience mild Li depletion early in their lifetime, they are observed to have abundances much higher than the current solar photospheric value. The expected degree of gravitational settling for Li is of of order of that found for He, and the degree expected for Be is smaller. A mild mixing mechanism, in addition to convection, is therefore required to explain the surface light element abundances, Models that include mixing induced by rotation are capable of reproducing the observed surface abundances of both Li and Be.

4.4.3 Rotation as a Function of Depth and Latitude The seismic inversions indicate that the surface differential rotation as a function of latitude persists throughout the surface convection zone. The rotation as a function of depth at a given latitude is nearly constant through the surface convection zone. There is a narrow region at the base of the surface

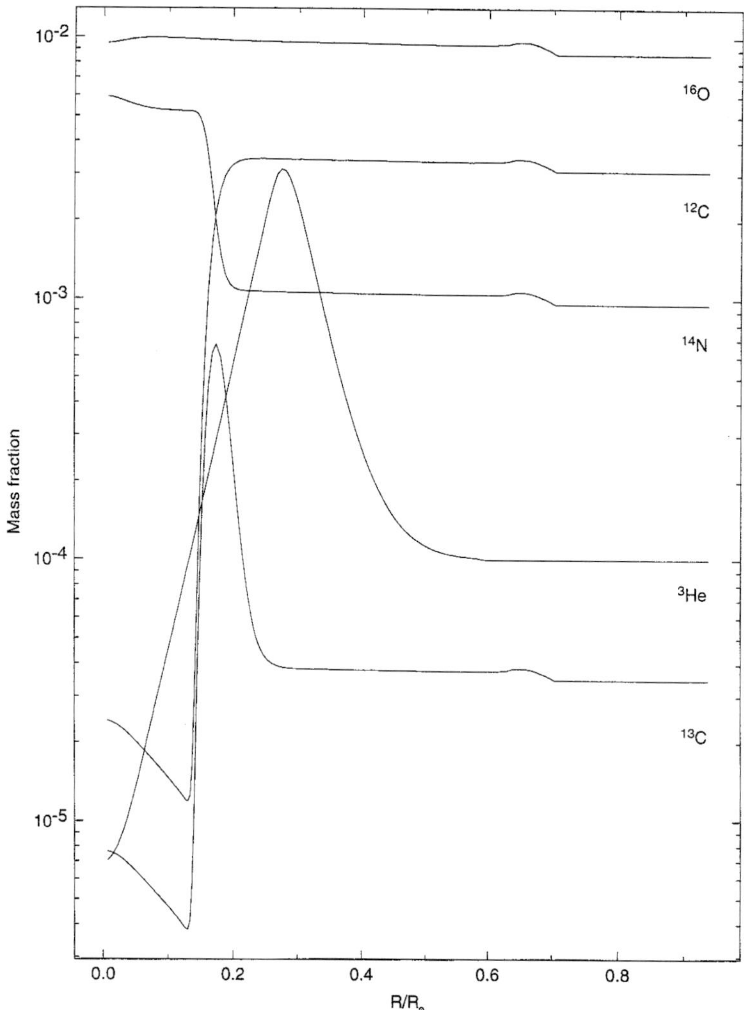

FIG. 7. Mass fractions of ^{3}He and the important C, N, and O isotopes on a logarithmic scale as functions of radius fraction in a theoretical solar model.

convection zone where differential rotation as a function of latitude vanishes; the depth of this region is currently unresolved. The layer below the surface convection zone is the likely seat of the solar dynamo, responsible for the sunspot cycle and the generation of the Sun's magnetic field.

4.5 Radiative Mantle

Below $0.713R_{\odot}$ the Sun is stable against convection. Nuclear energy generation does not become significant until about $0.2R_{\odot}$; the intervening region can be considered to be a radiative mantle surrounding the nuclear-burning core.

4.5.1 Evidence for Microscopic Diffusion and Mixing The depth of the solar convection zone and the surface He abundance inferred from helioseismology both require gravitational settling of helium in the solar envelope. Models with only diffusion, however, predict a sharp change in the mean molecular mass below the surface convection zone, which is not seen; this produces the discrepancy between the model and the Sun seen

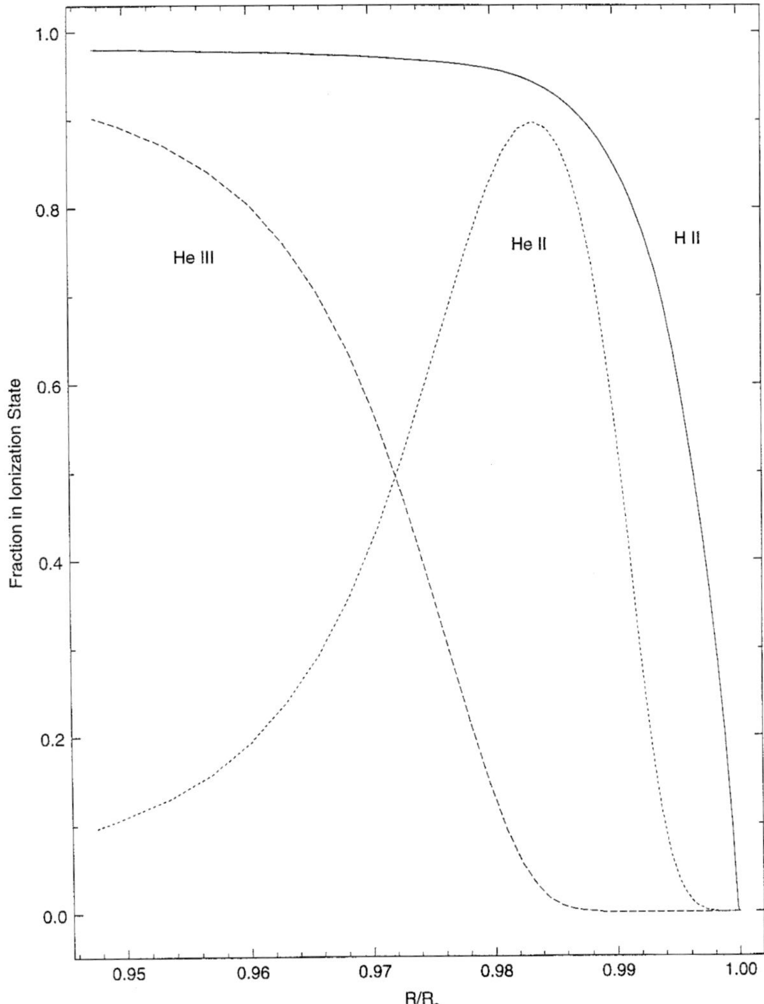

FIG. 8. The fractions of hydrogen that is fully ionized (H II, solid line), helium that is singly ionized (He II, short-dashed line) and helium that is doubly ionized (He III, long-dashed line) as functions of radius fraction in the outer layers of a theoretical solar model.

in Fig. 4(b). In addition, the surface abundances of lithium and beryllium require a slow mixing mechanism. The survival of these light elements places strong constraints on how much mixing can have occurred, and calculations indicate that mixing consistent with the surface abundances cannot entirely remove the effects of gravitational settling. Combined with the data on the solar rotation curve, this implies that the mantle of the Sun is the site of both slow separation and mixing processes not present in standard stellar models.

4.5.2 Rotation as a Function of Depth

There is general agreement that the rotation of the Sun depends neither on latitude nor on depth between a transition region near the base of the convection zone and $\approx 0.4R_\odot$. The shear at the base of the solar convection zone is a likely site for the solar dynamo that gives rise to the Sun's magnetic field. There have been contradictory claims, ranging from a mild decrease to rigid rotation to increased rotation in the region between $0.2R_\odot$ and $0.4R_\odot$. The most recent inversions of seismic data

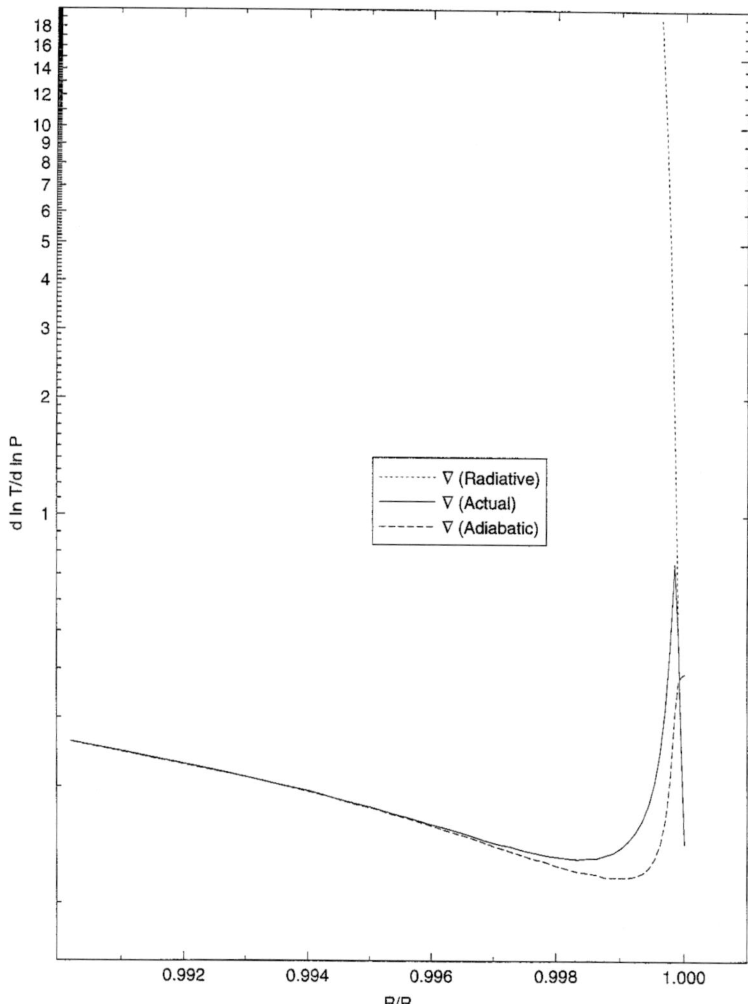

FIG. 9. An enlarged portion of Fig. 5, showing the behavior of the adiabatic, radiative, and actual temperature gradients near the solar surface as functions of radius fraction. The actual temperature gradient starts to depart significantly from the adiabatic gradient at 0.996; the radiative gradient is off the scale before 0.9985 and drops sharply with increased radius, becoming less than the adiabatic gradient just below the solar atmosphere.

favor essentially solid-body rotation down to $0.2R_{\odot}$ (Tomczyk *et al.*, 1995). Because the Sun is experiencing angular-momentum loss via the solar wind, this implies that the time scale for angular-momentum transport in the radiative core is less than the age of the Sun.

4.6 Nuclear-Burning Core

The deep core is the site of the Sun's energy generation. Conditions in this region are tested directly by the solar neutrino experiments, which are in disagreement with theory. The helioseismic data, however, are in much better agreement with the theoretical models in the core. Because large changes in the solar-model fluxes would required large changes in the thermal structure of the solar core, this severely restricts the allowed classes of solar-model solutions for the solar neutrino problem.

4.6.1 Thermal Structure Nuclear burning has altered the composition and temperature of the core of the Sun since its birth. To first order, the local sound speed scales as $(T/\mu)^{1/2}$. The agreement between observation and theory is at the 0.2% level in the core, which would require finely tuned changes in the temperature and composition to preserve the close observed match between the two.

4.6.2 Abundances and Energy Generation as Functions of Depth A series of different nuclear reactions produce energy in the Sun. Because nuclear reactions are strongly temperature sensitive, the reaction rates are usually a maximum at the center of the Sun. The differing temperature dependences of the individual reactions, however, imply that the degree of central concentration of the energy generation will differ from one reaction to the next. This in turn is reflected in the detailed abundance profiles seen in Figs. 6 and 7. In later phases of evolution the convection zone of Sunlike stars becomes much deeper and incorporates partially processed material from the core into their surface convection zone where it can be seen. Surface abundance changes from this dredge-up have been seen and are in fairly good agreement with theoretical predictions.

4.6.3 Predicted Neutrino Fluxes The neutrino fluxes from the different experiments are in contradiction with those predicted by theoretical models. The measured thermal structure is in good agreement with theoretical models; this implies that a large reduction in the neutrino fluxes predicted by these models is unlikely. The most likely explanation is that the Sun is producing neutrinos in the amount expected from the theoretical models, but they are being converted from electron neutrinos (which can be detected) into other kinds of neutrinos, which cannot be measured by the current generation of experiments. If there is a nonzero neutrino mass, electron neutrinos can be converted into other flavors between the region where they are produced and the Earth. The SNO experiment, which is currently under way, is sensitive to both neutral- and charged-current reactions. If the electron neutrino oscillates into a μ or τ neutrino, SNO should be able to provide direct experimental confirmation of this phenomenon. Below we compare the theoretical predictions and the observational data.

4.6.3.1 Kamiokande. The measured flux of ^{8}B neutrinos (with energies above 7.5 MeV) is $[3.0 \pm 0.41(\mathrm{stat}) \pm 0.35(\mathrm{syst})] \times 10^6\,\mathrm{cm^{-2}\,s^{-1}}$. This compares with a theoretical prediction of $6.6(1.0^{+0.14}_{-0.17}) \times 10^6\,\mathrm{cm^{-2}\,s^{-1}}$. The observed flux is thus only about 45% of the predicted value.

4.6.3.2 Homestake. The Homestake experiment is sensitive to both the higher-energy ^{8}B neutrinos and those from other sources, such as ^{7}Be and the pep reaction. It is not sensitive to the lowest-energy neutrinos arising directly from the pp reaction. The measured flux (in units of 1 SNU, or 10^{-36} interaction per target atom per second) is $2.55 \pm 0.17(\mathrm{stat}) \pm 0.18(\mathrm{syst})$ SNU; the theoretical prediction is $9.3^{+1.2}_{-1.4}$ SNU. This discrepancy is particularly difficult to explain with standard models in conjunction with the Kamiokande data discussed above. The measured Kamiokande flux can be converted into a minimum flux expected for the Homestake experiment, which leaves little or no room for any contribution from neutrinos produced by other reactions to which Homestake is sensitive.

4.6.3.3 GALLEX/SAGE. The GALLEX and SAGE experiments are sensitive to all of the neutrinos detected by the Homestake and Kamiokande experiments, plus the low-energy pp neutrinos. The measured fluxes from GALLEX and SAGE are, respectively, $79 \pm 10(\mathrm{stat}) \pm 6(\mathrm{syst})$ SNU and $69 \pm 10(\mathrm{stat}) \pm 6(\mathrm{syst})$ SNU. This compares with a theoretical prediction of 137^{+8}_{-7} SNU, which is relatively weakly model dependent. Of this total about 70 SNU arises directly from the pp reaction; the magnitude of this component is fixed by the measured energy output of the Sun.

5. SUMMARY

The Sun can be divided up into three distinct layers: the nuclear-burning core, a radiative mantle, and a surface convection zone. The speed of sound as a function of depth provides strong constraints on theoretical models, and the overall agreement between theory and observations is at a level (0.2% in the sound speed) not typically found in astronomy. The observed solar neutrino fluxes do not agree with those expected from theoretical models; the most likely explanation is a nonzero neutrino mass. A detailed comparison of model properties with the solar data provides evidence for element-separation processes and mild envelope mixing below the surface convection zone.

GLOSSARY

Adiabatic Temperature Gradient (∇_{ad}): The logarithmic derivative of temperature with respect to pressure for a fluid element that does not exchange heat with its surroundings. If the radiative temperature gradient exceeds the adiabatic temperature gradient in a given region, the region is unstable with respect to convection and energy is transported by mass motions rather than by photon diffusion.

CNO Cycle: A series of nuclear reactions in which carbon, nitrogen, and oxygen nuclei are used as catalysts for the conversion of hydrogen into helium. These reactions account for about 2% of the solar energy output.

Convection Zone: A region where energy is transported by mass motions; this occurs in the outer layers of the Sun.

g Mode: A mode of oscillation in the Sun where the restoring force of the wave is gravity; these modes have not been definitively identified in the Sun, but are predicted to have a characteristic period of 3 h.

Helioseismology: The study of the internal structure of the Sun deduced from the properties of nonradial oscillations observed at the solar surface.

Mixing-Length Theory: A simplified theory of convective energy transport used extensively in theoretical solar models, in which a characteristic length scale α is assumed and the characteristic velocities of fluid elements and the temperature gradient are then calculated. The value of the mixing length α is usually adjusted to reproduce the solar radius at the solar age.

Opacity: A measure of the mean free path of photons in the presence of matter; a typical length scale is 1 cm in the Sun. Astrophysical calculations use a weighted average over photon frequency, called the Rosseland mean opacity, and solve a diffusion equation to infer the temperature gradient needed to carry a given flux.

p Mode: An acoustic mode of oscillation in the Sun where the restoring force of the wave is pressure; these modes have a characteristic period of 5 min.

pp Chain: A series of nuclear reactions in which protons combine directly to form the intermediate species deuterium and helium-3, followed by three different branches that produce helium-4. These reactions produce 98% of the solar energy output.

Radiative Core: A region where energy is transported by photon diffusion; this occurs in the inner layers of the Sun.

Radiative Temperature Gradient (∇_{rad}): The logarithmic derivative of temperature with respect to pressure that would be required to carry the local flux by the diffusion of photons.

Works Cited

Anders, E., Grevesse, N. (1989), *Geochim. Cosmochim. Acta* **53**, 197–214.

Bahcall, J. N. (1989), *Neutrino Astrophysics*, Cambridge, UK: Cambridge Univ. Press.

Bahcall, J. N., Pinsonneault, M. H. (1992), *Rev. Mod. Phys.* **64**, 885–926.

Bahcall, J. N., Pinsonneault, M. H., Wasserburg, G. J. (1995), *Rev. Mod. Phys.* **67**, 781–808.

Cox, A. N., Livingston, W. C., Matthews, M. S. (Eds.) (1991), *Solar Interior and Atmosphere*, Tucson: Univ. of Arizona Press.

Iglesias, C. A., Rogers, F. J. (1996); *Astrophys. J.* **464**, 943–953.

Lang, K. R. (1991), *Astrophysical Data: Planets and Stars*, New York: Springer-Verlag.

Libbrecht, K. G., Morrow, C. A. (1991), in: A. N. Cox, W. N. Livingston, M. S. Matthews (Eds.), *Solar Interior and Atmosphere*, Tucson: Univ. of Arizona Press.

Richard, O., Vauclair, S., Charbonnel, C., Dziembowski, W. A. (1996), *Astron. Astrophys.* **312**, 1000–1011.

Rogers, F. J., Swenson, F. J., Iglesias, C. A. (1996), *Astrophys. J.*, **456**, 902–908.

Tomczyk S., Schou J., Thompson M. J., (1995). *Astrophys. J. Lett.*, **448**, L57–L60.

Further Reading

Bahcall, J. N. (1989), *Neutrino Astrophysics*, Cambridge, UK: Cambridge Univ. Press. A discussion of the solar neutrino problem, including the experiments and nuclear physics relevant to the solar problem. Updated review articles on the current status of the solar models are listed in Bahcall and Pinsonneault (1992) and Bahcall *et al.* (1995).

Clayton, D. (1983), *Principles of Stellar Evolution and Nucleosynthesis*, Chicago: Univ. of Chicago Press. A discussion of the physics appropriate for the interiors of stars.

Cox, A. N., Livingston, W. C., Matthews, M. S. (1991), *Solar Interior and Atmosphere*, Tucson: Univ. of Arizona Press. An extensive compilation of review articles on all aspects of the Sun, both observational and theoretical.

Kippenhahn, R., Weigart, A. (1990), *Stellar Structure and Evolution*, New York: Springer-Verlag. An introduction to stellar interiors, along with a discussion of the evolution of stars. A resource for putting the Sun in the broader context of astronomy.

SUPERFLUIDITY: LIQUID HELIUM SYSTEMS

SHAUN N. FISHER AND GEORGE R. PICKETT, *School of Physics and Chemistry, Lancaster University, Lancaster, United Kingdom*

INTRODUCTION

The two common isotopes of helium, ^{3}He and ^{4}He, are unique in remaining liquid down to absolute zero temperature. At low temperatures the liquid phases of both isotopes become superfluid. The most apparent property of a superfluid is its ability to flow without friction. This leads to the possibility of non-decaying persistent currents flowing indefinitely around loops. However, on a more fundamental level, the really unique aspects of superfluidity lie in the influence of the underlying quantum mechanics on the macroscopic behavior of the liquid.

Superfluidity was first observed in ^{4}He just before the second World War, and the effect was elucidated and named by Kapitza in Moscow in 1942. The superfluid transition temperature in liquid ^{4}He is close to 2 K. The superfluid transition in liquid ^{3}He does not occur until much lower temperatures, about 1 mK, and was first observed in 1971 by Osheroff, Richardson, and Lee, who received the Nobel Prize for this work in 1996.

The properties of the two helium isotopes are almost identical at ordinary temperatures; they are inert gasses with very weak interatomic forces. Nevertheless, their superfluid states are very different. The differences arise as a direct consequence of the nuclear spin, which is zero for ^{4}He and $\frac{1}{2}$ for ^{3}He. Even though the interatomic forces are almost identical for the two isotopes, quantum mechanics

ISBN 3-527-29308-6

dictates that liquids at very low temperatures will behave very differently depending on the net intrinsic spin of the constituent particles.

Superfluidity, unlike the analogous superconductivity seen in some metals and conducting ceramics, has as yet found very few practical applications outside low-temperature research. The real impact of the superfluid state is the simplicity with which it can be described. Quantum mechanics describes a superfluid, consisting of countless numbers of atoms, as a single macroscopic object (wave function). These systems therefore provide a very rich laboratory for testing ideas from other areas of physics, even extending as far as particle physics and cosmology.

1. WHAT IS SUPERFLUIDITY?

The two isotopes of helium, the common isotope ^{4}He and the lighter isotope ^{3}He, are so light and so weakly interacting that their behavior is dominated by quantum mechanics much more than that of heavier elements. As a result, helium in the liquid phase is often described as a quantum fluid.

The laws of thermodynamics dictate that a system becomes more ordered as the temperature is reduced. A typical example is the solidification of a substance (such as the freezing of water) as the temperature is reduced below the melting point. The level of order of the system is increased on solidification, since in the liquid the component atoms or molecules are randomly distributed in space, whereas in the solid they take up fixed positions relative to each other and in general form a regular ordered crystalline structure.

The property of superfluidity in helium follows from the fact that helium is unique in remaining liquid down to absolute zero temperature and will only solidify if a pressure of some 30 atm is applied. It is the unusually strong influence of quantum mechanics on helium that prevents solidification. To form a solid, each atom must be confinable in its place in the crystal. However, a particle may only be confined within a region of space at the expense of increasing its energy (the zero-point energy). For helium, which has weak interatomic forces, this energy is large enough to overcome the attractive forces between the atoms that would otherwise lead to solidification at a sufficiently low temperature.

While liquid helium cannot increase its order by becoming solid, there are other ways in which it may order. In fact, at some low temperature the atoms in the liquid begin to fall into the same-lowest energy quantum mechanical state. It is this process that leads to superfluidity. The term was first coined by Kapitza to describe the properties of liquid ^{4}He.

Once a significant fraction of the atoms occupy the same quantum mechanical state, the behavior becomes quite different from what we intuitively expect for normal liquids. For example, there is no viscosity. Viscosity implies the process whereby atoms are mutually scattered (with each atom changing its state) and information about velocity inhomogeneities is gradually transmitted through the liquid. If all the atoms are in the same state in the first place, then such scattering processes cannot occur; *i.e.* the viscosity is zero. With zero viscosity, flows of the liquid can be set up without any pressure gradient. Furthermore, in the absence of scattering processes, these currents may, in principle, continue indefinitely. It is this behavior that gave rise to the term superfluidity in analogy with the term superconductivity, which describes the closely related behavior of the electron gas in certain metals.

1.1 Bosons, Fermions

While both isotopes of helium (^{4}He and ^{3}He) become superfluid, the mechanism by which the ordering takes place is very different for the two isotopes.

All particles have an intrinsic angular moment or spin. Quantum mechanics distinguishes between two types of particle depending on the magnitude of their spin. These are known as fermions and bosons and have very different properties. Fermions are particles having a half-integral spin quantum number, whereas bosons have an integral spin quantum number. The building blocks of matter (neutrons, protons and electrons) are all fermions, whereas particles that carry information about forces—for example, photons (telling about the motion of charges)—are bosons. The distinction applies also to composite objects. An object made up of an even number of

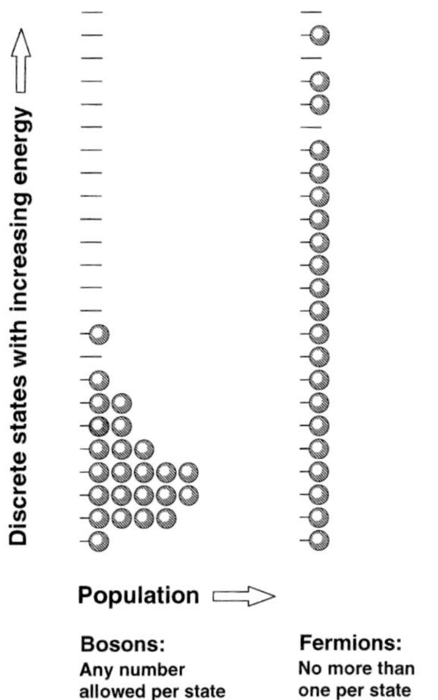

FIG. 1. The filling of energy states by bosons and fermions. Any number of identical bosons may fill any one state whereas only one fermion is allowed per state.

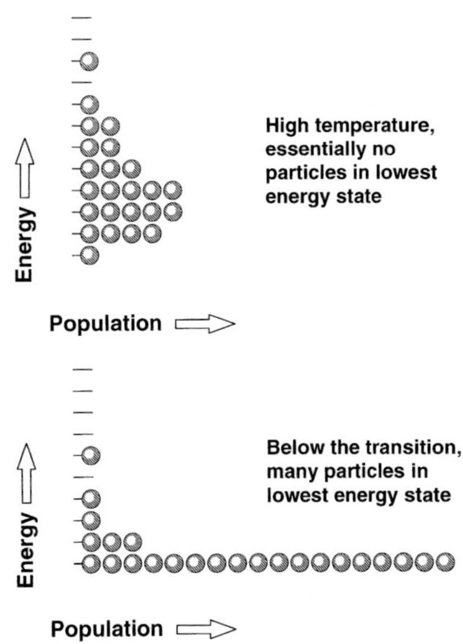

FIG. 2. Bose–Einstein condensation; below a critical temperature a macroscopic number of particles begins to populate the lowest energy state.

fermions becomes a boson (because the total spin in now integral), while an object constructed of an odd number of fermions remains a fermion.

A ^{4}He atom is made up of two electrons, two protons and two neutrons. The ^{4}He atom is therefore a boson (with spin 0). In contrast, ^{3}He is made from two electrons, two protons but only one neutron, and therefore the atom (with spin $\frac{1}{2}$) is a fermion.

The quantum mechanical distinction between these classes of particle comes into play when we have several particles occupying the same space. As shown schematically in Fig. 1, fermions are only allowed to occupy a quantum mechanical state singly, *i.e.*, no more than one particle per state. Bosons, on the other hand, face no such restriction, and there is no limit to how many particles may be in the same state with identical energies and momenta.

1.2 Bose–Einstein Condensation

At high temperatures, an assembly of bosons will be distributed over a large number of states. As the temperature is reduced the particles take up a distribution centered on lower and lower energies until at some temperature particles start filling the lowest state or ground state. This process is known as the Bose–Einstein condensation, and the temperature at which it begins is the Bose–Einstein condensation temperature (T_{BE}). This process is illustrated in Fig. 2. When a macroscopically large number of the particles occupy a single state the level of order in the system increases. In this case, unlike in a crystal, the order is not in terms of position but in terms of momentum, since the particles collect in the lowest state where the momentum is zero.

A system of bosons which are all in the same state displays some very unusual properties. If all the bosons are identical in every sense then they must behave as a single "object". It is meaningless to think of the system in terms of distinct individual particles since (according to quantum mechanics) it is fundamentally impossible to distinguish one particle from another. In consequence, while the bosons are governed by the laws of quantum mechanics only on the microscopic scale for temperatures higher than T_{BE}, at lower temperatures, once there is a macroscopic occupation of the lowest state, the bosons coalesce into a single macroscopic "object" represented by a single wave function. The system is still governed by quantum mechanics, but now operating on the entire macroscopic sample. Such an assembly of particles described by a common wave function is known as a coherent state.

1.3 Cooper Pairing

Direct Bose–Einstein condensation is impossible for a system of independent fermions since no more than one fermion may occupy a single state. Indeed, at moderately low temperatures a system of weakly interacting fermions has all the low-lying single-particle states singly occupied up to a cutoff energy called the Fermi energy. Nevertheless, liquid ^{3}He undergoes a transition into a superfluid state and also the conduction electrons of many metals undergo an analogous transition into a superconducting state. The superfluid state clearly exhibits properties associated with a coherent quantum state and therefore must represent an ordering related to Bose–Einstein condensation but in a system of fermions.

The solution to the problem was provided by Bardeen, Cooper, and Schrieffer (BCS) in 1957 (Bardeen *et al.*, 1957) who found that an attraction between two fermions with energies close to the Fermi energy will always lead to the formation of bound pairs at sufficiently low temperatures. These pairs are known as Cooper pairs. While the work of BCS was in the context of superconductivity, their ideas are equally applicable to ^{3}He atoms. The pairs prefer to form from fermions with equal and opposite momenta such that the combined momentum of the pair is zero. Furthermore, since an even number of fermions is involved, each pair is itself a boson. Therefore at sufficiently low temperatures, any attractive force between the fermions will lead to the production of Cooper-pair bosons, which are free to form a coherent ground state. This is, of course, a variant of Bose–Einstein condensation in which the bosons are formed from pairs of fermions.

The binding energy of the Cooper pairs ensures that the pairs have a lower energy than the constituent unbound particles. This means that there must be a minimum energy difference between an unpaired particle and those particles which are paired. This is known as the energy gap, usually represented by Δ. When a Cooper pair is broken, at least two unpaired particle excitations must be produced, needing a minimum energy of 2Δ.

2. SUPERFLUID ^{4}He

2.1 Cooling Methods

To reach the temperature at which liquid ^{4}He becomes superfluid at around 2 K requires only quite modest equipment. All that is needed is a simple insulating glass Dewar flask that may be filled with liquid ^{4}He and a modest vacuum pump to reduce the pressure of the vapor phase to reduce the temperature. A suitable system is shown in Fig. 3. It is particularly convenient for experimentalists that the superfluid transition can be reached simply by the reduction of the pressure over the liquid phase.

Liquid ^{4}He in equilibrium with its vapor at

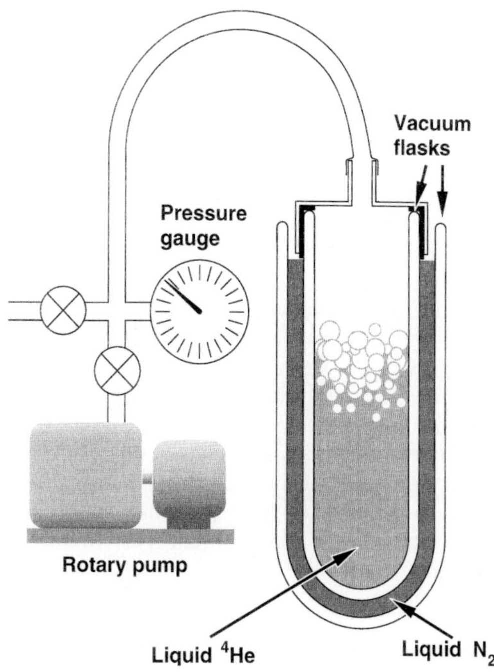

FIG. 3. A simple arrangement for reaching the superfluid transition in liquid ^{4}He.

one atmosphere sits at a temperature of 4.2 K. If the pressure above the liquid is slowly reduced the liquid will cool from the extraction of the latent heat of evaporation. At the modest vapor pressure of around 0.05 atm the temperature of the liquid–vapor mixture has fallen to 2.2 K, at which point the transition to superfluidity occurs.

The unique behavior of the superfluid state can already be appreciated with the naked eye even with such a modest arrangement. As the liquid is cooled to the transition temperature, it boils violently. A striking feature of the phase transition is the abrupt cessation of boiling once the superfluid state has been entered. The process of boiling arises from temperature gradients in the fluid. In a pan of boiling water, vapor bubbles will form in the hottest regions of the water at the heated surface of the pan. The abrupt disappearance of boiling in liquid ^{4}He below the transition suggests that temperature gradients must be greatly reduced in the superfluid phase. Clearly something very unusual has happened to the thermal properties, since the thermal conductivity of the superfluid phase must be extremely high. Vapor must still evaporate from the superfluid when the pressure is further reduced, but the vapor now forms at the free surface of the liquid rather than on the walls of the container. Many properties of superfluids are similarly unusual and quite at variance with what would be expected for a normal liquid.

2.2 The Superfluid Mechanism

The basic mechanism for superfluidity in ^{4}He is the Bose–Einstein condensation of the ^{4}He atoms. The calculated "ideal" Bose Einstein condensation temperature falls within a factor of two of the measured superfluid transition temperature. However, the real situation is more complex than the ideal case. The ^{4}He atoms cannot be considered weakly interacting, since the superfluidity occurs in the liquid phase where interactions are obviously important. They are in any case responsible for the liquid state itself. Therefore it is not surprising that an ideal Bose–Einstein condensation provides only a very qualitative description of superfluidity in ^{4}He.

Nevertheless, the superfluid properties of ^{4}He are certainly due to a boson condensate. The nonideality of the superfluid ^{4}He condensate is, in many ways, responsible for many of the more interesting superfluid properties. For example, neutron-scattering experiments (Campbell, 1983; Sears *et al.*, 1982) show that even at very low temperatures, the fraction of ^{4}He atoms in the zero-momentum state is only about 15% compared with the ideal Bose-gas prediction of 100%. However, as one consequence of the interatomic forces, the superfluid ground state incorporates a mixture of momentum states.

2.3 The Lambda Transition

Since the superfluid state is highly ordered, the entropy of the liquid must fall rapidly below the transition, and this should give rise to an anomaly in the heat capacity. The measured heat capacity plotted against temperature is shown in Fig. 4. The form of the heat capacity is reminiscent of the Greek letter λ.

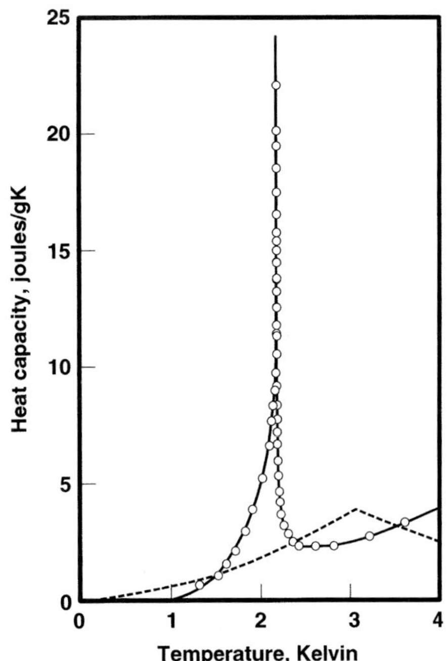

FIG. 4. The heat capacity of liquid ^{4}He showing the extremely steep lambda transition. The heat capacity for an ideal Bose–Einstein condensation in a gas of similar density is shown by the dashed line. (After Buckingham and Fairbank, 1961.)

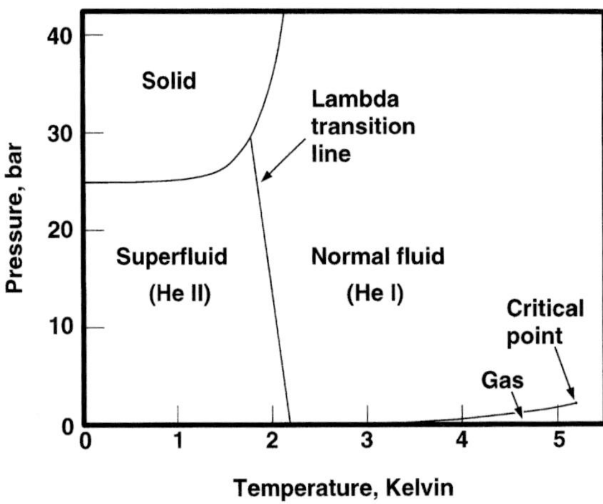

FIG. 5. The phase diagram for ^{4}He. Since liquid extends to zero temperature, there is no triple point. The designations He I for the normal liquid above the transition, He II for the superfluid phase, are often used.

Consequently the superfluid transition in ^{4}He is commonly known as the lambda transition. There is a logarithmic singularity at the transition in which the heat capacity rises dramatically and approaches infinity. This indicates the differences from the ideal Bose-gas case as indicated in Sec. 2.2 above, since the ideal Bose-gas heat capacity should display a cusp at the transition temperature as shown by the dashed curve in Fig. 4.

2.4 The Phase Diagram

The phase diagram for ^{4}He at low temperatures is shown in Fig. 5. Quite apart from the superfluidity itself, this is a very unusual system. We see, for example (as discussed in Sec. 1), that the liquid state extends all the way to zero temperature. This being so, there is clearly no triple point. Furthermore, the level of order in both the liquid and solid phases at low temperatures is virtually complete. In other words, the entropy of both phases is very small. The superfluid phase is often referred to as helium II, in contrast to the normal liquid, helium I.

2.5 The Two-Fluid Model

At absolute zero temperature, all the ^{4}He atoms may be considered to be in their lowest-energy state and therefore contribute to the superfluid properties, since, by definition, at absolute zero there are no thermal excitations. At higher temperatures there will be a finite number of excitations. In an ideal noninteracting boson system, these excitations simply represent those particles not in the ground state. However, in the case of liquid ^{4}He, which is strongly interacting, the excitations represent collective modes of the liquid involving many atomic states. These are phonons and, at higher energies, rotons, which are unique to liquid ^{4}He. At a microscopic level we cannot distinguish between atoms in the ground state and those contributing to the excitations. However, that having been said, it is very convenient to consider the liquid below the transition as consisting of a mixture of two freely interpenetrating fluid components, namely the superfluid and a normal fluid. The superfluid component has zero entropy and zero viscosity, whereas the normal component has a finite entropy and viscosity. This is the two-fluid model. The density (ρ) of the fluid can be separated into two terms:

$$\rho = \rho_s + \rho_n, \tag{1}$$

where ρ_s represents the superfluid density and ρ_n the normal-fluid density. At the lambda temperature, ρ_s is zero while $\rho_n = \rho$. As the temperature is reduced ρ_s increases and ρ_n decreases until at absolute zero $\rho_s = \rho$ and $\rho_n = 0$. Although we have noted that it is meaningless to label any particular atom as belonging to either component, nevertheless the two-fluid model, as we show below, is very successful in explaining many of the superfluid

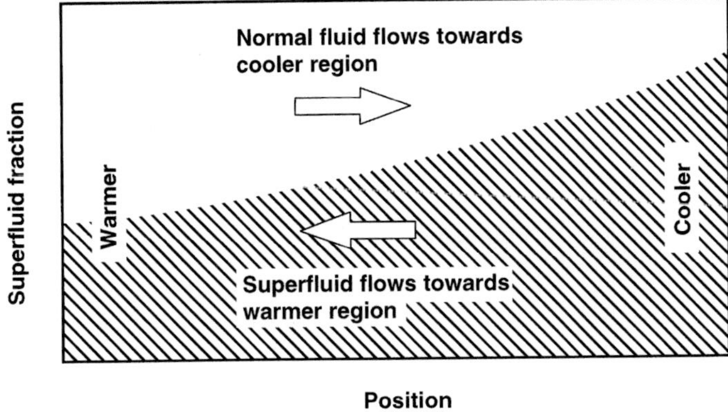

FIG. 6. The counterflow of normal and superfluid ^{4}He in a temperature gradient, giving a high thermal conductance just below the transition.

properties and is a very convenient concept for describing the behavior.

2.5.1 Hydrodynamic Heat Flow We can now explain the virtually infinite thermal conductivity of the superfluid state as a natural consequence of two-fluid behavior. The superfluid component has zero entropy and therefore cannot conduct heat. Since the ratio of the amounts of normal fluid and superfluid depends on temperature, a temperature gradient imposed on the superfluid phase must be associated with a gradient in the superfluid-to-normal fluid ratio such that there is a higher density of the normal fluid towards the warmer regions of the sample. The concentration gradient gives rise to an osmotic pressure, which drives the normal fluid towards lower temperatures and the superfluid towards higher temperatures (*i.e.*, the osmotic pressure acts to restore thermal equilibrium), as shown in Fig. 6. The counterflow of the two components leads to an extremely effective "hydrodynamic" heat flow in which the heat is carried bodily by the flow of normal fluid. The effective thermal conductivity is thus extremely high, in analogy with the heat transfer by the two-phase (but much less efficient) counterflow in a heat pipe.

2.5.2 Thermomechanical Effect The high thermal conductance is just one manifestation of an important consequence of two-fluid behavior, which is the linking of mechanical and thermal behavior. This has no analogies with the behavior of normal fluids. In these thermomechanical (and mechanothermal), effects a temperature gradient gives rise to a pressure gradient (and *vice versa*). Because of the very large thermal conductivity it is difficult to maintain a significant temperature gradient in the bulk fluid. However, it is easy to establish a temperature difference between two volumes of fluid that are connected by a device that only allows the passage of the superfluid component (analogous to a semipermeable membrane). Such a device, called a "superleak", consists of an array of very narrow passages, usually those threading a porous medium, such that the normal component is very effectively clamped by its finite viscosity, while the zero-viscosity superfluid component is free to pass. When a temperature difference is maintained between the two volumes, there must be a corresponding difference in the concentrations of the superfluid component on the two sides of the superleak. The osmotic pressure gradient will force some of the superfluid component through the superleak into the hotter of the two volumes leading to a hydrostatic pressure difference. The flow will continue until in equilibrium the hydrostatic pressure difference balances the osmotic pressure difference. Conversely, if a hydrostatic pressure gradient is applied across the superleak, superfluid is forced through the superleak thus "diluting" the normal component in one volume and concentrating it in the other leading to a temperature gradient.

A rather striking demonstration of such

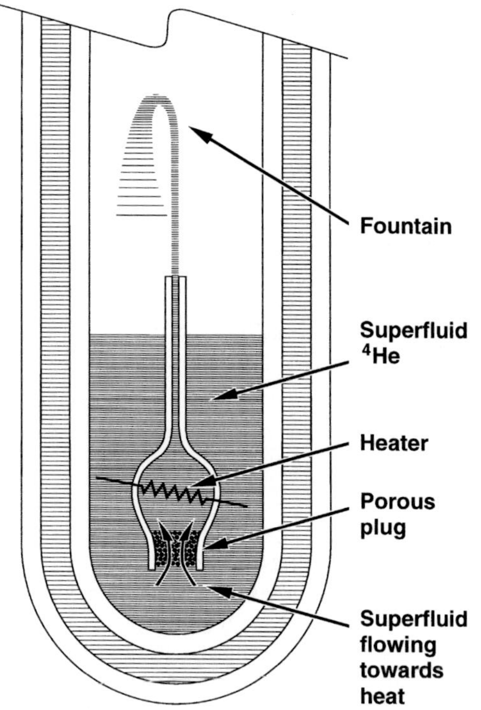

FIG. 7. The fountain effect. On heating the liquid in the container, superfluid flows in through the porous plug from the surrounding bath. The normal fluid cannot flow out to compensate. The excess liquid is therefore ejected upwards through the narrow tube as a fountain.

thermomechanical properties is provided by the fountain effect illustrated in Fig. 7. Here, a container with a narrow open neck at the top and a porous plug below is lowered into a reservoir of superfluid, such that the neck is above the liquid level. If the fluid inside the flask is then heated, the superfluid component is forced to flow from the reservoir into the flask via the pores which impede the flow of normal fluid. Since the only other outlet is the neck, the excess liquid is ejected from the neck of the flask in the form of a fountain. The effect is very dramatic: it is relatively easy to produce fountains more than 30 cm high. By this process we can pump superfluid ^{4}He around an apparatus simply by heating it.

2.5.3 Second Sound Ordinary sound consists of propagating density (or pressure) waves. Two-fluid behavior gives rise to the possibility of temperature waves in which the normal fluid and superfluid density oscillate in antiphase to each other, but such that the total density remains constant. This is called second sound and can be generated and detected in superfluid ^{4}He with a simple heater and thermometer arrangement.

2.6 Excitations and Superflow

Since the superfluid has zero viscosity, flow of the liquid must occur without the production of any excitations. The reason for this can be inferred from the energy-momentum spectrum (dispersion relation) for the excitations as shown in Fig. 8. In the superfluid phase there are two types of thermal excitation. Phonons are collective density oscillations of the liquid (sound waves) travelling at the speed of sound c_1, which is approximately constant at about

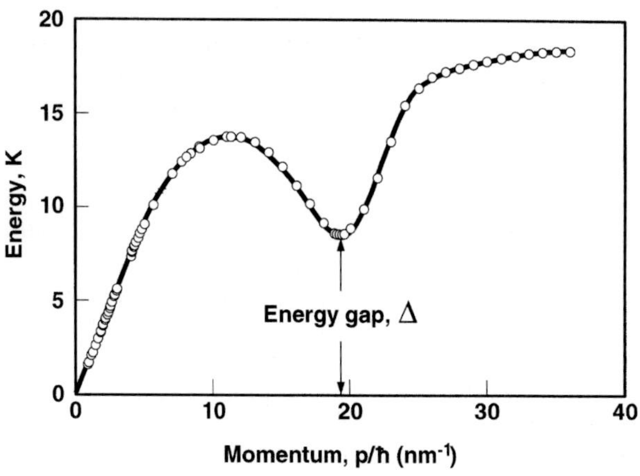

FIG. 8. The energy–momentum relationship (dispersion curve) for the excitations in superfluid ^{4}He. Phonons lie on the straight part of the curve near the origin, rotons around the minimum. The energy gap Δ to the minimum is indicated. (After Donnelly *et al.*, 1981.)

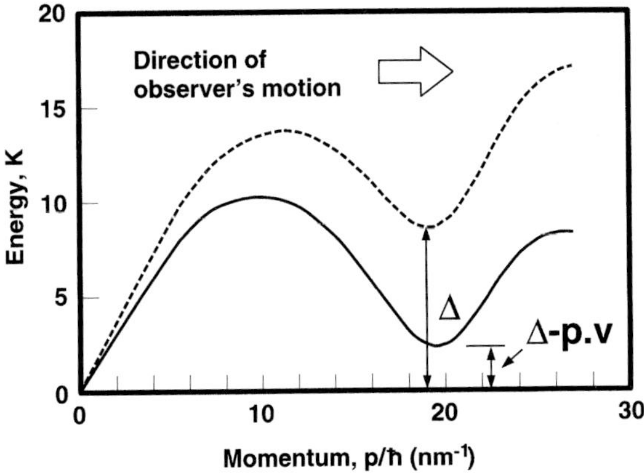

FIG. 9. The dispersion curve of Fig. 8 as seen in a reference frame moving at velocity $-\boldsymbol{v}$ relative to the superfluid. In this frame the roton minimum energy falls by $\boldsymbol{p}\cdot\boldsymbol{v}$. At the Landau velocity, when the minimum falls to zero energy, a scatterer moving at this velocity can freely create excitations.

240 m s^{-1}. Their dispersion relation is therefore approximately linear and described by the expression $E = c_1 p$. At higher energies and momenta the phonon excitations are gradually transformed into excitations called rotons, which have a minimum in the dispersion curve at an energy Δ and at a finite momentum p_0. In consequence, rotons that have momenta less than p_0 have a group velocity ($v_g = \mathrm{d}E/\mathrm{d}p$) which is in the opposite direction to their momenta; in other words they are "hole"-like excitations. Rotons with momenta greater than p_0 have group velocities in the same direction as their momenta and may be thought of as "particle"-like excitations. While the dispersion relation for rotons is well known, their microscopic nature remains a matter of conjecture.

The dispersion curve shown in Fig. 8 is valid in the stationary reference frame of the superfluid. Let us assume we have flow and that the superfluid is moving with velocity v_s with respect to some boundary wall. To a stationary observer in the rest frame of the wall the excitation energies will be shifted by an amount $-\boldsymbol{p}\cdot\boldsymbol{v}_s$ in accordance with the appropriate Galilean transformation. In other words, excitations with momenta approaching the observer, looking into the flow direction, are seen to have a higher energy and those retreating a lower energy. The resulting dispersion curve becomes canted as shown in Fig. 9. Since any scattering process must conserve energy in the rest frame of the scatterer (in this case the wall), new excitations can only be created at zero energy. As the figure shows, there are no such possibilities until the superflow exceeds the value $v_L = \Delta/p_0 \simeq 50\,\mathrm{ms}^{-1}$, at which velocity the dispersion-curve minimum reaches zero energy and excitations may be freely created. This process clearly extracts energy from the flow, *i.e.*, is dissipative. The velocity at which this process begins is known as the Landau critical velocity (Landau, 1941). From these arguments, superfluid ^{4}He should exhibit nondissipative superflow for velocities less than the critical value. It is interesting to note here that the dispersion curve for an ideal Bose gas follows the classical relation $E = p^2/2m$. In this case the Landau critical velocity is zero and there would be no regime of zero viscosity.

In practice it is found that the critical flow velocity for dissipation is very much smaller than the Landau velocity. Measured critical velocities vary according to the precise experimental geometry, but are generally of order 10–100 mm s^{-1}, or some three orders of magnitude less than the Landau value. These velocities correspond to those needed to generate quantized vortices in the superfluid (superfluid turbulence), which we describe in Sec. 2.9.

2.7 Persistent Currents

A firm experimental test of nondissipative flow is given by the demonstration of persistent currents. Persistent currents have been observed in "superfluid gyroscopes" (Clow and Reppy, 1972) consisting of a toroidal flow channel packed with fine powder. The powder is required to clamp the normal component and to prevent the motion of vortices, which would otherwise lead to mutual friction between the superfluid and normal components and thus to dissipation of the superflow. The superfluid is set into motion by rotating the torus about its axis of symmetry above a certain critical angular velocity. The torus is then held stationary while the superfluid continues to circulate around the torus with negligible dissipation. The superflow can be detected by tipping the torus, which induces a precession of the symmetry axis by the gyroscope effect. In such gyroscopes supercurrents have been observed to persist without noticeable loss for several months, demonstrating that the viscosity of the superfluid component is immeasurably small.

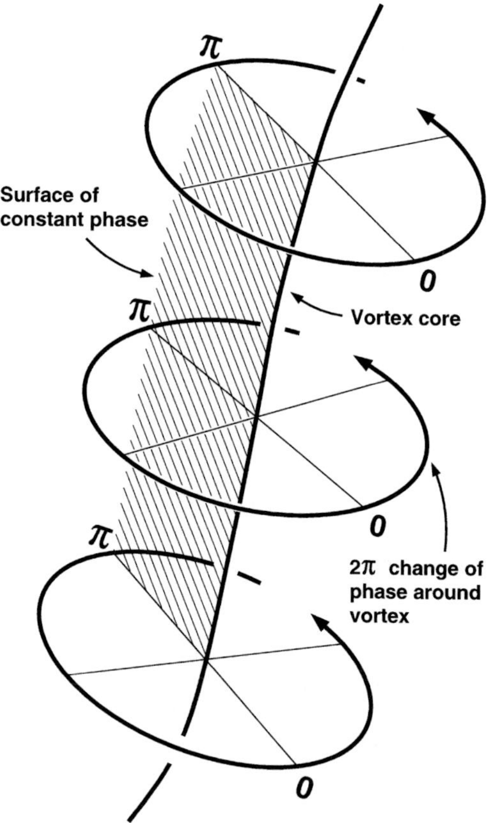

FIG. 10. The 2π change of the phase of the wave function around a vortex. Along the axis there must be a singularity where the phase cannot be defined.

2.8 The Wave Function and Phase Coherence

As discussed in Sec. 2.5, the ^{4}He superfluid component corresponds to the macroscopic occupation of the ground-state energy level in the fluid. All particles contributing to the superfluid component can thus be described by a single wave function of the form $\psi = \psi_0 \exp(i\theta)$. The magnitude of the wave function is determined by the density of the superfluid component, $\psi_0 = \sqrt{\rho_s}$. The value of the phase θ of the wave function is arbitrary; however, gradients of this phase determine the velocity of the superfluid component via the relationship $\hbar\nabla\theta = m_4\boldsymbol{v}_s$ where m_4 is the mass of a ^{4}He atom. The fact that a single phase factor describes all the atoms associated with the superfluid component is known as phase coherence.

2.9 Vortices

The existence of quantized vortex lines in superfluid ^{4}He (and the analogous flux lines in superconductors) is a direct consequence of the existence of a single macroscopic wave function describing the entire sample. Since a gradient in the phase of the wave function implies a flow of the superfluid, which must have some kinetic energy, the most energetically favorable configuration for a given volume of superfluid will be one with a slowly varying phase (small superflows) throughout; and for the special case of a stationary sample, the phase should be constant everywhere. However, in practice it is very difficult to realize this ideal configuration in any given experiment.

Since the wave function must be single-valued throughout the superfluid, the change in the phase around any closed path has to be an integer multiple of 2π. If we imagine such a path with a phase change of 2π around it, then we can shrink the path arbitrarily while still maintaining the 2π phase change, until the path becomes a single point. From Fig. 10 we see that there must be a line of such points through the liquid defining a linear topological defect. This represents the core of a vortex. The 2π phase change around the core leads to a circulating superflow with a velocity described by the expression

$$v_s = \hbar/m_4 r \tag{2}$$

where r is the distance from the core. The velocity thus decreases inversely with distance. The circulation of liquid, defined as $\kappa = \int v_s \cdot dl$ around a loop surrounding the core, is thus quantized in units of h/m_4.

We now appear to face a contradiction since, first, the phase of the wave function cannot be defined along the actual core and, second, adjacent to the line the gradient of the phase approaches infinity implying an infinite velocity of the superfluid. There is only one way of avoiding these contradictions: the magnitude of the wave function must vanish at the core of the vortex; *i.e.*, the core comprises normal fluid. The length scale over which the wave function is suppressed at the vortex core is called the coherence length and is of the order of an interatomic spacing. Since the velocity field extends to infinite distance from the core these vortices do not have a unique energy per unit length since the kinetic energy stored in the flow field depends on the extent of the container.

It is worth noting at this point that similar difficulties with infinite velocities are a feature of most types of vortex and are invariably resolved in the same way by the exclusion of the motion from the axis. For example, the vortex formed by water running out of the plughole of a bath has a cylindrical air core, which excludes the velocity field of the water from the axis.

2.9.1 Rotating Superfluid ^{4}He If the wave function of the superfluid is continuous (*i.e.*, contains no voids or vortex cores) then the change in phase of the wave function around any loop must be zero. This means that the superflow is irrotational ($\nabla \times \boldsymbol{v}_s = 0$). This has the important consequence that solid-body rotation of the superfluid is impossible. Furthermore, in the absence of vortices, the net rotation of the superfluid must be zero with respect to the fixed stars. If a bucket of normal liquid ^{4}He is set into rotation (with sufficiently small angular velocity) and then cooled below the lambda transition, the superfluid component will be stationary. This could in principle be seen from the shape of the free liquid surface. the surface would take up a parabolic profile in the normal state in response to the centrifugal force acting on the rotating helium. However as the liquid is cooled well below the transition temperature the surface should become flat since the superfluid component cannot rotate.

As the rotation speed of the container is increased, then eventually the relative velocity between the stationary superfluid and the moving walls will exceed the critical velocity for vortex creation. A small loop of vortex is then created which is immediately dragged around by the friction force of the rotating normal fluid. Here the Magnus force on the vortex from the moving superfluid comes into play. (The Magnus force is the sideways force acting on a rotating cylinder moving transversely in a fluid). The Magnus force causes the vortex line to grow outwards from the container wall and eventually take up a position along the rotation axis of the container as illustrated in Fig. 11. As the rotation is further increased, more vortex lines are produced, which move towards the center of the cell forming a regular lattice of vortex lines parallel to the rotation axis. Photographic images of the vortex line lattice have been produced by Yarmchuk *et al.* (1979). The vortex lattice is analogous to the flux-line lattice produced by a magnetic field applied to a type II superconductor. In this sense, rotation applied to a superfluid is exactly analogous to a magnetic

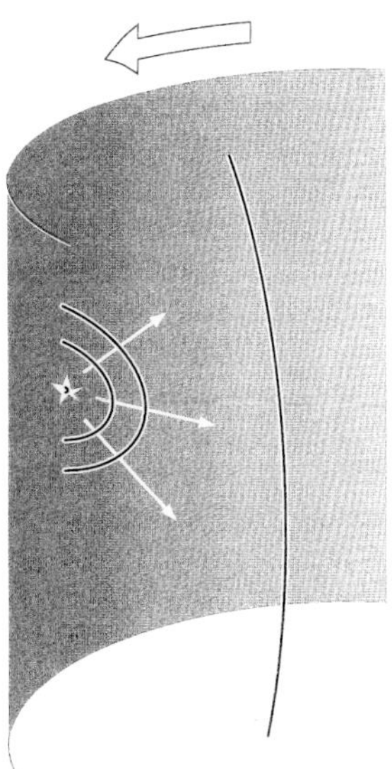

FIG. 11. In a rotating container, microscopic vortex loops created at the moving wall grow, under the influence of the Magnus force, and take up a position parallel to the axis of rotation.

field applied to a superconductor, since in the former case the rotation is quantized (circulation of mass) and in the latter flux is quantized (circulation of charge). The analogy may be taken further since another important property of a superconductor is the expulsion of magnetic field (the Meissner effect), which is analogous to the expulsion of solid-body rotation in a superfluid.

2.9.2 Measurement of Absolute Rotation with Superfluid ^{4}He Since superfluid ^{4}He is irrotational with respect to the fixed stars, a measurement of the superfluid rotation with respect to its container constitutes a measurement of the rotation of the container with respect to the fixed stars. In this manner, superfluid ^{4}He can be used to measure absolute rotation. This principle has already been tested by Schwab *et al.* (1997), who have recently measured the rotation of the Earth. The device they used is shown schematically in Fig. 12. A toroidal container incorporating a small aperture is filled with superfluid ^{4}He. The container is stationary with respect to the Earth and thus rotating with respect to the fixed stars. Since the superfluid cannot rotate, there must be a superflow with respect to the container which is magnified as the superfluid flows through the small aperture. The flow through the aperture may be measured, from which the rotation of the container (the Earth) is inferred. The device shown in Fig. 12 is actually the superfluid analog of a superconducting SQUID, which measures magnetic fields in a similar manner.

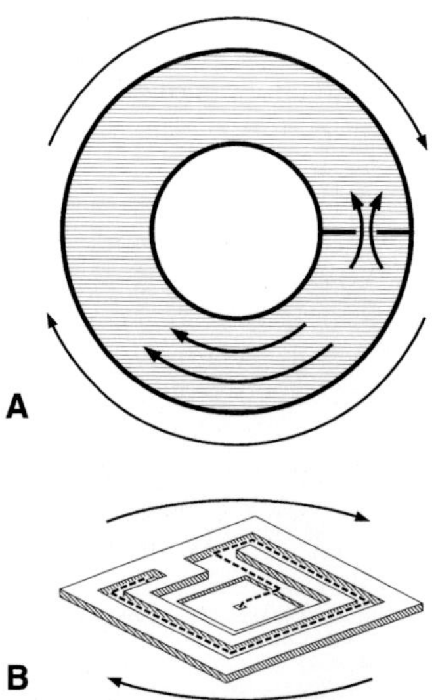

FIG. 12. Measurement of absolute rotation. (a) A rotating toroidal container incorporating a small aperture is filled with superfluid ^{4}He. Since the superfluid cannot rotate, there must be a flow through the aperture. This flow can be measured and the rotation of the toroid inferred. (b) In the actual device, the channel is formed in a thin slab. The flow loop, indicated by the dashed line, passes through the small aperture sited at the center. (After Schwab *et al.*, 1997.)

3. SUPERFLUID ^{3}He

Superfluid ^{3}He was first observed by Osheroff, Richardson, and Lee in 1971 (Osheroff *et al.*, 1972). Although the superfluid transition in ^{3}He had long been predicted, the observation of the superfluid transition was at first mistakenly associated with the solid phase. However, soon afterwards, the experiment was repeated with the use of nuclear magnetic resonance (NMR) techniques with spatial resolution that allowed the identification of two superfluid phases, designated the A phase and the B phase.

The arguments of Bardeen, Cooper, and Schrieffer for Cooper pairing discussed in Sec. 1.3 above were originally proposed in the context of superconductivity. The Cooper pairs that form in conventional superconductors are relatively simple, having zero spin, zero orbital angular momentum, and therefore no internal structure. (The form of the pairs in high-T_c materials is still under debate at the time of writing.) In contrast, the Cooper pairs in superfluid ^{3}He have spin unity and orbital angular momentum unity. The ^{3}He pairs thus have a more complex internal structure, which has a dramatic effect on the superfluid properties, not least in ensuring that there are several distinct superfluid phases.

3.1 Cooling Techniques

The superfluid transition temperature for ^{3}He varies between about 1 and 3 mK depending on the pressure. Even with current cooling techniques these temperatures are not straightforward to attain. The most usual methods in use rely on a continuously operating dilution refrigerator to produce a steady temperature in the millikelvin regime from which base a further cooling stage to reach the superfluid region is used. For this final stage nuclear cooling is currently almost universally used, but in the past Pomeranchuk cooling (see Sec. 3.1.1) has also been used successfully. A dilution refrigerator may be obtained commercially. However, while a modern commercial dilution refrigerator will typically have a base temperature of around 5 mK this is not as good as the best in-house produced machines, which currently can reach around 2 mK.

3.1.1 Pomeranchuk Cooling During the early 1970s before nuclear cooling had become established, the only other practical methods for cooling ^{3}He to low millikelvin temperatures were by the Pomeranchuk effect and, for work just reaching the superfluid transition, magnetic cooling using cerium magnesium nitrate. In fact Pomeranchuk cooling was the method employed by Osheroff *et al.* (1972) when in 1971 they first observed superfluidity in ^{3}He. The technique relies on an unusual property of the ^{3}He phase diagram shown in Fig. 13. At low temperatures, the pressure at which solidification occurs in liquid ^{3}He falls as the temperature rises. This follows from the unusual property of ^{3}He that the entropy of the solid phase is actually higher than that of the liquid phase below 0.3 K. In other words, the solid phase in more disordered than the liquid phase. While Fermi statistics ensure that the entropy of the liquid is low, the solid has almost complete disorder of the nuclear spin system above 1 mK. Consequently, an increase in pressure that causes solidification of the liquid will produce cooling from the latent heat of solidification, *i.e.*, will force the liquid–solid mixture to follow the solidification line to lower temperatures as the pressure is increased. This is the principle utilized in Pomeranchuk cooling.

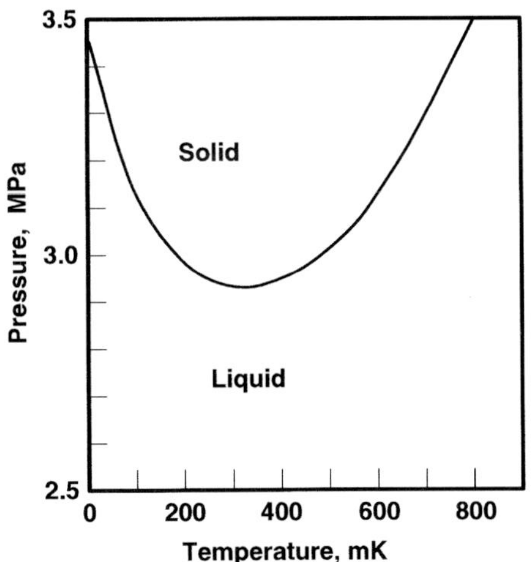

FIG. 13. The liquid–solid region of the ^{3}He phase diagram.

Pomeranchuk cooling has the advantage that the ^{3}He is itself the refrigerant, thus overcoming any problems of establishing thermal contact to the specimen at low temperatures. The disadvantages of this technique are

1. that the specimen will always contain a mixture of solid and liquid ^{3}He;
2. that the specimen is always at melting pressure, thus preventing studies at lower pressures; and
3. that the minimum temperature that can be achieved is limited by the magnetic ordering temperature of the solid (about 1 mK).

These limitations have meant that the Pomeranchuk technique has largely been replaced as a cooling method by the far more versatile technique of nuclear cooling.

3.1.2 Nuclear Cooling The principle of nuclear cooling is shown schematically in Fig. 14. A material containing weakly interacting spins (magnetic moments) is placed in a high magnetic field and then precooled to some temperature T_i. The high magnetic field B_i produces Zeeman splitting of the magnetic energy levels, and the precooling ensures a relatively high population of the lower energy levels as shown in the figure. The magnetic entropy of the system depends only on the relative populations of the energy levels, which in turn depend only on the ratio $\mu B_i/kT_i$ where μ is the magnetic moment associated with the spin. Now suppose that we can reduce the magnetic field adiabatically, *i.e.*, such that no thermal energy can be exchanged between the outside world and the spin system. Since no heat is exchanged the populations of the various levels must remain the same. Therefore, during "adiabatic demagnetization" the entropy of the spin system must remain constant,

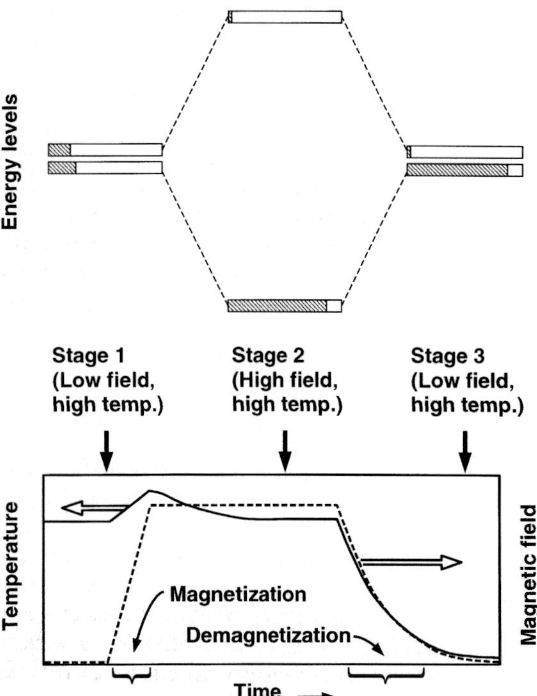

FIG. 14. Nuclear cooling. At the top is a schematic illustration of the energy levels and populations as the field is applied and removed. Below is shown schematically the variations of the magnetic field and the temperature during the process.

ensuring that the ratio $\mu B/kT$ is also constant. Therefore a reduction of the field to some final value B_f results in a lower temperature T_f as given by:

$$T_f = (B_f/B_i)T_i.$$

In principle the technique of nuclear cooling could not be simpler. When the field applied to a material containing nuclear magnetic moments is reduced the temperature falls in the same ratio. In practice, of course, it is not so straightforward.

The minimum temperature possible with nuclear cooling is ultimately limited to the temperature at which the nuclear spins spontaneously order as a result of the interactions between spins, which we neglected above. However, in the case of copper, the most widely used nuclear refrigerant, this ordering temperature lies in the nanokelvin regime and does not present any drawback for cooling liquid ^{3}He to the millikelvin and microkelvin regimes. In practice the minimum ^{3}He temperature achievable by this method is limited by a combination of heat leaks into the specimen and the very poor thermal contact between the specimen and the nuclear coolant. Since the nuclear heat capacity of copper is not very large compared with that of liquid helium, a volume of copper comparable to that of the liquid helium specimen is necessary to generate enough cooling power. Typical starting conditions for demagnetization are a starting temperature of 5–10 mK (well within the range of a dilution refrigerator) in a field of 6–8 T (readily generated by a commercial superconducting solenoid). By reducing the magnetic field by a factor of 1000, temperatures around 100 μK are attainable in the superfluid ^{3}He. This is already low enough that to achieve significantly lower temperatures we may need to consider shielding the specimen against the heating from cosmic rays. A practical system is shown schematically in Fig. 15.

3.2 A Multiplicity of Possible Phases

Since the Cooper pairs in superfluid ^{3}He have spin angular momentum $\boldsymbol{S} = 1$ and orbital angular momentum $\boldsymbol{L} = 1$, they have in-

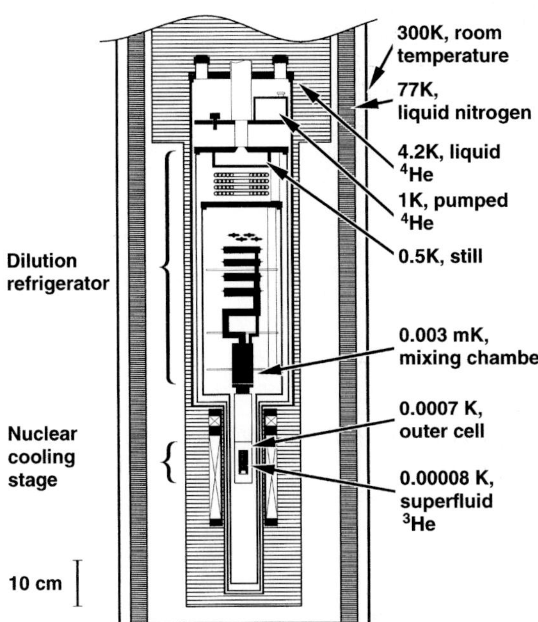

FIG. 15. A nuclear cooling cryostat for achieving superfluid temperatures in liquid ^{3}He.

ternal structure in the sense that there are two axes defined by the directions of the spin and orbital angular momenta. The internal degrees of freedom are specified by the spin projection quantum numbers $m_S = +1, 0, -1$ and orbital angular momentum projection $m_L = +1, 0, -1$. Therefore there are $3 \times 3 = 9$ different types of Cooper pair allowed in superfluid ^{3}He, and correspondingly the superfluid wave function has 9 complex components (18 parameters). This complex structure places superfluid ^{3}He in a class of its own. In contrast, superfluid ^{4}He can be described by a simple two-parameter wave function $\psi = \psi_0 \exp(i\theta)$ (see Sec. 2.8), which is also the case for conventional superconductors where the Cooper pairs have zero spin and zero orbital angular momentum and therefore have no internal structure.

One immediate consequence of the multiplicity of possible Cooper pair states in superfluid ^{3}He is the possibility of several distinct superfluid phases. It is a tribute to modern theoretical physics that several of the possible phases of systems with $\boldsymbol{S} = 1$ and $\boldsymbol{L} = 1$ Cooper pairs had been studied theoretically (albeit in the context of superconductivity) long before the discovery of superfluid ^{3}He. There are 13 possible phases but only three occur under normal experimental conditions. Which phase is preferred depends on temperature, pressure, and magnetic field. The phase diagram for the two phases that occur in zero magnetic field is shown in Fig. 16.

3.2.1 The A Phase As shown in Fig. 16, the preferred phase at high temperatures and pressures is known as the *A* phase. The Cooper-pair structure of this phase was first considered theoretically by Anderson, Brinkman, and Morel (ABM) (Anderson and Morel, 1961; Anderson and Brinkman, 1973). In this phase the Cooper-pair spin projection may only take the values $m_S = +1$ and $m_S = -1$. Furthermore, the orbital angular momentum of every pair points along a common direction called the orbital angular momentum vector (or $\boldsymbol{l}$ vector). The pairs are constructed from ^{3}He atoms close to the Fermi surface. If the component atoms are to couple to form pairs with angular momentum $\boldsymbol{L} = 1$ pointing in, say, the polar direction, then atoms with momenta lying in the equatorial plane can achieve this easily by mutually orbiting but atoms with momenta in the polar direction cannot mutually orbit to create an angular momentum also in this direction. In consequence this phase

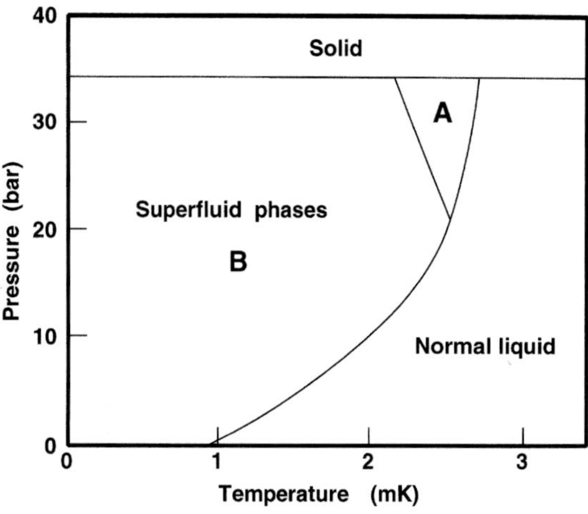

FIG. 16. Phase diagram of the superfluid phases of ^{3}He in zero magnetic field.

acts as an excellent superfluid for atoms with momenta in the equatorial plane but cannot at all form a superfluid for atoms with momenta along the polar direction. The superfluid energy gap Δ, which tells us the strength of the pairing, is thus anisotropic, vanishing along the $\boldsymbol{l}$ axis as shown in Fig. 17.

Since the pairs all orbit about a common axis, defined by the direction of the $\boldsymbol{l}$ vector, the nuclear magnetic moment associated with each ^{3}He nucleus experiences a changing magnetic field from its partner. This is equivalent to two parallel bar magnets orbiting about a common center. The magnetic (dipole–dipole) energy of such a system is reduced when the magnets are oriented perpendicular to the rotation axis. Therefore, in the absence of any other disturbing features such as walls or flow, this will align the spin perpendicular to the $\boldsymbol{l}$ vector.

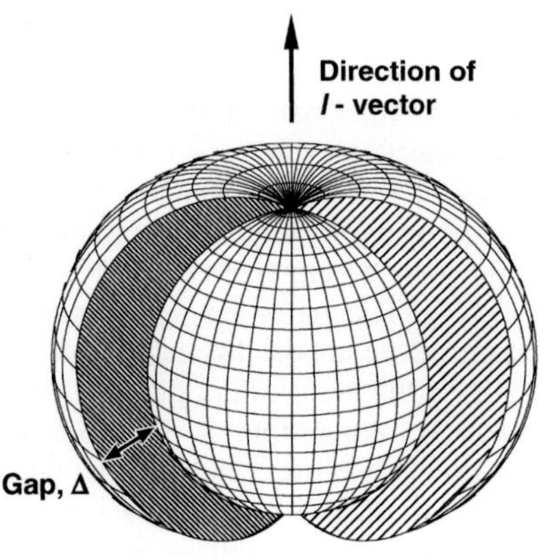

FIG. 17. The energy gap in the *A* phase. The gap is zero along the ***l***-vector axis and is at its maximum in the perpendicular directions.

For an isolated pair this energy is miniscule and would not cause any orientation at millikelvin temperatures. However, all the pairs are in the same quantum ground state acting as a single entity such that all the spin vectors acting together are orienting themselves relative to all the orbital moments acting together. The net result is a greatly enhanced spin–orbit coupling. This demonstrates that indeed we are dealing with a coherent ground state.

3.2.2 The B Phase The structure of the *B* phase of superfluid ^{3}He is somewhat more subtle than that of the A phase. In low magnetic fields the system shows a superposition of all possible m_S values and all possible m_L values, with the result that there is no net preferred spin or angular momentum direction in the liquid. As a consequence of this apparent higher degree of symmetry, the B-phase energy gap is isotropic. To first order, each pair attempts to make its total angular momentum (spin plus orbital) zero, $\boldsymbol{J} = \boldsymbol{S} + \boldsymbol{L} = 0$. Therefore pairs consisting of ^{3}He atoms with momenta parallel to the field are in the $m_S = 0$ state. Those formed from atoms with momenta perpendicular to the field have $m_S = +1$ and -1, with all three states contributing for other angles. However, the same magnetic dipole forces that orient the A phase also act here, forcing the spin direction to be rotated with respect to the orbital direction. However, unlike in the A phase where the total spin system of the pairs is rotated with respect to the total orbital momentum, in the B phase the rotation takes place for each pair individually, but about an axis common for all pairs, specified by a vector $\boldsymbol{n}$. The dipole–dipole coupling energy is minimized when this rotation angle is 104° (known as the Leggett angle). This leads to quite a subtle structure but does have the effect of producing a simple isotropic energy gap, at least in low magnetic fields.

The B phase corresponds to the theoretical BW state first examined by Balian and Werthamer (1963), long before this behavior was realized in superfluid ^{3}He. The original BW state had Cooper pairs with total angular momentum $\boldsymbol{J} = \boldsymbol{L} + \boldsymbol{S} = 0$, but the relative population of the m_S states was distributed around the Fermi surface as described above.

3.3 Broken Symmetries of the Superfluid Phases

3.3.1 The A Phase In Sec. 3.3.1 we introduced the $\boldsymbol{l}$ vector, which defines the axis along which the energy gap vanishes in the A phase. In the absence of any external fields and "foreign" objects such as container walls, it is clear that the direction of $\boldsymbol{l}$ must be arbitrary. Above the transition the direction of the $\boldsymbol{l}$ vector has no meaning. Thus as we pass through the transition the symmetry of the liquid is reduced, as there is now a definite (but arbitrary) direction for $\boldsymbol{l}$ whereas above the transition all directions were equivalent. This behavior is known as spontaneous symmetry breaking. Conventional superconductors and superfluid ^{4}He also exhibit spontaneous symmetry breaking, although here a more abstract symmetry is broken, known as gauge symmetry, which leads to the definite (but again arbitrary) phase factor in the superfluid/superconductor wave function. Since spatial gradients of the $\boldsymbol{l}$ vector give rise to superflows in a similar manner as spatial gradients of the phase factor give rise to superflows in superfluid ^{4}He, the broken gauge and orbital symmetries are linked in the A phase. This is described as a broken relative gauge-orbit symmetry. In addition, the directionality of the spin part of the wave function leads to broken spin rotation symmetry in ^{3}He-A. The spin part of the wave function in the A phase is described by a vector $\boldsymbol{d}$, which is defined to point along the direction of zero spin projection. In other words the $\boldsymbol{d}$ vector points perpendicular to the spin $\boldsymbol{S}$ and is common for all pairs.

3.3.2 The B Phase In superfluid ^{3}He-B gauge symmetry is broken (*i.e.*, the phase of the wave function takes a particular value). As discussed in Sec. 3.2.2, there is no independent directionality for the spin and orbital parts of the wave function as in the A phase. However, the orbital and spin vectors are coupled for each pair such that all the spin vectors may be rotated with respect to all the orbital vectors around a common axis specified by the vector $\boldsymbol{n}$. Thus all the directionality of the B phase (in low magnetic fields) is specified by this spin–orbit rotation axis and therefore the B phase exhibits a broken relative spin–orbit rotation symmetry.

3.4 The Phase Diagram

The phase diagram of superfluid ^{3}He has already been introduced in Fig. 16. In low magnetic fields, the B phase covers the majority of the *P-T* space while the A phase is confined to a region of high pressure and temperature. The phase transition between the normal state and the superfluid state is second order with a discontinuity in the heat capacity (in this sense very similar to conventional superconductors).

The phase transition between the A and B phases is, however, first order, and very substantial supercooling of the A phase has been observed (Hakonen *et al.*, 1985). The mechanism for the nucleation of the B phase from the A phase has been a matter of much debate. The first theoretical estimates of the nucleation time for the B phase suggested that it should be of order the age of the universe. The precise nucleation mechanism is as yet still uncertain; however, there is strong evidence that cosmic rays may be playing an important role (Leggett, 1984).

The phase diagram is transformed by the application of a magnetic field. Another phase, called the A1 phase, appears in a narrow temperature range separating the normal state from the A phase. The A phase becomes stable over a wider range of temperatures and pressures. For magnetic fields larger than about 0.5 T, the B phase is removed completely from the phase diagram, and the A phase becomes the low-temperature superfluid phase. The A1 phase is very similar to the A phase, the main difference being that the A1 phase contains only $m_S = +1$ (up spin) pairs; that is, the A1 phase is a magnetic superfluid. The transition from the A1 phase to the A phase simply corresponds to the phase transition for the $m_S = -1$ (down spin) Cooper pairs. The A phase is often considered as consisting of two interpenetrating superfluids, one for the down-spin pairs and one for the up-spin pairs.

3.5 Properties of Superfluid ^{3}He

Not surprisingly, the high level of structure of the Cooper pairs in superfluid ^{3}He leads to rather unusual properties. In spite of the complexity of the superfluid wave function, many of the superfluid properties were found theoretically prior to their discovery. The reasons for this are that

1. the basic interactions between ^{3}He atoms are known rather precisely;
2. the entire superfluid sample (typically consisting of some 10^{23} atoms) is described by a single, albeit rather complicated, wave function with a well defined structure in each superfluid phase; and
3. there are virtually no impurities in liquid ^{3}He at low temperatures.

The only element that has appreciable solubility in ^{3}He is ^{4}He, and the extrapolated solubility of ^{4}He at sub-millikelvin temperatures is around one ^{4}He atom in an astronomical volume of liquid ^{3}He. Superfluid ^{3}He is therefore an exceptionally clean medium, for which theories should be applicable with extreme accuracy.

Nevertheless, there are still new phenomena being found in superfluid ^{3}He that have, as yet, no theoretical explanation. The reason for this is that the system is so complex that even though the fundamental equations are known, all the solutions are not.

3.5.1 Thermal Excitations and Superflow

In order to describe the thermal excitations in the superfluid phases, it is first necessary to describe those in the normal phase of liquid ^{3}He. The normal phase is described well by Landau's phenomonological Fermi-liquid theory (Landau, 1956). Here, the main effect of the interatomic forces can be simply accounted for by introducing an effective ^{3}He mass, m^*. The modified atoms, called quasiparticles, can then be considered to be almost noninteracting, with energy levels given by $E = p^2/2m^*$. The main thermal excitations of the system correspond to either the occupation of energy levels above the Fermi energy (quasiparticle excitations) or empty energy levels below the Fermi surface (quasihole excitations). The excitation energies, given by $\xi = |p^2/2m^* - E_F|$, are shown by the dashed line in Fig. 18.

At absolute zero temperature, all the ^{3}He quasiparticles are paired and can be described by the superfluid wave function. At finite temperatures, some of these pairs will be broken by thermal agitation. The unpaired quasiparticle excitations are still influenced by the pairing interaction and their energies are raised with respect to the normal-state excitations as shown by the solid line in Fig. 18. The minimum quasiparticle excitation energy is equal to the superfluid energy gap Δ, as shown in the figure. The Cooper pairs may be considered to lie at the origin in Fig. 18, and in order to break a pair, two excitations must be created, requiring a minimum of 2Δ of energy.

The arguments presented in Sec. 2.6 relating dissipationless flow with the excitation dispersion curve for superfluid ^{4}He are equally applicable to superfluid ^{3}He. The dispersion curve shown in Fig. 18 will clearly allow the existence of dissipationless superflow below the Landau critical velocity $\boldsymbol{v}_L = \Delta/p_F$. In the B phase the measured critical velocity of the superfluid is indeed limited by the Landau velocity even in the presence of strong magnetic fields (Fisher *et al.*, 1991). (This is in contrast to the case of superfluid ^{4}He discussed in Sec. 2.6.) In the A phase the situation is far more complicated as a result of the extreme anisotropy of the energy gap. The Landau critical velocity for the A phase is dependent on direction, falling to zero for superflow along the $\boldsymbol{l}$ axis. The critical flow velocity for the A phase is in practice dependent on geo-

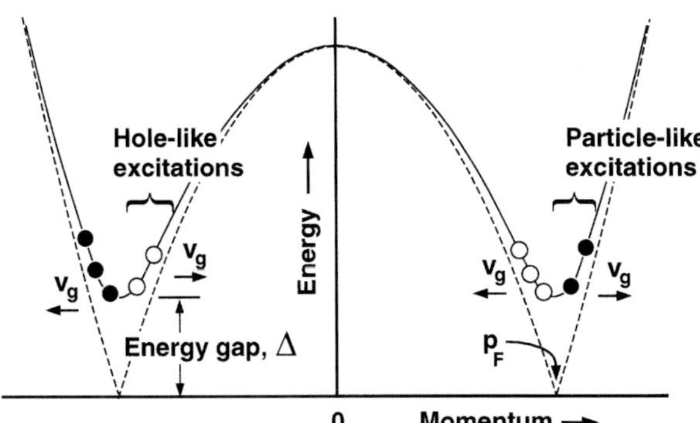

FIG. 18. The energy–momentum relations (dispersion curves) for the excitations in superfluid (solid line) and normal liquid ^{3}He (dashed line). The solid circles represent quasiparticles and the open circles quasiholes. The Fermi momentum p_F is indicated. The excitation energies plotted are measured from the Fermi energy.

metrical factors on account of the ***l*** textures which we describe in Sec. 3.5.3.

3.5.2 Thermal Properties of Superfluid ^{3}He The heat capacity of superfluid ^{3}He as measured by Greywall (1986) is shown in Fig. 19. The behavior at the transition is quite unlike that of ^{4}He. There is a clear second-order phase transition between the normal phase and the superfluid phases with no logarithmic singularity. The B-phase heat capacity is very similar to that of a conventional superconductor and is well described by the BCS theory with a slightly enhanced energy gap due to strong coupling corrections (the BCS theory is only applicable in the limit of weak pairing interactions). At low temperatures the heat capacity of the B phase falls exponentially with decreasing temperature because of the dominance of the Boltzmann factor $\exp(-\Delta/kT)$ governing the number of thermal quasiparticle excitations. The extremely small heat capacity of ^{3}He-B at temperatures of around 100 μK has been utilized in superfluid-^{3}He particle detectors (Bradley *et al.*, 1995) capable of measuring energy depositions of less than 1 keV.

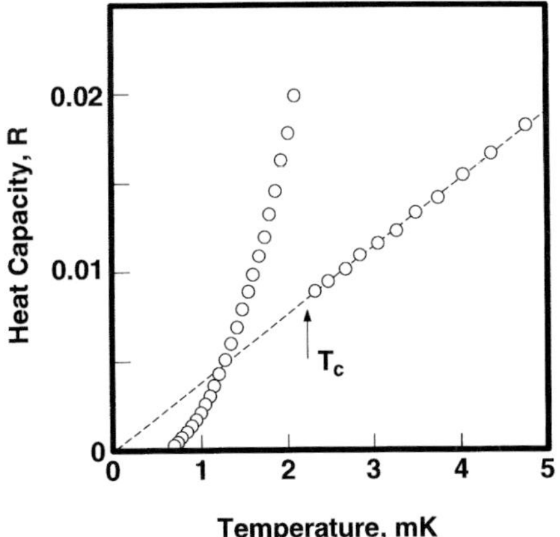

FIG. 19. The heat capacity of superfluid ^{3}He at 20.3 bar (after Greywall, 1986).

In contrast, the heat capacity of ^{3}He-A is expected to vary as T^3 at low temperatures on account of the nodes in the energy gap (see Sec. 3.3.1). However, as a significant field is needed to stabilize the A phase to low temperatures this behavior has not been experimentally verified.

The thermal transport properties of the superfluid phases are qualitatively similar to those of superfluid ^{4}He in that at high temperatures a counterflow of normal fluid and superfluid is set up by a temperature gradient and the thermal conductivity is very high. The situation is particularly interesting at very low temperatures where the normal-fluid density is very low, since the quasiparticle excitations making up the normal fluid travel ballistically. The theoretical mean free path of a quasiparticle in ^{3}He-B at a temperature of around $0.1\,T_c$ (the lowest temperature presently achievable) is about 1 km. This property has been utilized in quasiparticle beam experiments (Fisher *et al.*, 1992).

3.5.3 Textures The directional properties of the superfluid give rise to properties very reminiscent of those seen in liquid crystals. The directions of the ***l*** and ***d*** vectors in the A phase and the ***n*** vector in the B phase, which determine the dynamic behavior of the phases, are influenced both by external fields and by interfaces (including solid boundaries and superfluid phase boundaries). The arrangement of the various directionalities in a given situation is known as the texture.

In the A phase, a solid–superfluid boundary forces the $\boldsymbol{l}$ vector to lie normal to the surface. Put in simple terms, the pairs at the interface may orbit in the plane of the surface, whereas other directions will involve collisions of the component ^{3}He atoms with the boundary. In order to minimize bending energies, a slow variation in the direction of the $\boldsymbol{l}$ vector is preferable; however, in a simply connected volume this condition cannot be satisfied everywhere. There has to be at least one singularity (or textural defect). This is illustrated for a spherical cell in the first part of Fig. 20. If, on the other hand, we confine the A phase in a torus then the singularity can be avoided, as shown in the lower part of the figure. Since the spin direction can be influenced by a magnetic field, there is an interesting region near walls where there is a competition between the orienting effect of the wall on the orbital angular momentum and the orienting effect of the field on the spin angular momentum.

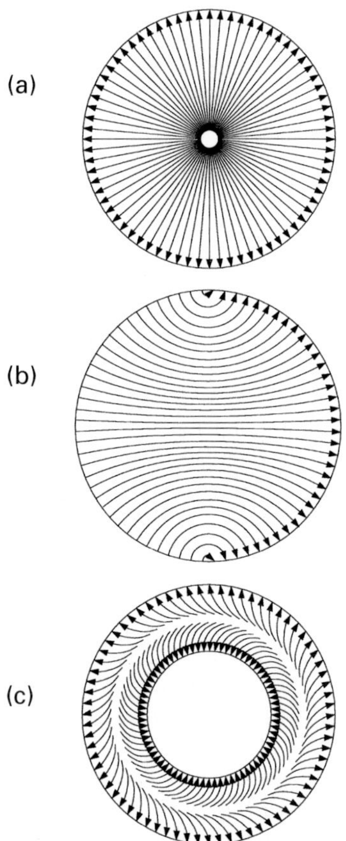

FIG. 20. (a), (b) The A-phase texture (the direction of the l-vector) in a spherical volume. As can be seen, for the direction of l to vary only slowly over the volume there must be either (a) a point singularity in the bulk or (b) a line singularity at the boundary. (c) The texture in a toroidal container can be accommodated with no singularities.

The effect of a solid boundary in the B phase is more complex. In the absence of walls the $\boldsymbol{n}$ vector prefers to lie parallel to the field. However, if we have a wall nearby, then the direction of $\boldsymbol{n}$ depends on the orientations of both the field and the wall.

Other orienting effects include electric fields, thermal conduction, and superflow. The combination of all these orienting effects in practice leads to very messy textures with many defects (somewhat in analogy with the structure of a deformed polycrystalline solid).

Textural effects in the A phase are obviously very strong as a result of the large anisotropy of the energy gap. In consequence virtually all the dynamic properties of the A phase (transport properties and NMR properties) are strongly affected by the texture. For example, at the lowest temperatures virtually all the quasiparticle excitations of the A phase have momenta along the $\boldsymbol{l}$ axis (where the energy gap is zero), so that in this sense the excitation gas is almost one dimensional. Spatial variations of the direction of $\boldsymbol{l}$ lead to confinement of the excitations, thus reducing all excitation transport properties by an enormous factor.

For low magnetic fields, the textures in the B phase are far weaker as far as excitations are concerned since the energy gap is isotropic. In higher magnetic fields (above 0.1 T, say) the energy gap becomes distorted thus strengthening the influence of the texture. The B-phase texture does, however, have an enormous effect on the NMR properties even in small magnetic fields (see Sec. 3.5.5.1).

3.5.4 Vortices The complexity of the superfluid ^{3}He wave function also reveals itself in the variety of superfluid ^{3}He vortices, which have been methodically studied and categorized by the Helsinki ultralow-temperature group (Krusius, 1993).

There is only one type of vortex possible in superfluid ^{4}He because of the simplicity of the wave function (which reflects the simplicity of the ordering). Put simply, in superfluid ^{4}He the phase of the wave function is allowed to change by a multiple of 2π around a vortex core. In superfluid ^{3}He, however, there are several properties that may change as we circle the core. The phase may change, but also the directions of the orbital and spin angular momentum vectors may change, and since the latter are coupled the number of possible structures is quite large.

The circulation κ of vortices in superfluid ^{4}He is quantized in units of h/m_4, as discussed in Sec. 2.9. Similarly, in superfluid ^{3}He vortices are quantized in units of $h/2m_3$ where m_3 is the bare ^{3}He atomic mass, the reason being that the superfluid is composed of Cooper-pair bosons consisting of two ^{3}He atoms (analogously, the flux carried by a flux-line vortex in a superconductor is quantized in units of $h/2e$).

In the B phase, the two most common types of vortex are the axisymmetric (V1) vortex and the nonaxisymmetric (V2 or dumbbell) vortex. Both of these are hard-core vortices, meaning that the structure of the wave function is modified within the vortex core over a length scale of the order of the superfluid coherence length $\xi \approx 10$–$100\,\mathrm{nm}$. The V1 vortex occurs at higher temperatures and carries a single quantum of circulation. The B phase is gradually transformed into the A phase on approaching the vortex axis. The V2 vortex has an exotic double-core structure consisting of two topologically confined half-quantum vortex cores, so that the core carries, in total, a single quantum of circulation. Single half-quantum vortices are topologically forbidden; their confinement within the V2 core is analogous to the confinement of quarks within nucleons.

The predominant vortices in the A phase are soft-core vortices, meaning that the structure of the superfluid is unchanged in the vortex core. They are in fact textural vortices in the sense that they are defined by the $\boldsymbol{l}$ texture (as mentioned in Sec. 3.3.1, a spatially varying $\boldsymbol{l}$ texture produces superflow). The size of the soft core, typically a few micrometers, is much larger than that of hard cores, and they carry two quanta of circulation. Other vortex types in the A phase include hard-core vortices and, remarkably, a vortex sheet consisting of multiple soft-core vortices locked together side by side.

3.5.5 Magnetic Properties Since the Cooper pairs in superfluid ^{3}He have spin $S = 1$ arising from the nuclear spin of the constituent ^{3}He atoms, they have a magnetic moment. This is the only superfluid/superconducting system which is currently known to have magnetic Cooper pairs. The absence of electrical charge in the ^{3}He pairs makes certain types of measurements (such as the measurement of superflow) rather difficult as compared with a similar measurement in a superconductor. However, this is more than compensated by the magnetic moment of the pairs, which allows them to be probed using the powerful technique of NMR, which is unique in that the technique allows us to look very directly at the Cooper-pair wave function. A part of the wave function concerns the direction of the nuclear spin (or magnetic moment). Not only can we see what this part of the wave function is doing by NMR, but we can also use NMR to manipulate it. In consequence, new phenomena can be observed such as spin supercurrents, spin vortices and coherent excited states of the Cooper-pair condensate.

3.5.5.1 Nuclear Magnetic Resonance Historically, the discovery of superfluid ^{3}He was made using NMR techniques, and the A and B phases of the superfluid were identified and then associated with the theoretical ABM and BW states primarily because of their NMR signatures.

The magnetic susceptibility of normal liquid ^{3}He is temperature independent at millikelvin temperatures. Since the A-phase Cooper pairs are formed with parallel spins the A-phase susceptibility is identical to that of the normal state. However, as mentioned above in Sec. 3.2.1, the pairs have the lowest magnetic (dipole) energy when the orbital and spin directions are perpendicular. In NMR we apply a magnetic field and then with a perpendicular rf field tip the magnetization from the equilibrium direction and follow the precession by observing the voltage output of a surrounding pickup coil. In the normal phase, the spins precess with a frequency determined only by the field and the value of the nuclear moment. However, in the A phase, since the spin direction is influenced by the orbital moment, if we tip the spins from their equilibrium direction there is a restoring force from the Cooper pairs themselves trying to return the spin direction to the perpendicular state. This gives an extra restoring force above the "bare" magnetic value, which means that the precession frequency is greater than that for ^{3}He atoms in the normal liquid. This is a unique signature of the A phase. In contrast, the B phase shows no such effect in the absence of other textural orienting forces, and the NMR frequency is similar to that of the normal fluid. However the susceptibility falls rapidly with decreasing temperature close to T_c and reaches a low-temperature value equal to about a third of the normal-state susceptibility. This is because in the normal liquid all the atoms have spin $\frac{1}{2}$, whereas in the B-phase superfluid the pairs have $S = 1$, $S = -1$, and $S = 0$. The $S = 0$ pairs do not contribute to the susceptibility, which therefore falls with the increasing superfluid fraction.

The dynamic magnetic properties of the superfluid phases are more interesting. Since the direction of the spins in the pairs is influenced not only by the magnetic forces but also the dipole–dipole interaction from the influence of the orbital moment, the spin dynamics becomes very complex. First, unlike the situation with free spins, we see the phenomenon of longitudinal NMR in which there are oscillations of the magnetization parallel to the steady field. The extra dipole–dipole restoring forces also give rise to subtle feedback mechanisms that lock the whole precessing system into one coherent state, as discussed in the next section.

3.5.5.2 Coherent Excited States in ^{3}He-B Any wave function has the property that a spatial gradient in any of its quantities is associated with an appropriate flow. Such gradients are responsible for some of the most fundamental properties of superfluids and superconductors. For example, gradients in the phase θ of the wave function in superfluid ^{4}He are associated with dissipationless mass flow (see Sec. 2.8). This can be thought of as the operation of Lenz's law. In an attempt to nullify the gradient the superfluid responds with a supercurrent. In superfluid ^{3}He-B we can set up (by NMR) gradients across a sample in the spin part of the wave function, by arranging for the spins to precess at different frequencies in different parts of the liquid. In this case the flow set up is a flow of spin, a spin supercurrent. Spin supercurrents are damped by thermal excitations (or the normal-fluid component), but nevertheless play a vital role in the spin dynamics and can give rise to extraordinary phenomena at very low temperatures where the normal-fluid damping becomes negligible.

The first coherent excited state discovered in ^{3}He-B is called the homogeneous precession domain (HPD) (Borovik-Romanov *et al.*, 1984; Fomin, 1984). It can be excited by both pulsed and continuous-wave (cw) NMR methods. The most striking property of the HPD, as observed with pulsed NMR, is that it is a very long-lived signal even in the presence of large gradients of the external field. In fact, a magnetic field gradient is required in order to stabilize the HPD signals.

The mechanism behind the HPD formation is illustrated in Fig. 21. Before the NMR tipping pulse, the net magnetization (spin) induced by the vertical magnetic field H is vertical throughout the whole volume. On application of a 90° NMR pulse, the net magnetization (spin) is tipped into the horizontal plane and then precesses around the steady field direction at the Larmor frequency $f_L = \gamma H/h$, where γ is the gyromagnetic ratio for ^{3}He and h is Planck's constant. If the cell is exposed to a vertical field gradient, the precession frequency is higher at the top of the cell than at the bottom. In normal liquid ^{3}He we would see the signal picked up by the NMR coil (reflecting the net transverse magnetization of the whole volume) rapidly decay as the spins precessing at different rates in different parts of the cell become out of phase. Under conditions we consider here this would limit the NMR signal time to about 1 ms.

In the superfluid B phase, however, the different precession frequencies of the spins in different parts of the cell leads to a gradient in the direction of the spin, which induces a spin supercurrent acting to counter the phase gradient. The vertical spin supercurrent so induced transports down-spin to the lower field regions and up-spin to the higher field regions.

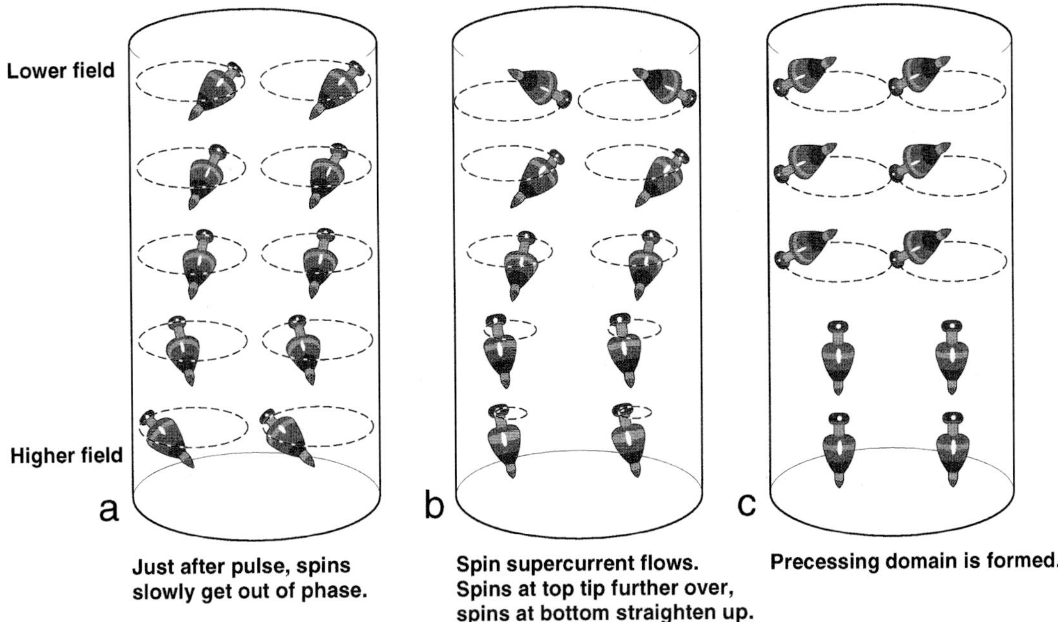

FIG. 21. Formation of the homogeneously precessing domain.

Thus the spin deflection gradually increases in the low-field region and decreases in the higher-field region. This continues until the angle of deflection reaches the critical deflection of 104°. Once this angle is exceeded the extra dipole–dipole restoring force cuts in. This has the effect of increasing the precession frequency above the local Larmor frequency.

Thus as the spin deflection in the low-field region (where the Larmor precession frequency is at its lowest) increases beyond 104°, the frequency increases. This then matches the frequency in the next highest field region; the gradient thus disappears, the supercurrent stops, and the two regions are locked together in frequency. After a short period all the spin deflection of the high-field region of the cell has been transported to the lower-field region and two domains remain, one with coherently precessing spins in the low-field region and the other of completely undeflected spin in the high-field region. This situation can be maintained for hundreds of milliseconds, and if the precessing domain can be trapped in a minimum in the texture away from the disturbing effect of walls, it can continue for tens of seconds (Bunkov *et al.*, 1992). This is in stark contrast to the 1-ms signal seen under the same conditions in the normal state.

The importance of these precessing states lies in the fact that they represent an excited but coherent state. The fact that we can study their evolution by NMR makes them not only interesting but very accessible to study.

4. OTHER SIMILAR SUPERFLUIDS

The superfluids are largely unique in that superfluidity is in general a low-temperature property and relies on the system being liquid. There are other potential superfluids, but phase transitions to solids intervene. However, superfluidity can occur in exotic circumstances. It is possible that some of the heavy-fermion metals that show superconductivity (see HEAVY-FERMION PHENOMENA) do so with Cooper pairs with spin 1. The exact structure of the pairs in high-temperature superconductors at present is a matter of conjecture.

The closest analogies to the helium superfluids may lie among the neutron and proton liquids that are believed to exist inside neutron stars. These exotic objects give rise to the very regular radio pulses that suggested the original name pulsars. They are the stellar bodies that remain after a supernova explosion. Although the temperature of these liquids in absolute terms is very high, they are actually low-temperature systems in the sense that the temperature is much lower than the appropriate Fermi temperatures ($T_F = E_F/k$). This suggests that the liquids are likely to be superfluid with neutron–neutron and proton–proton Cooper pairs as in superfluid ^{3}He. Since neutron stars rotate very rapidly, up to revolution rates approaching 1000 Hz, any internal superfluid must be traversed by a very dense lattice of vortices. It is possible that the starquakes observed when the very slowly decaying rotation rate makes very small (but easily detected) jumps are the result of sudden reorganizations of this vortex lattice.

The dilute gases of trapped cooled atoms in which Bose–Einstein condensations have recently been observed must also be superfluid below the condensation temperature. These systems are being intensively studied at present, and we expect a whole new range of superfluid phenomena to be observed. Superfluidity in these systems is quite different from that in a liquid in that the gas of particles is dilute and interactions are very small.

Finally, one further related system of great interest is provided by dilute solutions of ^{3}He in ^{4}He. At low temperatures, superfluid ^{4}He will dissolve ^{3}He up to around 10% depending on pressure. It is an intriguing thought to imagine what might happen if we took such a mixture to a temperature low enough for the dilute ^{3}He fraction to become superfluid while dissolved in an already superfluid ^{4}He. Such a mixture of two interpenetrating superfluids would certainly show unusual behavior. Interesting questions arise as, for example, in rotation how would the vortices in the two superfluids interact? Unfortunately, at the densities possible in the solution this ^{3}He transition may well need a temperature of fractions of a microkelvin, which is several orders of magnitude below our present capability.

GLOSSARY

Absolute Zero: The lowest possible temperature, at which the thermal energy and entropy are zero.

Adiabatic Demagnetization: Cooling method relying on the temperature reduction undergone by paramagnetic magnetic moments when the ambient magnetic field is reduced.

Boson: Any fundamental or composite particle whose total intrinsic spin is an even multiple of $\hbar/2$.

Bose–Einstein Condensation: The process which occurs at low temperatures in a collection of bosons when large numbers of the particles enter the ground state.

Circulation: The integrated scalar product of the velocity and distance around a closed loop.

Coherence Length: Minimum length scale on which the superfluid wave function may vary.

Cooper Pairs: The boson particles formed from a pair of fermions that allow fermion systems to become superfluid.

Dilution Refrigerator: Refrigerator operating in the millikelvin region, whose working fluid is ^{3}He dissolved in liquid ^{4}He.

Dipole–Dipole Interaction: The nonisotropic interaction between two magnetic moments. In superfluid ^{3}He this coupling determines the orientation of the (magnetic) nuclear spin relative to the (nonmagnetic) orbital angular momentum of the Cooper pairs.

Dispersion Relation: The relationship between the momentum and the energy of a particle.

Energy Gap: The minimum energy of a quasiparticle excitation in superfluid ^{3}He.

Fermion: Any fundamental or composite particle whose total intrinsic spin is an odd multiple of $\hbar/2$. Obeys Pauli's exclusion principle.

Fermi Momentum p_F, Fermi Energy E_F: Momentum and energy of a particle at the Fermi surface.

Fermi Surface: The surface marking the limit of filled states in momentum space for a system of fermions.

Gauge Symmetry: The invariance of a system in condensed matter physics under a global change of the phase of the wave function.

Homogeneous Precessing Domain: Domain that can be set up in superfluid ^{3}He in which the magnetizations of all the Cooper pairs precess coherently.

Landau Velocity: The critical velocity in a superfluid at which excitations may be freely created and the superfluidity breaks down.

Longitudinal NMR: Exotic nuclear magnetic resonance in which the exciting radiofrequency field is parallel to the static field.

Nuclear Cooling: Cooling method for the microkelvin region based on the adiabatic demagnetization of nuclear spins.

Persistent Current: A current in a superfluid that flows indefinitely because of the lack of dissipation.

Phonons, Rotons: Thermal excitations in superfluid ^{4}He.

Quasiparticle Excitations: Excited states of the system whose behavior is similar to propagating particles. These excitations may be either bosons or fermions. In the present article the term is used for the fermionic excitations, which may be particle-like or hole-like in superfluid ^{3}He.

Quantized Vortices: The vortices in a superfluid whose circulation is quantized since the wave function is limited to making a multiple of 2π change around the vortex axis.

Quantum Fluid: Fluid whose behavior is dominated on a macroscopic scale by quantum mechanics.

Spin Supercurrent: A flow of spin in superfluid ^{3}He in response to a gradient in the spin part of the wavefunction.

Spontaneous Symmetry Breaking: The lifting of a symmetry arising from ordering at a phase transition.

Superfluid: Fluid able to flow without dissipation; frictionless liquid.

Superleak: A restricted passage that allows the passage of superfluid but restricts the passage of normal fluid.

Texture: The "grain" in the structure of superfluid ^{3}He, governed ultimately by the directions of the orbital and spin angular momentum directions of the Cooper pairs.

Thermomechanical, Mechanothermal: Terms describing the linking of thermal and mechanical effects which occurs in superfluids.

Works Cited

Anderson, P. W., Brinkman, W. F. (1973), *Phys. Rev. Lett.* **30**, 1108–1110.

Anderson, P. W., Morel, P. (1961), *Phys. Rev.* **123**, 1991–1934.

Balian, R., Werthamer, N. R. (1963), *Phys. Rev.* **131**, 1553–1564.

Bardeen, J., Cooper, L. N., Schrieffer, J. R. (1957), *Phys. Rev.* **108**, 1175–1204.

Borovik-Romanov, A. S., Bunkov, Yu. M., Dmitriev, V. V., Mukharskiy, Yu. M. (1984), *Pis'ma Zh. Eksp. Teor. Fiz.* **40**, 256–259 [*JETP Lett.* **40**, 1033–1037.

Bradley, D. I., Bunkov, Yu. M., Cousins, D. J., Enrico, M. P., Fisher, S. N., Follows, M. R., Guénault, A. M., Hayes, W. M., Pickett, G. R., Sloan, T. (1995), *Phys. Rev. Lett.* **75**, 1887–1890.

Buckingham, M. J., Fairbank, W. M. (1961) *Prog. Low Temp. Phys.* **3**, 80–112.

Bunkov, Yu. M., Fisher, S. N., Guénault, A. M., Pickett, G. R. (1992), *Phys. Rev. Lett.* **69**, 3092–3095.

Campbell, L. J. (1983), *Phys. Rev. B* **27**, 1913–1915.

Clow, J. R. Reppy, J. D. (1972), *Phys. Rev. A* **5**, 424–438.

Donnelly, R. J., Donnelly, J. A., Hills, R. N. (1981) *J. Low Temp. Phys.* **44**, 471–489.

Fisher, S. N., Guénault, A. M., Kennedy, C. J., Pickett, G. R. (1991), *Phys. Rev. Lett.* **67**, 1270–1273.

Fisher, S. N., Guénault, A. M., Kennedy, C. J., Pickett, G. R. (1992), *Phys. Rev. Lett.* **69**, 1073–1076.

Fomin, I. A. (1984), *Pis'ma Zh. Eksp. Teor. Fiz.* **40**, 260–262 [*JETP Letters* **40**, 1037–1040].

Greywall, D. S. (1986), *Phys. Rev. B* **33**, 7520–7538.

Hakonen, P. J., Krusius, M., Salomaa, M. M., Simola, J. T. (1985), *Phys. Rev. Lett.* **54**, 245.

Krusius, M. (1993) *J. Low Temp. Phys.* **91**, 233–273.

Landau, L. D. (1941), *Zh. Eksp. Teor. Fiz.* **11**, 592–614.

Landau, L. D. (1956), *Zh. Eksp. Teor. Fiz.* **30**, 1058–1064 [(1957) *Sov. Phys. JETP* **3**, 920–925].

Leggett, A. J. (1984), *Phys. Rev. Lett.* **53**, 1096–1099.

Osheroff, D. D., Richardson, R. C., Lee, D. M. (1972), *Phys. Rev. Lett.* **28**, 885–888.

Schwab, K., Bruckner, N., Packard, R. E. (1997), *Nature (London)* **386**, 585–587.

Sears, V. F., Svensson, E. C., Martel, P., Woods, A. D. B. (1982), *Phys. Rev. Lett.* **49**, 279–282.

Yarmchuk, E. J., Gordon, M. J. V., Packard, R. E. (1979), *Phys. Rev. Lett.* **43**, 214–217.

Further Reading

Tilley, D. R., Tilley, J. (1974), *Superfluidity and Superconductivity*, London: Van Nostrand.

Wilks, J., Betts, D. S. (1987), *An Introduction to Liquid Helium*, Oxford: Oxford University Press.

SURFACES AND INTERFACES OF SOLIDS, STRUCTURE OF

FENG LIU, M. HOHAGE AND M. G. LAGALLY, *University of Wisconsin–Madison, Madison, Wisconsin, U.S.A.*

INTRODUCTION

Among the properties of surfaces and interfaces, structure is fundamental, as it determines or influences chemical, electrical, mechanical, and even magnetic and optical behavior. In describing structure, it is essential that a scale length be defined. For example, a surface may feel smooth or rough or it may scatter light specularly or diffusely, which are clearly macroscopic or mesoscopic measures of surface structure. On the other extreme, the atomic structure of the surface describes where the atoms are located relative to their positions in the bulk of the crystal. In the past half century, remarkable progress has been made in determining the atomic structure of the surface, driven initially by the development of ultrahigh vacuum, which permitted the preparation of clean surfaces and maintenance of the cleanliness for sufficient time to make detailed structural measurements. The subse-

ISBN 3-527-29308-6

quent invention of numerous structural characterization methods and the development of supercomputers and advanced computational algorithms have led to a broad understanding of surface atomic structure.

"Surface structure" is not just "where are the atoms". There are also larger-scale features for whose description the detailed positions of the atoms need not be known. These features are typically affected or determined by extended structural entities such as steps, terraces, dislocations, grain boundaries, and voids. The term "surface morphology" has come into use to describe such larger-range features. Their influence on surface properties can be significantly greater than that of the atomic structure. The analogous term for the bulk of materials is "microstructure". Looking down on a surface to observe its form and topology in the way one looks at the ground from an airplane illustrates why "morphology" is appropriate.

This article discusses surface atomic structure and surface morphology under the umbrella of "surface structure". It aims to provide an overview of basic concepts and fundamental principles related to structural properties of solid surfaces and interfaces and to illustrate systematic trends in structures of surfaces and interfaces in different classes of materials. Section 1 reviews surface thermodynamics, as a basis and a driving force for formation of surface structure. Section 2 introduces concepts and notations required to describe the structures of single-crystal surfaces. Section 3 overviews intrinsic structures of single-crystal surfaces and discusses surface relaxation and reconstruction, with examples chosen from semiconductor and metal surfaces. Section 4 deals with the morphology of real surfaces and interfaces, and the development of morphology through kinetic limitations or thermodynamic driving forces encountered during thin-film growth or surface treatment.

1. SURFACE THERMODYNAMICS

1.1 Gibbs Surface Model and Surface Tension

J. W. Gibbs (1928) laid the groundwork for surface thermodynamics. He introduced a no-

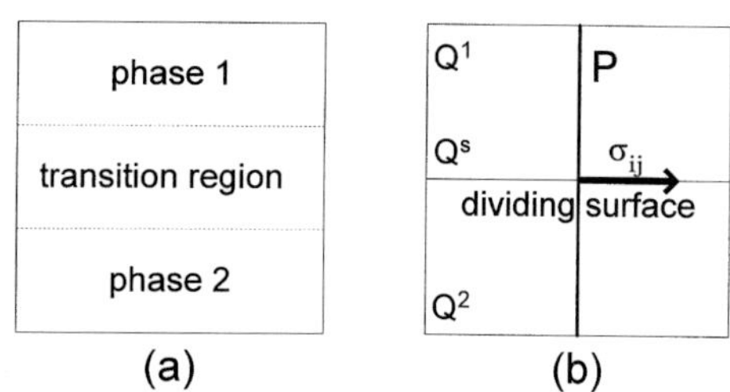

FIG. 1. The Gibbs surface model: (a) the real system; (b) the idealized system. The thin horizontal line in the middle marks the dividing surface; the thick vertical line marks the imaginary plane P used to derive the surface stress tensor σ_{ij}.

tion of a *dividing surface* and derived all surface thermodynamic properties by an *excess procedure*. A real system consisting of two homogeneous phases separated by a finite inhomogeneous transition region [Fig. 1(a)] is idealized as two homogeneous phases separated by a hypothetical geometrical surface, the dividing surface [Fig. 1(b)]. The dividing surface is chosen to be near the macroscopic observable physical interface and everywhere parallel to it; different choices of dividing surface may be used that are parallel to each other but displaced from each other in the direction of the surface normal. The excess procedure expresses the thermodynamic properties of the real system in terms of the thermodynamic properties of the ideal system and excess properties, which define the difference between the real and ideal values. Thus, all extensive surface thermodynamic functions can be expressed in terms of the dividing surface through the excess procedure. For the two-phase system that is represented by two homogeneous phases, phases 1 and 2, and a dividing surface, as shown in Fig. 1(b), any extensive property can be divided unambiguously into contributions from phase 1, phase 2, and the surface as

$$Q = Q^1 + Q^2 + Q^s, \qquad (1)$$

where Q is the extensive property of the whole system, which is the integration of the specific value of Q over the whole system including the inhomogeneous transition region; Q^1 and Q^2 are, respectively, the same extensive property for phases 1 and 2 that would apply as if they are homogeneous up to the (infinitely narrow) dividing surface; Q^s is the excess extensive property, *i.e.*, the difference between Q and $Q^1 + Q^2$, assigned to the surface. One intriguing problem with the excess extensive property is that it is usually not unique, depending on the choice (or location) of the dividing surface (see discussion below).

One defines surface tension γ as the reversible work required to create unit area of surface (by cleavage), at constant temperature (T) and pressure (p); this function is the partial derivative of Gibbs free energy (G) of the whole system with respect to area of surface formed (A), at constant T, p, and mole concentration of each component (n_i):

$$\gamma = \left(\frac{\partial G}{\partial A}\right)_{T,p,n_i}. \tag{2}$$

As $G = G(T, p, n_i, A)$, the differential of Gibbs free energy can be written as

$$dG = -S\,\mathrm{d}T + V\,\mathrm{d}p + \Sigma\mu_i\,\mathrm{d}n_i + \gamma\,\mathrm{d}A, \tag{3}$$

where S is the entropy, V the volume, and μ_i the chemical potential of molecular species i.

Alternatively, γ can be defined as the partial derivative of Helmholtz free energy (F) of the whole system with respect to A, at constant T, V, and n_i,

$$\gamma = \left(\frac{\partial F}{\partial A}\right)_{T,V,n_i}; \tag{4}$$

and the differential of Helmholtz free energy, $F = G - pV$, is

$$\mathrm{d}F = -S\,\mathrm{d}T - p\,\mathrm{d}V + \sum \mu_i\,\mathrm{d}n_i + \gamma\,\mathrm{d}A. \tag{5}$$

(Similarly, γ can be defined in terms of internal energy, $E = F + TS$, and grand canonical potential, $\Omega = F - \sum \mu_i n_i$.)

From Eqs. (1) and (3) and applying the first and second law of thermodynamics to the two homogeneous phases 1 and 2, the differential of surface Gibbs free energy is derived as

$$\mathrm{d}G^s = \sum \mu_i\,\mathrm{d}n_i^s + \gamma\,\mathrm{d}A. \tag{6}$$

Here n_i^s is the excess mole concentration of the ith species assigned to the surface. According to Euler's theorem, the surface Gibbs free energy is

$$G^s = \sum \mu_i n_i^s + \gamma A, \tag{7}$$

and similarly, the surface Helmholtz free energy is

$$F^s = pV^s + \sum \mu_i n_i^s + \gamma A \approx \sum \mu_i n_i^s + \gamma A, \quad \text{for } V^s \to 0. \tag{8}$$

Equations (7) and (8) give another definition for surface tension:

$$\gamma = g^s - \sum \mu_i \Gamma_i^s \approx f^s - \sum \mu_i \Gamma_i^s, \tag{9}$$

where g^s (f^s) is the Gibbs (Helmholtz) free energy per unit area of surface, *i.e.*, the specific surface Gibbs (Helmholtz) free energy, and Γ_i^s is the surface density n_i^s/A of the ith species.

In general, surface tension γ, which measures the change in Gibbs or Helmholtz free energy for the whole system in creating the surface, has two contributions: the change in Gibbs or Helmholtz free energy per unit area for the surface "phase", g^s or f^s, and the change per unit area of surface formed for the surrounding bulk phases. Note that g^s, f^s, and Γ_i^s are all *excess* surface properties that are dependent on the position of the dividing surface, but γ, a quantity associated with the whole system, is not, because the position of the dividing surface affects the Gibbs (or Helmholtz) free energy in such a way that the change in the surface term g^s (or f^s) is always balanced by the change in the bulk term $\Sigma\mu_i\Gamma_i^s$. For a one-component system, it is possible to choose the dividing surface so that n_1^s and, hence, Γ_1^s vanish; the specific surface Gibbs or Helmholtz free energy then becomes equal to surface tension. Consider a one-component system with solid and gas phases;

the atomic density assumes a constant value of n_{bulk} in the solid phase and a value of zero in the gas phase. In the interface (surface) region, the atomic density decreases gradually from the bulk value to zero as one goes from the bulk phase toward the gas phase. If the dividing surface is located inside the transition region, in the vicinity of the surface, there will be a thin region of "extra" atomic density (>0) in the gas phase and a thin region of "lost" atomic density ($<n_{\text{bulk}}$) in the solid phase. By choosing the position of the dividing surface so that the extra density in the gas phase equals exactly the lost density in the solid phase, there will be no excess surface density, *i.e.*, $n_1{}^{\text{s}}$ and hence $\Gamma_1{}^{\text{s}}$ vanish. For a multi-component system, however, one choice of the dividing surface that makes $n_i{}^{\text{s}}$ vanish will not make other $n_{j\neq1}{}^{\text{s}}$ vanish. Therefore, in general, surface tension γ should not be confused with the specific Gibbs (or Helmholtz) surface free energy, g^{s} (or f^{s}).

Surface tension is a very difficult quantity to measure experimentally, but it can in principle be calculated with a reliable energy functional, such as those based on first-principles (or *ab initio*) computational techniques. In an empirical way, surface tension (surface energy) of a solid surface has two contributions: the formation energy and the relaxation energy. The former reflects the breaking of bonds to make a solid surface (at the ideal bulk terminations); the latter reflects the tendency of a solid surface to distort because it is a quasi two-dimensional (2D) system and hence would like to assume an atomic structure and bonding configuration different from that of the bulk. To create a surface, it costs energy to break bonds, and the cost is partly recovered by the relaxation process involving rearrangement of atoms and bonds at the surface. The formation energy (positive) dominates the relaxation energy (negative). The energy of a single-crystal surface scales approximately with the cohesive energy of the bulk crystal. The greater the bulk cohesion, the stronger the interatomic bonds, the higher the surface energy. The weakly van der Waals–bonded rare-gas solids have the lowest surface energy; the strongly bonded semiconductors and metals have high surface energy. In contrast to that of an isotropic liquid surface, the energy of a solid surface is anisotropic, depending on the orientation of the surface, because a different number of bonds is broken to create different surface orientations.

1.2 Surface Stress and its Relation to Surface Tension

Imagine that there exists a plane P normal to the dividing surface [Fig. 1(b)]. One may derive the surface force acting on P by the same excess procedure discussed above. The total force acting across P, from the material on one side of P to the material on the other side, can be divided into contributions from the two homogeneous bulk phases and from the surface phase. The surface force per unit length of the line intersection of the plane P with the dividing surface plane is then defined as the surface stress.

Gibbs (1928) first pointed out the distinction, for the case of a solid, between surface tension and surface stress: the former measures the energy cost to create unit area of new surface; the latter measures the energy cost to deform the surface. The relationship between surface stress and surface tension can be derived as follows.

Suppose we deliberately deform the surface, at constant temperature and total amount of surface species, introducing an infinitesimal strain ε_{ij} to the surface. The work required to do so is

$$\delta W = A\sum \sigma_{ij}\varepsilon_{ij} \quad (i,j=1,2). \tag{10}$$

Here σ_{ij} denotes the surface stress tensor. This work is equal to the infinitesimal change of surface Gibbs or Helmholtz free energy, G^{s} or F^{s}, at constant T and $n_i{}^{\text{s}}$,

$$\begin{aligned}\delta G^{\text{s}} &= \delta F^{\text{s}} = \delta(\gamma A) = \gamma\delta A + A\delta\gamma \\ &= \gamma A\sum \varepsilon_{ii} + A\sum \frac{\partial\gamma}{\partial\varepsilon_{ij}}\varepsilon_{ij} \quad (i,j=1,2).\end{aligned} \tag{11}$$

Combining Eqs. (10) and (11), we have

$$\sigma_{ij} = \gamma\delta_{ij} + \frac{\partial\gamma}{\partial\varepsilon_{ij}} \quad (i,j=1,2), \tag{12}$$

where the Kronecker delta $\delta_{ij} = 1$ if $i = j$ and $\delta_{ij} = 0$ if $i \neq j$. Surface stress, σ_{ij}, is independent of the location of dividing surface, although it is an excess surface property, as indicated by its relation to surface tension γ in Eq. (12).

Equation (12) also implies that the surface stress tensor σ_{ij} is defined as the derivative of surface Gibbs or Helmholtz free energy (G^s or F^s) with respect to surface strain ε_{ij}, at constant temperature and total amount of surface species,

$$\sigma_{ij} = \frac{1}{A}\left(\frac{dG^s}{d\varepsilon_{ij}}\right)_{T,n_i^s} = \frac{1}{A}\left(\frac{dF^s}{d\varepsilon_{ij}}\right)_{T,n_i^s} \quad (i,j = 1,2). \tag{13}$$

Equivalently, surface stress may be defined in terms of other thermodynamic energy functions.

For a liquid, surface stress is equal to surface tension because liquid is a purely plastic medium and its surface tension is independent of strain, *i.e.*, the second strain-derivative term on the right-hand side of Eq. (12) vanishes. (For a one-component liquid, surface tension also equals the specific Gibbs or Helmholtz surface free energy. So, for this special case, the values of all three quantities, surface stress, surface tension and specific surface free energy, are the same, which has often caused confusion.) When a liquid surface is stretched, atoms rapidly enter the surface from the bulk; the existing surface atoms stay where they are and the new surface is formed of atoms from the bulk now occupying surface positions. Thus, expanding a liquid surface is equivalent to creating a new liquid surface. In contrast, solids are a purely elastic medium, at least at low deformations. When a solid surface is stretched, surface atoms are displaced from their minimum-energy positions, sitting at new sites for which the surface as a whole has a different energy. Thus, the surface tension of a solid surface depends on surface strain. While a liquid surface is always under tensile stress for γ to be positive, a solid surface can be under an overall stress that is either tensile or compressive because the second term in Eq. (12), the strain derivative of tension, can be either positive or negative and may have a magnitude larger than the first term, the tension contribution. Also, the strain-derivative term often makes the surface stress of a solid surface anisotropic, while the surface stress of a liquid surface is always isotropic.

2. SURFACE CRYSTALLOGRAPHY

2.1 Orientation of Crystalline Surfaces

Given a single crystal of a particular Bravais lattice (see CRYSTALLOGRAPHY), a lattice plane is defined to be any plane containing at least three noncollinear Bravais lattice points; it is denoted by Miller indices, which are the coordinates of the shortest reciprocal-lattice vector normal to the plane. A common way of creating a crystalline surface is to cut a bulk single crystal parallel to one of its low-index atomic planes (*e.g.*, {100}, {110} or {111} planes in a cubic single crystal). An ideal surface (one that has all atoms in bulk positions) is made by removing all the atoms lying on one side of a chosen atomic plane and keeping the positions of all remaining atoms intact. The orientation of this surface is then well defined, with its surface normal specified by the directional indices, [*hkl*], defined in the conventional way with respect to the three-dimensional (3D) unit cell of the bulk crystal (Ashcroft and Mermin, 1976). The surface is denoted as an (*hkl*) surface, using the Miller indices of the corresponding parallel set of atomic planes in the bulk crystal. The notation {*hkl*} denotes a set of equivalent planes or surfaces.

Low–Miller-index surfaces are widely used in surface science research and epitaxial growth of thin films because of their relatively greater stability and high symmetry. A real surface generally consists of a number of terraces of low-index surface planes of the same orientation separated by steps. In *nominal* {*hkl*} surfaces the normal to the average surface orientation coincides with the normal to the *singular* {*hkl*} terraces that constitute the surface. To achieve this condition, there must be on the average an equal number of "up" steps and "down" steps in a nominal {*hkl*} surface [see

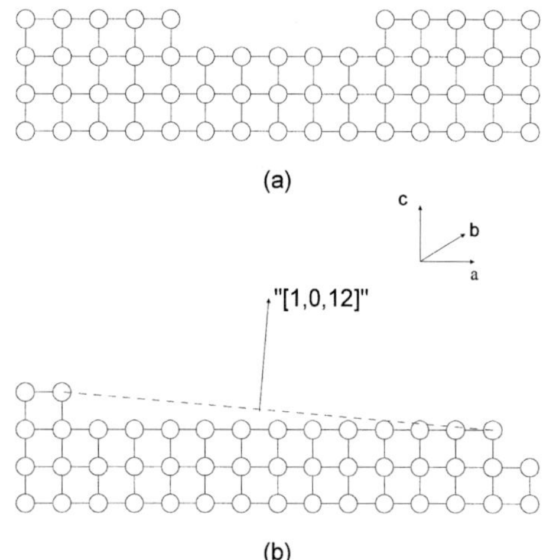

FIG. 2. Side views ([0$\bar{1}$0] projection) of (001) surfaces of a simple cubic lattice. (a) A nominal surface consisting of three singular (001) terraces separated by one up and one down single-atomic-height step. (b) A vicinal surface miscut by $\approx$4.8° toward [100]. The normal (arrow) to the average surface orientation (dashed line) is [1 0 12].

Fig. 2(a)]. In practice, it is very difficult to cut a crystal exactly parallel to a chosen low-index plane. Even if there is no average miscut, the uncertainties in cutting and polishing lead to a wavy surface, giving only a nominally low-index surface. A vicinal surface is made by a deliberate cut up to a few degrees away from a low-index plane toward a specific direction. For example, Fig. 2(b) shows a vicinal surface of a simple cubic lattice formed by a miscut $\approx 4.8^0$ away from [001] toward [100]. A vicinal surface may be specified either by the normal to its average surface orientation, which generally produces unreasonably large Miller indices [*e.g.*, the surface in Fig. 2(b) would be labeled as (1 0 12)], or by the indices of the constituent terraces plus the miscut angle and miscut orientation {*e.g.*, Si(001) miscut 2^0 toward [110]}. The latter choice makes it convenient to understand and discuss the properties of a small-miscut vicinal surface in terms of its low-index singular terraces and the effects of defects (steps). If the miscut angle is not too small and the index of the normal to the average surface orientation does not contain numbers that are too large, the surface is named with its appropriate Miller indices.

2.2 Surface Periodicity and Symmetry

The structure of a single-crystal solid is uniquely defined by its intrinsic symmetry (Ashcroft and Mermin, 1976), which includes translational symmetry (periodicity), symmetry of the Bravais lattice (primitive unit cell), and symmetry of the unit cell basis (atoms or molecules) in the unit cell. The translational symmetry in 3D is realized by 7 different *crystal systems*: cubic, tetragonal, orthorhombic, monoclinic, triclinic, trigonal, and hexagonal. Allowing more than one grid point per unit cell leads to 14 *Bravais lattices*, including types of primitive (p), body-centered (bc), face-centered (fc), and base-centered unit cells. Each lattice point may have more than one atom associated with it, reducing the symmetry of the original crystal system. There are 32 *crystallographic point groups* that a crystal structure can have, covering symmetry operations of rotation, reflection, and inversion, and combinations of them. The full symmetry of the crystal (including atomic arrangement) is described completely by 230 *crystallographic space groups*, with the addition of symmetry operations of screw axes and glide planes.

As a crystalline solid is truncated to make a surface, the translational symmetry of the solid perpendicular to the surface is removed, while the periodicity parallel to the surface remains. Because in the surface region each atom layer becomes intrinsically inequivalent to other layers, all the symmetry properties of a solid surface are two-dimensional. Consequently, surface structures, in analogy to 3D crystal structures, are classified by the 2D Bravais lattice (net) and symmetry groups of the net and of the atomic basis associated with the net. There are five different 2D Bravais nets: oblique, primitive (p) rectangle, centered (c) rectangle, square, and hexagonal, as depicted in Fig. 3; ten 2D point groups; and seventeen possible 2D space groups (Woodruff, 1981).

The 2D Bravais net consists of a grid of equivalent points; the atomic basis represents the arrangements of atoms at each net point. The translational symmetry of a real surface usually differs from that of an ideal bulk termination of the solid, *i.e.*, that of the corresponding atomic plane in the bulk, because of surface reconstruction (see discussion below). If $(\boldsymbol{a}_s, \boldsymbol{b}_s)$ and $(\boldsymbol{a}_i, \boldsymbol{b}_i)$ are the unit net vectors

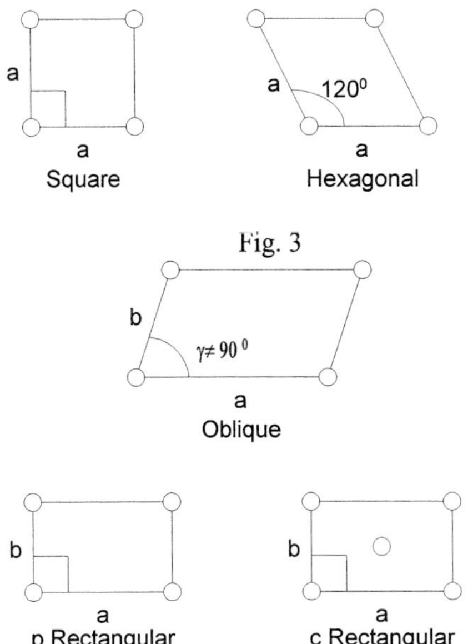

FIG. 3. The five two-dimensional Bravais nets: oblique, square, hexagonal, and primitive (p) or centered (c) rectangle.

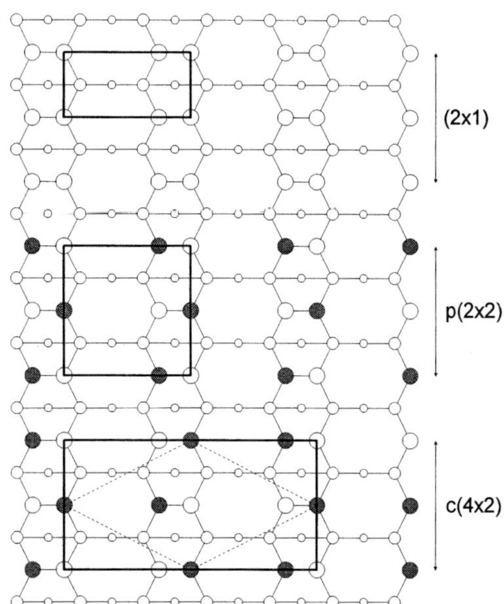

FIG. 4. Schematic top view of three possible reconstructions on Si(001). Top region: (2×1); middle: $p(2 \times 2)$; bottom: $c(4 \times 2)$. Open and solid circles mark atom positions; their size indicates different layers of atoms, with the largest circles in the outermost layer. In the middle and bottom regions, the slightly different heights of the two atoms in a dimer due to buckling are indicated by solid and open circles. The unit mesh of each reconstruction is depicted by a dark rectangle. For the $c(4 \times 2)$ structure, the primitive unit mesh is depicted by a dashed-line rhombus.

(the primitive translational vectors) of the real surface and of the ideal surface, respectively, the general translational vectors are

$$\boldsymbol{T}_s = n\boldsymbol{a}_s + m\boldsymbol{b}_s \tag{14}$$

in the real surface and

$$\boldsymbol{T}_i = n\boldsymbol{a}_i + m\boldsymbol{b}_i \tag{15}$$

in the ideal surface. Here n and m are integers. The relationship between the real surface net and ideal surface net is uniquely defined by a (2×2) matrix, $\boldsymbol{G}$ (Park and Madden, 1968),

$$\begin{pmatrix} \boldsymbol{a}_s \\ \boldsymbol{b}_s \end{pmatrix} = \boldsymbol{G}\begin{pmatrix} \boldsymbol{a}_i \\ \boldsymbol{b}_i \end{pmatrix}. \tag{16}$$

The determinant of $\boldsymbol{G}$, $|\boldsymbol{G}|$, equals the ratio of the areas of the real unit net and the ideal unit net. If $|\boldsymbol{G}|$ is an integer, the nets are *simply* related; if $|\boldsymbol{G}|$ is a rational number, the two nets are *rationally* related. When $|\boldsymbol{G}|$ is irrational, the real surface has an incommensurate structure relative to the ideal surface (substrate).

Although the matrix notation is exact, it is seldom used. Instead, a notation introduced by Wood (1964) has been widely adopted and is convenient for small commensurate unit meshes. This simple notation uses the ratio of the lengths of the real primitive net vectors to those of ideal primitive vectors plus an angle through which the real net is rotated with respect to the ideal net. Figure 4 illustrates the use of this notation to describe several possible reconstructions on Si(001). A detailed discussion of these reconstructions will be presented later.

2.3 Reciprocal-Space Representation and Diffraction

Conventionally, surface structures are detected through their representation in reciprocal space using diffraction-based surface probes. An important consequence of periodicity is the modification of the law of momentum conservation. As a wave (*e.g.*, electrons or photons) interacts with a periodic structure, its momentum is conserved plus or minus any reciprocal-lattice vector (lattice momentum). For a surface, the periodicity only exists parallel to the surface, so the modified law of momentum conservation applies only to the components of the wave vectors parallel to the surface,

$$\boldsymbol{k}_{||} = \boldsymbol{k}_{0||} + \boldsymbol{g}_{hk}. \tag{17}$$

$\boldsymbol{k}_{0||}$ and $\boldsymbol{k}_{||}$ are parallel to surface components of incident and diffracted-wave vectors. $\boldsymbol{g}_{hk}$ is a surface reciprocal-net vector (*i.e.*, a general translational vector in $\boldsymbol{k}$ space), which is expressed as

$$\boldsymbol{g}_{hk} = h\boldsymbol{a}^* + k\boldsymbol{b}^*, \tag{18}$$

where h and k are integers; $\boldsymbol{a}^*$ and $\boldsymbol{b}^*$ are primitive unit vectors of the reciprocal net, defined relative to the real-space net as

$$\boldsymbol{a}^* = (2\pi\boldsymbol{b}_s \times \boldsymbol{n})/A, \quad \boldsymbol{b}^* = (2\pi\boldsymbol{a}_s \times \boldsymbol{n})/A,$$

$$A = \boldsymbol{a}_s \cdot \boldsymbol{b}_s \times \boldsymbol{n}, \tag{19}$$

where $\boldsymbol{n}$ is the unit vector normal to the surface.

The conditions for diffraction are totally defined by the momentum conservation expressed in Eq. (17) together with an energy conservation equation,

$$k^2 = k_0{}^2 \text{ or } k_{||}{}^2 + k_{\perp}{}^2 = k_{0||}{}^2 + k_{0\perp}{}^2. \tag{20}$$

These conditions can be conveniently visualized using the Ewald sphere construction in $\boldsymbol{k}$ space, as illustrated in Fig. 5. One first draws a vector of the incident wave, $\boldsymbol{k}_0$, to the origin of the reciprocal space, and then draws a sphere of radius k_0 about the tail of the vector $\boldsymbol{k}_0$. In three dimensions, this (Ewald) sphere passes through certain reciprocal-lattice points in addition to the origin, and the diffracted beam $\boldsymbol{k}$ is simply given by a vector from the tail of $\boldsymbol{k}_0$ to those reciprocal-lattice points lying on the Ewald sphere. A surface can be approximated by a 2D net. It is straightforward to show that for a 2D net, the complete reciprocal space consists of a set of lines or "rods" that are infinite in extent, are

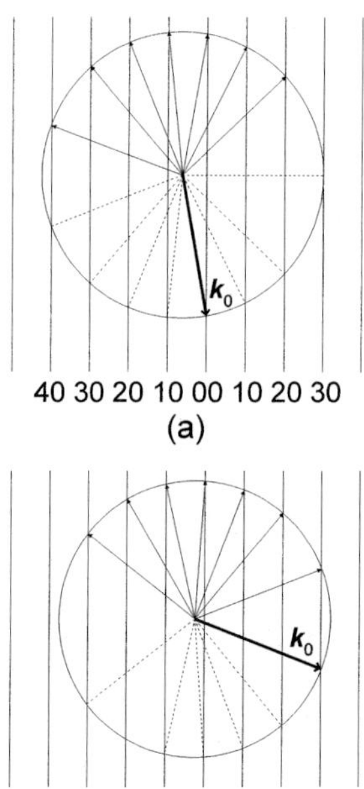

FIG. 5. Ewald sphere construction for a two-dimensional system, *e.g.*, a single layer. The incident beam is denoted by the vector k_0 and all other radial vectors ending on the Ewald sphere correspond to diffracted beams, k. (a) Near-normal incidence for LEED; (b) low-angle incidence for RHEED. Dashed lines indicate those diffracted beams penetrating into the bulk. For RHEED, k_0 is generally much larger, relative to the spacing of the reciprocal-lattice rods, than is shown in (b); for LEED, the relative magnitudes shown in (a) are appropriate.

perpendicular to the 2D net, and pass through the reciprocal-net points. Because those rods inside the Ewald sphere will always intersect the sphere, the conditions for the appearance of diffracted beams are greatly relaxed, relative to those for a 3D crystal, for which the reciprocal lattice consists of points. Therefore, a small change in energy of the incident wave, *i.e.*, in the length of $\boldsymbol{k}_0$ and the radius of Ewald sphere, will totally extinguish one set of diffracted beams in three dimensions. The same change for a surface will only shift slightly the direction of the diffracted beams.

Electron beams and x rays are typically used in diffraction-based surface probes. Figure 5 illustrates two common setups for electron diffraction: low-energy electron diffraction (LEED), generally performed at near-normal incidence, and reflection high-energy electron diffraction (RHEED), at glancing incidence. Most forms of radiation used for diffraction (electrons, x rays) have a mean free path in solids sufficiently large to penetrate some distance into the crystal. For example, for electrons this mean free path varies from approximately 5 Å up to thousands of ångströms, as the energy is changed. The actual penetration into the crystal can be limited by changing the angle of incidence (compare LEED and RHEED). The finite penetration of radiation into the crystal implies that a 2D net and perfect reciprocal-lattice rods to describe scattering from a surface form only a limiting approximation, *i.e.*, there is some 3D character to the reciprocal lattice. Thermal He and other noble-gas atoms, on the other hand, are scattered above the surface without penetrating the crystal because the energies for the wavelengths appropriate for crystal diffraction are much lower. For He scattering, a 2D net and pure reciprocal-lattice rods are a good approximation; consequently, of course, only the structure of the outer surface layer can be detected. Helium scattering is also very useful in studying "unstable" surface structures such as adsorbate layers because of the very low energy of the probe.

In addition to determining the periodicity (*i.e.*, long-range order) of the surface structure, diffraction can in principle also determine the positions of the atoms in the structure. This sensitivity to the atomic basis associated with the lattice occurs through the intensities of the diffracted beams. For electron diffraction, determination of atomic positions is complicated by the strong interaction of electrons with atoms, causing multiple scattering. The use, for example, of x rays (for which the interaction is much weaker) at grazing incidence avoids this problem.

As real-space surface structures are derived indirectly from their representations in reciprocal space with diffraction techniques, controversies often arise when different structural models are fitted to the diffraction data, especially when a large atomic basis is involved or when the relaxation from the ideal (bulk-termination) surface is small. The advance of a new generation of surface probes using scanning tunneling and force microscopies has allowed the direct observation of surfaces in real space, often with atomic resolution. In many ways, the "old" diffraction probes and the new scanned probes compensate each other: the former, working in reciprocal space, are sensitive to the long-range order, and the latter, working in real space, give local order with atomic detail. It is frequently much easier to construct a correct model to fit the diffraction data using scanned-probe real-space images as input. In modern laboratories, diffraction and real-space techniques are often combined to monitor and determine surface structures *in situ* and in real time.

3. SURFACE RELAXATION AND SURFACE RECONSTRUCTION

3.1 A General Discussion

Real surfaces do not retain the ideal bulk termination. Atoms of surface and near-surface layers are generally displaced from their ideal bulk positions, residing at new positions of minimum energy in response to a change of atomic and electronic environments in the surface regions. The atomic rearrangements can take place a few layers deep into the bulk, but the most dramatic structural changes usually occur at only the true surface layer. The structure of this outer layer dominates most surface properties. Surface *relaxation* involves only atomic displacements that do not change the translational symmetry of the surface (*i.e.*, the periodicity of the surface), *e.g.*, collective equal displacements of all atoms in the surface layer; surface *reconstruction* describes atomic displacements that change the periodicity of the surface. Surface relaxation should not be confused with atomic relaxations at defects (vacancies, adatoms, steps, *etc.*), at which the periodicity is already disturbed. A change in surface periodicity can also be caused by a change of atom number density in the surface layers. Because atoms in the surface layer are undercoordinated, missing all neighbors on the vacuum side of the surface, atoms may be added or deleted from the top layer in order to minimize the surface free energy, as has been discussed in Sec. 1.

The driving force for surface relaxation and reconstruction is minimization of surface (free) energy. Surfaces with high surface energy have a stronger tendency to relax and/or reconstruct than surfaces with low surface energy. For a given solid, higher-energy (generally higher-index) orientations of a surface usually relax and/or reconstruct more than lower-energy (lower-index) orientations [see, *e.g.*, the discussion in Sec. 3.3 of fcc metal (110), (100), and (111) surfaces]. It is, however, the reduction of surface energy rather than the surface energy itself that controls the tendency for a surface to relax and/or reconstruct. The nature of interatomic interactions in different classes of solids is the key factor controlling their respective surface structural properties. The surfaces of rare-gas solids, in which atoms are bonded with the weak van der Waals interaction, are most inert, exhibiting only small relaxations perpendicular to the surface; surfaces of metals that have isotropic metallic bonds most often do not reconstruct but display large relaxations; semiconductor surfaces with broken highly directional covalent bonds have the strongest tendency to reconstruct.

The lowering of surface energy can be achieved by reducing surface chemical energy and/or surface strain energy; the two often compete. The reduction of chemical energy, which dominates at the atomic level by optimizing bonding and/or electron density, often occurs at the expense of strain energy, which is caused by bond distortion [see, *e.g.*, the discussion of reconstruction on Si(001) in Sec. 3.2]. The reduction of strain energy becomes dominant in determining morphologies at the mesoscopic scale because elastic interactions extend to a much longer range than do chemical interactions.

The fundamental mechanisms for reducing surface chemical energy through surface relaxation/reconstruction depend strongly on the nature of chemical bonding in the underlying solid. In semiconductors, atoms bond with each other by sharing electron pairs in covalent bonds. The major effect of creating a surface is to break covalent bonds, introducing a large number of dangling bonds in the surface. Consequently, there exists a strong driving force for reconstructing semiconductor surfaces to remove dangling bonds on surface atoms. In metals, atoms bond with each other through the effective medium of an electron gas, and the major effect of creating a surface is the depletion of electron density in the vicinity of the surface. Surface atoms tend to rearrange themselves to reoptimize their surrounding electron density. Such atomic rearrangement in both semiconductor and metal surfaces involves, inevitably, changes of interatomic spacings and distortion of bond angles, so that the gain in chemical energy is usually achieved at the expense of strain energy.

The surface strain energy can be reduced by morphological changes at the mesoscopic scale, such as creation and modulation of steps and dislocations (see Sec. 4.2). Such structural and morphological changes give rise to a variation of surface stress over the surface, which generates a distribution of elastic force density. The strain relaxation energy equals the integral of force density times displacement over the whole surface area. Thus, the strain relaxation process is always accompanied by a redistribution of the surface stress field.

A large portion of research in surface science, especially in earlier years, focused on determination of the intrinsic surface structure—in particular, surface relaxation and surface reconstruction of clean crystalline solid surfaces. In the following we discuss some general features of relaxation and reconstruction in semiconductor and metal surfaces and present a few typical examples.

3.2 Semiconductor Surfaces

Semiconductors are generally made of atoms with s and p valence orbitals each containing on average two electrons (the half-filled s and p shells of an average atomic configuration yield the filled valence band and empty conduction band of the semiconducting bulk). In those semiconductors that crystallize in diamond (elemental semiconductors) and zincblende (binary compound semiconductors) structures, the s and p orbitals at each atomic site hybridize into an sp^3 configuration, which is characterized by four equivalent hybrid orbitals pointing to the corners of a tetrahedron with the atom at the center. Therefore, each atom is tetrahedrally coordinated, forming four covalent bonds with its four nearest neighbors, and each bond shares an electron pair, amounting effectively to a total of eight electrons, a very stable closed s and p shell, on each atom. In a Group-IV elemental semiconductor, each atom contributes one electron per covalent bond. In a binary compound semiconductor, each atom contributes $f = n_v/4$ electrons per covalent bond, where n_v is the number of valence electrons in a single atom; $f_c + f_a = 2$, where subscripts c and a denote cations and anions, respectively.

In creating a surface, some covalent bonds must be broken, leaving behind dangling bonds, hybrid orbitals with a single electron. Because of the high energy cost of dangling bonds, the surface energy of semiconductor surfaces is usually high, and there exists a strong tendency for semiconductor surfaces to relax and/or to reconstruct to reduce the number of dangling bonds or to reconfigure the dangling bonds into a lower-energy state.

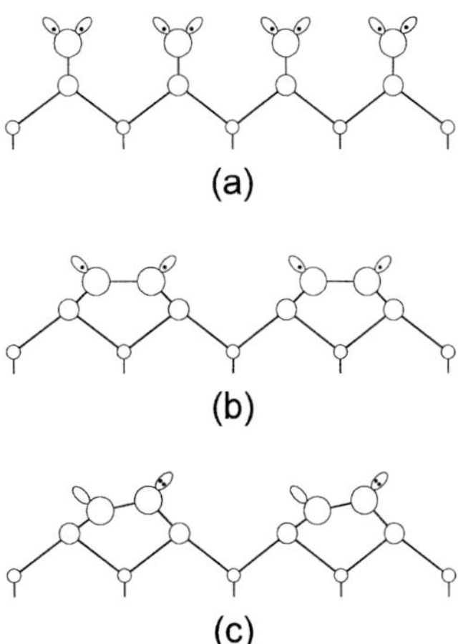

FIG. 6. Dangling-bond configurations on {100} surfaces of elemental semiconductors: (a) two dangling bonds (hybrids with a single electron) on each surface atom in an ideal bulk-terminated surface; (b) one dangling bond on each atom in a dimer. Each dangling bond is half-filled, leading to a metallic surface layer. (c) Charge transfer from downward atom to upward atom in a tilted ("buckled") dimer. The dangling bond on the lower atom is empty and the one on the higher atom is filled, leading to a semiconducting surface layer.

3.2.1 The {100} Surfaces The {100}-oriented surfaces of elemental and compound semiconductors have been the most extensively studied surfaces because of their technological importance. In particular, the Si(100) surface serves as the foundation of modern semiconductor technology. In an ideal {100} surface of a diamond or zincblende structure, each atom has two dangling bonds, caused by the loss of two of its nearest neighbors [see Fig. 6(a)]. The primary feature of a reconstructed {100} surface is dimerization, the rebonding of two neighboring surface atoms to form a dimer to eliminate one dangling bond per atom [Fig. 6(b)]. The rebonding, which bonds two second-nearest-neighbor atoms into a distance almost equal to the first-nearest-neighbor separation, requires a rather large distortion of bond angles. Thus, the decrease in electronic energy by rebonding is partly offset by an increase in strain energy. The minimization of strain energy drives the dimers to order into parallel rows, leading to a (2×1) reconstruction (see top region of Fig. 4). The dimerization of Si(001) and Ge(001) was first proposed by Schlier and Farnsworth (1959) as an explanation of their (2×1) LEED patterns observed at room temperature. The dimer rows have been "seen" directly by scanning tunneling microscopy (STM) on both Si(001) (Tromp *et al.*, 1985, Hamers *et al.*, 1986a) and Ge(001) (Kubby *et al.*, 1987), as well as on (001) surfaces of compound semiconductors, such as GaAs(001) (Biegelsen *et al.*, 1990a, 1990b).

Using a typical dimer bond length of 2.5 Å, it is estimated that the gain in electronic energy is about 4 eV per dimer, based on a tight-binding model (Harrison, 1981), and the cost in strain energy is about 2 eV per dimer, based on the Keating model (Keating, 1966); so, overall, the dimerization, or more precisely the (2×1) reconstruction, lowers the surface energy by ≈ 2 eV per dimer. This estimate is roughly the same for all the elemental and binary compound semiconductor {100} surfaces, and it is in very good agreement with *ab initio* calculations (Ramstad *et al.*, 1995) for Si(001).

3.2.1.1 Dimer Buckling in the {100} Surfaces of Elemental Semiconductors The energies of two dangling bonds are degenerate in a dimer reconstruction if the dimers consist of two identical atoms in identical positions. Such dimers are called symmetric or "unbuckled" dimers. In the symmetric configuration, the two degenerate dangling bonds are both half-filled, resulting in a metallic surface. Because such energy degeneracy is not a spin degeneracy, the symmetric structure is unstable against Jahn–Teller distortion, which will lower the free energy by lowering the symmetry of the dimer, *i.e.*, by removing the degeneracy of the dangling bonds in the dimer and changing the surface from metallic to semiconducting. An asymmetric dimer can be most easily achieved by a simple tilt, or buckling, as shown in Fig. 6(c).

The buckling of dimers induces rehybridization of surface bonds: the atom moving downward changes its back bonds more to an sp^2-like bonding (about equally bonded to its three nearest neighbors within a plane) and hence its dangling bond to more p-like; the atom moving upward changes its back bonds to more p-like (bond angles are reduced toward 90°) and its dangling bonds to more s-like. The two reconfigured dangling-bond states result in a charge transfer from the downward atom to the upward atom and a lowering of electronic energy. In other words, the buckling of a dimer splits two degenerate dangling-bond energy levels. The dangling-bond bonding level, of s character, lies below the maximum of the bulk valence band and will then be occupied; the dangling-bond anti-bonding level, of p character, lies above the maximum of the valence band and will be empty, making a metallic surface layer semiconducting. Chadi (1979), using tight-binding calculations, first showed that the symmetric dimers are unstable against buckling on Si(001). *Ab initio* calculations (for a list of these calculations, see, *e.g.*, Ramstad *et al.*, 1995) later confirmed this conclusion. The energy gain from buckling is ≈ 0.2 eV per dimer (Ramstad *et al.*, 1995). Similar results are obtained for Ge(001) (Needels *et al.*, 1987).

The buckling, like the dimerization, decreases electronic energy at the expense of increasing strain energy. The minimization of strain energy causes periodic arrays of buckling patterns that give rise to $c(4 \times 2)$ and $p(2 \times 2)$ reconstructions, which have been observed at low temperatures on Si(001) and Ge(001) by LEED (Kevan and Stoffel, 1984; Tabata *et al.*, 1987) and by x-ray diffraction (Lucas *et al.*, 1993). Low-temperature STM (Wolkow, 1992) of Si(001) shows that surface dimer buckling begins to "lock in" below ≈ 200 K, the number of buckled dimers that are locked into the buckled position increases with decreasing temperature, and the ordering of buckled dimers forms local $c(4 \times 2)$ and $p(2 \times 2)$ domains, with the $c(4 \times 2)$ domains dominating.

Although buckled (tilted) dimers are more stable than unbuckled (untilted) dimers, there seems to be a rather low energy (kinetic) barrier for a buckled dimer to switch from one buckling configuration (*e.g.*, with the left-side atom up) to another (with the left-side atom down). At sufficiently high temperature, surface dimers can switch their orientations so rapidly and independently of each other that the buckling configurations are in a dynamic disorder, leading to an averaged symmetric appearance in STM and to a (2×1) diffraction pattern. LEED (Kevan and Stoffel, 1984; Tabata *et al.*, 1987) and x-ray diffraction (Lucas *et al.*, 1993) have revealed a structural phase transition from the (2×1) to the $c(4 \times 2)$ phase as the temperature is decreased, corresponding to the freezing in of rocking dimers, in the temperature range between 150 and 250 K. Theoretical calculations (Ihm *et al.*, 1983; Saxena *et al.*, 1985; Zubkus *et al.*, 1991) predict a second-order phase transition from an ordered $c(4 \times 2)$ or $p(2 \times 2)$ structure to a disordered (2×1) structure at approximately 200–250 K, in good agreement with experiments.

3.2.1.2 The {100} Surfaces of Binary Compound Semiconductors and Electron Counting Covalent bonds in a binary compound semiconductor are partially ionic because the electronegativities are smaller for the metal atoms (cations) than for the nonmetal atoms (anions). In a zincblende structure, bulk {100} planes are alternately occupied by cations and anions, and so an ideal {001} surface will be completely terminated by either cations or anions, leading to a polar surface. Polar surfaces, especially the {100}-oriented ones, have been widely used in homo- and hetero-epitaxial growth of compound semiconductors because the sticking coefficient of anion molecules depends strongly on the cation surface concentration. For example, the sticking coefficient of As_2 molecules on a GaAs(001) surface is unity if the surface is Ga-terminated but zero if it is As-terminated (Arthur, 1974; Foxon and Joyce, 1977). During MBE growth, the surface composition can be controlled from anion- to cation-rich by adjusting growth conditions, *e.g.*, the substrate temperature, the growth rate, and the ratio of the respective fluxes. All

{100} surfaces of III–V and II–VI compound semiconductors display reconstructions depending on surface composition.

The GaAs(001) surface has attracted much attention because of its potential importance in optoelectronic devices. As the As-rich GaAs(001) surface is heated from room temperature to above 450 °C, the following sequence of reconstructions is observed (Cho, 1971; Drathen *et al.* 1978; Massies *et al.*, 1980):

$$\mathrm{c}(2\times 8)\xrightarrow{\approx 475\,^{\circ}\mathrm{C}} 1\times 6\longrightarrow \mathrm{c}(6\times 4)$$
$$\xrightarrow{\approx 575\,^{\circ}\mathrm{C}} 3\times 1\longrightarrow 4\times 1$$
$$\xrightarrow{\approx 650\,^{\circ}\mathrm{C}} \mathrm{c}(8\times 2). \qquad (21)$$

The sequence is correlated with the continuous decrease of the As concentration in the surface due to thermal desorption of As_2 molecules without evaporation of Ga (Arthur, 1974). A reverse sequence is observed (van Bommel *et al.*, 1978) when cooling down the Ga-rich surface:

$$\mathrm{c}(8\times 2)\xrightarrow{\approx 450\,^{\circ}\mathrm{C}}\mathrm{c}(6\times 4)\xrightarrow{\approx 350\,^{\circ}\mathrm{C}}\mathrm{c}(2\times 8), \qquad (22)$$

which is attributed to surface segregation of As from inside the bulk (Arthur, 1974; Neave and Joyce, 1978). In the As-rich surfaces, three different forms of (2×4) and/or $\mathrm{c}(2\times 8)$ reconstructions may occur depending on growth conditions, and they are denoted as (2×4)-α, -β, and -γ structures (Farrell and Palmstrøm, 1990). If an excess amount of As is present on the surface beyond one monolayer (ML) coverage, the surface displays a $\mathrm{c}(4\times 4)$ reconstruction (Neave and Joyce, 1978) and eventually becomes (1×1) when a thin film of As of a few monolayers thick is formed. All {100} surfaces of III–V and most of the II–VI compound semiconductors except ZnSe behave like GaAs(001) (Cornelissen *et al.*, 1988).

STM images of As-rich GaAs(001) surfaces (Biegelsen *et al.*, 1990a, 1990b) reveal that the basic features of the surface reconstructions are dimers and dimer vacancies (missing dimers). At different coverages, different reconstructions arise from the specific density, the arrangement, and the orientation of dimers and dimer vacancies. The three forms of $(2\times 4)/\mathrm{c}(2\times 8)$ reconstruction on the As-rich surface correspond to coverages of 0.5, 0.75, and 1 ML of As, respectively; the $\mathrm{c}(4\times 4)$ reconstruction corresponds to a As coverage of 1.75 ML.

The formation of dimers and dimer vacancies, and hence the reconstructions, can be understood by an electron-counting model (Chadi, 1987a; Pashley, 1989). As for elemental semiconductors, the dimerization on {100} surfaces of compound semiconductors reduces the number of dangling bonds. The energy of dangling-bond states can be reduced by charge transfer between two dangling bonds to fill one dangling bond with a lone electron pair while emptying the other one. On elemental {100} surfaces, this energy redistribution is achieved by intra-dimer charge transfer through dimer buckling, while on compound {100} surfaces, it is achieved by charge transfer between the second-layer atoms and surface dimers through the formation of dimer vacancy rows. Because dangling-bond energies are lower on anions than on cations, charge transfer from dangling bonds on cations to those on anions is greatly favored over the reverse. On a cation-terminated surface, all the dangling bonds on surface cation dimers can be emptied by transferring their electrons to the second-layer anions that are exposed at the dimer-vacancy sites; on an anion-terminated surface, all the dangling bonds on surface anion dimers can be filled by obtaining extra electrons from the second-layer cations that are exposed at the dimer-vacancy sites. Consequently, all compound-semiconductor {100} surfaces exhibit a general $(2\times N)$ reconstruction; the twofold periodicity results from the dimerization, and the N-fold periodicity results from the dimer vacancy row formation. The periodicity N defining the optimal density of dimer vacancies is determined by an electron-counting rule (Mönch, 1995).

Consider an anion-stabilized $(2\times N)$ structure with M_a anion dimers and $N-M_a$ missing dimers per unit mesh. Each dimer will contain six electrons: two in the dimer bond itself and two each in the dangling bonds at each of the dimer atoms. These electrons are supplied by the surface anions as well as by the second-layer cations that are exposed to the surface at the vacancy sites. Because each cation and anion contributes, respectively, f_c and f_a electrons per bond, counting the total number of electrons in the unit mesh, we have

$$6M_a = 4M_a f_a + 4(N - M_a) f_c. \tag{23}$$

Since $f_c = 3/4$ and $f_a = 5/4$ for III–V semiconductors, and $f_c = 2/4$ and $f_a = 6/4$ for II–VI semiconductors, we obtain from Eq. (23)

$$M_a(\text{III–V}) = 3N/4; \quad M_a(\text{II–VI}) = N. \tag{24}$$

Because the value of M_a, *i.e.*, the number of dimers in the unit cell, must be an integer, the relations in Eq. (24) predict that the smallest unit mesh on anion-stabilized {100} surfaces is (2 × 4) for III–V systems and (2 × 1) for II–VI systems. Figure 7 shows a schematic diagram of the (2 × 4) reconstruction on As-terminated GaAs(001). Similarly, for a cation-stabilized (N × 2) structure with M_c cation dimers, we have

$$2M_c + 8(N - M_c) = 4M_c f_c + 4(N - M_c) f_a, \tag{25}$$

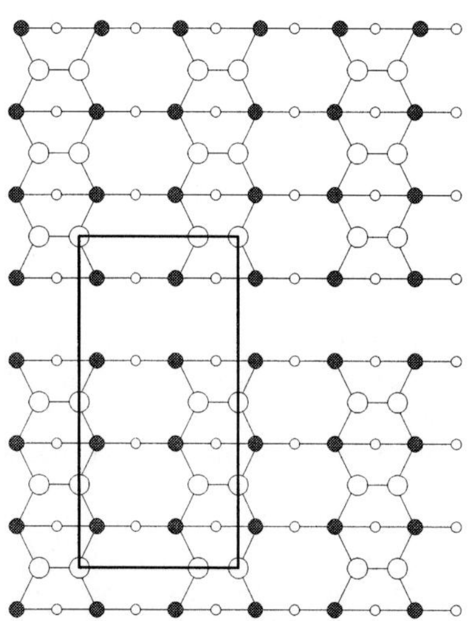

FIG. 7. Schematic top view of the (2 × 4)-β reconstruction on As-stabilized GaAs(001), induced by As vacancies. The large and small open circles denote As atoms in the first and third layers, respectively. The solid circles denote Ga atoms in the second layer. The solid rectangle marks the unit mesh.

and

$$M_c(\text{III–V}) = 3N/4; \quad M_c(\text{II–VI}) = N. \tag{26}$$

The smallest unit mesh on cation-stabilized {100} surfaces is (4 × 2) for III–V systems and (1 × 2) for II–VI systems. The acceptable values of N clearly depend on surface composition, which alters the number of electrons in the surface. So far, all the known reconstructions on the {100} surfaces of binary compound semiconductors satisfy the electron-counting rule (Mönch, 1995).

3.2.2 The {111} Surfaces The {111} surfaces of semiconductors, like the {100} surfaces, have also attracted much attention, the interest stemming from their complex structures, such as the well-known Si(111)-(7 × 7) reconstruction. In an ideally bulk-terminated {111} surface of diamond or zincblende structures, surface atoms have three nearest neighbors in the second layer and one half-filled dangling bond perpendicular to the surface [see Fig. 8(a)]. The elemental semiconductors Si and Ge cleave along {111} planes, and both as-cleaved surfaces exhibit a metastable (2 × 1) reconstruction. The ground-state structure of Si(111) is a (7 × 7) reconstruction and that of Ge(111) is a c(2 × 8) reconstruction. A simple rebonding (dimerization) mechanism like the one on Si(001) to reduce the number of dangling bonds is unlikely here because there is no easy direction for surface bonds to rotate. However, an adatom can effectively reduce the number of dangling bonds, two per adatom, by saturating three dangling bonds but introducing only one, a key ingredient on both the Si(111)-(7 × 7) and Ge(111)-c(2 × 8) reconstructions. For binary compound semiconductors, the {111} surfaces are fully terminated by either cations or anions, and the same (2 × 2) reconstructions are observed on different terminations, but with different structural origins.

3.2.2.1 The (2 × 1) Reconstruction on Cleaved Si(111) and Ge(111) and on Clean C(111) Both Si{111} and Ge{111} display a (2 × 1) reconstruction upon cleavage. The doubling of surface periodicity relative to the ideal surface (bulk {111} planes) occurs in

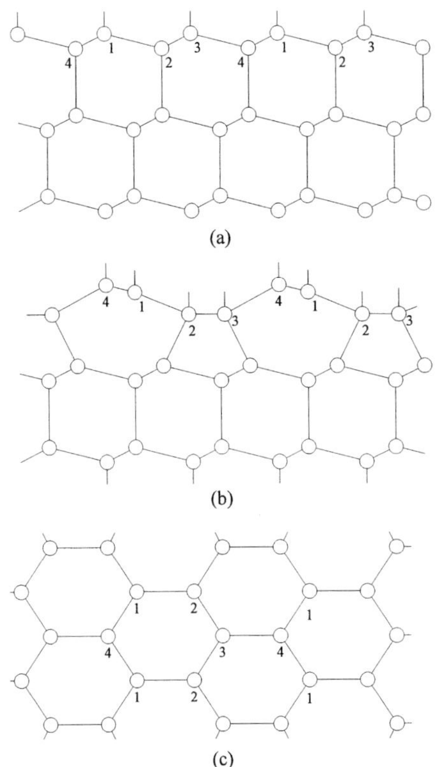

FIG. 8. π-bonded chain model of the (2×1) reconstruction on {111} surfaces of solids with the diamond structure. (a) Side view of ideally terminated surface. (b) Side view of the reconstructed surface. (c) Top view of the reconstructed surface.

⟨211⟩ directions, and the three equivalent ⟨211⟩ directions in the diamond lattice may give rise to three different orientations of (2×1) domains. In an early model (Haneman, 1961), the doubling of periodicity was thought to result from alternating rows of depressed and raised surface atoms, which was expected to lower surface energy by rehybridization of sp^3 orbitals on the surface atoms, inducing a charge transfer from the downward atoms with p-like dangling bonds to the upward atoms with s-like dangling bonds. This simple model, accepted until 1980, was found to be inconsistent with the measurements of surface electronic properties—in particular, the core-level shifts of surface atoms measured by photoemission (Himpsel *et al.*, 1980; Brennan *et al.*, 1980). A model proposed by Pandey (1982), which consists of zigzag π-bonded chains along ⟨110⟩ directions [Figs. 8(b) and 8(c)], was found to be consistent with these data. These chains are joined to the underlying bulk by five- and seven-member rings instead of the six-member rings that are characteristic of the diamond structure [see Fig. 8(b)]. The arrangement of surface atoms into such zigzag chains reduces surface energy because it conserves the total number of dangling bonds in the surface but makes them nearest neighbors and hence promotes some π-bonding among them rather than having them as second-nearest neighbors, as in the ideal or the earlier model surfaces. Surface energy is further reduced by tilting (or buckling) of chains, similar to buckling of dimers on Si(001) and Ge(001), as shown by both *ab initio* theories and various experiments (for a collection of theories and experiments, see e.g., Schlüter, 1988; Mönch, 1995).

The chains can be easily formed by simultaneously depressing a surface atom [atom 3 in Fig. 8(b)] into the second layer and raising a second-layer atom [atom 4 in Fig. 8(b)] into the surface layer, with a very small barrier of ≤ 0.03 eV (Northrup and Cohen, 1982). The reconstructed surface is about 0.3–0.4 eV per atom more stable than the ideal surface (Pandey, 1982; Northrup and Cohen, 1982). The small formation barrier explains why the chain structure can be readily formed during cleavage. The formation process may be mediated by generation of stacking faults that involves formation of five- and seven-member rings (Reichardt, 1991).

Clean diamond {111} surfaces, which are prepared by annealing polished samples in ultrahigh vacuum in the temperature range of 950–1200 °C, show a (2×2) LEED pattern (Marsh and Farnsworth, 1964; Lander and Morrison, 1966). The (2×2) pattern is interpreted as the superposition of diffraction patterns from more than one of the three possible (2×1) domains present in the surface. [Occasionally, a (2×2) pattern also occurs on Si(111) and Ge(111).] So, similar to the {111} surfaces of Si and Ge, the {111} surfaces of C display a (2×1) reconstruction, which can be explained by Pandey's chain model (Pandey, 1982). However, differences exist between the {111} surfaces of C and those of Si or Ge. In C(111), the chains are flat, as shown by total-energy calculations (for a list of theories, see

Mönch, 1995), and are purely stabilized by strong π-bonding between C dangling bonds. In Si(111) [or Ge(111)], however, the π-bonding between dangling bonds is much weaker, and it alone is not sufficient to stabilize the chain structure. Consequently, the chains are buckled and the structure is primarily stabilized by the buckling-induced charge transfer between surface atoms rather than by π-bonding (Badziag and Verwoerd, 1988).

3.2.2.2 The Si(111)-(7 × 7) and Ge(111)-c(2 × 8) Reconstructions The (2 × 1) reconstructions on the as-cleaved Si(111) and Ge(111) surfaces transform irreversibly into (7 × 7) and c(2 × 8) structures, respectively, after annealing at elevated temperature (Lander *et al.*, 1963), confirming the metastability of the (2 × 1) reconstruction. The calculated surface energy of Si(111)-(7 × 7) is indeed lower than that of Si(111)-(2 × 1) (Brommer *et al.*, 1992). The determination of atomic structure involving large surface unit meshes is generally difficult. It took almost thirty years to solve the Si(111)-(7 × 7) reconstruction after it was first observed (Schlier and Farnsworth, 1959). After many different experimental techniques had been applied and a variety of models had been proposed, a complete description of the structure was finally achieved by Takayanagi *et al.* (1985a,b), who used a comprehensive analysis of transmission electron-diffraction patterns and all the prior knowledge developed for this surface. They introduced the dimer–adatom–stacking fault (DAS) model, showing how three key ingredients involved in the reconstruction that had each been independently suggested earlier, the adatom, the stacking fault, and the dimer, contribute to the overall structure shown in Fig. 9.

The (7 × 7) unit cell consists of two triangular subunits; one of them (the left half in Fig. 9) contains a stacking fault, producing wurtzite-type stacking of the outermost two double layers. There are six adatoms in each of the two subunits, in a (2 × 2) arrangement. They are located directly over the second-layer atoms and bind to three first-layer atoms. The faulted and unfaulted subunits are separated by a triangular network of partial dislocations, and dangling bonds along these dislocations are partially saturated by the formation of dimers (three dimers at each triangle edge),

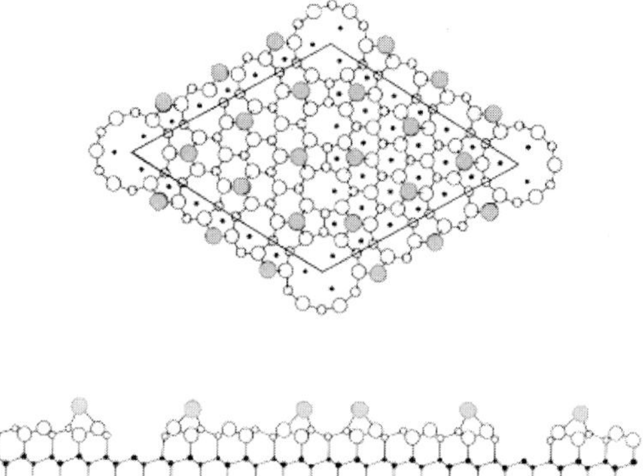

FIG. 9. (Top) Top view and (bottom) side view, taken along the long diagonal of the unit cell, of the dimer–adatom–stacking fault (DAS) model of the Si(111)-(7 × 7) reconstruction. The adatom layer (large shaded circles) and two double layers (open and solid circles) are indicated. The stacking fault is in the left half of the unit cell. (After Takayanagi *et al.*, 1985b.)

which are linked by eight-member rings. At the corners of the unit cell, *i.e.*, the crossing points of the dislocations, there are holes (corner holes) exposing large portions of the second double layer. Around the corner holes, atoms are connected by twelve-member rings.

The DAS model combines all the structural ingredients that had emerged from previous experimental studies. The adatom is the most effective way to reduce the dangling-bond density on a {111} surface of diamond structures (Harrison, 1976). However, incorporation of adatoms also introduces large strain in the surface. *Ab initio* theories (Northrup and Cohen, 1984; Northrup, 1986; Meade and Vanderbilt, 1989) show that the balance between the gain in electronic energy and the cost in strain energy produces, as the most stable arrangement of adatoms on Si(111), a (2 × 2) lattice in which adatoms sit on top of the second-layer atoms (in T_4 sites), with one rest atom (unsaturated surface atom) per (2 × 2) unit mesh. This arrangement has been seen directly by STM (Binnig *et al.*, 1983a; Becker *et al.*, 1985a; Hamers *et al.*, 1986b) in both triangle subunits of the (7 × 7) unit cell of Si(111). However, for Si(111), adatoms alone are unable to stabilize the surface. Additional structural changes must be incorporated; in particular, stacking faults (Bennett *et al.*, 1983) are introduced to create dimers at

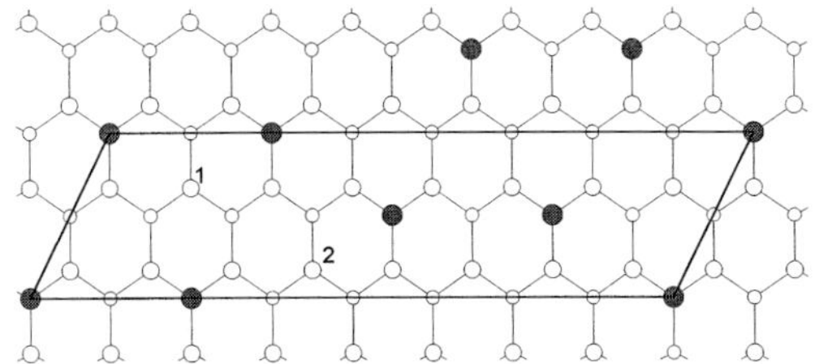

FIG. 10. Adatom model of the Ge(111)-c(2 × 8) reconstruction. Large and small open circles mark the first-layer and second-layer atoms, respectively. Shaded circles are Ge adatoms sitting on top of the second-layer atoms. Numbers 1 and 2 denote two inequivalent sites of rest atoms in the first layer. (After Becker *et al.*, 1989.)

the boundaries between faulted and unfaulted subunits (McRae, 1983; Himpsel, 1983) to reduce the dangling-bond density further. STM, equipped with spectroscopy (Tromp *et al.*, 1986), also reveals details of other structural features, such as the rest atoms, the corner holes, and the asymmetry between the two triangle subunits. Consequently, in the (7 × 7) structure, the overall electronic-energy gain (from the reduction of dangling bonds by a factor of 2.6 by creation of adatoms and dimers) offsets the strain-energy cost associated with creation of adatoms and with the formation of dimers and stacking faults, as shown by theoretical calculations (Qian and Chadi, 1987; Štich *et al.*, 1992; Brommer *et al.*, 1992).

The structure of the Ge(111)-c(2 × 8) surface, which involves only adatoms, is much simpler than that of the Si(111)-(7 × 7) surface. STM images (Becker *et al.*, 1985b, 1989) show that there are four adatoms and four rest atoms in the unit cell, as shown in Fig. 10. All adatoms are located at T_4 sites (van Silfhout *et al.*, 1990) while the rest atoms occupy two inequivalent positions (Hirschorn *et al.*, 1991). There is one dangling bond on each adatom and on each rest atom. The dangling bonds on adatoms are empty while those on rest atoms are filled (Becker *et al.*, 1989), indicating charge transfer from adatoms to rest atoms to reduce surface energy. The Ge dangling-bond energy is lower than that of Si. Also, Ge adatoms introduce less surface strain on Ge(111) than do Si adatoms on Si(111) because Ge is softer than Si (the elastic constants are approximately 40% larger for Si than for Ge). Consequently, for Ge(111), the introduction of adatoms alone is energetically favorable because the energy gain by reducing dangling-bond density by a factor of 2 more than offsets the adatom-induced energy cost in strain. The balance between the electronic-energy gain and the strain-energy cost is rather delicate in the (7 × 7) and c(2 × 8) reconstructions. Applying a compressive strain to a Ge(111) surface will convert the c(2 × 8) structure into the (7 × 7) structure, as demonstrated by the growth of thin Ge films on Si(111) (Gossmann *et al.*, 1985). Also, there exists a family of $[(2n-1) \times (2n-1)]$ DAS structures that employ the same underlying principle of energy balance as in the (7 × 7) structure. Depending on sample conditions and annealing temperatures, different domains with (5 × 5), (7 × 7), and (9 × 9) unit cells may appear on Si(111) (Becker *et al.*, 1986). The detailed mode of reconstruction can be controlled by varying the lateral stress (strain) in the surface. For example, the (5 × 5) reconstruction appears in the SiGe alloy surface under compressive stress as a thin SiGe alloy film is grown on a Si(111)-(7 × 7) substrate (Gossmann *et al.*, 1984), as well as in the Si(111) surface under tensile stress, obtained from a Si–Ge thin-film sandwich structures (Ourmazd *et al.*, 1986).

3.2.2.3 The {111} Surfaces of Compound Semiconductors The ideal {111}-(1 × 1) surfaces of binary compound semiconductors terminated by either cations or anions are conventionally denoted as (111) and $(\overline{1}\overline{1}\overline{1})$ surfaces, respectively. For III–V as well as II–VI compound semiconductors, both cation-terminated (111) and anion-terminated $(\overline{1}\overline{1}\overline{1})$ surfaces exhibit a (2 × 2) reconstruction (Ebina and Takahashi, 1982), but with different structural origins. The (2 × 2) structure of (111) surfaces arises from the formation of cation vacancies while that of $(\overline{1}\overline{1}\overline{1})$ surfaces arises from the formation of a regular array of anion trimers on top of the anion termination layer.

Most studies have been carried out on GaAs (111) and $(\overline{1}\overline{1}\overline{1})$ surfaces. In GaAs(111), each (2 × 2) unit mesh contains one Ga vacancy, exposing three As atoms in the second layer below the surface. Consequently, there are equal numbers (three) of Ga and As atoms in the unit mesh, and Ga dangling bonds are empty and As ones are filled as a result of the charge transfer from Ga to As. This structural and electronic configuration has been directly seen by STM (Haberern and Pashley, 1990). A LEED study (Tong *et al.*, 1984) shows that the exposed second-layer As atoms relax toward the center of Ga vacancies, leading to almost planar (on Ga atoms) and pyramidal (on As atoms) surface bonding configurations. The resultant atomic arrangement becomes similar to the tilted chains on nonpolar (110) surfaces of compound semiconductors (see discussion below). Total-energy calculations (Chadi, 1984) reveal that the formation of Ga vacancies on GaAs(111) is exothermic but the formation of As vacancies on GaAs($\overline{1}\overline{1}\overline{1}$) is endothermic. The vacancy-induced (2×2) reconstruction is also observed on similar systems, such as GaP(111), InSb(111) and GaSb(111) (Xu *et al.*, 1985; Bohr *et al.*, 1985; Feidenhans'l *et al.*, 1987).

Different models have been proposed for the GaAs($\overline{1}\overline{1}\overline{1}$)-$(2 \times 2)$ reconstruction. It cannot originate from As vacancies because of their large positive formation energy (Chadi, 1984). An anion-cluster model that consists of As trimers sitting on top of a complete terminating As layer was suggested from STM (Biegelsen *et al.*, 1990c). This structural model satisfies the electron-counting rule (Pashley, 1989) discussed in Sec. 3.2.1.2, and it possesses a lower surface energy than many other proposed models, as shown by an *ab initio* calculation (Kaxiras *et al.*, 1986). The anion trimer model is also supported by a TEM study of InSb($\overline{1}\overline{1}\overline{1}$) (Nakada and Osaka, 1991).

3.2.3 The {110} Surfaces The {110} surfaces of elemental semiconductors have complex structures, most notably superstructures involving steps. For example, a (16×2) reconstruction, which consists of 25-Å-wide stripes of singular terraces separated by up and down monatomic-height steps running along ⟨112⟩ directions, has been observed on both Si(110) (Yamamoto, 1994) and Ge(110) (Ichikawa *et al.*, 1995). A similar reconstruction with all the steps in the same sense (*i.e.*, all steps up or down) also forms on Si(110) (Yamamoto, 1994), leading to a vicinal facet of (17, 15, 1) structure.

Both III–V and II–VI semiconductors that crystallize in the zincblende structure cleave naturally along {110} planes, because these

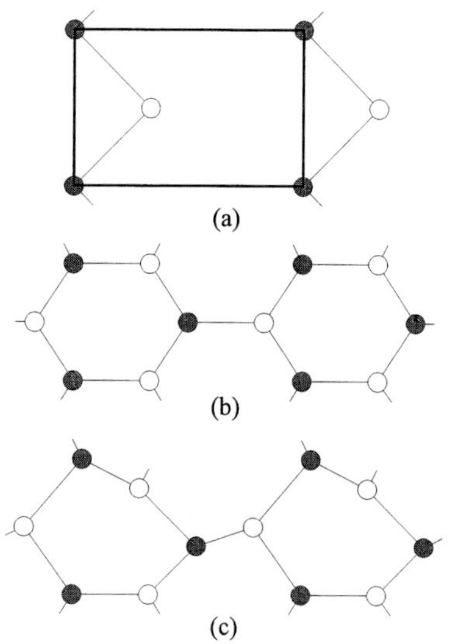

FIG. 11. Schematic atomic arrangements at {110} surfaces of zincblende-structure compound semiconductors. Open and shaded circles denote cations and anions, respectively. (a) Top view. (b) Side view of ideal bulk-terminated surface. (c) Side view of the relaxed surface.

planes contain an equal number of cations and anions and thus are intrinsically neutral. The ideal surface consists of planar zigzag chains of alternating cations and anions along the [$\overline{1}$10] direction; each surface atom remains threefold coordinated, with one dangling bond [Fig. 11(a)]. The surface displays no reconstruction but sizable relaxations. The simplest relaxation at the {110} surfaces of a zincblende crystal is the tilting (buckling) of the zigzag chains (MacRae and Gobeli, 1966; Levine and Freeman, 1970), as shown in Fig. 11(b). Because bond energies change less with changing bond angle than with changing bond length, the buckling of the chain involves mainly bond rotations with bond length being approximately conserved, as shown by a collection of experimental results on {110} surfaces of III–V, II–VI, and even I–VII compounds (Mönch, 1995). Similar to buckled dimers on Si(100) or Ge(100), the buckled chains reduce surface energy by charge transfer from the downward atoms to the upward atoms because the dangling bond on the lower atoms becomes more p-like (with backbonds more sp^2-like) while that on the upper atoms becomes more s-like (with backbonds more p-like). Also, the dangling-bond energy is lower on anions than on cations (see Sec. 3.2.2.3), so that all the cations on the {110} surfaces of compound semiconductors move downward and all the anions move upward [Fig. 11(b)], leading consistently to a charge transfer from the downward cations to upward anions. A correlation exists between buckling angle and ionicity of the underlying compound. For compounds with large ionicity, substantial charge transfer can be induced by the large difference in electronegativities between surface cations and anions, without a need for much buckling of chains. Thus, in general, the angle of buckling is found to decrease with increasing ionicity (Mönch, 1995).

3.3 Metal Surfaces

The nature of chemical bonding in metals differs from that in semiconductors. Covalent bonds are directional, while metallic bonds are more or less isotropic. Consequently, atoms in a semiconductor have to bond with each other in specific orientations, leading to a tetrahedral network and the diamond and zincblende structures; atoms in a metal, in contrast, bond as much as possible closely packed, forming face-centered cubic (fcc), hexagonal close-packed (hcp), or body-centered cubic (bcc) crystal structures, the three simple 3D lattices with highest atom number density. The fcc and hcp structures represent the closest packing of spheres in three dimensions, both having twelve nearest neighbors, while the bcc structure has eight nearest neighbors but six nearly as close second-nearest neighbors, with a separation only slightly larger than the first-nearest-neighbor distance. Such close packing maximizes the electron-density overlap. Different packing sequences result from different electronic structures of the constituent metal atoms.

The difference in bonding between semiconductors and metals also strongly influences the structural properties of the respective surfaces. As just described, in semiconductor surfaces, removal of atoms at the vacuum side produces a great number of dangling bonds, and thus a strong tendency for reconstruction to reduce their number. This driving force for reconstruction disappears in metal surfaces because there are no dangling bonds. Metal surfaces generally exhibit less tendency to reconstruct than semiconductor surfaces.

An empirical estimate of the magnitude of surface relaxation and the tendency to reconstruct for systems of close-packed spheres can be obtained by comparing the interatomic spacing to the diameter of a free atom. For example, rare-gas crystals can be approximated as the close packing of 'hard' spheres. The atoms bond with each other *via* the weak van der Waals interaction, and the atomic radii hardly change in the solid. Consequently, there is very little relaxation (less than a few percent) and no reconstruction of rare-gas crystal surfaces. Metallic bonds are much stronger than the van der Waals interaction; consequently interatomic spacings in a metal are less than the diameters of free metal atoms and metal surfaces have a stronger tendency to relax. Some metal surfaces also reconstruct, optimizing the interatomic spacings and/or the coordination numbers of surface atoms in order to minimize the surface free energy.

A more detailed understanding of the structure of metal surfaces than the sphere packing model requires knowledge of the electronic properties of metals. They can be described in the simplest limit by a jellium model, in which a periodic array of positive ions is embedded in the uniform density of a free-electron gas. Two theoretical approaches in the spirit of the jellium model, the embedded-atom method (EAM; see, *e.g.*, Foiles, 1987) and the effective-medium theory (EMT; see, *e.g.*, Jacobson and Nørskov, 1988) have been widely used to determine many properties of metals and metal surfaces. At equilibrium, metal atoms are located at sites of most favorable electron density to minimize their free energy. At the surface the electron density is reduced because of the removal of neighbors on the vacuum side. The most intuitive (but not always correct) conclusion one draws is that the atoms in the outermost surface layer can reduce their energy by relaxing inward to a higher electron density. Such inward relaxation frequently does occur. For example, the spacing of the two outermost layers of unreconstructed fcc (110) metal surfaces is reduced between 5.1% in Pd(110) (Schottke *et al.*, 1987) and 17.1% in Pb(110) (Breuer *et al.*, 1990) relative to the bulk value.

A relaxation in the opposite direction, *i.e.*, an increase in the separation of the two outer layers, is also possible, as observed, *e.g.*, in the Pt(111) surface (Materer *et al.*, 1995). Metals are frequently not well described by a jellium model, *i.e.*, the electron density does not decrease uniformly at the surface. Electronic surface states can introduce a local maximum in the electron density at a position outside the regular 3D lattice position of the outermost layer of atoms. The lowest-energy state of the surface may then involve an outward relaxation.

The electron density at the surface can also be optimized by reducing the interatomic spacing within the surface layer or by increasing the coordination number of the surface atoms, either by incorporating extra atoms into the surface layer or by rearrangement of the atoms. These effects produce reconstructions. A reconstruction may be caused by large atomic displacements or by atomic displacements so small that the reconstruction is difficult to observe. We restrict ourselves here to reconstructions that are unambiguously identified.

The desire of the surface atoms to reduce their lateral spacing leads to tensile stress in not yet reconstructed surfaces (this means that the surface energy would be reduced if the crystal were compressed). For example, *ab initio* calculations (Feibelman, 1995) have shown an excess surface stress in Pt(111) of about $0.25\,\text{eV}\,\text{Å}^{-2}$. Whether the excess surface stress is strong enough to induce a reconstruction depends on its value and on the energy cost in creating the reconstruction. For example, atomic rearrangement, including the incorporation of additional atoms within the surface layer, displaces surface atoms from their regular lattice positions. Because atoms in the topmost layer will lose some of their nearest neighbors in the second layer, the interaction between the topmost layer and the second layer is weakened. Whether the surface shows a reconstruction depends on the balance between the energy gain from increasing the atom number density in the topmost layer and the energy loss from weakening of the interaction between the topmost and second layer.

Most metal surfaces exhibit large relaxation but only some surfaces, including some orientations of the bcc metals W and Mo and of the fcc metals Au, Pt, and Ir, reconstruct. Intuitively, the higher the atom number density in a surface layer, the lower is the tendency for the surface to reconstruct, because a higher coordination implies a more bulk-like environment. The atom number density of a surface plane depends on the surface orientation (see Fig. 12). Nevertheless, there are surfaces with highest atom number densities [*e.g.*, Au(111)] that do reconstruct.

Among bcc metals, the W(100) surface is unreconstructed at room temperature, but during cooling to 250 K W(100) reconstructs spontaneously to a $\sqrt{2} \times \sqrt{2}$ R45° structure (Fig. 13) (see, *e.g.*, Pendry *et al.*, 1988). Figure 13(b) shows that the lateral displacement of the atoms in the first layer leads to zigzag chains. The atoms in the second layer are also slightly displaced, by about 20% of the displacement of the atoms in the topmost layer (Altmann *et al.*, 1988). Additionally the topmost layer shows an inward relaxation of about 6%. *Ab initio* calculations (Singh and Krakauer, 1988) have identified the gain in energy due to an increase in the effective coordination of the surface atoms as the driving force towards the reconstruction. Mo(100) exhibits a similar reconstruction at lower temperatures, but with a seven times larger unit cell (Daley *et al.*, 1993).

The fcc 5d transition metals (Au, Pt, Ir) display a variety of reconstructions on different low-index surfaces. In the following we discuss, as an example, the general trend of surface relaxation and reconstruction on these 5d metals.

3.3.1 The fcc (110) Surface Many experiments (for a list of these experiments, see, *e.g.*, Rous, 1995) have shown that the (110) surface of fcc 5d metals has a strong tendency to reconstruct. Figure 14(a) shows a schematic diagram of the unreconstructed surface. Atoms in the topmost layer have only seven nearest neighbors, a great reduction from the bulk value of twelve. Atoms in the second layer also are missing one nearest neighbor, which would sit on top of them. Even more importantly, an atom located in the (110) plane has only two in-plane nearest neighbors along the [110] direction. In the perpendicular direction there are no in-plane nearest neighbors, leading to a very weak interaction between atoms in different [110] chains. Chains of atoms in such a configuration can be rearranged easily without losing nearest-neighbor bonds.

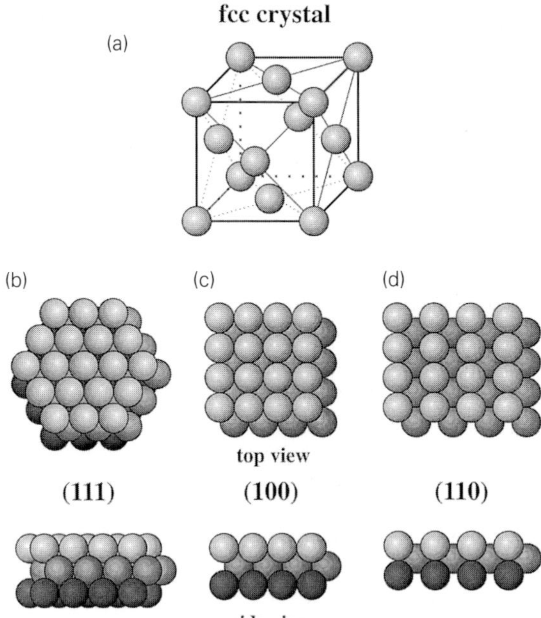

FIG. 12. Surfaces of fcc crystals. (a) Model of the fcc crystal. (b)–(d) Top and side views of the (111), (100), and (110) surfaces, respectively.

The large potential for reconstruction leads to the missing-row structure, observed, *e.g.*, on Au(110) and Pt(110). Figure 14(c) shows a schematic view of the missing-row reconstruction. Counting all atoms exposed to vacuum in the top, second, and third layers as surface atoms, the reconstruction changes neither the average coordination of surface atoms nor the surface atom-number density. However, the distribution of coordination of surface atoms has changed. In the unreconstructed surface [Fig. 14(b)], half of the surface atoms (in the topmost layer) lose five neighbors, and the other half (in the second layer) lose one neighbor; in the reconstructed surface [Fig. 14(d)], a fourth of the surface atoms (in the topmost layer) lose five neighbors, one half (in the second layer) lose three neighbors, and another fourth (in the third layer) lose one neighbor. The missing-row reconstruction transforms half of the atoms missing five or one neighbors into ones missing three neighbors by breaking some second-nearest-neighbor bonds between rows. Calculations have shown that this more uniform coordination leads to a reduction in free energy. The free-energy reduction for the missing-row reconstruction can be visualized by recognizing that the reconstruction is equivalent to forming small (111) facets (Binnig *et al.*, 1983b), the most stable surface of fcc crystals.

Surfaces of fcc (110) metals exhibit a systematic correlation between formation of a missing-row reconstruction and electronic properties. Whereas all 5d fcc transition metals reconstruct, the 4d fcc metals and the 3d transition metals show no tendency to reconstruct. The trend can be understood by comparing the energy cost of forming the missing-row reconstruction with the energy gain. The cost of

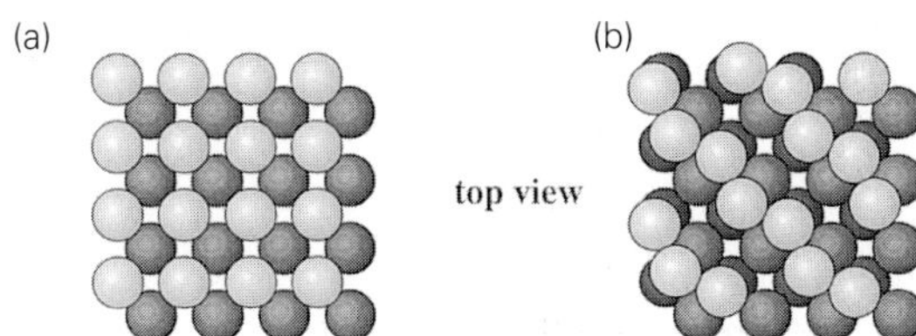

FIG. 13. Top view of the W(100) surface: (a) unreconstructed and (b) $\sqrt{2} \times \sqrt{2}$ reconstructed.

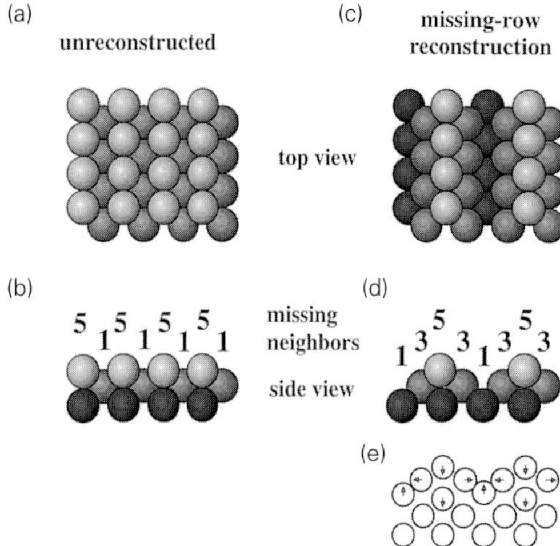

FIG. 14. Reconstruction of fcc (110) surfaces. (a), (b) Top and side views of the unreconstructed surface same as Fig. 12(d). (c), (d) Top and side views of the 2×1 missing-row reconstructed surface. (e) A schematic side view of the relaxation of the Au(110)-(2×1) missing-row reconstruction as proposed by Moritz and Wolf (1985).

breaking the second-nearest-neighbor bonds between neighboring rows is balanced by the gain produced by a more uniform distribution of the number of nearest neighbors. Thus, the gain is related to a more localized effect (the nearest-neighbor bonds) than is the cost (the second-nearest-neighbor bonds). The electrons of the 5d metals are very localized, so that the first-nearest-neighbor effect dominates over the second-nearest-neighbor effect, favoring the reconstruction. As one moves from 5d to 4d and 3d metals, electrons become less and less localized. As a result, the second-nearest-neighbor effect becomes dominant, preempting the reconstruction. This trend is reproduced qualitatively by many calculations (see Woodruff, 1994, for a selection of theoretical publications).

An interesting question is why the (110) surface does not form larger (111) facets, leading to ($N \times 1$) ($N > 2$) instead of (2×1) missing-row reconstructions. The simple model described above would lead to larger facets because the energy gain would be greater. The cause lies in the atomic relaxation. As shown in Fig. 14(e), the relaxation involves atomic displacements both perpendicular and parallel to the surface (Fery *et al.*, 1988). For larger (111) facets the relaxation would have to be mainly a uniform displacement of atoms perpendicular to the (111) facet. In certain cases, such as Au(110), calculations have shown that there is only a slight energy advantage for the 2×1 missing-row reconstruction relative to ($N \times 1$) structures with larger (111) facets [about 10 meV per atom (Ho and Bohnen, 1987)]. In agreement with these calculations, the Au(110) surface has an intrinsic amount of disorder involving (3×1) and (4×1) reconstructed regions (Binnig *et al.*, 1983b; Gritsch *et al.*, 1991). On Ir(110) a (3×1) reconstruction has been observed (Naumovets *et al.*, 1975).

3.3.2 The fcc (100) Surface In the (100) surface, atoms have eight nearest neighbors, four of them within the surface plane. The (100) surface of metal crystals is expected to have a smaller tendency to reconstruct than the (110) surface. Nevertheless all three 5d transition metals have reconstructed (100) surfaces (for a list of relevant experiments, see, *e.g.*, Rous, 1995). In contrast to the reconstruction of the (110) surface, the reconstruction of the (100) surface is controlled by gain and cost both related to nearest-neighbor bonds. The details of the underlying mechanisms are not yet known.

Both Pt(100) and Au(100) form a hexagonal surface layer on top of the square (100) lattice (Fig. 15) (Abernathy *et al.*, 1992). The

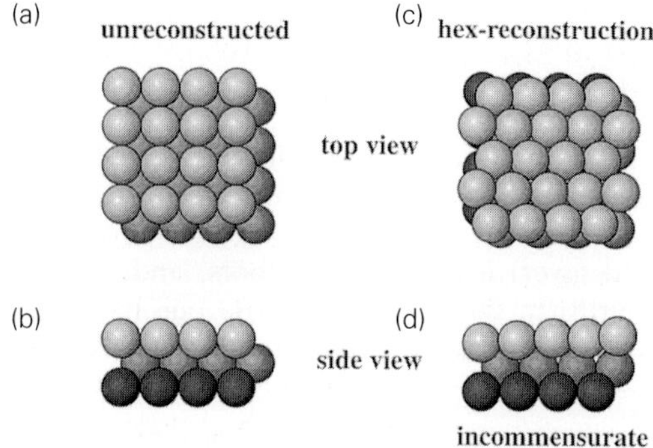

FIG. 15. Reconstruction of fcc (100) surfaces. (a), (b) Top and side views of the unreconstructed surface same as Fig. 12(c). (c), (d) Top and side views of the hexagonally reconstructed surface. The hexagonal layer has a periodicity incommensurate with that of the substrate lattice.

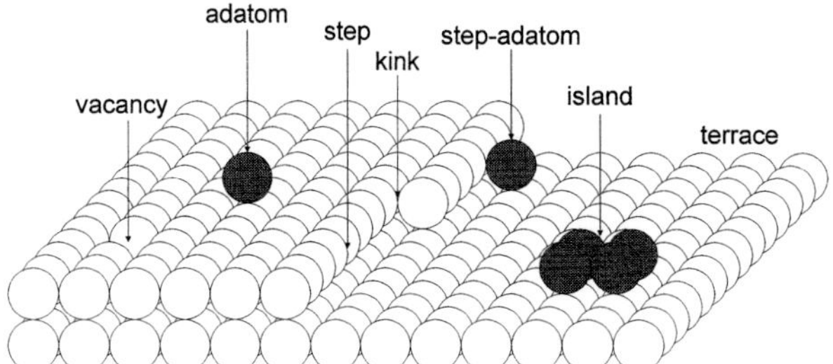

FIG. 16. The TSK model of a surface defined for a simple cubic crystal. Open circles are substrate atoms. Shaded circles are adatoms deposited on the substrate. There are two terraces (the upper terrace is on the left) separated by a step containing a single-atom kink.

atom number density in this hexagonal outermost layer is 8% higher than in a regular (111) plane. This enhanced atom number density lowers the surface free energy. Such a layer is generally higher-order commensurate or incommensurate with the underlying bulk. The energy cost of this type of reconstruction is the reduced interaction with the second layer, because the incommensuration leads to a lower average coordination with the second layer.

Ir(100) reconstructs by forming a quasihexagonal (1×5) reconstruction (Lang *et al.*, 1983). In one close-packed direction, $\langle 110 \rangle$, the interatomic spacing remains unchanged, whereas along the other direction the structure is highly compressed. The surface layer remains commensurate with the second layer. Because more atoms are located at (or near) their regular lattice sites, the weakening of interactions between the topmost layer and the second layer is not as great as in the incommensurate structures of Pt(100) and Au(100). On the other hand, the bonds within the outermost layer are not as optimized as those in an incommensurate structure.

3.3.3 The fcc (111) Surface The (111) surface is the most stable and inert surface of fcc crystals, in which surface atoms have nine nearest neighbors, three fewer than in the bulk. The reconstructions on Au(100) and Pt(100), with their hexagonal outer layers having about 8% higher atom number density than a regular (111) plane, suggest that a (111) surface reconstruction that increases the outer-layer atom number density would lower the free energy.

Au(111) reconstructs by forming the $(22 \times \sqrt{3})$ herringbone structure that allows the surface to increase the atom number density by about 5% (Perderau *et al.*, 1974; Barth *et al.*, 1990). Pt(111) reconstructs by forming a similar structure in equilibrium at high temperatures (≈ 1300 K) (Sandy *et al.*, 1992). The reconstruction of Pt(111) is also observed during homoepitaxial growth at lower temperatures (Hohage *et al.*, 1995). Calculations (Needs *et al.*, 1991) show that the excess surface stress causes a general tendency for the fcc 5d metals to reconstruct in this manner.

4. MORPHOLOGICAL FEATURES OF REAL SURFACES, INTERFACES AND THIN FILMS

4.1 The Terrace–Step–Kink (TSK) Model

Real surfaces are not perfect but contain defects, such as steps, kinks at steps, adatoms, and vacancies, which are generally described in terms of the terrace–step–kink (TSK) model (Burton *et al.*, 1951), as shown in Fig. 16. These morphological features not only influence surface properties but also play a fundamental role in thin-film growth (see THIN-FILM DEPOSITION). The STM has allowed the direct visualization of the TSK model (Swartzentruber *et al.*, 1990; Wang *et al.*, 1990) and has made possible the quantitative determination of kinetic and thermodynamic properties of morphological features ranging from a single adatom to micrometer-sized terraces.

At thermal equilibrium, the concentration of surface defects is determined by minimization of free energy, *i.e.*, the competition between their formation energy and their configurational entropy. The formation energy of a particular type of defect can be determined by measuring its concentration as a function of temperature. For example, surface steps meander by forming kinks. Step energies and kink energies on Si(001) have been determined by STM from the distribution function of kinks along the steps (Swartzentruber *et al.*, 1990). Beyond a critical temperature, the surface roughening temperature, surface steps form spontaneously, as the entropy term, which increases linearly with temperature, becomes equal to the step formation energy.

Surface defects may induce both local structural and long-range morphological changes. For example, steps and dimer vacancies on Si(001) "freeze" the motion of buckled dimers (see Sec. 3.2.1.1) between the two equivalent positions at room temperature, as a means to release surface strain energy (Wolkow, 1992). Step–step interactions (*e.g.*, entropic repulsion) influences kink formation on steps; the meandering of an individual step is independent of that of other steps only when steps are far separated (Swartzentruber *et al.*, 1990; Wang *et al.*, 1990; Zandvliet *et al.*, 1992). The kinetics of step motion (Kitamura *et al.*, 1993a) and vacancy diffusion (Kitamura *et al.*, 1993b) have also been measured with STM at high temperatures. A novel mechanism for the vacancy diffusion, involving a rolling motion at surface and subsurface atoms, has been proposed (Zhang *et al.*, 1993).

4.2 Equilibrium Step Configurations

4.2.1 Stress-Domain Structure Surface stress plays an important role in determining surface structure and morphology. Theories (Marchenko, 1981; Alerhand *et al.*, 1988) predict that if the intrinsic surface stress is anisotropic, a single-domain surface is unstable against formation of alternating degenerate stress domains. For example, the (2×1) reconstruction of Si(001) introduces a highly anisotropic surface stress: the stress $\sigma_{||}$ along the dimer bond is tensile (*i.e.*, surface atoms would like to be closer together along this direction than they are); the stress $\sigma_{\perp}$ along the dimer row is consequently compressive (or at least less tensile than $\sigma_{||}$). Because of the tetrahedral bonding configuration in the diamond structure, the dimer directions are orthogonal on terraces separated by monatomic steps, giving rise to orthogonal (2×1) domains on adjacent terraces, which are referred to as (2×1) and (1×2). The intrinsic stress anisotropy, $F = \sigma_{||} - \sigma_{\perp}$, causes a morphological instability (Marchenko, 1981; Alerhand *et al.*, 1988): a flat Si(001) surface is unstable against formation of monatomic steps as domain boundaries separating alternating (2×1) and (1×2) domains. Such stress domains are quite generally possible in systems that lower their surface energy by reconstruction.

The discontinuity of stress at the step introduces a force monopole, giving rise to an elastic step–step interaction that lowers the surface energy with a logarithmic dependence on step separation (terrace width, L) (Marchenko, 1981; Alerhand *et al.*, 1988), $E_s = C_1 - C_2 \ln(L/a)$, where C_1 denotes the step-formation energy per unit length, C_2 reflects the strength of the interaction, which is related to the stress anisotropy and elastic constants, and a is a microscopic cutoff length on the order of a lattice constant. The surface assumes the lowest-energy configuration with an optimal step separation of $L_0 = \pi a \exp(C_1/C_2 + 1)$. Therefore, a surface with sufficiently low step density could reduce its energy by introducing extra steps. However, the spontaneous formation of extra steps to reduce the size of the stress domain predicted by theory is not observed on nominally singular Si(001). Instead, the existing steps are observed to become wavy in the low–step-density regions (Tromp and Reuter, 1992). Step undulations lower the surface energy, just as creation of new steps does, by effectively reducing the size of the stress domains (Tersoff and Pehlke, 1992). Step undulations preempt the formation of extra steps because the undulations are kinetically preferred and are compatible with step flow.

4.2.2 Steps on Vicinal Surfaces Vicinal surfaces—in particular, vicinal Si(001)—play important roles in epitaxial growth and device fabrication, as templates for growing smooth thin films. The step morphology is frequently related to the vicinality of the surface. The behavior of steps on vicinal Si(001) demonstrates the range of morphological phenomena related to steps that can be encountered. Most effort has been dedicated to surfaces miscut toward a $\langle 110 \rangle$ direction, because in these directions, the dimer rows on the surface are either parallel or perpendicular to the steps. On vicinal

Si(001) surfaces with a small miscut (<1.5°) toward ⟨110⟩, the misorientation produces only monatomic steps, which separate alternating (2×1) and (1×2) domains constituting the stress-domain structure. By convention, the (2×1) notation is chosen to describe the domains that have the dimer rows perpendicular to the steps. If the surface is nominally singular (miscut by less than a few tenths of a degree), the equilibrium populations of these two domains are about equal (Men *et al.*, 1988). When an external stress is applied to such a surface, the relative populations of the two domains can be changed by straining the surface at a temperature sufficiently high so that the step mobility is high (Men *et al.*, 1988). The domain compressed along the dimer bond is favored. The experiment is well understood by the theoretical model (Alerhand *et al.*, 1988) mentioned in the last section. The response of the relative domain population to an external stress provides a unique way to determine the intrinsic surface stress anisotropy (Webb *et al.*, 1990, 1991). For Si(001), the anisotropy is $0.07 \pm 0.01\,\mathrm{eV\,Å^{-1}}$, as obtained from LEED and STM measurements (Webb *et al.*, 1990, 1991). *Ab initio* calculations (Garcia and Northrup, 1993; Dabrowski *et al.*, 1994) have produced a value in good agreement with the experiment.

Step configurations on high-miscut vicinal Si(001) surfaces are more complicated and their behavior is less well understood. As the miscut angle toward ⟨110⟩ increases, the populations of (2×1) and (1×2) domains begin to differ, with the population of (2×1) domains increasing gradually with increasing miscut angle (Tong and Bennett, 1991; Pehlke and Tersoff, 1991; Swartzentruber *et al.*, 1992). At sufficiently high miscut (≈4°) double-atomic-height steps start to form (Wierenga *et al.*, 1987; Swartzentruber *et al.*, 1992). The formation energy of double-atomic-height steps is lower than that of the monolayer-high steps (Chadi, 1987b), but, at a sufficiently small miscut angle, single-atomic-height steps may be stabilized by their elastic step–step interactions (Vanderbilt *et al.*, 1989). Several model calculations (Alerhand *et al.*, 1990; Poon *et al.*, 1990; Pehlke and Tersoff, 1991) predict the existence of a first-order phase transition from the single-atomic-height-step surface to the double-atomic-height-step surface as the miscut angle increases. Experimentally, however, no indication of a first-order phase transition has been observed. Instead, the step concentration changes continuously with both miscut angle (Swartzentruber *et al.*, 1992) and temperature (de Miguel *et al.*, 1991). Experiments (de Miguel *et al.*, 1992) also indicate that at high temperatures and/or large miscut angles, a simple 1D elastic model becomes unsatisfactory.

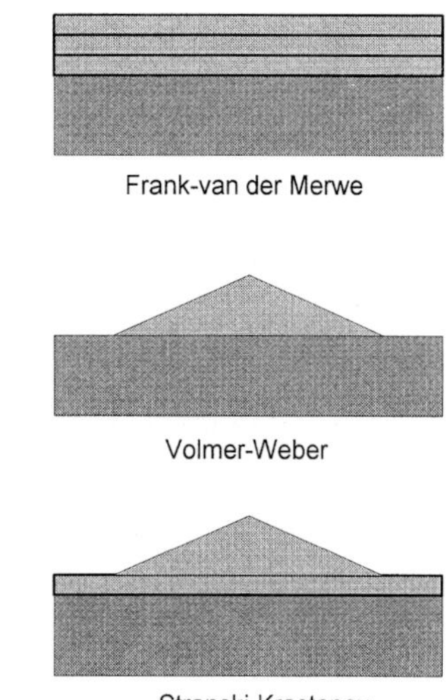

FIG. 17. Three equilibrium growth modes in heteroepitaxy.

4.3 Surface/Interface Morphology of Epitaxial Thin Films

4.3.1 Equilibrium Growth Modes The surface morphology of a thin film grown on a substrate by vapor deposition is described, at thermal equilibrium, by three different growth modes, as depicted in Fig. 17. In the Frank–van der Merwe mode, growth proceeds layer by layer, maintaining a smooth film; in the Volmer–Weber mode, 3D islands form immediately and the film does not cover the whole substrate surface; in the Stranski–Krastanov mode, the film first grows layer by layer—*i.e.*, "wets" the surface—followed by the formation of 3D islands. The realization of different growth modes, and hence different thin-film morphologies, is determined again by minimization of surface free energy, in this case for the combined substrate/film system. The surface energy of the substrate ($\gamma_{\mathrm{substrate}}$), the surface energy of the film (γ_{film}), and the interface energy between the film and substrate ($\gamma_{\text{interface}}$) (Bauer, 1958) enter into this minimization. For heteroepitaxy, *i.e.*, the growth of material A on a different substrate B, the surface energy of the film (A) will differ from that of the substrate (B). If $\gamma_{\text{substrate}} < \gamma_{\text{film}} + \gamma_{\text{interface}}$, the deposited film tends to nucleate 3D islands, leaving the low-energy substrate exposed (the Volmer–Weber mode). The condition of $\gamma_{\text{substrate}} > \gamma_{\text{film}} + \gamma_{\text{interface}}$ defines the layer-by-layer or Frank–van der Merwe growth mode. However, γ_{film} increases with increasing film thickness because of misfit strain, which leads to Stranski–Krastanov growth: the film first grows layer by layer as long as $\gamma_{\text{substrate}} > \gamma_{\text{film}} + \gamma_{\text{interface}}$, forming a finite thickness of wetting layer, followed by formation of 3D islands as $\gamma_{\text{substrate}} < \gamma_{\text{film}} + \gamma_{\text{interface}}$ to relieve the misfit strain built up in the film. In general, if A wets B, then B will not wet A. For homoepitaxy, *i.e.*, the growth of material A on a substrate of the same material A, $\gamma_{\text{substrate}} = \gamma_{\text{film}} + \gamma_{\text{interface}}$. The equilibrium film grows layer by layer in the Frank–van der Merwe mode.

Stranski–Krastanov growth has attracted considerable interest recently because of its potential for fabricating 3D spatially confined structures (Liu and Lagally, 1997a). One typical example is the growth of Ge or SiGe alloy on Si. Because Ge has a lower surface energy than Si, Ge initially grows layer by layer, wetting the Si substrate. As the film gets thicker, the strain energy in the film increases, and eventually 3D islands start to form to relieve the 4% lattice mismatch between Ge and Si. The equilibrium surface morphology of a thin Ge film grown on Si(001) is rather complex (Liu *et al.*, 1997). For pure Ge films grown on Si(001), the thickness of the wetting layer is extended up to 3 ML by strain-relaxation mechanisms involving only 2D surface roughening processes before 3D islanding takes place (Mo *et al.*, 1990; Tersoff, 1991). The misfit strain is partially relaxed by the formation of dimer vacancy lines, the $(2 \times N)$ reconstruction (Chen *et al.*, 1994). The reconstruction also induces changes in surface (step) morphology and stress field, reversing the relative roughness of the two types of steps on the surface (Wu *et al.*, 1995) as well as the surface stress anisotropy (Wu and Lagally, 1995; Liu and Lagally, 1996; Liu *et al.*, 1997). The 3D islands form in two stages. Initial islands are small and coherent with the substrate lattice, *i.e.*, without dislocations (Eaglesham and Cerullo, 1990; Mo *et al.*, 1990). Larger dislocated islands form later.

4.3.2 Kinetics-Limited Growth Morphology Equilibrium thin-film morphologies for different film systems can be achieved by growth at high temperature and low deposition rate. Very often, growth is carried out far from equilibrium, and consequently, the evolving surface morphology during thin-film growth is kinetically limited (Liu and Lagally, 1997b). The ability to control kinetic parameters to manipulate thin-film morphology allows us to grow novel artificial film structures with desirable properties. Surface diffusion is the most important kinetic parameter controlling growth morphology. Because the rate of diffusion is an exponential function of temperature, different regimes of growth that lead to different thin-film morphologies can be achieved by tuning the growth temperature, even for systems that belong to the Frank–van der Merwe growth mode, e.g., homoepitaxial growth.

Real surfaces have imperfections such as steps. As adatoms are deposited onto a surface, they are preferentially incorporated at steps, particularly at kink sites where they can bind to the most neighbors (Burton *et al.*, 1951). If all deposited adatoms are attached at steps, growth will proceed with advancement of steps, so called step-flow growth. Step-flow growth requires a high mobility of the adatoms so that they reach the existing steps before meeting with each other to nucleate islands. It is achieved, in homoepitaxy, by growing at high temperature, at low deposition rates, or on high–step-density surfaces. Step-flow growth, representing a situation close to thermal equilibrium, produces smooth films for homoepitaxial film growth. However, morphological instabilities, *e.g.*, the bunching of substrate steps, may arise from the additional kinetic barrier for adatoms crossing the steps (Ehrlich and Hudda, 1966; Schwoebel and Shipsey, 1966) or from complex dynamics of step flow (Mullins and Hirth, 1963; Kandel and Weeks, 1995). For step-flow growth of a strained film, a step-bunching instability caused by the elastic attractions between steps (Tersoff *et al.*, 1995) or by strain-modified adatom diffusion (Duport *et al.*, 1995) can occur.

At lower temperature, at high deposition rates, or on surfaces with a low step density, adatoms may meet each other in the middle of terraces before reaching the existing steps and nucleate stable 2D islands. As the deposition continues, the nucleated islands grow by incorporating new arriving adatoms, and more islands form by new nucleation events. Eventually, the 2D islands meet and coalesce. This regime is referred to as two-dimensional island growth. The 2D island growth mode is used for fundamental studies of kinetic mechanisms of homoepitaxial growth and to calibrate deposition rates in layer-by-layer growth, but serves little technological function.

The nucleation and growth of 2D islands can be described by phenomenological rate equations (Venables *et al.*, 1984). The properties of islands (*e.g.*, island number density, island compactness, and island shape) are largely determined by surface diffusion. The surface diffusion coefficient can be quantitatively determined by STM measurement of the island number density as a function of temperature at the initial stage of island growth (Mo *et al.*, 1991; Bott *et al.*, 1996; Liu and Lagally, 1997b). The compactness of islands (fractal-like versus close-packed) is controlled by how fast an adatom diffuses along island edges and crosses corners where two edges meet (Zhang, *et al.*, 1994; Hohage *et al.*, 1996; Zhang and Lagally, 1997). In general, growth at relatively lower temperatures produces less compact (more fractal-like) islands, because adatoms diffuse more slowly and thus do not reach their equilibrium positions before they become trapped by newly arriving atoms. On reconstructed surfaces, 2D island growth becomes more complex. For example, on (2×1) reconstructed Si(001), the large anisotropy in adatom sticking coefficients to two different edges (steps) of an island creates kinetics-limited islands with long and thin shapes. Reconstruction can also introduce additional roughness to the growth front. Two islands nucleating at different locations in the same terrace can form antiphase domain boundaries when they meet. For example, in Si(001) islands have a 50% chance to align their dimer rows out of phase (Hamers *et al.*, 1989), roughening the growth front. The GaAs(001) (4×2) structure has four different possibilities of island positions with only 1/4 of the islands meeting without antiphase boundaries. The larger the unit mesh, the more possibilities that exist.

Islands may nucleate not only on terraces but also on top of those islands that are already formed, when these become large enough so that atoms falling on them can no longer reach the edge. In addition to surface diffusion, the barrier for an adatom to jump off an island (*i.e.*, crossing a step) now becomes a key kinetic parameter in determining the film roughness, even for homoepitaxial films. The roughness of the film increases as the film grows thicker if atoms cannot cross the steps.

In heteroepitaxy, misfit strain can lead to various 2D morphological instabilities that roughen the surface, even in the absence of kinetic roughening, and are due to energetic (thermodynamic) driving forces. For example, as Ge is deposited on Si(001), the surface first roughens by formation of dimer vacancies, giving rise to the $(2 \times N)$ reconstruction. The dimer vacancy lines reduce adatom diffusion (Lagally, 1993), leading subsequently to additional roughness induced by new kinetic limitations (Mo and Lagally, 1991; Lagally, 1993; Wu and Lagally, 1996).

4.3.3 Interface Morphology The quality of the interface—in particular, the smoothness of the interface of a thin-film structure—frequently influences other properties, and may determine the usefulness of the film, for example, for electronic and magnetic applications. A strained film is inherently unstable; strain energy increases as the film grows thicker. Upon reaching the "equilibrium critical thickness" (van der Merwe, 1963; Matthews and Blakeslee, 1976), a strained film will relax back to its equilibrium lattice constant, forming dislocations at the interface, if the kinetics permit it. Films can be grown beyond the equilibrium critical thickness by growing at low enough temperatures so that the kinetics are so small that dislocation formation is suppressed. For homogeneous nucleation of dislocations, the kinetic barrier for nucleating a dislocation decreases with increasing strain (LeGoues *et al.*, 1993), so that eventually, as the misfit strain increases, dislocations always will form. A structural defect at the surface or interface may activate the nucleation of dislocations.

It is difficult to create a smooth interface,

even without forming dislocations. Interdiffusion roughens the interface by creating a transition region in which a compositional gradient exists. In general, interdiffusion lowers the interface energy by increasing the configurational entropy. In those cases when $2E_{A-\mathrm{B}} > E_{A-A} + E_{\mathrm{B-B}}$, where E denotes the bond energy, there will also be a strong tendency for interdiffusion to form interface alloy compounds. Substantial interdiffusion can occur even at relatively low temperatures, at which bulk diffusion is negligible, through surface diffusion-assisted atom place exchange in the outer few layers, driven by surface segregation of the thermodynamically more stable component and/or by preferential occupation of different components at different lattice sites induced by surface reconstruction. For example, Ge surface segregation causes a decrease of Ge concentration at the interface when a SiGe alloy film is grown on Si and a nonabrupt interface when Si is grown on Ge. The (2×1) reconstruction on Si(001) introduces a nonuniform stress distribution in the surface layers (LeGoues *et al.*, 1990; Liu and Lagally, 1996), leading to preferential occupations of Si or Ge at different sites and formation of an ordered SiGe phase at the interface when SiGe alloy is deposited on Si(001) (LeGoues *et al.*, 1990).

Pre-existing surface roughness generally remains at the interface after overgrowth. In the growth of a multilayer thin film, the situation is more complex because the growth of B on A can be quite different from that of A on B. As stated earlier, if B wets A, A will not wet B. In general, the roughness of a surface or interface can be quantitatively characterized by its rms roughness, lateral correlation length, and fractal exponent (Sinha *et al.*, 1988). An x-ray diffraction technique has been developed to measure the magnitude and correlation of interfacial roughness in a multilayer film (Savage *et al.*, 1991; Phang *et al.*, 1994). The interfacial roughness is frequently correlated among different layers (Savage *et al.*, 1991; Phang *et al.*, 1994). In particular, strain-induced interaction between islands in different layers leads to a self-organization of islands, with progressively improved uniformity in island size, shape, and spacing, as the multilayer thickness increases (Tersoff *et al.*, 1996; Teichert *et al.*, 1996; Liu *et al.*, 1999), which provides an attractive route to fabricating arrays and superlattices of quantum dots (Liu and Lagally, 1997a).

GLOSSARY

Adatom: A single atom adsorbed on a surface (terrace) bonded only to the surface atoms.

Basis: A group of atoms, molecules, or ions that constitute a single lattice point of the Bravais lattice.

Bravais Lattice: An infinite array of discrete points with an arrangement and orientation that appears the same from whichever of the points the array is viewed.

Crystal Systems: Point groups of the Bravais lattice (the symmetry of the basis is excluded).

Crystallographic Point Groups: Point groups of the general crystal structure including the symmetry of the basis.

Crystallographic Space Groups: Space groups of the general crystal structure.

Dangling Bond: A hybridized electron orbital on an atom without bonding on account of missing neighbors.

Dislocation: An extended line defect resulting from missing half of an atomic plane in the solid or from missing half of an atomic row in the surface.

Dividing Surface: A hypothetical geometrical surface separating two homogeneous phases to replace the real physical transition region between the two phases.

Epitaxial Growth: Growth of a film on a substrate with lattice structure of the film confined to that of the substrate.

Excess Surface Properties: The difference between the extensive thermodynamic property of a real system and that of an ideal system consisting of two homogeneous phases separated by the dividing surface.

Heteroepitaxy: Epitaxial growth of a material A on a substrate of different material B.

Homoepitaxy: Epitaxial growth of a material A on a substrate of the same material A.

Jahn–Teller Distortion: Lattice distortion that lowers the symmetry, removing the degeneracy in electronic levels and hence lowering the total energy.

Kink: A jog of any length in a step terminated by one inside corner and one outside corner.

Miller Indices: The indices of an atomic (lattice) plane that are coordinates of the shortest reciprocal lattice vector normal to the plane.

Misfit Strain: Strain induced by different equilibrium lattice constants (misfit) between the film and the substrate during heteroepitaxial growth.

Primitive Unit Cell: A volume (area) of space that, when translated through all the vectors in a Bravais lattice, just fills all the space without either overlapping itself or leaving a void.

Reciprocal Lattice: The set of all wave vectors $\boldsymbol{K}$ that yield plane waves with the periodicity of a given Bravais lattice.

Stacking Fault: A single plane where the stacking differs from the usual stacking sequence of atomic planes inside the crystal.

Stacking-Fault Region: A finite region of a stacking sequence of atomic planes different from that in the surrounding lattice.

Surface Reconstruction: Atomic displacements at a surface that change the surface periodicity relative to that of an ideally bulk-terminated surface.

Surface Relaxation: Atomic displacements at a surface that do not change the surface periodicity relative to that of an ideally bulk-terminated surface.

Surface Stress: (I) The reversible work per unit area required to deform a surface. (II) The force per unit length on a line cut in the surface.

Surface Tension: The reversible work required to create a new unit area of surface.

Vicinal Surface: A crystal surface formed by a miscut a few degrees away from a low-index atomic plane.

Works Cited

Abernathy, D. L., Mochrie, S. G., Zehner, D. M., Grübel, G., Gibbs, D. (1992), *Phys. Rev. B* **45**, 9272–9291.

Alerhand, O. L., Vanderbilt, D., Meade, R. D., Joannopoulos, J. D. (1988), *Phys. Rev. Lett.* **61**, 1973–1976.

Alerhand, O. L., Berker, N. A., Joannopoulos, J. D., Vanderbilt, D., Hamers, R. J., Demuth J. E. (1990), *Phys. Rev. Lett.* **64**, 2406–2409.

Altmann, M. S., Estrup, P. J., Robinson, I. K. (1988), *Phys. Rev. B* **38**, 5211–5214.

Arthur, J. R. (1974), *Surf. Sci.* **43**, 449–461.

Ashcroft, N. W., Mermin, N. D. (1976), *Solid State Physics*, London: Holt, Rinehart and Winston.

Badziag, P., Verwoerd, W. S. (1988), *Surf. Sci.* **201**, 87–96.

Barth, J. V., Brune, H., Ertl, G., Behm, R. J. (1990), *Phys. Rev. B* **42**, 9307–9318.

Bauer, E. (1958), *Z. Kristallogr.* **110**, 372–394.

Becker, R. S., Golovchenko, J. A., McRae, E. G., Swartzentruber, B. S. (1985a), *Phys. Rev. Lett.* **55**, 2028–2031.

Becker, R. S., Golovchenko, J. A., Swartzentruber, B. S. (1985b), *Phys. Rev. Lett.* **54**, 2678–2680.

Becker, R. S., Golovchenko, J. A., Higashi, G. S., Swartzentruber, B. S. (1986), *Phys. Rev. Lett.* **57**, 1020–1023.

Becker, R. S., Swartzentruber, B. S., Vickers, J. S., Klitsner, T. (1989), *Phys. Rev. B* **39**, 1633–1647.

Bennett, P. A., Feldman, L. C., Kuk, Y., McRae, E. G., Rowe, J. E. (1983), *Phys. Rev. B* **28**, 3656–3659.

Biegelsen, D. K., Swartz, L.-E., Bringans, R. D. (1990a), *J. Vac. Sci. Technol.* **A8**, 280–283.

Biegelsen, D. K., Bringans, R. D., Northrup, J. E., Swartz, L.-E. (1990b), *Phys. Rev. B* **41**, 5701–5706.

Biegelsen, D. K., Bringans, R. D., Northrup, J. E., Swartz, L.-E. (1990c), *Phys. Rev. Lett.* **65**, 452–455.

Binnig, G., Rohrer, H., Gerber, Ch., Weibel, E. (1983a), *Phys. Rev. Lett.* **50**, 120–123.

Binnig, G., Rohrer, H., Gerber, Ch., Weibel, E. (1983b), *Surf. Sci.* **131**, L379–L384.

Bohr, J., Feidenhans'l, R., Nielsen, M., Toney, M., Johnson, R. L., Robinson, I. K. (1985), *Phys. Rev. Lett.* **54**, 1275–1278.

Bott, M., Hohage, M., Morgenstern, M., Michely, Th., Comsa, G. (1996), *Phys. Rev. Lett.* **76**, 1304–1307.

Brennan, S., Stöhr, J., Jaeger, R., Rowe, J. E. (1980), *Phys. Rev. Lett.* **45**, 1414–1418.

Breuer, U., Prince, K. C., Bonzel, H. P., Oed, W., Heinz, K., Schmidt, G., Müller, K. (1990), *Surf. Sci.* **239**, L493–L497.

Brommer, K. D., Needels, M., Larson, B. E., Joannopoulos, J. D. (1992), *Phys. Rev. Lett.* **68**, 1355–1358.

Burton, W. K., Cabrera, N., Frank, F. C. (1951), *Philos. Trans. R. Soc. London Ser. A* **243**, 299–358.

Chadi, D. J. (1979), *Phys. Rev. Lett.*, **43**, 43–46.

Chadi, D. J. (1984), *Phys. Rev. Letters* **52**, 1911–1914.

Chadi, D. J. (1987a), *J. Vac. Sci. Technol.* **A5**, 834–837.

Chadi, D. J. (1987b), *Phys. Rev. Lett.* **59**, 1691–1694.

Chen, X., Wu, F., Zhang, Z. Y., Lagally, M. G. (1994), *Phys. Rev. Lett.* **73**, 850–853.

Cho, A. Y. (1971), *J. Appl. Phys.* **42**, 2074–2081.

Cornelissen, H. J., Cammack, D. A., Dalby, R. J. (1988), *J. Vac. Sci. Technol.* **B6**, 769–772.

Dabrowski, J., Pehlke, E., Scheffler, M. (1994), *Phys. Rev. B* **49**, 4790–4793.

Daley, R. S., Felter, T. E., Hildner, M. L., Estrup, P. J. (1993), *Phys. Rev. Lett.* **70**, 1295–1298.

de Miguel, J. J., Aumann, C. E., Kariotis, R., Lagally, M. G. (1991), *Phys. Rev. Lett.* **67**, 2830–2833.

de Miguel, J. J., Aumann, C. E., Jaloviar, S. G., Kariotis, R., Lagally, M. G. (1992), *Phys. Rev. B* **46**, 10257–10261.

Drathen, P., Ranke, W., Jacobi, K. (1978), *Surf. Sci.* **77**, L162–L166.

Duport, C., Nozières, P., Villain, J. (1995), *Phys. Rev. Lett.* **74**, 134–137.

Eaglesham, D. J., Cerullo, M. (1990), *Phys. Rev. Lett.* **64**, 1943–1946.

Ebina, A., Takahashi, T. (1982), *J. Cryst. Growth* **59**, 51–64.

Ehrlich, G., Hudda, F. G. (1966), *J. Chem. Phys.* **66**, 1039–1049.

Farrell, H. H., Palmstrøm, C. J. (1990), *J. Vac. Sci. Technol.* **B8**, 903–907.

Feibelman, P. J. (1995), *Phys. Rev. B* **51**, 17867–17875.

Feidenhans'l, R., Nielsen, M., Grey, F., Johnson, R. L., Robinson, I. K. (1987), *Surf. Sci.* **186**, 499–510.

Fery, P., Moritz, W., Wolf, D. (1988), *Phys. Rev. B* **38**, 7275–7286.

Foiles, S. M. (1987), *Surf. Sci.* **191**, L779–L786.

Foxon, C. T., Joyce B. A. (1977), *Surf. Sci.* **64**, 293–304.

Garcia, A., Northrup, J. E. (1993), *Phys. Rev. B* **48**, 17350–17353.

Gibbs, J. W. (1928), in *Collected Works*, Vol. 1, New York: Longmans, Green, p. 55.

Gossmann, H.-J., Bean, J. C., Feldman, L. C., Gibson, W. M. (1984), *Surf. Sci.* **138**, L175–L180.

Gossmann, H.-J., Bean, J. C., Feldman, L. C., McRae, E. G., Robinson, I. K. (1985), *Phys. Rev. Lett.* **55**, 1106–1109.

Gritsch, T., Coulman, D., Behm, R. J., Ertl, G. (1991), *Surf. Sci.* **257**, 297–306.

Haberern, K. W., Pashley, M. D. (1990), *Phys. Rev. B* **41**, 3226–3229.

Hamers, R. J., Tromp, R. M., Demuth, J. E. (1986a), *Phys. Rev. B* **34**, 5343–5357.

Hamers, R. J., Tromp, R. M., Demuth, J. E. (1986b), *Phys. Rev. Lett.* **56**, 1972–1975.

Hamers, R. J., Köhler, U. K., Demuth, J. E. (1989), *J. Vac. Sci. Technol.* **A8**, 195–200.

Haneman, D. (1961), *Phys. Rev.* **121**, 1093–1100.

Harrison, W. A. (1976), *Surf. Sci.* **55**, 1–19.

Harrison, W. A. (1981), *Phys. Rev. B* **24**, 5838–5843.

Himpsel, F. J. (1983), *Phys. Rev. B* **27**, 7782–7785.

Himpsel, F. J., Heimann, P., Chiang, T.-C., Eastman, D. E. (1980), *Phys. Rev. Lett.* **45**, 1112–1115.

Hirschorn, E. S., Lin, D. S., Leibsle, F. M., Samsavar, A., Chiang, T. C. (1991), *Phys. Rev. B* **44**, 1403–1406.

Ho, K. M., Bohnen, K. P. (1987), *Phys. Rev. Lett.* **59**, 1833–1836.

Hohage, M., Michely, Th., Comsa, G. (1995), *Surf. Sci.* **337**, 249–267.

Hohage, M., Bott, M., Morgenstern, M., Zhang, Z., Michely, Th., Comsa, G. (1996), *Phys. Rev. Lett.* **76**, 2366–2369.

Ichikawa, T., Sueyosi, T., Sato, T., Iwatsuki, M., Udagawa, F., Sumita, I. (1995), *Solid State Commun.* **93**, 541–545.

Ihm, J., Lee, D. H., Joannopoulos, J. D., Xiong, J. J. (1983), *Phys. Rev. Lett.* **51**, 1872–1875.

Jacobson, K. W., Nørskov, J. K. (1988), in: J. K. van der Veen, M. A. Van Hove (Eds.) *The Structure of Solid Surfaces II*, Berlin: Springer.

Kandel, D., Weeks, J. D. (1995), *Phys. Rev. Lett.* **74**, 3632–3635.

Kaxiras, E., Bar-Yam, Y., Joannopoulos, J. D., Pandey, K. C. (1986), *Phys. Rev. Lett.* **57**, 106–109.

Keating, P. N. (1966), *Phys. Rev.* **145**, 637–645.

Kevan, S. D., Stoffel, N. G. (1984), *Phys. Rev. Lett.* **53**, 702–705.

Kitamura, N., Swartzentruber, B. S., Lagally, M. G., Webb, M. B. (1993a), *Phys. Rev.* **B48**, 5704–5707.

Kitamura, N., Lagally, M. G., Webb, M. B. (1993b), *Phys. Rev. Letters* **71**, 2082–2085.

Kubby, J. A., Griffith, J. E., Becker, R. S., Vickers, J. S. (1987), *Phys. Rev. B* **36**, 6079–6093.

Lagally, M. G. (1993), *Jpn. J. Appl. Phys.* **32**, 1493–1501.

Lander, J. J., Gobeli, G. W., Morrison, J. (1963), *J. Appl. Phys.* **34**, 2298–2306.

Lander, J. J., Morrison, J. (1966), *Surface Sci.* **4**, 241–247.

Lang, E., Müller, K., Heinz, K., Van Hove, M. A., Koestner, R. J., Somorjai, G. A. (1983), *Surf. Sci.* **127**, 347–365.

LeGoues, F. K., Kesan, V. P., Iyer, S. S., Tersoff, J., Tromp, R. (1990), *Phys. Rev. Lett.* **64**, 2038–2041.

LeGoues, F. K., Mooney, P. M., Tersoff, J. (1993), *Phys. Rev. Lett.* **71**, 396–399.

Levine, J. D., Freeman, S. (1970), *Phys. Rev. B* **2**, 3255–3272.

Liu Feng, Davenport S. E., Erans, H. M., Lagally M. G. (1999), *Phys. Rev. Lett.* **82**, 2528–2531.

Liu Feng, Lagally, M. G. (1996), *Phys. Rev. Lett.* **76**, 3156–3159.

Liu Feng, Lagally, M. G. (1997a), *Surf. Sci.* **386**, 169–181.

Liu Feng, Lagally, M. G. (1997b) in: D. A. King, D. P. Woodruff (Eds.), *The Chemical Physics of Solid Surfaces*, Vol. 8, Amsterdam: Elsevier, p. 258–296.

Liu Feng, Wu Fang, Lagally, M. G. (1997), *Chem. Rev.* **97**, 1045–1061.

Lucas, C. A., Dower, C. S., McMorrow, D. F., Wong, G. C. L., Lamelas, F. J., Fuoss, P. H. (1993), *Phys. Rev. B* **47**, 10375–10382.

MacRae, A. U., Gobeli, G. W. (1966), in: R. K. Willardson, A. C. Beer (Eds.), *Semiconductors and Semimetals*, Vol. 2, New York: Academic, p. 115.

Marsh, J. B., Farnsworth, H. E. (1964), *Surface Sci.* **1**, 3–21.

Massies, J., Etienne, P., Dezaly, F., Linh, N. T. (1980), *Surf. Sci.* **99**, 121–131.

Materer, N., Starke, U., Barbieri, A., Döll, R., Heinz, K., Van Hove, M. A., Somorjai, G. A. (1995), *Surf. Sci.* **325**, 207–222.

Marchenko, V. I. (1981), *Pis'ma Zh. Eksp. Teor. Fiz.* **33**, 397–399 [*JETP Lett.* **33**, 397–399].

Matthews, J. W., Blakeslee, A. E. (1975), *J. Cryst. Growth* **29**, 273–280.

Matthews, J. W., Blakeslee, A. E. (1976), *J. Cryst. Growth* **32**, 265–273.

McRae, E. G. (1983), *Surf. Sci.* **124**, 106–128.

Meade, R. D., Vanderbilt, D. (1989), *Phys. Rev. B* **40**, 3905–3913.

Men, F. K., Packard, W. F., Webb, M. B. (1988), *Phys. Rev. Lett.* **61**, 2469–2471.

Mo, Y. W., Lagally, M. G. (1991), *J. Cryst. Growth* **111**, 876–881.

Mo, Y. W., Savage, D. E., Swartzentruber, B. S., Lagally, M. G. (1990), *Phys. Rev. Lett.* **65**, 1020–1023.

Mo, Y. W., Kleiner, J., Webb, M. B., Lagally, M. G. (1991), *Phys. Rev. Lett.* **66**, 1998–2001.

Mönch, W. (1995), *Semiconductor Surfaces and Interfaces*, 2nd ed., Berlin, New York: Springer-Verlag, Chap 7.

Moritz, W., Wolf, D. (1985), *Surf. Sci.* **163**, L655–L665.

Mullins, W. W., Hirth, J. P. (1963), *J. Phys. Chem. Solids* **24**, 1391–1404.

Nakada, T., Osaka, T. (1991), *Phys. Rev. Lett.* **67**, 2834–2837.

Naumovets, A. G., Fedorus, A. G. (1975), *Zh. Eksp. Teor. Fiz.* **68**, 1183 (1975) [*Sov. Phys. JETP* **41**, 587–589].

Neave, J. H., Joyce, B. A. (1978), *J. Cryst. Growth* **44**, 387–397.

Needels, M., Payne, M. C., Joannopoulos, J. D. (1987), *Phys. Rev. Lett.* **58**, 1765–1768.

Needs, R. J., Godfrey, M. J., Mansfield, M. (1991), *Surf. Sci.* **242**, 215–221.

Noonan, J. R., Davis, H. L. (1984), *Phys. Rev. B* **29**, 4349–4355.

Northrup, J. E. (1986), *Phys. Rev. Lett.* **57**, 154–157.

Northrup, J. E., Cohen, M. L. (1982), *Phys. Rev. Lett.* **49**, 1349–1352.

Northrup, J. E., Cohen, M. L. (1984), *Phys. Rev. B* **29**, 1966–1969.

Ourmazd, A., Taylor, D. W., Bevk, J., Davidson, B. A., Feldman, L. C., Mannaerts, J. P. (1986), *Phys. Rev. Lett.* **57**, 1332–1335.

Pandey, K. C. (1982), *Phys. Rev. Lett.* **49**, 223–226.

Park, R. L., Madden, H. H. (1968), *J. Surf. Sci.* **11**, 188–202.

Pashley, M. D. (1989), *Phys. Rev. B* **40**, 10481–10487.

Pehlke, E., Tersoff, J. (1991), *Phys. Rev. Lett.* **67**, 465–468, 1290.

Pendry, J. B., Heinz, K., Oed, W., Landskron, H., Müller, K., Schmidtlein, G. (1988), *Surf. Sci.* **193**, L1–L6.

Perderau, J., Biberian, J. P., Rhead, G. E. (1974), *J. Phys. F* **4**, 798–806.

Phang, Y. H., Teichert, C., Lagally, M. G., Peticolas, T. J., Bean, J. C., Kasper, E. (1994), *Phys. Rev. B* **50**, 14435–14445.

Poon, T. W., Yip, S., Ho, P. S., Abraham (1990), *Phys. Rev. Lett.* **65**, 2161–2164.

Qian G.-X., Chadi, D. J. (1987), *Phys. Rev. B* **35**, 1288–1293.

Ramstad, A., Brocks, G., Kelly, P. J. (1995), *Phys. Rev. B* **51**, 14504–14523.

Reichardt, D. (1991), *Prog. Surf. Sci.* **35**, 63–66.

Rous, P. J. (1995) in: F. R. de Boer, D. G. Pettifor (Eds.), *Cohesion and Structure of Surfaces*, Amsterdam: Elsevier, vol. 4, 1.

Sandy, A. R., Mochrie, S. G. J., Zehner, D. M., Grübel, G., Hwang, K. G., Gibbs, D. (1992), *Phys. Rev. Lett.* **68**, 2192–2195.

Savage, D. E., Schimke, N., Phang, Y. H., Lagally, M. G. (1991), *Phys. Rev. B* **71**, 3283–3293.

Saxena, A., Gawlinski, E. T., Gunton, J. D. (1985), *Surf. Sci.* **160**, 618–640.

Schlier, R. E., Farnsworth, H. E. (1959), *J. Chem. Phys.* **30**, 917–926.

Schlüter, M. (1988) in: King D. A. and Woodruff, D. P. (Eds.), *The Chemical Physics of Solid Surfaces*, Vol. 5, Amsterdam: Elsevier, p. 64.

Schottke, M., Behm, R. J., Ertl, G., Penka, V., Moritz, W. (1987), *J. Chem. Phys.*, **87**, 6191–6198.

Schwoebel, R. L., Shipsey, E. J. (1966), *J. Appl. Phys.* **37**, 3682–3686.

Singh, D., Krakauer, H. (1988), *Phys. Rev. B* **38**, 3999–4006.

Sinha, D. T., Sirota, E. B., Garoff, S., Stanley, H. B. (1988), *Phys. Rev. B* **38**, 2297–2311.

Štich, I., Payne, M. C., King-Smith, R. D., Lin, J.-S., Clarke, L. J. (1992), *Phys. Rev. Lett.* **68**, 1351–1354.

Swartzentruber, B. S., Mo, Y.-W., Kariotis, R., Lagally, M. G., Webb, M. B. (1990), *Phys. Rev. Lett.* **65**, 1913–1916.

Swartzentruber, B. S., Kitamura, N., Lagally, M. G., Webb, M. B. (1992), *Phys. Rev. B* **47**, 13432–13441.

Tabata, T., Aruga, T., Murata, Y. (1987), *Surf. Sci.* **179**, L63–L70.

Takayanagi, K., Tanishiro, Y., Takahashi, M., Takahashi, S. (1985a), *J. Vac. Sci. Technol.* **A3**, 1502–1510.

Takayanagi, K., Tanishiro, Y., Takahashi, M., Takahashi, S. (1985b), *Surf. Sci.* **164**, 367–392.

Teichert C., Lagally M. G., Peticolas L. J., Bean, J. C., Tersoff J. (1996), *Phys. Rev. B* **53**, 16334–16337.

Tersoff, J. (1991), *Phys. Rev. B* **43**, 9377–9380.

Tersoff, J., Pehlke, E. (1992), *Phys. Rev. Lett.* **68**, 816–819.

Tersoff, J., Phang, Y. H., Zhang, Z., Lagally, M. G. (1995), *Phys. Rev. Lett.* **75**, 2730–2733.

Tersoff, J., Teichert, C., Lagally, M. G. (1996), *Phys. Rev. Lett.* **76**, 1675–1678.

Tong, S. Y., Xu, G., Mei, W. N. (1984) *Phys. Rev. Lett.* **52**, 1693–1696.

Tong, X., Bennett, P. A. (1991), *Phys. Rev. Lett.* **67**, 101–104.

Tromp, R. M., Hamers, R. J., Demuth, J. E. (1985), *Phys. Rev. Lett.* **55**, 1303–1306.

Tromp, R. M., Hamers, R. J., Demuth, J. E. (1986), *Science* **234**, 304–309.

Tromp, R. M., Reuter, M. C. (1992), *Phys. Rev. Lett.* **68**, 820–822.

van Bommel, A. J., Crombeen, J. E., Dirschot, T. G. J. (1978), *Surf. Sci.* **72**, 95–108.

van der Merwe, J. H. (1963), *J. Appl. Phys.* **34**, 117; 123–127.

van Silfhout, R. G., van der Veen, J. F., Norris, C., MacDonald, J. E. (1990), *Faraday Discuss. R. Soc. Chem.* **89**, 169–180, 204–205.

Vanderbilt, D., Alerhand, O. L., Meade, R. D., Joannopoulos, J. D. (1989). *J. Vac. Sci. Technol.* **B7**, 1013–1016.

Venables, J. A., Spiller, G. D. T., Hanbücken, M. (1984), *Rep. Prog. Phys.* **47**, 399–459.

Wang, X.-S., Goldberg, J. L., Bartelt, N. C., Einstein, T. L., Williams, E. D. (1990), *Phys. Rev. Lett.* **65**, 2430–2433.

Webb, M. B., Men, F. K., Swartzentruber, B. S., Kariotis, R., Lagally, M. G. (1990), in: M. G. Lagally (Ed.), *Kinetics of Ordering and Growth at Surfaces*, New York: Plenum.

Webb, M. B., Men, F. K., Swartzentruber, B. S., Kariotis, R., Lagally, M. G. (1991), *Surf. Sci.* **242**, 23–31.

Wierenga, P. E., Kubby, J. A., Griffith, J. E. (1987), *Phys. Rev. Lett.* **59**, 2169–2172.

Wolkow, R. A. (1992), *Phys. Rev. Lett.* **68**, 2636–2639.

Wood, E. A. (1964), *J. Appl. Phys.* **35**, 1306–1312.

Woodruff, D. P. (1981), in: D. A. King, D. P. Woodruff (Eds.), *The Chemical Physics of Solid Surfaces and Heterogeneous Catalysis*, Vol. 1, Amsterdam: Elsevier, p. 175.

Woodruff, D. P. (1994), in: D. A. King, D. P. Woodruff (Eds.), *The Chemical Physics of Solid Surfaces*, Vol. 7, Amsterdam: Elsevier, p. 513.

Wu, F., Lagally, M. G. (1995), *Phys. Rev. Lett.* **75**, 2534–2537.

Wu, F., Chen, X., Zhang, Z. Y., Lagally, M. G. (1995), *Phys. Rev. Lett.* **74**, 574–577.

Xu, G., Hu, W. Y., Puga, M. W., Tong, S. Y., Yeh, J. L., Wang, S. R., Lee, B. W. (1985), *Phys. Rev. B* **32**, 8473–8476.

Yamamoto Y. (1994), *Surf. Sci.* **313**, 155–167.

Zandvliet, H. J., Elswijk, H. B., van Leonen, E. J. (1992), *Surf. Sci.* **272**, 264–268.

Zhang, Z. Y., Chen, H., Bolding, B. C., Lagally, M. G. (1993), *Phys. Rev. Lett.* **71**, 3677–3680.

Zhang, Z. Y., Chen, X., Lagally, M. G. (1994), *Phys. Rev. Lett.* **73**, 1829–1832.

Zhang, Z., Lagally, M. G. (1997), *Science*, **276**, 377–383.

Zubkus, V. E., Kundrotas, P. J., Molotkov, S. N., Zatarsi, V. V., Tornau, E. E. (1991), *Surf. Sci.* **243**, 295–302.

Further Reading

Bauer, E. (1958), *Z. Kristallogr.* **110**, 372–395.

Burton, W. K., Cabrera, N., Frank, F. C. (1951), *Philos. Trans. R. Soc. London Ser. A* **243**, 299–358.

Ertl, G., Küppers, J. (1974), *Low Energy Electrons and Surface Chemestry*, Weinheim: Verlag Chemie.

Herring, C. (1953), in: R. Gomer, C. S. Smith, (Eds.), *Structure and Properties of Solid Surfaces*, Chicago: The University of Chicago Press, p. 5.

King, D. A., Woodruff, D. P. (Eds.) (1981), *The Chemical Physics of Solid Surfaces and Heterogeneous Catalysis*, Vol. 7, Amsterdam, Elsevier.

King, D. A., Woodruff, D. P. (Eds.) (1998), *The Chemical Physics of Solid Surfaces*, Vol. 8, Amsterdam: Elsevier.

Liu, Feng, Wu, Fang, Lagally, M. G. (1997), *Chem. Rev.* **97**, 1045.

Mönch, W. (1995), *Semiconductor Surfaces and Interfaces*, 2nd ed., Berlin, New York: Springer-Verlag.

Yang, H. N., Wang, G. C., Lu, T. M. (1993), *Diffraction from Rough Surfaces and Dynamic Gorwth Fronts*, Singapore: World Scientific.

Zhang, Z., Lagally M. G. (1997), *Science* **276**, 377–383.

Addenda

ATMOSPHERIC STRUCTURE

(An Addendum to *Encyclopedia of Applied Physics*, Volume 2, pages 201–224.)

Raymond G. Roble, *High Altitude Observatory, National Center for Atmospheric Research, Boulder, Colorado, U.S.A.*

The original article Atmospheric Structure describes the basic vertical structure of the Earth's atmosphere, with a brief description of latitudinal and seasonal variation and the response of the atmosphere to solar variability. This Addendum describes new insights into the atmospheric structure in regions that were poorly known when the original article was written but have been measured recently by new satellite- and ground-based instruments. A brief summary of the nature of the variability found in each atmospheric layer is also given, starting from the upper atmosphere downward to the Earth's surface.

1. THERMOSPHERE AND IONOSPHERE

Within the thermosphere and ionosphere most of the variability about the basic state is controlled by the solar radiative and plasma outputs. Since the original article was written there have been only a few new measurements of the sun's spectral irradiance in the extreme ultraviolet portion of the spectrum (10 to 120 nm), primarily from rockets. However, there have been extensive measurements of the solar UV spectrum (120 to 400 nm) and total solar irradiance that have been made from various satellites. These measurements have defined the range of variability of the solar radiative output over a solar cycle and also have shown variability associated with the 27-d solar rotation period, and shorter-term variability, such as during solar flares. In general, the solar radiative output over a solar cycle varies by about 10% near 200 nm, 1% near 300 nm, and 0.1% for the total solar irradiance. The older solar EUV measurements show considerable day-to-day variability that influences the basic temperature and compositional structure of the thermosphere and ionosphere as discussed in the original article.

There have also been new measurements from satellites and new analysis techniques that better define the energy and momentum inputs into the thermosphere and ionosphere from the aurora. The aurora is driven by the interaction of the solar-wind plasma with the Earth's magnetic field, and images of the aurora made by the NASA POLAR satellite in the UV and visible range have observed a tremendous source of variability introduced into the Earth's atmosphere especially during geomagnetic storms. During these storms the energy and momentum inputs into the Earth's upper atmosphere can vary by over two orders of magnitude. As a result, the high-latitude thermosphere and ionosphere are subjected to a wide range of dynamic phenomena such as large horizontal scale waves that travel globally, circulation changes to the basic wind structure, temperature and compositional perturbations that occur at high latitudes and propagate equatorward, and many others. Schematics showing the response to the aurora have been presented in the original article, and the new data have helped to better define the forcing that drive the variability and the small-scale structure.

In addition to the variability introduced into the thermosphere and ionosphere by solar and auroral variability, numerical models that couple the lower and upper atmosphere show

ISBN 3-527-29308-6

that another important source of variability in the thermosphere and ionosphere consists of gravity waves, tides, and other large-scale disturbances generated by weather systems in the lower atmosphere. These waves and disturbances propagate upward and deposit their energy and momentum in the thermosphere and are a major source of variability, especially during quiet auroral conditions and steady solar EUV output. During these conditions waves from the lower atmosphere are the major source of day-to-day variability that has been observed over a given station or by a satellite in orbit.

2. LOWER THERMOSPHERE AND MESOSPHERE

For many years the atmospheric structure and dynamics in the upper mesosphere and lower thermosphere between about 60 and 200 km altitude has not been measured on a global basis. Most data came from rockets and remote sensing from ground stations. The Upper Atmosphere Research Satellite (UARS) that was launched in 1991 had instruments on board that could measure the global temperature and winds between 30 and 500 km altitude. These measurements not only defined the basic structure of the region, but also illustrated the range of variability that exists in the upper atmosphere. Analysis of the data indicated that the equatorial upper atmosphere below about 110 km is strongly influenced by the diurnal tide that is forced by solar radiation being absorbed by water vapor in the troposphere and propagating upward to the 100-km altitude region before dissipating by molecular viscosity. At mid and high latitudes the tidal structure is dominated by a semidiurnal tide that is excited not only in the troposphere but also by *in situ* absorption of solar radiation primarily by ozone in the middle atmosphere near 50 km. During the summer months a strong signature of an intermittent two-day wave was found to dominate the winds in the upper mesosphere and lower thermosphere between 70 and 100 km altitude. The UARS satellite also found equatorially trapped fast Kelvin waves with periods of 3 to 5 d and regions of strong atmospheric gravity-wave activity that influenced the basic wind, temperature, and compositional structure of the upper mesosphere and lower thermosphere profoundly. In fact, the gravity-wave activity in this region of the upper atmosphere is so strong that it primarily controls the temperature and dynamic structure and drives it away from a radiative equilibrium structure that would result if only absorption of solar radiation acted upon the region. An example of the complex wind structure observed by the UARS satellite is shown in Fig. 1, which illustrates the zonal (positive eastward) and meridional (positive northward) winds for March equinox and December solstice at 1200 local time. These are the first global observations of the wind structure between 60 and 200 km, and they shown that the diurnal tide propagates to about 110 km at low latitudes and that there is a major transition in the dynamic structure above 110 km. This structural change occurs because the atmosphere is so thin above 100 km that the waves in the lower atmosphere are damped and strong solar EUV forcing in the thermosphere gives rise to a different circulation pattern. These figures show the changing wind structure for both equinox and solstice conditions near 100 km altitude. These data filled a critical gap in our knowledge of the wind structure in the region between 60 and 200 km altitude. These data, with the additional new data from ground stations such as LIDARs, radars, and airglow instruments, are greatly improving our understanding of the upper-atmosphere structure and dynamics.

In addition to the wind and temperature measurements, there have been composition measurements of the upper atmosphere. Briefly, the new measurements indicate that current aeronomical theory predicts too much hydroxyl, OH, that catalytically destroys ozone in the mesosphere. The ozone presently calculated by upper-atmosphere models is about 20% to 30% lower than measurements indicate. It is important to resolve this discrepancy because ozone is such an important constituent for absorbing damaging solar UV radiation before it reaches the Earth's surface. The current problem suggests that the laboratory rate coefficients for certain chemical reactions need to be re-evaluated because they may be different for upper-atmosphere environmental conditions. There are also problems

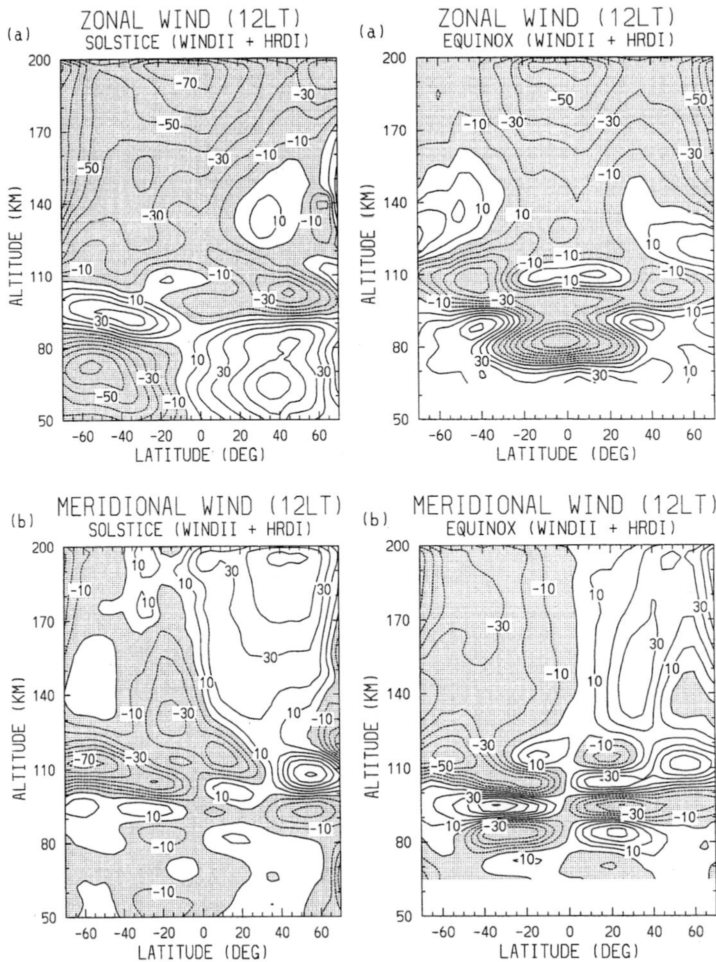

FIG. 1. Combined zonally averaged mesospheric and thermospheric winds for December solstice period (left panels) and March equinox period (right panels) at 1200 h local time measured by the Upper Atmosphere Research satellite. The upper frames are the zonal and the lower frames are the meridional components of the wind. A 10-m/s contour interval is employed. Shading is used for westward winds in the upper panels and for southward winds in the lower panels (McLandress *et al.*, 1996). The acronym WINDII represents the WIND Imaging Interferometer, and HRDI, the High Resolution Doppler Imager, instruments used to measure the winds from the Upper Atmosphere Research Satellite (UARS).

between theory and observations for other chemical species such as nitric oxide and carbon monoxide, indicating that there is still considerable work needed before we are able to understand the current compositional structure of the upper atmosphere.

A very important problem that governs the temperature structure of the upper mesosphere and lower thermosphere between 80 and 150 km altitude is the magnitude of the O–CO_2 vibrational exchange coefficient. This coefficient governs the infrared radiative cooling of the upper atmosphere. Current theory, laboratory measurements, and observational analysis of satellite infrared emissions indicate a factor of 6 uncertainty in the magnitude of this important rate coefficient, which governs the temperature structure of the upper atmosphere. Not only does this problem need to be resolved for understanding current-day atmospheric structure, but it is important for determining the magnitude of possible future global change.

There have been observations that suggest

that the upper atmosphere above about 50 km altitude has been cooling during the last decade. This secular trend has been identified in satellite, LIDAR, and rocket data, but the cause of the cooling is not known. Studies of the upper-atmosphere response to global change and a CO_2 doubling from present-day levels indicate that the upper atmosphere will cool because of increased infrared radiation to space. But the major cooling is expected to occur slowly during the next century, and it is generally believed to be too small at the present time to account for the observed cooling. Another possible cause may be an association with the observed ozone decrease caused by chlorofluorocarbons (CFC's) affecting the ozone layer, or it may be a natural variability of the ocean–atmosphere system that is not well understood.

3. STRATOSPHERE

Ozone is one of the greenhouse gases that modify the radiative structure of our atmosphere. A change in ozone amount and distribution can influence climate. In addition, ozone protects the biosphere from harmful solar ultraviolet radiation. Ozone is continually being produced and destroyed by solar ultraviolet radiation. Additionally ozone is destroyed by catalytic reactions of nitrogen, hydrogen, chlorine, and bromine oxides. In the early 1970s it was predicted that human activity will lead to a diminishing ozone layer; and since the detection of the Antarctic ozone hole, the downward trends in the column total ozone have been well documented for the regions both inside and outside Antarctica by satellite, ground-based, and aircraft observations. Decadal trends have been observed, with the greatest ozone loss occurring in the Antarctic and the smallest in the equatorial region. Rates as large as −3%/yr have been observed in spring over Antarctica and −0.8%/yr in northern-hemisphere spring at mid-latitudes (Stolarski *et al.*, 1991). The ozone loss has also been linked to an observed cooling of the lower stratosphere. These observations all indicate that the basic structure described in the original article is subject to considerable variability and that these changes may be caused by a combination of natural variability, man's activities, changes in the solar radiative and plasma output, and changes associated with atmosphere–ocean and atmosphere–biosphere interactions. These complex interactions all contribute to a possible change in the vertical structure of the atmosphere.

Some of the major sources of atmospheric variability in the stratosphere are global oscillations associated with the quasi-biennial oscillation (QBO), an oscillation with an irregular period of about 26 months where the equatorial winds in the stratosphere oscillate in an easterly/westerly direction. A seasonal variation occurs with interannual variability superimposed and there are solar-cycle variations that have been detected within the stratosphere.

4. TROPOSPHERE WEATHER AND CLIMATE

It is well known that the Earth's climate has undergone considerable change over geologic history. In its early history the atmosphere was much warmer because of higher CO_2 levels. There were also ice ages and interglacial periods, all indicating that the basic structure of the atmosphere has undergone considerable change with time. Not only have the surface temperatures changed with time but most likely the entire atmospheric vertical structure has changed.

Variability in the lower atmosphere is governed by many factors. The most prominent that has emerged in recent years consists of the major changes to the climatological structure caused by El Niño. The period 1991 to 1995 was one of the longest El Niño events in recent times, and the major event that is presently underway in 1997 is expected to cause major changes in weather patterns. It is well documented that El Niño can affect the weather near the surface, but it is not clear whether there are associated changes that occur in the upper atmosphere. Interannual variability in the lower atmosphere is large and the major causes of this variability are uncertain but are subject to intensive study. Much of the variability is associated with atmosphere–ocean interactions, but there is also considerable variability introduced by nonlinear interactions of the large-scale dynamical processes.

Atmospheric chemistry is also having a strong influence on regional climate. The atmosphere is not only subjected to increasing amounts of important atmospheric trace gases such as CO_2, CH_4, and NO_x, but aerosol-forming gases give rise to cloud seeding and aerosol layers that affect the solar radiative balance of the lower atmosphere. All of these chemical processes exert some influence on the climatological balance within the lower atmosphere that in turn influences the structure of the entire atmosphere.

The limits of variability in the lower atmosphere have been quantified by use of historical records, but each year the record highs or lows for an individual station may be broken, indicating that the true range of atmospheric variability has not been determined. With possible global change occurring in the next century, computer models are suggesting that the range of variability may increase, with periods of prolonged drought and intense storms occurring in future climate scenarios. There is considerable effort being expended to identify and understand possible global change caused by increasing CO_2 concentrations and other greenhouse gases.

Works Cited

McLandress, C. M., Shepherd, G. G., Solheim, B. H., Burrage, M. D., Hays, P. B., Skinner, W. R. (1996), "Combined mesosphere/thermosphere winds using WINDII and HRDI data from the Upper Atmosphere Research Satellite," *J. Geophys. Res.* **101**, 10441–10455.

Stolarski, R. S., Bloomfield, P., McPeters, R. D., Herman, J. R. (1991), "Total ozone trends deduced from NIMBUS 7 TOMS data," *Geophys. Res. Lett.* **18**, 1015–1018.

Further Reading

Brekke, A. (1997), *Physics of the Upper Polar Atmosphere*, New York: Wiley.

Brasseur, G., Solomon, S. (1986), *Aeronomy of the Middle Atmosphere*, Dordrecht: D. Reidel Publishing.

Salby, M. L. (1996), *Fundamental of Atmospheric Physics*, San Diego: Academic Press.

BIOLOGICAL EFFECTS OF ELECTROMAGNETIC AND PARTICLE RADIATION

(An Addendum to *Encyclopedia of Applied Physics*, Volume 2, pages 365–401.)

JÜRGEN KIEFER, *Strahlenzentrum der Justus-Liebig-Universität, Giessen, Germany*

INTRODUCTION

It is impossible to review here all new developments that took place after the publication of the original article, and so an admittedly subjective choice was made. Quite a few larger publications appeared, the most important being the 1990 recommendations of the International Commission for Radiological Protection (ICRP) (ICRP 60, 1991), a new report of the UN Scientific Committee on the Effects of Atomic Radiations (UNSCEAR) (UNSCEAR, 1994) and the proceedings of the 10th International Congress of Radiation Research (Hagen *et al.*, 1995). Particularly the last volume is a valuable source to be consulted for an overview of the present state of the art in the field.

1. DEPOSITION OF ENERGY

Energy deposition plays an important role in the understanding of the action of ionizing radiations; with non-ionizing radiations the actual spatial pattern depends more on the structure of the target than on the property of the radiation. These questions have so far received less attention and will not be treated here. Classical microdosimetry is limited to micrometer dimensions. For even smaller sizes one has to recur to Monte-Carlo calculations. They have now been performed for protons and alpha particles. Presently it is attempted to determine energy deposition in molecules of biological interest, particularly DNA, to understand the initial processes of lesion formation and link them to the biological effective-

ISBN 3-527-29308-6

ness of different radiations. It has been found that the yield of DNA double-strand breaks as measured by conventional techniques does not depend to a great extent on radiation type. Chromosomal aberrations, cell killing, and mutation induction, on the other hand, show a clear dependence. This is interpreted by assuming that the interaction between primary lesions and their proximity plays an important role. Densely ionizing particles produce strand breaks close to each other forming "clusters" not resolvable by the usual experimental approaches. They presumably give rise to errors in repair. An understanding of this phenomenon is essential to assessment of the risk of, e.g., neutrons and alpha particles (Rossi and Zaider, 1996).

2. CELLULAR EFFECTS

2.1 Delayed Reproductive Death, Apoptosis, Genomic Instability, and Low-Dose Responses

If surviving cells as defined by the colony-forming assay are further cultivated, they display a decrease in plating efficiency that depends on the dose originally given. It appears that they retain some kind of "cryptic" damage expressed at later stages. In many cases these cells die in a characteristic manner that has been termed *apoptosis* and constitutes some kind of a cellular "suicide program". The cells do not just disintegrate as in necrosis but follow a given scheme of interphase death: At first the cells shrink and their mean density increases; this is followed by changes in their morphological appearance. Most characteristically the chromatin is fragmented into nucleosome sizes, i.e., about 200 nucleotide pairs. Apoptosis is an active cellular process requiring active protein synthesis, mediated by enzymes and under genetical control. It can also be directly triggered by ionizing radiation and seems to play an important role particularly at low doses (Hendry *et al.*, 1995).

Cells surviving the exposure to ionizing radiation display also a phenomenon called *genomic instability*, which was first demonstrated after alpha-particle irradiation: As discussed in the original article, dicentric chromosomal aberrations can be scored after one or two replication rounds. Their number decreases with time because the affected cells die. After several generations there is another increase that is due to newly formed aberrations, indicating fragility of the genome caused by the previous irradiation. At the same time, mutations may arise that were not present in the population before. All these findings have an important bearing on the understanding of radiocarcinogenesis: it is now well established that it proceeds in a multistage fashion, and that more than one (presumably around 5) mutational events are required. It is extremely unlikely that they are all caused directly by radiation; more probably genomic instability is responsible for the later steps (Little, 1994).

The response of cells to low doses of ionizing radiation has received increased attention. It is now clear that gene activity and its regulation can be influenced by x rays as well as by UV. Studies using single-cell observations suggest that mammalian cells may be more sensitive to killing by low doses of x rays than anticipated from an extrapolation from the overall survival curve. At the same time it has been reported by several groups that the yield of chromosomal aberration after high doses is reduced if the cell received "priming doses" in the order of a few hundred milligrays several hours before. This "adaptive response" depends on an active cellular metabolism and may be related to inducible repair processes, but there is presently no proof for this supposition. Studies related to these questions can be found in a special issue of *Radiation Research* (Skov and Marples, 1994) and is also summarized in UNSCEAR (1994).

2.2 Repair Processes

The last years have brought impressive progress in the understanding of cellular repair of radiation-induced damage (Friedberg *et al.*, 1995). Most of the relevant genes have been cloned, both in microorganisms and in mammalian cells. Excision repair was shown to work preferentially in active genes via a coupling with transcription. Apart from this, "general" excision can also be found. A number of genes responsible for the repair of radiation-induced double-strand breaks (DSB) in mammalian cells could be identified. It is now clear that DSB repair in mammalian cells operates through an "end joining" mechanism, in contrast to the case in yeast where homologous recombination plays the most important role. A major breakthrough was the cloning of the *Ataxia telangiectasia* (AT) gene, although its role remains still elusive.

2.3 Radiofrequency Electromagnetic Radiation

Because of the wide-spread public interest in the possible adverse effects of power lines, e.g., the induction or promotion of cancer (see below), effects on the cell and tissue level have to be analyzed. There are a number of mechanisms that could be invoked: The quantum energy of rf (radio-frequency) electromagnetic radiation is much too low to cause any direct damage at the molecular level and also well below the "thermal noise": it is 2.5×10^{-13} eV at 60 Hz while the mean molecular kinetic energy ($\frac{3}{2}kT$) is 0.04 eV. Forces on charged particles by a field of an electric field of 1000 V/m or a magnetic field of 0.1 mT, ranging between 10^{-7} and 10^{-10} pN, are even smaller than those generated when a single hair cell in the inner cochlea of the ear is activated (about 1 pN), the most sensitive interaction known in biological systems. An interaction with isolated components can thus be ruled out and some kind of cooperative mechanism has to be invoked. Electromagnetic fields may alter the lifetime of radicals, which could cause biological effects, but the levels required are much larger than created from external electricity. Resonance phenomena have frequently been suggested, but a careful analysis reveals that they are not plausible in real biological systems. The main argument is that the medium causes rapid damping so that with the external fields applied resonance cannot build up. The sensitivity to the pickup of even very weak signals might be increased if a number of sensors cooperate in a coherent way. This has been found, e.g., in sharks, who are able to sense even 10^{-5} V/m, but no such arrangement has ever been found or even suggested in higher mammals; and it would also be extremely difficult to link it with carcinogenesis. The authors of a very careful and rigorous study (Valberg *et al.*, 1997) come to the conclusion that there is no plausible mechanism to invoke a harmful action of rf electromagnetic radiation on biological systems; and, particularly, no clue to support either tumour induction or tumour promotion can be provided (see also Malyapa, 1997a, 1997b).

3. ORGAN AND WHOLE-BODY EFFECTS: CANCEROGENESIS

UNSCEAR (1994) has reviewed the available epidemiological data concerning cancer induction by ionizing radiations by exposure both from external and from internal sources. The most important information still stems from the survivors of the atomic bombs in Hiroshima and Nagasaki, which is also the core of the new ICRP recommendations (see below). They are supported by quite a number of studies with medical radiation applications and occupational exposures. On the basis of the available data risk estimates for various cancer sites were computed; these are summarized in Table 1.

A new dosimetry system for the atomic bomb survivors was introduced in 1986 (Roesch, 1987), which resulted in a significant reduction of the neutron component compared to the older estimates. In the meantime new figures based on measurements of isotopes produced by neutron reactions became available, casting some doubt on the correct assessment of thermal neutrons. A compilation of the data (Straume *et al.*, 1992) suggests that particularly at larger distances from the epi-

Table 1. Risk of exposure-induced deaths for various cancer sites, estimated on the basis of data for the bomb survivors and cancer incidences in the Japanese population (UNSCEAR 1994). Figures are given in %/Sv and relate to a single acute exposure.

Cancer site	Male	Female	Both
Leukemia	1.3	0.9	1.1
Esophagus	0.3	0.7	0.5
Stomach	0.9	2.0	1.4
Colon	0.5	0.6	0.6
Liver	2.2	0.3	1.2
Bladder	0.4	0.2	0.3
Lung	1.8	3.1.	2.5
Breast	—	2.0	2.0
Ovary	—	0.5	0.3
Others	4.3	2.0	3.1
Total	**11.7**	**12.3**	**12.0**

centre the dose by thermal neutrons is underestimated by factors between at least 2 to 10. This does not drastically change the overall risk factor but may have implications for neutron weighting factors (see below).

4. APPLIED RADIOBIOLOGY

4.1 New Developments in Biodosimetry

4.1.1 Fluorescence *in Situ* Hybridization for Chromosomal Translocations (FISH). Chromosomal aberrations were discussed in the original article. Apart from their fundamental relevance they are an important tool of biodosimetry. If lymphocytes taken from exposed individuals are cultivated *in vitro*, aberrations can be scored and used as an indicator of dose. Classically, dicentric chromosomes ("dics") are used for this purpose since they are comparatively easy to recognize. The lowest dose limit resolvable is about 0.1 Gy with x or gamma rays. Dics interfere with further cellular proliferation; 50% of the cells carrying them will die at every division. This means that these "unstable" aberrations will rapidly fade out so that reliable dose estimates are impossible after months or years. Translocations, on the other hand, are stably transmitted to the progeny, allowing in principle biodosimetry at any time, even after decades. Technical problems of their scoring prevented the use until recently. The introduction of the so-called FISH technique has now led to an improvement: Monoclonal antibodies coupled to fluorescent dyes can be raised against human chromosomes so that every chromosome can be selectively "painted". Translocations are then visualized by colour changes along the chromosomal structures. Studies have shown that they persist indeed much longer than dics. Their dose dependence is similar to that of dics, and their detection limit lies in the same range (about 100 mGy for gamma rays). Developments for automatic scoring are already well under way (Lucas *et al.*, 1992; Yang, 1997).

4.1.2 ESR Dosimetry. Ionizing radiation induces the formation of free radicals in the exposed tissue, but their lifetime is usually very short. This is not the case if they are created in a solid matrix where their movability is severely restricted. In this case they may persist for very long periods and may thus be detected by electron spin resonance (ESR), e.g., in teeth or bone. The method has practical applications both in radiation protections and in radiotherapy (Kudynski *et al.*, 1993).

4.2 Developments in Radiation Therapy

The "classical" radiation therapy of tumours uses quantum or electron radiation provided commonly by linear accelerators. The energy is adjusted in such a way that the dose maximum lies at the tumour site. By careful treatment planning using multiple beams and beam shaping by custom-made absorbers, the tumour volume can be delineated and the healthy tissue spared to a great extent. This technique has now reached a high degree of sophistication. More recently particle therapy has also gained some importance. The most important is based on protons. Since they deliver most of their energy just before the end of their range, a very good dose discrimination can be achieved. There is typically a dose factor of 5 between the so-called "plateau" and the "peak" regions. As the dose maximum is very sharp, small tumours can be very effectively treated with an optimal sparing of the surrounding areas. This technique has been applied especially to ocular melanomas; several thousand patients have so far been treated with excellent results. Not only could the tumour be locally cured, but also eyesight could be saved to a large extent in most cases. While in the beginning treatments were performed at accelerators built for research in physics (the forerunner being Boston, Massachusetts, U.S.A.), more recently "dedicated" machines have been established in specialized hospitals. The largest is situated in Loma Linda, California, U.S.A., commissioned in 1990; others are located in Boston, Massachusetts, U.S.A., Nice, France, and Villigen, near Zürich, Switzerland. Also more deep-seated tumours are now treated, which requires, of course, higher proton energies, which now go up to about 250 MeV (Miller, 1995).

Protons possess a very favourable dose distribution, but their biological effectiveness is not much different from that of gamma or x rays in consequence of their comparably low linear energy transfer (LET). In order to deliver an even higher "biological" dose and

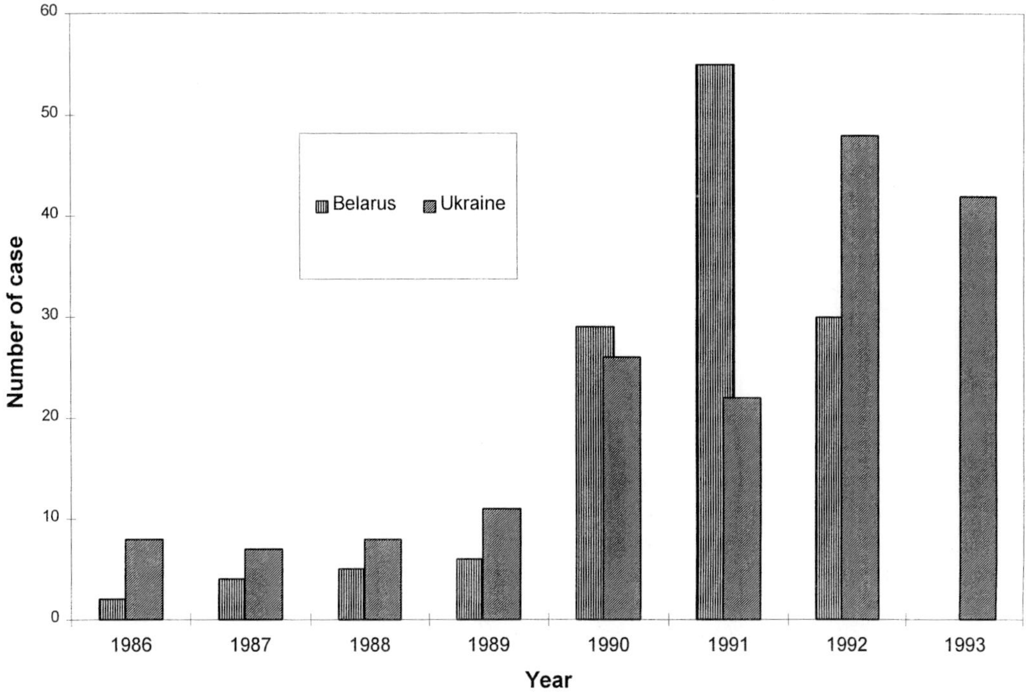

FIG. 1. Thyroid cancers in children up to age 14 in Belarus (Kazakov *et al.*, 1992) and Ukraine (Likhtarev *et al.*, 1995) as a consequence of the Chernobyl disaster. The 1992 value for Belarus is only for half a year.

overcome complications like the "oxygen effect" (see the original article), heavier ions are required. Studies along this line started many years ago in Berkeley, California, U.S.A., but had to be terminated because of the shutdown of the accelerator. Meanwhile two new studies were started, one in Chiba, Japan, where a dedicated heavy-ion accelerator was built, and in Darmstadt, Germany, which makes use of the existing heavy-ion accelerator and is still in the preclinical study phase. A new method for treatment planning has here been developed where the beam energy is modulated in such a way that the treatment volume matches as closely as possible the tumour ("raster scan") (Amaldi and Larsson, 1994).

5. RADIOECOLOGY, RADIATION PROTECTION, AND GUIDELINES

5.1 Effects of the Chernobyl Accident

The 1986 accident in the Soviet reactor in Chernobyl created a widespread radioactive contamination affecting more than 100 000 people. Because of uncertain dose assessment, firm epidemiological data are not readily available, but a few conclusions can now be drawn:

1. Contrary to common perceptions, there is **no** increase in childhood leukaemia, both in the Ukraine and in Belarus (Ivanov *et al.*, 1996; Drozdova *et al.*, 1996).
2. A dramatic increase in thyroid cancers was found in children up to age 14, as shown in Fig. 1.

5.2 Ultraviolet Radiation

All types of skin cancer show a world-wide dramatic increase. This is not only the case for the comparatively harmless basal- and squamous-cell carcinoma but also for the very aggressive melanoma. Figure 2 summarizes data obtained in British Columbia illustrating this alarming development. It is certainly not due to the reduction in stratospheric ozone but

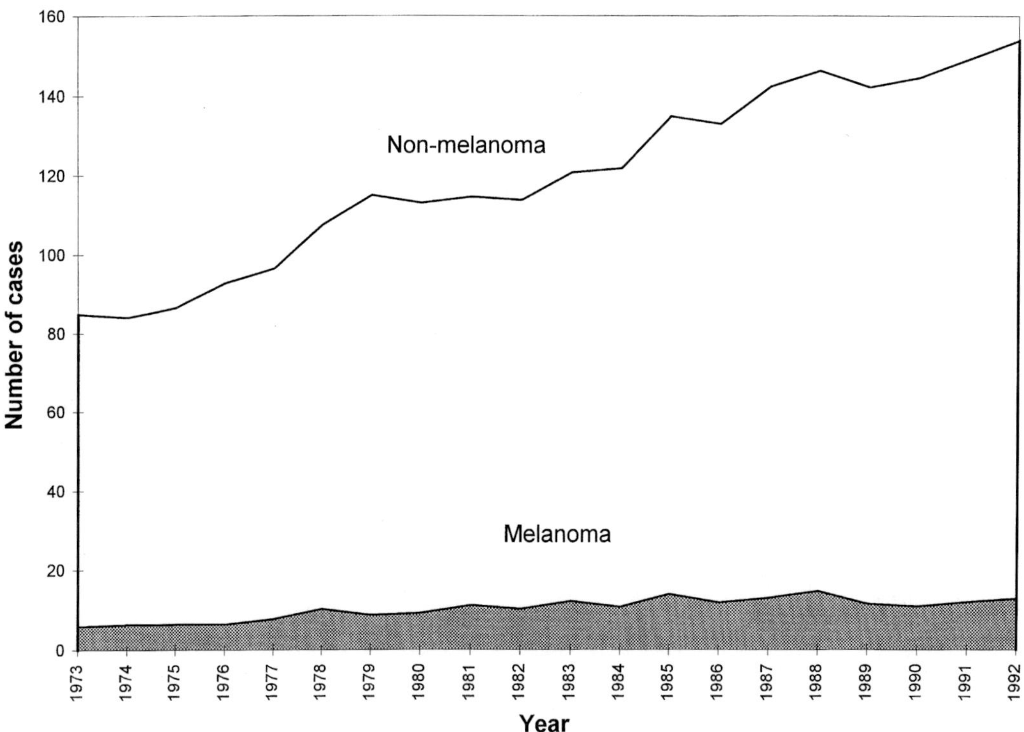

FIG. 2. Increase in non-melanoma and melanoma skin cancer in British Columbia from 1973 to 1992 (Philips *et al.*, 1996).

presumably to changed leisure habits and possibly also the increased use of sunbeds and other tanning devices.

There is a still on-going debate whether near UV (UV-A) has a carcinogenic action. Unfortunately there is not yet an action spectrum available for melanomas in mammals, but action spectra for other types of skin cancer obtained in hairless mice strongly indicate that also long-wavelength UV has clearly a carcinogenic activity.

Ozone depletion will add to this situation. Evidence is mounting that this is not only a theoretical possibility. It can now be seen world-wide, although the most dramatic effect has been registered in Antarctica. Future trends are difficult to predict and depend essentially on the use of chlorofluorocarbon compounds. It has to be kept in mind, however, that atmospheric chemistry operates at long time scales. A comprehensive review of all problems involved can be found in SSK (1996).

5.3 Radiofrequency Electromagnetic Radiation

The US National Research Council (NRC, 1997) has reviewed in a careful and comprehensive study all reports dealing with the question of a possible relationship between power lines and cancer. The result is best summarized by a sentence taken from the Executive Summary:

"Specifically, no conclusive and consistent evidence shows that exposures to residential electric and magnetic fields produce cancer, adverse neurobehavourial effects, or reproductive and developmental effects."

The stress is on "conclusive and consistent," which does not rule out single, as-yet-unexplained findings. Taken together with biophysical considerations reviewed above it may be safely stated that scientific judgement does not provide any possibility for a link between normal environmental levels of electric and magnetic fields and human diseases.

Table 2. Radiation weighting factors (ICRP 60, 1991).

Type and energy range	Radiation weighting factor w_R
Photons, all energies	1
Electrons and muons, all energies	1
Neutrons, energy < 10 keV	5
10–100 keV	10
100 keV–2 MeV	20
2–20 MeV	10
> 20 MeV	5
Protons, other than recoil, energy > 2 MeV	5
Alpha particles, fission fragments, heavy nuclei	20

5.4 Principles and Fundamentals of Radiation Protection

The International Commission on Radiological Protection (ICRP) issued in 1990 a new set of fundamental recommendations (ICRP 60, 1991) that are likely to have also considerable impact on national legislation. More formally, the term "dose equivalent" has been changed to "equivalent dose" but also to some extent redefined. It is given by the product of dose and radiation weighting factors that replace the "quality factor". Radiation weighting factors are listed in Table 2.

A functional relationship with LET is no longer recommended, to avoid the impression of an accuracy that is inherently not possible. The step function for neutrons is somewhat artificial and commonly replaced by a smooth approximation.

Equivalent dose is measured over a tissue or an organ. To estimate a whole-body equivalent it has to be multiplied by "tissue weighting factors" w_T. This concept had already been used before but the actual values and organs to be considered have been altered, taking into account new medical results from the atomic-bomb survivors.

It has to be emphasized that the system described is only devised for application in occupational radiation protection. Its value in assessing the risk of medical radiation usage is obviously limited.

The ICRP has again reviewed all available evidence for the quantification of radiation carcinogenesis and models for risk projection. As a result new estimates for the expected occurrence of fatal cancer were produced, which must not be taken as firm evidence but rather as upper limits in order to limit possible risks. The numbers have increased considerably over the years, essentially as a result of the application of altered projection models. Consequently also new upper limits are recommended (Table 3). The data from the survivors of Hiroshima and Nagasaki are still the main source of information. Since they received acute doses in a short time, and as there is evidence that lowering the dose rate reduces the effect level, the ICRP introduced a "Dose and Dose Rate Effectiveness Factor" (DDREF) of 2 that is applicable in the common occupational situation. Extended studies with nuclear workers that became recently available are not at variance with the ICPR estimates.

Table 3. Recommended dose limits (ICRP 60, 1991).

	Occupational	Public
Effective dose (Whole body)	20 mSv/year averaged over 5 years	1 mSv/year
Eye lens (Equivalent tissue dose ETD)	150 mSv/year	15 mSv/year
Skin (ETD)	500 mSv/year	50 mSv/year
Hands or feet (ETD)	500 mSv/year	

ETD: Equivalent tissue dose

GLOSSARY

Apoptosis: "Programmed cell death": disintegration of a dying cell characterized by a sequence of distinct metabolic processes.

***Ataxia telangiectasia*:** Recessive hereditary disease in humans characterized by extreme sensitivity to ionizing radiations.

Chromosomal Aberrations: Change in the number of chromosomes (numerical aberration) or in their structure (structural aberrations). Radiation produces mainly the latter.

Dicentric Chromosomes: Chromosomal aberration where two chromosomes are joined in such a way that there are two centromeres (attachment sites for the spindle during mitosis).

DNA: Abbreviation for *deoxyribonucleic acid*, the macromolecule that carries the genetic information in cells.

Dose Equivalent: Formerly, the product of the (physical) dose and the quality factor for different radiation types. It is now replaced by "equivalent dose," which is the product of dose and the newly introduced "radiation weighting factor". If dose is measured in grays (Gy), the equivalent dose is given in sieverts (Sv). The old unit was "rem" with 100 rem = 1 Sv.

Double Strand Breaks (DSB): Rupture of the phosphate–sugar backbone of DNA in both strands opposite to each other. DSB are assumed to be the most important lesions caused by ionizing radiations.

ICRP: International Commission on Radiological Protection, an international non-government organisation that reviews the scientific literature and issues recommendations for radiation protection.

Linear Energy Transfer (LET): Energy imparted per unit path length by a charged particle.

Necrosis: Disintegration of a dying cell.

Recombination: Exchange of parts of genetic material between DNA molecules or chromosomes.

Translocations: Exchange of parts between different chromosomes without loss of genetic material.

Works Cited and Further Reading

Amaldi, U., Larsson, B. (Eds.) (1994), *Hadrontherapy in oncology*, Amsterdam: Elsevier, 755 pp.

Drozdova, V. P., Moroz, G. I., Kiriyewa, S. S., Kuzmina, S. G., Ivanova, I. A. (1996), "Descriptive epidemiology of leukemias in children in the regions of Ukraine caused by the Chernobyl APS accident," *Exp. Oncol.* **18**, 128–131.

Friedberg, E. C., Walker, G. C., Siede, W. (1995), *DNA repair and mutagenesis*, Washington D.C.: ASM Press, 698 pp.

Hagen, U., Harder, D., Jung, H., Streffer, C. (Eds.) (1995), *Radiation Research 1895–1995, 10th ICRR Congress*, Würzburg: 10th ICRR Society, 1236 pp.

Hendry, J. H., Potten, C. S., Merritt, A. (1995), "Apoptosis induced by high- and low-LET radiations," *Radiat. Env. Biophys.* **34**, 59–62.

ICRP 60 (1991), "1990 recommendations of the International Commission on Radiological Protection," *Ann. ICRP* **21**, 1–3.

Ivanov, E. P., Tolochko, G. V., Shavaeva, L. P., Becker, S., Nekolla, E., Kellerer, A. M. (1996), "Childhood leukemia in Belarus before and after the Chernobyl accident," *Radiat. Env. Biophys.* **35**, 75–88.

Kazakov, V. S., Demidchik, E. P. Astakhova, I. N. (1992), "Thyroid cancer after Chernobyl," *Nature* **359**, 21.

Kudynski, R., Kudynski, J., Buckmaster, H. A. (1993), "The application of ESR dosimetry for radiotherapy and radiation protection," *Appl. Radiat. Isotopes* **44**, 903–906.

Likhtarev, I. A., Sobolev, B. G., Kairo, I. A., Tronko, N. D., Bogdanova, T. I., Oleinic, V. A., Epshtein, E. V., Beral, V. (1995), "Thyroid cancer in the Ukraine," *Nature* **375**, 365.

Little, J. B. (1994), "Changing views of cellular radiosensitivity," *Radiat. Res.* **140**, 299–311.

Lucas, J. N., Awa, A., Straume, T., Poggensee, M., Kodama, Y., Nakano, M., Ohtaki, K., Weier, U., Pinkel, D., Gray, J., Littlefield, G. (1992), "Rapid translocation frequency analysis in humans decades after exposure to ionising radiation," *Int. J. Radiat. Biol.* **62**, 53–63.

Malyapa, R. S., Ahern, E. W., Straube, W. L., Moros, E. G., Pickard, W. F., Roti Roti, J. L. (1997a), "Measurement of DNA damage after exposure to 2450 MHz electromagnetic radiation," *Radiat. Res.* **148**, 608–617.

Malyapa, R. S., Ahern, E. W., Straube, W. L., Moros, E. G., Pickard, W. F., Roti Roti, J. L. (1997b), "Measurement of DNA damage after exposure to electromagnetic radiation in the cellular phone communication frequency band (835.62 and 847.74 MHz)," *Radiat. Res.* **148**, 618–627.

Miller, D. W. (1995), "A review of proton beam radiation therapy," *Med. Phys.* **22**, 1943–1953.

NRC (1997), *Possible health effects of exposure to residential electric and magnetic fields*, Washington, DC: National Research Council National Academy Press, 356 pp.

Philips, N. A., Gallagher, R. P., Coldman, A. J., Band, P. R. (1996), "Skin cancer incidence: what has happened and what can we expect," in: *Environmental UV-radiation, risk of skin cancer and primary prevention*, Veröffentlichungen der Strahlenschutzkommission, Band 34, Stuttgart: G. Fischer-Verlag, pp. 323–332.

Roesch, W. C. (Ed.) (1987), *U.S.–Japan Reassessment of Atomic Bomb Radiation Dosimetry in Hiroshima and Nagasaki—Final Report*, Vols. 1 and 2, Hiroshima: Radiation Effects Research Foundation.

Rossi, H. H., Zaider, M. (1996), *Microdosimetry and its applications*, Berlin: Springer, 333 pp.

Skov, K. A., Marples, B. (Eds.) (1994), "Molecular, cellular and genetic basis of radiosensitivity at low doses: a case of inducible repair?," *Radiat. Res. Suppl.* **138**, S1–S131.

SSK (1996), *Environmental UV-radiation, risk of skin cancer and primary prevention*, Veröffentlichungen der Strahlenschutzkommission Band, 34, Stuttgart: G. Fischer-Verlag, 443 pp.

Straume, T., Egbert, S. D., Woolson, W. A., Finkel, R. C., Kubik, P. W., Grove, H. E., Sharma, P., Hoshi, M. (1992), "Neutron discrepancies in the DS86 Hiroshima dosimetry system," *Health Phys.* **63**, 421–426.

UNSCEAR (1994), *Sources and effects of ionizing radiation*, Geneva: United Nations, 272 pp.

Valberg, P. A., Kavet, R., Rafferty, C. N. (1997), "Can low-level 50/60 electric and magnetic fields cause biological effects?," *Radiat. Res.* **148**, 2–21.

Yang, T. C (Ed.) (1997), "Space radiation damage and biodosimetry," *Radiat. Res. Suppl.* **148**, S1–S114.

BIOLOGICAL EFFECTS OF SOUND AND ULTRASOUND

(An Addendum to *Encyclopedia of Applied Physics*, Volume 2, pages 403–420.)

WESLEY L. NYBORG, *Department of Physics, University of Vermont, Burlington, Vermont, U.S.A.*

INTRODUCTION

Since the original article (hereafter cited as A) was prepared, there has been considerable research activity into biological effects of sound, especially ultrasound, and its applications. Much of this has been motivated by medical applications, both to diagnosis, for which safety guidelines are desired, and to therapy, where the goal is to optimize treatment conditions. For either medical purpose, and for other applications, it is recognized as advantageous to know the mechanisms for the biological effects, whether they are adverse or beneficial. This brief review of selected findings and developments in the past six years is organized according to the mechanisms, both thermal and nonthermal, that are believed responsible for the various kinds of effects under consideration.

Theory for these mechanisms was discussed in A, and is not repeated here. In this Addendum the emphasis is on advances in experimental findings and applications. References are made to the literature for recent developments in theory.

1. BIOLOGICAL EFFECTS OF THERMAL ORIGIN

1.1 Safety Concerns

At frequencies in the megahertz range used in most medical ultrasound, acoustic absorption coefficients for mammalian tissues are high enough to allow temperature elevations produced in medical applications to be significant, even in some applications of diagnostic ultrasound. In tissues where cells divide rapidly, as in the mammalian embryo or fetus, a temperature rise of 2 °C or less can produce damage if prolonged sufficiently. For a wide range of teratogenic effects, the time t in minutes required for exposure to temperature T (in degrees Celsius) to produce the effect has been equated, approximately, to

$$t = 4^{43-T}, \tag{1}$$

when T is not greater than 43 °C (Miller and Ziskin, 1989; NCRP, 1992); Eq. (1) was

ISBN 3-527-29308-6

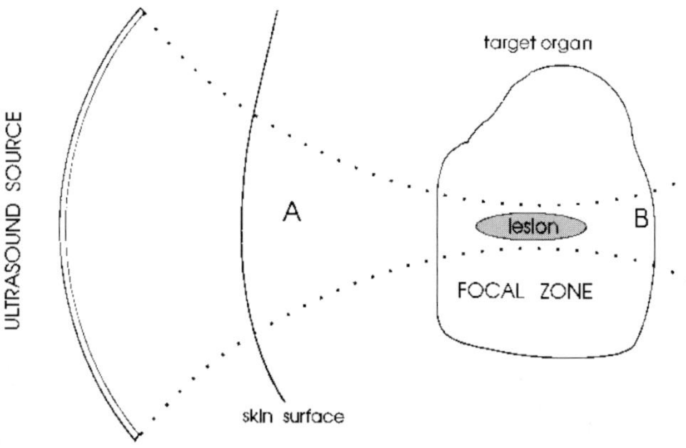

FIG. 1. Schematic diagram of arrangement used in surgical applications of focused ultrasound. A lesion (a region of severely damaged tissue that the body will remove by natural physiological processes) is produced in the focal region, while microscopic examination reveals no damage to tissue lying between the source and the target volume (A) or beyond it (B). Adapted from ter Haar (1995).

adapted from a more general equation given by Sapareto and Dewey (1984). According to Eq. (1), damage may occur after 4 min of exposure to a temperature of 42 °C, and after 16 min of exposure to 41 °C.

Among fetal tissues, the absorption coefficient is highest in bone, and the potential for heating increases with gestational age. Safety criteria derived from algorithms for calculating estimates of the maximum temperature rise that would be produced by ultrasonic exposure of bone and other tissues have been formulated during the past few years (NCRP, 1992; AIUM/NEMA, 1992). These criteria are expressed in terms of *thermal indices*, which are becoming familiar to users of diagnostic ultrasound equipment; many manufacturers of such equipment (in satisfying U.S. regulatory requirements) now display on the imaging screen a "real-time thermal index", i.e., an indicator of the maximum temperature rise being produced in the body of a patient during an ultrasound examination with current instrument settings.

Investigations of the accuracy of algorithms for temperature estimation, while not extensive, have included experimental measurements using special phantoms, as well as measurements made in living animals. The general conclusion is that the thermal indices, while based on estimates of temperature elevation that are subject to error, provide the user with information that is much more relevant to patient safety than was available before the indices were available; errors are usually in the direction of overestimating the temperature rise. It is hoped that in the future the accuracy of the indices can be improved, especially for the more critical situations.

1.2 Thermal Therapy Using Ultrasound

First introduced in the 1940's and 1950's, focused ultrasound has received increasing attention in recent years as a mean of modifying tissue in selected parts of the body; special emphasis has been given to its capability for elevating tissue temperature in a controlled manner. In *hyperthermia* applications, tissues are exposed to ultrasound at moderate intensity levels so that the temperature is maintained at a temperature of 43 to 45 °C for 10 to 30 minutes. In *noninvasive surgery* much higher intensities are used and the temperature raised to the range 60 to 90 °C for 0.1 to 10 seconds. The subject was reviewed recently by ter Haar (1995) and Sanghvi *et al.* (1996), and is featured in a special journal issue (Ebbini, 1996). Ultrasound hyperthermia and surgery are being considered in many fields of medicine, their significant applications at this time being in the fields of ophthalmology and urology. In Fig. 1 is a simplified diagram showing the basic features of an arrangement used in surgery for generating a beam of focused sound and using it to create a *lesion*, i.e., an ellipsoidally shaped volume of damaged tissue in soft tissue. Here the source is a spherical shell of piezoelectric ceramic a few centimeters in diameter, the frequency lies in the range 1 to 8 MHz, and the maximum intensity in the focal region is 300 to 2000 W/cm^2.

In other implementations the single source is replaced by an array of smaller ones, each with its own amplitude and phase; such an array can readily be steered, focused and programmed, so that multiple lesions can be produced in desired configurations. This facilitates the treatment of tumors, for example, by providing the capability of damaging the tissue in a controllable manner throughout regions of various sizes and shapes. Treatment design is greatly assisted by calculations of the expected temperature distribution, with computational or analytical solutions of the bio-heat transfer equation (A; NCRP, 1992; Ebbini, 1996; Sanghvi *et al.*, 1966) or, for very short exposures where the influence of blood flow can be neglected, the equation for thermal diffusion. Ultrasound and magnetic-resonance imaging methods are used to define the region to be treated and to visualize changes that the treatment has produced in the tissue.

2. BIOLOGICAL EFFECTS MEDIATED BY ACOUSTIC CAVITATION

2.1 Sonoluminescence

A major advance in experimental studies of cavitation came from the discovery (Gaitan and Crum, 1990; Gaitan *et al.*, 1992) that a continuous sound field of frequency 22 kHz and pressure amplitude 0.12 MPa can cause a single air bubble in liquid to undergo repeated "collapse events," i.e., implosions, at a regular rate synchronized with the sonic frequency for an indefinitely long period of time, and to generate a short flash of light during each im-

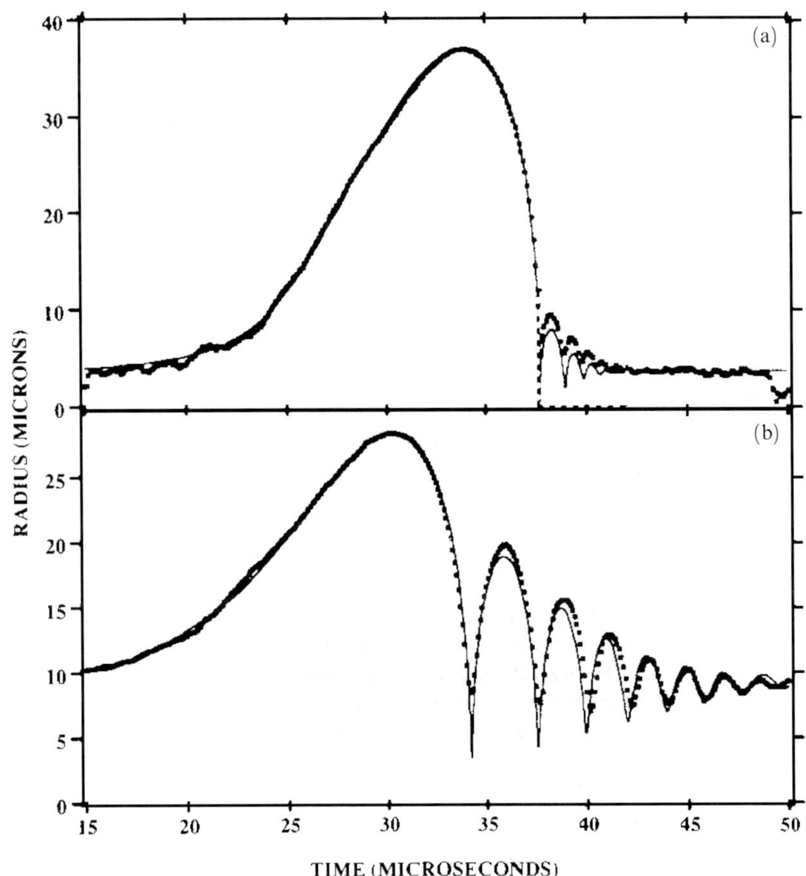

FIG. 2. Bubble radius as a function of time, repeated during each cycle, for a spherical air bubble in water driven by a sound field of frequency 26.5 kHz. The points are experimental; the solid curve is from theory. Sonoluminescence occurs under the conditions of (a), but not under those of (b), the only di¤erence being that the pressure amplitude in the driving field was 0.133 MPa for (a) and 0.108 MPa for (b). From Barber and Putterman (1992).

plosion. These findings caused considerable interest because of the short duration of the flash (often less than 50 ps), the high regularity of the phenomenon, and the nature of the light spectrum. Optical techniques were used to determine the bubble radius R as a function of time t. A typical R *vs* t curve for one cycle of the motion for a frequency of 26.5 kHz is shown in Fig. 2; the points are experimental and the solid curves are from theory. The latter is a solution of the nonlinear equation of motion [similar to Eq. (13) of A]; the excellent agreement gives strong support for the theory. Flynn (1964) has pointed out that the violent collapse events come about because of the inertia of the liquid that rushes inward as the bubble radius decreases. The activity in and near a bubble undergoing the kind of motion illustrated in Fig. 2 has become known as *inertial cavitation* (known earlier as "transient cavitation"). The light production, known as *sonoluminescence* (*q.v.*), is critically dependent on the pressure amplitude in the driving sound field. For example, there was copious light production under conditions of Fig. 2(a) and none under those of Fig. 2(b), yet the conditions were the same except that the pressure amplitude was 0.133 MPa for the former and 0.108 MPa, about 20% less, for the latter. Interesting features of single-bubble sonoluminescence have been discussed by Crum (1994) and Putterman (1995).

Sonoluminescence produced by ultrasound in water or other liquids has been known for many years, and apparently emanates from activity of small bubbles that, in turn, owe their existence to pre-existing microscopic bodies of gas, stabilized against diffusion, called *cavitation nuclei* (see A). Such nuclei are ubiquitous in most liquids unless the latter have been very carefully prepared. An ultrasound field of sufficiently high pressure amplitude may cause free bubbles to emerge from these nuclei and, further, may produce inertial cavitation by causing the bubbles to expand quickly, then contract violently. The optical spectrum of "multibubble sonoluminescence" that appears when ultrasound is applied to water in a test tube or other vessel contains discrete lines characteristic of free radicals, especially O and OH; these are produced by dissociation of water molecules during brief periods of high temperatures (typically, several thousands of kelvins) following the violent contractions. Discrete lines are not seen in the spectra of single-bubble sonoluminescence; instead, a plot of spectral density *vs* wavelength published by Hiller *et al.* (1992) for the light generated by a single air bubble in water is a smooth curve which is fitted well over the 300 to 700 nm wavelength range by a 25000-K blackbody spectrum. Crum (1994) suggests that the conditions of single-bubble inertial cavitation, involving symmetry of the driving pressure field about the bubble center, can lead to production of much higher temperatures in the bubble (which is compressed with little loss of heat to the liquid) than are produced in the more typical experimental conditions where many bubbles are active. In the latter situation, because of high-speed bubble motions and interactions between bubbles, the latter probably do not maintain their spherical shapes as they contract. Instead, available information suggests that the motion during collapse is complicated by surface deformation accompanied by the formation of liquid jets and microbubbles (see A; SONOLUMINESCENCE). It can be expected that, because of the energy used in forming new surface area and because of increased heat transfer from the gas, the pressures and temperatures reached during collapse will be reduced.

There have been many studies in which ultrasound is applied to suspensions of biological cells *in vitro* with the consequence that inertial cavitation occurs and the cells are lysed or otherwise damaged. Sonoluminescence may have occurred, but biological effects have not been attributed directly to the light (even though its spectrum extends into the ultraviolet); instead, the effects have usually been attributed to mechanical or chemical aspects of inertial cavitation (Miller *et al.*, 1996). For example, Miller *et al.* (1995) showed that inertial cavitation generated by ultrasound in suspensions of Chinese hamster ovary (CHO) cells caused H_2O_2 to be produced, which entered surviving cells and caused single-strand DNA breaks. Kim *et al.* (1996) have shown that cavitation produced by continuous 1-MHz ultrasound with pressure amplitude about 400 kPa can make mammalian cells temporarily permeable to large molecules, and appears to provide a promising method of gene transfer. (Note also related findings for lithotriptor-produced shock waves, reported below.)

2.2 Extracorporeal Shock-Wave Lithotripsy

Much has been learned from experiments in which the source is an extracorporeal lithotriptor, a device used medically for breaking up kidney stones by means of focused ultrasound pulses of short duration and high pressure amplitude that are generated outside the patient's body and propagated into it. These high-amplitude pulses are often called *shock waves* and the procedure for applying them is *extracorporeal shock wave lithotripsy* (ESWL). Conditions are favorable for the production of inertial cavitation in water, in the field of a typical lithotriptor, because of the high acoustic pressures (up to 120 MPa) that exist in the focal region during a shock and because of the frequency spectrum, which includes considerable energy in the range 200 kHz and below.

Delius (1995) has reviewed a wide variety of effects produced by ESWL fields on cell suspensions, concluding that these are caused by shear forces resulting from cavitation. A novel finding (Gambihler *et al.*, 1994) is that mouse tumor cells can be made temporarily permeable, allowing large molecules to enter them, after which the cells remain viable. On the basis of such findings, Delius *et al.* (1994) have proposed that ESWL techniques be developed for application to patients as a mean of gene therapy. (Note related findings for ultrasound, reported above.)

Evidence has been obtained that inertial cavitation occurs in the body of patients during ESWL procedures. Coleman *et al.* (1995, 1996) used B-scan ultrasound imaging to examine a region exposed to an ESWL field while the latter was focused on stones in the kidney. In a region (possibly a fat layer) just outside the kidney, through which the shock wave is transmitted into the kidney, increases in image brightness, i.e., echogenicity, appeared when the output setting of the lithotriptor was raised to values above threshold. This increased echogenicity was attributed to gas-filled cavities that had been produced by the field. From hydrophone measurements of the acoustic pressure in water, it was determined that threshold values for the cavity formation were about 15 MPa for the positive peak pressure and 3.5 MPa for the negative peak during passage of a shock.

2.3 Pulsed Ultrasound

Acoustic cavitation, both inertial and noninertial, capable of causing biological effects, is readily produced by ultrasound in water and in cell suspensions, even with pulsed ultrasound such as is used in medical diagnosis, where frequencies are higher (usually 2 MHz or above) than for a lithotriptor and the focal dimensions are smaller (Deng *et al.*, 1996; Miller *et al.*, 1996).

Even in some mammalian tissues, pulsed ultrasound, similar to that used in diagnostic applications, produces cavitation activity of some kind, though its nature is still under investigation; these are tissues where there are known gas-filled spaces, as in the mammalian lung and gastrointestinal system. The biological endpoint is usually structural damage, made evident by the leakage of erythrocytes from small blood vessels and capillaries on the inner surface of the lung and gastrointestinal tract. Systematic studies of lung damage produced by pulsed focused ultrasound have yielded information on thresholds for damage, in terms of the peak negative pressure p. All threshold values lie within an order of magnitude of 1 Mpa, but there is variation with age and species of the animal, and with frequency and pulse duration of the ultrasound. See, for example, Carstensen *et al.* (1992), Baggs *et al.* (1996), and Holland *et al.* (1996). Surprisingly, the thresholds for damage to mammalian lung are in the same range as thresholds for damage to fruit-fly larvae and to leaves of the plant *Elodea* (AIUM, 1993). The common element to all of these biological systems is the presence of stabilized gas-filled cavities; i.e., lung in the mammal, respiratory channels in the larvae, and intracellular gas-filled channels in the plant. For the last, the primary mechanism for damage appears to be viscous stress from acoustic microstreaming produced near vibrating intracellular channels (Miller, 1987; A). The detailed mechanisms for damage caused by ultrasound to mammalian lung or to insect larvae are, as yet, unknown.

2.4 Contrast Agents

A recent development in diagnostic ultrasound procedures is the increasing use of contrast agents (de Jong, 1996). These agents take various forms, but those being considered here are particulates that, when injected into the blood stream of a patient, cause small bodies of gas to exist there. These gas-filled cavities are surrounded by fluid, or are encapsulated by elastic shells or by viscoelastic films; they are very effective as scatterers of the incident ultrasound (at moderate pressure amplitudes) or as emitters of cavitation-generated harmonics (at higher pressure amplitudes), and thus add information to an ultrasound scan.

Contrast agents also can contribute to biological effects of ultrasound. It was found in experiments with the commercial product Albunex® that these gas-containing particles can serve as nuclei for cavitation and, when added to a cell suspension, can greatly increase the destruction of cells when the latter are exposed to pulsed ultrasound (Brayman *et al.*, 1996) or when they are exposed to lithotriptor shock waves (Miller and Thomas, 1996). Studies are in progress on the extent to which damage occurs to living mammals upon exposure to ultrasound when contrast agents are present in the circulatory system. The mechanisms for the action of contrast agents are not known in detail, but evidently include various aspects of inertial and noninertial cavitation.

2.5 Safety Considerations

Theoretical and experimental results have led to safety criteria, analogous to those developed for thermal mechanisms (NCRP, 1992), which define conditions under which tissue damage might occur as a result of cavitation. A criterion adopted by the Output Display Standard (AIUM/NEMA, 1992) is expressed in terms of a *Mechanical Index* (MI), the latter being defined as the ratio of the peak acoustic pressure, defined in a specified way and expressed in megapascals, to the square root of the center frequency, the latter expressed in megahertz. A standard formula is used for calculating the peak negative acoustic pressure in the patient's body, given the results of measurements in water. It is assumed that damage does not occur if the MI is less than 1.0 (or other assigned number). Many manufacturers provide for display of the "real-time MI", i.e., the MI appropriate for current instrument settings, on the screen of ultrasound imaging equipment, in satisfying U.S. regulatory requirements.

3. CONCLUSIONS

Biological effects of ultrasound are receiving increased attention, partly because of their relevance to safety of medical procedures, and partly because of the diverse therapeutic applications which are possible because of the ability of ultrasound to change the structure and function of cells and tissues. For continued progress, it will be necessary to achieve much better understanding of the mechanisms involved in the applications described above, and in others which it has not been possible to discuss in this brief review.

GLOSSARY

Acoustic Cavitation: The mechanical response of one or more gas-filled cavities to a sound field; the cavitation may be inertial or noninertial.

Cavitation Nuclei: Small bodies, usually gas-filled cavities, that can serve as sites for acoustic cavitation.

Extracorporeal Shock-Wave Lithotripsy (ESWL): A procedure used for destruction of kidney stones, involving production of short ultrasound pulses (with very high acoustic pressures) outside the body of a patient and focusing them on the kidney.

Hyperthermia: A procedure for killing tumor cells by raising their temperature to a critical value.

Inertial Cavitation: A class of acoustic cavitation involving growth and violent collapse of one or more cavities.

Lesion: A region of damaged tissue, such as can be produced by focused ultrasound under specific conditions.

Mechanical Index: A number calculated according to standards accepted in the U.S.A., used as a safety index for alerting the user of diagnostic ultrasound equipment to conditions under which damaging acoustic cavitation might occur in the body of a patient.

Noninvasive Surgery: A procedure, such as an application of focused ultrasound, for creating damaged tissue, especially in a tumor, which will subsequently be removed by natural processes.

Sonoluminescence: Production of light by sound in water or other liquid, usually as a consequence of inertial cavitation.

Thermal Index: A number calculated according to standard algorithms, used as a safety index for alerting the user of diagnostic ultrasound equipment to conditions under which damaging temperature elevation might occur in the body of a patient.

List of Works Cited

AIUM (1993), *Bioeffects and Safety of Diagnostic Ultrasound*, Laurel MD: American Institute of Ultrasound in Medicine.

AIUM/NEMA (1992), *Standard for Real-Time Display of Thermal and Mechanical Indices on Diagnostic Ultrasound Equipment*, Laurel, MD: American Institute of Ultrasound in Medicine, and Washington, DC: National Electrical Manufacturers Association.

Baggs, R., Penney, D. P., Cox, C., Child, S. Z., Raeman, C. H., Dalecki, D., Carstensen, E. L. (1996), *Ultrasound Med. Biol.* **22**, 119–128.

Barber, B. P., Putterman, S. J. (1992), *Phys. Rev. Lett.* **69**, 3839–3842.

Brayman, A. A., Azadniv, M., Cox, C., Miller, M. (1996), *Ultrasound Med. Biol.* **22**, 927–938.

Carstensen, E. L., Duck, F. A., Meltzer, R. S., Schwartz, K. Q., Keller, B. (1992), *Echocardiography* **6**, 605–623.

Coleman, A. J., Kodama, T., Choi, M. J., Adams, T., Saunders, J. E. (1995), *Ultrasound Med. Biol.* **21**, 405–417.

Coleman, A. J., Choi, M. J., Saunders, J. E. (1996), *Ultrasound Med. Biol.* **22**, 1079–1087.

Crum, L. A. (1994), *Phys. Today* **47** (9), 22–29.

de Jong, N. (1996), *IEEE J. Eng. Med. Biol.* **15**, 72–82.

Deng, C. X., Xu, Q., Apfel, R. E., Holland, C. K. (1996), *Ultrasound Med. Biol.* **22**, 939–948.

Delius, M. (1994), *Shock Waves* **4**, 55–72.

Delius, M., Hofschneider, P.-H., Lauer, U., Messmer, K. (1995), *Lancet* **345**, 1377.

Ebbini, E. S. (1996), *IEEE Trans. Ultrason. Ferroelec. Freq. Control* **43**, 989–1129.

Flynn, H. G. (1964), "Physics of acoustic cavitation in liquids", in: W. P. Mason (Ed.), *Physical Acoustics*, Vol. 1B, New York: Academic Press, pp. 57–172.

Gaitan, D. F., Crum, L. A. (1990), in: M. F. Hamilton, D. T. Blackstock (Eds.), *Frontiers of Nonlinear Acoustics: Proceedings of 12th ISNA*, London: Elsevier Science Publishers pp. 459–463.

Gaitan, D. F., Crum, L. A., Church, C. C., Roy, R. A. (1992), *J. Acoust. Soc. Am.* **91**, 3166–3183.

Gambihler, S., Delius, M., Ellwart, J. W. (1994), *J. Membrane Biol.* **141**, 267–275.

Hiller, R., Putterman, S. J., Barber, B. P. (1992), *Phys. Rev. Lett.* **69**, 1182–1184.

Holland, C. K., Deng, C. X., Apfel, R. E., Alderman, J. L., Fernandez, L. A., Taylor, K. J. W. (1996), *Ultrasound Med. Biol.* **22**, 917–925.

Kim, H. J., Greenleaf, J. F., Kinnick, R. R., Bronk, J. T., Bolander, M. E. (1996), *Human Gene Therapy* **7**, 1339–1346.

Miller, D. L. (1987), *Ultrasound Med. Biol.* **13**, 443–470.

Miller, D. L., Thomas, R. M. (1996), *Ultrasound Med. Biol.* **22**, 1089–1095.

Miller, D. L., Thomas, R. M., Buschbom, R. L. (1995), *Ultrasound Med. Biol.* **21**, 841–848.

Miller, M. W., Ziskin, M. C. (1989), *Ultrasound Med. Biol.* **15**, 707–722.

Miller, M. W., Miller, D. L., Brayman, A. A. (1996), *Ultrasound Med. Biol.* **22**, 1131–1154.

NCRP (1992), *Exposure Criteria for Medical Diagnostic Ultrasound*, Report No. 113, National Council on Radiation Protection and Measurements, Bethesda, MD: NCRP Publications.

Putterman, S. J. (1995), *Sci. Am.* **272** (2) 46–51.

Sanghvi, N. T, Hynynen, K., Lizzi, F. L. (1996), *IEEE J. Engin. Med. Biol.* **15**, 83–92.

Sapareto, S. A., Dewey, W. C. (1984), *Int. J. Radiat. Oncol. Biol. Phys.* **10**, 787–800.

ter Haar, G. (1995), *Ultrasound Med. Biol.* **21**, 1089–1100.

Further Reading

AIUM (1993), *Bioeffects and Safety of Diagnostic Ultrasound*, Laurel, MD: American Institute of Ultrasound in Medicine.

Crum, L. A. (1994), *Phys. Today*, **47** (9) 22–29.

Ebbini, E. S. (Ed.) (1996), Special Issue on Therapeutic Ultrasound, *IEEE Trans. Ultrason. Ferroelec. Freq. Control* **43**, 989–1129.

Leighton, T. G. (1994), *The Acoustic Bubble*, London: Academic Press.

NCRP (1992), *Exposure Criteria for Medical Diagnostic Ultrasound*, Report No. 113. National Council on Radiation Protection and Measurements, Bethesda, MD: NCRP Publications.

Nyborg, W. L. (1991), "Biological effects of sound and ultrasound," in: G. L. Trigg (Ed.), *Encyclopedia of Applied Physics*, Vol. 2, New York: VCH, pp. 403–420.

CHARACTERIZATION AND ANALYSIS OF MATERIALS

(An Addendum to *Encyclopedia of Applied Physics*, Volume 3, pages 215–240.)

JOHN P. SIBILIA, *Sibilia Associates Inc., Livingston, New Jersey, U.S.A.*

1. OPTICAL MICROSCOPY AND IMAGE ANALYSIS

Considerable strides have been made in recent years in developing quantitative image analysis (QIA) (Russ, 1990). Faster and greater processing-capacity computers and improvements in software have made the acquisition, digitization, storage, and image display processes much more efficient. For example, one of the most widely used applications is in the analysis of particle size distributions and particle shapes (Li *et al.*, 1995). The improved computational developments have resulted in improved resolution by enhancing the contrast between particles and displaying a variety of particles in a mixture by discrete colors. These techniques have been applied to the analysis of polymer composites, metal alloys, ceramic mixtures, and various combinations of polymer, metal, and ceramic composites (Sibilia, 1996). In addition to information on the particle sizes and shapes, QIA also provides information on the degree of mixing of the components, their ability to "wet", and their packing efficiency. Contiguity, a parameter which measures the degree of interaction between neighboring particles, is also determinable by QIA.

QIA has also been applied in the medical and biological sciences. The various features of cell structures have been much more clearly elucidated and quantified by image analysis. Furthermore, the technique has been applied to the analysis of virus-containing tissues and the micromolecular diffusion in skin.

2. MOLECULAR SPECTROSCOPY

2.1 Infrared Spectroscopy

The infrared dispersive spectrometers have essentially been replaced by Fourier-transform infrared (FTIR) spectrometers. These instruments utilize an interferometer in place of a monochromator to achieve the separation of absorbed frequencies when infrared light is passed through a sample (Ferraro and Krishnan, 1990). The advantages of FTIR are several. They have much higher source-radiation throughput, increased signal/noise ratio, reduced measurement times, and higher wavelength accuracy. These advantages have allowed infrared spectroscopy to be used in a number of new applications, as well as the conducting of more standardized experiments in a more efficient manner (Coleman, 1993).

One of the areas where FTIR has widened the scope of applications in infrared spectroscopy is in surface analysis. Infrared spectroscopy in general can be used to provide information about the molecular structure of the surfaces of materials. There are three available

ISBN 3-527-29308-6

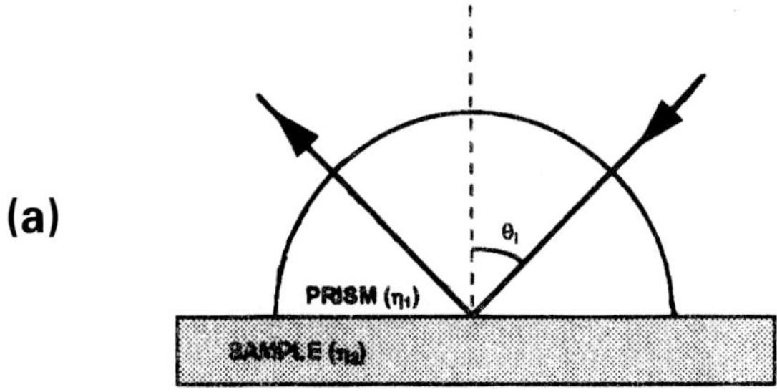

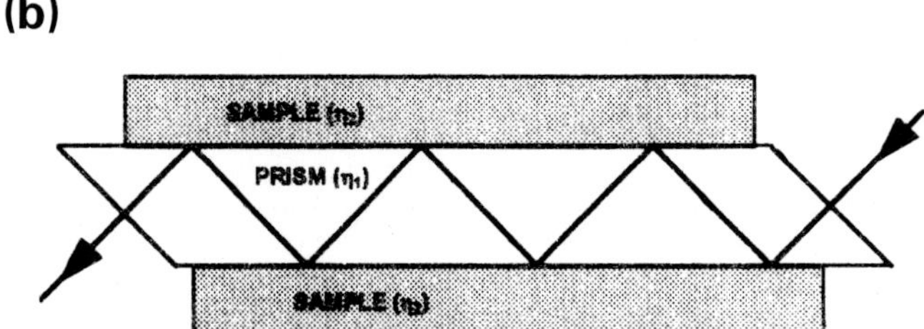

FIG. 1. Schematic of sampling apparatus for ATR. (a) Single-reflection hemispherical element; (b) multiple-reflection parallelogram element. From Sibilia (1996), p. 20, Fig. 2.1.1.

techniques, which provide information at three different depth levels in materials and for different kinds of materials. The most common of these FTIR surface analytical techniques is attenuated total reflectance (ATR), which probes the depth in the range of about 0.5 to 3 μm. In this technique the sample is held in contact with a prism as shown in Fig. 1. In Fig. 1(a), a single-reflection hemispherical prism is shown while in 1(b), a multiple-reflection prism is used with the sample pressed against both prism surfaces. Light entering the prism (usually KRS-5, or some other high–refractive-index, infrared-transparent inorganic crystal) undergoes internal reflection at the prism surfaces. At the interface, the light penetrates a small distance into the sample. The depth of penetration depends on the angle of incidence of the light beam and the difference between the index of refraction of the prism (n_1) and that of the sample (n_2) (Fig. 1). The light exiting the prism is attenuated at specific frequencies because of absorption by the sample. The resulting spectrum is similar to that obtained by transmission infrared spectroscopy. The sensitivity of the ATR technique is enhanced if the multiple-pass configuration [Fig. 1(b)] is used rather than single-pass mode shown in 1(a). Applications of the ATR technique include the surface analysis of thin films, adhesives, surface contaminants, surface compositions of materials, and migration of substances to the surface of materials.

Another infrared spectroscopic surface technique is photoacoustic spectroscopy (PAS) (*q.v.*). The sampling depth of PAS is about 1 to 10 μm. In this technique the incident infrared energy that is absorbed by the sample is converted into heat. Gas, which is in contact with the sample surface, is heated by the sample and expands. The expansion of the gas produces an acoustic signal when the incident beam intensity is modulated at a frequency in the acoustic range. Microphonic detection of this signal can be made to produce an IR spectrum when it is processed by the detector amplification electronics of an FTIR spectrometer. PAS is particularly useful for examining samples as received without any further preparation. For example, powders, pellets, films, opaque materials, and rough-surfaced materials can all have their surfaces analyzed by PAS.

The third infrared surface analytical technique, which is capable of examining ultrathin (monolayer) films, is grazing incidence reflectance (GIR) spectroscopy. In this methodology the IR beam intersects the surface at a high angle of incidence. This technique is highly sensitive but has been limited to the examination of only metallic surfaces.

Another infrared technology which has been expanded through the use of FTIR is infrared microspectroscopy. Infrared spectra, with a spatial resolution of about 10 μm, can be achieved by coupling a high-quality optical microscope with an FTIR spectrometer. Spectra in both transmission and reflectance modes can be obtained. It is possible to perform localized surface analyses *in situ* using specialized ATR and GIR objectives at lower spatial resolution (about 25 μm). Mapping of individual components can also be achieved by using an *x-y* translation stage. Infrared microspectroscopy has been applied in analysis of contaminants or defects on surfaces of ceramic, metallic, and polymer substrates; gels and defects in pharmaceutical tablets and packages; and electronic failures; and in forensic analysis of fibers, pigments, dirt etc.

Another 'combination' technique that has been extensively applied is chromatography-FTIR. Components in the microgram and subnanogram range have been identified by this combination technique. In the GC-IR mode, compounds can be identified in the gas phase in real time. Techniques such as liquid chromatography/IR or gel permeation chromatography/IR require that the sample first be separated and collected on a disc and then analyzed.

The study of kinetic systems and transient species has also been enhanced through the application of FTIR techniques. Spectra can be obtained in times as short as 1/20 of a second with the use of conventional FTIR. If thc spectrometer is operated in a step-scan mode, microsecond and even nanosecond time domains can be achieved. This allows for real-time studies of chemical reactions to detect unstable intermediates, establish reaction mechanisms, and determine kinetic parameters.

2.2 Raman Spectroscopy

Both dispersive and Fourier-transform spectrometers are used routinely for Raman measurements. A choice of incident laser frequencies, ranging from the ultraviolet to visible regions of the electromagnetic spectrum, is usually available. FT-Raman spectrometers use the same principles as FTIR but operate in the near-IR region of the spectrum and use a near-IR laser at 1.06 μm. The advantage of the FT-Raman spectrometers is that sample fluorescence can be reduced. Sample fluorescence is a common problem in Raman spectroscopy; it results in masking of the spectrum and sometimes creates totally unusable spectra.

One recent application of Raman spectroscopy is in the analysis of carbon/graphite-containing materials. Infrared spectroscopy is of little use in the analysis of such materials since they absorb too strongly throughout the IR region and provide little structural information. Raman spectroscopy, on the other hand, can distinguish between the different forms of graphite (determined by their different degrees of graphitization) and diamond.

2.3 Nuclear Magnetic Resonance Spectroscopy

Significant advances have been made in multi-dimensional high-resolution NMR. Two-dimensional NMR (2D NMR) is a methodology that provides molecular structural information in a simple visual display. High-resolution one-dimensional NMR spectra of even relatively low–molecular-weight compounds can be complex on account of the overlapping array of signals arising from a multitude of different chemical shifts and spin–spin interactions. When a second frequency dimension (2D NMR) or in some cases even 3D or 4D is utilized, structural information becomes easier to extract because the interactions between specific nuclei in the molecule become more evident.

Advances in the application of cross-polarization (CP) and high-power decoupling (DD) have also made solid-state NMR a more valuable technique. These and other advances have accelerated the development of multidimensional solid-state NMR to applications in polymers (Mathias, 1991) as well as small-molecule organic and inorganic compounds. Magnetic resonance imaging of materials is another field that is also rapidly developing in solid-state NMR.

2.4 Mass Spectrometry

Two relatively new techniques in the field of structural mass spectrometry have been used to analyze the structures of high-molecular weight polymers. They are plasma desorption mass spectrometry (PDMS) and matrix assisted laser desorption mass spectrometry (MALDI). Both techniques utilize a pulse of energy to create a thermal spike in the sample, which in turn results in the ejection of both positively and negatively charged ions into the gas phase. Time-of-flight mass analyzers are now generally used for both PDMS and MALDI. Time-of-flight mass spectrometers have only moderate mass-resolving power but have no constraints on mass range. They are also ideally suited for this application because they have the ability to detect all the ions ejected by the ion source.

3. CHROMATOGRAPHY

3.1 Supercritical Fluid Chromatography

Supercritical fluid chromatography (SFC) is yet another type of chromatography in a growing field that is more appropriately known as separation sciences. The technique requires that the sample to be analyzed must be soluble in a solvent for dilution and injection purposes and must also be soluble in the mobile phase used in the SFC method. The mobile phase is a supercritical fluid, i.e., a "gas" above its critical pressure and above its critical temperature. Most analyses are carried out at room temperature or slightly higher but rarely above 100 °C. Sensitivities by this methodology are in the parts-per-million range or better depending on the method of detection used and the nature of the sample. The principle of the method is similar to that of other chromatographies where components are separated by their ability to partition between the mobile phase and the stationary phase, which is a component of the column. Some of the supercritical fluids used in SFC are carbon dioxide, ammonia, monochlorodifluoromethane, and dichlorodifluoromethane. SFC has advantages over GC and HPLC in certain analytical applications, but it does not share the wide range of applicability of these two well established techniques.

3.2 Capillary Electrophoresis

High-performance capillary electrophoresis (HPCE) (see CAPILLARY ELECTROPHORESIS in this Supplement) is a technique that combines the strength of high-performance liquid chromatography (HPLC) and conventional electrophoresis to yield rapid, precise, automated, and highly efficient analysis of complex mixtures (Wenclawiak, 1992). It is particularly useful for analyzing both organic and inorganic compounds in aqueous media. Since HPCE utilizes such small sample sizes (nanoliters and less), it is most effective as an analytical rather than a preparative technique.

The basic principles of HPCE are rather simple. A small plug of sample solution is introduced into one end of a fused-silica capillary tube that contains an appropriate electrolyte. Upon application of a large electrical field to the solution flowing through the capillary column, the molecules in the sample migrate at different rates through the column. A number of different detectors such as ultraviolet/visible, fluorescence, conductivity, or mass spectrometry are used to detect the effluent molecules. The technique has found wide use in a broad range of applications, especially in the pharmaceutical industry.

4. SCANNING PROBE MICROSCOPY (SCANNING TUNNELING AND ATOMIC FORCE MICROSCOPY)

Scanning probe microscopy (SPM) (see TUNNELING MICROSCOPY AND SPECTROMETRY) is the term associated with a variety of techniques which measure the topography of surfaces at the atomic level. Scanning tunneling microscopy (STM) and atomic force microscopy (AFM) are two of the earlier SPM methodologies that were described in the original article. In Fig. 2 are schematic drawings showing the basic operations of an STM and an AFM. In STM the piezoelectric scanner probe passes over the surface of a relatively flat conducting sample. If a potential differ-

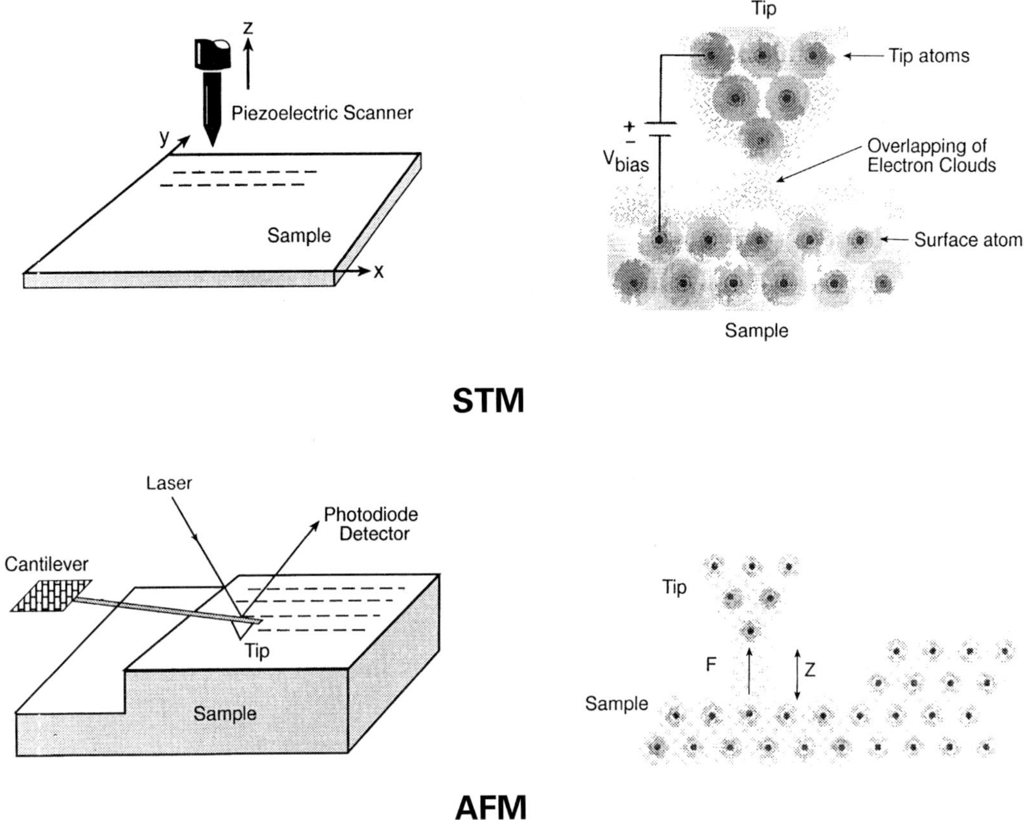

FIG. 2. Schematic of the STM and AFM setups. From Sibilia (1996). p. 190, Fig. 8.18.

ence is maintained between the tip and sample surface as illustrated in Fig. 2, a current will be generated if the tip passes very close to the sample surface. The current is generated by electrons "tunneling" their way through the surface of the sample and through the air gap to the probe tip. As the tip is traversed over the surface it will experience an up-and-down motion corresponding to the topography of the surface at the atomic level. The motion of the probe can be sensed by a transducer and in turn can be converted to a contour map of the surface. This is only one mode of operation, and it involves feedback to the (vertical) positioning mechanism based on the magnitude of the tunneling current. In the AFM technique, the electronic forces between the probe tip and the sample surface cause a deflection in the cantilever arm. The deflections can be detected by means of a laser beam reflecting off the cantilever and into a photodiode detector. A mapping of the cantilever deflections as the probe is traversed over the sample surface results in a map of the atomic level surface topography. Basically, all the SPM techniques consist of a probe which scans the surface of a material and measures the surface topography and the specific surface interaction, which is dependent on the nature of the probe and sample type of surface being studied. Three more recent techniques (in addition to STM and AFM) to join this expanding field of SPM are magnetic force microscopy (MFM), scanning thermal microscopy (SThM), and scanning near-field optical microscopy (SNOM).

MFM is a variation of AFM that uses a ferromagnetic probe to detect the forces exerted on the tip by the stray magnetic fields of the sample. While AFM operates in a contact mode, the tip does not contact the surface in the MFM mode. Both surface topography signals, which dominate at close distances to the surface, and magnetic signals, which are more prevalent at greater distances from the surface, can be detected by moving the probe closer to or further away from the surface. This is important to establish the relationship between the topographic and magnetic data. Resolution down to 50 nm and the non-contact/nondestructive nature of MFM make this a useful addition to the SPM arsenal.

SThM uses the sample's thermal conductivity characteristics as a contrast mechanism in imaging microscopic features (Dinwiddie *et al.*, 1994). There are several variations of the thermal probe. It can be made with two wires to form a thermocouple junction at the scanning tip or it can comprise an integrated resistive thermal probe with the AFM cantilever. The probe is operated in a constant-force mode as in an AFM. SThM can collect and display topographic and thermal (conductivity or temperature) data simultaneously. Like the MFM technique, which provides both topographic and surface magnetic information, SThM provides both topographic and basic information about the conductivity/temperature characteristics of the sample surface.

SNOM is a scanning probe technique that provides optical microscopic information at a resolution better than 100 nm. The optical image is generated by means of a subwavelength optical probe that is held in close proximity to the sample (i.e., in the near field). The near-field interaction, which is based on fiber-optic waveguide theory, is monitored while the probe is scanned across the sample surface. The SNOM technique, which does not use lenses, has improved resolution over conventional optical microscopy since the diffraction limitations due to the use of lenses are eliminated. The technique has been finding application in the semiconductor, bioscience and magneto-optical device fields.

List of Works Cited

Coleman, P. B. (1993), *Practical Sampling Techniques for Infrared Analysis*, Boca Raton: CRC Press.

Dinwiddie, R. B., Pylkki, R. J., West, P. E. (1994), *Thermal Conductivity* **22**, 668–670.

Ferraro, J. R., Krishnan, K. (1990), *Practical Fourier Transform Infrared Spectroscopy*, New York: Academic Press.

Li, C.-W., Lui, S.-C., Goldacker, J. (1995), *J. Am. Ceram. Soc.* **78** (2), 449–450.

Mathias, L. J. (1991), *Solid State NMR of Polymers*, New York: Plenum Press.

Russ, J. C. (1990), *Computer-Assisted Microscopy, The Measurement and Analysis of Images*, New York: Plenum Press.

Sibilia, J. P. (Ed.) (1996), *A Guide to Materials Characterization and Chemical Analysis*, 2nd ed., New York: Wiley-VCH, Chapters 8 and 9.

Wenclawiak, B. (1992), *Analysis with Supercritical Fluids: Extraction and Chromatography*, New York: Springer.

CHEMICAL ANALYSIS

(An Addendum to *Encyclopedia of Applied Physics*, Volume 3, pages 307–343.)

KENNETH A. RUBINSON, *The Five Oaks Research Institute, Cincinnati, Ohio, U.S.A.*

INTRODUCTION

Steady improvement in practically every area of chemical analysis has occurred since the original article appeared in this Encyclopedia. However, some areas have progressed so fast that little is similar over that time span. This update focuses on four major changes that have occurred:

1. in spectrometry, where a larger fraction of the light available for analyses can be collected through the use of multichannel monitoring;
2. in electrophoresis, where running the separations in capillary tubes has resurrected this old method of separations;
3. in mass spectrometry, where molecules with molecular weight up to 100 000 can be brought into the gas phase; and
4. in sample preparation, where microwave heating has helped to speed sample preparation to match the ever greater throughput of advanced instruments.

1. PARALLEL COLLECTION OF ANALYTICAL INFORMATION

A quiet revolution in detectors has occurred with the advent of silicon solid-state charge-transfer devices (CTDs), which include charge-coupled devices (*q.v.*) (CCDs). Contemporary CCDs now outperform photomultiplier tubes in nearly every application. Each photon that hits the CCD produces one (in the near-IR to visible range) or more (in the UV and x-ray ranges) electron-hole pairs, which are stored inside the volume under the location where the photon hit the detector. These charges are stored until the packets of charge from each pixel are read out individually and amplified. Figure 1 shows the difference between using a single detector such as a photomultiplier in conjunction with a monochromator in a scanning instrument and a CCD array in a dispersive, nonscanning instrument (a polychromator).

The scanning instrument gets it name because over time it scans the colors past a slit that selects a narrow range of the whole to produce a spectrum (a plot of light power *vs.* wavelength). However, the scanning instrument makes inefficient use of the light entering it. Most of the light—that not at the specific wavelength falling on the detector—is wasted. However, when the entire spectrum can be recorded simultaneously, far more signal is recorded per unit time. This inherently better performance is designated the multichannel advantage. Closely related is the multiplex advantage of infrared spectrometry.

Polychromators are used to great benefit in atomic emission spectrometry. Early polychromators had up to ten photomultiplier tubes set to record the lines from ten elements simultaneously. In contrast, contemporary instruments can quantitate essentially all the common elements in a sample in about a minute through a combination of the CCD-based polychromator and contemporary small computers. Improvement in the signal-to-noise ratio over the older methods can be up to a factor of 100.

ISBN 3-527-29308-6

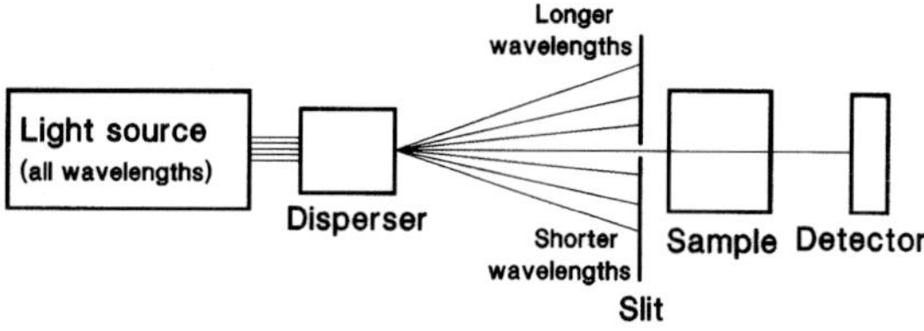

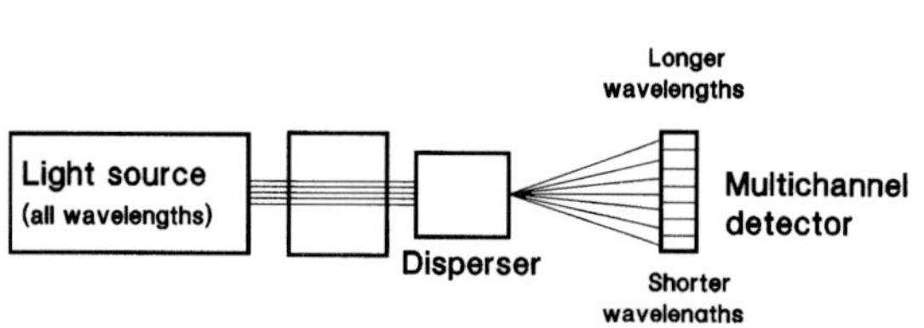

FIG. 1. Diagram of the general differences between a scanning spectrometer (top) and a polychromator system (bottom). The disperser consists of one or more gratings. The scanning spectrometer scans sequentially through the wavelengths. Here, it is shown in the middle of the range. The multichannel spectrometer has optics that are fixed in position.

Separations provide another application where polychromators are rapidly being applied. Compounds are separated in time (and volume) by column liquid chromatography (see CHROMATOGRAPHY) and then pass sequentially through a detector. The volumes containing the individual compounds are called bands. If the band holds a single compound, the spectrum is the same throughout. However, if a band is not pure, but composed of two compounds, it is likely that one is enriched at the leading edge and the other is enriched at the trailing edge. Two different spectra are seen at the edges and a combination of the two in the middle. The benefits of multiplexing are so great that multidimensional detectors have become the method of choice, since no more time is necessary for the chromatographic separation to be run. After all, why should all that extra data not be collected if it can be used to answer any questions that arise about the purity, identity, and properties of the compounds in each band?

2. ELECTROPHORESIS

Separations of different species based on their dissimilar behavior under the influence of an applied electric field are classified generally as electroseparations. There are six phenomenological rules that describe the factors that underlie the various types of electroseparations.

1. Negative ions tend to migrate toward the higher (positive) voltage and positive ions tend to migrate toward the lower (negative) voltage.
2. The average rate of migration of an analyte species is proportional to the average charge on the ion.
3. The average rate of migration of an analyte species is proportional to the average voltage applied.
4. The average rate of motion of an ion decreases as its average cross section increases. (The cross section is the area of the ion that lies perpendicular to the direction of the motion.)

The ability to separate slightly different molecules, such as deoxyribonucleic acids (DNAs) that differ by one base unit, is severely compromised by convection, however. Convection is the macroscopic mixing of liquids due to differences in density, and heating causes much convection in solutions. So,

5. the quality of electroseparations in liquids improves as the convection can be suppressed.

Heating of a liquid in electroseparations *cannot* be avoided since the electric current passing through the solution heats it. Finally,

6. the quality of electroseparations in liquids improves when the temperature can be held constant over the time of the experiment.

There are two ways to suppress convection in a liquid when gravity is present:

A. running the electroseparation in a gel, where the molecular strands of the gel stop macroscopic convection by trapping the liquid between the strands, and

B. running the electroseparation inside a small ($\sim 100\,\mu m$) capillary, where the walls of the capillary have the same beneficial effect as the gel molecules. In addition, the small size of the capillary lowers the temperature differences that can arise since the heat can leave relatively easily across the short distances to the surface of the tube.

The analytical methods that fall under the classification of electroseparations in solution are used to separate and quantitate analytes ranging from small molecules (formula weight ~ 100) to the largest macromolecules (formula weight up to 10^8) such as DNA chains or the largest proteins.

2.1 Gel Electrophoresis

The most commonly used methods of electrophoresis are done by carrying out the separation in a polymeric gel (as in *gelatin*). The gels are composed of a three-dimensional web of polymer strands having the spaces between the strands filled with liquid. Two different separating mechanisms exist: electrophoresis, that separates by the charge-to-size ratio (that is, the total electric charge on the molecule divided by its effective cross section), and sieving, that separates mostly by size. The sieving effect is caused by the macromolecules' migration being slowed by colliding with the gel molecules; larger molecules are slowed more than smaller ones. Especially prominent among the applications of gel electrophoresis is the analysis of DNA and RNA fragments, which forms the analytical basis of contemporary molecule biology of thc gene.

Conditions for electrophoresis can be divided into two classes—denaturing and nondenaturing. To *denature* means to make the molecules biologically inactive. The molecules are still polymers with the monomers covalently linked in their original order, but the polymer is no longer folded into the form that it has when it is active. Detergents such as sodium dodecyl sulfate [SDS, $CH_3(CH_2)_{10}CH_2OSO_3^-Na^+$] and other nondetergent agents (*e.g.*, urea) are known to cause such denaturing. The names of the methods, such as *SDS polyacrylamide gel electrophoresis*, relate the type of gel and that it is run under denaturing conditions.

2.2 Capillary Electrophoresis

In an electrophoretic separation, both the speed of the separation and the resolution of the components improve as the applied electric field is increased. The reason that the voltage cannot be increased at will is that the heating increases to the point where the convection ruins the quality of the separation. The heating effects can be reduced drastically by running the electrophoresis inside a capillary tube having an inside diameter less than 0.1 mm and a length about 50 cm to 1 meter. Naturally, the technique is called *capillary electrophoresis* (CE) (see CAPILLARY ELECTROPHORESIS in this update volume). Another name is capillary zone electrophoresis (CZE). One reason why running in a capillary is effective follows from its electrical properties. The electrical resistance is so high that the current remains low (less than about $10\,\mu A$). As a result, the heating is reduced.

The apparatus for capillary electrophoresis is illustrated in Fig. 2. The technique has great resolving power, and methods have been developed so that electrophoresis can even be done for neutral, nonionic molecules. (This latter method is called micellar electrokinetic chromatography, abbreviated MECC.)

The analyte zones separated by capillary electrophoresis can be detected in their path through the capillary as they pass a detector, since the solution contained inside the tube move while the separation is occurring. This solution movement is called electroosmosis or electroendoosmosis.

Using a narrow capillary does involve a tradeoff: the samples must be about a nanoliter or less, and, after separation, the analytes are each in a similar volume. With such small liquid samples, only a few methods of detection can be used effectively. For example, light absorption is not routinely used to detect concentrations below $1\,\mu M$ in such small volumes. As a result, analytes are required to be at least $10\,\mu M$ to 0.1 mM. Nevertheless, nanogram to picogram quantities can be detected and quantitated, and it is this progress in detection that

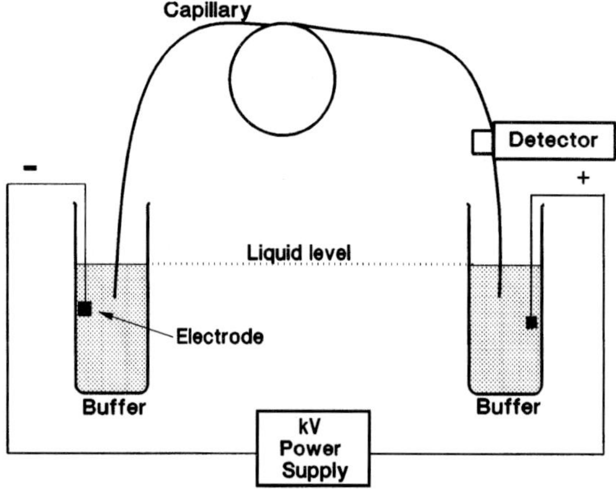

FIG. 2. Diagram of the basic parts of a capillary electrophoresis apparatus with spectroscopic detection. The sample is put in at the left side by placing the end of the capillary in the sample solution and raising its level so that the solution flows in. To run, the end is placed in a buffer set at the same height as the right end. While running, the levels of the two buffer solutions must be the same (as indicated by the dotted line) to avoid further bulk flow. The entire unit shown is held at a constant temperature.

has led to the revival of electrophoresis in this newer, capillary, form.

In running the separation, when the anode (+) is on the injection side of the capillary, this is called running in the normal mode. When the cathode (−) end is on the injection side, it is the reversed polarity mode. Voltages up to $1000\,V\,cm^{-1}$ may be applied.

3. MASS SPECTROMETRY

Mass spectrometry (MS) is one of the most versatile and powerful tools of chemical analysis. Its broad usefulness is due mostly to the wide variety of capabilities possible for each of the three sections of a mass spectrometer: the mass source, the mass discriminator, and the transducer/detector. These were illustrated in the original article. The gas-phase ions are separated according to their differences in mass-to-charge ratio, abbreviated m/z.

Since its inception, mass spectrometry has been used effectively to analyze small molecules and atoms. However, a variety of sources have been devised to vaporize and ionize materials that are not initially gases, which has allowed mass spectrometry to analyze virtually all types of samples. Because of these newer vaporization/ionization methods, mass spectrometry is now firmly established in biochemical, polymer, and inorganic analysis. For molecules greater than 10 000 daltons [1 dalton (Da) = 1 unified atomic mass unit (u)], electrospray ionization (ESI), matrix-assisted laser desorption ionization (MALDI), and plasma desorption (PD) can be used. ESI has an advantage over the others in that multicharged ions (up to the range of 20 charges) are produced, which brings these heavy molecules into the m/z range of 500 to 2000. As a result, electrospray ionization permits mass spectrometry up to the 100 000-Da range on commonly available instruments. The limitation, then, is only whether such large molecules can be brought intact into the gas phase, and, apparently, many can.

Mass spectrometry also has become one of the two techniques of choice for the simultaneous analysis of many elements. The source of the ions of the elements is the inductively coupled plasma torch. This torch is a plasma formed from argon that is heated to about 3000 K with high-power radio waves. The sample in solution passes into the torch, where ions of the elements form and are passed into a mass spectrometer for analysis.

4. SAMPLE PREPARATION

Contemporary instrumentation has increased our ability to measure the contents of samples more quickly, and the bottleneck in productive analyses in many cases has become the relatively slow preparation of samples. For example, one might require a liquid solution for analysis, but the sample is soil from a field.

The conversion from a solid to the solution involves dissolution of soluble components and bringing otherwise insoluble components into solution. This latter process is called digestion.

A relatively new addition to accelerate sample dissolution and digestion is the microwave oven. A number of such ovens are specifically designed for chemical sample digestions with heating compartments lined with relatively inert materials and a means to remove corrosive vapors. Their power levels are far more precisely controlled than those of household microwave ovens, and monitors of the sample temperatures and pressures are provided. Gram-size samples can be heated to a given temperature from 10 to 100 times faster than by conventional thermal heating and with far better control of the speed of heating and the temperature. Microwave assistance also facilitates digestions in sealed containers. These containers are most often made of a type of Teflon^{TM} called (perfluoroalkoxy)ethylene (or Teflon PFA). Digestion in closed containers obviously is a superior way to ensure that analyses will be accurate and precise, since the loss of analyte is limited and introduction of interferents is easier to avoid. Both characteristics are especially important in trace and ultratrace analyses. In addition, the closed containers allow digestion to be accelerated since the boiling point can rise above the normal boiling point of the acid at atmospheric pressure. As an example, concentrated (constant boiling) nitric acid boils in an open container (at 1 atmosphere pressure) at 120.5 °C, while in a microwave pressure vessel, the temperature can be held at about 185 °C. A typical digestion reaction could be accelerated about 800-fold with this temperature difference. The limits of microwave digestion arise from the pressure and temperature limits of the containers and the electrical and chemical properties of the samples.

Further Reading

Bilhorn, R. B., Epperson, P. M., Sweedler, J. V., Denton, M. B. (1987), "Spectrochemical Measurements with Multichannel Integrating Detectors," *App. Spectrosc.*, **41**, 1127–1136.

Baker, D. R. (1995), *Capillary Electrophoresis*, New York: Wiley.

Johnstone, R. A. W., Rose, M. E. (1996), *Mass Spectrometry for Chemists and Biochemists*. Cambridge, U.K.: Cambridge Univ. Press.

Kingston, H. M., Jassie, L. B. (1988), *Introduction to Microwave Sample Preparation*, Washington, D.C.: American Chemical Society.

CYCLOTRONS

(An Addendum to *Encyclopedia of Applied Physics*, Volume 4, pages 427–462.)

R. A. Baartman, R. E. Laxdal, and G. H. Mackenzie, *TRIUMF, Vancouver, British Columbia, Canada*

INTRODUCTION

The cyclotron was invented in 1931, and from its inception into the 1960s installations were designed chiefly for research in nuclear physics and radiochemistry. Several laboratories developed strong affiliated programs in nuclear medicine, radiation oncology, and materials science, and these became the primary functions of some laboratories during the 1960's. Since then ever more cyclotrons have been built to service research in nonnuclear fields and for industrial and medical applications. This situation was described in the original article Cyclotrons. That article also gave the history of cyclotrons and described their design features in some detail. This Addendum introduces new fields of application, some speculative, and gives illustrations of recent activity in some older ones. It describes some new cyclotron technology and contains a discussion of the factors that may limit the intensity of the beam from cyclotrons.

The most recent overview of the field can be found in the Proceedings of the 14th Conference on Cyclotrons and their Applications (Cornell, 1996). There are presently more than 130 cyclotrons operating commercially and about 75 more in research institutions. About 30 projects involving new machines or major expansions of existing facilities are underway.

1. APPLICATIONS OF CYCLOTRONS

1.1 Cyclotrons as Sources of Neutrons

It is becoming harder to convince the public to license new reactors or to replace older ones. Accelerators are being perceived as a safer source of neutrons for the production of power by nuclear fission and for other traditional reactor applications. They function by directing an ion beam onto a target with a

ISBN 3-527-29308-6

high atomic number. Spallation reactions, and sometimes fission reactions, produce a copious number of secondary particles including fast neutrons. These last may then be slowed in a moderator to give the velocity spectrum suited to the application. The optical source of the neutrons is less diffuse than in a reactor and much less total power is dissipated.

Examples of spallation neutron sources are ISIS at the Rutherford-Appleton Laboratory, near Oxford, United Kingdom, and LANSCE at the Los Alamos National Laboratory, New Mexico. ISIS is based on a 200-μA, 800-MeV proton synchrotron and LANCSE on a linac and storage ring handling 100-μA, 800-MeV protons. The nature of operation of these types of accelerator is to produce neutrons with a pulsed time structure that can be useful for time-of-flight spectroscopy. The most powerful proton beam at present is at the Paul Scherrer Institute (PSI), Villigen, Switzerland, where 850 μA of the 1.5 mA of 590-MeV protons extracted from the c.w. PSI cyclotron are directed toward the SINQ cold neutron source to produce $10^{14}\,n/\mathrm{s\cdot cm^2}$; this is comparable to a medium-flux reactor. Liquid-deuterium moderators can be placed near the maximum neutron flux, and the continuous beam of very cold, 25-K or 2-meV, neutrons will match that of the current best source, the reactor at the Institut Laue-Langevin (ILL), Grenoble, France.

Articles on NEUTRON SCATTERING and NEUTRON DIFFRACTION describe how physicists, chemists, and biologists use neutron beams to elucidate complicated molecular structures. Neutron therapy and neutron radiography are other established applications that use beam from cyclotrons with $k = 40$ to 70. Two proposed new applications are described below. Neutron sources based on cyclotrons have the advantage that they are smaller and more economical to operate than those using linacs or synchrotrons, although linacs may eventually produce more beam power.

1.1.1 Radioisotope Production by Neutrons. The most widely used medical radioisotope is $^{99\mathrm{m}}$Tc with tens of millions of diagnoses per year. Since it has a 6-h half-life it is supplied in the form of a generator using the 66-h half-life ^{99}Mo parent. This isotope is obtained from reactors as a fission byproduct of ^{235}U. There have been proposals to make $^{99\mathrm{m}}$Tc or ^{99}Mo by proton bombardment of an enriched ^{100}Mo target; however, technetium produced directly would have to be used within a few hours of production, and there may be difficulties in separating ^{99}Mo from the ^{100}Mo target. Ion Beam Applications (IBA), Louvain, Belgium, has proposed a facility to make ^{99}Mo by fission, as at present, but to create the neutrons by bombarding a lead target with a 1.5-mA beam of 150-MeV protons. About one fast neutron would be produced per proton. The target would be surrounded by a water moderator, which would contain the secondary production targets of ^{235}U and thermalize the neutrons. The system would operate safely below the critical level and the calculated flux of $2 \times 10^{14}\,n/\mathrm{cm}^2 \cdot \mathrm{s}$ should produce 2×10^{14} Bq/week of ^{99}Mo, which if efficiently extracted and distributed would match world demand.

1.1.2 Production of Thermal or Electrical Power. In the so-called energy amplifier (Rubbia *et al.*, 1994) a beam from an accelerator would initiate a fission cascade in a target placed inside a calorimeter. The fission products would slow down in the calorimeter and their energy would be transformed into heat. Electrical energy transformed in turn from the heat would exceed that required to operate the facility and the excess, as electricity or heat, could be delivered to external users. Thorium is proposed as the target; it is more abundant than uranium, fissile isotopes can be "bred" from it through nuclear interactions, and its use would result in a smaller amount of radioactive waste. Such ideas have been proposed before but the required technology seems closer now.

A schematic energy-generating complex is illustrated in Fig. 1. In this proposal (Fiétier *et al.*, 1996) two injector cyclotrons deliver 6 mA of 10-MeV H^- ions, and two separated-sector cyclotrons accelerate a combined 12 mA of protons first to 120 MeV, then to 1 GeV, for injection into an amplifier module. Interestingly, one 10-MeV beam would be extracted by stripping and the other using a deflector channel; the two would be overlapped in the injection line to reduce space-charge forces. Each amplifier module would produce 1500 $\mathrm{MW_{TH}}$, of which 30 MW would be returned to operate the accelerators. The neutron flux in the target would be in the range of $10^{14}/\mathrm{cm}^2 \cdot \mathrm{s}$, well below criticality so that en-

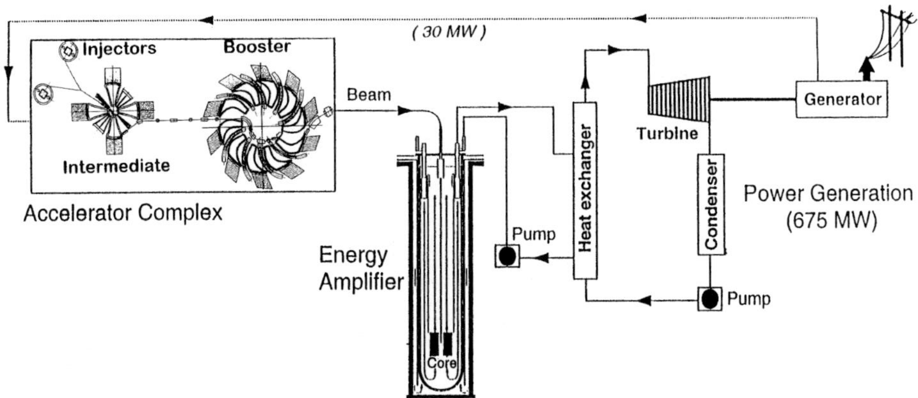

FIG. 1. The major components of an energy-producing complex. The cyclotron chain consumes 30 MW to produce 12 MW of 1-GeV protons. These initiate fission reactions in the core of the energy amplifier. About 1500 MW of thermal energy would be produced, which could be transformed to 675 MW of electrical energy. (Courtesy of P. Mandrillon, Cyclotron Laboratory, Nice, France.)

ergy production will stop when the incident beam of protons is shut off, thus offering an extra measure of safety. A plant might consist of several such generators in parallel.

1.2 Production of Radioactive Ion Beams (RIB)

Nuclear physicists have used separated isotopes in their targets and ion sources for a number of years; the isotopes could be unstable provided half-lives were reasonably long. The current interest is in the production on line of isotopes with short half-lives (minutes or less) and energies from ~ 1 keV/amu to many MeV/amu. Nuclear physicists are interested in the structure of nuclei far from the line of stability in the mass (A)–nuclear charge (Z) diagram. Astrophysicists as well as nuclear physicists are interested in the reaction mechanisms of these nuclei, since they can be part of the chain of reactions that breed heavy elements in stars. Some experiments with the less easily produced isotopes can be performed by confining them in neutral-atom traps. Ions are neutralized by first implanting the beam in a surface and then re-evaporating them as atoms. These are cooled and confined by interchanging their momentum with crossed laser beams. The final ensemble may have the atomic shell structure elucidated by probing with lasers of a different frequency, and nuclear properties can be inferred from their effects on the atomic shells. 6×10^3 ^{38m}K atoms have been constrained to a volume of 2 mm^3 in such a trap; the half-life of the metastable isotope is about 1 second.

There are two chief methods to generate beams of short-lived isotopes. The first directs an energetic beam onto a thin target; the unstable ions are made by direct nuclear reactions and emerge with significant energy according to the laws of kinematics. They are captured in a large-acceptance beam-transport system and led to an experimental area. This method was used at the Bevalac high-energy synchrotron (now shut down) at Lawrence Berkeley Laboratory (LBL), California, and is presently employed using cyclotrons at Michigan State University (MSU), East Lansing, Michigan, and GANIL, Caen, France. Most new RIB facilities, including an upgrade at GANIL, will use a second method in which a high-power primary beam from an accelerator hits a thick target to produce stable and unstable isotopes of many different species by spallation reactions. These continually diffuse out of the target, which is usually hot, into an ion source. Ions emerge at the source potential, and a mass separator selects the species of interest, which is then transported to an experimental area or to a second accelerator. Such an on-line isotope separator, ISOLDE, at CERN, Switzerland, has been producing unstable-ion beams of ~ 1 keV/amu for 20 years using first a proton beam from a synchrocyclotron and more recently a 1-μA, 1.2-GeV proton beam from a synchrotron. Isochronous cyclotrons offer a 10- to 100-fold increase in primary beam power and are the foundation of facilities under construction at TRIUMF at Vancouver, Canada, GANIL, INFN-LNS at Catania, Italy, and RIKEN at Saitama, Japan. One quite different proposal, PIAFE, would place the target and ion source, biased at 30 kV, within the pile of the reactor at ILL. Ions of the neutron-rich isotopes produced would be transported 400 m to the cyclotrons at the nearby Institut des Sciences Nucléaires for acceleration.

1.3 Ion Beams in Industrial and Materials Research

There are now two institutes dedicated to research into condensed matter that use beams from cyclotrons. These are the Hahn–Meitner Institute (HMI), Berlin, Germany, and the Institute for Advanced Materials (IAM), Ispra, Italy.

1.3.1 Thin-Layer Activation. This process and its applications were discussed in CYCLOTRONS, and a recent example of its utility is the investigation of the spallation of the oxide scale that protects alloys from corrosion (Stroosnijder, 1996). This scale often has a different coefficient of thermal expansion from the underlying layer, and thermal cycles can cause cracks to appear and material to spall off. The conventional procedure for measuring resistance to corrosion is to subject a sample to several cycles of the industrial process and then to measure the change in mass. The difference in weight of the sample gives the material lost, which may be only a small fraction of the sample mass. The amount of material spalled away can be measured much more accurately by first producing radioactivity in the surface layer by bombarding a specimen with a proton or deuteron beam, then subjecting the sample to cycling, and finally measuring the activity in the spalled material only. IAM achieves a precision of ng/cm^2. One may choose to irradiate only a part of a sample; for example, one need not irradiate regions of high curvature in a scaled-down model since these will not exist in the full-scale item.

The radioisotope to be produced by the beam for thin-layer studies must have an activity that can be easily measured, usually γ rays. It must have a half-life matched to the duration of the tests and may have to interact chemically with the surface, e.g., in the formation of an oxide scale. In some cases no suitable isotope can be made, or the material may be damaged by the beams producing them. A solution may be to implant radioactive ions directly. Not every element has an isotope with a suitable half-life or γ activity, however, and the range of the heavier ions may be too short. ^{7}Be has been used at MSU and at the Kurchatov Institute, Russia, to irradiate ceramic, polyamid, or viton machine parts and measure their rate of wear during operation of the machines. There is a proposal at TRIUMF to measure wear of replacement hip joints by implanting ^{7}Be, or possibly ^{22}Na.

1.3.2 Constructive Processes in Materials. Evidence is emerging that ion beams can cause constructive processes, i.e., those that create order. Ion beams passing through matter cause electronic excitation, ionization, and displacement of atoms. These processes usually destroy molecules, but it is well known that electron beams can cause cross linking of molecules in plastics. Constructive processes involving ions appear as the energy density transferred to excitation and ionization increases (Fink *et al.*, 1994), and at sufficiently high energy densities the constructive yield can surpass the destructive. The passage of a heavy ion produces a dense cloud of positive charges within a few nanometers of the track, and this cloud in turn is surrounded by a penumbra of knocked-on electrons that excite nearby atoms. The ion thus produces a transient pulse with a high energy in a small volume that quickly quenches once the ion has passed. Such a sequence is suitable for the synthesis of buckminsterfullerene. C_{60} has been found within the tracks of 5-MeV $^6Li^+$ in polyimide at HMI and of 130 MeV/amu $^{161}Dy^{22+}$ on pyrolytic graphite at the UNILAC Laboratory, Darmstadt, Germany. Other examples are given in the article by Apel (1996).

1.3.3 RIB complementing μSR Techniques. Research laboratories have used muon beams to investigate solid state structure, magnetism, chemical reaction rates etc. These procedures are described in MUON SPIN ROTATION/RELAXATION/RESONANCE. Unstable heavy ions also can be used in these fields. Polarized beams of β^- emitters can be stopped in a layer of material $<10\,\mathrm{nm}$ thick, the isotope precesses in a magnetic field, and the observation of the modulation of the decay rate by a β detector will yield information analogous to μSR. The precision may be better than μSR for some of these cases, since the sample is thinner and because more precession cycles may be observed since β decay lifetimes are much longer than muon lifetimes.

1.4 Medical Applications

1.4.1 Radioisotope Production by Ions. The most popular accelerator for isotope production remains the variable-energy H^- cyclotron with 0.1–1-mA beam and a maximum energy in the range 18 to 30 MeV; over 100 have been installed in the last decade. They are economic and reliable in operation and have a high beam power for relatively low capital cost. The TRIUMF laboratory and Nordion Inc., Canada, have jointly demonstrated extracted beams of 30 kW of 30-MeV protons. Also their small size requires a smaller volume of shielding than some competing accelerators. Nine very small 12-MeV H^- superconducting OSCAR cyclotrons have been sold by Oxford Instruments Corp., United Kingdom.

Other continuing developments include the production of generators to supply short-lived isotopes to institutions without production facilities. These generators are isotopes that decay into the short-lived species required for imaging and have a lifetime long compared with transportation times; e.g., Sr^{82} (25 d) $\rightarrow$ Rb^{82} (6 h and 1 min). There are proposals also to use focused beams of therapeutic radioisotopes (RIB) to accumulate high activities in a small volume.

1.4.2 Proton Therapy. The vast majority of medical treatments with particle beams are intended to destroy cancerous tissue while attempting to spare neighboring healthy tissue. Ion beams are well suited to this since they have a well-defined range (see Fig. 15 in CYCLOTRONS) and can be focused to spots in tissue within a diameter of a few millimeters. Proton beams have a lower magnetic rigidity than heavy-ion beams with the same range and thus the accelerator and beam-transport system will be smaller and cheaper. There are 19 proton irradiation facilities at present, and the number continues to increase.

The equipment for a 70-MeV beam to treat ocular melanoma is shown in Fig. 2. Deeper tumors require higher particle energies. 230-MeV protons would have sufficient range for most people but an accelerator may have to provide 250 MeV or even 270 MeV to compensate for energy lost in the beam-preparation system. Tumors with large area can be treated either by incorporating a second scatterer into the standard passive system illustrated in Fig. 2(b) or by steering, usually termed scanning, a small-diameter beam over the tumor area. The second scatterer is designed to broaden the center of the Gaussian distribution produced by the first and to remove the penumbra. The scatterers absorb some of the beam energy and may produce a small background of secondary particles. Scanning techniques avoid these effects and are thought to match tumors of irregular shape better, but they are more complicated. Each scan treats a certain depth of tumor, and the energy of the beam entering the patient is changed to treat "slices" at different depths. The beam current must be constant to a few percent for scanning and to accommodate refinements such as synchronization with breathing. The beam from a cyclotron is well suited as it shows little modulation on a millisecond and longer time scale.

It can take an hour to change the energy of a beam from a variable-energy proton cyclotron. Consequently the treatment facilities often operate with a fixed extraction energy and use energy-absorbing material followed by a magnetic momentum-selection system to obtain the treatment energy. The radiation background and residual activity in the degrader and momentum collimator are acceptable. Both are well shielded, the beam currents used are $\leq 300\,\mathrm{nA}$, and the beam is usually on for less than 10% of the working day since most of the time is taken in setting up the patients for treatment. Cyclotrons for therapy occupy a space smaller than most treatment rooms and are less expensive and easier than synchrotrons to build and operate. IBA have designed a compact 235-MeV proton cyclotron, beam-preparation system, and gantry specifically for therapy. The magnet of this cyclotron is not superconducting but does operate with highly saturated iron; the hill field

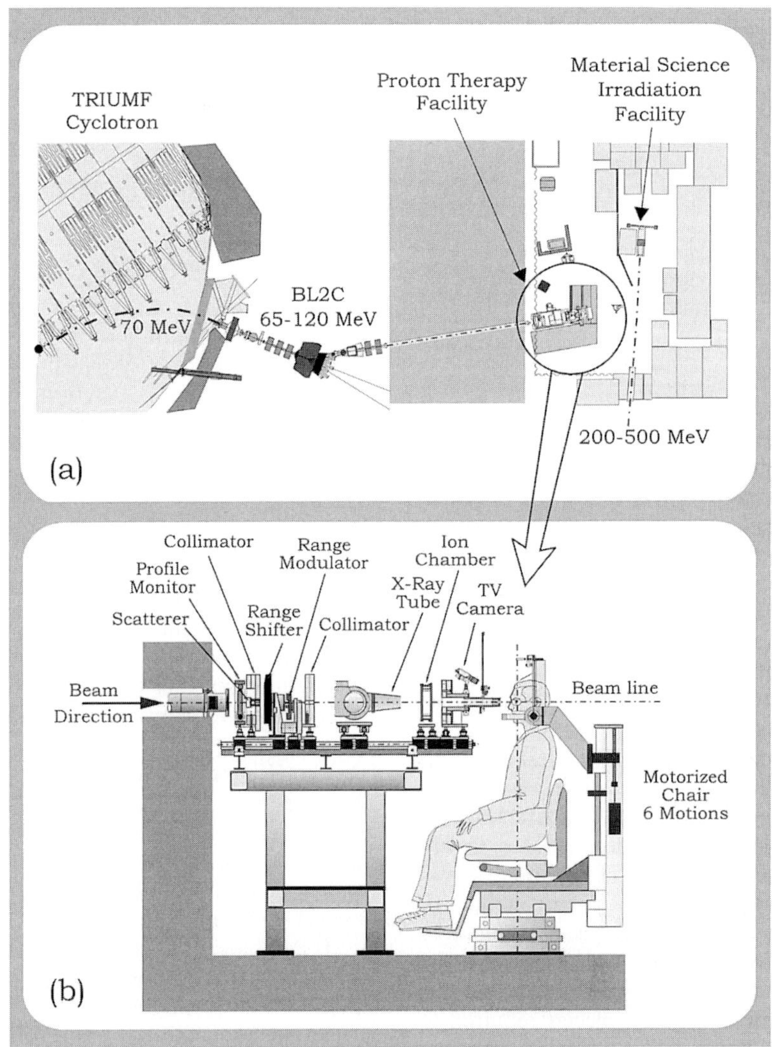

FIG. 2. (a) A portion of the H^- beam circulating in the TRIUMF cyclotron is stripped at 70 MeV, at approximately half of the maximum orbit radius. The protons curl out of the cyclotron into a beam line, which focuses and directs them into a shielded patient treatment area. (b) The equipment used to treat tumors of the eye at TRIUMF. The scatterer introduces a broad Gaussian distribution in the angular divergence of the 70-MeV proton beam extracted from the cyclotron. The collimators trim the sloping sides of this distribution and reduce the flux from the ~1 nA extracted to a treatment level of about 10^7 protons $cm^{-2} \cdot sec^{-1}$. The shape of the final, patient, collimator is matched to the tumor being treated. The material in the range shifter reduces the proton energy to match the distance to the rear of the tumor; the variable-thickness range modulator spreads the dose over the tumor depth. The ion chamber measures the dose of ionizing radiation delivered. The x-ray tube is inserted to confirm the tumor position with respect to the proton beam line. (Courtesy of E. Blackmore, TRIUMF, Canada.)

is 2.3 T and the outer diameter of the yoke is about 4.5 m. This results in a cyclotron of relatively small size without the complication of having to provide a supply of liquid helium in a hospital. Two have been sold, the first to Massachusetts General Hospital, Boston, U.S.A., the second to the National Cancer Center, Kishiwa, Japan.

1.4.3 Neutron Therapy. Salivary-gland and advanced prostate tumors are examples of the sites that respond best to direct irradiation by neutron beams, and there are perhaps 10 cyclotrons attached to hospitals performing such treatments. Boron-capture therapy is a new modality that incorporates ^{10}B into molecules that enter tumors easily; this is followed by irradiating the patient with neutrons. The reaction $^{10}B + n \rightarrow {}^{7}Li + \alpha$ has a cross section of hundreds of barns for thermal neutrons, and the reaction products are highly ionizing and destructive over a short range. At the moment trials are being carried out at reactors and hospitals such as The Department of Radiation Oncology, University of Washington, Seattle, that have cyclotrons that can produce 10^8 slow neutrons/sec$\cdot$cm^2 thermalized in the patient. Ideally about $10^9\,s^{-1}\cdot cm^{-2}$ thermal neutrons are required at the tumor, and if the method proves successful an optimized cyclotron and neutron preparation system will be required for hospital use.

2. ADVANCES IN CYCLOTRON DESIGN

2.1 The Separated-Orbit Cyclotron

A most innovative but technically challenging machine is the superconducting separated-orbit cyclotron (SOC) Tritron at Munich, Germany. It combines some of the features of a cyclotron and synchrotron. The typical isochronous fixed-frequency cyclotron with $\beta \cong r\omega/c$ accelerates a continuous stream of beam bunches but with a turn separation that unfortunately decreases as $(\beta\gamma^3)^{-1}$. Tritron is also isochronous, with a fixed field and accelerating frequency, but in addition the turn separation is constant. This last requires that the accelerating voltage increase proportionately with radius. Ions in Tritron follow a fixed spiral path in the field of superconducting magnets, each turn occupying its own channel in the iron, Fig. 3. Efficient superconducting rf accelerating cavities produce a radial turn separation of 40 mm, much larger than the beam width. Particles leading or lagging in phase with respect to the reference particle will deviate in energy and eventually in radius. The field gradient dB/dR in each separate channel can be chosen to provide a longitudinal (phase) restoring behavior as well as transverse focusing provided that the beam bunches lie at negative phases where $dT/d\varphi > 0$ (Trinks, 1991). This characteristic resembles a synchrotron; it takes about 4 turns in Tritron to complete one synchrotron, or phase, oscillation. The existence of longitudinal focusing allows operation at 170 MHz with harmonic numbers $h = 14$ to 55, larger than the 1 to 10 typical of more conventional cyclotrons. All components are now in place and a beam of $^{32}S^{14}$ ions has been accelerated over several turns.

2.2 Injection from Radiofrequency Quadrupole Linacs

Linear accelerators usually operate with a fixed rf frequency and a fixed particle-velocity profile, which results in a fixed output energy per nucleon (see ACCELERATORS, LINEAR). This is a limitation to their use as injectors to variable-frequency cyclotrons. Some years ago a variable-frequency drift-tube linac, RILAC, was built at the RIKEN Laboratory, Saitama, Japan. The drift tubes were attached to the inner end of coaxial $\lambda/4$ lines, and the frequency was altered by varying the electrical length of these lines by means of a sliding short. These Wideröe-type linacs still require an injected beam of several hundred keV. Radiofrequency quadrupole linacs (RFQ's) accept beam at a few keV/amu, bunch it, and accelerate it with high efficiency, but again they usually operate at a fixed frequency, and moreover their emittance is not well matched to a cyclotron acceptance. This situation is changing. A variable-energy and -frequency 4-rod RFQ injector linac is being built by a group at Frankfurt University to match the dee frequencies, 80 to 110 MHz, of the $K = 134$ MeV separated-sector cyclotron at HMI. It will replace the present 8-MV tandem injector. Another 4-rod RFQ has successfully operated at 18 to 40 MHz at RIKEN; it will be used to inject from a high-power ion source into RILAC. The efficiency of an RFQ in turning electrical power into particle energy is greater at high frequencies. RFQs could operate at harmonics of the cyclotron dee frequency, but the ion beam itself must be bunched to the lower, cyclotron, frequency in

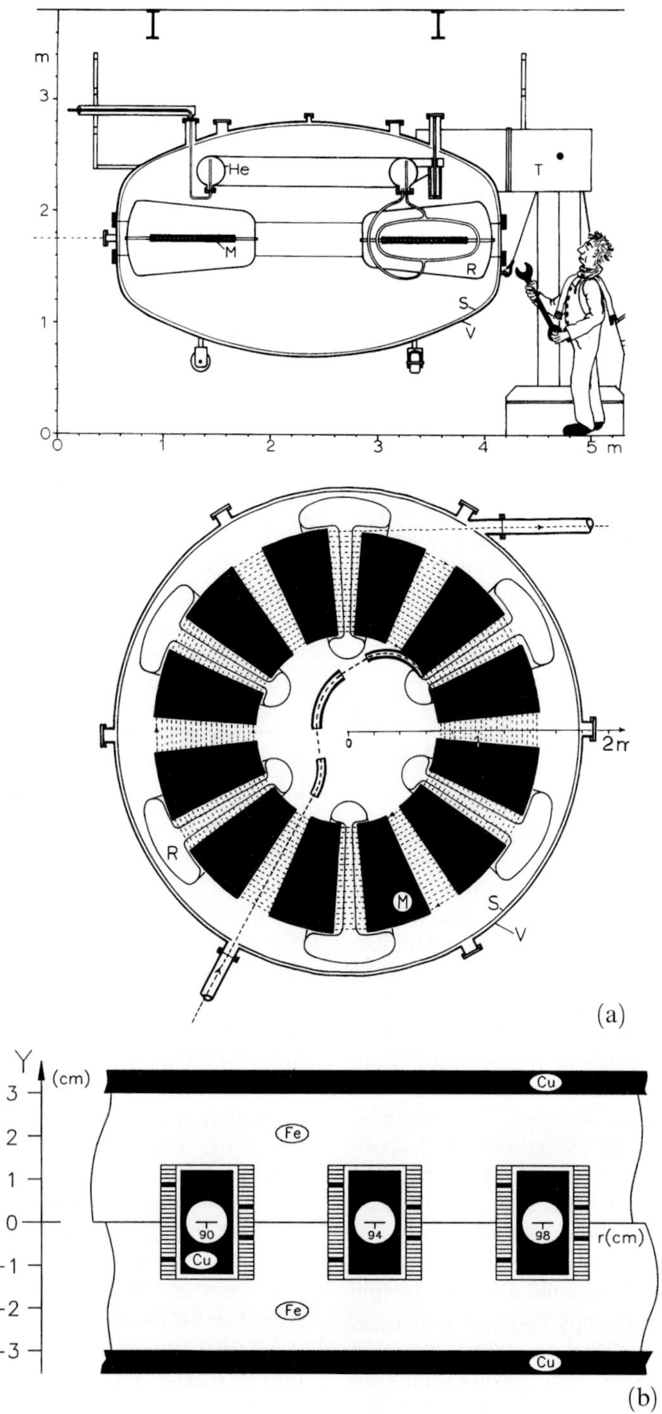

FIG. 3. (a) Cross sections of the Tritron cryostat. The rf cavity R and magnet sectors M are cooled by liquid helium from the reservoir He. S is a shield at 80°K and V the vacuum vessel. (b) A cross section of three beam channels in a magnet sector. The superconducting coils exciting the field have two separate windings in each channel to produce the field gradient that provides focusing. (Courtesy of U. Trinks.)

order to capture most of the ions leaving the ion source.

2.3 Altering the Beam Microstructure and Macrostructure

The typical cyclotron operates in cw mode, producing a train of beam bunches each a few nanoseconds long at the frequency of the accelerating field. These frequencies range from ~ 1 MHz for heavier ions to ~ 100 MHz for protons. There has been no macrostructure, i.e., modulation at lower frequencies, unless it was imposed by chopping the beam with a pulsed deflector or by pulsing either the ion source or the rf voltage, and these methods reduce the total charge delivered. Recently cavities operating at some harmonic h of the main rf have demonstrated modulation of the time structure without loss of beam.

Consider such a cavity to be installed approaching the outer radius of a cyclotron and to provide an additional energy gain per turn ΔT_h that can be phased to either augment or oppose the energy gain ΔT_1 from the main rf. The phase Φ_r of some particle at radius r is obtained from

$$\Delta T_1(r_0)\sin\Phi_0 = \Delta T_1(r)\sin\Phi_r + [k\Delta T_h(r)/h]\sin(h\Phi_r), \quad (1)$$

where Φ_0 is the starting phase at radius r_0. The parameter $k = 1$ when the additional cavity is phased to assist the main cavity in accelerating the beam. In this case the width of the phase band decreases during acceleration from r_0 to r and phase compression is said to occur. The phase width is expanded when the additional cavity opposes acceleration—k is set equal to -1 in Eq. (1). In the latter case the phase band can be split into two when $\Delta T_h > \Delta T_1$. The results of an experiment to confirm this are shown in Fig. 4(a). The beam extracted from TRIUMF was a single bunch 3 ns wide at half maximum. This split into two bunches 8.5 ns apart when an additional cavity was turned on and its phase adjusted to oppose acceleration. Figure 4(b) shows how further splitting is possible with the addition of extra cavities. This effect may possibly be exploited to match a cyclotron with a post-accelerator of higher frequency.

Large betatron oscillations can be excited by a kick from a pulsed deflector or by a field oscillating at the betatron frequency. If an extraction system—stripping foil or dc deflector—is placed at the point of maximum excursion, the beam extracted will have the time structure of the driving term. For example, calculations have shown that a stripping foil placed at a radius where Q_z is 0.25 and positioned above a deflector pulsed to impart a vertical kick every fourth turn to drive this resonance can provide a beam with three unpopulated rf buckets followed by one four times as bright as normal.

2.4 Limitations to Beam Intensity from Cyclotrons

The most important advantage of cyclotrons for many ion-beam applications is the high beam current that they can deliver. It is useful to consider the factors that may limit beam intensity.

2.4.1 Space-Charge Limits. Coulomb forces between the constituent ions of the beam being accelerated oppose the focusing forces provided by the magnetic and electric fields. The most common limitation to beam current arises at low energy when this transverse space-charge force increases the height of the beam to the limit of the acceptance. This may occur in an axial injection system, in an inflector, or over the first few turns in the cyclotron where the limiting proton current, in amperes, is given by

$$I_{\max} = \beta B_f (Q_z b\,\omega_{ion}/c)^2 [7.8\times 10^6] \quad (2)$$

In Eq. (2) B_f is the bunching factor, defined as the ratio of average to local current, and b is the half-height of the accepting aperture. The other terms are defined in CYCLOTRONS. A 1-MeV test-bed cyclotron with a solid pole, an intense external cusp H^- source, a spiral inflector, and an injection energy of only 25 keV has achieved 2.5 mA of H^- ions at TRIUMF.

The ring cyclotron at PSI, which holds the beam power record, has delivered 1.5 mA, or 0.9 MW, of 590-MeV protons. Beam is injected into the ring at 72 MeV, and low-energy

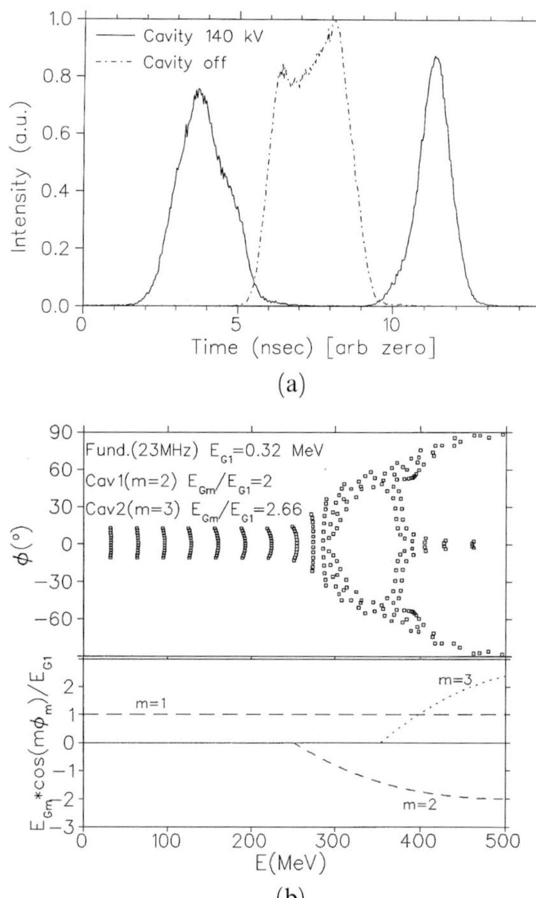

FIG. 4. (a) An experimental confirmation of beam splitting using one 300-keV/turn, $h = 4$ cavity at TRIUMF. 1 ns is equivalent to 8.3° of phase for the main 23-MHz rf. (b) A computer simulation illustrating the splitting of a single phase band into three by means of two cavities operating at the second and third harmonics of the main accelerating rf frequency.

space-charge problems are handled by the specially designed injector cyclotron. Within the ring itself 10^{-4} of a 1-mA current is lost, and this loss scales as $1/N^3$, where N is the number of turns from injection to extraction. This is consistent with the limiting factor being longitudinal space charge, which tends to accelerate the front of the beam bunch and decelerate the rear. The bunch twists in the horizontal plane, increasing the overlap of turns and reducing the extraction efficiency. An extra rf cavity, oscillating at a harmonic of the main accelerating cavities, may ameliorate this situation by providing accelerating forces opposing those from space charge.

Cyclotrons normally operate in a steady state, and dynamic feedback between a beam signal and a subsystem is unnecessary. The rf cavities in the PSI ring cyclotron are illustrated in Fig. 5 of CYCLOTRONS. Four 50-MHz cavities accelerate the beam, and one 150-MHz cavity opposes their action in order to flatten the 50-MHz waveform over the length of a bunch. This flattening results in all protons having about the same energy in any given turn and thus increases the separation between turns and the extraction efficiency. The 0.9 MW absorbed by the 1.5-mA beam is comparable to the power dissipated in the walls of the 50-MHz cavities. The 150-MHz cavity decelerates the beam and thus absorbs power from it, and at 1.5 mA this absorbed power also compares with the cavity losses but has the opposite phase. The net phasor (voltage and phase) of the power supplied to this cavity depends strongly, therefore, on the beam current being delivered at that instant. PSI had to develop a fast feed-forward control system for the 150-MHz power supply in order to operate stably at these higher currents. The system uses the signal from a beam-current sensor to adjust the drive signal to the amplifier. In addition an external load was added to the cavity to absorb both forward and reflected power at the cost of increasing the total power required (Sigg *et al.*, 1966).

2.4.2 Lifetime of Foils Used for Charge Exchange. Foils are used to extract a beam from a cyclotron by stripping (CYCLOTRONS, Sec. 2.5), or to increase the charge state of an ion in order to make more effective use of an accelerating voltage. The useful life of a foil is affected by the energy dissipated within it. Power dissipated by the passage of beam ions and stripped electrons can raise the temperature to a point where evaporation is important. Also the structure of the foil material may alter at high temperatures, resulting in dimensional changes and curling or tearing of the foil; for example, pyrolytic graphite may be transformed into polycrystalline carbon. The carbon foils and monofilaments used to extract proton beams by stripping circulating H^- ions take 1 to 3 mAh/mm^2, or (2 to 7) $\times 10^{19}$ protons/mm^2, before curling. Lattice changes can also occur at normal temperatures; in fact this is the basis of some applications in the modification of materials. Heavy ions are most effective in displacing foil atoms because of their mass and short range. It was found at GANIL that carbon foils 150 to 400 µg/cm^2 thick take only 3.6×10^{15}/mm^2 of 13.6-MeV/amu argon ions before foil thickening renders unacceptable the emittance of the outgoing beam. For very thin foils Livingston *et al.* (1978) obtained

$$\tau(\mathrm{p\mu A \cdot min \cdot mm^{-2}}) = A \times T^{1.15}(\mathrm{MeV/amu}) \tag{3}$$

where the beam intensity is expressed in particle microamps (pµA) and the empirical constant A is 20 for Argon. The rupture lifetime–intensity product τ may be extended >10-fold by more sophisticated manufacturing methods or by mounting the foils in an oscillating frame. Pure metal foils should be more resistant to lattice deformations than carbon; however, they must be thin to reduce scattering and to be able to handle the thermal power from the beam. The ISIS synchrotron uses 200-µg/cm^2 Al_2O_3 foils successfully.

ERRATA

Here are some corrections to the original article CYCLOTRONS.

p. 432, between Eqs. (14) and (15), replace ν_z and ν_r with Q_z and Q_r.

p. 434, Sec. 1.3, replace "section" with "sectors" to read, "The iron sectors in these superconducting cyclotrons are saturated at all working fields."

p. 441, above Fig. 10, insert "kW" to read, "Cyclotrons tend to operate ... and rf power requirements range from 15 kW to 1 MW".

p. 446, Sec. 3, the second sentence should begin "This is because a single-stage cyclotron that provides a large energy increase ..."

p. 461, the definition of "Particle Microamp" should read "Cyclotron beam intensities are usually expressed as an electric current, e.g., 1 µA being 6.24×10^{12} electronic charges/s. For an ion of charge qe the intensity may be expressed in particle microamperes where 1 pµA is 6.24×10^{12} particles/s, the electrical current being q times larger.

p. 461, the definition of betatron frequency, replace $\omega_{z,r}/\omega_{\mathrm{rot}}$ with $f_{z,r}/f_{\mathrm{rot}}$.

Works Cited

Apel, P. Yu. (1996), in: J. C. Cornell (Ed.), *Proceedings of the 14th International Conference on Cyclotrons and their Applications*, Singapore: World Scientific Publishing, pp. 136–143.

Cornell, J. C. (Ed.) (1996), *Proceedings of the 14th International Conference on Cyclotrons and their Applications*, Singapore: World Scientific Publishing.

Fiétier, N., Mandrillon, P., Rubbia, C. (1996), in: J. C. Cornell (Ed.), *Proceedings of the 14th International Conference on Cyclotrons and their Applications*, Singapore: World Scientific Publishing, pp. 598–605.

Fink, D., Chadderton L. D., Hosoi, F., Omichi, H., Sasuga, T., Schmoldt, A., Wang, L., Klett, R., Hillenbrand, J. (1994), *Nucl. Instrum. Methods* **B91**, 146–150.

Livingston, A. E., Berry, H. G. Thomas, G. E. (1978), *Nucl. Instrum. Methods* **148**, 125–127.

Rubbia, C., Mandrillon, P., Fietier, N. (1994), in: V. Suller, Ch. Petit-Jean-Genaz (Eds.), *Proceed-*

ings of the 4th European Particle Accelerator Conference, Vol. 1, Singapore: World Scientific Publishing, pp. 270–272.

Sigg, P, Cherix, J., Fetze, H. R., Frei, H., Märki, M., Schryber, U., (1996), in: J. C. Cornell (Ed.), *Proceedings of the 14th International Conference on Cyclotrons and their Applications*, Singapore: World Scientific Publishing, pp. 161–168.

Stroosnijder, M. F. (1996), in: J. C. Cornell (Ed.), *Proceedings of the 14th International Conference on Cyclotrons and their Applications*, Singapore: World Scientific Publishing, pp. 144–147.

Trinks, U., (1991) *Nucl. Instrum. Methods* **A306**, 27–35.

Further Reading

Many of the larger laboratories maintain Web pages which describe the laboratory facilities and give links to the experimental programs. Some examples are the following:

European Centre for Nuclear Research, Geneva, Switzerland, http://www.cern.ch.

Grand Accélérateur National d'Ions Lourds, Gaen, France, http://ganinfo.in2p3.fr.

Institute of Nuclear Studies and KEK, Tokyo, Japan, http://www.ins.u-tokyo.ac.jp.

Indiana University Cyclotron Facility, Bloomington, IN, http://www.iucf.indiana.edu.

Laboratorio Nazionale del Sud, Catania, Italy, http://www.lns.infn.it.

National Superconducting Cyclotron Laboratory, Lansing, MI, http://www.nscl.msu.edu.

Paul Scherrer Institute, Villigen, Switzerland, http://www.psi.ch.

Institute of Physical and Chemical Research, Saitama, Japan, http://www.riken.go.jp.

Tri-University Meson Facility (TRIUMF), Vancouver, Canada, http://www.triumf.ca.

DISPLAY TECHNOLOGY

(An Addendum to *Encyclopedia of Applied Physics*, Volume 5, pages 101–126.)

DIETMAR THEIS, *Corporate Technology, Siemens AG, Munich, Germany*

INTRODUCTION

Since the beginning of the last decade of this century the technical and economic importance of electronic displays has dramatically increased. It is predicted that the ever-growing application of displays will continue over many years to come. We are now only at the threshold of the "information age," which inevitably will stimulate new applications and new technical developments in the field of visual man-machine interface. In recent years we have already witnessed the appearance of flat-panel displays as daily commodity items. The market penetration of display-based products and systems occurs in an environment where digitized hardware and networks, advances in display technologies, and user-friendly software are rapidly developing. The information age provides new modalities like multimedia, high-definition television, and virtual reality on an internetworked global scale. Advanced storage and data-compression technologies allow for the provision of huge image data bases at one's fingertip. All these developments pave the way for an increased demand for user-friendly light and flat displays.

The intense market penetration of flat-panel displays is nearly completely to be credited to the products of liquid-crystal display (LCD) technology. In 1995/96 the production volume of high-information LCDs was two orders of magnitude larger than the sum of all other high-information content flat-panel technologies.

This Addendum will cover some of the new developments in the mainstream LCD technology; in the active light-emitting display technologies like field emission, plasma, electroluminescence, visible-light–emitting diodes; and in projection display technologies.

1. LIQUID-CRYSTAL DISPLAYS

1.1 Tailored Nematic Liquid-Crystal Materials

Liquid crystals appear in several "mesophases," which exist between the isotropic liquid phase and the highly ordered crystalline structure. Technically, the most important mesophases are the "nematic" (string-like) and the "smectic" (soapy) mesophases. Nematic liquid-crystal molecules consist of a flexible component such as an alkyl group and rigid components such as a biphenyl group and a cyano group. The cyano group serves to pro-

ISBN 3-527-29308-6

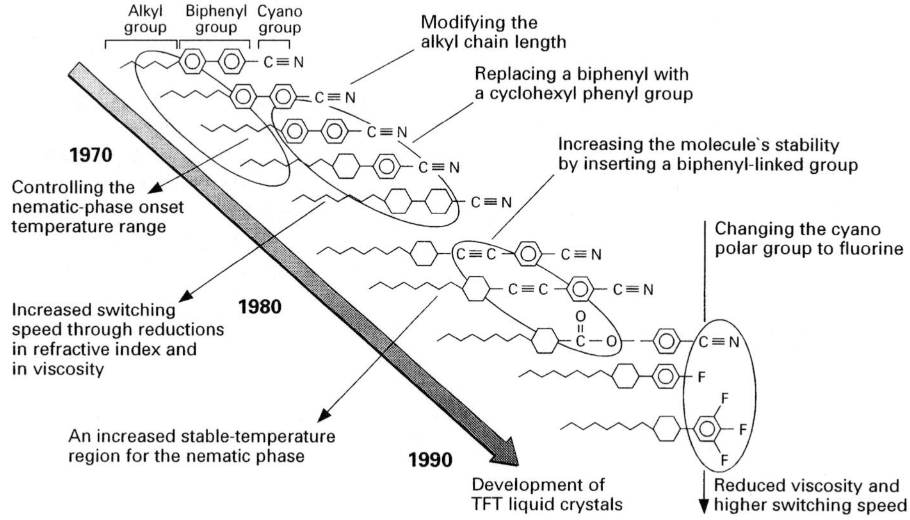

Fig. 1. Molecular design for liquid-crystal molecules.

duce positive dielectric anisotropy (the difference between the dielectric constant in the direction of the long axis of the liquid crystal molecule and the dielectric constant at a right angle to the long axis). Molecular design efforts in recent years have involved replacing or modifying the three parts of a liquid crystal in order to synthesize new liquid crystals. This is illustrated in Fig. 1, which shows the main results of chemical modifications on the device performance (Yamazaki *et al.*, 1996). Properties that cannot be produced by molecular design are achieved by mixing various liquid crystal materials, as is being done to increase the practical temperature range.

When the liquid-crystal molecules are oriented by a surface alignment layer or by an electric field, the bulk sample has all the optical properties of an anisotropic crystal. LCDs operate by electronically modulating this anisotropy between two distinct optical states. The anisotropy is made visible with the aid of crossed polarizers.

1.2 Novel Addressing Schemes for Direct-Addressed Supertwist LCDs

In LCD matrix displays the driving circuitry sequentially selects the address lines, while the data information is sent simultaneously to all pixels in the selected line. Liquid crystals have relatively slow turn-on times; they respond to the time-average voltage (rms) instead of the instantaneous voltage. A selected pixel, which spends most of its time with its driving voltage below threshold voltage, has only a slightly higher rms voltage than a non-select pixel. All contrast needs to be produced with only this small difference in voltage. This is why a highly nonlinear response is required as is provided by supertwist (STN) LCDs.

A drawback of STN is the inherent comparatively long response time (300 to 100 ms) and the chromaticity of the cell. To get rid of the chromaticity, double supertwisted nematic cells (D-STN) have been developed. Here a second LC panel containing a helix of the opposite orientation is placed in front of the first device. This technique later gave way to film-compensated STN (F-STN) in which the undesired coloration was compensated for by a phase-difference film.

The inherent response time of the display must be many times longer than the refresh time to satisfy the rms response condition. This generally rules out displays with response times short enough to show moving images at video rates. Indeed, if the display uses a "slow" LC mixture, the difference between OFF and ON states never gets large enough to generate good contrast. If the display uses a "fast" responding LC, then the contrast of an individual row begins to fade before the next refresh cycle begins, again deteriorating contrast. This phenomenon is called frame response—instead of responding to the average voltage applied over several frames, the LCD reacts to the relatively high-voltage row pulse within a frame.

The problem can be avoided by a completely different approach (Scheffer and Clifton, 1992: "Active Addressing™"; Ihara *et al.*, 1992; "Multiple Line Selection"). Instead of sending the high-voltage pulses sequentially down the rows, a set of predetermined signals are applied to multiple rows simultaneously and a unique calculated analog voltage is applied to the columns, depending on the state of the pixels in that column. The key is to determine which rows need which signal and to calculate the proper column voltage within the time frame available. One still has to provide an LCD cell that is capable of switching at video rates, i.e., decrease the cell thickness and use fast material.

The ultimate success of these novel addressing schemes will depend on the added cost of making these enhancements and the competition with active-matrix LCDs in terms of cost and performance.

1.3 Reflective LCDs for Low-Power Systems

Reflective color LCD without back light would be the enabling technology for multifunctional portable information terminals by which one could receive and send any information any time at any place. Such displays do not need heavy batteries and represent the most promising approach to "electronic paper" today.

A wide variety of reflective LC devices have been developed so far. A comprehensive description of the basic LC modes and design concepts has been given by Uchida (1996).

1.4 Chiral Nematic and Chiral Smectic Liquid Crystals

Both nematic and smectic liquid crystals forming a helical (chiral) structure are optically active. Optically active (chiral) nematic liquid crystals are also called cholesteric liquid crystals. They have the property of selectively reflecting or passing the light of a specific wavelengh if the pitch of their helical structure is equal to the wavelengh λ. This is the condition for constructive interference of light waves occurring if identical regions in the sample are separated by a distance of $\lambda/2$. Cholesteric liquid crystals can form one of three major textures depending on boundary conditions and applied fields: planar, focal conic, or homeotropic (Fig. 2). With increasing field strength the planar structure first transforms into a focal conic structure, then into a homeotropic aligned nematic state—which is highly transparent. The planar texture is highly reflective at a preselected wavelength and band-

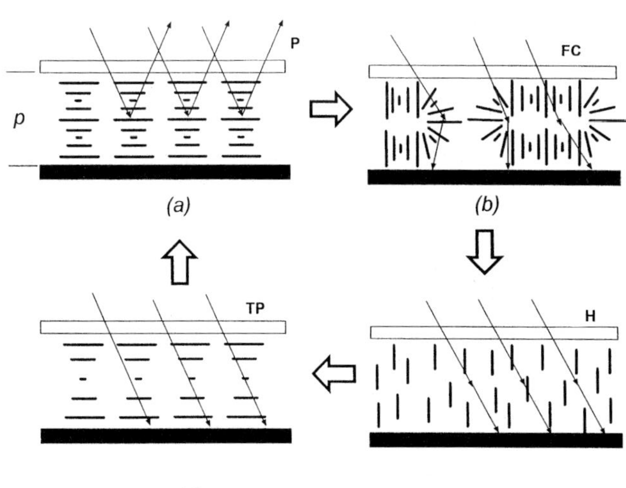

Fig. 2. Schematic textures in cholesteric liquid-crystal material: (a) light-reflecting planar texture; (b) nearly transparent focal-conic texture; (c) transparent homeotropic texture; (d) transient planar texture.

width. With the introduction of a polymer network or with suitable surface treatment of the cell substrates, a domain structure is introduced. This broadens the viewing angle, stabilizes the focal conic texture to create bistable switching, and introduces a gray-scale regime. The ability of the display to maintain pixels in both the ON and OFF states once they are written (without refresh) allows for unlimited resolution on a low-cost passive matrix without sacrificing contrast. Since a recent publication on a full-page size prototype (Pfeiffer *et al.*, 1995), renewed interest in this material is growing.

In the first engineering samples a total reflection around 40% was observed, substantially more than an ordinary reflective twisted nematic LCD. The achievable contrast is determined by transparency of the focal conic structure, which can exhibit a weak scattering. Different driving schemes are still under development. Real-time television has already been viewed on a cholesteric display cell at Kent State University (Doane and Stefanov, 1996). The technology has unique features like high resolution, low power due to bistability, and low cost, and seems to be very attractive for portable and handheld devices. The absence of polarization makes it possible to use low-cost plastic substrates—their inherent birefringence does not interfere with device performance. Principally polymer-stabilized cholesteric-texture devices could be one of the premier enabling technologies for electronic books, newspapers, or document viewers.

Optically active (chiral) smectic liquid crystals have a permanent electric dipole moment in the absence of an electric field, and for this reason they are called also ferroelectric liquid crystals (FLCs). The intrinsic polarization of these molecules, which is perpendicular to their long molecular axis (director), has the consequence that when subjected to an external field the molecules respond to the direction as well as the magnitude of the field. The helical structure would average out the direction of the spontaneous dipole, thus suppressing the ferroelectric behavior. Rubbed polymer surface alignment layers, similar to those in supertwist LCDs, remove the helical structure. The spacing between the two glass substrates should be about 2 μm. This spacing is also required for the ferroelectric liquid crystal cell to act as a half-wave plate. The operation of a FLC display depends on the switching of the FLC's dipole, whose orientation depends on the polarity of the addressing pulse (Fig. 3).

The FLC molecule cannot simply rotate

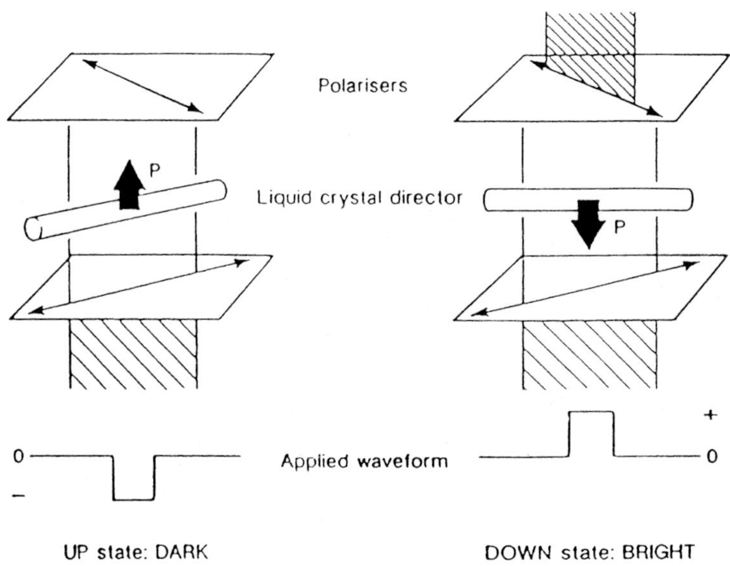

Fig. 3. An FLC display operates by changing the orientation of the LC molecule in the plane of the display.

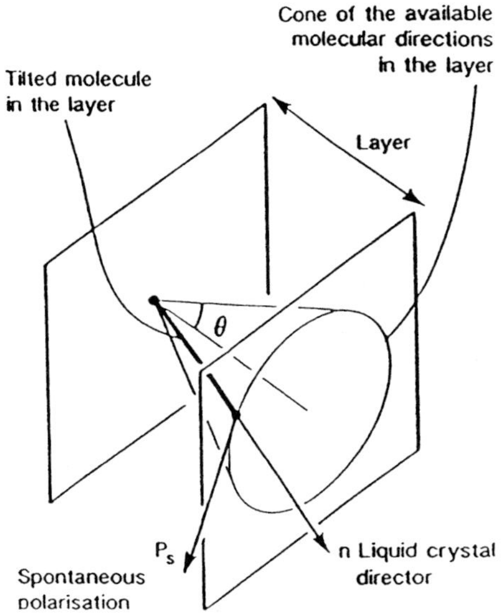

Fig. 4. The FLC molecule is constrained to rotate about an imaginary cone.

about its own axis; instead it is constrained to rotate about an imaginary cone (Fig. 4). If the cone half-angle θ, which is equal to the tilt angle of the molecules in the smectic layer, is approximately equal to 22.5°, then the change in the orientation of the FLC molecules, 2θ, will be 45°. This is the angular change that produces the optimum values of transmission and contrast ratio (Mosley, 1997).

The two main attributes of FLC displays are very fast switching—typically 100 μs per row—and bistability. This property can be made use of to produce a nonvolatile display where the image is maintained at zero volts. The molecules switch in the plane of the display rather than perpendicular to it (as in supertwist and active-matrix LCDs). This leads to a very good viewing angle. The main obstacles in the commercialization of ferroelectric LCDs have been production issues (very small distance between substrate plates), the absence of gray scale, and susceptibility to mechanical shock, although it is said that there are solutions to these problems. In addition to ferroelectric LCDs there are also antiferroelectric LCDs. Yamada *et al.* (1995) have developed a display with gray scale and have demonstrated high-quality video images on a $125 \times 80\text{-mm}^2$ display area.

1.5 Active-Matrix–Addressed Liquid-Crystal Displays (AMLCDs)

Active-matrix addressing has been the most broadly pursued technology to enhance liquid-crystal display performance. In 1996 the world market for AMLCDs was estimated to be about US\$$6.1 \times 10^9$; a global turn-over in the order of US\$$10 \times 10^9$ is expected around the year 2000.

Active-matrix addressing leaves the fundamental twisted nematic operation unchanged and uses pixel-external nonlinear devices to enhance the contrast of large multiplexed displays. The devices do not respond to the rms time-average voltages but rather to the instantaneous peak voltage. Two- or three-terminal thin-film devices act as switches, electrically isolating a pixel until its input voltage exceeds a certain specified threshold. Pixels in active-matrix devices are driven at duty cycles close to 100%. The voltage at the pixel remains practically constant during each frame, as if the pixel were directly driven. The pixel acts as a capacitor storing the charge and maintaining the voltage during the frame time. Thin-film transistors (TFTs) formed by sophisticated deposition and patterning technologies are the most frequently used switches in AMLCDs commercially available. The semiconductor material used for the TFT in most cases is amorphous silicon because it can be fabricated at relatively low temperatures and generates high image quality. The use of polysilicon material would allow the integration of driver circuits on the same glass plate because of the much higher mobility achievable. On account of high process temperatures polysilicon is not compatible with normal glass substrates. Laser or furnace annealing are key processes for the advancement of polysilicon AMLCD technology. For details of the fabrication the reader is referred to Morozumi (1996).

AMLCD manufacturers now are moving into the third level of development with significant improvement issues. The coming wave of AMLCDs is intended to at least partially replace CRT monitors. This means that competitive size, cost, and performance of the flat displays will be of paramount importance. In 1995 the AMLCD mass product was the VGA 10.4-in. (26.4-cm) panel. Now the trend has shifted to 11.3-in. (28.7-cm) and 12.1-in. (30.7-

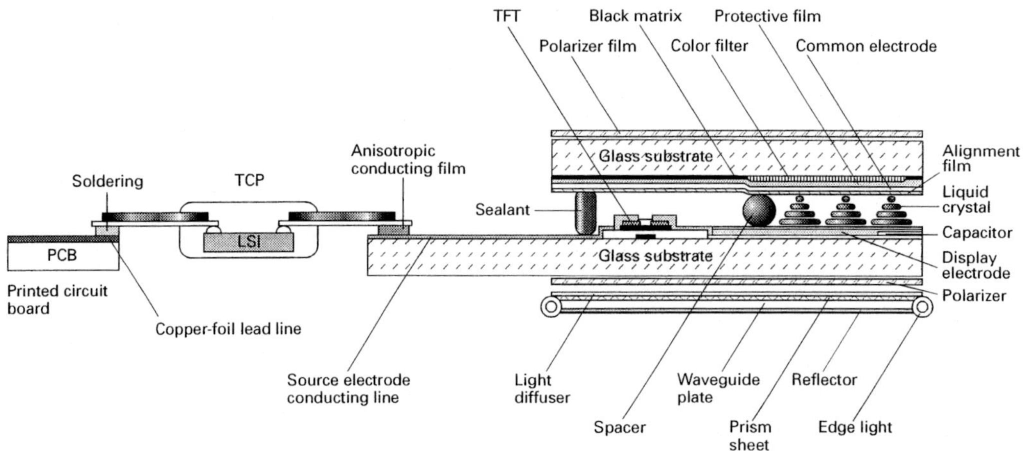

Fig. 5. Cross section of a color AMLCD and connection to the addressing circuitry (after Yamazaki *et al.*, 1996).

cm) SVGA panels. The size enlargement will continue to 13.3 in. (33.8-cm) in XGA resolution with future options in the 15- to 17-in. (38.1- to 43.2-cm) range. Manufacturers are shifting their $390 \times 470\text{-mm}^2$ or $400 \times 500\text{-mm}^2$ substrate sizes to $550 \times 650\,\text{mm}^2$ and will ultimately be targeting even 1 m diagonal size. They will simplify their array processes by lowering the number of photo masks and eliminate rubbing by photo alignment technologies. They will have to develop means to lower the resistance of bus lines. At the same time they are striving to widen the viewing angle and reduce power consumption of the panels. Figure 5 shows a cross section of a color AMLCD module and indicates how the driving circuitry is connected to the glass substrate carrying the TFTs. Mostly anisotropic conducting films are used to connect the ITO pads on the glass to the driver LSIs. Another connection method would be chip-on-glass.

Several technologies have been proposed to improve the originally rather narrow and nonuniform viewing characteristic of single-domain twisted nematic cells. Two of the major technologies are the formation of multi-domain cells with reversed LC orientations in each domain (by multiple rubbing and photolithographic processes) as suggested, e.g., by Kaneko *et al.* (1993) and the in-plane switching as first devised by Kiefer *et al.* (1992) and implemented by Kondo *et al.* (1996).

Standard AMLCD have five electrodes: three for gate, source, and drain of the transistor, one for the pixel electrode, and one common electrode. In most cases the common electrode is formed on the counter-substrate. As shown in Fig. 6, in-plane switching uses an interdigital electrode structure. The five electrodes are arranged in two layers on the TFT substrate side. The lower layer carries the gate and common electrodes, the upper one the source and drain electrodes, which are co-used as pixel electrodes. This approach avoids the indium-tin-oxide transparent electrode on the counter-substrate. The electric field is no longer vertically but horizontally oriented as indicated by the arrows in Fig. 6. Under the influence of this field the optical axis of the nematic LC molecules deviates from the polarization axis and the transmittance of incident light through crossed polarizers is gradually increased. Horizontally and vertically 70° viewing angle in both directions has been achieved at a contrast ration of 10.

The overall transmissivity of AMLCDs is determined by the transmissivities of the glass, 95%, the polarizers, 35%, the color filters, 25%, and the cell aperture (that portion of the cell which is transparent to light), which depending on design can vary between 40% and 65%. As a result the overall transmissivity of the cell is not more than 3.3% to 5.0%! One of the major research issues for AMLCDs is to improve transmissivity to reduce power consumption. The two other areas to work on for further

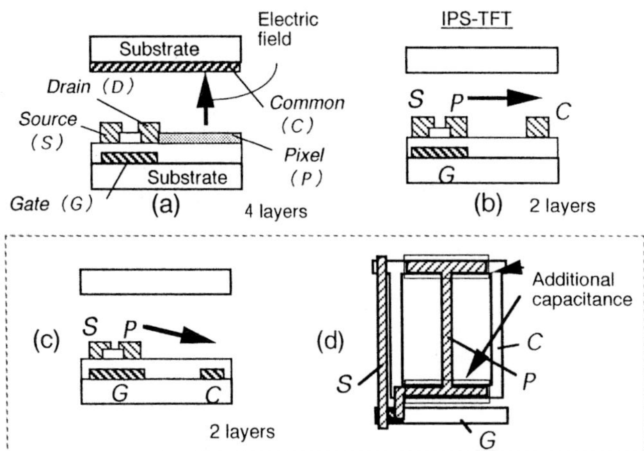

Fig. 6. (a) Cross section of conventional AMLCD and (b) in-plane-switching. (IPS-AMLCD). (c) In the IPS version the common electrode is located in the gate electrode layer. (d) The planar arrangement of (c) (after Kondo *et al.*, 1996).

power reduction are improvement of the efficiency of backlights and reduction of the power consumption of the drive-circuit unit. As a result of a comprehensive optimization the power consumption of a 10.4-in. (26.4-cm) SVGA AMLCD with a luminance of 70 cd/m^2 could be reduced from 3.5 W to 1.5 W (57% reduction) or with the same power consumption the luminance could be raised to 120 cd/m^2 (+72%) (Ohshita, 1996).

1.6 Plasma-Addressed Liquid-Crystal Displays

Despite their immense success in the portable display area and the anticipated large market share they might conquer in the monitor market in the near future, it is not predicted that AMLCDs will be available as meter-sized panels for consumer HDTV in the next ten years. The scaling difficulties of photolithography and metallization to sizes larger than 25 in. (63.5 cm) can elegantly be bypassed by a new approach called plasma-addressed liquid crystal (PALC) (Buzak, 1990). PALC uses the electrical switching properties of an ionized gas to address the liquid-crystal pixels actively. In the integrated LCD/plasma structure of the PALC, each scan line is defined by a plasma channel. The cathodes in the channels are sequentially scanned by applying a plasma discharge voltage that produces a full plasma within several microseconds. The analog video signals are fed to each column electrode. Except for the one row where the gas is ionized, the analog voltage has no effect on the light transmissivity of the liquid crystal since there is no electrical connection or conductor on the bottom side of the liquid crystal. Along the line above the ionized channel, however, the pixel capacitance can charge to the value determined by the data voltage. In effect, the plasma completes the electrical circuit between the data electrodes and the grounded channel electrode. Kakizaki *et al.* (1996) have presented a 25-in. (63.5-cm) PALC display for wall-hanging TV with 1.8 lm/W light efficiency, 250 cd/m^2, and high contrast even in bright ambient.

2. ACTIVE DISPLAYS

Active, self-emitting displays do not need any backlight. Every pixel emits light (and uses power) only as long as this is required for the image buildup. In the backlit LCDs every pixel including the OFF pixels receives full backlight power as long as the display is operating. In terms of energy management active displays have a conceptual advantage. On the other hand the separation of the functionalities “pixel addressing” and “light emission” allows a separate optimization of both. In the case of AMLCD this freedom in design has led to an overall energy consumption which is lower than in any active light-emitting device.

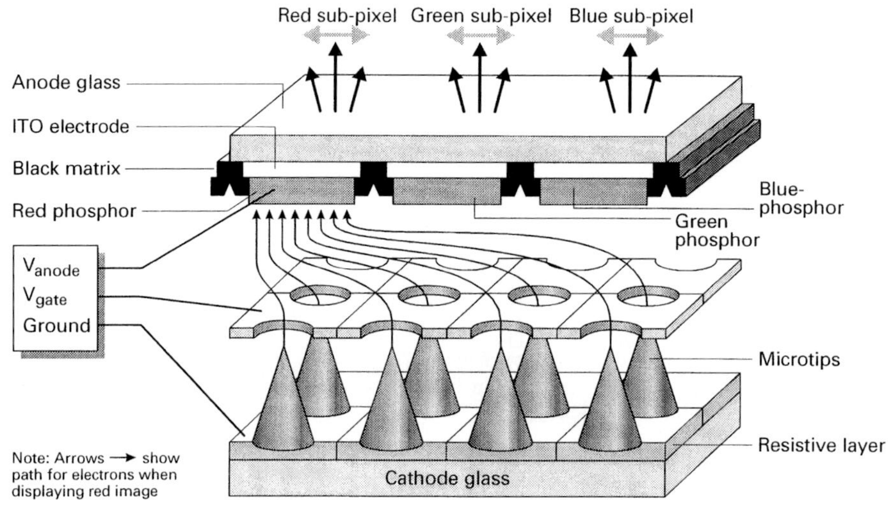

Fig. 7. Basic structure of field-emission displays (courtesy of PixTech).

2.1 Field-Emission Displays (FEDs)

The operating principle of these newly emerged vacuum electron devices is comparable to vacuum-fluorescent displays with the important difference that the electrons are not emitted by thermal activation from hot filaments but rather by field emission from a cold cathode. Electrons escaping from the cathode are accelerated towards a phosphor screen, where their energy is partly transformed into light (Fig. 7). In FEDs electrons are generated by a voltage difference between the cathode and closely positioned gates according to the Fowler–Nordheim tunneling equation for field emission.

Field enhancement induced by tips is the most widely investigated method to obtain useful currents at moderate macroscopic fields. One way of fabricating microtips is to make use of the wet etching properties of *n*-silicon (Jiang and White, 1993). This approach limits the display size to the available diameters of Si wafers. Meyer (1990) and his group have developed a fabrication process for molybdenum microtip cathodes on glass. The process also provides a gate electrode with a 1.4-μm self-aligned hole around each cathode tip. This original design was refined through the addition of a ballast resistive layer to average out random fluctuations of current yield between tips. The negative electron affinity on diamond-like carbon surfaces has fueled research in diamond emitters as a low-cost manufacturing alternative (Kumar *et al.*, 1994). The lwo-cost cathode architecture allows only the design of diode-type FEDs with the inherent disadvantage that gray scale is difficult and expensive to obtain (Courreges, 1996).

A big challenge in FED construction is the anode technology. The anode acceleration voltage determines the penetration depth of electrons into the phosphor grains. At low acceleration voltages, below 1000 V, the RGB phosphors are deposited on a conductive transparent electrode on the anode glass. If the phosphors are deposited on electrically isolated comb-like ITO structures, switched-anode FEDs become possible. Here the cathode is time multiplexed with a sequential addressing of color anodes as shown in Fig. 7. As a result of small electron penetration depth in the low-voltage approach, the grain surface largely determines the efficiency and the ultimate brightness. At higher acceleration voltages, above 5 kV, the phosphors can be covered by a thin layer of aluminum to provide surface conduction and avoid charging up. Here efficiency and brightness come close to those of CRTs. Low-voltage architecture avoids the need for focusing and allows for a low-cost spacer technology in the 200-μm cell gap. For high voltage the cell gap has to be increased at least ten times, which leads to

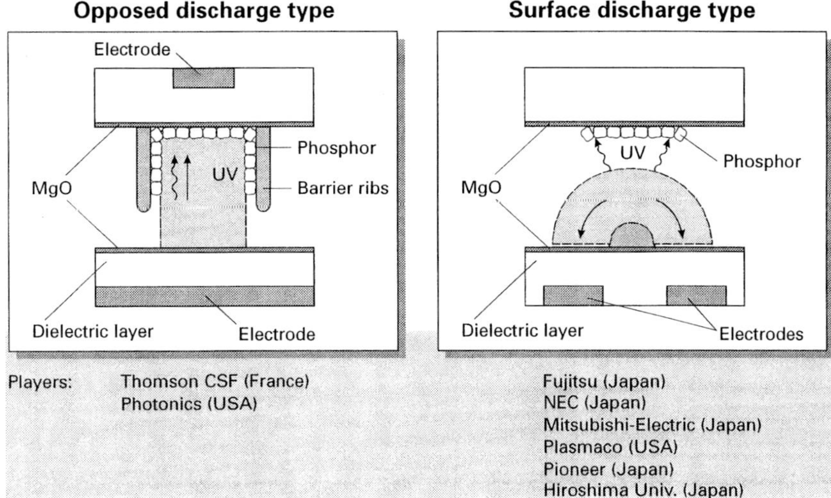

Fig. 8. Two designs for color ac plasma displays.

complex solutions for the focusing and spacer technologies to suppress arcing and sputtering effects. The future will show whether the attractive properties of FEDs will prevail in terms of cost/performance ratio.

2.2 Plasma Displays (PDPs)

Plasma-display technology has gained a strong momentum in recent years. Demonstrations of exceptional large-area image quality have generated new interest in this 25-year-old technology. Plasma is now regarded as the leading direct-view flat-panel technology for large-area (meter-sized) displays. After the successful prototype demonstrations of 42-in. (107-cm) diagonal plasma displays, the major challenge now is to achieve high-volume manufacturing at affordable cost.

Like LCD panels, PDPs consist of two glass plates with helium or neon gas with an addition of 3 to 10% of xenon gas sealed between them. Electrodes are formed at the inner surface of the glass plates. To avoid catastrophic arc the current in the gas discharge must be limited: dc plasma displays use a resistor, and have the electrodes in intimate contact with the gas discharge; ac plasma displays limit the current with an internal glass dielectric covering the electrodes, which are capacitively coupled to the gas discharge.

In color PDPs the Xe component of the gas mixture emits ultraviolet light in the 147-nm range. In a line-at-a-time scanned-matrix display the overall brightness decreases with increasing number of scanned lines. It is very helpful in this respect that for both ac and dc plasma displays an inherent memory function is available. RGB light emission is achieved via the photoluminescence of phosphors excited by the UV light.

Two ac plasma-display electrode configurations are shown in Fig. 8.

The single-substrate structure separates the discharge cathode areas from the phosphor by applying the sustain voltage only the lower electrodes while the phosphor is on top. This approach avoids direct bombardment of the phosphor by energetic ions. The basic structure of single-substrate ac plasma displays is shown in Fig. 9.

The front glass carries the two parallel transparent sustain electrodes for each pixel. On the rear glass substrate the RGB phosphors are deposited on top of the address electrodes, separated by barrier ribs to avoid color cross-talk. The two glass plates are aligned orthogonally, forming pixels automatically at the intersection of the orthogonal electrodes. In PDP production the fabrication of the barrier ribs with their high aspect ratio (50 μm wide and 130 μm high) is a formidable

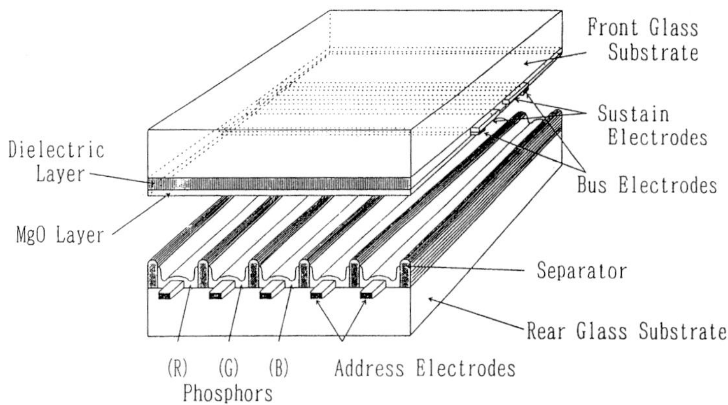

Fig. 9. Structural details of Fujitsu's ac plasma television screen.

problem. For low and medium resolution, multiple-screen printing can be used. For high-resolution displays this process is not suitable. For high-density barrier-rib formation, sand blasting and photosensitive molding of glass paste are being investigated.

In current generations of PDPs lamp phosphors like $(Y,Gd)BO_3:Eu$ are used for red, $BaMgAl_{14}O_{23}:Eu^{2+}$ for blue, and $Zn_2SiO_4:Mn$ for green emission. Phosphor aging is an important issue but seems to be solved. Probably the greatest challenge for plasma-display engineers is to increase the luminous efficiency of the devices. 0.7 lm/W is the highest efficiency reported so far for a commercial full-color plasma display (Shinoda, 1993). To increase this value improvements in the discharge efficiency, better optical coupling, and higher phosphor efficiencies have to be achieved.

To obtain gray scale in the memory mode of ac and dc plasma displays, the time that the pixel is ON in a given frame rate is modulated. This means that the pixels must be addressed multiple times per frame. In a binary chop technique, e.g., a 256-intensity gray scale would require 8 addresses per pixel per frame (Weber, 1996).

2.3 Electroluminescent Displays (ELDs)

2.3.1 Electroluminescence in Inorganic Materials. The performance of inorganic color thin-film electroluminescent (TFEL) displays has made significant progress in recent years due to the improvements of color phosphor materials with blue and broadband emissions. The reader is referred to the comprehensive article ELECTROLUMINESCENCE, together with its Addendum.

The classical EL phosphor is the yellow-emitting ZnS : Mn. This large–band-gap semiconductor is sandwiched between two insulating thin films and operated with ac voltage. The power dissipated in generating light is 8 mW/cm^2. Accordingly the power required to generate light from a 10-in. (25.4-cm) VGA display is just over 1 W; the luminous efficiency is 6 lm/W.

Two parallel roads have been taken to achieve color EL:

1. the development of an efficient white (broadband) phosphor that can be filtered and
2. the development of efficient red, green, and blue primary-color EL phosphors.

Tuenge and Karne (1991) have shown that an efficient red emission (1 lm/W) can be achieved with the combination of the standard yellow ZnS : Mn phosphor and a red filter. The green ZnS : Tb phosphor also has an efficiency near 1 lm/W (Härkönen *et al.*, 1990). In the blue part of the spectrum for some time the efficient SrS : Ce phosphor was available. Actually this phosphor has a blue-green chromaticity and requires filtering to achieve a true blue color. Barrow *et al.* (1993) achieved an important breakthrough by showing that a group of ternary sulfides (thiogallates:

MGa_2S_4 where M stands for Ca, Sr, or Ba) activated with Ce were efficient blue-emitting phosphors.

White EL phosphors were first based on the broad emission of the system SrS : Ce,Eu, but the achieved brightness did not satisfy the needs of high-information displays. White emission from stacked layers of ZnS : Mn and SrS : Ce phosphors has achieved the luminance level required for filtered color ELDs. A white luminance of 340 cd/m^2, and red, green, and blue filtered luminance values of 39, 90, and 18 cd/m^2 at 60 Hz have been obtained by Velthaus *et al.* (1994) using a multilayer structure with 9 layers. The peak luminance could be further increased to 470 cd/cm^2 by dual layers grown by atomic-layer epitaxy (Soininen *et al.* 1994).

2.3.2 Electroluminescence in Organic Materials. Electroluminescence can also be obtained from numerous organic materials. Following the pioneering work of Tang and Van Slyke (1987) and Burroughes *et al.* (1990), research efforts have been directed at developing efficient organic EL devices based on thin films, driven typically below 10 V. Substantial improvements in efficiency, color, and stability of the devices have been achieved.

Organic EL devices generally consist of a hole-transport layer (HTL) and an electron-transport layer (ETL) sandwiched between current-injecting electrodes. Potential barriers at the interface of the two layers localize electrons and holes and thus enhance the probability of their recombination in this area. Also, the channel for radiative recombinations can be further optimized by introducing fluorescent centers near this interface. The two-layer structure is thus extended to include a thin emitting layer (EML) between HTL and ETL. The three-layer structure is particularly useful in tuning the EL colors (Tang, 1996).

The current in the organic layers in the EL device is principally space-charge or injection limited. Space-charge limitation is avoided by using organic films only a few hundred nanometers thick. As ETL organic materials, often a complex of 8-hydroxyquinoline with a *tris* ligand configuration around the central Al ion (called Alq) is used, while the EML contains a small amount of fluorescent molecules and the HTL is made of an aromatic amine material. Another successful approach is the use of light-emitting polymer material for the fabrication of flexible full-color displays. Low-barrier contacts are designed by matching the organic layers with electrode materials of proper work functions like Mg : Ag or Al : Li as cathode material and ITO for the anode. Red, green, and blue light-emitting devices have been obtained with efficiencies above 1 lm/W. Depending on the luminance output, operational half-life on the order of several thousand hours is achievable from orgainc EL emitters.

2.4 Visible-Light–Emitting Diodes (LEDs)

Visible LEDs have been on the market since the 1960's. Only in the last few years, however, have high-performance LEDs been available in all colors, allowing the fabrication of full-color flat-panel displays viewable in direct sunlight. Before 1985, when LED performance was generally in the range of less than 2 lm/W, LEDs were mostly limited to indicator-type applications. Since 1985 the performance has increased to as high as 50 lm/W for the highest performance type of red-emitting AlInGaP LEDs, With the development of GaInN devices, the range of colors has expanded to include the violet, blue, and pure green such that high-performance LEDs are available throughout the color spectrum. With an efficiency that exceeds many types of conventional lighting sources, LEDs can now be used in a variety of new application areas such as vehicle lighting, traffic signals, road signs, and large-area displays for advertisement. Red-light–emitting high-performance devices in AlGaAs technology were first introduced to the market around 1983. In 1991 red, orange, yellow, and green light-emitting diodes based on AlInGaP came up (Kuo *et al.*, 1990; Sugawara *et al.*, 1991). In 1993 this was followed by the successful commercialization of a new materials system for the green and blue part of the visible spectrum: the InGaN system (Nakamura *et al.*, 1995). The luminous efficiency vs. emission wavelength for commercial LEDs is shown in Fig. 10.

The AlInGaP system shows the highest performance in the red and amber region; the efficiency degrades sharply towards the green region. Nevertheless green AlInGaP LEDs can

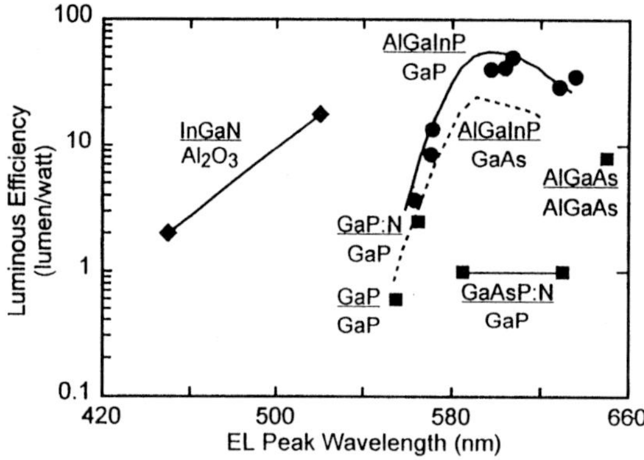

Fig. 10. Luminous efficiency of different materials systems vs. peak wavelength (after Craford, 1996).

be produced with an efficiency still doubling that of the conventional GaP : N green devices. The wavelength, however, being 560 nm or greater, is perceived as a yellowish green and therefore is not suitable for many full-color applications. The efficiency degradation towards the green is due to the fundamental transition in the band structure of the semiconductor alloy, which changes from "direct type" to "indirect type" with much less light-generation efficiency. The direct-to-indirect change is also the reason why AlGaAs LEDs work well in the infrared and red but not in the green part of the spectrum. The InGaN material system is a direct-type semiconductor all the way from the red to the ultraviolet. The best devices have been made in the blue and green regions. In the yellow there is a sharp falloff due to problems with materials growth and a bad match in atomic spacing between the different layers of the device structure.

3. PROJECTION DISPLAYS

In 1995 the total market for projection displays was in the order of US $ 4.6 billion. 58% of this market was covered by CRT-based products, and 42% by LCD-based products. The range of these products includes projectors for video home theater applications, projectors for overhead presentations, head-up displays for automotive systems, and many more.

At the heart of the projector is the light engine, which converts the electrical signal to an optical image (Stupp, 1996). The light engine can be either light emissive or light modulating. In the light-emissive group we find the CRT projectors and systems under development like field-emission or laser-based projection systems. Light-modulator systems employ either light-valve (LV) or light-amplifier (LA) systems to modulate the output light spatially.

The typical CRT projector uses three CRTs and three lenses—one for each of the primary colors. Consumer projectors are mostly CRT-based rear-projection systems with the light passing through a viewing screen to the viewer. The screen consists of a Fresnel lens and a double lenticular lens to minimize color separation in the image resulting from the angular separation of the sources.

LV or LA projector systems can be based on one or more (usually three) LVs. The systems based on multiple light valves often use dichroic mirrors and filters to separate the white light into color bands and to recombine the spatially modulated bands into a single converged image. Color information in single-LV systems can be obtained either by color-field sequence, by microfilters on the panel, or by angular color separation. The typical light source for LV systems is a high-intensity discharge lamp or occasionally a tungsten halogen lamp. The efficiency of light collection and utilization is determined by the size of the light arc, the area of the LV, and the usable angles of illumination. Since the peak white intensity is continuously generated, the maximum contrast achievable is determined by the attenuation capability of the LV. Active-matrix LCDs (frequently in the polysilicon version) are the most commonly used LVs in LCD-based projection systems. Polymer-dispersed LCDs are of specific interest, because they do not need polarizers with their inherent optical losses.

Another LV technology entering commercial application is based on micromechanics (Sampesell, 1993). The Digital Mirror Device™ (DMD™) consists of an array of square mirrors ($17 \times 17\,\mu m^2$) torsionally suspended over a surface of monocrystalline silicon containing SRAM and other CMOS circuitry. The mirrors are tilted to $\pm 10°$ under the influence of voltages from the underlying address circuitry. When the mirror is in the $+10°$ position, the light from the lamp enters the projection lens and is transmitted to the screen. In the $-10°$ position, the light misses the lens and is internally absorbed. Gray scale is achieved by time-multiplexing the position of the mirror. The switching speed is sufficient to achieve 8 bits per color in a 60-Hz frame time.

In LA devices the signal "writing" is done optically rather than electrically with an active matrix. The devices are multilayered. The input side consists of a photoconductor such as amorphous Si coupled to a reflective liquid-crystal cell that forms the output stage. A CRT or other light source is imaged onto the input photoconductor, generating an electrical image, which is transferred to the LC cell. An intense illumination onto this output stage is modulated by the cell, as with an LV. The input and output sides of the LA are separated by an efficient dielectric mirror that is part of the reflective LC cell and by a light-blocking layer to keep the output illumination from degrading the response of the input photoconductor.

Another emerging technology is AMLCDs fabricated on crystalline silicon wafers. These reflective-mode devices offer the potential of small pitch and high aperture ratio. In transmissive LVs, the aperture ratio is limited by the address lines and the TFTs. In reflective cells these can be fully hidden under the reflective back electrode. As with the DMD™, this produces small interpixel spacings with high aperture ratios.

Works Cited

Barrow, W. A., Coovert, R. C., Dickey, E., King, C. N., Laakso, C., Sun, S. S., Tuenge, R. T., Wentross, R. (1993) , *SID 93 Digest*, 761–764.

Burroughes, J. H., Bradley, D. D. C., Brown, A. R., Marks, R. N., MacKay, K., Friend, R. H., Burn, P. C., Holmes, A. B. (1990), *Nature* **347**, 539–543.

Buzak, T. S. (1990), *SID 90 Digest*, 420–423.

Courreges, F. (1996), *SID 96 Digest*, 45–48.

Craford, M. G. (1996), in: *HBF International Forum "Blue Light Emission for Future Imaging Technology,"* Tokyo: Hoso-Bunka Foundation, pp. 147–154.

Doane, J. W., Stefanov, M. E. pp. (1996), *Inf. Displ.* **12** (12), 18–21.

Härkönen, G., Härkönen, K., Törnqvist, R. (1990), *SID 90 Digest*, 232–235.

Ihara, S., Sugimoto, Y., Nakagawa, Y., Kuweta, T., Koh, H., Hasebe, H., Ruckmongathan, T. N. (1992), *SID 92 Digest*, 232–235.

Jiang, J. C., White, R. C. (1993), *SID 93 Digest*, 596–598.

Kakizaki, T., Tanamachi, S., Hayashi, M. (1996), *SID 96 Digest*, 915–918.

Kaneko, S., Hirai, Y., Sumiyoshi, K. (1993), *SID 93 Digest*, 265–268.

Kiefer, R., Weber, B., Windscheid, F., Baur, G. (1992), *J. Displ.* **92**, 547–550.

Kondo, K., Kinugawa, K., Konishi, N., Kawakami, H. (1996), *SID 96 Digest*, 82–84.

Kumar, N., Schmidt, H. K., Clark, M. H., *et al.* (1994), *SID 94 Digest*, 43–46.

Kuo, C. P., Fletcher, R. M., Osentowski, T. D., Lardizable, M. C., Craford, M. G. (1990), *Appl. Phys. Lett.* **57**, 2937–2939.

Meyer, R. (1990), *Eurodisplay* **90**, 374.

Morozumi S. (1996), *SID 96 Seminar Lecture Notes*, M8/1–35.

Mosley, A. (1997), "The Structure, Performance and Future of Passive LCDs," in: L. W. MacDonald, A. L. Lowe (Eds.), *Display Systems*, Chichester: John Wiley, pp. 173–189.

Nakamura, S., Senoh, M., Iwasa, N., Nagahama, S., Yamaka, T., Mukai, T. (1995), *J. Appl. Phys.* **34**, L1332–L1335.

Ohshita, S. (1996), *Displ. Dev.* **13**, 28–30.

Pfeiffer, M., Yang, D. K., Doane, J. W. (1995), *SID 95 Digest*, 706–709.

Sampsell, J. B. (1993), *SID 93 Digest*, 1012–1015.

Scheffer, T. J., Clifton, B. (1992), *SID 92 Digest*, 228–231.

Shinoda, T., Wakitami, M., Nanto, T., Yoshikawa, K., Ohtsuka, A., Hirose, T. (1993), *SID 93 Digest*, 161–164.

Soininen, E., Leppänen, M., Törnqvist, R., Viljanen, J. (1994), *7th International Workshop on EL, Beijing 1994*, pp. 97–100.

Stupp, E. (1996), *Inf. Displ.* **12** (10), 20–24.

Sugawara, H., Ishikawa, M., Hatakoshi, G. (1991), *Appl. Phys. Lett.* **58**, 1010–1012.

Tang, C. W. (1996), *SID 96 Digest*, 181–184.

Tang, C. W., Van Slyke, S. A. (1987), *Appl. Phys. Lett.* **151**, 913–915.

Tuenge, R. T., Kane, J. (1991), *SID 91 Digest*, 279–281.

Uchida, T. (1996), *SID 96 Digest*, 31–34.

Velthaus, K. O., Troppenz, U., Hüttl, B., Herrmann, R., Mauch, R. H. (1994), *International Display Research Conference 1994*, pp. 346–348.

Weber, L. F. (1996), *SID 96 Seminar Lecture Notes*, M6/1–41.

Yamada, Y., Yamamoto, N., Nakamura, K., Koshobu N. *et al.* (1995), *SID 95 Digest*, 789–792.

Yamazaki, T., Kawakami, H., Hori, H. (1996), in: SEMI Standard FDP Technology Group (Ed.), *Color TFT Liquid Crystal Displays*, Mountain View, CA: Semiconductor Equipment and Materials International.

EARTH, INTERIOR STRUCTURE OF THE

(An Addendum to *Encyclopedia of Applied Physics*, Volume 5, pages 127–148.)

SHUN-ICHIRO KARATO, *Department of Geology and Geophysics, University of Minnesota, Minneapolis, Minnesota, U.S.A.*

EIJI OHTANI, *Institute of Mineralogy, Petrology and Economic Geology, Tohoku University, Sendai, Japan*

1. EARTH STRUCTURE AND COMPOSITION

Significant progress has occurred in seismic tomography. The resolution in detecting velocity heterogeneity has been improved, and it is now well established, on the basis of both P-wave and S-wave tomography, that some subducted slabs penetrate down to the bottom of the lower mantle (North and South America), although others appear to be deflected near the 660-km discontinuity (e.g., in the western Pacific) (e.g., van der Hilst *et al.*, 1997). These latter slabs, however, also are likely to reach to the bottom of the lower mantle after a certain time, judging from the correlation of high-velocity anomalies at the bottom of the mantle and the history of subduction. Variation of the behavior of subducted slabs appears to correlate with the velocity of trench migration and the temperature of slabs: slab penetration deep into the lower mantle occurs where the velocity of trench migration is small and/or the temperature of the slabs is high (young and/or fast subducting slabs). Thus, the style of mantle convection appears to be neither a simple whole-mantle convection nor strictly layered convection at the present time and during the recent past.

Seismic tomography has also shown large-scale low-velocity anomalies at the bottom of the mantle. They are located beneath the south Pacific and Africa. The presence of partial melting is suggested (Williams and Garnero, 1996). These regions are the likely source for plumes that result in ocean island volcanoes. However, plumes with shallower sources (transition zone) are also inferred (e.g., in Europe).

The presence of anisotropic structures in the inner core has been established both by the analysis of body-wave travel-time anomalies and by the analysis of anomalous splitting of free oscillation peaks. The anisotropy is rather large ($\sim 3\%$) and has nearly axial symmetry, with the symmetry axis slightly ($\sim 10°$) tilted with respect to the current rotation axis (e.g., Song, 1997). The anisotropy is best explained by the preferred orientation of anisotropic iron crystals in the inner core, although neither the physical mechanisms of alignment nor the crystal structure of iron in the inner core is well understood. By analyzing old seismic records, two groups of seismologists have detected the migration of the pole of anisotropy, suggesting rotation of the inner core with respect to the mantle (Song and Richards, 1996; Su *et al.*, 1996). It is remarkable that this "super-rotation" of the inner core was predicted by theoretical modeling of the geodynamo as caused by magnetic coupling between the inner and outer cores (Glatzmaier and Roberts, 1995).

In contrast to the inner core, the bulk of the Earth's mantle is nearly isotropic except for the near-surface layer (lithosphere and asthenosphere), the D$''$ layer (the bottommost layer of the mantle), and possibly the layer near the 660-km discontinuity. Anisotropy in the mantle is likely caused by lattice-preferred orientation of elastically anisotropic minerals such as olivine or perovskite or by laminated structures. Boundary layers of convection are the likely place to cause seismic anisotropy be-

ISBN 3-527-29308-6

cause of highly concentrated deformation at high stresses and the accumulation of chemically distinct materials that tend to form laminated structures. Thus the presence of strong anisotropy in the top and the bottom (D″) layers indicates that they are the boundary layers of convection, and also that the possible presence of an anisotropic layer around the 660-km discontinuity suggests the presence of a mid-mantle boundary layer, implying that convection is somewhat layered. It is surprising that the bulk of the lower mantle is nearly isotropic despite the fact that its constituent minerals are highly anisotropic. The most likely explanation is to assume that the deformation mechanism in the lower mantle is grain-size–sensitive diffusion creep. In this case, rheology is linear (rate of deformation increases linearly with stress) and the grain-size reduction due to phase transformations in downgoing slabs will cause significant rheological weakening.

New seismological observations on gross Earth structure include the evidence of the 520-km discontinuity and a discontinuity around ~1000 km in some areas (e.g., Kawakatsu and Niu, 1994). These discontinuities are rather subtle as compared to those at ~410 km and at ~660 km, which are more distinct and are global features. The 520-km and ~1000-km discontinuities may be regional as opposed to global. The 520-km discontinuity may be attributed to a phase transformation in Ca-rich silicate to $CaSiO_3$ perovskite, but the origin of a discontinuity around 1000 km is uncertain. No phase transformations in major minerals are known at conditions equivalent to this depth. A chemical layering caused by subduction of oceanic lithosphere is a possible mechanism.

The chemical composition of the Earth's mantle has continuously been a subject of extensive studies. Advances in high-pressure technology combined with high-energy x-ray sources from synchrotron radiation now allow us to measure directly the molar volume of minerals under conditions equivalent to the shallow lower mantle (to ~1000 km depth). Experimental techniques to measure seismic-wave velocities under deep-Earth conditions (to ~500 km deep so far) have also been developed. With these techniques, the density, compressibility, and thermal expansion of $MgSiO_3$ perovskite have been determined under shallow lower-mantle conditions (Funamori *et al.*, 1996). Also, seismic-wave velocities of the α (olivine) and β (modified spinel) phases of $(Mg,Fe)_2SiO_4$ have been measured near conditions equivalent to the 410-km depth. The results of these studies have been compared with seismological observations to place constraints on the chemical composition of the Earth's mantle. The main issue here is if the Earth's mantle is chemically layered or not. Despite much progress in high-pressure mineral physics, the issue is far from resolved. The major difficulty is the trade-offs among various parameters (e.g., temperature, iron/silicon content). For example, seismological data can be fitted to either a chemically homogeneous model or a chemically layered model with some iron and/or silicon enrichment in the lower mantle, when a significant temperature increase around the 660-km discontinuity is assumed. However, in view of the strong evidence of penetration of some downgoing slabs into the lower mantle, a strictly chemically layered mantle structure appears to be ruled out.

One important discovery in mineral physics is a dramatic effect of Al on the crystal structure of $(Mg,Fe)SiO_3$ perovskite. It was found, from high-pressure experiments using a multi-anvil apparatus, that a small amount of Al changes the valence state and the crystallographic site of Fe in perovskite (Wood and Rubie, 1996). Fe appears to occur mostly as ferric iron and occupies six-coordinated sites (similar to Si) when a small amount of Al is present. This behavior affects the partitioning of Fe between perovskite and magnesiowüstite and also is likely to change other physical properties including electrical conductivity and elastic properties. Discussions on the chemical composition of the lower mantle will have to take this point into account.

Another potentially important source of information to place constraints on the chemical composition of the deep interior of the Earth is the chemical composition of rocks that are carried from the deep interior of the Earth. During the last few years, several rocks have been discovered that are considered to have originated in the transition zone or the lower mantle of the Earth on the basis of their mineralogy. They include peridotites from South Africa containing unusually silica-rich garnets, a peridotite from the Swiss Alps containing olivine grains with exotic Ti-rich inclusions, and diamonds from Brazil that contain a pair of minerals (magnesiowüstite and pyroxene) that could have been in chemical equilibrium in the lower mantle. These rock specimens provide direct information as to the chemical composition of the deep interior of the Earth. However, the relevance of their compositions to the composition of average Earth's mantle is not always clear, because these samples may have been carried from some unusual regions.

The sharpness and the regional variation in depth of seismic discontinuities have been studied in detail. Many of the seismic discontinuities appear to be rather sharp (major changes in physical properties occur within a few kilometers interval in most cases). Regional variation in the depths of these discontinuities has also been detected. The most marked one is the variation in the depth of "660-km" discontinuity. The depth of the "660-km discontinuity" (boundary between upper and lower mantles) appears to vary on a regional scale (Shearer and Masters, 1992). It is significantly deeper than average in the regions where the temperature is considered to be lower than average as a result of the subduction of oceanic lithosphere. This is consistent with the results of mineral physics, which indicate that the phase transition from upper-mantle minerals to lower-mantle minerals occurs at higher pressures at lower temperatures. The depth of the 410-km discontinuity does not vary as much as that of the 660-km discontinuity. Because the temperature sensitivity of the depth of the phase transformation corresponding to the 410-km discontinuity is as large as that of the 660-km discontinuity (but with a different sign), this observation suggests that lateral variation in temperature around the 410-km discontinuity is not as large as that around the 660-km discontinuity.

2. MANTLE DYNAMICS

The study of mantle convection through computer simulation has made major progress during the last few years. The increasing capability of computation allows incorporation of more and more "realistic" physical properties in numerical modeling. The most important result in this area is the demonstration that the density contrast associated with the 660-km discontinuity has an important effect in modifying the convection pattern. The lateral density contrast near the 660-km boundary as inferred from the undulation of the 660-km discontinuity tends to prevent vertical mass transfer across this boundary. The results of numerical modeling during the early 1990s showed that this effect is large enough to change convection pattern to intermittently layered convection when the contrast in viscosity between descending currents and the surroundings is small (e.g., Tackley, 1995).

In contrast to the effects of density variation, the effects of viscosity structures on the style of convection are less well understood. The viscosity profile in the Earth's mantle is much less constrained than the density profile, but viscosity could change very dramatically both radially and laterally (e.g., Karato, 1997). Recent analysis of vertical crustal movement associated with the post-glacial rebound combined with the analysis of Earth's rotation and geoid observations yields a unified model characterized by a significantly higher viscosity in the deep lower mantle than in the upper mantle (Forte and Mitrovica, 1996). Such a viscosity profile provides a natural explanation for the fixed position of hotspot volcanoes, which may be caused by plumes originating in the deep portions of the lower mantle. Lateral viscosity variation, particularly the viscosity of subducting lithosphere, has been shown to significantly affect convection patterns. The significant role of the density contrast associated with the 660-km discontinuity in changing the convection pattern is important only when the descending current does not have much higher viscosity than the surrounding mantle. The observed significant deflection of subducted oceanic lithosphere in the western Pacific therefore suggests that subducted lithosphere there is rather weak. Likely causes of weakness include grain-size reduction associated with phase transformation(s) (Riedel and Karato, 1997).

Deformation in deep subducted oceanic lithosphere often occurs through seismic faulting. The origin of these deep earthquakes has been an enigma, because brittle fracture that is responsible for shallow earthquakes is impossible under high confining pressures. The role of phase transformations in causing instabilities in deformation has been demonstrated in laboratory studies, which provides a potential mechanism for deep earthquakes (Green, 1994). However, the new data obtained from the Bolivia and the Fiji deep earthquakes during 1994 showed much wider fault widths than predicted from simple models. Active research is ongoing to understand better the role of phase transformations in deep earthquakes.

3. CORE DYNAMICS AND GENERATION OF THE GEOMAGNETIC FIELD

The last few years witnessed significant progress in the understanding of the origin of magnetic field generation in the Earth's core. The first successful modeling was made in U.S.A. and in Japan in which both the equation of motion of a rotating viscous fluid and the electromagnetic induction equation are simultaneously solved to show the generation of a nearly dipole magnetic field (Glatzmaier and Roberts, 1995, 1997; Kageyama and Sato, 1997). The driving forces in these models include both thermal and chemical buoyancy forces, the latter being due to the solidification of nearly pure iron from iron-rich alloy. The important effects of the inner core are also noted in these models. These models were able to show a stable nearly dipole magnetic field, but some instabilities causing reversals of magnetic field were also seen. One of the important outcomes of this type of model is the prediction of a "superrotation" of the inner core as a result of magnetic coupling between outer and inner cores due to a strong toroidal field (Glatzmaier and Roberts, 1995). "Superrotation" of the inner core may also be attributed to viscous coupling.

Observationally, one of the most fascinating observations is the suggestion that the virtual geomagnetic pole (VGP: the pole of magnetic dipole field projected to the Earth's surface) follows specific paths during the reversal. This notion was inferred from the analysis of magnetic records in sediments, and the results show that the VGP follows a band in the Earth along which seismic-wave velocities are anomalously high at the bottom of the mantle (Laj *et al.*, 1991). This suggests that the temperature anomalies at the core–mantle boundary caused by mantle convection may influence the convection pattern in the outer core that in turn causes the variation in magnetic field (Kuang and Bloxham, 1997).

List of Works Cited

Forte, A. M., Mitrovica, J. X. (1996), *Geophys. Res. Lett.* **33**, 1147–1150.

Funamori, N., Yagi, T., Utsumi, W., Kondo, T., Uchida, T. (1996), *J. Geophys. Res.* **101**, 8257–8269.

Glatzmaier, G. A., Roberts, P. H. (1995), *Phys. Earth Planet. Inter.* **91**, 63–75.

Glatzmaier, G. A., Roberts, P. H. (1997), *Nature* **377**, 203–209.

Green, H. W., II. (1994), *Sci. Am.* **271** (3), 64–71.

Kageyama, A., Sato, T. (1997), *Phys. Rev.* **E55**, 4617–4626.

Karato, S. (1997), in: D. J. Crossley (Ed.), *Earth's Deep Interior*, London: Gordon and Breach, pp. 223–272.

Kawakatsu, H., Niu, F. (1994), *Nature* **371**, 301–305.

Kuang, W., Bloxham, J. (1997), *Nature* **389**, 371–374.

Laj, C., Mazaud, A., Weeks, R., Fuller, M., Herro-Brevera, E. (1991), *Nature* **351**, 447.

Riedel, M. R., Karato, S. (1997), *Earth Planet. Sci. Lett.* **148**, 27–44.

Shearer, P. M., Masters, T. G. (1992), *Nature* **355**, 791–796.

Song, X., Richards, P. G. (1996), *Nature* **382**, 221–224.

Song, X. (1997), *Rev. Geophys.* **35**, 297–313.

Su, W. J., Dziewonski, A. M., Jeanloz, R. (1996), *Science* **274**, 1883–1887.

Tackley, P. J. (1995), *Rev. Geophys. Suppl.* (U.S. Natl. Rep. Int. Union Geo. Geophys., 1991–1994), 275–282.

van der Hilst, R. D., Widiyantoro, S., Engdhal, E. R. (1997), *Nature* **386**, 578–584.

Williams, Q., Garnero, E. J. (1996), *Science* **273**, 1528–1530.

Wood, B. J., Rubie, D. C. (1996), *Science* **273**, 1522–1524.

Further Reading

Lay, T., Wallace, T. C. (1995), *Modern Global Seismology*, San Diego: Academic Press. Introduction to global seismology with basic concepts and some recent results.

Crossley, D. J. (Ed.) (1997), *Earth's Deep Interior*, London: Gordon and Breach. A collection of review papers related to Earth's interior. Topics discussed include seismic tomography, mineral physics, mantle convection, and the geodynamo.

FUSION, INERTIAL CONFINEMENT

(An Addendum to *Encyclopedia of Applied Physics*, Volume 6, pages 545–573.)

BARRETT H. RIPIN, *The American Physical Society College Park, Maryland, U.S.A.*

INTRODUCTION

Developments in the inertial-controlled fusion (ICF) program occurring in the mid to late 1990's have raised expectations that ignition of an ICF pellet in the laboratory may be demonstrated as early as 2005.

1. OVERVIEW OF INERTIAL CONFINEMENT FUSION (ICF)

The basic physics principles of the inertial-confinement fusion concept are described in detail in the original article. As a brief reminder of the concept, recall that the objective is to implode a small pellet containing deuterium (D) and tritium (T), heavy isotopes of hydrogen, to high enough density and temperatures that a substantial fraction of the fuel nuclei undergo fusion reactions and yield bursts of energy. Calculations predict that the fusion energy release in this process can be of the order of 10^2 times the energy driving the implosion. The ratio of released energy to driver energy is called gain. If high-gain ICF can be demonstrated, then it raises hopes that it might be further developed for use in civilian electric-power generation.

To achieve the conditions for high-gain ICF, the outside surface of a pellet of about 1 mm diameter that encloses a layer of cryogenic DT fuel on the shell's inner surface is irradiated by an intense (10^{14} to 10^{15} W/cm^2) energy pulse from a directed energy source, such as a laser. The energy deposition on the pellet surface causes it to ablate and heat, become a hot plasma, and blow away from the surface at high velocity. Pressures of tens of megabars created by the hot ablating plasma surface material drives the remaining shell and interior fuel inward, compressing the DT fuel to over 10^3 times solid density. The driver pulse characteristics, and hence the ablation pressure, are staged such that the bulk of the compressed fuel remains on a low isentrope (dense but cold material) except for a small fuel region in its center. The drive pulse is shaped in time so that the central region, called the ignitor, is heated to temperatures of about 10 keV. This configuration is shown schematically in Fig. 1. If the hot-spot density-radius product is high enough (>0.3 g/cm) to stop the fusion-produced alpha particles at this temperature, then self-heating from the nuclear fusion reactions will initiate a fusion "burn" front that propagates outward throughout the main fuel volume. If all this is done properly, the energy yield of the explosion will greatly exceed the incident energy, yielding net energy. Although most ICF research to date has used a large laser driver, in principle, particle beams from accelerators could also be used.

Figure 2 illustrates the two general ap-

ISBN 3-527-29308-6

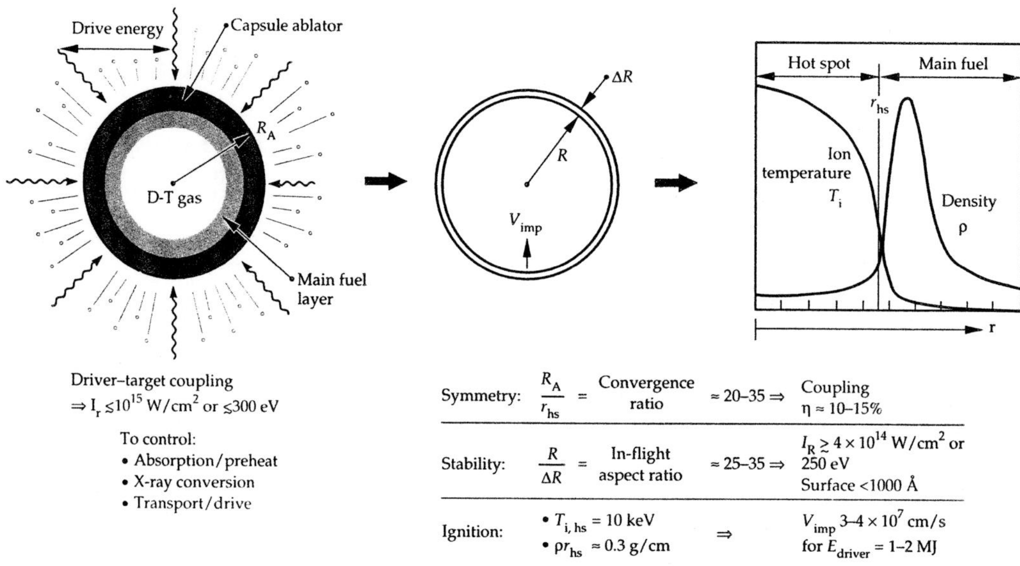

FIG. 1. Elements of inertial confinement fusion. (Left) A schematic of an irradiated pellet, (middle) aspect ratio ($R/\Delta R$) of imploding pellet, and (right) density–temperature profile of hot spot (ignitor) and main compressed DT fuel at point of ignition. Tables summarize constraints on drive symmetry, hydrodynamic stability requirements, and ignition parameters. [From LLNL (1995).]

proaches to ICF. When the driver energy beam directly irradiates the pellet, the ICF approach is called "direct-drive." If the driver energy first strikes high-Z material inside of an enclosure, a hohlraum, that converts the beam energy into intense x rays that subsequently strike, ablate, and implode a pellet, the approach is called "indirect drive," "x-ray drive," or the "classified approach." There was a major declassification and disclosure of the x-ray drive approach to ICF in 1993. Most of what follows is a description of the

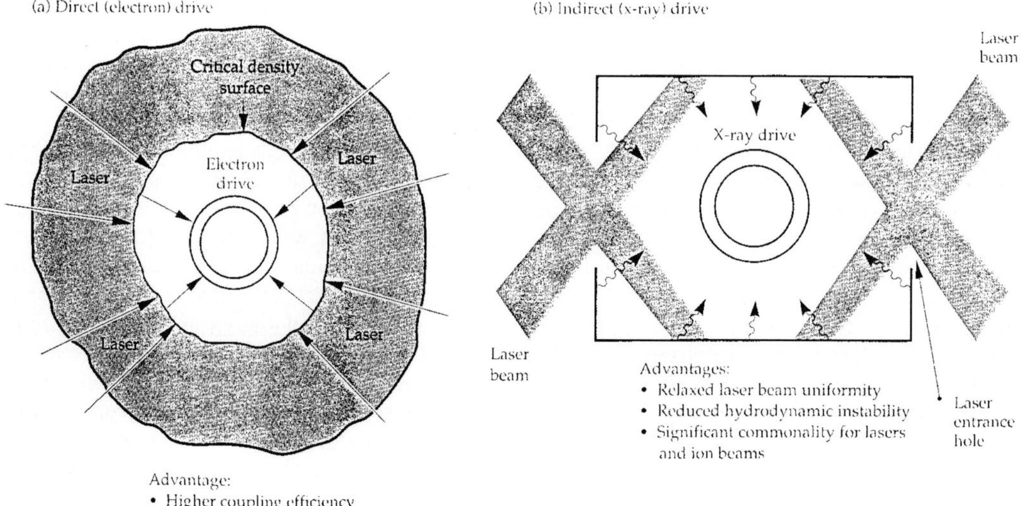

FIG. 2. Two approaches to ICF: (left) direct drive and (right) indirect (x-ray) drive [From LLNL (1995).]

x-ray drive approach that was omitted in the original article because of classification restrictions, along with more recent advances.

There has also been a significant shift in the programmatic rationale for ICF since 1993. Most of the funding for ICF research in the United States comes from the Department of Energy's defense programs, ostensibly as a laboratory means to gain general information related to nuclear weapons design, operation, and effects. With the breakup of the Soviet Union and the advent of a comprehensive nuclear test ban, the ICF program has become a key component in the U.S. stockpile stewardship maintenance program (SSMP) in the mid-1990's. The purpose of SSMP is to maintain a nuclear deterrent capability in the absence of underground nuclear testing.

2. PRINCIPLES OF X-RAY DRIVE

The x-ray drive approach has actually been the field's principal effort since 1976, although little was known about it in the general scientific community because of U.S. Department of Energy (DOE) secrecy restrictions. The formerly classified information release, combined with recent advances in the scientific basis of the ICF research program and in driver and target fabrication technologies, has raised expectations that an x-ray–driven ICF target could achieve ignition in the laboratory in the not too distant future. A comprehensive review of the development of x-ray drive ICF approach can be found in Lindl (1995).

The distinctive element of x-ray drive is the hohlraum, which is a radiation-filled enclosure made of high-Z material (typically gold) that surrounds a pellet (also known as a 'capsule'), such as depicted in Fig. 2. Laser beams enter the hohlraum by focusing through small entrance holes. Copious x rays are produced where the beam spots strike the interior hohlraum walls. Most x rays are absorbed by and heat the remainder of the hohlraum walls, which in turn, reradiate x rays forming a bath of x rays with an equilibrium distribution. Typical beam–to–x-ray energy conversion efficiencies are reasonably high (70% to 80%). Laser beams are usually arranged in two clusters. Hohlraums are usually cylindrically shaped with an entrance hole in each end cap to accept the beam clusters. The optimal hohlraum design depends upon details of the driver type and pellet design. The beams are arranged to strike the walls such that the x-ray flux falling on the pellet surface is a uniform mixture of x-rays emitted directly from the laser spots and the somewhat lower intensity x-ray albedo from other portions of the hohlraum. The exact beam placement required to achieve the required uniformity and symmetry for high-convergence ratio implosions is an area of active research.

The pressure P_a applied at the ablation surface of the pellet is what determines the implosion dynamics of an ICF pellet, whether that pressure be generated by direct-drive or indirect-drive process. The ablation pressure accelerates the remaining shell material inward at a velocity given by the spherical version of the rocket equation, $V_{imp} = V_{pl}\ln(m_0/m)$. Here m_0 and m are respectively the original and final mass/surface areas of the pellet shell. The velocity of the plasma ablating away from the pellet, V_{pl}, is given by $V_{pl} = P_a/[dm/dt]$).

X rays are well suited to generation of high ablation pressures because they deposit their energy directly into the high-density regions of the pellet shell. Dense material deposition results in a higher mass ablation rate (dm/dt) and an ablation velocity that is lower and better matched to the peak efficiency point (where $V_{imp} \approx V_{pl}$) of the rocket equation. X rays inside a hohlraum can be characterized by a blackbody temperature, at least within the relevant wavelength regime containing most of the energy. The x-ray intensity impinging on the ablation surface is given approximately by the relation $I_a \approx T_r{}^4$. By way of calibration, a hohlraum temperature of $T_r = 300\,\text{eV}$ has the equivalent x-ray intensity of about $I_a = 10^{15}\,\text{W/cm}^2$.

For typical ignition target designs, the effective radiation temperatures are in the range 225 to 300 eV. The high-temperature limit is set primarily by the need to keep incoming laser-beam irradiances low enough to avoid stimulating parametric instabilities in the hohlraum plasma. The low-temperature limit comes from the need to achieve the minimum shell velocity for ignition, which is about 3×10^7 cm/s. Here the consensus is that the pellet shell in-flight aspect ratio (IFAR $= R/\Delta R$) is limited to 30 or less (i.e., this is the thinnest feasible shell) because of excessive hydrodynamic instability.

Capsules need to converge in radius by a factor of 25 to 35 in order to form the small hot-spot region needed to ignite the pellet at peak compression. For a pellet to converge this far and remain spherical requires x-ray fluxes uniform to within 1 to 2%. x-ray uniformity improves with larger hohlraums for a given pellet size, but with a loss of efficiency. Alternatively, optimal placement of beam deposition patterns and careful balance between them can reduce x-ray asymmetries inside a smaller hohlraum.

In addition to good overall x-ray symmetry impinging on the pellet surface, the pellet shell must resist excessive hydrodynamic instability. One of the most critical regions for hydrodynamic instability in indirect-drive pellets is at the inner surface of the main fuel. This is because hydrodynamic instability will tend to mix cold dense fuel into the hot-spot region, cooling it and quenching ignition. The most important factor affecting inner-surface mix is the initial inner-surface roughness of the fuel layer. The upper limit for the inner-surface roughness of the cryogenic DT in typical NIF (National Ignition Facility; see Sec. 4) ignition target designs is about 1 μm when a CH ablator shell is employed, and a factor of two or three higher if a beryllium ablator shell can be used. Hydrodynamic instability causing mix is fed by areal mass nonuniformities. Beryllium is a more tolerant material than CH because its density is lower so that the effect of a given roughness is reduced. Also, a Be ablator shell would be about twice as thick as a plastic (CH) shell, which further reduces the growth rate of the hydrodynamic instability that leads to mix.

Although there is a small loss of efficiency in the process of converting the driver energy into x rays, x-ray drive has a number of potential advantages over the direct drive approach.

1. The pellet finds itself immersed in an x-ray bath that tends to be naturally uniform on account of the broad pattern of the x-ray emission process plus absorption and reradiation by the hohlraum walls. The x-ray emission is relatively insensitive to nonuniformities in the driver beams and, to some degree, to the geometric pattern of incoming beam clusters.
2. X rays penetrate and directly deposit their energy in the vicinity of the pellet ablation surface ($\sim 10^{24}$ electrons/cm^3), which is a few times solid density. This is a more efficient process than the case of direct laser irradiation in which initial laser-light absorption occurs at much lower densities (near or below the critical density, where the laser-light frequency equals the plasma frequency, $\sim 10^{22}$ electrons/cm^3 for 0.35-μm light), requiring the energy then to diffuse up a density ramp to the ablation surface.
3. x-ray deposition creates a softer density gradient in the vicinity of the ablation region that reduces the shell's susceptibility to hydrodynamic instability over that of the direct-drive case.
4. The principal proponents of the x-ray approach, DOE weapons laboratories, have considerable experience, infrastructure, and data base upon which to draw in this area.

This is not to claim that the x-ray ICF approach is ideal or a sure thing. For example, only about 5% of the driver energy actually ends up as useful hydrodynamic implosion energy. This is a relatively inefficient use of high-quality energy from the driver beam. The losses take the form of direct conversion loss of beam energy into x-rays, loss of x-ray energy out the holes in the hohlraum and by reabsorption in the hohlraum walls, and waste heat carried away by the ablating plasma.

Scale sizes of plasmas generated inside a hohlraum can be large, creating a favorable environment for laser parametric instabilities, such as Brillouin or Raman backscatter instabilities. Sizable levels of backscatter would reduce beam energy available to generate x rays and thereby lower efficiency. Avoidance of parametric instabilities tends to cap the maximum laser-beam irradiances that can be utilized.

Finally, there always exists a small residual asymmetry in the x-ray albedo falling on the pellet surface due to the beam deposition geometry. Although the x-ray radiation pattern falling on the pellet is made nearly spherically symmetric by multiple absorption and reradiation of x rays by the hohlraum walls, there can still be enough of a residual asymmetry to affect the implosion symmetry adversely. Excessive asymmetry can seed hydrodynamic instabilities during pellet implosion that prevent formation of a small ignitor region, which, in turn, would increase the fuel needed to be brought to 10 keV and, consequently, the driver energy required for ignition.

For energy applications, the presence of extraneous mass, particularly high-Z materials typically used in hohlraums, is a practical disadvantage. This is because some of the accumulated material from the millions of capsules exploded in an energy reactor will become activated by the fusion neutron flux and must be extracted and disposed of.

Experiments on Nova and elsewhere, as well as computer simulation projections, have demonstrated that most of these potential problems can be averted or made tolerable, and that current x-ray–driven pellets designed for use in the NIF are likely to reach ignition yields.

In spite of the advantages of the indirect-drive approach, direct-drive research is also being pursued vigorously. If direct-drive ICF can be demonstrated, then it might permit ignition and high gain with lower-energy (smaller) drivers, since direct drive does not have energy losses associated with the conversion to x rays and other hohlraum losses. Direct-drive pellets would be simpler in design and cheaper, again since they do not require a carefully crafted hohlraum. Therefore, direct-drive targets are more desirable for energy reactor applications. Since either approach may yet encounter formidable, but presently unknown, problems as they scale up into new parameter regimes, the current ICF program strategy is to pursue both vigorously until it is clear which is superior.

3. CURRENT STATUS OF ICF RESEARCH

3.1 Laser Facilities

Two new ICF-laser facilities designed to pursue the direct-drive ICF approach, under construction in 1993, have been completed and put into operation. The Omega glass laser, at the University of Rochester's Laboratory for Laser Energetics, underwent a complete upgrade to deliver over 40 kJ of UV (0.35 μm wavelength) in each nanosecond-duration pulse in sixty uniform, balanced, and flexible beams. Figure 3 shows a photo of the Omega Upgrade target chamber. Although Omega Upgrade was designed primarily to do integrated direct-drive pellet implosion research, it is also being used to investigate aspects of x-ray–drive physics in preparation for ignition experiments on the NIF. An overview of the Omega Upgrade experimental program may be found in Soures *et al.* (1995). The other new laser facility, Nike, is a relatively small (4 kJ) krypton-fluoride gas laser built at the Naval Research Laboratory. Nike was designed with excellent beam-smoothing features to allow precision planar-geometry hydrodynamic-instability experiments that may help determine the upper yield limit of direct-drive pellet implosions.

FIG. 3. Omega Upgrade target facility at the Laboratory for Laser Energetics, University of Rochester [From University of Rochester (1996).]

3.2 Target Experiments

The U.S. ICF program has made significant progress in reducing uncertainties in most technical areas required for ignition. Experiments on the Nova laser at Lawrence Livermore National Laboratory (LLNL) have provided a better physical understanding of ignition physics, especially those that may affect target performance with the NIF. The status of some of those investigations is provided in the following.

Currently, gas-filled hohlraums are the baseline targets for use on NIF. The purpose of the gas is to impede the flow of plasma from the hohlraum wall into its volume, thereby reducing the environment for parametric instabilities of the incoming beams. Early Nova experiments indicated levels of laser light backscattered from instabilities and dynamics shifts in beam pointing inside hohlraums large enough to affect adversely the x-ray drive effectiveness. It was conjectured that these effects resulted from high-intensity peaks in the fairly nonuniform Nova beams. Beam-smoothing techniques, such as use of phase plates, and an active beam-fluctuation averaging technique, such as smoothing by spectral dispersion (SSD), were applied to Nova's beams in 1996. Experiments and modeling then confirmed that use of smooth beams largely eliminated these potential problems and brought about a corresponding rise in x-ray radiation temperature and symmetry. These favorable results scale to expected NIF conditions according to current theory and models.

The Omega laser at the University of Rochester has also been used to do precision hohlraum x-ray–drive experiments. Excellent agreement was obtained on Omega with equivalent Nova experiments and LASNEX calculations (Decker *et al.*, 1997). Omega has also been used to test beam configurations planned for the NIF facility. About half of Omega's 60 beams have been arranged into a similar geometry to the NIF point-target design, i.e., two outer and two inner rings of beam irradiation spots inside the hohlraum. Experiments to exercise precise temporal and spatial control over the symmetry of x-ray emission by use of beam delays and pulse shapes were performed. Again good agreement between simulation and experiments was found, showing that similar beam phasing on NIF should provide adequate drive symmetry for ignition implosions.

3.3 Predictive Capability

A key component of successful extrapolation from Nova to NIF, and from NIF to a high-yield facility, is possession of reliable predictability through an accurate computer simulation capability. Most of the experimental results are used either to validate predictive codes, or to help add or correct the physics models required to do so. Increasingly, it has been found that two-dimensional (2D) codes are not sufficient to model implosion dynamics and that three-dimensional (3D) simulations are needed. In the mid-1990s, a radiation-hydrodynamics code, HYDRA, which includes full fusion burn physics, was developed. HYDRA is now routinely used for modeling ICF capsules. In particular, the surface-finish smoothness requirements for the inner surface of the deuterium-tritium ice layer in NIF ignition targets have been studied. HYDRA showed that a DT-containing shell made with a plastic ablator can tolerate up to 1-μm rms DT layer roughness, whereas the corresponding shell made of beryllium ablator can tolerate up to three times more roughness.

3.4 NIF Ignition Target Point Design

The baseline ignition capsule (pellet) for use on the NIF has a cryogenic DT-ice shell surrounding a central DT-gas region that resides inside a gas-filled gold hohlraum, as depicted in Fig. 4. The DT ice is surrounded by either a plastic or a beryllium shell that both forms a vessel to hold the DT and acts as the ablator material. With an expected NIF input of 1.8 MJ in a 500-TW pulse shaped as shown in Fig. 4, yields of order 15 MJ are predicted. This leaves about a safety margin of close to 10 above the ignition level for nonideal performance. There are over a dozen target designs that reach ignition levels, most of which have moderate robustness for nonideal behavior. As mentioned previously, the most sensitive parameter that can spoil performance is the surface finish of the inner DT-ice surface, which affects the level of hydrodynamic instability causing mix in the hot-spot region. Since significantly higher DT-ice roughness is tolerated with beryllium ablators than with plastic ablators, considerable research is being

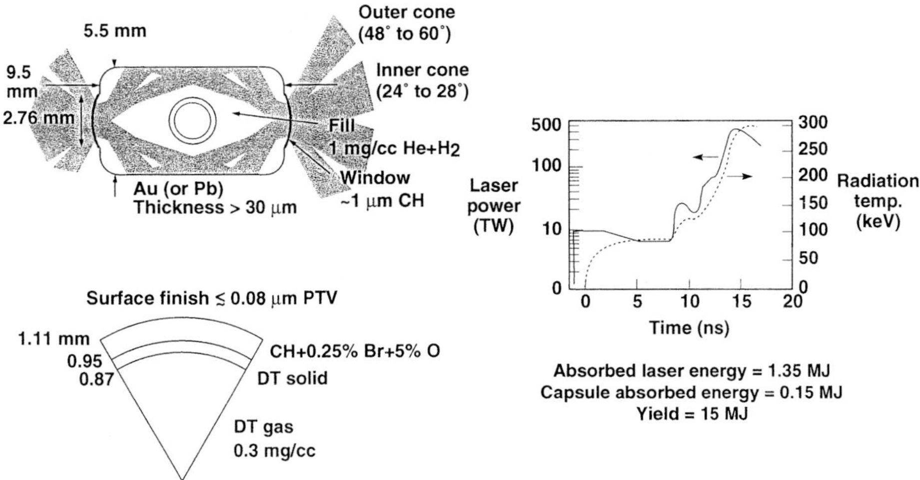

FIG. 4. Baseline ignition target design for the NIF (Courtesy of LLNL).

performed on fabricating suitable Be shells. A more complete description of ignition target design can be found in Haan *et al.* (1995).

3.5 Target Fabrication

Plastic shells are the mainline target for NIF ignition because their fabrication and characterization has been demonstrated. The relaxed specifications on surface finishes of cryogenic DT layers in Be targets have motivated efforts to develop the means to fabricate Be shells with the proper uniformity. Both machining and sputter-coating methods are being evaluated for this purpose. Although good progress has been achieved, the fabrication of beryllium shells with both sufficient uniformity and a means to load them with DT fuel needed further development as of 1998. It is expected that beryllium will indeed become the *de facto* ablator material, with plastic kept as the fallback option.

4. NATIONAL IGNITION FACILITY

Among the facility proposals under consideration in the early 1990s was a "Nova Upgrade," the purpose of which was to increase the existing 10-beam Nova laser's energy to a level that could approach pellet ignition. Subsequently, a very large glass laser facility of this type, called the National Ignition Facility (NIF), was designed and approved for construction in the United States. Figure 5 shows a sketch of the NIF as it is expected to look when completed. The new name reflects both the fact that the NIF is an entirely new facility and that it is being viewed as a national resource for use by all research participants in the U.S., and perhaps worldwide. The NIF is viewed as a key element of the U.S. stockpile stewardship management program. An ignition facility very similar to the NIF is also planned for construction in France.

29 May 1997 was the groundbreaking for the National Ignition Facility at Lawrence Livermore National Laboratory. When operational in the 2005 time frame, NIF will be the world's largest laser, about twenty times more powerful than Nova, the LLNL laser it is replacing. The heart of the system is an Nd laser with 192 beams that will produce in excess of 1.8×10^6 J of ultraviolet (0.35 μm) laser light energy in 3 ns (about 500 TW).

The NIF was designed to be very flexible, incorporating beams with excellent uniformity, interbeam balance, and pulse shaping. It was also constructed to allow relatively easy reconfiguration into direct-drive geometry facility. The size (energy) of the NIF was chosen to

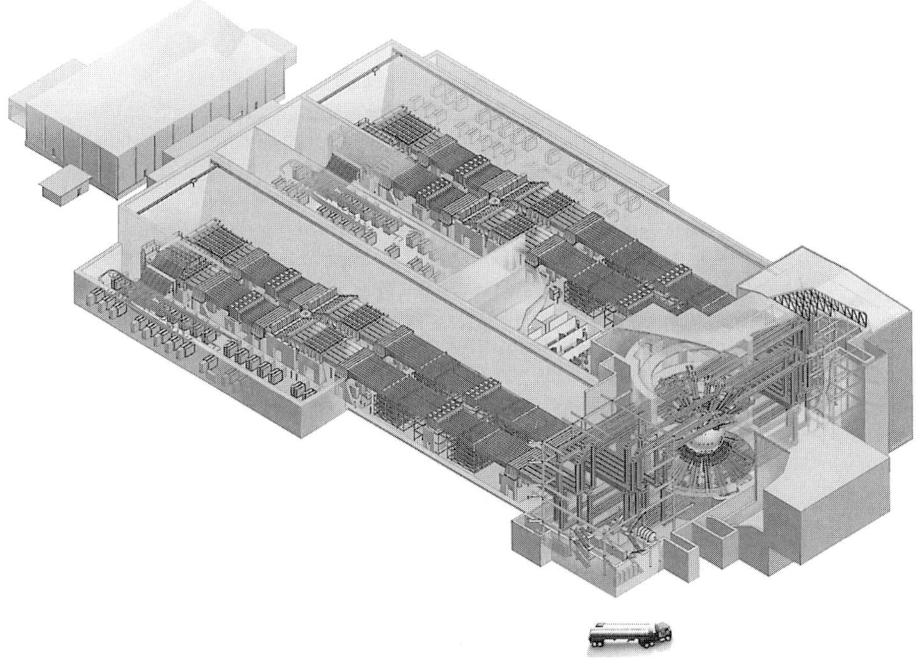

FIG. 5. The National Ignition Facility (artist's sketch courtesy of LLNL).

have a likelihood to achieve ignition level yields, yet small enough to contain costs. The completed facility is expected to cost about 1.2×10^9 (1997) dollars, including the laser, associated experimental and target fabrication facilities, and infrastructure. The NIF will consist of about 8000 large (over 1 ft^2) optical components and more than 30 000 smaller ones. The operational lifetime of the NIF is expected to be about 15 years.

Although located at LLNL, the NIF was designed, built, and operated as a national facility through a partnership with Los Alamos National Laboratory, Sandia National Laboratories, and the University of Rochester Laboratory for Laser Energetics, in addition to LLNL. General Atomics, Inc., is a partner in developing the target fabrication technology to be used in NIF experiments.

The NIF is scheduled to be completed in several phases, with the first set of eight beam lines ready for testing in about 2001. Beamlet, a prototype of one NIF beam, began operating at LLNL in 1994. The Beamlet prototype was necessary to validate the laser architecture used in NIF, which is much different from previous-generation lasers.

The light from NIF's beams will be directed into a 10-m-diam spherical target chamber and focused onto the relatively small (millimeter) targets. The target chamber will be loaded with sophisticated diagnostic instruments to measure the performance of the pellet performance. The entire NIF facility is being designed to accept either indirect-drive geometry, in which the beams enter the chamber primarily in clusters on the top and bottom of the target chamber, or direct-drive geometry, in which the beams enter the chamber more symmetrically. Although the indirect-drive configuration will be implemented first, the system will be reconfigured should direct-drive experiments at Omega Upgrade prove more promising.

4.1 Science-Based Stockpile Stewardship

Establishment of the science-based stockpile stewardship (SBSS) program, a subset of the comprehensive stockpile stewardship maintenance program (SSMP), was a consequence of the cessation of the cold war and President Clinton's order for a moratorium on U.S.

nuclear weapon testing in 1993. (Subsequently, in 1996, the U.S. and over 90 other nations signed a Comprehensive Test Ban Treaty.) The SBSS represents DOE's strategy to address the challenge of retaining nuclear capability in a test-free environment. The 1994 National Defense Authorization Act called for "a stewardship program to ensure the preservation of the core intellectual and technical competencies of the United States in nuclear weapons." The SSMP is directed to do this in a manner consistent with the broad arms-control and nonproliferation goals of the United States and most other nations.

The SBSS is structured to maintain a continuing capability to anticipate, detect, and correct problems related to aging of the nuclear stockpile. In addition to accumulating detailed knowledge about nuclear technologies, understanding of high–energy-density physics, reliable computer simulation capability, and the like, it requires that there continue to be a number of excellent scientists and engineers trained and knowledgeable in weapons related areas.

The role that ICF, and NIF in particular, play in the SBSS program is to explore the physics of matter and radiation at high energy density and the physics of thermonuclear fusion in a laboratory setting. Ignition of an ICF pellet in NIF is a major goal. The goals of SBSS and NIF are also fully consistent with the goals of both the defense and energy ICF applications. Many people believe that it was only the confluence of favorable events, such as of the end of the cold war, test bans, and nonproliferation policy, concurrent with readiness of glass-laser technology and advanced and positive ICF research results, that allowed the NIF to be authorized in the tight federal fiscal environment of the mid-1990s.

5. HEAVY-ION DRIVER DEVELOPMENT

Although the most advanced and utilized drivers for ICF pellet implosion research to date are lasers, there has been some progress in developing particle-beam accelerators for ICF driver use. Particle-beam driver concepts currently assume use of the x-ray drive approach (although direct-drive schemes are also possible in principle). Particle-beam energy is first converted in a hohlraum into x rays that, in turn, irradiate and drive pellet implosions. The technical feasibility of the development of a heavy-ion induction linear accelerator continues at Lawrence Berkeley Laboratory, but at a slow pace due to low funding levels ($\$7 \times 10^6$ level in FY97 out of a total ICF funding of $\$250 \times 10^6$). The reader is directed to the original article for further details and references that describe several heavy-ion driver approaches.

6. *Z*-PINCH DRIVER

There has been a program, from the early days of the ICF program, to determine if electric-pulse–power devices could be used as a driver. Pulse-power devices store large amounts of electrical energy in capacitor banks or inductive units and use switching elements to discharge the electrical energy rapidly into a load. Discharge-pulse parameters are in the tens of nanoseconds time scale with voltages in the multimegavolt range and with multimegampere currents.

In the 1970s, the approach used was to generate an electron beam in the load, frequently called a diode, to drive the pellet implosion. The electron-beam approach was dropped when the multi-MeV electrons from the driver beam were found to penetrate and preheat the pellet interiors to unacceptable levels. In the 1980s and early 1990s, the diodes were configured to have reverse polarity, and efforts were shifted to accelerating light ions (hydrogen or lithium) for pellet driver application. After encouraging initial results, light-ion driver ICF development stagnated for a number of years because no way was found to focus the beam to sufficiently high energy densities. This resulted in hohlraum temperatures much too low to be useful. The light-ion–beam accelerator approach was essentially abandoned in the mid-1990s.

In 1996, Sandia National Laboratory made a major advance in the generation of x rays with a different pulse-power approach. They rejuvenated and refined an old pulse-power device, called a *Z*-pinch, to generate x rays directly, without use of a particle beam inter-

FIG. 6. Wire array load for Sandia *Z*-pinch x-ray source. The diameter of the cylindrical array of 300 tungsten wires is about 2 cm. The pulse-power voltage is applied across the top and bottom support rings. (Courtesy of Sandia National Laboratory.)

mediary. This *Z*-pinch consists of a wire array load that gets vaporized, is made into a plasma, and is pinched to high density and temperature by the magnetic field from the huge currents flowing through it. In the process, the pinched plasma emits copious x rays. Sandia produced a record 1.9 MJ of x-ray energy in a 200-TW pulse with the 300-wire array load pictured in Fig. 6. Using this x-ray source, they produced a 130-eV black body temperature inside of a 2.5-cm-diam hohlraum in 1997. While the hohlraum temperatures demonstrated are still at least a factor 2 to 3 too low to drive ignition-level pellets, the results are viewed as encouraging. Sandia proposed to build a larger version of this device, called X-1, that might be used for ignition-level experiments in the late 1990's. A *Z*-pinch ICF driver is best suited for defense applications that require single-shot capability (e.g., one shot per day or less), since the *Z*-pinch load region is destroyed with each firing. It is difficult to imagine a *Z*-pinch being compatible with the high repetition rate needed for a civilian power plant.

7. ALTERNATE CONCEPT: FAST IGNITOR

If inefficiencies and waste heat could be eliminated, high-gain fusion yield could be achieved with as little as 100 kJ energy input. This is approximately the energy required to compress low-temperature main pellet fuel to the high densities required to support propagating fusion burn. What is then needed, in principle, is only a few tens of kilojoules delivered to the central small ignitor region that raises its temperature enough to initiate propagating fusion burn. In conventional approaches (direct and x-ray drive) ignition is achieved by carefully programming in a set of shocks converging though the cold, dense main fuel region that coalesce at the right instant to heat and ignite the central region. This is a rather inefficient process, as only a small fraction of the shock energy transfers to the ignitor, even if done optimally.

The fast-ignitor concept would employ an ultrahigh-intensity laser to burrow through the corona surrounding the compressed fuel and deliver energy directly to an ignitor region in the form of energetic electrons, as depicted in Fig. 7. By way of an example, to heat a DT ignitor region of density 300 g/cm^3 to about 10-kV temperature using energetic electrons requires focal intensities of at least 10^{20} W/cm^2 with 1-μm light in a pulse duration of 1 ps (i.e., 10 kJ). Any laser–matter inefficiencies will increase the associated laser requirements. A laser prepulse of order 10^{19} W/cm^3 is needed over a few tens of picoseconds to create the immense pressures required to push the cold fuel aside and form a low-density channel through which the much more intense ignition pulse can be delivered.

The fast-ignitor concept is considered to be very speculative approach, since it requires specific (favorable) physics behavior through large extrapolations in unexplored, highly nonlinear portions of parameter space. Nonetheless, if it can be demonstrated to work, the costs to achieve high-yield ICF and the engineering requirements for the reactor chamber plummet and the variety of driver options decrease.

A petawatt laser (10^{15} W) has been developed at LLNL to test some of the physics basis of the fast-ignitor concept. The petawatt laser produces about 1 kJ of 1.053-μm light at irradiances above 10^{21} W/cm^2 with a pulse length in the 1-ps regime. The laser starts with short chirped pulses (broad bandwidth) that are stretched to about 1-ns duration with a dif-

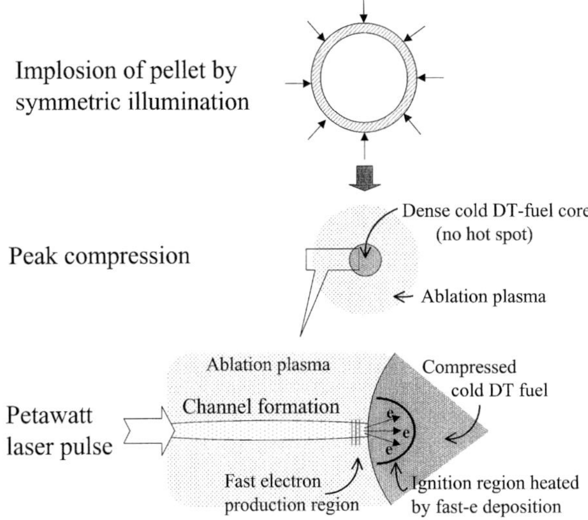

FIG. 7. Fast-ignitor concept. The pellet containing DT fuel is compressed without forming a hot spot. A picosecond laser pulse bores a hole through the dense plasma leading up to the compressed core and creates an intense burst of energetic (fast) electrons near the core that heat a small region to ignition.

fraction grating. The stretched pulse length is well matched for efficient amplification using a slightly modified Nova amplifier chain, followed by recompression of the pulse duration down to the picosecond regime using large recompression diffraction gratings. Proof-of-principle experiments of the critical fast-ignitor physics issues began in 1997 using the petawatt laser. A more complete description of the fast ignitor concept may be found in Tabak (1994).

8. INERTIAL FUSION ENERGY

Efforts directed specifically to produce a civilian energy source based on ICF, called inertial fusion energy (IFE), continue, but constitute a minor portion of the total ICF program. However, the essential stage of demonstrating ICF ignition in the laboratory is common to both energy and defense applications; hence the ICF program resources have been concentrated on this goal. A critical requirement for civilian energy use of ICF is a driver that is capable of being cycled with a repetition rate of about 1 shot/second with good efficiency and reliability, long lifetime, and compatibility with a reactor environment. Heavy-ion accelerators appear to be capable of meeting these requirements. Two laser technologies that might potentially satisfy the energy driver requirements, high–repetition-rate KrF-gas laser and laser-diode–pumped glass lasers, are also being investigated at low levels.

9. PROGNOSIS

The inertial-confinement fusion program has progressed to the point where the scientific basis and technology for the next major step is in place, that of demonstrating ICF ignition in the laboratory. When complete in about 2005, the National Ignition Facility will have this capability. Although the primary impetus for the program is defense related, this phase is also necessary for the development of inertial fusion as an electrical power source. Obviously, the next decade will be an exciting and telling period.

List of Works Cited

Decker, C. *et al.* (1997), "Hohlraum radiation drive measurements on the Omega laser," *Phys. Rev. Lett.* **79**, 1491–1494.

Haan, S. *et al.* (1995), "Design and modeling of ignition targets for the National Ignition Facility," *Phys. Plasmas* **2**, 2480–2087.

Lindl, John (1995), "Development of the indirect-drive approach to inertial confinement fusion and the target physics basis for ignition and gain," *Phys. Plasmas* **2**, 3933–4024.

LLNL (1995), *ICF Quart. Rep.* **5** (July–September).

Soures, J. *et al.* (1995), "Direct-drive laser fusion experiments with the OMEGA, 60-beam, >40 kJ, ultraviolet laser," *Phys. Plasmas* **3**, 2108–2112.

Tabak, Max *et al.* (1994), "Ignition and high gain with ultrapowerful lasers," *Phys. Plasmas* **1**, 1626–1634.

University of Rochester (1996), *LLE Rev.* **68** (July–September).

Further Reading

Berk, Herbert (1998), "Fusion, Magnetic Confinement," in: G. L. Trigg (Ed.), *Encyclopedia of Applied Physics*, Vol. 6, New York: VCH, pp. 575–607. An up-to-date review of magnetic confinement fusion research, the principle alternative approach to ICF.

Kruer, William (1988), *The Physics of Laser Plasma Interactions*, Reading, MA: Addison-Wesley. An excellent and physically intuitive description of the interaction physics between plasma and intense laser light.

Lindl, John (1995), "Development of the indirect-drive approach to inertial confinement fusion and the target physics basis for ignition and gain," *Phys. Plasmas* **2**, 3933–4024. A comprehensive technical review of the indirect-drive ICF approach.

Ripin, Barrett H. (1993), "Fusion, Inertial Confinement," in: G. L. Trigg (Ed.), *Encyclopedia of Applied Physics*, Vol. 6, New York: VCH, pp. 545–573. A summary of the basic physical principles of inertial confinement fusion.

Soures, J. *et al.* (1995), "Direct-drive laser fusion experiments with the OMEGA, 60-beam, >40 kJ, ultraviolet laser," *Phys. Plasmas* **3**, 2108–2112. A description of the current status of the direct-drive ICF research approach.

Tabak, Max *et al.* (1994), "Ignition and high gain with ultrapowerful lasers," *Phys. Plasmas* **1**, 1626–1634. A good general description of the fast-ignitor concept.

FUSION, MAGNETIC CONFINEMENT

(An Addendum to *Encyclopedia of Applied Physics*, Volume 6, pages 575–607.)

H. L. Berk, *Institute for Fusion Studies, The University of Texas at Austin, Austin, Texas, U.S.A.*

INTRODUCTION

In the 1990s there has been significant progress in the development of the tokamak concept and serious study by an international research group [the International Thermonuclear Experimental Reactor (ITER) team] on the feasibility and design of an experiment to demonstrate fusion power production. In this Addendum we present an overview of the status of tokamak research in relation to achieving this goal.

1. POWER-PRODUCING TOKAMAK EXPERIMENT

1.1 Magnetic Field

Recall that a tokamak is a toroidally shaped plasma in a magnetic field as shown in the schematic diagram of Fig. 1. Its cross section is shown in Fig. 2, with dimensions envisioned for a 1500-MW power-producing experiment. The dominant magnetic field is in the toroidal direction and is established by toroidal field (TF) coils that poloidally loop the plasma. A smaller, poloidally directed magnetic field arises from toroidally directed current flowing in the plasma itself and toroidal currents flowing in poloidal field coils that are outside and parallel to the plasma (currents external to the plasma control the shape and position of the plasma). All these currents produce magnetic field lines that circulate in the toroidal and poloidal directions and form magnetic flux surfaces. The q value (a parameter that will be frequently referred to) of a surface is the number of times a field line circulates in the toroidal direction in one poloidal circuit. The surfaces are represented by closed surfaces in Fig. 2.

1.2 Central Solenoid

The central solenoid is used to supply the toroidal voltage that sustains the plasma current. The voltage arises from the time rate of change of the entrained magnetic flux in the coil of the central solenoid. However, without other current drive methods being imposed, the establishment of the plasma equilibrium and the sustainment time of the discharge are limited by the time integral of the emf that is obtained by reversing this magnetic flux. This time is expected to be about 1000 s in a fusion-producing plasma.

ISBN 3-527-29308-6

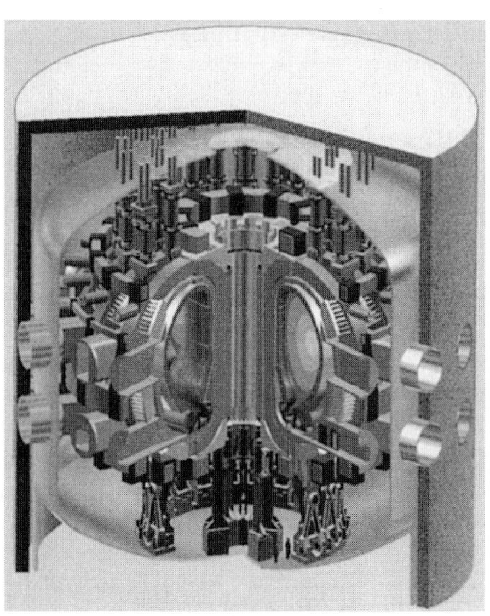

FIG. 1. An artist's conception of a tokamak reactor, with cutout showing plasma discharge.

1.3 Fusion Burn

Charged particles are trapped in continuous spirals that closely follow the entrapped field lines. This containment allows the particles to be heated to temperatures where fusion reactions occur. For a deuterium-tritium mixture, this temperature is in the range 10 to 25 keV where we use temperature in units of energy (where $1\,\text{eV} = 1.16 \times 10^4\,\text{K}$). When deuterium and tritium fuse ($d + t \Rightarrow {}^4\text{He} + n + 17.6\,\text{MeV}$), the 14-MeV neutrons leave the plasma and are absorbed by the walls. Their energy is to be used to produce steam that drives an electrical power plant. In addition the neutrons are needed to breed tritium (tritium is radioactive with a half-life of 12 years and is available only in limited amounts from fission reactors). Hence it will be necessary to surround a fusion reactor with a lithium blanket (see FUSION TECHNOLOGIES) to breed tritium ($n + {}^6\text{Li} \Rightarrow t + {}^4\text{He} + 4.8\,\text{MeV}$; $n +$

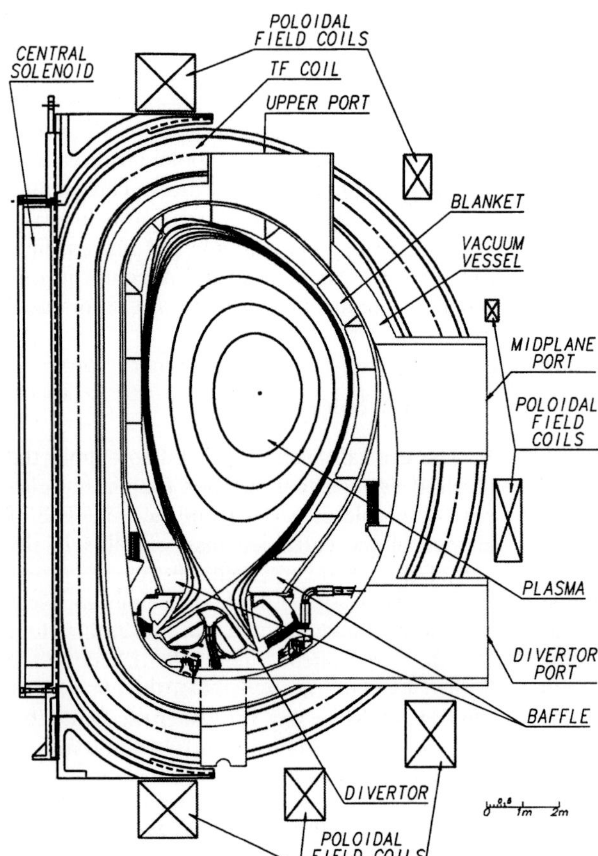

FIG. 2. Cross-section of a typical design of a tokamak for a fusion-producing experiment which has been produced by the ITER team.

$^7Li \Rightarrow t + {}^4He + n' - 2.5\,MeV$). The 3.5-MeV charged alpha particles (^4He) should be retained in the tokamak and their energy transferred to the background plasma by collisions. Then, even without external heating, a "burn" is established, whereby the plasma fuel remains hot while losing energy by bremsstrahlung, atomic radiation, and convection of heat to the "cold" solid walls. Ignition occurs when the fusion-produced alpha particles supply the required heating instead of an external source.

1.4 Divertor

Figure 2 also illustrates a single divertor configuration where there is a special magnetic flux surface called the separatrix (the one where the poloidal field lines appear to cross). Field lines beyond the separatrix (the scrape-off layer) impinge on structure called divertor plates designed to handle large escaping heat loads (charged particles that diffuse outside the separatrix strike here). Gas formed from the recombined plasma is to be pumped away from the surrounding divertor chamber into a closed tritium-recovery system.

2. OPERATION OF A BURNING PLASMA

2.1 Lawson Condition

In order for fusion reactions to be thus sustained, the confinement properties at fusion temperatures must satisfy the Lawson criterion, $n\tau_E \approx 2 \times 10^{20}\,m^3\,s$ (here n is the d-t ion density and τ_E the overall energy lifetime). The energy lifetime is determined by transport arising from particle collisions and plasma-induced short-wavelength turbulence. At fusion temperatures, the density is limited by magnetohydrodydamic (MHD) stability. This sets a limit on the parameter β, the ratio of the average pressure (the particle density times the temperature) to magnetic field pressure. In standard tokamak operation, where the current density is largest in the plasma's center, there is a further MHD limitation that prevents the q value (which is inversely proportional to the current density) from being too much less than unity. In the standard case the β is limited to about 5%. In addition there exists a less well understood empirical density limitation, perhaps associated with charged-particle recombination, known as the Greenwald limit, which seems to limit the maximum density to a level proportional to the mean current density of the discharge.

2.2 Magnetic Field Limits

Another limitation is set by the magnitude of the magnetic field that can be imposed. With pulsed operation ($\sim 10\,s$), TF coils with about a 20-T peak field (producing a 13-T field at the plasma center) have been designed. However, superconducting magnets are generally considered to be needed for a fusion experiment designed to operate in quasisteady conditions of order 1000 s. Then peak fields are limited to 12 T, which allows a 6-T field at the plasma center.

2.3 Empirical Scaling

The energy lifetime of a typical tokamak discharge is uncertain as it is primarily governed by incompletely understood turbulent processes. Thus the extrapolation of tokamak parameters to larger-size reactor regimes is typically based on empirical scaling to the experimental database, obtained from many different installations. A reasonable fit to many experiments is achieved by assuming that the energy lifetime is proportional to algebraic powers of the physical parameters defining the experiment (such as plasma current, magnetic field, plasma heating rate, etc.), with the value of each power the same for all the experiments. A general conclusion from this procedure is that confinement improves with increasing plasma current but degrades with increasing heating power. When this law is applied to an ignition experiment, the extrapolated lifetime is compatible with achieving ignition, but with some uncertainty due to the variance in the data.

2.4 Heat Load

The dimensions shown in Fig. 2 arise as a result of conforming to the constraints mentioned above. They lead to a total fusion-power production in the gigawatt regime (Post and Uckan, 1992). One fifth of this power (from alpha particles) directly heats the plasma. The heat exhausts on plasma-facing material with a rather modest average power load of about 0.2 MW m^{-2}. However, should most of this heat impinge on the divertor plate, an unacceptable 40-MW m^{-2} heat load would arise.

Operation with a divertor is viewed as essential. It has been found to be extremely important to control the impurities that enter the plasma, to design efficient pumping methods, and to establish the conditions for optimum plasma confinement in the so-called *H*-mode. The *H*-mode of operation improves the energy confinement time a factor of 2 or more compared to an alternative mode of operation, the so-called *L*-mode (see below for further discussion).

A fix to this heat exhaust problem is to radiate away most of the exhaust energy. For example, if 1500 GW of fusion power is being produced, then the plasma is directly heated by the 300 MW in alpha particle production. Roughly 100 MW radiates as bremsstrahlung. Impurity seeding at the plasma edge and diverter throats is required to radiate away about 150 MW so that the remaining 50 MW of heat flux in the escaping plasma allows a tolerable power-density load of less than 10 MW m^{-2} at the divertor plates. Favorable results are now available in several tokamak experiments that demonstrate that the heat load on the divertor plates is indeed reduced by such impurity radiation shielding, in a so-called semidetached mode of operation, while maintaining relatively good purity of the plasma. An important associated issue is whether optimal *H*-mode confinement properties can be achieved in semidetached operation. Most experiments observe confinement degradation, but there have been results that demonstrate, in semidetached operation, confinement properties close to that in typical *H*-mode operation.

2.5 Pumping

Pumping in the divertor region is extremely important in reactor operation. The helium accumulating in the core must be pumped away. Otherwise the burn quenches after ignition is achieved, as helium ash is replacing plasma fuel. Experiments have shown that the ratio of helium to energy lifetime is about 5, and with pumping and *d-t* gas puffing the helium accumulation can be limited to 10% of the plasma density, an acceptable level. The pumping in the divertor system is a principal control of the plasma impurities as well. The density of seeded impurities needs to be peaked near the scrape-off layer but diffuse in the plasma bulk. Experiments have demonstrated that this can be achieved.

2.6 Ignition Physics

There are several intrinsic physics issues associated with achieving ignition and burn. One is the control of the parameters of the plasma. There can be an ignition-induced instability, where once the fusion burn dominates, the plasma heating rapidly increases the plasma pressure. The increased pressure can cause the system to enter unstable MHD regimes where confinement deteriorates or is even lost. The conditions for such excursions are believed to be understood and in principle can be avoided. Ultimately the appropriate operation must be experimentally demonstrated.

Another burn issue is the physics associated with alpha-particle confinement. To achieve burn the alpha kinetic energy must be absorbed by the plasma. It is possible for energetic alpha particles to diffuse to the plasma edge and then embed themselves into the plasma-facing material. It is estimated that a limit of only 5% of the generated energetic alpha particles can be lost in this way if serious wall deterioration is to be averted.

Energetic alpha particles near the edge are primarily lost as a result of complicated motion arising from the presence of toroidal magnetic field ripple resulting from the spacing of a finite number of TF coils in the machine. It has been shown that if only the energetic alpha particles born near the edge are lost, there is a tolerable loss with a reasonable number of TF coils. However, there is still a concern that alpha particles can be fed into the edge region from plasma instabilities. One possibility is that Alfvén-wave instabilities can be spontaneously generated with the alpha particles themselves the instability source. Another possibility is that alpha particles are lost through MHD activity of the plasma, which either directly drives the alpha particles into the loss region or indirectly triggers the Alfvén instability by causing a rearrangement of the alpha-particle pressure profile. Such scenarios, with loss, have been observed in experiments that use energetic particles arising from neutral-beam injection or ion-cyclotron rf heating. In the TFTR tokamak, where up to 10 MW of fusion power has been produced, the alpha-particle–driven Alfvén wave has been observed, but in these cases the observed saturation level is too low to cause significant energetic particle loss. Theory predictions indicate that an alpha-particle–driven Alfvén instability is unlikely to arise for many operating parameters, but it may prevent some otherwise desirable set of plasma parameters from being achieved.

3. PLASMA CONFINEMENT ISSUES

3.1 *H*-mode

Clearly energy-confinement time is a critical parameter in determining the feasibility and design of fusion-power production. The plasma properties can be significantly altered if turbulence conditions are changed. This appears to be the case when there is a transition from *L*-mode to *H*-mode operation. In *H*-mode a pedestal in the temperature appears at the edge of the plasma with a radial thickness typically of an ion orbit width. In this layer there is a flowing plasma with radial shear. A widely accepted explanation of why *H*-mode is established (Itoh and Itoh, 1996), is that the escaping plasma causes a torque to be generated that drives a shearing flow. The shear in turn suppresses fluctuations in the pedestal region by either "ripping" apart the coherence of the fluctuations so that they saturate at a low level, or preventing unstable fluctuations from appearing. It is experimentally observed that a threshold power level is needed to establish an *H*-mode, and this observation is consistent with the hypothesis for the cause of shear flow. The determination of this level is still uncertain; there is no generally accepted quantitative theory, and empirical studies have too much scatter to predict whether, in the reactor operation described above, there is enough power flowing through the edge to establish an *H*-mode. This is an important issue that is to be resolved in ongoing studies.

3.2 Transport Barriers

The studies of the *H*-mode have been crucial in the discovery of new ways to improve energy confinement (Burrell, 1997). It has been found in many tokamaks that by applying torques in either narrow or wide regions of the plasma, by use of either neutral beams or rf sources, transport barriers can be established in the core of the plasma. High temperatures have been established in the central region, with a large drop in the temperature taking place in the transport barrier. It has been further found that establishing this barrier is aided by having a hollow current profile, which leads to a reversed q profile that has a local maximum at the plasma center and a minimum q value, typically ~ 2, further out radially. The thermal barrier is generally established near the minimum-q position. Stability theory indicates that in the region where q is radially decreasing, the strength of the diamagnetic-drift– and magnetic-curvature–driven instabilities (see FUSION, MAGNETIC CONFINEMENT) are reduced. Then the flow shear that is established is enough to stabilize the modes completely. Some experiments report that the residual transport can be explained by transport due only to collisional effects.

The major issue with modes that have hollow current profiles is that they are transient, unless a tailored emf can be applied from auxiliary-current drive methods involving rf or neutral beam sources. It needs to be demonstrated that power requirements for such a drive are compatible with reactor operation.

Another benefit from the hollow current profiles is that they lead to plasmas where a large fraction of the current is in the form of bootstrap current (see FUSION, MAGNETIC CONNEMENT), which does not decay in absence of voltage. If this can indeed be established, then true steady-state operation in a reactor with non-Ohmic current-drive power is feasible as it leads to acceptable power requirements for driving the nonbootstrap current component. Steady-state operation is extremely desirable. It avoids what may be delicate plasma control issues associated with startup and shutdown, as well as thermal stresses associated with repetitive operation that lead to material deterioration.

3.3 Transport Theory

There has been considerable development in theory to achieve transport predictions based on plasma-physics principles. The theories used are still in a post-processing phase, i.e., they explain the temperatures achieved in a large number of experiments but have not predicted the temperature of any experiment before it was performed. One reason is that a theory for energy transport is easier to develop than a theory for particle transport. Hence to determine temperature, a density profile has to be taken from experiment. In addition the transport processes in the core of the plasma are thought to differ from the processes at the edge, and typically the various predictive methods make allowances for less understood edge transport processes either by using experimentally observed temperatures at the edge or by adjusting edge transport coefficients to match empirical observation. As a result many theory-based transport calculations produce good correlation of the predicted and observed energy lifetimes for a variety of experiments (Connor *et al.* 1997). However, when the theories are extrapolated to reactor-type plasmas, significant deviations are predicted. One theory predicts that gigawatt fusion power operation can be expected only if a high edge temperature, 4 keV at the inner edge of the pedestal, can be achieved, while another theory indicates that such fusion power levels can be achieved even with a modest edge temperature. In such a case the former theory would predict a fusion power output that is about a factor of 2 larger than the external heating input, which would give a few hundred megawatts of fusion power when about 100 MW input power is available.

The issue of attaining a predictive theory is of high priority, and tests to resolve existing discrepancies are still in progress. Nonetheless, it is clear that there has been a great deal of basic progress in the understanding of the turbulent transport problem, and we can expect the theoretical reliability of this approach to improve rapidly.

3.4 MHD Considerations

If the plasma's MHD stability characteristics can be improved it can lead to better reactor properties. At a given temperature, a greater density can be achieved, making it easier to satisfy the Lawson criterion. One of the important parameters controlling MHD stability is the plasma shape. Note the vertical elongation and the triangular shape of the outer flux surfaces in Fig. 2. The decreasing pressure profile at the pedestal is a destabilizing effect on the outer part of the surface and is a stabilizing effect on the inner part of the surface. The triangularity allows stronger weighting from the stabilizing inside than would arise without this type of shaping. Then the temperature drop at the pedestal where the H-mode formed can be increased to a limit apparently determined by MHD oscillations known as elms. As one example we note that in 1996 the best bulk confinement parameter $n\tau_E T = 1.5 \times 10^{21}\,\mathrm{m}^{-3}\,\mathrm{s\ keV}$ was achieved in the JT-60U experiment, in part as a result of shape optimization.

There are many experiments that achieve higher β values, but they frequently have MHD fluctuations that reduce the energy-confinement time or cause a disruption where the plasma discharge rapidly disintegrates. One route to a disruption is through a so-called locked mode, where a perturbed wave "attaches" to an imperfection of the stationary magnetic field, which can cause deterioration of the discharge. Locked modes are avoided with a sufficiently large plasma rotation, and hence enough rotation needs to be generated in a fusion power-producing experiment.

There is a great deal of optimism that discharges with reversed q profiles in the center can attain significantly increased β values, particularly when a close-fitting conducting wall is used. However, in this case slowly growing modes still arise on account of the finite wall resistivity. Such modes can in principle be controlled by feedback with low-power sensing circuitry. More experiments on this topic are needed, and appropriate reactor designs still need to be developed.

3.5 Reliability

The demonstration of reliability is another challenge for tokamak operation. In conventional operation, pulsations, such as sawtooth oscillations and elms, are common. Typically the sawteeth arise when the q value of the discharge is somewhat below unity and the pressure profile peaks too strongly within the $q = 1$ surface. The time interval between sawtooth pulses is typically a fraction of the overall current diffusion time (the time interval is of order 100 ms in the larger present-day machines and is expected to be about 1 s in a tokamak like that described in Fig. 2). The pulsation flattens the pressure profile within the $q = 1$ surface, but it also has a beneficial effect of preventing further steepening of the central current profile, a destabilizing tendency. These pulsations generally cause rather small perturbations at the edge so that ordinary sawteeth by themselves are not likely to induce heat-flow problems at the divertor plates. However, in some operations energetic particles induce a change of plasma behavior that causes the sawtooth oscillations to be suppressed for long time intervals. Unfortunately, at times, a giant sawtooth pulse appears that disturbs edge conditions and can even disrupt the entire plasma discharge. Such behavior must be controlled in reactor operation.

The elms, mentioned earlier, are a plasma edge perturbation that cause enhanced heat pulses to outflow at regular time intervals. There is an acceptable type of elm operation, so-called grassy elms, where the pulses are frequent but low level, which leads to heat loads that are readily handled, good global confinement characteristics, and an observed advantageous property of keeping impurities from accumulating in the main part of the plasma. However, elms arise that give larger heat pulses and less favorable plasma confinement. A further difficulty is that a large elm pulse can cause termination of H-mode operation, which will then lead to a significant reduction of the fusion output unless active counter measures to restore H-mode operation are found.

Large disruptions are known to arise in tokamaks, leading to the loss of the plasma discharge. Frequently these disruptions are uncontrolled and are more likely to arise for tokamaks that operate at the limits of acceptable operation. Of concern is that large disruptions can cause damage to plasma-facing wall components due to localized heat deposition, the formation of so-called halo currents where the plasma current penetrates the surrounding conducting wall, and the formation and subsequent wall bombardment of relativistic "runaway" electrons accelerated from the inductive electric field.

Studies are in progress to develop reliable ways that disruptions can be guided to a "soft" landing, where the heat is distributed benignly over the entire surface of plasma-facing walls (for example, from injection of so-called killer pellets, that cause the heat to radiate), and the current in the remaining cold plasma, together with its associated magnetic field energy, is allowed to dissipate on the time scale of a second, a time interval long enough so that the inductively produced electric field is too low to cause runaway electrons.

4. ALTERNATIVE APPROACHES

4.1 Alternative Tokamak Approaches

On the basis of the knowledge known today there is a high confidence level that a fusion power-producing experiment can be designed to operate, at a minimum, in a fusion-power amplification mode where several hundred megawatts of power are produced. A more challenging goal is to achieve, reliably, quasi-steady burn conditions or a high enough energy amplification level to produce gigawatt power levels. Such success requires maintaining optimal tokamak performance (such as H-mode operation, steady alpha-particle heating, limited parameter excursions from pulsations, etc.) in the standard-type discharge discussed above.

Alternative directions of research will proceed either in conjunction with, or as an alternative to, the direct implementation of a fusion power experiment. Perhaps the most exciting development in the mid-1990s is the demonstration that turbulence levels can be dramatically diminished, leading to enhanced confinement, as a result of the combination of plasma flow shear and establishing q profiles that decrease radially over much of the discharge. The future challenge in the operation of this type of discharge is to demonstrate that such conditions can be steadily maintained with modest external power control that extrapolates properly to reactor-size plasmas. The intrinsic bootstrap current in such discharges is expected to be high enough that the power needed to maintain true steady-state reactor operation can be achieved. Significantly, non-inductive current-drive experiments in the Tore-Supra and in other tokamaks have already proven that plasma discharges can be sustained indefinitely, but the needed efficiency is still to be demonstrated.

Future research will also focus on achieving higher plasma pressure operation by implementing conducting-wall stabilization with feedback to control slowly growing resistive wall instabilities. The new methods of obtaining enhanced confinement may also lead to alternative designs for plasma shaping. Other studies are examining whether plasma confinement properties can be improved by increasing a tokamak's minor-to-major radius or major-to-minor radius ratio. Preliminary arguments and data give supportive reasons why such approaches can be beneficial.

There is extensive interest in building an experiment that demonstrates ignition and a short burn, but that does not address long-time reliability issues (Coppi *et al.*, 1994). By using the highest possible magnetic fields, plasmas can be established that are well within intrinsic parameter limits. Hence in such an experiment the likelihood of achieving a proof-of-principle controlled burn is high. Future improvements in the development of large superconducting magnets to the 20-T range would have a dramatic impact on the progress of controlled magnetic fusion. Then long-time burn experiments can be performed at high fields, allowing engineering issues associated with power production to be tested.

Another alternative is to attempt to obtain net fusion-power operation when the *d-t* temperature is greater than the electron temperature. It is significant that such operation characterizes experiments that have achieved some of the highest Lawson parameters in JT-60U, JET, and TFTR and the highest fusion power (16 MW in JET). If the $n\tau_E$ parameter can be increased another factor of 3 to 4 in the former two machines, the condition of a relatively compact reactor operation would nearly be established. There is a problem with maintaining this mode, because as confinement improves the electron and ion temperatures tend to approach each other, and in the process degrade the favorable features of this type of confinement. A novel mechanism known as energy channeling has been suggested, where through the controlled stimulation of rf waves, the alpha-particle energy directly heats the *d-t* fuel rather than the alpha particles primarily heating the electrons (as generally occurs). The technical feasibility of this idea is still under study.

4.2 Other Concepts

The most developed alternative to tokamak research is the stellarator concept (see the original article), where charged particles are confined by the three-dimensional magnetic field established by currents in external coils. This concept is intrinsically steady state; some plasma confinement regimes have been shown to be comparable to similar tokamak confinement regimes, and experiments have been free of disruptions. Of principal concern is that the achieved β values in stellarators are generally lower than in tokamaks, and in most designs a large class of energetic particles are not contained in the magnetic fields. A positive recent development has been the design of so-called quasisymmetric fields, which improves confinement of energetic particles. However, this design still needs to be made compatible with the achievement of reasonably large β limits. Another advantage of a stellarator is its typical property of a radially increasing q profile. This may allow easier access into regimes demonstrating enhanced confinement. A new large stellarator in Japan, the LHD, began operation in 1998, and it will further test the prospects for achieving high-grade plasma confinement suitable to reactor operation.

Research with different types of pinched plasmas has made significant progress during the 1990s. The simplest in conception is the *z*-pinch (see the original article), where a current is directed in the *z* direction, producing a magnetic field that winds poloidally around this direction. This concept was one of the first for which the controlled-fusion problem was investigated in the 1950s. The conclusion was that the configuration is unstable. Nonetheless, in recent years there has been appreciable progress in the study of such pinches. When they are produced from thin fibers, extremely dense and hot plasmas are produced that are useful as x-ray sources and for understanding the interaction of plasmas with solids. By using many parallel strands of fiber, it is found that for short time intervals (of the order of microseconds) uniformity in the pinch is maintained. Because of the high densities and temperatures that have been achieved, proposals have now been made for demonstrating fusion power production with this concept.

Another pinch concept, the field-reversed pinch (FRC; see the original article) is also a candidate for demonstrating fusion power production. In the latter case, it is proposed to compress an FRC to high density and temperature using an imploding metal liner that produces magagauss magnetic fields. However, both types of pinch experiments are intrinsically pulsed, and it is now too premature to tackle the difficult problem of obtaining a duty cycle that would allow the consideration of commercial fusion power production.

The reversed-field pinch (RFP; see the original article) is a pinch more like a tokamak, in that it forms a toroidal plasma with toroidal and poloidal magnetic fields. In the 1990s, significant improvement of its parameters of operation and in the understanding of its physical mechanisms of turbulence have been made. This concept has also contributed to the understanding of the plasma dynamo problem of interest in astrophysics and to the understanding of how to control locked modes, which is crucial to understand in tokamaks as well as the RFP.

4.3 Final Statement

As a result of a great deal of progress, a plan for the demonstration of fusion-power production in a tokamak has been made. A detailed design for a fusion power-producing experiment has been completed, but cost considerations are likely to cause postponement of its implementation or a downscaled design. Many scientific issues remain to be clarified and other alternative paths are likely to develop. The physics issues that need to be resolved and the goal of a long-term energy source remain challenges to the technical creativity of mankind.

GLOSSARY

Alfvén Wave: A plasma wave that may be excited by alpha particles in a fusion plasma.

Burn: The self-sustained heating of a plasma arising from the fusion reaction.

Divertor: The edge region of a tokamak where field lines divert escaping plasma away from the immediate vicinity of the main discharge.

Energy Channeling: A proposed process to allow charged fusion products, with assistance from radio-frequency excitation, to heat ions rather than directly heating electrons.

Halo Current: Currents that are partially carried by the plasma but pass through the surrounding walls.

Ignition: The process whereby plasma heating transfers from being externally supplied to being internally supplied from the fusion reactions.

Magnetic-Field Ripple: The spatial modulation in the magnitude of the magnetic field due to the discrete number of toroidal field coils.

Pedestal: A region of plasma adjacent to the edge, where the plasma temperature abruptly rises.

Semidetached: A divertor plasma condition where a large fraction of the escaping plasma has recombined into neutral atoms.

Shear Flow: Flow of plasma that varies with minor radius.

Thermal Barrier: A region in a plasma where local transport properties are suppressed.

List of Works Cited

Burrell, K. (1997), *Phys. Plasmas* **4**, 1499.

Connor, J. W. *et al.* (1997), in: *Plasma Physics and Controlled Nuclear Fusion Research, 1996*, Proceedings of the 16th Conference, Montreal, Vol. 2, Vienna: International Atomic Energy Agency, pp. 935–944.

Coppi, B., the Ignitor Project Group, (1994), *J. Fusion Energy* **13**, 111.

Itoh, K., Itoh, S., (1996), *Plasma Phys. Control. Fusion* **38**, 1.

Post, D. E., Uckan, N. A., (1992), *Fusion Technol.* **21**, 1427.

Further Reading

Aymar, R., the ITER Team (1997), in: *Plasma Physics and Controlled Nuclear Fusion Research, 1996*, Proceedings of the 16th Conference, Montreal, Vol. 2, Vienna: International Atomic Energy Agency, pp. 3–17 and 737–1002.

MODULATORS AND DEMODULATORS, ELECTRICAL

(An Addendum to *Encyclopedia of Applied Physics*, Volume 10, pages 379–391.)

KOHJI HOHKAWA, *Faculty of Engineering, Kanagawa Institute of Technology, Atsugi Kanagawa, Japan*

INTRODUCTION

With the advent of the Internet, communication systems are making an evolutional impact on people's lives all over the world. A large quantity of information in various categories, such as telephone, telegraph, computer data, and broadcast, and various services, including bi-directional moving pictures, are required to be transmitted through communication networks. Extremely high-capacity and flexible communication networks are required in order to serve these requirements. In these networks, pulse modulation schemes capable of transmitting a large quantity of data through a restricted communication band play an important role, as does the digital signal processing. This Addendum introduces modulators and demodulators widely used in up-to-date digital communication systems.

1. PULSE MODULATION METHODS FOR DATA-COMMUNICATION SYSTEMS

This section introduces advanced digital pulse modulation schemes and modulators and demodulators. Device and circuit technology are in common for various kinds of modulation schemes. The functions relating modulation and demodulation process, such as line filtering and timing extraction, represented in Sec. 1.2 are also used for other modulation schemes.

1.1 Advanced Modulation Method in Quadrature Phase-Shift Keying Scheme

Binary phase-shift keying (BPSK) and quadrature phase-shift keying (QPSK) modulation schemes have the highest performance and, therefore, are extensively used for various kinds of communication systems, such as sat-

ISBN 3-527-29308-6

ellite communication, mobile communication, and personal handy telephone systems. This section discusses modulators and demodulators for two types of advanced schemes based on QPSK:

1. offset quadrature phase-shift keying (OQPSK) modulation, which reduces signal distortion caused by nonlinear effects, and
2. quadrature quadrature phase-shift keying (Q^2PSK) modulation, which improves signal throughput.

1.1.1 Offset Quadrature Phase-Shift Keying Modulation (OQPSK). In the modulated signal using BPSK and QPSK schemes, there are abrupt phase transitions with maximum values of π at the boundaries of data. The large phase transition tends to increase signal distortion and to increase the outer band spectrum of the modulated signals when such signals pass through a relatively high-efficiency C-class power amplifier with saturation characteristics. There are several kinds of phase-shift keying modulation schemes to overcome the degradation by reducing the phase transition of the carrier at the data boundary. Offset quadrature phase-shift keying modulation (OQPSK) is the basic method among many kinds of modulation schemes with the small phase transition.

As shown in Fig. 1, the OPQSK modulator has about the same circuit structure as that for BPSK except that data in the quadrature channel are delayed by $T/2$, where T is the duration of the BPSK-modulated signal of in-phase and quadrature channels. Similarly, demodulation of OQPSK and MSK signals is done using a QPSK demodulator. In an OPQSK-modulated signal, the phase transition occurs in every $T/2$, but the signal spectrum does not increase because the operation is a simple summation. The sum signal at the phase transition point is produced by a vector sum of a continuous signal and a signal with $\pi/2$ (or $-\pi/2$) difference in phase at the transition point. Therefore, the maximum value of the phase transition reduces to $\pi/2$.

The degradation due to the phase transition at the boundary is minimized if we can realize a continuous phase condition at the data boundary. In the frequency-shift keying scheme, it is simply realized by selecting frequencies f_i to meet the orthogonal condition under which the total phase shift on each datum $(2\pi f_i T)$ is equal to a multiple of 2π (this scheme is called CPFSK and extensively used). In the quadrature phase-shift keying scheme, it is realized by shaping each signal bit in both the in-phase channel (I channel) and the quadrature channel (Q channel) by weighting as shown in Fig. 1(c). The shaping is carried out by any of several weighting functions such as sinusoidal or Gaussian. The modulation scheme using a sinusoidal weighting function is called minimum-shift keying (MSK).

Reduction of the abrupt phase transition of OQPSK makes it possible to use the high-efficiency C-class power amplifier. This feature is effective for downsizing and light-weighting battery-assisted receivers.

1.1.2 Quadrature Quadrature Phase-Shift Keying Modulation (Q^2PSK). This modulation scheme is a focus of attention because of its effective utilization of the frequency resources and is now being applied in various communication systems. Figure 2 illustrates the structure of a Q^2PSK modulator. The scheme uses quadrature phase and an orthogonal shaping function as is used for the MSK. However, it encodes four bits per symbol, compared to two for conventional QPSK, and twice the data per modulation interval is accommodated. On account of the existence of an abrupt change of phase in the modulating signal as shown in waveform (*d*), this scheme requires an increase of 1.6 dB transmission power in order to provide an identical bit error rate for the same bandwidth. However, this scheme can obtain twice the throughput of QPSK.

A Q^2PSK modulated signal has both a phase transition and an amplitude variation in the overlap wave shape. In order to reduce the degradation due to nonlinear effects due to the power amplifier, a linearized saturation power amplifier, which has a saturation characteristic to the peak value without degradation on linearity against a wide range of input signal level, has been developed. Basically, this characteristic is attained by an A-class or AB-class amplifier whose gain is controlled according to the input signal level by some means, such as feed-forward technology, predistortion, and postdistortion. The linearized amplifier tends to be inefficient from the point of view of power consumption, but is now essential for a communication system having an abrupt

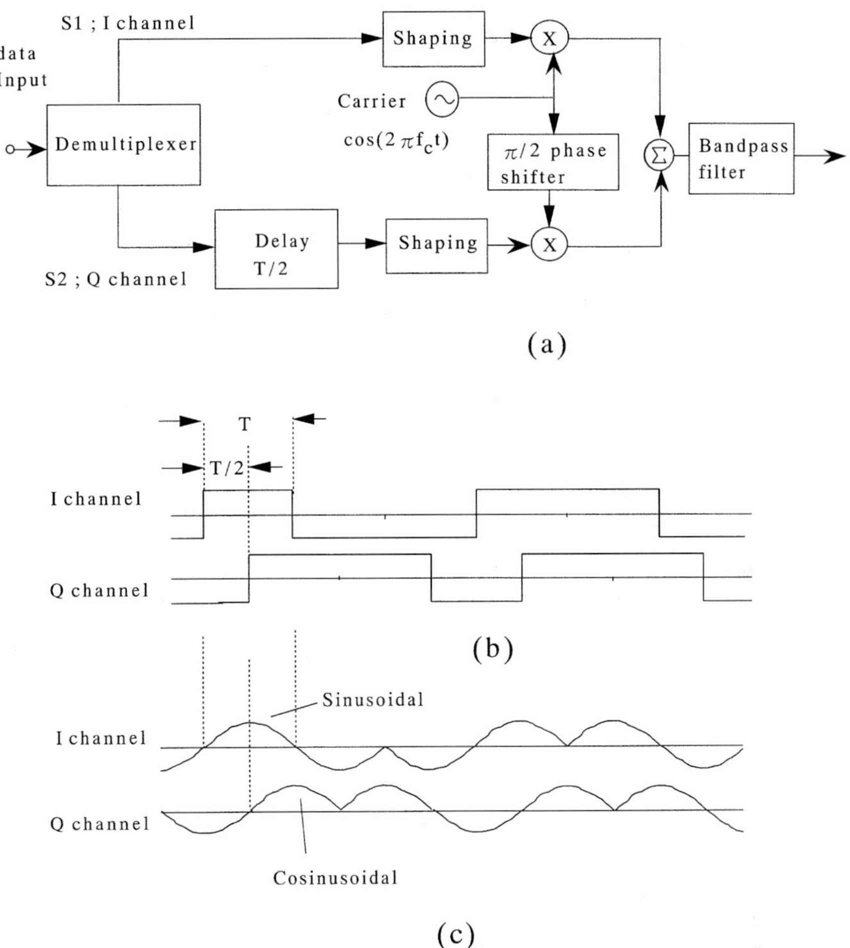

FIG. 1. OQPSK modulation scheme: (a) modulator, (b) OQPSK signal wave forms, (c) MSK signal wave forms. In (a), ordinary OQPSK needs no shaping function. OQPSK with shaping using a half-cycle sinusoidal weighting function is called MSK. (b), (c) Wave forms for input data with the duration time of $T/2 = T_b$ and sequence "11010011110". The I-channel signal is modulated by a cosinusoidal carrier and the Q-channel by a sinusoidal one.

phase shift or a nonuniform amplitude envelope in its modulated signals. This technology is also applied on systems with modulation schemes presented in the following sections.

1.2 Digital Pulse Amplitude Modulation (PAM-DSB, PAM-SSB, and PAM-VSB)

This section introduces efficient digital modulation schemes for transmitting binary data through a restricted communication channel by using digital pulse amplitude modulation (PAM).

1.2.1 Modulator and Demodulator. Figure 3 illustrates a basic circuit configuration of a PAM modulator and demodulator. At the sending end, the input data sequence is subdivided and mapped onto a corresponding amplitude level of a multiple-level (M-ary) digital signal. The modulated signal is fed to the filter and transmitted through a baseband communication channel or modulated to form a bandpass data signal with a carrier using the conventional analog amplitude modulator: double sideband (DSB), single sideband (SSB), or vestigial sideband (VSB).

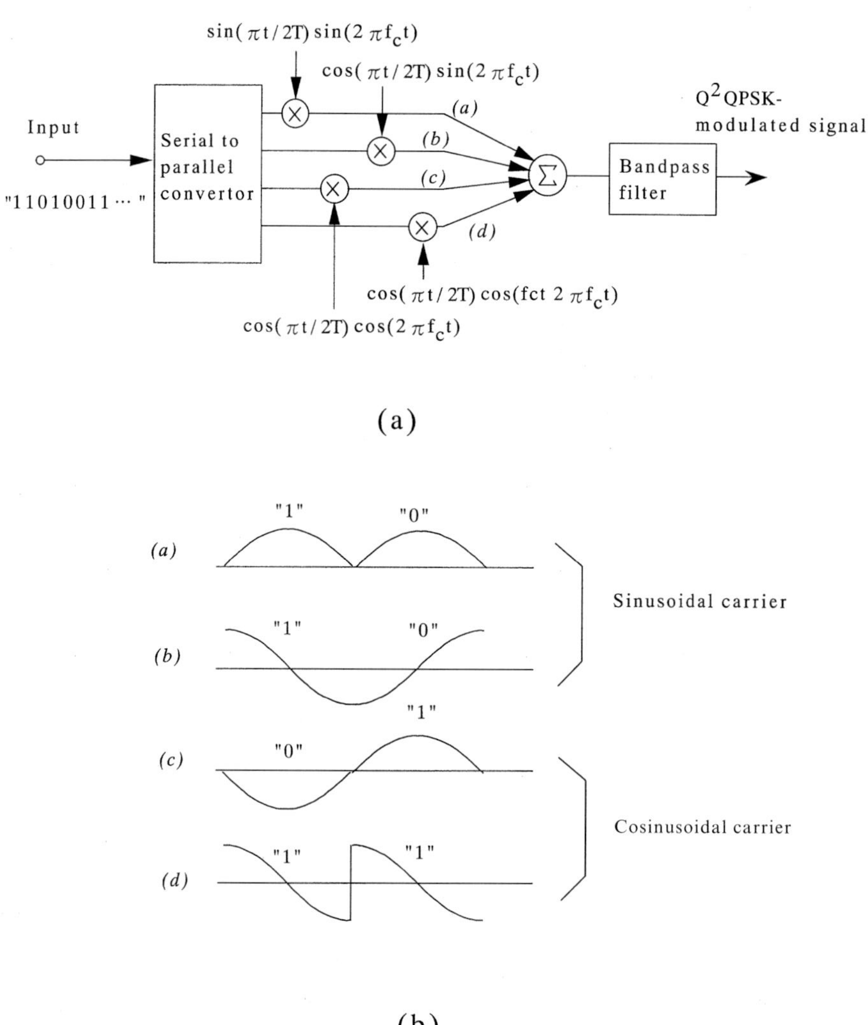

FIG. 2. Modulator for quadrature quadrature phase-shift keying.

1.2.2 Timing Extraction. At the receiving end, the received signal must be sampled periodically at the symbol rate and the signal level must be detected by a decision circuit. We need a clock signal to perform the periodic sampling. One method for extracting the timing signal is for the transmitter to transmit simultaneously the clock frequency $1/T$ or a multiple of $1/T$ along with the information signal. The receiver may simply extract the clock signal by employing a narrow bandpass filter tuned to the transmitted clock frequency. This method has the advantage of being simple to implement but has several disadvantages, one of which is that the transmitter must allocate some of its available power and bandwidth to the transmission of the clock signal. There are a number of methods for extracting the clock from the received signal. Figure 4 shows typical circuits: In (a), the clock signal is detected directly using harmonics caused by nonlinear elements. In (b) the clock signal is generated by a VCO whose phase is controlled by the position of the input signal.

1.2.3 Equalization Process. The received signal is distorted by the communication channel by several factors, such as nonlinear amplitude and phase characteristics of the channel, or echo signal reflected at the discon-

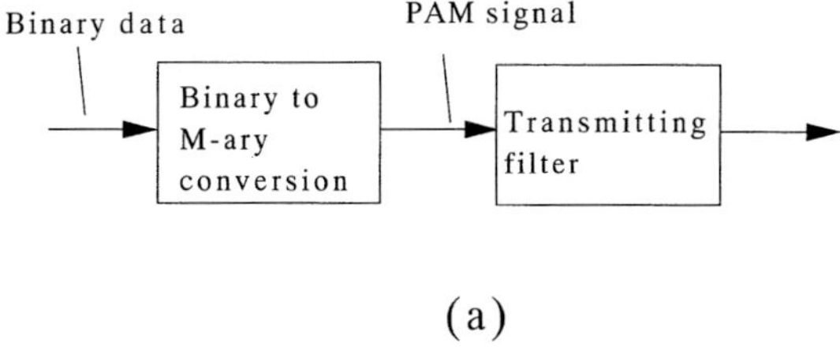

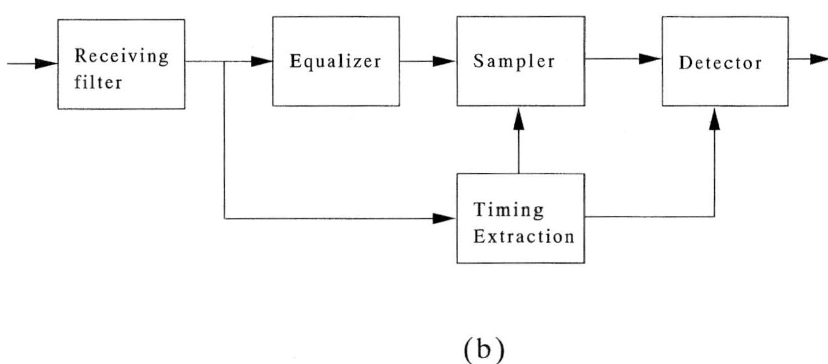

FIG. 3. (a) PAM modulator, (b) PAM demodulator.

tinuity. This results in intersymbol interference at the output of the demodulator and leads to an increase in the probability of error at the detector. A channel equalizer is used for correcting or undoing the distortion. If the channel is precisely set and its characteristics do not change with time, we can design the equalizer easily. One example is a linear equalizer, which uses a finite–impulse-response filter or a transversal filter, as shown in Fig. 5. We often encounter channels whose characteristics are unknown or change with time. In these circumstances, the channel equalizer tracks the variations of the channel and adopts its coefficients to reduce the intersignal interference dynamically.

1.3 Quadrature Amplitude Modulation (QAM)

This modulation scheme is also called quadrature amplitude-shift keying. It has high efficiency and is extensively used in the modem in computer networks and various kinds of data communication systems including mobile radio, satellite communication, and the information superhighway. In this method, a single pulse in modulated signals contains multiple bits of information in the form of phase and amplitude. This scheme may be viewed as a form of combined digital amplitude and digital phase modulation.

The QAM signal employs two quadrature carriers, $\cos 2\pi f_c t$ and $\sin 2\pi f_c t$, each of which is modulated by an independent sequence of information bits to form a digital PAM signal. Adapting N-level PAM, we can transmit $M = N \times 4$ signal points with one PAM symbol. Therefore, the bandwidth required can be reduced to $1/(2 + \log_2 N)$ times that required for the ordinary BPSK system. A QAM system with up to 256 data points M is being used in an existing system.

1.3.1 Symbol Constellation and Mapping. In this scheme, information bits are mapped onto a two-dimensional complex signal space. Figure 6 illustrates examples of allocation of data points in signal space for an $M = 16$ QAM. These figures are called *symbol constellations*, and *mapping* is the procedure for assigning data symbols onto data points of the

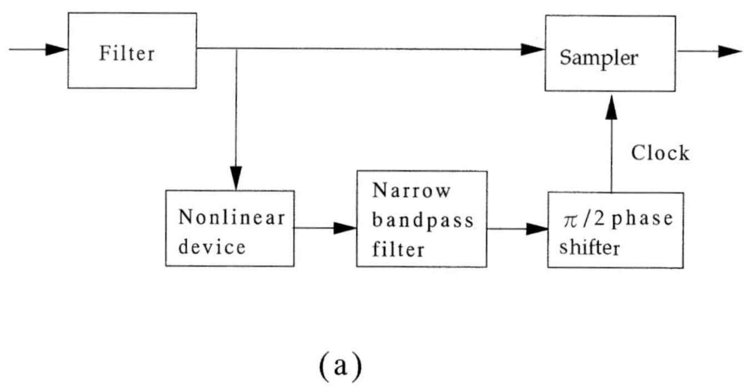

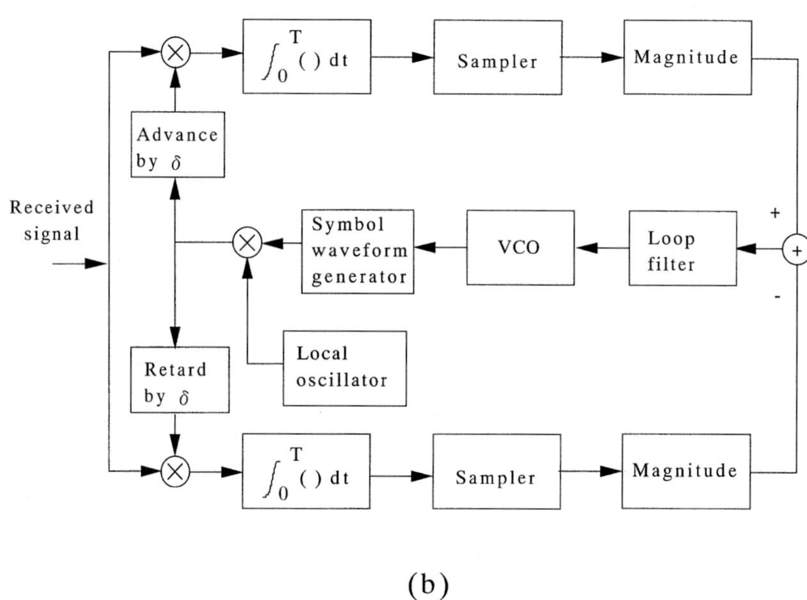

FIG. 4. Timing extraction for PAM. (a) Spectral-line method. Since the received signal has no power spectrum at the clock frequency, it is fed to the nonlinear circuit to generate second-harmonic components of $1/T$ (Hz) and higher harmonics. The clock signal is extracted by a narrow-bandpass filter. (b) Early-late gate synchronizer. In this method, the input signal is correlated by two correlators with a reference signal over the symbol interval $1/T$; one of the correlators starts integration δT early relative to the estimated optimum sampling time and the other starts it δT late relative to the estimated optimum sampling time. The optimal sample point is found by adjusting the phase of the voltage-controlled oscillator that generates the estimated optimum timing of the reference signal so as to make the peak values of the two correlated wave forms coincide with each other.

constellation. These procedures strongly affect system performance.

On designing a constellation, one must consider the following factors:

1. The Euclidian distance between phasors (coordinates on the complex plane) should be minimized in order to realize the characteristic noise immunity of the scheme.
2. The minimum phase rotation between phasors, which determines phase jitter immunity, should be sufficient to stand up against clock recovery imperfection and channel phase rotation.

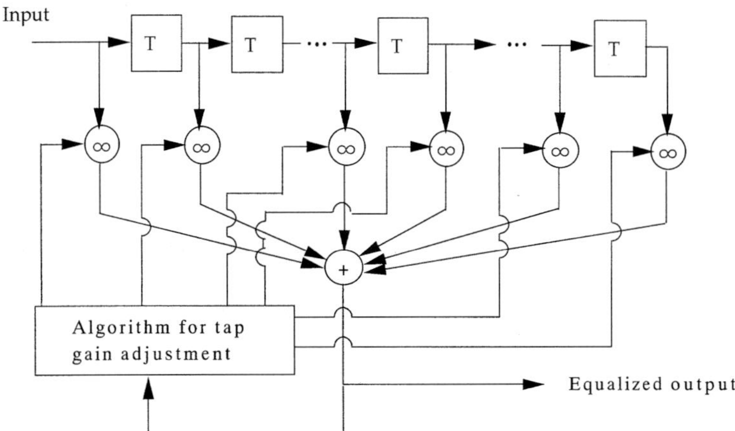

FIG. 5. Equalizer based on a transversal filter. In this circuit, tap coefficients are adjusted so as to realize the inverse frequency response of the channel.

3. The ratio of peak to average phasor power should be a minimum to ensure robustness against nonlinear distortion.

The rectangular constellation with Gray code mapping procedure as shown in Fig. 6(a) has merits with respect to power and signal-to-noise ratio. Systems using this scheme with the numbers of points up to 256 have been developed and are widely used as the standard modulation schemes for communication systems, such as modems. Figure 6(b) shows a circular (or star) constellation. This type of constellation does not have the equidistance condition between phasors. However, it has the advantage of being capable of employing a noncoherent demodulation process, which does not require complex carrier regeneration and AGC circuits. This scheme is extensively used for radio communication systems with a Layleigh fading channel, in which undesired multiple-echo signals with time-varying characteristics are superposed onto the desired receiving signal.

1.3.2 Modulator and Demodulator. Figure 7 shows a block diagram of a modulator and demodulator for a QAM signal with a rectangular constellation. In the modulator, the input data sequence is subdivided and mapped into two digital signal sequences whose amplitudes are modulated according to the input binary sequence, using the same circuit as that used for the PAM modulator. The two PAM signals produced are then modulated by sinusoidal carrier signals whose phases differ by $\pi/2$, using the same circuit as that used in a quadrature phase-shift keying (QPSK) modulator. In the demodulator, the received signal is divided into two PAM signals by multiplying the quadrature carriers in the same way as in the QPSK demodulation circuit. In the QAM demodulator, the same circuits or circuits based on the same principles as those of QAM and QPSK demodulators are employed to perform basic processes, such as synchronization and equalization.

In the circular constellation for 16-point QAM, the modulation process is performed through the encoding process shown in Fig. 8. The demodulation process is a differential decoding process of the modulated signal: The first signal is detected by comparing the signal amplitude at the data timing and that at the previous data timing. If there is a large change in amplitude then the first signal b1 is defined to be "1", and otherwise "0". This process is done by two adaptive threshold comparators. The remaining three signals are decoded by detecting the phase difference between the data and the previous data.

2. MODULATOR AND DEMODULATOR FOR SPREAD-SPECTRUM COMMUNICATION SYSTEMS

Spread-spectrum (SS) signals for digital communication were originally developed and

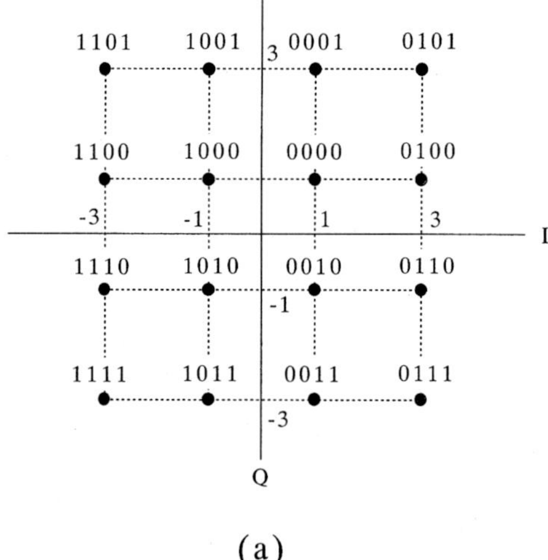

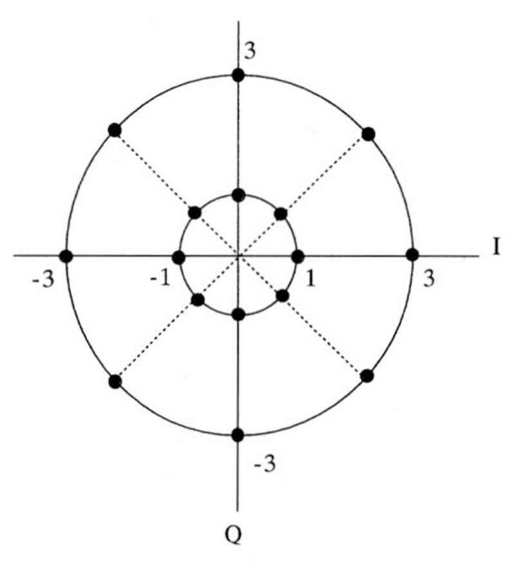

FIG. 6. Symbol constellation for QAM signal. (a) Rectangular constellation, so called from its shape in the signal space. This constellation has equidistant constellation points arranged in such a way that the average energy of the phasors is maximized. This scheme provides the higher signal-to-noise ratio to achieve the same bit error rate (BER). From the point of view of mapping, this example employ so-called Gray code mapping, in which the Hamming distance between a phasor and its closest neighbor is always 1. This procedure ensures that whenever a transmitted phasor is corrupted by noise that is incorrectly identified as a neighboring constellation point, the demodulator chooses a phasor with a single bit error. This minimizes the error probability. (b) Circular constellation. An example of the mapping procedure is illustrated in Fig. 8.

used for military communication to provide resistance to jamming, or to hide the signal by transmitting it at low power and thus making it difficult for the unintended listeners to detect its presence in noise. They are also useful for code-division multiple-access communications (CDMA) when two or more transmitters use the same common channel to transmit information, and they allow reliable communication through extremely noisy communication channels in which systems with conventional modulation schemes are useless. In addition to the robustness against interference, this modulation scheme has the excellent property of efficient usage of the limited bandwidth. Because of its features, this technology is expected to be useful for, and is now beginning to be applied for, a variety of civilian communication systems including satellite, mobile vehicle, and inter- and intra-office communication.

To overcome the problem of intentional or unintentional interference, the bandwidth of a transmitted signal is increased in this modulation scheme so that the bandwidth expansion factor W/R is much greater than unity (where W is the transmission bandwidth and R is the baseband data bandwidth). In addition, the information signal at the modulator is spread in bandwidth by means of a code (channel code) that is independent of the information sequence. The code has a pseudorandom property, so that it appears to be random to receivers other than the intended receiver, which uses the code to demodulate the signal. The usage of the channel codes allows great freedom in the choice of low interuser-interference codes. Proper choices of channel codes make it possible for a large number of users to communicate simultaneously through a limited bandwidth.

There are many kinds of modulation schemes in which the direct sequence scheme and the frequency-hopping scheme are generally used.

2.1 Direct-Sequence Spread-Spectrum Communication

A modulator for accomplishing the spreading in the direct-sequence (DS) scheme is shown in Fig. 9. Here, we assume that the binary information signal is at a rate R $(= 1/T_{\mathrm{b}},$

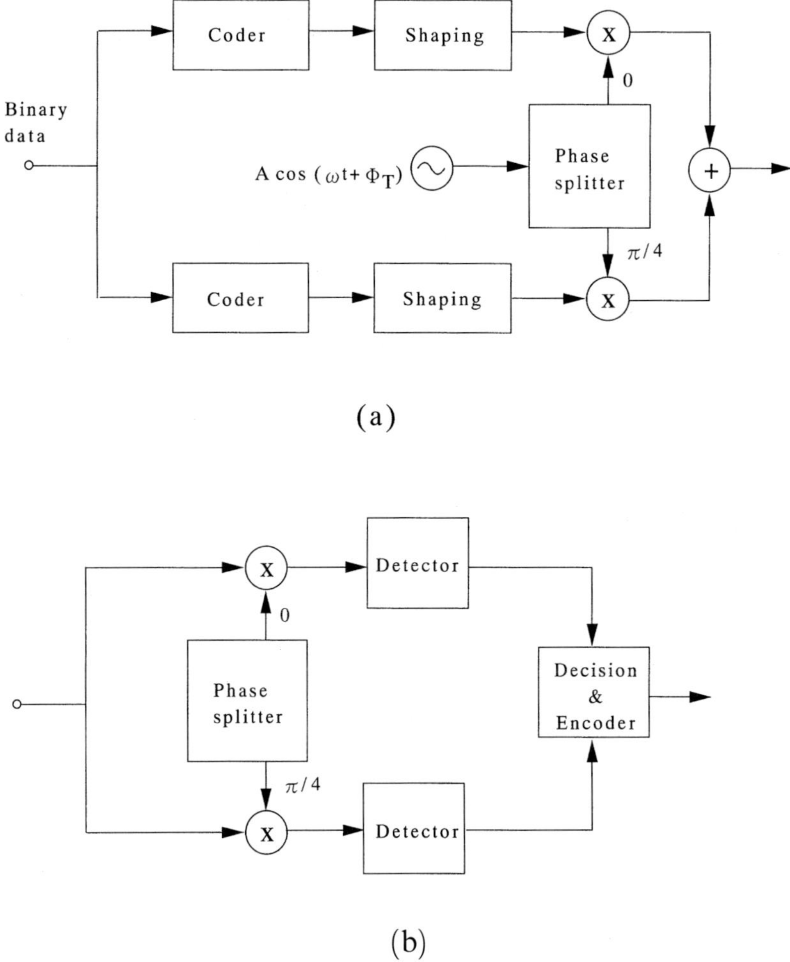

FIG. 7. (a) Modulator and (b) demodulator for QAM with rectangular constellation.

where T_b is the bit interval) bits per second and the available channel bandwidth is B_c Hz, where $B_c \gg R$. At the modulator, the information signal (a) is modulated with a bipolar type of pseudonoise [or pseudorandom noise, PN; this term is commonly used in this field as the same category of pseudorandom binary sequence (PRBS)] pulse sequence with its bit interval of $T_c(= 1/B_c)$ using the product modulator. The bandwidth of the information signal is expanded to $W = B_c$ Hz by shifting the phase of the carrier pseudorandomly at a rate of W times per second. The DC-SS signal is produced by modulating the expanded signal as shown in (c) with a sinusoidal carrier to form the BPSK signal and is transmitted through a communication channel.

We usually call a pulse in the PN code sequence the *chip*, T_c the *chip interval*, and $R_c = B_c$ the *chip rate*. The ratio of the bit interval to the chip interval $L_c = T_b/T_c$ is usually selected to be equal to the number of chips in the PN code sequence per information bit. The PN code sequence mentioned here is a code sequence of 1s and 0s whose autocorrelation has properties similar to those of white noise; the wave form of the autocorrelated signal has a steep peak at only one time point and at any other time position has a side lobe with a small noise amplitude. In the DS-SS

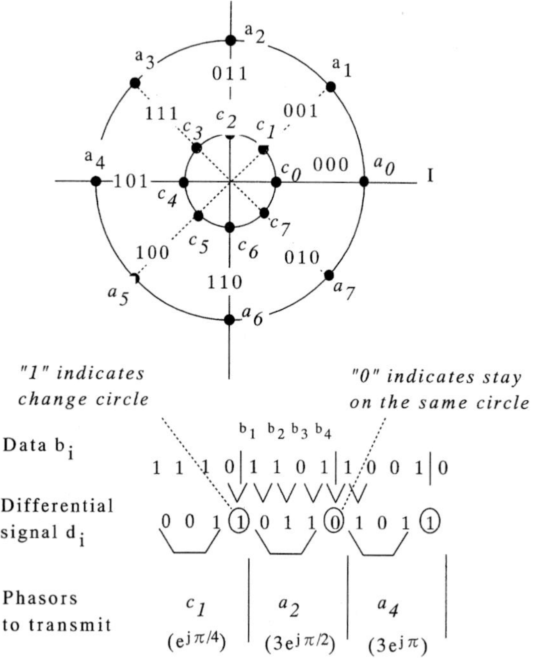

FIG. 8. Modulation process for 16-point QAM with a circular constellation. Of 4 bits in each symbol, b_1, b_2, b_3, and b_4, the first signal b_1 is differentially encoded onto the QAM phasor amplitude so that "1" causes a change to the amplitude circle that was not used in the previous signal and "0" causes the current symbol to be transmitted at the same amplitude. The remaining three signals are differentially Gray encoded on the circle with the same amplitude as that of the first signal.

system, the binary sequence with element $\{0, 1\}$ is mapped into a corresponding binary sequence with element $\{-1, 1\}$. The most widely known PN code sequence is the maximum-length sequence, which has a length $2^m - 1$ bits and is generated by an m-stage shift register as shown in Fig. 9(b).

The demodulation is the process of *despreading* of the spread signal. This is done by the circuits shown in Fig. 10. That in Fig. 10(a) uses direct cross correlation using a matched filter, such as a SAW correlator or convolver. This method has the advantage that demodulation can be performed with an extremely simple circuit and does not require a complicated synchronization process.

In the coherent demodulation process shown in Fig. 10(b), signal synchronization is critical to maintaining stable communication. The synchronization step can be separated into two distinct phases. The usual procedure for establishing initial synchronization is for the transmitter to send a known pseudorandom sequence to the receiver. The receiver is continuously in a search mode looking for this sequence to establish initial synchronization. In principle, matched filtering or cross correlation [as used in the circuit of Fig. 10(a)] are optimum methods for establishing initial synchronization in the presence of additive Gaussian noise. The matched filter establishes the initial synchronization in the same way as that of the modulator using a matched filter, and the demodulator enters the "data receive" mode. Examples of circuits for initial synchronization and tracking are shown in Fig. 11. Figure 11(a) shows an example of the initial synchronization method called a sliding correlator. The tracking is usually performed by means of a tracking loop called a delay-locked loop (DLL) as shown in Fig. 11(b). This implementation is equivalent to the early-late gate bit tracking synchronization described in Sec. 1.2.

2.2 Frequency Hopping

An example of the signal pattern of FH is shown in Fig. 12. In this scheme, the available channel bandwidth W is subdivided into a large number of nonoverlapping frequency slots. In any signaling interval the transmitted signal occupies one or more of the available

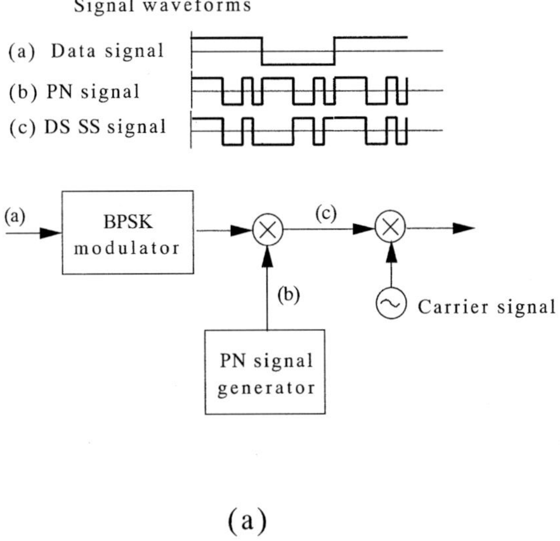

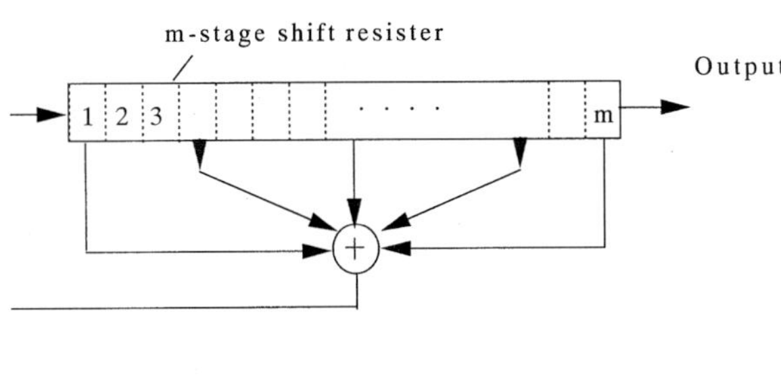

FIG. 9. Modulator for a direct sequence SS signal. (a) Block diagram, (b) maximum-sequence PN code generator.

frequency slots. The selection of the frequency slot in each signal interval is made pseudorandomly according to the output from a PN generator.

A block diagram of the modulator and demodulator for a FH-SS system is shown in Fig. 13. The modulation is either binary or M-ary FSK (MFSK). For example, if binary FSK is used, the modulator selects one of two frequencies, say f_0 or f_1, corresponding to the transmission of a 0 or a 1. The resulting binary FSK is translated in frequency by an amount that is determined by the output sequence from the PN generator, which used to select the frequency f_c that is synthesized by the frequency synthesizer. This frequency is mixed with the output of the output FSK modulator and the resultant frequency-translated signal is transmitted over the channel. For example, by taking m bits from the PN generator, we may specify $2^m - 1$ possible carrier frequencies.

At the receiver, there is an identical PN sequence generator, synchronized with the received signal, which is used to control the output of the frequency synthesizer. Thus, the pseudorandom frequency translation introduced at the transmitter is removed at the demodulator by mixing the synthesizer output with the received signal. The resultant signal is then demodulated by means of an FSK demodulator.

A signal for maintaining synchronization of the PN sequence generator with the FH received signal is usually extracted from the re-

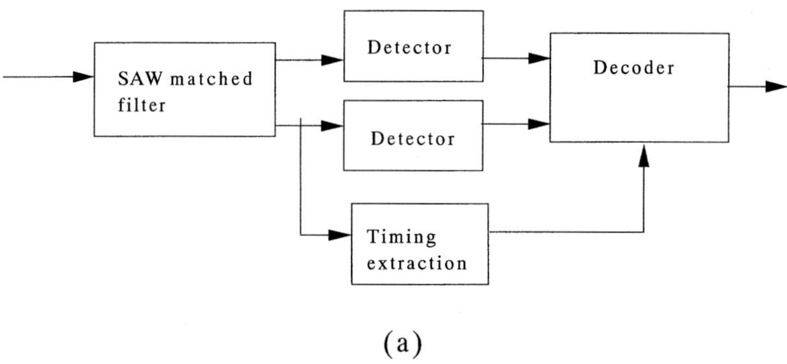

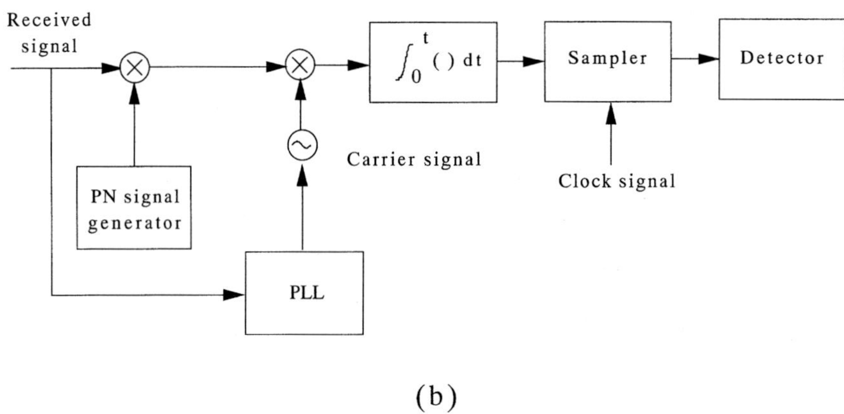

FIG. 10. Demodulator for a direct-sequence SS signal. (a) DS demodulation using SAW matched filter, (b) coherent demodulation. In (a), the received signal is autocorrelated by a matched filter and a compressed signal is generated directly by detecting the peak values from matched filters; thus the data will detected. In (b), the input signal is multiplied by the same PN sequence and is cross-correlated with a rectangular pulse of duration T_b. The data decision is performed by detecting whether the level of the output signal from the sampler exceeds thresholds.

ceived signal. Nonsynchronous detection is usually used for FH SS systems because it is difficult to maintain phase coherence both in the synthesis of the frequencies used in the hopping patterns and also in the propagation of the signal over the channel as the signal is hopped from one frequency to another over a wide bandwidth.

In the FH method, the initial acquisition process is to synchronize the known PN code sequence generated at the receiver that controls the hopping frequency pattern. The same procedures used in the DS signal synchronization procedure, such as serial search and sliding correlator, are useful for FH signals. However, it requires a long searching time. By employing multiple correlators in parallel, the searching time can be reduced at the expense of a more complex and costly implementation. For the tracking, the same technology as that used in DS-SS is applied.

3. OFFSET FREQUENCY-DIVISION MULTIPLEX

There exist some other modulation schemes with high efficiency in both transmission power and frequency resources. Among them,

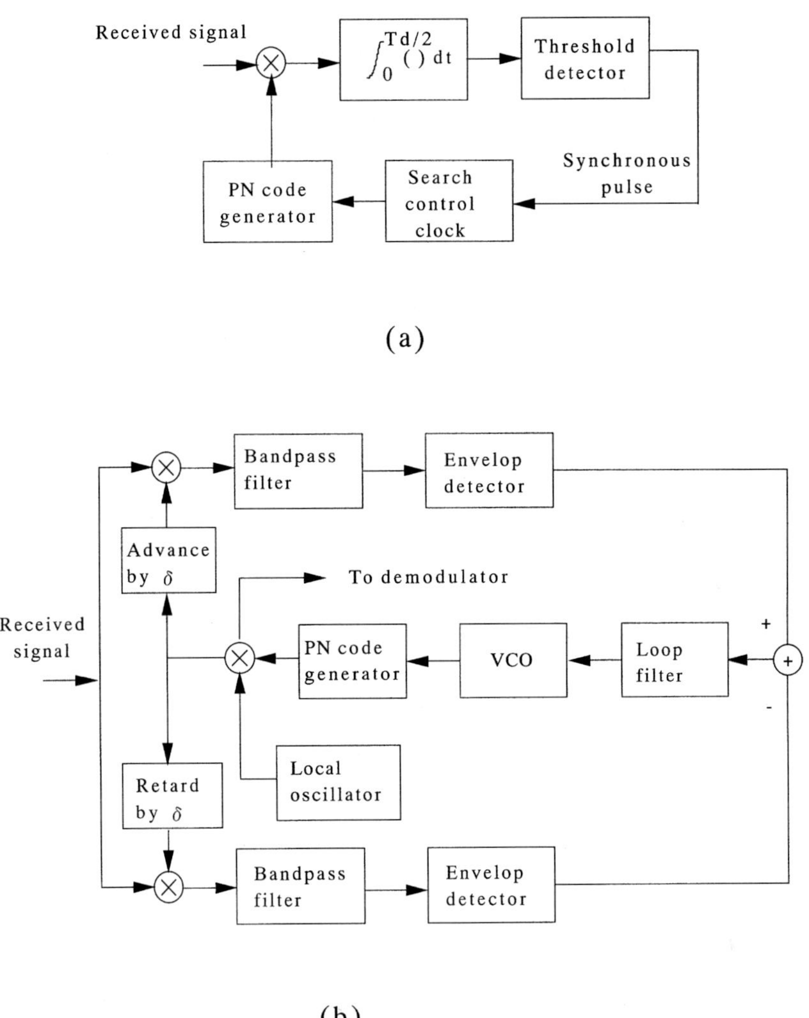

FIG. 11. Methods for synchronization. (a) Sliding correlator for DS signal acquisition. The received signal is cross correlated with a rectangular pulse of $T_c/2$ seconds or less. Then integration is performed over the time interval NT_c, and the peak value of the output is compared with a threshold to determine if the known signal sequence is present. If the known threshold is not exceeded, the known reference signal is advanced by $T_c/2$ seconds and the correlation process is repeated. These operations are performed until a signal is detected or until the search has been performed over the uncertain interval. (b) Tracking using delay-locked loop (DLL). In this loop, the received signal is applied to two multipliers, where it is multiplied by two outputs from the local PN code generator, that are delayed relative to each other by an amount $2\delta < T_c$. The product signals are bandpassed and detected, and then subtracted. They are then applied to the loop filter that drives the clock VCO. The VCO output serves as the clock for the PN code-signal generator. In this process if the synchronization is not exact, the filtered outputs from two correlators will be different from each other and the phase of the VCO will be appropriately advanced or delayed. At the equilibrium point, the two filtered outputs will be equally displaced from the peak value, and the PN code output fed to the demodulator will be exactly synchronized to the received signal.

an excellent scheme is the offset frequency-division multiplex (OFDM) modulation. The operation principle of this system is that the original bandwidth is divided into a large number of narrow sub-bands in which the mobile channel can be considered to be non-

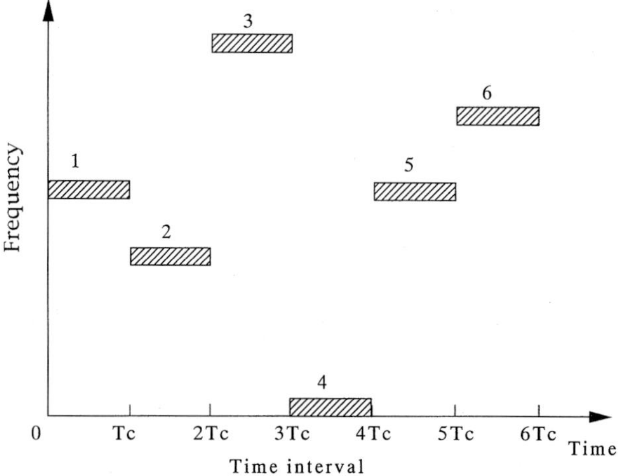

FIG. 12. An example of an FH signal.

dispersive. In other words, by setting the duration time of the modulated signal larger than the delay time of echo signals, it makes an undesirable echo no disturbance on the desired signal.

The modulation is performed by transforming serial time-sequence pulse data to parallel long-duration pulse signals with orthogonal carrier frequencies. This process is

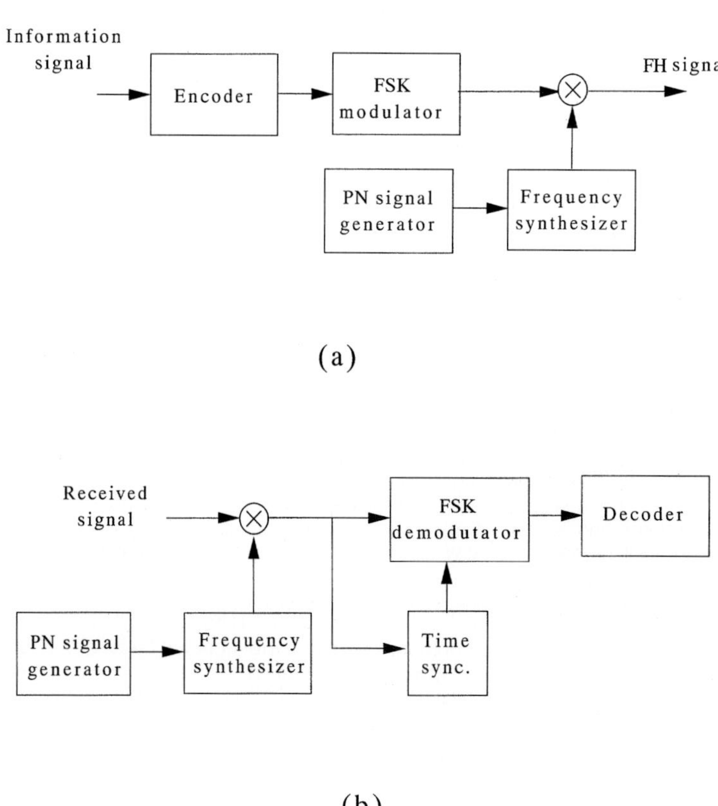

FIG. 13. (a) Modulator and (b) demodulator for FH-SS signal.

simply performed by a fast Fourier transformer (FFT) [or a discrete Fourier transformer (DFT)]. At the receiving end, the demodulation is also done by performing inverse discrete Fourier transform using an FFT. This scheme has the advantages of being implemented without a complex channel equalizer. Since the echo signal within a data interval contributes to an increase in the received signal power, this scheme has an advantage also in power efficiency, in addition to the robustness against multipath fading. Because of these features, this scheme is adopted for various digital wireless communication systems such as audio and digital TV broadcasting systems.

4. CONCLUSIONS

The basic concepts for modulation schemes stated here were proposed in relatively old days (several tens of years ago, or more). However, recent progress in device technology and global needs for communication systems as an infrastructure of social life accelerated the research and development of systems with these schemes. Technologies to improve system performances have also developed—for example, in DAM schemes, the development of linearized power amplifiers, adoption of error correction, and pilot-signal–assisted modulation—with the results that these systems are now in practical use in a wide area. As is represented by the information-superhighway concept, the importance of communication technologies is increasing still more, so that further development of technology of modulators and demodulators can be expected.

ABBREVIATIONS

DFT: Discrete Fourier transform.
FFT: Fast Fourier transform.
DLL: Delay-locked loop.
DS: Direct sequence.
DSB: Double sideband.
FH: Frequency hopping.
I-channel: In-phase channel.
Q-channel: Quadrature-phase channel.
MSK: Minimum-shift keying.
MFSK: *M*-ary frequency-shift keying.
OQPSK: Offset quadrature phase-shift keying.
PN: Pseudoranom noise.
PRBS: Pseudorandom binary sequence.
QAM: Quadrature amplitude modulation.
Q^2PSK: Quadrature quadrature phase-shift keying.
SAW: Surface acoustic wave.
SS: Spread spectrum.
SSB: Single sideband.
VCO: Voltage-controlled oscillator.
VSB: Vestigial sideband.

Further Reading

Alisouskas, V. F. A., Tamsi, W. (1985), *Digital and Data Communications*, Englewood Cliffs, NJ: Prentice-Hall.

Campbell, C. (1989), *Surface Acoustic Wave Devices and Their Signal Processing Applications*, London: Academic Press.

Dixon, R. C. (1976), *SpredSpectrum Systems*, New York: Wiley.

Gilcic, S. G., Leppanen, P. A. (Eds.) (1995), *Code Division Multiple Access Communications*, London: Kluwer Academic Publishers.

Proakis, J. G., Salhi, M. (1994), *Communication Systems Engineering*, Englewood Cliffs, NJ: Prentice-Hall.

Webb, W. T., Hanzo, L. (1995), *Modern Quadrature Amplitude Modulation—Principles and Applications for Fixed and Wireless Channels*, London: Pentech Press Publishers, and New York: IEEE PRESS.

CONTENTS OF VOLUMES 1–23*

VOLUME 1

*Complete addresses can be found at the beginning of each volume.

VOLUME 2

VOLUME 3

VOLUME 4

COMBUSTION
Charles K. Westbrook, *Lawrence Livermore National Laboratory*

COMPOSITE MATERIALS
Charles P. Beetz, Jr., *Advanced Technology Materials, Inc.*

COMPRESSIBLE FLOWS
J. E. Shepherd, *Rensselaer Polytechnic Institute*

COMPUTER-AIDED DESIGN IN ELECTRONICS
Antun Domic, *Digital Equipment Corporation*

COMPUTER-AIDED MANUFACTURING (CAM)
Kazuaki Iwata, *Osaka University*

COMPUTER DATABASES
H. Behrens and J. W. Tepel, *Fachinformationszentrum Karlsruhe*

COMPUTER GRAPHICS
Bruce Naylor and Arun N. Netravali, *AT&T Bell Labs*

COMPUTER HARDWARE
Kenneth H. Slater, *Digital Equipment Corporation*

COMPUTER PROGRAMMING LANGUAGES
Herbert J. Bernstein, *Bernstein + Sons*

COMPUTERS
John W. Carr, III, *University of Pennsylvania*

CONFORMATIONAL ANALYSIS
S. Scott Zimmerman, *Brigham Young University*

CONSTANTS, FUNDAMENTAL
Barry N. Taylor, *National Institute of Standards and Technology*

COPPER
Mary W. Covington,* *Consultant*
Nathan L. Church and William H. Dresher, *International Copper Association*
Bernard R. Cooper, *West Virginia University, Morgantown*

*Deceased.

CRITICAL PHENOMENA
Cyril Domb, *Bar-Ilan University*

CRYOGENICS
Hans M. Bozler and Christopher M. Gould, *University of Southern California*

CRYSTAL GROWTH
Erich Schönherr, *Max-Planck-Institut für Festkörperforschung*

CRYSTALLINE STATE
A. I. Goldman, *Iowa State University*

CRYSTALLOGRAPHY
W. B. Pearson and C. Chieh, *University of Waterloo*

CVD (CHEMICAL VAPOR DEPOSITION)
Bernard S. Meyerson, *IBM T. J. Watson Research Center*

CYCLOTRONS
G. H. Mackenzie, P. W. Schmor, and H. R. Schneider, *TRIUMF, Vancouver*

DATA-ACQUISITION SYSTEMS
Walter N. Gekelman, *University of California, Los Angeles*

DETECTORS, PARTICLE, CALORIMETRIC
Christian W. Fabjan, *CERN*

DETECTORS, SCINTILLATION
Eric B. Norman, *Lawrence Berkeley Laboratory*

DETECTORS, SEMICONDUCTOR
John Walter, *IntraSpec Inc.*

DIALYSIS
Robert H. Barth, M. D., *Department of Veterans' Affairs Medical Center and State University of New York, Health Science Center at Brooklyn*

DIAMAGNETISM
Leopoldo M. Falicov, *University of California, Berkeley, and Lawrence Berkeley Laboratory*

VOLUME 5

VOLUME 6

VOLUME 7

VOLUME 8

VOLUME 9

VOLUME 10

*Deceased.

VOLUME 11

VOLUME 12

VOLUME 13

*Deceased.

VOLUME 14

*Deceased.

VOLUME 15

VOLUME 16

VOLUME 17

VOLUME 18

VOLUME 19

*Deceased.

VOLUME 20

VOLUME 21

VOLUME 22

VOLUME 23

LIST OF RECOMMENDED UNITS AND SYMBOLS

RECOMMENDED UNITS AND CONVERSION FACTORS

The SI system provides six basic units: meter m, kilogram kg, second s, ampere A, kelvin K, candela cd, and mole mol.

Some important derived units are also allowed and bear special names, e.g.:

1 N (newton) = $1\,\mathrm{kg\,m\,s^{-2}}$
1 J (joule) = 1 N m = $1\,\mathrm{kg\,m^2\,s^{-2}}$
1 W (watt) = $1\,\mathrm{J\,s^{-1}} = 1\,\mathrm{kg\,m^2\,s^{-3}}$
1 Pa (pascal) = $1\,\mathrm{N\,m^{-2}} = 1\,\mathrm{kg\,m^{-1}s^{-2}}$

For mass, the gram g, or the metric ton t which equals 1000 kg, may be used instead of kilogram kg.

The so-called "long ton" (UK) and "short ton" (US) have been abandoned and will not be used in the *Encyclopedia of Applied Physics.*

For pressure, the bar (name and symbol alike) may be used, and for temperature, the degree celsius °C.

From all of these, decimal multiples or fractions can be derived:

Power of ten	Prefix	Symbol	Power of ten	Prefix	Symbol
10	deca	da	10^{-1}	deci	d
10^2	hecto	h	10^{-2}	centi	c
10^3	kilo	k	10^{-3}	milli	m
10^6	mega	M	10^{-6}	micro	μ
10^9	giga	G	10^{-9}	nano	n
10^{12}	tera	T	10^{-12}	pico	p
10^{15}	peta	P	10^{-15}	femto	f
10^{18}	exa	E	10^{-18}	atto	a

For mass, multiples or fractions are derived from g, not from kg (since the latter already contains a prefix), e.g., mg.

Units with the prefixes are considered as one entity and can, therefore, be raised to any power, e.g., $\mathrm{cm^3}$.

The liter is now considered as synonymous with $\mathrm{dm^3}$ (which does not hold in the older literature!). Please use the capital letter L as unit symbol (following a recent IUPAC recommendation). The use of the Ångström unit is discouraged; it should be replaced by fractional meters (1 Å = 100 pm = 0.1 nm).

SELECTED QUANTITIES, UNITS, AND SYMBOLS

Name of quantity[a]	Symbol[b]	SI unit[c]	Name of unit	Other units
Space and time				
length*	l	m	meter[d]	
breadth, width	b	m		
height	h	m		
radius	r	m		
thickness	d	m		
area	A, S	m^2		a (are) h (hectare)
volume	V	m^3		L, l (liter)[e]
plane angle	$\alpha, \beta, \gamma,$ ϑ, φ	1, rad	radian	° (degree) ′ (minute) ″ (second)
solid angle	ω, Ω	1, sr	steradian	
wavelength	λ	m		
wave number	σ, ν	m^{-1}		
time*	t	s	second	min (minute) h (hour) d (day)
frequency	ν, f	s^{-1}		Hz (hertz)[f]
relaxation time	τ	s		
velocity	u, v	$m\ s^{-1}$		km/h
acceleration	a	$m\ s^{-2}$		
Mechanics				
mass*	m	kg	kilogram	g (gram) t (tonne)[g]
(mass) density	ρ	$kg\ m^{-3}$		g/cm^3
momentum	$\mathbf{p}$	$kg\ m\ s^{-1}$		
angular momentum	$\mathbf{L}$	$kg\ m^2\ s^{-1}$		
force	$\mathbf{F}$	N	newton[f]	
moment of force	$\mathbf{M}$	N m		
weight	G, W	N		
pressure	p	Pa	pascal	bar (bar)[h]
energy	E, W	J	joule	W h (watt hour)[i] eV (electron volt)[j]
work	W, A	J		
power	P	W	watt	J/s, V A[k]
Molecular Physics and Thermodynamics				
thermodynamic temperature*	T	K	kelvin	°C (degrees Celsius)
Celsius temperature	ϑ, t			°C
number of entities	N			
Avogadro constant[l]	N_A, L	mol^{-1}	(particles) per mole	
Boltzmann constant[m]	k, k_B	$J\ K^{-1}$		
Planck constant[n]	h	J s		
(molar) gas constant[o]	R	$J\ mol^{-1}\ K^{-1}$		
(quantity of) heat	Q	J		
entropy[p]	S	$J\ K^{-1}$		
internal energy[p]	U	J		
Helmholtz function,[p] (Helmholtz) free energy, Helmholtz energy	F, A	J		
enthalpy[p]	H	J		

Name of quantity[a]	Symbol[b]	SI unit[c]	Name of unit	Other units
Gibbs function,[p] (Gibbs) free energy, Gibbs energy	G	J		
heat capacity[p]	C_p, C_v	$\mathrm{J\,K^{-1}}$		
Chemical Physics				
amount of substance*	n	mol	mole	
relative atomic mass	A_r	1		
relative molecular mass	M_r	1		
atomic mass constant	m_u	kg		u (atomic mass unit)[q]
mass of a portion (of substance B)	$m_B, m(\mathrm{B})$	kg		g (gram)
molar mass (of substance B)	$M_\mathrm{B}, M(\mathrm{B})$	$\mathrm{kg\,mol^{-1}}$		
concentration (of substance B)	$c_\mathrm{B}, c(\mathrm{B})$	$\mathrm{mol\,m^{-3}}$		mol/L
mole fraction[r] (of substance B)	$\kappa_\mathrm{B}, \kappa(B)$	1		
mass fraction (of substance B)	$\omega_\mathrm{B}, \omega(\mathrm{B})$	1		%, ‰,ppm,ppb
volume fraction (of substance B)	$\varphi_\mathrm{B}, \varphi(\mathrm{B})$ $\phi_\mathrm{B}, \phi(\mathrm{B})$	1		%, ‰,ppm,ppb
mass concentration	ρ	$\mathrm{kg\,m^{-3}}$		g/L
molality		$\mathrm{mol\,kg^{-1}}$		mmol/kg
volume concentration[s]	σ	1		
molar volume	V_m	$\mathrm{m^3\,mol^{-1}}$		L/mol
molar heat capacity	C_m	$\mathrm{J\,mol^{-1}\,K^{-1}}$		
molar conductivity	Λ_m	$\mathrm{S\,m^2\,mol^{-1}}$	(S: siemens)	
Faraday constant[t]	F	$\mathrm{C\,mol^{-1}}$		
Electricity and Magnetism				
quantity of electricity[u]	Q	C	coulomb	
charge density	ρ	$\mathrm{C\,m^{-3}}$		
electric potential	ϕ, V	V	volt	
electric potential difference, voltage	$U, \Delta\phi, \Delta V$	V		
electric dipole moment	$\mathbf{p}, \mathbf{p}_\mathrm{c}$	C m		
electric current*	I	A	ampere	
electric current density	j	$\mathrm{A\,m^{-2}}$		
electric field strength	$\mathbf{E}$	$\mathrm{V\,m^{-1}}$		
electric displacement	$\mathbf{D}$	$\mathrm{C\,m^{-2}}$		
capacitance	C	F	farad	
permittivity	ε	$\mathrm{F\,m^{-1}}$		
relative permittivity	ε	1		
dielectric polarization	$\mathbf{P}$	$\mathrm{C\,m^{-2}}$		
electric susceptibility	χ_c	1		
polarization (of a particle)	α	$\mathrm{m^2\,C\,V^{-1}}$		
magnetic flux	Φ	Wb	weber	
magnetic flux density	$\mathbf{B}$	T	tesla	
magnetic field strength	$\mathbf{H}$	$\mathrm{A\,m^{-1}}$		
permeability	μ	$\mathrm{H\,m^{-1}}$, $\mathrm{N\,A^{-2}}$	(H: henry)	
relative permeability	μ_r	1		
magnetization	M	$\mathrm{A\,m^{-1}}$		
magnetic susceptibility	χ	1		
molar magnetic susceptibility	χ_m	$\mathrm{m^3\,mol^{-1}}$		

Name of quantity[a]	Symbol[b]	SI unit[c]	Name of unit	Other units
(electrical) resistance	R	Ω	ohm	
(electrical) conductance	G	S	siemens	
(electrical) resistivity	ρ	Ω m		
(electrical) conductivity	κ, o	S m^{-1}		
self-inductance	L	H	henry	
Radiation				
radiant energy	Q, W, Q_c	J		
luminous intensity*	I	cd	candela	
radiant intensity	I_c	W sr^{-1}, W		
emissivity, emittance	ε	1		
absorptance	α	1		
reflectance	ρ, R	1		
transmittance	τ	1		
absorption coefficient:				
linear (decadic)	a	m^{-1}		
molar (decadic)	ε	m^2 mol^{-1}		
refractive index	n	1		
molar refraction	R_m	m^3 mol^{-1}		
angle of optical rotation	α	1, rad		
Transport Properties				
flux of quantity X	J_X, J	(varies)		
mass flow rate	$q_m, \dot{m}$	kg s^{-1}		
volume flow rate	$q_v, \dot{V}$	m^3 s^{-1}		
heat flow rate	Φ	W		
thermal conductivity	κ, k, λ	W m^{-1} K^{-1}		
coefficient of heat transfer	h	W m^{-2} K^{-1}		
thermal diffusivity	a	m^2 s^{-1}		
diffusion coefficient	D	m^2 s^{-1}		
thermal diffusion coefficient	D_T	m^2 s^{-1}		
viscosity	η, μ	Pa s		
kinematic viscosity	υ	m^2 s^{-1}		

[a] SI base quantities are marked by asterisks (*)
[b] Recommended by IUPAC
[c] SI base units as well as derived and supplementary units are listed; all are to be used with prefixes as needed
[d] Do not use "metre"
[e] Do not use "litre"; 1 L = 10^{-3} m^3
[f] 1 Hz = 1 s^{-1}; 1 N = 1 kg m s^{-2}
[g] Formerly metric ton; 1 t = 10^3 kg
[h] 1 bar = 10^5 Pa
[i] 1 W h = 3.6×10^3 J
[j] 1 eV = $1.602\,189 \times 10^{-19}$ J
[k] 1 W = 1 J/s = 1 VA
[l] $N_A = 6.022\,136\,7 \times 10^{23}$ mol^{-1}
[m] $k = 1.380\,658 \times 10^{-23}$ J K^{-1}
[n] $h = 6.626\,075\,5 \times 10^{-34}$ J s
[o] $R = 8.314\,510$ J mol^{-1} K^{-1}
[p] Molar quantities can be distinguished from the quantity of a system by adding the subscript m; e.g., molar internal energy U_m, in J mol^{-1}
[q] 1 u = $1.660\,565\,5 \times 10^{-27}$ kg
[r] A more accurate, but rather uncommon, name is "amount-of-substance fraction"
[s] σ refers to the total volume of a mixture, whereas the volume fraction φ relates the volume of a substance to the volume of several components before mixing
[t] $F = 9.648\,530\,9 \times 10^4$ C mol^{-1}
[u] Also called electric charge